An Invitation to Combinatorics

Active student engagement is key to this classroom-tested combinatorics text, boasting 1200+ carefully designed problems, ten mini-projects, section warm-up problems, and chapter opening problems. The author – an award-winning teacher – writes in a conversational style, keeping the reader in mind on every page. Students will stay motivated through glimpses into current research trends and open problems as well as the history and global origins of the subject. All essential topics are covered, including Ramsey theory, enumerative combinatorics including Stirling numbers, partitions of integers, the inclusion–exclusion principle, generating functions, introductory graph theory, and partially ordered sets. Some significant results are presented as sets of guided problems, leading readers to discover them on their own. More than 140 problems have complete solutions and over 250 have hints in the back, making this book ideal for self-study. Ideal for a one-semester upper undergraduate course, prerequisites include the calculus sequence and familiarity with proofs.

Shahriar Shahriari is Professor of Mathematics at Pomona College. He has over 50 publications in mathematics including two books: *Approximately Calculus* (AMS 2006) and *Algebra in Action: A Course in Groups, Rings, and Fields* (AMS 2017). His book *Approximately Calculus* was chosen as an American Library Association's Choice Outstanding Academic Title of 2007, and he won the Mathematical Association of America's Carl B. Allendoerfer Award for expository writing in 1998. Shahriari was awarded the Mathematical Association of America's Haimo National Teaching Award in 2015, the Southern California-Nevada Section of the Mathematics Association of America's Teaching Award in 2014, and Pomona College's collegewide student-voted Wig Distinguished Teacher award five different times.

CAMBRIDGE MATHEMATICAL TEXTBOOKS

Cambridge Mathematical Textbooks is a program of undergraduate and beginning graduate-level textbooks for core courses, new courses, and interdisciplinary courses in pure and applied mathematics. These texts provide motivation with plenty of exercises of varying difficulty, interesting examples, modern applications, and unique approaches to the material.

A complete list of books in the series can be found at www.cambridge.org/mathematics
Recent titles include the following:

Chance, Strategy, and Choice: An Introduction to the Mathematics of Games and Elections, S. B. Smith
Set Theory: A First Course, D. W. Cunningham
Chaotic Dynamics: Fractals, Tilings, and Substitutions, G. R. Goodson
A Second Course in Linear Algebra, S. R. Garcia & R. A. Horn
Introduction to Experimental Mathematics, S. Eilers & R. Johansen
Exploring Mathematics: An Engaging Introduction to Proof, J. Meier & D. Smith
A First Course in Analysis, J. B. Conway
Introduction to Probability, D. F. Anderson, T. Seppäläinen & B. Valkó
Linear Algebra, E. S. Meckes & M. W. Meckes
A Short Course in Differential Topology, B. I. Dundas
Abstract Algebra with Applications, A. Terras
Complex Analysis, D. E. Marshall
Abstract Algebra: A Comprehensive Introduction, J. W. Lawrence & F. A. Zorzitto
Algebra, P. Aluffi
An Invitation to Combinatorics, S. Shahriari

An Invitation to Combinatorics

Shahriar Shahriari

Pomona College, California

CAMBRIDGE
UNIVERSITY PRESS

CAMBRIDGE
UNIVERSITY PRESS

University Printing House, Cambridge CB2 8BS, United Kingdom

One Liberty Plaza, 20th Floor, New York, NY 10006, USA

477 Williamstown Road, Port Melbourne, VIC 3207, Australia

314–321, 3rd Floor, Plot 3, Splendor Forum, Jasola District Centre, New Delhi – 110025, India

103 Penang Road, #05–06/07, Visioncrest Commercial, Singapore 238467

Cambridge University Press is part of the University of Cambridge.

It furthers the University's mission by disseminating knowledge in the pursuit of education, learning, and research at the highest international levels of excellence.

www.cambridge.org
Information on this title: www.cambridge.org/9781108476546
DOI: 10.1017/9781108568708

© Shahriar Shahriari 2022

First published 2022

Printed in the United Kingdom by TJ Books Limited, Padstow, Cornwall, 2022

A catalogue record for this publication is available from the British Library.

Library of Congress Cataloging-in-Publication Data
Names: Shahriari, Shahriar, author.
Title: An invitation to combinatorics / Shahriar Shahriari.
Description: New York : Cambridge University Press, 2021. |
Series: Cambridge mathematical textbooks | Includes bibliographical references and index.
Identifiers: LCCN 2021012365 (print) | LCCN 2021012366 (ebook) |
    ISBN 9781108476546 (hardback) | ISBN 9781108568708 (epub)
Subjects: LCSH: Combinatorial analysis. | BISAC: MATHEMATICS / Discrete
    Mathematics | MATHEMATICS / Discrete Mathematics
Classification: LCC QA164 .S495 2021 (print) | LCC QA164 (ebook) | DDC 511/.6–dc23
LC record available at https://lccn.loc.gov/2021012365
LC ebook record available at https://lccn.loc.gov/2021012366

ISBN 978-1-108-47654-6 Hardback

Additional resources for this title at www.cambridge.org/combinatorics

To my sons, nieces, and nephews: Tascha, Kiavash, Neema, Noosha, Borna, and Shooka

Contents

The plate section can be found between pages 248 and 249

Preface

Combinatorics is a fun, difficult, broad, and very active area of mathematics. Counting, deciding whether certain configurations exist, and elementary graph theory are where the subject begins. There are a myriad of connections to other areas of mathematics and computer science, and, in fact, combinatorial problems can be found almost everywhere. To learn combinatorics is partly to become familiar with combinatorial topics, problems, and techniques, and partly to develop a can-do attitude toward discrete problem solving. This textbook is meant for a student who has completed an introductory college calculus sequence (a few sections require some knowledge of linear algebra but you can get quite a bit out of this text without a thorough understanding of linear algebra), has some familiarity with proofs, and desires to not only become acquainted with the main topics of introductory combinatorics but also to become a better problem solver.

Key Features

- **Conversational Style.** This text is written for students and is meant to be read. Reading mathematics is difficult, but being able to decipher complicated technical writing is an incredibly useful and transferable skill. The discussions in between the usual theorems, proofs, and examples are meant to facilitate the reader's (ad)venture into reading a mathematics textbook.
- **Problem-Solving Emphasis.** One advantage of combinatorics is that many of its topics can be introduced without too much jargon. In fact, you can turn to almost any chapter in this book and find problem statements that are understandable regardless of your background. Initially, you may not know how to do a problem or even where to begin, but gaining experience in passing that hurdle is at the heart of becoming a better problem solver. For this to happen, you, the reader, have to get actively involved, and learn by solving problems. Getting the right answer is not really the objective. Rather, it is only by trying to solve a problem that you will really understand what the problem is asking and what subtle issues need to be considered. It is only after you have given the problem a try that you will appreciate the solution. This text gives you ample opportunity to get actively involved. In addition to over 1200 problems, we highlight the following features.
  - **Collaborative Mini-projects.** The mini-projects – there are 10 of these scattered throughout the text – are meant to be projects for groups of three or four students to collaboratively explore. They are organized akin to a science lab. A few preliminary

problems are to be done individually. The collaborative part of the project is meant to take up the good part of an afternoon, and it is envisioned that a project report – really a short mathematics paper – will be the result. The mini-projects are of two types. In one type (Mini-projects 1, 6, 7, 8, 9, and 10), the project explores new material that is not covered elsewhere in the text. The second type (Mini-projects 2, 3, 4, and 5) are meant to be done by the students *before* the relevant topic is covered in class. (This should explain their curious placement in the text.) It has been my experience that much learning happens if the students first work on a topic, in a guided and purposeful manner, on their own, followed by class discussion/lecture.

– **Guided Discovery through Scaffolded Projects.** In addition to collaborative mini-projects, quite a number of other topics are organized as a sequence of manageable smaller problems. I often assign these problems in consecutive assignments, so that I can provide solutions, and allow time for discussion before proceeding to subsequent steps. In some other problems, students are guided toward a solution through an explicit sequence of steps. Some of the proofs are presented in this format with the hope that an engaged reader, with a paper and pencil at hand, will fill in the details. As such, these problems are aimed at training the reader in the art of reading terse mathematical proofs.

– **Warm-Up Problems and Opening Chapter Problems.** Nearly every section starts with a warm-up problem. These are relatively straightforward problems, and I use them for in-class group work as a prelude to the discussion of a topic. In contrast, every chapter starts with a more subtle opening problem. My guess is that, for the most part, you, the student, will not be able to do these opening problems before working through the chapter. The purpose is to give you a glimpse of what we are going to do in the chapter, and motivate you to make it through the material. Each opening chapter problem is solved somewhere in the chapter, and all of the warm-up problems have a short answer in Appendix A.

– **Selected Hints, Short Answers, Complete Solutions.** The appendices have hints, short answers, and complete solutions for selected problems. A problem number in *italics* – there are 142 of these – indicates that the problem has a complete solution in Appendix D. Try a problem first without looking at the solution, but when you are stuck – and hopefully you will get stuck often and learn to cherish the experience – then first look at the hint section, then later at the short answer section, and finally at the complete solutions section. The complete solutions serve two purposes. They provide further examples of how to do problems, and they model how to write mathematics in paragraph style. By contrast, the short answers are not particularly helpful in telling you how to do a problem. Instead, after you are done with a problem, the short answer can either reassure you or send you back to the drawing board.

- **Open Problems and Conjectures.** A number of the chapters end with a section highlighting a few easily stated open problems and conjectures. Some of these are important unsolved problems, while others are mere curiosities. This is not meant to be a guide to current cutting-edge research problems. Rather, the modest aim is to whet the reader's appetite by convincing her that even some seemingly innocent-looking problems remain unsolved, and that combinatorics is an active area of research.
- **Historical Asides.** I am not a historian and the historical comments and footnotes barely scratch the surface. Even so, combinatorial problems and solutions are a wonderful example of the international nature of mathematics. In addition, mathematics is created by humans who are affected by, participate in, and sometimes, for good or bad, help shape the communities and societies that they are a part of. I will declare success even if just a few of you become curious and further investigate the historical context of the mathematics. I have also chosen to highlight the international nature of combinatorics by naming some well-known mathematical objects differently. See the discussion later in this preface for a bit more context.

Coverage and Organization

The text has more than enough material for a one-semester course in combinatorics at the sophomore or junior level at an American university. The sections of the book that I do not cover in my classes, and that I consider optional, are marked by a *. Induction proofs and recurrence relations will be used throughout the book and are the subject of Chapter 1. Counting problems – so-called *enumerative combinatorics* – take up more than half of the book, and are the subject of Chapters 3 through 9. In Chapter 3, we introduce a slew of "balls and boxes" problems that serve as an organizing principle for our counting problems. Chapters 3, 4, and 5 cover the basics of permutations and combinations as well as a good dose of exploration of binomial coefficients. Chapters 6 and 7 are on Stirling numbers and integer partitions. Two substantial chapters on the inclusion–exclusion principle and generating functions conclude our treatment of enumerative combinatorics. Graph theory is about one-third of the book and is covered in Chapters 2 and 10. The basic vocabulary of graphs is introduced early in Section 2.2, but, for the most part, the material on graph theory is independent of the other chapters. I start the course with Ramsey theory since I want to make sure that all students are seeing something new, and that they are not lulled into thinking that the class is going to be only about permutations and combinations. But Ramsey theory is difficult and could be postponed to much later. Alternatively, you could start with Chapter 10, and do graph theory first. Finally, Chapter 11 brings together material on partially ordered sets, a favorite of mine. Matchings in bipartite graphs is also covered in this chapter, since I wanted to bring out the close connection between the two frameworks.

The instructor may want to augment the usual fare of introductory combinatorics with one or two more substantial results. Among the topics offered here are the Chung–Feller theorem, Euler's pentagonal number theorem, Cayley's theorem on labeled trees, Stanley's theorem on acyclic orientations, Thomassen's five-color theorem, Pick's formula, the Erdős–Ko–Rado theorem, Ramsey's theorem for hypergraphs, and Möbius inversion.

The Global Roots of Combinatorics and the Naming of Mathematical Objects

Most "new" mathematical ideas and concepts have antecedents and precursors in older ones, and, as a result, the search for the "first" appearance of this or that mathematics is never-ending and often futile. As such the naming of mathematical objects and results sometimes – possibly always – is a bit arbitrary.[1] However, when you look at the totality of the common names of mathematical objects in combinatorics – Pascal's triangle, Vandermonde's identity, Catalan numbers, Stirling numbers, Bell numbers, or Fibonacci numbers – a remarkable and seemingly non-random pattern emerges. All the names chosen are from the European tradition. Undoubtedly, European mathematicians contributed significantly – and, in many subareas of mathematics, decisively – to the development of mathematics. However, this constellation of names conveys to the beginning student that combinatorial ideas and investigations were limited to Europe. In the case of combinatorics, nothing could be further from the truth. Mathematicians from China, Japan, India, Iran, northern Africa, the wider Islamic world, and the Hebrew tradition, to mention a few, have very much worked on these topics. (For some of this history, see Wilson and Watkins 2013.) Certainly, the later European scholars have taken some topics further, but this does not take away from the international character of mathematics in general and combinatorics in particular. For this reason, we have tried to use a more inclusive set of names for at least some of the familiar objects. It is completely understandable to want to be familiar with the more common – often universally accepted – names for various objects, and those are pointed out in the text as well. The author does not claim expertise in the history of combinatorics, and it is quite possible that very good historical arguments can be made in support of other attributions or against the ones suggested here. If such a discussion ensues, we will all be better for it.

---

[1] "It takes a thousand men to invent a telegraph or a steam engine, or a phonograph, or a telephone, or any other important thing – and the last man gets the credit and we forget the others. He added his little mite – that is all he did. These object lessons should teach us that ninety-nine parts of all things that proceed from the intellect are plagiarisms, pure and simple; and the lesson ought to make us modest. But nothing can do that." Mark Twain, *Letter to Helen Keller*, Riverdale-on-the-Hudson, St. Patrick's Day 1903.

Acknowledgments

Like many mathematicians, I was introduced to combinatorics by having to use it in other areas of mathematics. I actually did not take a combinatorics course in college or in graduate school. When I got the opportunity to teach combinatorics at Pomona College, I relied heavily on the many wonderful texts available. Some of my favorites, from which I learned a lot – and much of that can be seen on the pages of this book – are Brualdi (2010) (I used the first edition to teach my first combinatorics class), Tucker (1995), Stanley (2012), van Lint and Wilson (2001), Anderson (2002), and Wilf (2006). In teaching that first combinatorics class, the attempt to loosen the assumptions in one particular homework problem led me to my first research paper in combinatorics (Shahriari (1996)). I also realized that combinatorics is a fertile area for involving undergraduates in research. As a result, and over time, I shifted my primary research area from finite group theory to combinatorics. In addition to the authors of the numerous texts that I have relied on over the years, I also thank both my research collaborators and my students. It has been exciting to do research in combinatorics, and it has been fun to share that excitement with my students. My colleagues Vin de Silva and Ghassan Sarkis, who used earlier versions of the text, have been generous with their suggestions and comments. The staff and editors at Cambridge University Press – Katie Leach, Maggie Jeffers, Rachel Norridge, and John King – have been instrumental in getting the book completed. They have been helpful, professional, and most importantly patient. Finally, my immediate family, Nanaz, Kiavash, and Neema, have supported, motivated, and endured me all through this project. Thank you.

Introduction

Accurate reckoning. The entrance into the knowledge of all existing things and all obscure secrets.

–The Ahmes–Rhind Papyrus

What is Combinatorics?

Combinatorics is a collection of techniques and a language for the study of (finite or countably infinite) discrete structures. Given a set of elements (and possibly some structure on that set), typical questions in combinatorics are:

- Does a specific arrangement of the elements exist?
- How many such arrangements are there?
- What properties do these arrangements have?
- Which one of the arrangements is maximal, minimal, or optimal according to some criterion?

Unlike many other areas of mathematics – e.g., analysis, algebra, topology – the core of combinatorics is neither its subject matter nor a set of "fundamental" theorems. More than anything else, combinatorics is a collection – some may say a hodgepodge – of techniques, attitudes, and general principles for solving problems about discrete structures. For any given problem, a combinatorist combines some of these techniques and principles – e.g., the pigeonhole principle, the inclusion–exclusion principle, the marriage theorem, various counting techniques, induction, recurrence relations, generating functions, probabilistic arguments, asymptotic analysis – with (often clever) ad hoc arguments. The result is a fun and difficult subject.

In today's mathematical world, in no small part due to the power of digital computers, most mathematicians find much use for the tool box of combinatorics. In problems of pure mathematics, often, after deciphering the layers of theory, you find a combinatorics problem at the core. Outside of mathematics, and as an example, combinatorial problems abound in computer science.

Typical Problems

To whet your appetite, here is a preliminary sample of problems that we will encounter in the course of this text.

- How many sequences $a_1, a_2, \ldots, a_{12}$ are there consisting of four 0's and eight 1's, if no two consecutive terms are both 0's?
- A bakery has eight kinds of donuts, and a box holds one dozen donuts. How many different boxes can you buy? How many different boxes are there that contain at least one of each kind?
- A bakery sells seven kinds of donuts. How many ways are there to choose one dozen donuts if no more than three donuts of any kind are used?
- Determine the number of n-digit numbers with all digits odd, such that 1 and 3 each occur a *positive* even number of times.
- We are trying to reconstruct a word that is made from the letters A, B, C, D, and R. We are given a frequency table that shows the number of times a specific triple occurs in the word:

triple	frequency
ABR	2
ACA	1
ADA	1
BRA	2
CAD	1
DAB	1
RAC	1

For example, ABR occurs twice while ACA appears once. We want to know all words with the same triples and with the same frequency table. The answer may be that there are no such words. Note that by a word we mean an ordered collection of letters and we are not concerned with meaning.

- A particular signaling network consists of six pieces of communications equipment:

$$x_1, x_2, y_1, y_2, z_1, z_2$$

We can choose various pairs of these and link each pair through an intermediate facility (e.g., microwave towers, trunk groups). Intermediate facilities are expensive to build but if one fails, then all the links through them become inoperative. So, if we build just one intermediate facility and route all of our connections through it, then its failure will disconnect everything. However, the design specification requires that if one intermediate facility fails, then there will remain at least one link between at least one of the x's and one of the y's and between one of the x's and one of the z's, as well as between one of the y's and one of the z's. What is the minimum number of facilities that we need and how should the connections be designed?

- A soccer ball is usually tiled with 12 pentagons and 20 hexagons. Are any other combinations of pentagons and hexagons possible?

How Do We "Count"?

Counting the number of configurations of a certain type is an important part of combinatorics. In all of the examples in the previous section, it is clear what kind of an answer we are looking for. We want a specific numerical answer or an example of a specific configuration.

However, in many problems, it may be possible to present a solution that is satisfactory in many ways but is not quite a direct answer. We look at several examples.

(a) Let $[n] = \{1, 2, \ldots, n\}$, and let $f(n)$ be the number of subsets of $[n]$. Then $f(n) = 2^n$.

Proof. For any particular subset of $[n]$, each element of $[n]$ is either in that subset or not. Thus, to construct a typical subset, we have to make one of two choices for each element of $[n]$. Furthermore, these choices are independent of each other. Hence, the total number of choices – and consequently the total number of subsets – is

$$\underbrace{2 \times 2 \times \cdots \times 2}_{n} = 2^n.$$
□

This proof gives a closed formula for the answer, and, in fact, gives more than was asked. It also tells us how to construct all the subsets.

(b) Assume n people give their n hats to a hat-check person. Let $f(n)$ be the number of ways that the hats can be returned, so that everyone has one hat, but no one has their own hat. If we list the hats on the left and their owners on the right, then the figure below shows three ways of returning hats to four people so that none of them gets their own hat back.

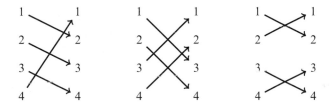

You should check that $f(1) = 0, f(2) = 1$, and $f(3) = 2$. You should also try to find $f(4)$.

We will show in Section 8.3 that

$$f(n) = n! \sum_{i=0}^{n} \frac{(-1)^i}{i!}.$$

This is a formula – and you should check your answers for $n = 1, \ldots, 4$ using it – but we would have preferred a "nice" closed formula. While you may not find this formula pleasing, it does work, and will become even more meaningful when we understand the significance of each term in the sum.

It is also possible to show that $f(n)$ is the nearest integer to $\frac{n!}{e}$. This is, of course, easier to use. But this formula may not have a combinatorial significance, and hence, it may be argued, that it gives us less insight. It is, however, fascinating that in answering a question about hats, the number e would make an appearance.

(c) Let $[n] = \{1, \ldots, n\}$. Let $f(n)$ be the number of subsets of $[n]$ that do not contain two consecutive integers. For example, if $n = 4$, then the subsets of $\{1, 2, 3, 4\}$ that do not contain two consecutive integers are

$$\emptyset, \{1\}, \{2\}, \{3\}, \{4\}, \{1,3\}, \{1,4\}, \{2,4\}.$$

Thus $f(4) = 8$. You should also check that $f(1) = 2, f(2) = 3$, and $f(3) = 5$.

We will show in Example 1.18 (and again in Example 9.34) that

$$f(n) = \frac{1}{\sqrt{5}}(\tau^{n+2} - \overline{\tau}^{n+2}), \text{ where } \tau = \frac{1}{2}(1 + \sqrt{5}), \overline{\tau} = \frac{1}{2}(1 - \sqrt{5}).$$

Again, it is unclear how irrational numbers got involved in counting a discrete phenomenon. This formula can actually be used but seems to give little insight into the problem. Sometimes, there are alternatives to finding a closed formula. For this problem, we can prove the following recurrence relation:

CLAIM: $f(n) = f(n-1) + f(n-2)$.

Proof. All "good" subsets of $[n]$ either have n or don't have n. The ones that don't have n are exactly the "good" subsets of $[n-1]$. The "good" subsets of $[n]$ that include n are exactly the "good" subsets of $[n-2]$ together with n. Thus $f(n) = f(n-1) + f(n-2)$. $\square$

Seeing the recurrence relation, we know that the sequence $f(1), f(2), \ldots$ is the Piṅgala–Fibonacci sequence (see Section 1.2.1), and we can use the recurrence relation to generate as many values of f as we want. In fact, the closed formula quoted above can be derived from this recurrence relation.

(d) We have a sequence $a_0 = 1, a_1, a_2, \ldots$ such that, for all $n \geq 1$,

$$\sum_{k=0}^{n} a_k a_{n-k} = 1.$$

We want to find a_{47}.

For example,

$$a_0 = 1$$
$$a_0 a_1 + a_1 a_0 = 1 \qquad \Rightarrow \qquad a_1 + a_1 = 1 \qquad \Rightarrow \qquad a_1 = 1/2$$
$$a_0 a_2 + a_1 a_1 + a_2 a_0 = 1 \quad \Rightarrow \quad a_2 + \left(\tfrac{1}{2}\right)^2 + a_2 = 1 \quad \Rightarrow \quad a_2 = 3/8.$$

We see that we could continue and, step by step, calculate the terms of the sequence. This is quite tedious. An alternative way to approach this problem is through generating functions. If we want to find a formula for some function $f(n)$ where n is a natural number, then we can form the generating function of $f(n)$:

$$\sum_{n \geq 0} f(n)x^n = f(0) + f(1)x + f(2)x^2 + \cdots + f(n)x^n + \cdots.$$

Here we are concerned with formal power series, and questions of convergence do not come up (at least for elementary applications). Sometimes this power series has a nice closed form and then we can manipulate this function and get information about $f(n)$.

So for our problem, let

$$F(x) = a_0 + a_1 x + a_2 x^2 + \cdots.$$

Now

$$F(x)F(x) = (a_0 + a_1x + a_2x^2 + \cdots)(a_0 + a_1x + a_2x^2 + \cdots)$$
$$= a_0^2 + (a_0a_1 + a_1a_0)x + (a_0a_2 + a_1a_1 + a_2a_0)x^2 + \cdots$$
$$= 1 + x + x^2 + \cdots$$
$$= \frac{1}{1-x}.$$

We have $(F(x))^2 = \frac{1}{1-x}$ and so

$$F(x) = \frac{1}{\sqrt{1-x}}.$$

Now a_n is the coefficient of x^n in the Taylor series expansion of $F(x)$. So, we can use a symbolic algebra software such as SageMath®, Maple®, or Mathematica® to find any desired value of a_n.

As an example, in Maple, we first define the function by $> F := \frac{1}{\sqrt{1-x}}$, and then get the coefficient of x^{47} in the Taylor polynomial expansion of F at $x=0$ by `>coeftayl(F,x=0,47)`. We get

$$a_{47} = \frac{50803160635786570329644235}{6189700196426901374495621112}.$$

It is amazing that calculus can help in solving such a discrete problem. In fact, the generating function $F(x)$ can actually be used to get

$$a_n = \frac{1 \cdot 3 \cdot 5 \cdots (2n-1)}{2^n n!}.$$

However, it is not clear that this formula is any better than the generating function.

As the examples show, we will not only use a myriad of techniques for solving counting problems, but we will also refine our sense of what a good solution should look like. This all will (hopefully) become clear as we get our hands dirty and start solving problems.

1   Induction and Recurrence Relations

On a large square piece of paper, I draw **100** *straight lines that start from one side of the square and end on another side. Each two of the lines intersect but no three (or more) lines go through the same point. Into how many regions do the lines split the piece of paper? If instead of* **100**, *I had drawn only* **3** *lines, the answer would have been* **7** *as illustrated in Figure 1.1.*

Induction is a powerful method of proof and immensely useful in combinatorics. We suspect that most readers already have some experience with induction, and so the first two sections of this chapter should provide a quick review and some additional practice. Recurrence relations are ubiquitous in combinatorics and provide another powerful tool in analyzing counting problems. This will be important through the text, but Section 1.3 gives you experience constructing and using recurrence relations. Often recurrence relations and induction provide a one-two punch. You are interested in a sequence of integers – maybe a sequence that arises from a counting problem – so you first find a recurrence relation for the sequence, then you use it to generate data, and finally you use induction to prove any patterns that you find. If you have prior experience with induction and recurrence relations, then you should try some of the problems and move quickly to the subsequent chapters. However, gaining experience with setting up recurrence relations by doing problems – maybe concurrently as you work on later chapters – is highly recommended. The optional Section 1.4 introduces you to two possible methods for attacking recurrence relations.

1.1   Induction

Warm-Up 1.1. *You have* 100 *briefcases numbered* 1 *through* 100. *You are told that, for any positive integer n, if a briefcase numbered n holds cash, then so does the briefcase numbered* $n + 3$. *You open up briefcase numbered* 55 *and it has a stuffed animal in it. Can you conclude anything about any of the other briefcases? What?*

In philosophy and in the sciences, the inductive method refers to the process of starting with observations and then looking for general laws and theories. Mathematical induction – which we just call induction – is quite different. For us, induction is a method of proof. When we have an infinite sequence $P_1, P_2, \ldots, P_n, \ldots$ of (related) mathematical statements that need a proof, then instead of proving each one of these statements, we may be able to just prove that whenever one of the statements is true, then so is the next statement (in other words, if P_k is true, then so is P_{k+1}). If we manage such a feat, then proving P_1, the first statement,

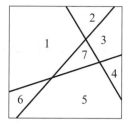

Figure 1.1 Three intersecting lines that do not all go through one point split the plane into seven regions.

starts a domino effect resulting in all of the statements being true. We state this principle and then give an example.

Principle of Mathematical Induction.

Given an infinite sequence of propositions

$$P_1, P_2, P_3, \ldots, P_n, \ldots,$$

in order to prove that all of them are true, it is enough to show two things:

- The base case: P_1 is true.
- The inductive step: For all positive integers k, if P_k is true, then so is P_{k+1}.

Example 1.2. Prove that for any integer $n \geq 1$, we have

$$1 + 2 + 3 + \cdots + n = \frac{n(n+1)}{2}.$$

Proof. Even though the statement that we want to prove is one formula, in actuality, we are trying to prove an infinite number of statements. The first few are:

For $n = 1$ $1 = \frac{1 \cdot 2}{2}$

For $n = 2$ $1 + 2 = \frac{2 \cdot 3}{2}$

For $n = 3$ $1 + 2 + 3 = \frac{3 \cdot 4}{2}.$

These first few statements can be checked directly and are true. This may be assuring and tempt us to think that the pattern must continue, but that is not a proof. It is possible that the pattern breaks down only when n is very large. This is exactly the kind of situation where proof by induction may come in handy (for this particular problem, there are a multitude of other methods of proof that could also be used). To do a proper proof by induction, we need to check the first case of our statement (e.g., when $n = 1$) and then also prove the general statement that, no matter what k is, *if* we know that the statement is true for $n = k$, then it must also be true for $n = k+1$. In other words, we show that if one of these statements is true, then so is the next one. It follows that since the first statement is true, then so is the second statement, and since the second statement is true, then so is the third statement, and so on. In a proof by induction, we have to be clear what exactly is the statement that we are trying to prove and what is the parameter that we are inducting on. In our case, we are to prove

$$P_n : 1 + 2 + \cdots + n = \frac{n(n+1)}{2}$$

for n any positive integer. We proceed to prove this statement by induction on n (this means that n is the parameter that is being incremented). The first statement (i.e., for $n = 1$) is

$$P_1 : 1 = \frac{1(2)}{2},$$

which is clearly correct. Now for the inductive step we can *assume* that P_k is true and we have to prove (using the fact that P_k is known to be true) that P_{k+1} is also true. In other words we assume that

$$P_k : 1 + 2 + \cdots + k = \frac{k(k+1)}{2}$$

is true, and we have to prove that

$$P_{k+1} : 1 + 2 + \cdots + k + (k+1) = \frac{(k+1)(k+2)}{2}$$

is also true.

We now start with $1 + 2 + \cdots + k + (k+1)$ and try to show that it is equal to $\frac{(k+1)(k+2)}{2}$. To begin with, $1 + 2 + \cdots + k + (k+1) = [1 + 2 + \cdots + k] + (k+1)$ and we had declared that we can assume that $1 + 2 + \cdots + k = \frac{k(k+1)}{2}$ (this does feel like cheating, but it is not, since as part of the inductive argument, our aim is not to directly prove P_{k+1} but to prove that P_k implies P_{k+1}). So we have

$$1 + 2 + \cdots + k + (k+1) = [1 + 2 + \cdots + k] + (k+1)$$
$$= \frac{k(k+1)}{2} + (k+1) = \frac{k(k+1) + 2(k+1)}{2}$$
$$= \frac{(k+1)(k+2)}{2},$$

and the proof is complete.   □

Remark 1.3. The sequence of integers

$$1, \ 3 = 1 + 2, \ 6 = 1 + 2 + 3, \ 10 = 1 + 2 + 3 + 4, \ 15 = 1 + 2 + 3 + 4 + 5, \ \ldots$$

are called *triangular numbers* since they count the number of dots in progressively larger triangles (see Figure 1.2). We will continue to run into this sequence of numbers throughout the text. In Example 1.2, we proved, by induction, that the nth triangular number is $n(n+1)/2$. We didn't have to use induction, and, in fact, we could have discovered the formula by first writing the sum $1 + 2 + \cdots + n$ twice, as follows:

$$\begin{array}{ccccccccccc}
1 & + & 2 & + & 3 & + & \cdots & + & (n-1) & + & n \\
n & + & (n-1) & + & (n-2) & + & \cdots & + & 2 & + & 1
\end{array}$$

Now, the sum of the two numbers in each column is $n + 1$ and there are n columns. So the total sum of all the numbers listed is $n(n + 1)$. It follows that $1 + 2 + \cdots + n$ is half of this number and equal to $n(n + 1)/2$.

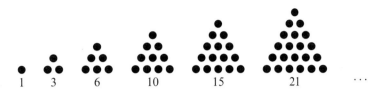

Figure 1.2 The triangular numbers.

Remark 1.4. One drawback of induction is that to use it, you already have to know the pattern and the answer. Mathematical induction – unlike the inductive method in science – does *not* help in finding the pattern. One strength of induction is that it allows you to use P_k to prove P_{k+1}. This is a big advantage – it almost seems like cheating – since often P_k looks very much like P_{k+1}.

Remark 1.5. We write mathematics in paragraphs consisting of English as well as mathematical sentences. If you write $2^{n+1} - 1$, then this is *not* a sentence since it does not have a verb. On the other hand,

$$1 + 2 + 2^2 + \cdots + 2^n = 2^{n+1} - 1$$

is a sentence. The equality symbol $=$ reads as "is equal to" and provides the verb for the sentence. This is not quite a *complete* sentence, since it is not clear what n is. Could n be the number π? Could it be an apple? If we write

$$1 + 2 + 2^2 + \cdots + 2^n = 2^{n+1} - 1, \text{ where } n \text{ is a positive integer,}$$

then we have a complete mathematical statement (or sentence). Such a statement may be true or false. If we believe that it is true, we look for a proof. If we believe it false, then we try to come up with a counter-example.

Problems

P 1.1.1. Problem 79 in the Ahmes–Rhind Papyrus[1] is translated as "An estate's inventory consists of 7 houses, 49 cats, 343 mice, 2401 spelt plants, and 16807 units of heqat. List the items

---

[1] Named after the scribe Ahmes who copied it circa 1650 BCE from an older, now lost, text, or more commonly after Henry Rhind who purchased it in 1858 in Luxor, Egypt. The papyrus contains 87 mathematics problems and is an indispensable source for the study of ancient Egyptian mathematics. The majority of this papyrus is now kept at the British Museum in London (it was object #17 in the Museum's much-recommended radio series and podcast "A History of the World in 100 Objects"), with a few small fragments at the Brooklyn Museum in New York City.

in the estate's inventory as a table, and include their total."[2] For this problem, one of the solutions in the papyrus is merely to find $7 \times 2801 = 19{,}607$. Is this solution correct? Write a general mathematical statement that captures this solution.

P 1.1.2. You notice that

$$1 + 5 = \frac{5^2 - 1}{5 - 1}$$

$$1 + 5 + 5^2 = \frac{5^3 - 1}{5 - 1}$$

$$1 + 5 + 5^2 + 5^3 = \frac{?}{5 - 1}.$$

(a) Is there a continuing pattern? If so, express it as a complete mathematical sentence, and prove it using induction.

(b) Can we replace 5 with an arbitrary real number a? If so, state the more general result. Does your proof for $a = 5$ generalize?

(c) Does your identity work for $a = 1$?

P 1.1.3. Let a be an arbitrary real number other than 1, and n a non-negative integer. Let

$$S = 1 + a + a^2 + \cdots + a^n.$$

(a) What is $aS - S$?

(b) Can you use your result for the previous part to find S?

(c) Is what you did related to Problem P 1.1.2? Do you need to prove your formula for S (using induction, for example) or have you already proved it?

P 1.1.4. Recall that, for n a positive integer, $n! = n(n-1)(n-2) \cdots 3 \times 2 \times 1$. Calculate the following:

$$1(1!) + 1$$

$$1(1!) + 2(2!) + 1$$

$$1(1!) + 2(2!) + 3(3!) + 1,$$

find a pattern, express it in mathematical notation, and prove it.

P 1.1.5. Calculate the following:

$$3 [1 \cdot 2]$$

$$3 [1 \cdot 2 + 2 \cdot 3]$$

$$3 [1 \cdot 2 + 2 \cdot 3 + 3 \cdot 4]$$

---

[2] This problem is related to the St. Ives nursery rhyme (circa 1730). A later version reads:

As I was going to St. Ives,
I met a man with seven wives,
Each wife had seven sacks,
Each sack had seven cats,
Each cat had seven kits:
Kits, cats, sacks, and wives,
How many were there going to St. Ives?

Do you see a pattern? Does it continue? Express your conjecture as a complete mathematical sentence, and prove it.

P 1.1.6. Let n be a positive integer. We want to discover and prove a formula for

$$1^2 + 2^2 + \cdots + n^2.$$

Do so by using the following steps.

STEP 1: Do Problem P 1.1.5, and use the result to find and prove a formula for

$$\sum_{k=1}^{n} k(k+1).$$

STEP 2: Is $\sum_{k=1}^{n} k(k+1) = \sum_{k=1}^{n} k^2 + \sum_{k=1}^{n} k$?

STEP 3: Use Steps 1 and 2, as well as Example 1.2, to prove a formula for $1^2 + 2^2 + \cdots + n^2$.

P 1.1.7. Calculate the following:

$$\frac{1}{1 \cdot 2}$$

$$\frac{1}{1 \cdot 2} + \frac{1}{2 \cdot 3}$$

$$\frac{1}{1 \cdot 2} + \frac{1}{2 \cdot 3} + \frac{1}{3 \cdot 4}.$$

Find a pattern, express it as a proper mathematical sentence, and prove it.

P 1.1.8. Calculate the following:

$$\frac{1}{1 \cdot 3}$$

$$\frac{1}{1 \cdot 3} + \frac{1}{3 \cdot 5}$$

$$\frac{1}{1 \cdot 3} + \frac{1}{3 \cdot 5} + \frac{1}{5 \cdot 7}.$$

Find a pattern, express it as a mathematical sentence, and prove it.

P 1.1.9. Calculate the following:

$$\frac{1}{2 \cdot 5}$$

$$\frac{1}{2 \cdot 5} + \frac{1}{5 \cdot 8}$$

$$\frac{1}{2 \cdot 5} + \frac{1}{5 \cdot 8} + \frac{1}{8 \cdot 11}.$$

Find a pattern, express it as a mathematical sentence, and prove it.

P 1.1.10. In Problem P 1.1.6, you were asked to find a formula for the sum of the squares of integers. What about the alternating sum of squares? To explore this question, calculate the following:

$$1^2$$
$$1^2 - 2^2$$
$$1^2 - 2^2 + 3^2$$
$$1^2 - 2^2 + 3^2 - 4^2.$$

Find a pattern, express it in suitable mathematical notation, and prove it.

P 1.1.11. In Example 1.2 we found the sum of the first n integers. What about the alternating sum of the first n integers? Calculate

$$1$$
$$1 - 2$$
$$1 - 2 + 3$$
$$1 - 2 + 3 - 4$$

and find a pattern. To express the pattern as a mathematical sentence, you may want to use the ceiling function. If r is a real number, then $\lceil r \rceil$ is defined to be the closest integer to r that is also bigger or equal to r (in other words, $\lceil r \rceil$ is the smallest integer greater than or equal to r). After expressing your pattern as a mathematical sentence, prove it using induction.

P 1.1.12. Calculate the following:

$$(1 - 1/2)$$
$$(1 - 1/2)(1 - 1/3)$$
$$(1 - 1/2)(1 - 1/3)(1 - 1/4).$$

Use the data to make a conjecture. Then prove your conjecture.

P 1.1.13. Calculate the following:

$$(1 - 1/4)$$
$$(1 - 1/4)(1 - 1/9)$$
$$(1 - 1/4)(1 - 1/9)(1 - 1/16).$$

Use the data to make a conjecture. Then prove your conjecture.

P 1.1.14. We want to find an identity for

$$1^2 + 3^2 + 5^2 + \cdots + (2n - 1)^2.$$

In a dream, I am told to calculate the following:

$$\frac{1}{4}\left[1 + 3(1^2)\right]$$

$$\frac{1}{4}\left[2 + 3(1^2 + 3^2)\right]$$

$$\frac{1}{4}\left[3 + 3(1^2 + 3^2 + 5^2)\right].$$

Is there a pattern? Even if there is, can you use it to find an identity for the original expression? Can you prove the identity?

P 1.1.15. A skeptical friend offers a proof by induction that all people have the same favorite color. To be precise, she states the result as the following:

For n a positive integer, P_n is the statement "In a set of n people, everyone has the same favorite color."

We want to prove an infinite set of statements $P_1, P_2, \ldots, P_n, \ldots$, and we proceed by using mathematical induction. We induct on n.

The base case P_1 is clearly true since in any set consisting of just one person, all the people in that set—namely that one person—will have the same favorite color.

For the inductive case, we get to assume P_k and prove P_{k+1}. Hence, we know it to be true that if a group of k people are singled out, then they will all have the same favorite color. So consider a group of $k + 1$ people. Line them up in a row, and ask the first k people to step forward. This is a set of k people and so by the inductive hypothesis, they all have the same favorite color. Now ask the last k people in the line to step forward. By the inductive hypothesis, this k people will have the same favorite color as well. But there are some people who are in both sets of k people (the first k and the last k) and so the two sets must have had the same favorite color. We conclude that all $k + 1$ people have the same favorite color. The proof is now complete.

Please comment.

P 1.1.16. Consider the following sequence of numbers: $4, 12, 32, 80, 192, \ldots$, where the formula for the nth number is $(n + 1)2^n$. For $n = 1, 2$, and 3, we find the average of the first n integers in this sequence:

$$\frac{4}{1} = 4$$

$$\frac{4 + 12}{2} = 8$$

$$\frac{4 + 12 + 32}{3} = 16.$$

Could it be that the averages are always a power of 2? Make a precise conjecture and prove it.

P 1.1.17. We calculate

$$5 + 2 = 7 \qquad\qquad 5^2 + 2^2 = 29$$
$$5^3 + 2^3 = 7 \times 19 \qquad\qquad 5^4 + 2^4 = 641$$
$$5^5 + 2^5 = 7 \times 11 \times 41 \qquad\qquad 5^6 + 2^6 = 29 \times 541$$

and notice that every other one is divisible by 7. Does this pattern continue? Prove your assertion.

P 1.1.18. In Problem *P 1.1.17*, we notice that 29 occurred twice as a prime factor. Being curious, we calculate

$$5^{10} + 2^{10} = 9{,}766{,}649 = 29 \times 61 \times 5521.$$

Make a conjecture and prove it.

P 1.1.19. Let n be a positive integer. We want to compare 2^n and n^2. Which one is bigger? We calculate that $2^3 < 3^2$ but $2^6 > 6^2$. Make a precise conjecture and prove it using induction.

P 1.1.20. Let n be a positive integer. We want to compare 3^n and $n!$. Which one is bigger? Make a precise conjecture and prove it using induction.

P 1.1.21. Bernoulli's Inequality. Let n be a positive integer and x a real number greater than -1. Prove that

$$(1 + x)^n \geq 1 + nx.$$

P 1.1.22. Let n be a positive integer, and let $f(x)$ be an infinitely differentiable function, and let $g(x) = xf(x)$. What is the nth derivative of $g(x)$ (with respect to x) in terms of the derivatives of $f(x)$? By calculating the answer for small values of n, find a pattern, and prove it by induction.

P 1.1.23. The function $y = e^x$ is familiar to you from calculus. The domain of this function (the values of x that you are allowed to plug in) can be the set of all real numbers. If you study complex analysis, then you are introduced to a definition of this exponential function even if x is a complex number. You then prove (or actually you could even take this as the definition) that $e^{ix} = \cos(x) + i\sin(x)$, providing a surprising expression for the exponential function as a combination of two periodic functions. You also prove that this new expanded exponential function continues to satisfy the usual rules of exponents and so $\left(e^{ix}\right)^n = e^{inx}$, and as a result

$$(\cos(x) + i\sin(x))^n = \cos(nx) + i\sin(nx).$$

Give a direct proof of this last identity using induction.

P 1.1.24. In a certain government there are n cabinet ministers. Each of these has a senior aide. Exactly k of these n aides leak important information to the press. None of the ministers knows whether their own aide leaks information or not. However, because of interdepartmental rivalries, every minister has many sources in the other ministries, and thus is aware of all the leaks outside of her ministry.

There is an unspoken law in this government that as soon as any minister has irrefutable proof that their aide leaked information, they must fire the aide at their first press conference. Every ministry has one press conference per day at exactly 8:00 a.m.

For a long time no one is fired, because no minister is aware of the leaks, if any, made by their own aide. One afternoon a newspaper runs a story, and announces that news has been leaked by some of the top aides. Exactly k days later, all the guilty aides are fired.

What happened?

1.2 Strong Induction

Warm-Up 1.6. *You again have* 100 *briefcases numbered* 1 *through* 100, *and you are again told that, for any positive integer n, if a briefcase numbered n holds cash, then so does the briefcase numbered n + 3. This time you open up the briefcases numbered* 1, 8, *and* 15, *and they all have cash in them. What can you conclude about the other briefcases? Why?*

In the original formulation of mathematical induction, given a sequence P_1, P_2, ..., P_n, ... of statements, we would attempt to show that P_{k+1} is true given that P_k – the statement immediately preceding it – was already proven true. However, in proving P_{k+1}, we don't have to rely just on the truth of P_k. If need be, we can use *any* of the earlier statements. This more general method of proof is called strong induction. Here is one formulation:

Principle of Strong Induction.

Given an infinite sequence of propositions

$$P_1, P_2, P_3, \ldots, P_n, \ldots,$$

in order to demonstrate that all of them are true, it is enough to know two things:

- The base case: P_1 is true.
- The inductive step: For all integers $k \geq 1$, if $P_1, P_2, P_3, \ldots, P_k$ are true, then so is P_{k+1}.

Example 1.7. In the game of Nim, there are two players and two piles of matches. The players take turns choosing a pile and then taking as many matches as they want from that pile. Each player at each turn must take at least one match. The player who takes the last match wins.

CLAIM: In the case when the two piles have an equal (positive) number of matches, the second player can always win.

Proof. Whenever the first player takes a number of matches from a pile, the second player counters by taking the same number of matches from the other pile. We claim that this is a winning strategy for the second player.

Let n be the number of matches in each pile. We prove our claim by (strong) induction on n.

For the base case, if $n = 1$, then the first player has to take the one match in one of the piles. The second player reciprocates and wins the game.

For the inductive step, assume that the claim is true for all $n \leq k$, and consider the case $n = k+1$. In the first round of the game, the first player takes a non-zero number of matches from one pile, and the second player reciprocates by taking exactly the same number of matches from the other pile. At this point, the two piles have an equal number of matches. But this number is less than $k + 1$. If the number of remaining matches is zero, then the second player has already won. Otherwise, we have a game of Nim with equal piles and with a smaller n. By strong induction, the second player's strategy will win.   □

Note that in Example 1.7 we had to use strong induction since, after the first move by each player, we would not know how many matches were left in each pile. To show that Player 2's strategy is successful with piles of $k + 1$ matches, it would not have been enough to only consider the case with k matches in each pile.

The Base Case for Strong Induction

The base case for an inductive proof is crucial. Without it the domino would not start. In the formulation of strong induction that we presented earlier, you prove one base case P_1 and then prove that P_k is true assuming that every one of $P_1, P_2, \ldots, P_{k-1}$ is true. This certainly works, since the truth of P_1 implies the truth of P_2, and the truth of P_1 and P_2 then implies the truth of P_3, and so on. However, there are other variations on strong induction. For example, in your proof of the inductive step, you may always need three prior cases to prove the new case. In other words, to prove P_k, maybe you use every one of P_{k-3}, P_{k-2}, and P_{k-1}. Such an argument would not apply to P_3 or P_2, and you would need the truth of each of P_1, P_2, and P_3 to conclude that P_4 is true. Thus, in such a case, for the base case of induction, you would have to prove P_1, P_2, and P_3, and not just P_1. Given the many varieties of arguments in inductive proofs, there is no general rule, and you have to look at your inductive argument to see exactly which statements are needed to get it started.

Problems

P 1.2.1. In a game of football, a touchdown (with conversion) is worth 7 points and a field goal is worth 3 points. Let n denote the total number of points in a football game. What are the possible values of n if only touchdowns (with conversion) and field goals are used?

P 1.2.2. I have an infinite supply of 4 and 5 cent stamps. I want to combine my 4 and 5 cent stamps to get exactly n cents worth of stamps. For which values of n can I succeed?

P 1.2.3. A specific statement about the positive integer n is denoted by $P(n)$. We can prove that, whenever $P(k)$ is true, then $P(k+1)$ is also true. It is also known that $P(47)$ is false. Given only this information, what is the strongest conclusion that follows?

1.2.1 The Piṅgala–Fibonacci and Lucas Numbers

Define a sequence of positive integers as follows: $F_0 = 0$, $F_1 = 1$, and for $n = 2, 3, \ldots$ we have

$$F_n = F_{n-2} + F_{n-1}.$$

Thus, for example, $F_2 = F_0 + F_1 = 1$ and $F_3 = F_1 + F_2 = 2$. This sequence is almost universally called the *Fibonacci sequence*, but we choose to call it the *Piṅgala–Fibonacci* sequence.[3] The first few terms of the Piṅgala–Fibonacci sequence starting with F_0 are

$$0, 1, 1, 2, 3, 5, 8, 13, 21, \ldots$$

If we use the same (recursive) rule but change the initial values, we, of course, will get a different sequence. For example, if $L_0 = 2$, $L_1 = 1$, and $L_n = L_{n-2} + L_{n-1}$, then we get the *Lucas numbers*:[4]

$$2, 1, 3, 4, 7, 11, 18, 29, 47, \ldots$$

Problems (continued)

P 1.2.4. We notice that the Lucas numbers are mostly different than the Piṅgala–Fibonacci numbers. Is there any relation? For example, $7 = 2 + 5$ and $11 = 3 + 8$. Can we write every Lucas number in terms of Piṅgala–Fibonacci numbers? Prove your assertion(s).[5]

P 1.2.5. Let $F_0, F_1, \ldots, F_n, \ldots$ denote the Piṅgala–Fibonacci sequence. We notice that

$$
\begin{aligned}
F_1 &= 1 = F_2 \\
F_1 + F_3 &= 3 = F_4 \\
F_1 + F_3 + F_5 &= 8 = F_6.
\end{aligned}
$$

Make a conjecture and prove it.

P 1.2.6. Let F_n be the nth Piṅgala–Fibonacci number. We consider the sequence of Piṅgala–Fibonacci numbers together with their squares:

n	0	1	2	3	4	5	6	7	...
F_n :	0	1	1	2	3	5	8	13	...
F_n^2 :	0	1	1	4	9	25	64	169	...

---

[3] Generations of commentators in India starting with Ācārya Piṅgala—circa third to second century BCE—studied the sequence of integers that we are calling the Piṅgala–Fibonacci numbers. (See Problem P 1.3.5.) According to Parmanand Singh, "Ācārya Piṅgala is the first authority on the metrical sciences in India whose writings exhibit a knowledge of the so-called Fibonacci numbers." (Singh 1985). In Europe, Leonardo of Pisa—he started to be known as Fibonacci only in the nineteenth century—in his thirteenth-century work *Liber Abaci*, included a problem on rabbits that leads to the same sequence. (See Problem P 1.3.6.)

[4] Named after the French mathematician Édouard Lucas (1842–1891). Lucas was an artillery officer in the Franco-Prussian War (1870–1871). After the French defeat, he became a professor of mathematics.

[5] Lucas numbers are taken up again in Problem P 1.4.1.

Consider $F_3^2 = 4$. Of course, $4 = 2^2 = F_3^2$, but 4 is also close to $1 \times 3 = F_2 F_4$. Likewise, $F_4^2 = 9$ is close to $F_3 \times F_5 = 10$. Is there anything to this observation? Can you write a formula for F_n^2 in terms of other Piṅgala–Fibonacci numbers? Prove your assertion.

P 1.2.7. Let $F_0, F_1, F_2, \ldots, F_n, \ldots$ denote the Piṅgala–Fibonacci sequence. Let n be a non-negative integer. For each of the following, conjecture a general formula (in terms of Piṅgala–Fibonacci numbers) and then prove it:

(a) $F_0 + F_1 + \cdots + F_n$

(b) $F_0^2 + F_1^2 + \cdots + F_n^2$.

P 1.2.8. Let $F_0, F_1, F_2, \ldots, F_n, \ldots$ denote the Piṅgala–Fibonacci sequence. For each of the following conjecture a general formula and then prove it:

(a) $F_0 + F_2 + \cdots + F_{2n}$

(b) $F_2 - F_3 + F_4 - \cdots - F_{2n+1}$.

P 1.2.9. For a non-negative integer n, let F_n denote the nth Piṅgala–Fibonacci number. Prove, using strong induction, that, for $n \geq 2$, we have $F_{n+2} < \left(\dfrac{7}{4} \right)^n$.

P 1.2.10. For a non-negative integer n, let F_n denote the nth Piṅgala–Fibonacci number.

(a) Calculate

$$F_0 F_4 - F_2 F_3$$
$$F_1 F_5 - F_3 F_4$$
$$F_2 F_6 - F_4 F_5.$$

Do you see a pattern? Express it as a mathematical statement and prove it.

(b) Two positive integers are said to be *relatively prime* if their greatest common divisor is 1. Prove that two consecutive Piṅgala–Fibonacci numbers are relatively prime.

P 1.2.11. Start with 0 and 1 and take their average. You get $1/2$. Take the average of 1 and $1/2$, and get $3/4$. Now take the average of $1/2$ and $3/4$ to get $5/8$. In this manner create a sequence that starts with 0 and 1 and every term (after the second term) is the average of the two numbers immediately preceding it. Thus the sequence begins as

$$0, 1, \frac{1}{2}, \frac{3}{4}, \frac{5}{8}, \frac{11}{16}, \frac{21}{32}, \ldots$$

(a) Formulate a conjecture about the nth term of the sequence, and then prove it.

(b) Does this sequence converge? If it does, what is the limit of the sequence?

P 1.2.12. We have an 8×8 chess board with 64 squares. We want to tile this board with pieces that are made of three squares as shown:

The three squares on the above piece are identical to the squares on the board, and we want each piece to cover exactly three squares of the board. Because 3 does not divide

64, we cannot completely tile the board. So we cut one square out of the board, leaving 63 squares. Can we now tile the board with 21 pieces? Does the answer depend on which square we cut out? Make as precise a conjecture as possible.

Can you generalize your conjecture and prove it?

1.3 Recurrence Relations

Warm-Up 1.8. *We have n dollars. Every day we buy exactly one of the following products: Mustard (1 dollar), Mint (2 dollars), Marjoram (2 dollars). Let $f(n)$ be the number of possible ways of spending all the money. For example, $f(3) = 5$, since the possible ways of spending 3 dollars are: Mustard-Mustard-Mustard, Mustard-Mint, Mustard-Marjoram, Mint-Mustard, and Marjoram-Mustard. What is $f(1)$, and $f(2)$? Which one(s) (if any) of the following are true?*

(a) $f(n) = 2f(n-1) + f(n-3)$

(b) $f(n) = f(n-1) + \frac{(n-1)}{2}[3 + (-1)^n]$

(c) $f(n) = f(n-1) + 2f(n-2)$

(d) $f(n) = 2f(n-1) - f(n-2)$.

Give adequate and complete reasoning for your answer.

In the warm-up problem, you are asked to find a formula for $f(n)$ but not the usual kind of formula of $f(n)$ in terms of n. What you are asked to do is to find a formula that gives $f(n)$ in terms of $f(n-1)$, $f(n-2)$, and possibly other values of the f function. Such a relation is called a *recurrence relation* and, while it does not give a direct closed formula for $f(n)$, it provides an efficient way for computing specific values of the function f. We illustrate the construction and use of recurrence relations by restating and solving the opening problem of the chapter.

Problem 1.9. *On a large square piece of paper, I draw 100 straight lines that start from one side of the square and end on another side. Each two of the lines intersect but no three (or more) lines go through the same point. Into how many regions do the lines split the piece of paper? If instead of 100, I had drawn only 3 lines, the answer would have been 7, as illustrated in Figure 1.1.*

Solution. This problem is an instance of a sequence of problems. Let n be a positive integer, and let $f(n)$ denote the number of regions created by n straight lines assuming that each pair of lines intersects and no three lines go through the same point. The problem is asking for $f(100)$, while Figure 1.1 shows that $f(3) = 7$. It is easy enough to see that $f(1) = 2$ and $f(2) = 4$. It is already a bit tedious to find $f(4)$. In fact, it is not even clear that the problem is well posed. In other words, how do we know that the number of regions does not depend on the configuration of the lines? Maybe if we draw the lines in a different relation to each other, the number of regions will change. For $n = 1$ and 2, it is clear that the number of regions is always 2 and 4, but for larger n, we need some kind of an argument.

Instead of trying to find a formula for $f(n)$ or even trying to prove that $f(n)$ is independent of the position of the lines, we try to determine $f(n)$ in terms of $f(n-1)$. Not only will this allow us to inductively start with $f(1)$ and $f(2)$ and find other values of $f(n)$, but it could also show that $f(n)$ is well defined. We do a thought experiment. Assume that $n-1$ straight lines split the plane into $f(n-1)$ regions. How many additional regions are created when we add one more line?

If we just zoom onto this new line and follow its path, it will start from one side of the square, one by one cross the $n-1$ other lines, and end at another side of the square. Hence, it will go through n regions created by the original $n-1$ lines. (The first region is the one our line is traversing before it hits the first line, the second region is between the first and the second, and so on until the $(n-1)$th region – that is, between the $(n-2)$th and $(n-1)$th lines. Finally, the nth region is after the $(n-1)$th line.) Our new line will split each of these n regions into two, and as a result will add n regions to what was there before. We conclude that, for $n > 1$,

$$f(n) = f(n-1) + n.$$

This allows us to calculate further values of $f(n)$. For example, $f(4) = f(3) + 4 = 11$ and $f(5) = f(4) + 5 = 16$. The relationship also proves that $f(n)$ does not depend on the configuration of the lines, and that it is well defined. (After all, no matter how the lines are configured, if $f(n-1)$ is well-defined, then $f(n)$ must be $f(n-1)+n$. Since $f(1)$ is well defined, by induction all values of $f(n)$ are well defined.) Note that we could have even started with the case $n = 0$. Clearly, $f(0) = 1$, and our formula now gives $f(1) = f(0) + 1 = 2$. We list the first few terms of this sequence:

n	1	2	3	4	5	6	7	8	9	10
$f(n)$	2	4	7	11	16	22	29	37	46	56

At this point, we could, either patiently by hand or more efficiently by a simple computer program, find $f(100)$. Sometimes we can do better. We may notice that these numbers are just one more than the triangular numbers (see Remark 1.3).[6] Hence, we claim:

CLAIM: $f(n) = 1 + \frac{n(n+1)}{2}$.

But how do we know this claim to be true? We prove it by induction!

We have already checked the base case, so assume that $f(k) = 1 + \frac{k(k+1)}{2}$ is known to be true, and use it to prove that $f(k+1) = 1 + \frac{(k+1)(k+2)}{2}$. We argue as follows (note that we have already proved that, for $n > 1, f(n+1) = f(n) + n$):

$$f(k+1) = f(k) + (k+1) = 1 + \frac{k(k+1)}{2} + (k+1)$$

$$= 1 + \frac{k(k+1) + 2(k+1)}{2} = 1 + \frac{(k+1)(k+2)}{2},$$

---

[6] Problem P 1.4.11 suggests techniques, other than guessing the pattern, for completing this problem.

and the proof is complete. We can finally answer the original question:

$$f(100) = 1 + \frac{100 \times 101}{2} = 5{,}051.$$

$\square$

Remark 1.10. Our solution to Problem 1.9 illustrates a common approach to some combinatorial problems.

Often a given combinatorial problem is an instance of a sequence of problems indexed by a positive integer n, where n is a natural parameter for the problem. If we let $f(n)$ denote the answer to the nth instance of our problem, ideally we may want a closed-form formula for $f(n)$. Sometimes, instead, we can find a "recurrence relation" for f. This means finding a formula for $f(n)$ in terms of $f(n-1), f(n-2), \ldots$ (For example, in the previous problem, the recurrence relation was $f(n) = f(n-1) + n$. Another example of a recurrence relation is $f(n) = f(n-1) + f(n-2)$, the defining relation for the sequence of Piṅgala–Fibonacci numbers; see Section 1.2.1.) Usually, the recurrence relation itself is *not* proved by induction. Rather, we often use a "thought experiment." What are the possibilities for the "first" or "last" step of the problem? In Problem 1.9, we focused on the *last* line drawn. Sometimes, for example in Warm-Up 1.8, it is useful to concentrate on the first step.

After we have a recurrence relation and a few base cases, we can easily generate data and record many values for $f(n)$. At this point, there are a number of possible approaches:

- Use a simple computer program to find any particular value of $f(n)$ that you need.
- Look at the values of $f(n)$ for small n and try to guess the pattern. If your guess is correct, sometimes you can translate it to a closed formula, and often you can prove your conjecture using induction. Just as in the solution to Problem 1.9, you can use the recurrence relation in your proof by induction. In fact, recurrence relations are especially suited to help with proofs by induction.
- Some classes of recurrence relations (e.g., so-called linear recurrence relations) can be solved systematically (see Section 1.4).
- Sometimes you can "unwind" the recurrence relation (see Section 1.4).
- Sometimes you can use "generating functions" to get information about the sequence $f(n)$ (see Section 9.2.2).

Guessing the pattern from the starting values of an infinite sequence is not easy. For one thing, there are many distinct infinite sequences that agree in the first few terms – after all, there are not that many very small integers, and, in contrast, there are many infinite sequences of integers. If you have the beginning of a sequence of numbers, and are wondering what the pattern could be, one very fun tool is the *Online Encyclopedia of Integer Sequences* at `https://oeis.org/`. Try it for the sequence in Problem 1.9.

A recurrence, for the integer valued function f, of the form $f(n) = \alpha_1 f(n-1) + \alpha_2 f(n-2) + \cdots + \alpha_k f(n-k) + g(n)$, where $\alpha_1, \ldots, \alpha_k$ are scalars and g is a function of n, is called a *linear recurrence relation*. If you are lucky and the recurrence relation is linear, then the most fruitful method – this is expanded in Section 1.4 – is to first ignore the initial conditions and find, basically by an educated guess, a set of functions that satisfy the recurrence relation, and then use the initial conditions to choose a function that satisfies both the initial conditions and the recurrence relation. The key observation – which will be used often – is that if two functions have the same initial values and satisfy the same recurrence conditions, then they will have to be the same for all later inputs. We will pursue this in Section 1.4, where we explore both linear recurrence relations and the idea of unwinding a recurrence relation.

Remark 1.11. The problems in this section are meant to give you experience in setting up a recurrence relation for a counting problem. This skill will be used throughout the book. In some problems, you are also asked to use the recurrence relation, generate some data, guess the pattern, and prove it by induction. Some of these problems will be revisited in Section 1.4. The two techniques of Section 1.4 – solving linear recurrences and unwinding – when applicable give a better alternative to "guess-the-pattern" of the current section. Yet a fourth method – using generating functions – will be discussed in Chapter 9.

Problems

P 1.3.1. We are given the following recurrence relation:

$$a_1 = 2, \text{ and } a_n = 3a_{n-1} + 2 \text{ for } n = 2, 3, \ldots$$

Find the values of a_n for small n. Do you see a pattern? Make a conjecture. Prove your conjecture using induction.[7]

P 1.3.2. Let n be a non-negative integer. We know that $f(0) = -3$, and for $n = 1, 2, \ldots$ we have

$$f(n) = 3f(n-1) + 10.$$

Generate some data, and use the data to conjecture a closed formula for $f(n)$. Prove your conjecture using induction.

P 1.3.3. Let h_n denote the number of ways of covering a $2 \times n$ array with 1×2 dominoes (see Figure 1.3). Find h_1, h_2, h_3, and a recurrence relation for h_n. Use these to find h_8.

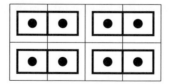 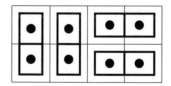

Figure 1.3 Two possible ways—there are others—of covering a 2×4 array with 1×2 dominoes.

P 1.3.4. Let n be a positive integer. Consider a $1 \times n$ strip of cardboard. We have (a large number of) three types of pieces: 1×2 red pieces, 1×4 yellow pieces, and 1×4 blue pieces. Let $f(n)$ be the number of ways that we can tile the strip of cardboard with our pieces. For example,

---

[7] Problem P 1.4.10 revisits this problem.

$f(3) = 0$ since you cannot tile a 1×3 board with the available pieces, while $f(6) = 5$ since the possibilities are YR, RY, RB, BR, RRR. Find $f(14)$. (Give an actual number.)

P 1.3.5. Generations of Indian commentators, apparently beginning with Piṅgala in the third to second century BCE, were interested in the number of rhythmic patterns having a total of n beats. Assuming that we use only one-beat and two-beat rhythms, how many different patterns can we make that have a total of n beats? For example, if $n = 4$, we can have $2 - 2, 2 - 1 - 1, 1 - 2 - 1, 1 - 1 - 2$, and $1 - 1 - 1 - 1$ for a total of 5 possibilities. If we denote the answer by $r(n)$, then find a recurrence relation for $r(n)$. Is $r(n)$ related to the Piṅgala–Fibonacci numbers?

P 1.3.6. In his 1202 book *Liber Abaci*, Leonardo of Pisa (known as Fibonacci since the nineteenth century) poses the following problem:

> A certain man put a pair of rabbits in a place surrounded on all sides by a wall. How many pairs of rabbits can be produced from that pair in a year if it is supposed that every month each pair begets a new pair which from the second month on becomes productive?

Give a solution. Is this related to the Piṅgala–Fibonacci numbers? How?

P 1.3.7. Assume that, for $n \geq 3$, the sequence $a_1, \ldots, a_n, \ldots$ satisfies the recurrence relation $a_n = a_{n-1} + a_{n-2}$. Assume further that we know that $a_2 = 3$ and $a_{50} = 300$. Find $a_1 + a_2 + \cdots + a_{48}$ and justify your answer.

P 1.3.8. You work at a car dealership that sells three models: A pickup truck, an SUV, and a compact hybrid. Your job is to park the vehicles in a row. The pickup trucks and the SUVs take up two spaces while the hybrid takes up one space. Let n be a positive integer and let $f(n)$ be the number of ways of arranging vehicles in exactly n spaces. Find a recurrence relation for $f(n)$ and use it to find $f(10)$, the number of ways of arranging vehicles if you have 10 parking spaces.

P 1.3.9. Continuing with the assumptions and the notation of Problem P 1.3.8.
   (a) Find $f(1), \ldots, f(5)$.
   (b) Find and prove a one-step recurrence relation for $f(n)$ (i.e., one that only depends on $f(n-1)$).
   (c) Find and prove a formula for $f(n)$ of the form
   $$f(n) = \frac{1}{?}(2^{n+1} + ?).^8$$

P 1.3.10. At a dinner party on the spaceship Enterprise, three life forms are present: Humans, Klingons, and Romulans. The dinner table is a long $1 \times n$ board, and the life forms sit on one side of it, next to one another. From each life form there are more than n individuals present, and so only a total of n sit at the table. The only problem is that no two humans want to sit next to each other. Let h_n denote the number of different ways that n individuals can be seated at the dinner table. Assume that all humans look alike, as do all Klingons and all Romulans.

---

[8] Problems P 1.4.2 and P 1.4.14 revisit this problem.

(a) What is h_1? What is h_2?

(b) Which one(s) (if any) of the following are true, and which are false?

    (i) $h_n = 3h_{n-1} - h_{n-2}$

    (ii) $h_n = 2h_{n-1} + 2h_{n-2}$

    (iii) $h_n = 3h_{n-1} - (n-1)!$

    (iv) $h_n = h_{n-1} + 3h_{n-2} + 2h_{n-3}$.

Give adequate and complete reasoning for your answers.[9]

P 1.3.11. Using only the digits 1, 2, and 3, how many integers can you construct in a way that the sum of the digits is 9? Examples would be 333, 1323, 2133, and 22221.

P 1.3.12. I have 12 identical irises and 4 distinct flowerpots. All the irises are to be planted, and I want to plant 2, 3, or 4 irises in each pot. I am interested in finding out the number of ways that this can be done. We first generalize the question. Given n identical irises and k distinct pots, let $F(n, k)$ be the number of ways that we can distribute the irises among the pots if each pot gets 2, 3, or 4 irises.

(a) Find $F(5, 2)$.

(b) Find a recurrence relation for $F(n, k)$. Explain your reasoning.

(c) Find $F(n, k)$ for the given values of n and k in Table 1.1.

(d) What is $F(12, 4)$? Why?

Table 1.1 The values of *F(n,k)* in Problem P 1.3.12.

$n \backslash k$	1	2	3
1			
2			
3			
4			
5			
6			
7			
8			
9			
10			

P 1.3.13. Two thick panes of glass are adjacent to each other. Light that enters from one side can be reflected by the internal faces. In other words, after entering the medium, the light may go straight through or be reflected back and forth by the three internal faces, as shown in Figure 1.4. Let a_n be the number of different ways that the light can be reflected n times.

---

[9] Problem *P 1.4.3* revisits this problem.

Figure 1.4 shows that $a_0 = 1$, $a_1 = 2$, and $a_2 = 3$. Find a recurrence relation for a_n. Use the recurrence relation to find a_7.[10]

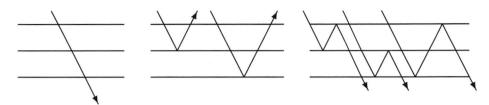

Figure 1.4 The possibilities, respectively, with 0, 1, or 2 reflections for Problem P 1.3.13.

P 1.3.14. Let n be a positive integer, and let $h(n)$ be the number of sequences

$$a_1, a_2, \cdots, a_n,$$

with the conditions that for $i = 1, \ldots, n$, each a_i is 0, 4, or 7, and each 4 in the sequence is *immediately* followed by a 7.
(a) Find $h(1)$, and $h(2)$.
(b) Find a recurrence relation for $h(n)$. Use it to find $h(5)$.
(c) I entered the first few terms of the sequence into the *Online Encyclopedia of Integer Sequences*, and it told me that a formula for $h(n)$ could be

$$h(n) = \frac{(1 + \sqrt{2})^{n+1} - (1 - \sqrt{2})^{n+1}}{2\sqrt{2}}.$$

Is this true? Prove your assertion.[11]

P 1.3.15. Let $h(0) = 1$ and let $h(n)$ be the number of sequences

$$a_1, a_2, \cdots, a_n,$$

with the conditions that, for $i = 1, \ldots n$, each a_i is 0, 4, or 7, and if both 4 and 7 occur in the sequence then the first 4 occurs before the first 7.
(a) Find $h(1)$ and $h(2)$.
(b) Find a recurrence relation for $h(n)$.
(c) What is $h(5)$?[12]

P 1.3.16. You are interested in finding the number of sequences

$$a_1, a_2, \ldots, a_n,$$

such that each a_i is 0, 1, 2, 3, or 4 *and* the number of 3's in the sequence is even. Denote the number of such sequences by $E(n)$.
(a) Find $E(1)$ and $E(2)$.

---

[10] Adapted from Moser and Wyman 1963.
[11] Problem P 1.4.4 revisits this problem.
[12] This problem is continued in Problem P 1.4.16.

(b) Let $O(n)$ be the number of sequences $a_1, a_2, \ldots, a_n$ such that each a_i is from the set $\{0, 1, 2, 3, 4\}$ *and* the number of 3's in the sequence is odd. How is $O(n)$ related to $E(n)$?

(c) Find a recurrence relation for $E(n)$ and state clearly your reasoning for why it works.

(d) Find $E(4)$.[13]

P 1.3.17. We want to divide $275,000 among four people. We have to divide all the money, and we can only divide the money in $25,000 increments. In how many ways can this be done? Make a conjecture.

To make the question clear, let us consider a simpler example. If we had $100,000 to divide among four people, then we could give all of it to one person, give $50,000 to two people, give $75,000 to one person and $25,000 to another, give $50,000 to one and $25,000 to two others, or finally give $25,000 to each of the four people. Thus we can divide $100,000 among four people in five different ways. Note that we consider the four people indistinguishable. If it mattered to us who got how much, then the number of ways of dividing the money would have been considerably higher.[14]

P 1.3.18. **The Gamma Function.** Let x be a positive real number. Define the function $\Gamma(x)$ by

$$\Gamma(x) = \int_0^\infty t^{x-1} e^{-t} \, dt.$$

The integral defining $\Gamma(x)$ is an improper integral—since one of the end points is ∞—and yet it converges for all $x > 0$. (This is not hard to prove using standard calculus techniques, and you are welcome to try, but for the purposes of this problem, you can assume that the integral converges for all $x > 0$. In fact, interesting things happen when one extends the definition to complex numbers.)

(a) What is $\Gamma(1)$?

(b) Use integration by parts, and the fact that, for a fixed x, $\lim_{t \to \infty} \frac{t^x}{e^t} = 0$, to prove the following recurrence relation for $\Gamma(x)$:

$$\Gamma(x + 1) = x\Gamma(x).$$

(c) Use the recurrence relation to find $\Gamma(2), \ldots, \Gamma(5)$. If n is a positive integer, what is $\Gamma(n)$? Prove your assertion using induction.

(d) The Encyclopedia Britannica entry for the Gamma function [Encyclopedia Britannica 2019] begins: "Gamma function, generalization of the factorial function to nonintegral values introduced by the Swiss mathematician Leonhard Euler in the 18th century." Explain the relation of the Gamma function to the factorial function.

P 1.3.19. Define a *selfish set* to be a set which has its own cardinality (number of elements) as an element.[15] A set is a *minimal selfish set* if it is selfish, and no proper subset of it is selfish. So, for example, $\{2, 47\}$ is a minimal selfish set, $\{2, 3, 5\}$ is selfish but not minimal, while $\{4, 5, 6\}$ is not selfish at all. Let n be a positive integer, and let f_n be the number of subsets

---

[13] Problem P 1.4.15 revisits this problem.
[14] Problem P 7.1.2 revisits this problem.
[15] Adapted from Problem B-1 of the 1996 Putnam Mathematical Competition.

of $[n]$ which are minimal selfish sets. Find a recurrence relation for f_n. The following steps may be helpful:

STEP 1: Find f_1, f_2, and f_3.

STEP 2: Which minimal selfish sets contain 1?

STEP 3: Is the size of a minimal selfish subset of $[n]$ ever n?

STEP 4: What can you say about the number of minimal selfish subsets of $[n]$ that do not include n?

STEP 5: Assume a minimal selfish subset of $[n]$ includes n. If you subtract 1 from each of the elements of the subset, and then throw out $n - 1$, do you get a minimal selfish set?

STEP 6: Write down, with proof, a recurrence relation for f_n.

1.4   Linear Recurrence Relations; Unwinding a Recurrence Relation*

In this optional section, we consider two methods for attacking recurrence relations. When they work – and there are many common recurrence relations that can be solved using these methods – they work well and give a closed formula for the underlying function. However, there are also plenty of recurrence relations that defy both of these methods.

1.4.1   Linear Recurrence Relations

Warm-Up 1.12. *Let $f(n) = \alpha n + \beta$, where α and β are two mystery real numbers. Find α and β such that the resulting function f satisfies the recurrence relation $f(n) = 5f(n-1) - 6f(n-2) + n$.*

Recurrence relations can be thought of as the discrete analogs of differential equations, and as such they are sometimes called *difference equations*. The idea is that for a function defined on the integers, you cannot find the usual derivative defined as $\lim_{h \to 0} \frac{f(n+h) - f(n)}{h}$ since the domain of the function is only the integers. Hence, the best you can do is to let $h = 1$ and find the difference $f(n+1) - f(n)$ or equivalently $f(n) - f(n-1)$. If you have a formula for this latter difference, then you really have a recurrence relation for $f(n)$ in terms of $f(n-1)$. From this point of view, a recurrence relation should behave like a differential equation, and recurrence relations that involve $f(n-2)$ in addition to $f(n-1)$ are "similar" to second-order differential equations. Solving differential equations in general is hard, but solving *linear differential equations with constant coefficients* is not. If you have already studied differential equations, then you will immediately recognize the parallels with our brief presentation here. We illustrate the ideas with an example.

Example 1.13. Define a function f on non-negative integers as follows: $f(0) = 0, f(1) = 1$, and for $n \geq 2, f(n) = 3f(n-1) + 28f(n-2)$. So, $f(0) = 0, f(1) = 1, f(2) = 3, f(3) = 37$, $f(4) = 195, f(5) = 1621, f(6) = 10323, \ldots$ We want to find a closed formula for f, and guessing a pattern for the generated data does not seem promising. Here, we first ignore the initial conditions (and the data that we generated) and just look for a function of the form a^n, for some non-zero real (or complex) number a, that satisfies the recurrence relation. The function $f(n) = a^n$ would satisfy the recurrence relation if and only if

$$a^n = 3a^{n-1} + 28a^{n-2} \quad \Leftrightarrow \quad a^2 = 3a + 28 \quad \Leftrightarrow \quad a = -4 \text{ or } 7.$$

Thus both $g(n) = (-4)^n$ and $h(n) = 7^n$ satisfy the recurrence relation. However, neither one satisfies the initial conditions. Here is where "linearity" comes in. If $g(n)$ and $h(n)$ satisfy the linear recurrence relation of this example, then we claim that any function of the form $f(n) = \alpha g(n) + \beta h(n)$ also satisfies the recurrence relation.[16] We show the claim as follows.

$$
\begin{aligned}
f(n) = \alpha g(n) + \beta h(n) &= \alpha \left(3g(n-1) + 28g(n-2)\right) + \beta \left(3h(n-1) + 28h(n-2)\right) \\
&= 3\left(\alpha g(n-1) + \beta h(n-1)\right) + 28\left(\alpha g(n-2) + \beta h(n-2)\right) \\
&= 3f(n-1) + 28f(n-2).
\end{aligned}
$$

This means that every function of the form $f(n) = \alpha(-4)^n + \beta(7)^n$ satisfies the recurrence relation. Now, we use the initial conditions to find α and β. From $f(0) = 0$ and $f(1) = 1$, we get

$$
\begin{cases} \alpha + \beta & = 0 \\ -4\alpha + 7\beta & = 1 \end{cases}
\quad \Rightarrow \quad
\begin{cases} \alpha & = -\frac{1}{11} \\ \beta & = \frac{1}{11}. \end{cases}
$$

Hence, $f(n) = \frac{1}{11}\left((-1)^{n+1}4^n + 7^n\right)$.

Remark 1.14. In Example 1.13, we found one function that satisfied the recurrence relation and the initial conditions of the original function f. Could there be others? Assume that f and g are two functions that are defined on the non-negative integers. If f and g satisfy the same initial conditions and if the later values of f and g can be generated from the same recurrence relation, then, by necessity, $f(n) = g(n)$ for all non-negative integers n. In other words, $f = g$, and we have a unique solution. In the case of differential equations, the situation can be more complicated. You can prove that certain initial-value problems (differential equations together with initial conditions) have a unique solution while others do not. Such existence and uniqueness theorems play an important role in the theory of differential equations.

We now try to distill Example 1.13 to see when its method works.

Definition 1.15. Let $f(0), f(1), \ldots, f(n), \ldots$ be a sequence of real (or complex) numbers. Assume that

$$f(n) = \alpha_1 f(n-1) + \alpha_2 f(n-2) + \cdots + \alpha_k f(n-k), \tag{1.1}$$

for some integer k and some real numbers $\alpha_1, \alpha_2, \ldots, \alpha_k$. We then say that f satisfies a *homogeneous linear recurrence relation* (with constant coefficients). If

$$f(n) = \alpha_1 f(n-1) + \alpha_2 f(n-2) + \cdots + \alpha_k f(n-k) + g(n), \tag{1.2}$$

---

[16] In the parlance of linear algebra, linear combinations of solutions are also solutions. This means that the set of functions that are solutions to $f(n) = 3f(n-1) + 28f(n-2)$ form a vector space.

for some integer k, some real numbers $\alpha_1, \alpha_2, \ldots, \alpha_k$, and some non-zero function g, then we say that f satisfies a (non-homogeneous) *linear recurrence relation* (with constant coefficients).

Example 1.16. A sequence $\{a_n\}_{n=0}^{\infty}$ with the recurrence $a_n = a_{n-1} + a_{n-2}$ satisfies a homogeneous linear recurrence, while a sequence $\{b_n\}_{n=0}^{\infty}$ with the recurrence relation $b_n = b_{n-1} + b_{n-2} + n^2$ satisfies a non-homogeneous linear recurrence.

We first formalize the "linearity" property of *linear* recurrence relations already observed in Example 1.13.

Lemma 1.17. (a) *Assume that the functions $f_1, f_2, \ldots, f_m$ satisfy the homogeneous linear recurrence of Equation 1.1. Then, for scalars $\beta_1, \ldots, \beta_m$, every linear combination*

$$\beta_1 f_1 + \beta_2 f_2 + \cdots + \beta_m f_m$$

also satisfies the homogeneous linear recurrence of Equation 1.1.

(b) *If functions h_1 and h_2 satisfy the non-homogeneous linear recurrence Equation 1.2, then their difference $h_1 - h_2$ satisfies the homogeneous linear recurrence of Equation 1.1.*

Proof Sketch. We use the recurrence relations that are satisfied by $f_1, \ldots, f_m$ as well as h_1 and h_2 to show that both $\beta_1 f_1 + \beta_2 f_2 + \cdots + \beta_m f_m$ and $h_1 - h_2$ also satisfy Equation 1.1. Writing this down involves keeping track of the notation, and is very similar to what we did in Example 1.13. You are asked to complete the proof in Problem P 1.4.22. $\square$

Given the homogeneous linear recurrence of Equation 1.1 – $f(n) = \alpha_1 f(n-1) + \alpha_2 f(n-2) + \cdots + \alpha_k f(n-k)$ – we can (as we did in Example 1.13) see if a function $f(n) = a^n$, for some yet-to-be-discovered $a \neq 0$, satisfies the recurrence relation. Plugging this into the recurrence relation, we get $a^n = \alpha_1 a^{n-1} + \alpha_2 a^{n-2} + \cdots + \alpha_k a^{n-k}$. Dividing through by a^{n-k}, we have

$$a^k - \alpha_1 a^{k-1} - \alpha_2 a^{k-2} - \cdots - \alpha_k = 0.$$

Hence, any root $x = a$ of the polynomial $x^k - \alpha_1 x^{k-1} - \cdots - \alpha_k$ – called the *characteristic polynomial* of the recurrence relation – will give a function a^n that satisfies the original recurrence relation. If the characteristic polynomial is of degree k, then it can have up to k distinct roots (it may have fewer as it may have roots with higher multiplicity). Each one of these roots gives a different solution to the recurrence relation, and, by Lemma 1.17(a), every linear combination of the different solutions is also a solution. Hence, if $a_1, a_2, \ldots, a_m$ are the distinct roots of the characteristic polynomial, then the function

$$f(n) = \alpha_1 a_1^n + \alpha_2 a_2^n + \cdots + \alpha_m a_m^n, \tag{1.3}$$

where $\alpha_1, \ldots, \alpha_m$ are scalars, satisfies the recurrence relation. Now, to pinpoint the scalars $\alpha_1, \ldots, \alpha_m$, we need to have initial conditions. If $f(n)$ depends on $f(n-1)$ through $f(n-k)$ – as we have been assuming – then we need to know k consecutive values of f to be able to generate all the subsequent values of f. Hence, maybe $f(0), \ldots, f(k-1)$ are given. We plug these into Equation 1.3, and get k equations and m unknowns.

In the case that $k = m$ – i.e., if all the roots of the characteristic polynomial are distinct and there are no multiple roots – one can prove that this system of k equations and k unknowns will always have a unique solution, thereby giving a unique solution to the recurrence relation with initial conditions. The proof – see Problem P 1.4.24 – amounts to showing that the matrix of coefficients of the system of linear equations is invertible. This is done by showing that the determinant of this matrix – which is known as the Vandermonde matrix – is non-zero (see Problem P 1.4.23). In the case when $k > m$ – i.e., when some of the roots of the characteristic polynomial are repeated – it is possible that the system of linear equations (that we used to find the scalars) has no solution. At that point, you need to find solutions other than those of the form a^n for the recurrence relation. It turns out – akin to what happens to the corresponding cases for differential equations – that if a is a root of the characteristic equation with multiplicity ℓ (that is, if the characteristic equation has a factor of $(x-a)^\ell$), then each of

$$a^n, \, na^n, \, n^2 a^n, \, \ldots, \, n^{\ell-1} a^n$$

is a solution to the recurrence relation. We will not give a proof of this fact (Problem P 1.4.25 walks you through the proof of a special case, but that proof could be generalized to cover the more general case), but, in the case of specific recurrence relations, even without a proof, you can use this as a guide for trying functions that may satisfy the relation. By Lemma 1.17(a), again every linear combination of solutions continues to be a solution as well. You can then use the initial conditions to find the scalars in the linear combination. (See Problems P 1.4.6– P 1.4.8.)

Example 1.18. Assume $F_n = F_{n-1} + F_{n-2}$, $F_0 = 0$, and $F_1 = 1$. Then the characteristic polynomial of this homogeneous linear recurrence is $x^2 - x - 1$ and the roots of this polynomial are $\frac{1\pm\sqrt{5}}{2}$. Hence, if $\mu = \frac{1+\sqrt{5}}{2}$ and $\nu = \frac{1-\sqrt{5}}{2}$ (in fact, $\mu\nu = -1$ and so $\nu = -1/\mu$), then $F_n = \alpha\mu^n + \beta\nu^n$ satisfies the recurrence relation. We now plug in the initial conditions and get

$$\begin{cases} \alpha + \beta & = 0 \\ \alpha\mu + \beta\nu & = 1 \end{cases} \Rightarrow \begin{cases} \alpha & = \frac{1}{\mu-\nu} \\ \beta & = \frac{-1}{\mu-\nu}. \end{cases}$$

and so (surprisingly) a closed formula that gives the Piṅgala–Fibonacci numbers is

$$F_n = \frac{\mu^n - \nu^n}{\mu - \nu} = \frac{1}{\sqrt{5}}\left(\frac{1+\sqrt{5}}{2}\right)^n - \frac{1}{\sqrt{5}}\left(\frac{1-\sqrt{5}}{2}\right)^n.$$

Turning to the non-homogeneous linear recurrence relations, we again initially ignore the initial conditions, and look for functions that satisfy the recurrence relation. Lemma 1.17(b) will be very helpful. We, by hook or by crook, first find one particular solution to the non-homogeneous linear recurrence relation. This one solution will most likely not be the one we are looking for, since it may not satisfy the initial conditions. Lemma 1.17(a) says that the difference between the solution we found and *any* other solution of the recurrence relation is a solution to the corresponding homogeneous linear recurrence. So, if we add all possible solutions of the homogeneous recurrence to the one solution of the non-homogeneous recurrence, then we get all possible solutions of the non-homogeneous linear recurrence. In summary, to solve a non-homogeneous linear recurrence relation, we find a *general* solution to the corresponding homogeneous linear recurrence relation as well as one *particular* solution to the non-homogeneous recurrence, and add them. We then use the initial conditions to determine all the scalars. The difficulty in this scheme is finding that one particular solution to the non-homogeneous relation. This is done by judiciously guessing functions that may work and checking if they actually do. We pay particular attention to the function $g(n)$ in Equation 1.2, and our guess for the particular solution is often based on the form that $g(n)$ takes. In particular, if $g(n)$ is a polynomial in n of degree ℓ, then we try a polynomial of degree ℓ as our choice for $f(n)$. In other words, we write $f(n) = \beta_\ell n^\ell + \cdots + \beta_1 n + \beta_0$ with yet-to-be-determined coefficients $\beta_0, \ldots, \beta_\ell$. Plugging this into Equation 1.2 and equating like terms will give us a system of equations that we may be able to solve for the β's. If $g(n) = c^n$ is an exponential function, then we try $f(n) = \beta c^n$, an exponential as well. (For an example where this simple heuristic doesn't quite work, see Problem P 1.4.11.)

Example 1.19. Assume $a_0 = 0$, $a_1 = 1$, and, for $n \geq 2$, $a_n = a_{n-1} + a_{n-2} + n^2$. Hence, the sequence starts as

$$0, 1, 5, 15, 36, 76, \ldots$$

This is a non-homogeneous linear recurrence and the corresponding homogeneous linear recurrence is $a_n = a_{n-1} + a_{n-2}$. In Example 1.18, we already found that $c\mu^n + d\nu^n$, where $\mu = \frac{1+\sqrt{5}}{2}$ and $\nu = \frac{1-\sqrt{5}}{2}$, is the general solution to this homogeneous recurrence. We now look for a particular solution to the non-homogeneous recurrence by trying a second-degree polynomial (since n^2 is a second-degree polynomial). Hence, let $f(n) = \alpha n^2 + \beta n + \gamma$. If $f(n)$ were going to satisfy the original recurrence, then we would have

$$\alpha n^2 + \beta n + \gamma = f(n) = f(n-1) + f(n-2) + n^2$$
$$= \alpha(n-1)^2 + \beta(n-1) + \gamma + \alpha(n-2)^2 + \beta(n-2) + \gamma + n^2.$$

Simplifying and gathering like terms, we get

$$(-\alpha - 1)n^2 + (6\alpha - \beta)n + (-5\alpha + 3\beta - \gamma) = 0.$$

Hence,

$$\begin{cases} -\alpha - 1 & = 0 \\ 6\alpha - \beta & = 0 \\ 5\alpha + 3\beta - \gamma & = 0 \end{cases} \Rightarrow \begin{cases} \alpha & = -1 \\ \beta & = -6 \\ \gamma & = -13. \end{cases}$$

So, $-n^2 - 6n - 13$ is a particular solution to the non-homogeneous recurrence, and the general solution to the non-homogeneous linear recurrence is

$$c\mu^n + d\nu^n - n^2 - 6n - 13.$$

We now use the initial conditions to find c and d.

$$\begin{cases} c + d - 13 & = 0 \\ c\mu + dv - 20 & = 1 \end{cases} \Rightarrow \begin{cases} c & = \frac{65+29\sqrt{5}}{10} \\ d & = \frac{65-29\sqrt{5}}{10}. \end{cases}$$

We conclude that a closed formula for a_n is

$$a_n = \frac{65 + 29\sqrt{5}}{10}\left(\frac{1+\sqrt{5}}{2}\right)^n + \frac{65 - 29\sqrt{5}}{10}\left(\frac{1-\sqrt{5}}{2}\right)^n - n^2 - 6n - 13.$$

This is a complicated formula, and, it is sort of amazing that it produces our original sequence. Even though this is a closed formula, if you have any questions or conjectures about this sequence, then the recurrence relation itself is most likely a more transparent tool for analyzing the sequence.

Remark 1.20. An alternative but related method for solving homogeneous linear recurrences is to use matrix algebra techniques after translating the problem into a problem of finding powers of matrices. Problem P 1.4.26 illustrates this method, and Problems P 1.4.27 and P 1.4.28 give additional examples.

1.4.2   Unwinding (or Back-Substitution)

Warm-Up 1.21. *Assume $a_1, a_2, \ldots, a_n, \ldots$ is a sequence of numbers, and, for $n \geq 2$, we have $a_n = 3a_{n-1}+2$. Does this mean that $a_{n-1} = 3a_{n-2}+2$ and so $a_n = 3(3a_{n-2}+2)+2 = 9a_{n-2}+8$? Find a similar expression for a_n in terms of a_{n-3}, and then another expression for a_n in terms of a_{n-4}.*

Given a recurrence relation for a function f, we can sometimes, by the repeated use of the relation, find a formula for f in terms of its initial conditions. Often this formula will be in the form of a sum, which we can attempt to write in a closed form. If we need to be formal about our procedure, then we use induction to prove our final result. We illustrate the method with two examples.

Example 1.22. Let n be a non-negative integer. Assume that $f(0) = -3$, and for $n = 1, 2, \ldots$ we have

$$f(n) = 3f(n-1) + 10.$$

In Problem P 1.3.2 you were asked to generate some data, guess a formula for $f(n)$, and then prove it using induction. This is also a non-homogeneous linear recurrence and so we can use the general method for solving such recurrences. Here, to illustrate how to use unwinding, we present a third approach. Since the recurrence relation holds for all n, we not only have $f(n) = 3f(n-1) + 10$ but we also have $f(n-1) = 3f(n-2) + 10, f(n-2) = 3f(n-3) + 10,$

and so on. We "unwind" the recurrence relation as follows:

$$f(n) = 3f(n-1) + 10 = 3\left[3f(n-2) + 10\right] + 10$$
$$= 3^2 f(n-2) + 3 \times 10 + 10 = 3^2 \left[3f(n-3) + 10\right] + 3 \times 10 + 10$$
$$= 3^3 f(n-3) + 3^2 \times 10 + 3 \times 10 + 10 = 3^3 \left[3f(n-4) + 10\right] + 3^2 \times 10 + 3 \times 10 + 10$$
$$= 3^4 f(n-4) + 3^3 \times 10 + 3^2 \times 10 + 3 \times 10 + 10.$$

The pattern now stands out clearly. For every k with $1 \le k \le n$, we have, for $n \ge 0$,

$$f(n) = 3^k f(n-k) + 3^{k-1} \times 10 + 3^{k-2} \times 10 + \cdots + 3 \times 10 + 10.$$

Recall that $a^m + a^{m-1} + \cdots + a + 1$ is a geometric series and equal to $\frac{a^{m+1}-1}{a-1}$ (see Problem *P 1.1.3*). Now, in the equation for $f(n)$, if we replace k with n (in other words, we unwind the recurrence relation n times), we get

$$f(n) = 3^n f(0) + 10\left[3^{n-1} + \cdots + 3 + 1\right]$$
$$= -3^{n+1} + 10\frac{3^n - 1}{3 - 1} = -3 \times 3^n + 5 \times 3^n - 5$$
$$= 2 \times 3^n - 5.$$

If we trust the pattern that we got from unwinding the recurrence relation, then we know our final result to be true as well. Sometimes, as in this problem, it is pretty clear that the pattern will continue, but sometimes you may not be so sure. The rigorous approach would be to take our final answer as a conjecture, and then prove it with induction. Also note that when we were unwinding the recurrence relation, we did *not* multiply the constants out. If we had, it would have been harder to find the pattern.

Our second example is a bit more complicated. We use the notation $\lg(n)$ for the logarithm of n in base 2. So if $n = 2^k$, then $k = \lg(n)$.

Example 1.23. Assume that n is a power of 2 and the function T satisfies the following recurrence relation and has the given initial value:

$$T(n) = 2T(n/2) + n, \quad \text{and} \quad T(1) = 3.$$

Note that n is a power of 2 and so repeated divisions by 2 continue to give us an integer and, in fact, a power of 2. Now, the relation $T(n) = 2T(n/2) + n$ is true no matter what n is – as long as it is a power of 2 – and so, plugging in $n/2$ for n, we have $T(n/2) = 2T(n/4) + n/2$. We can now plug this back into the original formula to get a relation between $T(n)$ and $T(n/4)$. This "unwinding" can be continued all the way down. Let $n = 2^k$, so that $k = \lg(n)$. Proceed as follows:

$$T(n) = n + 2T(n/2) = n + 2[n/2 + 2T(n/4)]$$
$$= 2n + 2^2 T(n/2^2) = 2n + 2^2[n/2^2 + 2T(n/2^3)]$$
$$= 3n + 2^3 T(n/2^3)$$
$$\vdots$$
$$= kn + 2^k \underbrace{T(n/2^k)}_{1} = kn + 3 \times 2^k$$
$$= n \lg(n) + 3n.$$

To be sure that the pattern is actually correct, we use a proof by induction.

CLAIM: $T(n) = n \lg(n) + 3n$ for n a power of 2 greater than or equal to 1.

Proof. Induct on n. For $n = 1$, we have $T(1) = 3 = 1 \underbrace{\lg(1)}_{0} + 3$. Assume the result is true for n, to prove it for $2n$:

$$T(2n) = 2T(n) + 2n = 2(n \lg(n) + 3n) + 2n$$
$$= 2n \lg(n) + 8n = 2n(\lg(n) + \underbrace{1}_{\lg(2)}) + 6n$$
$$= 2n \lg(2n) + 3(2n).$$

Note that since n is a power of 2, we used the statement for $T(n)$ to prove the statement for $T(2n)$.

Remark 1.24. This kind of reasoning is often used in the analysis of computer algorithms. The only difference is that in such analysis we are not concerned with precise values of functions but rather their asymptotic behavior, and hence lower-order terms are ignored. For this purpose we use the so-called Θ notation. Let f and g be functions with positive values. We say $g(n) = \Theta(f(n))$ if, for large n, there exist constants c_1 and c_2 such that $0 \leq c_1 f(n) \leq g(n) \leq c_2 f(n)$. Roughly, this means that, asymptotically, f and g are of similar orders of magnitude. As an example, in the analysis of the merge-sort algorithm, if the worst-case running time of the algorithm, for an input of size n, is denoted by $T(n)$, then you can argue that $T(n) = 2T(\lceil n/2 \rceil) + \Theta(n)$ and $T(1) = \Theta(1)$. (For a real number r, $\lceil r \rceil$ denotes the *ceiling* of r and is the smallest integer greater than or equal to r.) This is basically because the merge-sort works by splitting a problem of size n into two problems of half the original size (hence the term $2T(\lceil n/2 \rceil)$) and uses $\Theta(n)$ time to "merge" the solutions of the two subproblems. From a calculation similar to the one in the example above, one concludes that the worst-case running time of merge-sort is $\Theta(n \lg(n))$.

Problems

P 1.4.1. Recall that the sequence L_n of Lucas numbers is defined (see Section 1.2.1) by

$$L_0 = 2, \ L_1 = 1, \text{ and, for } n \geq 2, \ L_n = L_{n-1} + L_{n-2}.$$

Find a closed formula for the Lucas numbers.

P 1.4.2. Let $f(1) = 1, f(2) = 3$, and, for $n \geq 3, f(n) = f(n-1) + 2f(n-2)$. Find a closed formula for $f(n)$.[17]

P 1.4.3. Assume that $h_1 = 3, h_2 = 8$, and, for $n \geq 3$, we have $h_n = 2h_{n-1} + 2h_{n-2}$.[18] Find the roots of the characteristic polynomial of this recurrence relation, and find a closed formula for h_n.

P 1.4.4. Let n be a positive integer. In Problem P 1.3.14, we defined $h(n)$ to be the number of sequences $a_1, a_2, \cdots, a_n$ with the conditions that for $i = 1, \ldots, n$, each a_i is 0, 4, or 7, and each 4 in the sequence is *immediately* followed by a 7. Using the linear recurrence for $h(n)$, find a close formula for it.

P 1.4.5. Assume that $a_0 = 1, a_1 = -1, a_2 = 0$ and, for $n \geq 3$, we have $a_n = 6a_{n-1} - 11a_{n-2} + 6a_{n-3}$.
(a) Find a_3 and a_4.
(b) Find a closed formula for a_n.

P 1.4.6. Let $f(0), \ldots, f(n), \ldots$ be a sequence of numbers that satisfies $f(n) = 4f(n-1) - 4f(n-2)$.
(a) What is the characteristic polynomial of this recurrence relation?
(b) Is there a function of the form $f(n) = a^n$ that satisfies the relation? What about one of the form $f(n) = na^n$? Prove your assertion.
(v) Find a function $f(n)$ that satisfies the relation as well as the initial conditions $f(0) = 4$ and $f(1) = 7$.

P 1.4.7. For the sequence of numbers $a_0, a_1, a_2, \ldots, a_n, \ldots$, we have $a_0 = a_1 = a_2 = 1$, and, for $n \geq 3, a_n = 9a_{n-1} - 27a_{n-2} + 27a_{n-3}$.
(a) Use the recurrence relation to find a_3, a_4, and a_5.
(b) Find a closed formula for a_n.

P 1.4.8. For the sequence of numbers $a_0, a_1, a_2, \ldots, a_n, \ldots$, we have $a_0 = a_1 = a_2 = 1$, and, for $n \geq 3, a_n = 3a_{n-1} - 9a_{n-2} + 27a_{n-3}$.
(a) Use the recurrence relation to find a_3, a_4, and a_5.
(b) Find the characteristic polynomial of the recurrence relation. Check that $x = 3$ is a root. Use long division to factor the polynomial, and then find the other (complex) roots.
(c) Find a closed formula for a_n.
(d) If your formula doesn't look real, then plug in $n = 0, 1$, and 5 to check that it actually works. Anything surprising?

---

[17] We encountered this sequence of integers in Warm-Up 1.8 and Problem P 1.3.8. A more convoluted approach for finding a closed formula was pursued in Problem P 1.3.9.
[18] See Problem P 1.3.10.

(e) By considering four cases ($n = 4k, \ldots, 4k+3$), simplify your formula so that it contains no complex numbers.

P 1.4.9. Assume that $f(0) = 1, f(1) = 2$, and, for $n \geq 2, f(n) = 5f(n-1) - 6f(n-2) + n$.

(a) Use the recurrence relation to find $f(2)$ and $f(3)$.

(b) Ignoring the initial conditions, find a general solution for $f(n) = 5f(n-1) - 6f(n-2)$, the corresponding homogeneous linear recurrence.

(c) Finish (or repeat) Warm-Up 1.12 and, again ignoring the initial conditions, find a particular solution to the original recurrence.

(d) Find a closed formula for $f(n)$ (satisfying the recurrence relation as well as the initial conditions).

(e) Is it obvious that your formula for $f(n)$ will produce the correct values for $f(2)$ and $f(3)$? Check that it actually does.

P 1.4.10. Assume $a_1, a_2, \ldots, a_n, \ldots$ is a sequence of numbers, $a_1 = 2$, and, for $n \geq 2$, we have $a_n = 3a_{n-1} + 2$.

(a) Continue Warm-Up 1.21, unwind the recurrence relation, and find a formula for a_n.

(b) Re-do the problem using the general method for solving non-homogeneous linear recurrence relations.

(c) Compare the approaches here with your solution to Problem P 1.3.1.

P 1.4.11. Assume that $f(1) = 2$, and, for $n \geq 2, f(n) = f(n-1) + n$. In Problem 1.9, we guessed a closed form for $f(n)$ and proved our conjecture using induction. We want alternative methods for finding a closed formula for $f(n)$ using the techniques of this section (general method for solving non-homogeneous linear recurrences and unwinding).

(a) What is the general solution to $f(n) = f(n-1)$, the corresponding homogeneous recurrence?

(b) Show that no function of the form $an + b$ satisfies the non-homogeneous recurrence relation. Find a particular solution to the non-homogeneous equation of the form $n(an + b)$.

(c) Using the previous parts, find a closed form for $f(n)$.

(d) Find a closed form for $f(n)$ by unwinding the recurrence relation.

P 1.4.12. Let n be a non-negative integer. Assume that $f(0) = 4, f(1) = 7$, and, for $n \geq 2, f(n) = 4f(n-1) - 4f(n-2) + 3n + 1$. Find a formula for $f(n)$.

P 1.4.13. Let n be a non-negative integer, and assume $a_0 = 0$, $a_1 = 2$, and, for $n \geq 2$,

$$a_n = -a_{n-1} + 6a_{n-2} - 3^n + 2^{2n-2}.$$

(a) Find a_2, a_3, and a_4.

(b) Find a closed formula for a_n.

P 1.4.14. If $f(1) = 1$ and, for integers $n > 1$, we have $f(n) = 2f(n-1) + (-1)^n$, find a formula for $f(n)$ by unwinding the recurrence relation. Compare the solution to your solution to Problem P 1.3.9.

P 1.4.15. Define $E(1) = 4$, and, for $n > 1$, assume $E(n) = 3E(n-1) + 5^{n-1}$. Unwind the recurrence relation to find a formula for $E(n)$. Compare to your solution to Problem P 1.3.16.

P 1.4.16. Let n be a positive integer. In Problem P $1.3.15$, we defined $h(n)$ to be the number of sequences of length n where the terms of the sequence are either 0, 4, or 7, with the added condition that if both 4 and 7 occur in the sequence, then the first 4 occurs before the first 7. Unwind the recurrence relation for $h(n)$ to find an explicit formula for $h(n)$.

P 1.4.17. Assume n is a power of 2. We have a function T for which we know

$$T(n) = 8T(n/2) + n^2, \quad \text{and} \quad T(1) = 6.$$

Unwind this relation to find a formula for $T(n)$. Prove your result using induction.

P 1.4.18. Assume n is a power of 2. We have a function T for which we know

$$T(n) = 7T(n/2) + n^2, \quad \text{and} \quad T(1) = 6.$$

Unwind this relation to find a formula for $T(n)$.

Remark: A similar recurrence appears in the analysis of Strassen's matrix multiplication algorithm.

P 1.4.19. For a real number r, the *floor* of r, denoted by $\lfloor r \rfloor$, is the largest integer that is less than or equal to r. We have a function T defined on positive integers. We know that

$$T(n) \leq T(\lfloor n/5 \rfloor) + T(\lfloor 3n/4 \rfloor) + 2n, \quad \text{for} \quad n > 50.$$

$T(1), T(2), \ldots, T(50)$ are arbitrary positive integers. Prove that there exists a constant c such that

$$T(n) \leq cn \quad \text{for all } n > 0.$$

P 1.4.20. Let n be a power of 2 and assume that

$$T(n) = \begin{cases} 5 & \text{if } n = 1 \\ 2T(n/2) + n \lg(n) & \text{if } n > 1. \end{cases}$$

Unwind T to find a formula for $T(n)$.

P 1.4.21. Assume that $T(0) = 0$, $T(1) = T(2) = \cdots T(7) = 2$, and, for $n > 7$, we have

$$T(n) = T(n-7) + T(7) + n.$$

Unwind T to find a formula for $T(n)$.

P 1.4.22. Write down a complete proof of Lemma 1.17.

P 1.4.23. **The Vandermonde[19] determinant.**[20] Let $a_1, \ldots, a_k$ be k distinct real numbers, and define the $k \times k$ matrix V (the Vandermonde matrix) by

$$V = \begin{bmatrix} 1 & a_1 & a_1^2 & \cdots & a_1^{k-1} \\ 1 & a_2 & a_2^2 & \cdots & a_2^{k-1} \\ \vdots & \vdots & \vdots & \vdots & \vdots \\ 1 & a_k & a_k^2 & \cdots & a_k^{k-1} \end{bmatrix}.$$

Show that

$$\det(V) = \prod_{1 \le i < j \le k} (a_j - a_i) = (a_2 - a_1)(a_3 - a_1)(a_3 - a_2) \cdots (a_k - a_{k-1}) \neq 0.$$

You may find the following steps useful.

STEP 1: As an example, and as a point of reference, find $\det(V)$ in the case when $k = 3$.

STEP 2: Using general properties of determinants, convince yourself that if you replace a_2 with a_1 in the matrix V, then the determinant of the resulting matrix will be zero.

STEP 3: You can think of $\det(V)$ as a multivariable polynomial with variables $a_1, \ldots, a_k$. For example, for the case $k = 3$, one term of this polynomial is $a_2 a_3^2$. The degree of a term is defined to be the sum of the exponents of the variables. Thus the degree of $a_2 a_3^2 = a_1^0 a_2^1 a_3^2$ is $0 + 1 + 2 = 3$. For an arbitrary k, what is the degree of each of the terms of $\det(V)$?

STEP 4: The polynomial $x^2 - 48x + 47$ has 1 as a root. As a result we know that the polynomial has a factor of $x - 1$. Indeed $x^2 - 48x + 47 = (x - 1)(x - 47)$. Think of $\det(V)$ as a polynomial in a_2 (with $a_1, a_3, \ldots, a_k$ thought of as parameters and not variables). Using Steps 2 and 3, show that $a_2 - a_1$ is a factor of $\det(V)$.

STEP 5: Generalizing Step 4, show that

$$P = \prod_{1 \le i < j \le k} (a_j - a_i) = (a_2 - a_1)(a_3 - a_1)(a_3 - a_2) \cdots (a_k - a_{k-1})$$

is a factor of $\det(V)$.

STEP 6: Show that the number of parentheses of the form $(a_j - a_i)$ in the product P is $1 + 2 + \cdots + (k - 1)$. P and $\det(V)$ are both multivariable polynomials and P is a factor of $\det(V)$. By comparing the degrees of the terms of both show that $\det(V) = cP$ where c is some non-zero constant.

STEP 7: One term in $\det(V)$ is the product of the diagonal elements of V. By considering this term in $\det(V)$ and P show that $c = 1$, and so $\det(V) = P \neq 0$.

---

[19] Alexandre-Théophile Vandermonde (1735–1796) was a violinist, and turned to mathematics only at age 35. In his work on music theory, he—somewhat unexpectedly, coming from a mathematician—argued that musicians should ignore music theory and only focus on their trained ears for judging music. In mathematics, he was one of the first mathematicians studying the solvability of algebraic equations by looking at the effect of permuting the roots, and he also wrote an important paper on determinants. However, it seems that what is called the Vandermonde determinant never appeared in his works, and the naming of this determinant for him amounts to a misunderstanding. Vandermonde was French and an ardent supporter of the French Revolution.

[20] This problem involves linear algebra.

P 1.4.24. **Linear Recurrences with Distinct Roots for the Characteristic Polynomial.** Let $a_1, \ldots, a_k$ be k distinct real numbers. Consider the homogeneous linear recurrence of Equation 1.1, and assume that $f_1(n) = a_1^n, \ldots, f_k(n) = a_k^n$ are functions satisfying the recurrence relation. Show that there exists a function f that satisfies the recurrence of Equation 1.1 as well as the following initial conditions:

$$f(0) = \beta_0, \; f(1) = \beta_1, \; \ldots, \; f(k-1) = \beta_{k-1}, \tag{1.4}$$

where $\beta_0, \ldots, \beta_{k-1}$ are arbitrary real numbers. You may find the following steps useful.

STEP 1: Let $\alpha_1, \ldots, \alpha_k$ be yet-to-be-determined scalars, and let $f = \alpha_1 a_1^n + \alpha_2 a_2^n + \cdots + \alpha_k a_k^n$. Does f satisfy Equation 1.1? Why?

STEP 2: For f to satisfy the initial conditions of Equation 1.4, $\alpha_1, \ldots, \alpha_k$ have to be the solutions to a system of k linear equations and k unknowns. Write down the linear system of equations.

STEP 3: Show that the coefficient matrix of the system in Step 2 is the transpose of the Vandermonde matrix V of Problem *P 1.4.23*.

STEP 4: If the determinant of the coefficient matrix V is non-zero, then will we be able to find the required $\alpha_1, \ldots, \alpha_k$? Why?

STEP 5: Using the result of Problem *P 1.4.23*, complete the proof.

P 1.4.25. **Double Roots of the Characteristic Polynomial.** Prove that if a is a double root of the characteristic polynomial of the homogeneous linear recurrence of Equation 1.1, then the function $f(n) = na^n$ satisfies the recurrence relation. You may find the following steps useful.[21]

STEP 1: Show that we need to prove that

$$na^n - (n-1)\alpha_1 a^{n-1} - (n-2)\alpha_2 a^{n-2} - \cdots - (n-k)\alpha_k a^{n-k} = 0.$$

STEP 2: Recall that if a is a double root of a polynomial $h(x)$, then $h(x) = (x-a)^2 \ell(x)$ for some polynomial $\ell(x)$. Show that if a is a double root of a polynomial $h(x)$, then a is also a root of $h'(x)$, the derivative of h.

STEP 3: Let $g(x) = x^k - \alpha_1 x^{k-1} - \cdots - \alpha_k$ be the characteristic polynomial of the homogeneous linear recurrence of Equation 1.1, and assume that a is a double root of $g(x)$. Is a also a double root of $h(x) = x^{n-k} g(x)$?

STEP 4: Write out the polynomial $h(x)$, find its derivative $h'(x)$, and find a polynomial expression for $xh'(x)$.

STEP 5: Use Step 2 to argue that a is a root of $xh'(x)$. Plug a into the expression for $xh'(x)$ and complete the proof using Step 1.

P 1.4.26. **Solving Homogeneous Linear Recurrences via Matrix Diagonalization.** For n a non-negative integer, define $h(n)$ by

$$h(0) = 3, \quad h(1) = -2, \quad h(2) = 6,$$

---

[21] Adapted from Bennet 2014.

and, for $n \geq 2$,

$$h(n) = 2h(n-1) + h(n-2) - 2h(n-3).$$

Let

$$A = \begin{bmatrix} 0 & 1 & 0 \\ 0 & 0 & 1 \\ -2 & 1 & 2 \end{bmatrix}, \text{ and } v_n = \begin{bmatrix} h(n) \\ h(n+1) \\ h(n+2) \end{bmatrix}.$$

Our goal is to use A to solve the recurrence relation.

(a) What is Av_0? What about $A^2 v_0$?

(b) How is $A^n v_0$ related to the sequence $\{h(n) \mid n = 0, 1, \ldots\}$?

(c) Is A diagonalizable? If so, find a matrix P such that $D = P^{-1}AP$ is a diagonal matrix.

(d) How is char(A), the characteristic polynomial of A, related to the characteristic polynomial of the recurrence relation defined in the paragraph following Lemma 1.17.

(e) Show that $A^n = PD^n P^{-1}$, and then find A^n.

(f) Give a formula for $h(n)$.

P 1.4.27. The sequence $\{k(n) \mid n = 0, 1, 2, \ldots\}$ satisfies the same recurrence relation as the sequence $h(n)$ of Problem P 1.4.26, but has different initial conditions:

$$k(0) = 0, \quad k(1) = 0, \quad k(2) = 6.$$

Find a formula for $k(n)$.

P 1.4.28. Use the method of Problem P 1.4.26 to find a formula for the nth term of the sequence a_0, $a_1, \ldots$, where $a_0 = 4$, $a_1 = -7$, and, for $n \geq 2$, we have $a_n = a_{n-1} + a_{n-2}$.

1.5 Open Problems and Conjectures

Given a recurrence relation for a function, it is not always straightforward to find all the information that we may want. An example of this is the famous $3n + 1$ problem.

Conjecture 1.25 (The $3n + 1$ Conjecture). *Let n_0 be a positive integer, and define a sequence of numbers $a_0, a_1, \ldots, a_n, \ldots$ by letting $a_0 = n_0$, and, for $n > 0$,*

$$a_{n+1} = \begin{cases} \dfrac{a_n}{2} & \text{if } a_n \text{ is even} \\ 3a_n + 1 & \text{if } a_n \text{ is odd.} \end{cases}$$

Then, regardless of the value of n_0, the sequence will eventually become

$$\ldots, 4, 2, 1, 4, 2, 1, 4, 2, 1 \ldots$$

Clearly, if the sequence ever hits 4, then its subsequent terms are going to be 2, 1, 4, 2, 1, and so on. The conjecture says that, no matter where you start, you will eventually always make your way to $4, 2, 1$. For example, if you start with 3, the sequence becomes

$$3, 10, 5, 16, 8, 4, 2, 1, \ldots$$

On the other hand, if you start with 47, then the sequence becomes

$$47, 142, 71, 214, 107, 322, 161, 484, 242, 121, 364, 182, 91, 274, 137, 412,$$
$$206, 103, 310, 155, 466, 233, 700, 350, 175, 526, 263, 790, 395, 1186, 593,$$
$$1780, 890, 445, 1336, 668, 334, 167, 502, 251, 754, 377, 1132, 566, 283, 850,$$
$$425, 1276, 638, 319, 958, 479, 1438, 719, 2158, 1079, 3238, 1619, 4858, 2429, 7288,$$
$$3644, 1822, 911, 2734, 1367, 4102, 2051, 6154, 3077, 9232, 4616, 2308, 1154, 577, 1732,$$
$$866, 433, 1300, 650, 325, 976, 488, 244, 122, 61, 184, 92, 46, 23, 70,$$
$$35, 106, 53, 160, 80, 40, 20, 10, 5, 16, 8, 4, 2, 1, 4, 2, 1, 4, 2, 1, \ldots$$

If the conjecture were true, then you could have made a finite number of mistakes in your calculations and the sequence – since its long-term behavior is independent of its starting point – would end up in the same place anyway! The $3n + 1$ conjecture is also known as the Collatz[22] conjecture (or the Syracuse problem, the Ulam conjecture, the Kakutani problem, the Hasse algorithm, or the Thwaites conjecture), and, despite its simplicity, it remains completely open. In fact, Paul Erdős once remarked: "Mathematics may not be ready for such problems." By 2017, the conjecture had been verified for all starting values up to 87×2^{60}, and while many partial results have been obtained (see Lagarias 2010) no proof for the conjecture has been found.

As we have seen, recurrence relations can also be thought of as the discrete analogs of differential equations. From this point of view, given a recurrence relation, we can ask questions similar to those in differential equations. Is the solution function always bounded? As n goes to infinity, are the values of the function approaching a limit and does this limit depend on the initial condition? Many such problems have been posed and answered. Many have applications in applied mathematics and engineering. But there are many – including the following problem – that remain open.

Problem 1.26 (See Kulenović and Ladas 2002). *Consider a sequence $a_0, a_1, \ldots, a_n, \ldots$ that satisifes the recurrence relation*

$$a_{n+1} = -1 + \frac{a_{n-1}}{a_n}.$$

The values of this sequence depend on the initial values of a_0 and a_1. If, for example, $a_0 = a_1 = 1$, then $a_2 = 0$, but a_3 is not defined (since dividing by zero is not allowed). Find all real numbers a_0 and a_1 for which the above sequence is well defined for all $n > 0$.

---

[22] Named after Lothar Collatz (1910–1990) a German mathematician who introduced the idea in 1937.

2 The Pigeonhole Principle and Ramsey Theory

In an ancient cave, you see a seemingly arbitrary list of **100** *positive integers carved in a row. You add up all the integers and the sum is* **152.** *Show that, regardless of what the integers are, among them you can locate an unbroken block of adjacent integers that add up to exactly* **47.**

2.1 The Pigeonhole Principle

Warm-Up 2.1. *A bag contains* 100 *apples,* 100 *bananas,* 100 *oranges, and* 100 *pears. If, every minute, I randomly pick one piece of fruit out of the bag, how long will it be before I am assured of having picked at least a dozen pieces of fruit of the same kind?*

We begin with an almost trivial fact.

Theorem 2.2 (The Pigeonhole Principle). *If we put* $n + 1$ *pigeons into n pigeonholes, then at least one pigeonhole contains two or more pigeons.* [1]

Proof. This principle is so obvious that it really doesn't need a proof. You could write down a proof by contradiction: assume that each pigeonhole has at most one pigeon, then the total number of pigeons would be no more than $\underbrace{1 + 1 + \cdots + 1}_{n} = n$. But there were $n + 1$ pigeons.

The contradiction proves the claim. $\qquad\square$

It is surprising that this obvious fact can have many applications. But, as we shall see, clever uses of this principle can yield unexpected results. In this chapter, in addition to applications of the pigeonhole principle, we will consider Ramsey theory, which can be considered as a vast generalization of this principle.

This whole chapter does give a reasonable view of contemporary combinatorics. We often start with common-sense observations and small examples, and then, through a mixture of theory and clever ad hoc arguments, create effective tools for dealing with a host of problems.

The pigeonhole principle can be reformulated to sound more mathematical. If X is a finite set, then $|X|$ denotes the number of elements of X.

Theorem 2.3 (The Pigeonhole Principle, Reformulated). *Let P and H be finite sets. Let* $f : P \to H$ *be a function. If* $|P| > |H|$, *then f is not 1-1.*

---

[1] The pigeonhole principle is also called the *Dirichlet principle*.

Proof. Call the elements of P "pigeons" and the elements of H "pigeonholes." Then the map f assigns the pigeons to pigeonholes, and the conclusion follows from the (original) pigeonhole principle.   □

Remark 2.4. There are other forms of the pigeonhole principle as well. For example, if you have a big basket of oranges, apples, and bananas, and if you randomly pick 21 pieces, then you will either have at least 7 oranges, or at least 5 apples, or at least 11 bananas. You can prove this by contradiction, since if you had at most 6 oranges, 4 apples, and 10 bananas, then you had picked at most 20 pieces of fruit. You can think of this as a strong form of the pigeonhole principle where the pigeonholes have some predetermined capacity (as opposed to just 1 pigeon), and if you have one more pigeon than the sum of the capacities, then at least one of the pigeonholes will go over capacity. (See Problem P 2.1.4.)

Applying the Pigeonhole Principle

The idea of the pigeonhole principle is pretty straightforward. Even so, we can prove many unexpected results using it. Here, we give a few examples of its use (including the opening problem for this chapter) and you will find a number of other examples in the problems at the end of this section.

Example 2.5. At a party, some of the people present shook hands with some of the others. Assume that at least two people were at the party. Show that there exist at least two people who shook hands with the same number of people.

Proof. Assume there are n people at the party. Ask each one how many hands they shook. You will get answers:

$$a_1, a_2, a_3, \cdots, a_n.$$

Each of these numbers could be

$$0, 1, 2, \cdots, n - 1.$$

At this point, since we have n numbers and n possibilities, it is possible that every one of the answers is different.

INSIGHT: If one of the answers is zero, then someone didn't shake hands with anyone and so, in that case, none of the answers can be $n - 1$.

   Thus either the numbers are from among

$$0, 1, 2, \cdots, n - 2$$

or from among

$$1, 2, \cdots, n - 1.$$

In either case, we have n numbers (pigeons) and only $n - 1$ possibilities (pigeonholes), and so at least two of the answers will be the same!   □

Example 2.6. In an ancient cave, you see a seemingly arbitrary list of 100 positive integers carved in a row. You add up all the integers and the sum is 152. Show that, regardless of what the, integers are, among them you can locate an unbroken block of adjacent integers that add up to exactly 47.

Proof. Let the list of integers be $a_1, a_2, \ldots, a_{100}$. Let $s_1, s_2, \ldots, s_{100}$ be the sequence of the partial sums. In other words,

$$s_1 = a_1,$$
$$s_2 = a_1 + a_2,$$
$$\vdots$$
$$s_i = a_1 + a_2 + \cdots + a_i,$$
$$\vdots$$
$$s_{100} = a_1 + a_2 + \cdots + a_{100}.$$

Note that we know $s_{100} = 152$, and $1 \leq s_1 < s_2 < \cdots < s_{100} = 152$. Now just add 47 to each s_i to get a new integer t_i. In other words,

$$t_1 = s_1 + 47 \qquad\qquad\qquad\qquad t_2 = s_2 + 47$$
$$\cdots \qquad\qquad\qquad\qquad t_{100} = s_{100} + 47.$$

Now, $t_{100} = 152 + 47 = 199$, and we have $48 \leq t_1 < t_2 < \cdots < t_{100} = 199$. Consider a list of 200 integers consisting of *both* the s's and the t's:

$$1 \leq s_1, s_2, \ldots, s_{100}, t_1, t_2, \ldots, t_{100} \leq 199.$$

We have 200 integers, and each is an integer no smaller than 1 and no larger than 199. The pigeonhole principle says that two of them must be the same. The s's can't be equal to each other and the t's can't be equal to each other either. So, at least one of the s's must be equal to one of the t's. Hence, we have

$$s_j = t_i = s_i + 47,$$

for some $1 \leq i < j \leq 100$. This means that $s_j - s_i = 47$, but

$$47 = s_j - s_i = (a_1 + \cdots + a_j) - (a_1 + \cdots + a_i)$$
$$= a_{i+1} + a_{i+2} + \cdots + a_j.$$

We conclude that the sum of the $(i + 1)$th integer all the way to the jth integer is exactly 47. $\square$

The Erdős–Szekeres Theorem

Our next example of the use of the pigeonhole principle is the Erdős–Szekeres theorem of 1935 (Erdős and Szekeres 1935). We could state and give a formal proof of this theorem right away and with no extra song and dance, but maybe it is more interesting to see how one may discover such a theorem and its proof. In many parts of combinatorics, one starts with a somewhat straightforward – you could say simple-minded – question, then looks at concrete examples. Analysis of concrete examples gives the researcher a "feel" for the question and what may be subtle about it. Often at this point, based on the examples, one modifies the question, or one realizes that there are prior questions to answer, or one comes up with conjectures. Then one tries to prove the conjectures, but, even then, trying to prove a conjecture for concrete examples – i.e., coming up with rigorous arguments that explain the particular example – paves the way for more general arguments. Of course, experience with combinatorial methods and techniques can help and even be indispensable, but even very seasoned researchers often try their hands at concrete examples first. We begin with a definition and an example.

Definition 2.7 (Monotone Sequences). Let $a_1, a_2, \ldots, a_k, \ldots$ be a (finite or infinite) sequence of numbers. The sequence is called *(strictly) increasing* if $a_1 < a_2 < a_3 < \ldots < a_k < \ldots$ It is called *non-decreasing* (or *weakly increasing*) if equalities are allowed and $a_1 \leq a_2 \leq a_3 \leq \ldots \leq a_k \leq \ldots$ Likewise, if $a_1 > a_2 > a_3 > \ldots > a_k > \ldots$, then the sequence is *(strictly) decreasing*, while if $a_1 \geq a_2 \geq a_3 \geq \ldots \geq a_k \geq \ldots$, then the sequence is *non-increasing* (or *weakly decreasing*). A sequence is *monotone* (or *monotonic*) if it is either non-decreasing or non-increasing. The *length* of a finite sequence is the number of terms of the sequence.[2]

Example 2.8. Consider the sequence

$$4, 2, 5, 7, 3, 1, 9, 10, 6, 8.$$

This sequence is not monotone, but 4, 5, 7, 9, and 10 is a monotonic subsequence of length 5, while 7, 3, 1 is a monotonic subsequence of length 3.

Question 2.9. In Example 2.8, we started with a sequence of 10 numbers and the longest monotonic subsequence was of length 5. What can we say about the length of the longest monotonic subsequence in an arbitrary sequence of length 10?

Of course, the sequence itself could be monotone – 1, 2, ..., 10, for example – and so it is possible to have a monotonic subsequence of length 10. We can also just mess the monotonicity slightly – for example, 10, 1, 2, ..., 9 – so as to get a sequence whose longest monotonic subsequence is of length 9. You can continue this process and come up with sequences of length 10 whose longest monotonic subsequence is of length 8, 7, ..., or 5. For example, the length of the longest monotonic subsequence of 6, 7, 8, 9, 10, 1, 2, 3, 4, 5 is 5. Example 2.8 was another case of a sequence where the longest monotonic subsequence was of length 5.

---

[2] While this vocabulary is common, it is not universal. Some authors use "increasing" for what we call "non-decreasing." The problem with the latter expression is that it requires a moment's thought to realize that we are basically talking about an increasing sequence. The expressions "strictly increasing" and "weakly increasing" avoid most pitfalls, but the latter is not that common.

Can we find a sequence of length 10, where the longest monotonic subsequence is of length 4? What about 3? We urge the reader to attempt an answer before proceeding.

A related question is to ask for a good algorithm for finding the longest monotonic subsequence of a given sequence. This is a question of interest to computer scientists and is discussed in courses on algorithms.

Table 2.1 A sequence of length 10 whose longest monotonic subsequence has length 4.

	a_1	a_2	a_3	a_4	a_5	a_6	a_7	a_8	a_9	a_{10}
The sequence	5	3	7	1	10	8	2	6	4	9
Length of longest non-increasing subsequence starting at a_i	3	2	3	1	3	3	1	2	1	1
Length of longest non-decreasing subsequence starting at a_i	4	4	3	4	1	2	3	2	2	1

It is not too difficult to come up with a sequence of 10 integers whose longest monotonic subsequence is of length 4. The sequence 5, 3, 7, 1, 10, 8, 2, 6, 4, 9 is one such sequence, while 3, 2, 1, 6, 5, 4, 9, 8, 10, 7 is another. To check the size of the longest monotonic subsequence manually, we may record for each number in the sequence the longest non-increasing and the longest non-decreasing subsequence starting at that number. For the first sequence, we get Table 2.1. Now, try as hard as you may, and you will not find a sequence of length 10 whose longest monotonic subsequence has length 3. Why would that be true? In fact, we note that $3, 2, 1, 6, 5, 4, 9, 8, 7$ shows that there are sequences of length 9 whose longest monotonic subsequence is of length 3. But there don't seem to be sequences of length 10 with that property. Table 2.1 gives a clue for the proof of this fact.[3]

Under each term of the sequence, we have two numbers. One is the length of the longest non-increasing subsequence starting at that term, the other is the length of the longest non-decreasing subsequence starting at that term. For the specific sequence in that table, these integers are no smaller than 1 and no bigger than 4. If we had a sequence whose longest monotonic subsequence had length 3, then these integers would be from the set $\{1, 2, 3\}$. How many pairs (x, y) of integers can we make from this set? We have three choices for x and three choices for y and so the total number of possible choices is nine; that is, $(1, 1)$, $(1, 2)$, $(1, 3)$, $(2, 1)$, $(2, 2)$, $(2, 3)$, $(3, 1)$, $(3, 2)$, and $(3, 3)$. But our sequence has 10 terms. By the pigeonhole principle, two terms of our sequence will have to have identical integers attached to them. In other words, there is a pair a_i and a_j with $i < j$ in our sequence such that the length of the longest non-decreasing subsequence starting at a_i is the same as the length of the longest non-decreasing subsequence starting at a_j, *and* the length of the longest non-increasing subsequence starting at a_i is the same as the length of the longest non-increasing subsequence starting at a_j.

---

[3] This particular proof is due to Seidenberg (Seidenberg 1959).

But is that possible? Since $i < j$, we know that the element a_i appeared before a_j in our sequence. We ask if $a_i \leq a_j$ or $a_i \geq a_j$? If $a_i \leq a_j$, then you can add a_i to the beginning of the longest non-decreasing subsequence starting at a_j and get a longer non-decreasing subsequence. Likewise if $a_i \geq a_j$, we can get a longer non-increasing subsequence starting at a_i compared to the ones starting at a_j. So the numbers associated with a_i and a_j cannot be both the same. The contradiction proves that we could not have limited the lengths of the longest monotonic subsequences to 1, 2, and 3.

What was so special about 10? In the proof, we needed 10 to be more than $3 \times 3 = 9$. Hence, the same proof as above shows the following:

Any sequence of $n^2 + 1$ real numbers contains a monotonic subsequence of length at least $n + 1$. This is the Erdős–Szekeres theorem.

We now present a formal proof of the Erdős–Szekeres theorem. The argument is exactly the same as the more informal discussion of the example and of Table 2.1. In this sense, if you have followed the discussion above, then you don't need to read this proof. However, often in mathematics, you are presented with cleaned-up versions of proofs. To understand such a proof, you have to decipher it. You need paper and pencil and you may have to go through the proof by applying it to a specific example to see what it says. In essence, you may have to reconstruct the concrete arguments in our example to be able to understand the proof. So we urge you to read the more formal proof and to think about how it formalizes the same argument as the one we gave before. We first introduce an oft-used notation.

Definition 2.10. Let n be a positive integer. Then the notation $[n]$ will denote the set $\{1, \ldots, n\}$. In addition, for convenience, $[0]$ will denote the empty set.

Also recall that the *cartesian product* of two sets A and B is denoted by $A \times B$ and consists of pairs of elements whose first entry is from A and second entry is from B. In other words, $A \times B = \{(a, b) \mid a \in A, b \in B\}$. If A and B are finite sets, the size of $A \times B$ is the size of A times the size of B.

Theorem 2.11 (Erdős–Szekeres, 1935). *Let n be a positive integer. Any sequence of $n^2 + 1$ real numbers contains a monotonic subsequence of length at least $n + 1$. Furthermore, there exist sequences of n^2 numbers that do not have a monotonic subsequence of length $n + 1$.*

Proof. Let a sequence $a_1, a_2, \ldots, a_{n^2+1}$ of $n^2 + 1$ terms be given. Assume that there is no monotonic subsequence of length at least $n + 1$. Define a function $f : [n^2 + 1] \longrightarrow [n] \times [n]$ by $f(i) = (u(i), d(i))$, where $u(i)$ is the length of the longest non-decreasing subsequence starting with a_i and $d(i)$ is the length of the longest non-increasing subsequence starting with a_i. The function f is well defined since we are assuming that $1 \leq u(i), d(i) \leq n$ for all i. But the codomain $[n] \times [n]$ has n^2 elements while the domain $[n^2 + 1]$ has $n^2 + 1$ elements. So, by the pigeonhole principle, there exists $i < j$ with $f(i) = f(j)$. (In other words, f is not 1-1.) We now focus on this special i and j where $i < j$ but $f(i) = f(j)$.

Can $a_i \geq a_j$? No, since otherwise the longest non-increasing sequence starting at a_i has length at least 1 more than the one starting at a_j. Can $a_i \leq a_j$? No, since otherwise the longest non-decreasing sequence starting at a_i has length at least 1 more than the one starting at a_j. The contradiction proves the claim that there must have been a monotonic subsequence of length at least $n + 1$.

The sequence

$$n, n - 1, n - 2, \ldots, 1,$$
$$2n, 2n - 1, \ldots, n + 1,$$
$$3n, 3n - 1, \ldots 2n + 1,$$
$$\vdots$$
$$n^2, n^2 - 1, \ldots, n^2 - n + 1$$

shows that there exist sequences of length n^2 with monotonic subsequences of length no more than n. □

Problems

P 2.1.1. The last four digits of a US Social Security number can be any combination of four digits (including 0's) except 0000. You sort a database of 10,000 registered voters according to the last four digits of their Social Security numbers. Show that the last four digits of the Social Security number are identical for at least two of the voters in your database.

P 2.1.2. In the supermarket, I want to buy jars of various herbs and spices. I am only interested in turmeric, tarragon, and thyme. I start picking jars (of turmeric, tarragon, and thyme) randomly from the shelf. How many jars shall I pick, before I am assured of having picked at least five jars of the same kind?

P 2.1.3. A large number of Humans, Klingons, and Romulans have lined up to enter your restaurant. What is the minimum number of life forms that you can allow in and still be assured that at least 13 of the same kind have come in?

P 2.1.4. **The Strong Form of the Pigeonhole Principle.** Let $q_1, q_2, \ldots, q_n$ be positive integers. Prove that if $q_1 + \cdots + q_n + 1$ objects are put in n boxes, then either the first box contains at least $q_1 + 1$ objects, or the second box contains at least $q_2 + 1$ objects, ..., or the nth box contains at least $q_n + 1$ objects.

P 2.1.5. A round-robin tournament is one where every player competes exactly once against every other player. In a round-robin tournament of 47 players, no one lost all their matches. Assuming no ties are allowed, show that two players must have won exactly the same number of matches. Is there something special about the number 47? Was the condition that "no one lost all their matches" necessary for the conclusion?

P 2.1.6. You have a 3×3 square, and you throw 10 darts at it. Show that no matter where the darts land, you can always find two darts whose separation is at most $\sqrt{2}$.

P 2.1.7. At a gathering of 47 people, a total of 93 pairs of people shook hands. Show that there is at least one person who shook fewer than four hands.

P 2.1.8. Let k be a positive integer. A function $f: X \rightarrow Y$ is called *k-to-one* if for every element $y \in Y$ there are at most k elements $x \in X$ such that $f(x) = y$. Assume X and Y have 189 and 47 elements, respectively. What is the largest positive integer k, so that f is guaranteed *not* to be k-to-one?

P 2.1.9. A runner has four weeks before a one-mile race. She has decided to run a whole number of miles each day, at least one mile per day, and no more than 12 miles in any one week. Show that there exists a succession of days during which the runner will have run *exactly* 7 miles. Would the conclusion be still correct if we replace 7 with 6? What about if we replaced 7 with 8?

P 2.1.10. You are planning an event, and there are 47 days left until the event. You estimate that you will need 70 hours of work to prepare. You are going to somehow do exactly a total of 70 hours of work in the next 47 days, but you only know that on each day you will work a positive whole number of hours.

(a) Show that no matter how many hours per day that you work, there will be a succession of days during which you will have worked exactly 20 hours.

(b) What other integers, other than 20, would make the previous statement true?

P 2.1.11. You worked 64 hours on a combinatorics problem over a 14-day period. If each day you worked a whole number of hours, show that on some pair of consecutive days, you worked at least 10 hours.

P 2.1.12. I have a book with pages numbered 1 to 100. Every page has a whole number of words. Show that there exists two pages m and n, such that $1 \leq m \leq n \leq 100$ and the total number of words from the beginning of page m to the end of page n is a multiple of 100.

P 2.1.13. I pick nine positive integers and line them up in a row. I happen to find an unbroken block of adjacent integers whose sum is divisible by 9:

$$71 \quad 7 \quad \underbrace{219 \quad 86 \quad 47 \quad 93 \quad 14}_{\text{sum}=459=9\times51} \quad 61 \quad 35.$$

Is this a coincidence? Can you generalize?

P 2.1.14. You have 47 candy bars and 11 friends. Can you distribute all of the candy bars among your friends (you are allowed to give a friend 0 candy bars but can't give any one a fraction of a candy bar or a negative number of candy bars) in a way that no two get the same number of candy bars? What if you had 12 friends or 10 friends?

P 2.1.15. There are 47 airports in some region. There are direct flights between some of these airports. (If there is a direct flight from airport A to airport B, there is also a direct flight from airport B back to airport A.) For each airport, you list the number of airports that you can get to with a direct flight. You notice that your list consists of only positive even numbers. Prove that on your list there is at least one integer that is repeated at least three times. Would this still be true if the total number of airports was 46 and all the numbers on your list were odd integers?

P 2.1.16. I pick 136 distinct integers from the set $\{1, 2, \ldots, 270\}$.

(a) Prove that you can find (at least) two integers from my collection that add up to 271.

(b) Either give a counter-example or prove that you can always find (at least) two integers from my collection whose difference is 135.

(c) What is so special about 270 and 136? Can you generalize?

P 2.1.17. I again pick 136 distinct integers from the set $\{1, 2, \ldots, 270\}$. Prove that you can find (at least) two integers from my collection such that one is exactly one more than the other.

P 2.1.18. I pick 136 distinct integers from the set $\{2, \ldots, 271\}$. Either prove that you can find (at least) two integers from my collection such that their greatest common divisor is one, or give a counter-example.

P 2.1.19. I continue to pick 136 distinct integers from the set $\{1, 2, \ldots, 270\}$. Prove that you can find (at least) two (different) integers from my collection such that one exactly divides the other.

P 2.1.20. I am interested in pairs of integers such as 47 and 153 whose sum is divisible by 100. I choose 52 random integers from a bag. Show that if I can't find a pair whose sum is divisible by 100, then, for sure, I can find a pair whose difference is divisible by 100.

P 2.1.21. Can you position the numbers 1 through 47 around a circle in such a way that the sum of any four consecutive numbers is less than 96? Justify your answer.

P 2.1.22. I have 51 rectangular pieces of cardboard, each of which has integer length ≤ 100 and integer width ≤ 100. (Note that squares are allowed—they are special types of rectangles.) Prove that among my rectangles I can find two, such that one will completely hide the other when placed on top of it.[4]

P 2.1.23. Let $a_n = 2^n - 1$ be the nth term of the sequence of positive integers that are one less than a power of 2:

$$1, 3, 7, 15, 31, \ldots, 2^n - 1, \ldots$$

Let m be an arbitrary odd positive integer. Does m have to divide one of the terms of this sequence? We note that (the notation $m \mid n$ means that m divides n)

$$1 \mid a_1, \ 3 \mid a_2, \ 5 \mid a_4, \ 7 \mid a_3, \ 9 \mid a_6 = 63, \ 11 \mid a_{10} = 1023.$$

(a) What is the smallest n such that a_n has a factor of 13?

(b) Is it true that m (an arbitrary odd positive integer) will have to divide at least one of $a_1, a_2, \ldots, a_m$? Either prove this or give a counter-example.

P 2.1.24. A mathematician and her partner go to a party at which there are four other couples. Some of the 10 people at the party shook hands, but no one shook hands with their own partner, and so the maximum possible number of hands that any individual could have shaken is 8.

The mathematician asks all the other nine people at the party how many hands each of them shook, and receives the nine answers:

$$0, 1, 2, 3, 4, 5, 6, 7, 8.$$

How many hands did the mathematician's partner shake?

P 2.1.25. Assume that you are given 17 arbitrary points in $\mathbb{R}^2$ (the usual Euclidean plane). Show that you can always choose five of these points $P_1, \ldots, P_5$, in such a way that the slopes of the

---

[4] Adapted from the 1989–1990 University of Wisconsin-Madison Talent Search for high school students.

line segments P_1P_2, P_2P_3, P_3P_4, and P_4P_5 are either all non-negative or all non-positive (∞ and $-\infty$ are considered non-negative and non-positive, respectively).

P 2.1.26.   Give an example of 16 points in $\mathbb{R}^2$ in such a way that the conclusion of Problem P 2.1.25 does not hold.

P 2.1.27.   Can you find a sequence of 12 real numbers that has neither a monotonically non-decreasing subsequence of length 5 nor a monotonically non-increasing subsequence of length 4?

P 2.1.28.   Assume that a finite sequence of real numbers of length 13 is given. Assume that the sequence does not have a non-decreasing subsequence of length 5. Does it have to have a non-increasing subsequence of length 4? Prove your assertion.

P 2.1.29.   **Erdős–Szekeres Generalized.** Assume that a finite sequence of real numbers of length $(r-1)(s-1)+1$ is given. Prove that either you can find a monotonically non-decreasing subsequence of length r or a monotonically non-increasing subsequence of length s.

P 2.1.30.   **The Chinese Remainder Theorem.** Let n and m be two positive integers. Assume that n and m are relatively prime (in other words, no prime divides both n and m). Let a be an integer with $0 \leq a \leq n-1$, and, likewise, let b be an integer with $0 \leq b \leq m-1$. Prove that there exists an integer L such that the remainder of L when divided by n and m is, respectively, a and b. In other words, as long as n and m are relatively prime, we can find integers that when divided by m and n have any combination of remainders that we desire. You may find the following steps helpful:

STEP 1: Consider the sequence of integers

$$a, n+a, 2n+a, \ldots, (m-1)n+a. \tag{2.1}$$

How many integers are in this sequence? What is the remainder of any of these when divided by n?

STEP 2: Assume that for positive integers i, j, k, and ℓ we have $(i-j)n = (k-\ell)m$. Use the fact that m and n are relatively prime to show that either $i = j$ or the difference between i and j is at least m. Conclude that if $0 \leq i, j \leq m-1$, then $i = j$.

STEP 3: Assume that two of the integers in the list 2.1 had the same remainder when divided by m. Use Step 2 to arrive at a contradiction.

STEP 4: If you divide an integer by m, how many possible remainders are there? Use Step 3 to show that one of the integers in the list 2.1 has remainder b when divided by m.

STEP 5: Complete the proof and rewrite it concisely.

P 2.1.31.   **Approximating Irrational Numbers.** Let $r > 0$ be an *irrational* number. Of course, there are rational numbers that are very close to r. But are there any "nice" rational numbers that are close to r? Intuitively, we will consider a rational number "nice" if it can be written as a ratio of two integers and the integer in the denominator is not "too big." Show that there are positive integers m and n such that

$$\left| r - \frac{m}{n} \right| < \frac{10^{-10}}{n}.$$

The fraction on the right is not only very small, it gets smaller the bigger the n is. In other words, in approximating r with the fraction m/n, we are imposing a less stringent requirement, the smaller the n is.

P 2.1.32. **Divisors of Piṅgala–Fibonacci Numbers.** Let F_0, F_1, ..., F_n, ... denote the sequence of Piṅgala–Fibonacci numbers (see Section 1.2.1). Let m be an arbitrary positive integer. Show that there exists an integer n with $1 \leq n \leq m^2$ such that F_n is a multiple of m.[5] The following steps may be helpful:

STEP 1: Consider the set $\mathcal{S}$ of pairs of consecutive Piṅgala–Fibonacci numbers defined by

$$\mathcal{S} = \{(F_i, F_{i+1}) \mid i = 0, 1, \ldots, m^2\}.$$

Use the pigeonhole principle and remainders when dividing by m to show that there exist two pairs (F_j, F_{j+1}) and $(F_\ell, F_{\ell+1})$ in $\mathcal{S}$ such that $j < \ell$, and both $F_j - F_\ell$ and $F_{j+1} - F_{\ell+1}$ are divisible by m.

STEP 2: Show that $F_{j-1} - F_{\ell-1}$ is also divisible by m.

STEP 3: Repeat Step 2 to show that $F_{j-k} - F_{\ell-k}$ is also divisible by m for $k = 2, \ldots, j$.

STEP 4: Conclude that $F_{\ell-j}$ is divisible by m.

2.2   Multisets and Graphs

Warm-Up 2.12. *Mojdeh, Mehrdokht, Māmak, Marjān, Mehrnāz, Mahshid, and Marzieh are on the same social media platform. Some of them are friends with some of the others. For each we have a list of their friends.*

Mojdeh	*Marjān, Mamāk, Marzieh*
Mehrdokht	*Māmak, Mehrnāz, Mahshid*
Māmak	*Mojdeh, Mehrdokht*
Marjān	*Mojdeh, Marzieh*
Mehrnāz	*Mehrdokht*
Mahshid	*Mehrdokht, Marzieh*
Marzieh	*Mojdeh, Marjān, Mahshid*

If two people are friends, we say that their distance is 1. If two people are not friends, but have a friend in common, we say that their distance is 2, and so on. We are interested in questions such as "What is the distance between Mehrnāz and Marjān?" and "What is the largest distance between any two individuals?" Suggest a visual way to present the data that makes answering such questions easier.

---

[5] Adapted from Jameson (Jameson 2018).

In this section, we digress, to introduce some notation and vocabulary that will be used throughout the text. For n a positive integer, we have already introduced (see Definition 2.10) the notation $[n]$ for the set $\{1, 2, \ldots, n\}$. This set will be our prototypical example of a finite set with n elements. Likewise, $[0]$ and $\emptyset$ will both denote the empty set. We will also be working with multisets. A *multiset* is like a set except that its members need not be distinct. Slightly more formally,

Definition 2.13 (Multisets). A *multiset* is a set together with a function that assigns a positive integer or ∞ – called a *repetition number* (or a *multiplicity*) – to each member of the set.

Example 2.14. $M = \{a, a, a, b, c, c, d, d\} = \{3 \cdot a, b, 2 \cdot c, 2 \cdot d\}$ is a multiset with members a, b, c, and d and with repetition numbers 3, 1, 2, and 2, respectively.

Remark 2.15. A set can be thought of as a multiset with all repetition numbers equal to 1. In an infinite multiset, some members could have infinite repetition numbers. In other words, we allow for the possibility that a multiset contains an infinite number of copies of a certain element.

Definition 2.16 (The $|X|$ Notation). If X is a finite set, then $|X|$ will denote the number of elements in X. If X is a finite multiset, then $|X|$ will denote the sum of its repetition numbers.

Graph theory is the subject of (the long) chapter 10, but graphs are such a good way of organizing certain kinds of information and formulating problems that we introduce the basic definitions here. Intuitively, a graph is a set of dots – called vertices – where some of the dots are connected to some of the other dots. The connections are called edges. We usually like to depict graphs visually. Figure 2.1 is an example of a graph. Each edge of a graph connects just two vertices and so it can be represented as a pair of vertices. For example, in Figure 2.1, there is an edge between vertices 2 and 7. This edge is referred to as $\{2, 7\}$, i.e., as a pair of vertices. More formally,

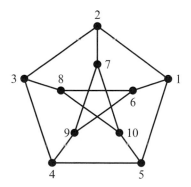

Figure 2.1 Intuitively, a graph is a bunch of vertices where some pairs of vertices are joined through edges. The graph in this figure – called the *Petersen graph* – has 10 vertices and 15 edges. Vertices numbered 6 and 9 are joined through an edge while vertices 6 and 3 are not.

Definition 2.17 (Graphs and Subgraphs). A *simple graph* (or just a *graph*) G is a pair of sets (V, E) where V is a non-empty set called the set of *vertices* of G, and E is a (possibly empty) set of unordered pairs of distinct elements of V. The set E is called the set of *edges* of G. If the number of vertices of G is finite, then G is a *finite graph* (or a *finite simple graph*).

If $G = (V, E)$ and $H = (V', E')$ are both graphs, then H is a *subgraph* of G if $V' \subseteq V$ and $E' \subseteq E$.

Example 2.18. Let $G = (V, E)$, where

$$V = \{1, 2, 3, 4\} \text{ and } E = \{\{1, 2\}, \{2, 3\}, \{3, 4\}, \{1, 4\}\}.$$

We can represent G with any of the drawings in Figure 2.2.

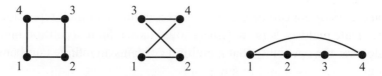

Figure 2.2 Three drawings for the graph of Example 2.18.

We could have drawn the picture in many other ways as well. All that matters is what the vertices are and which vertices are connected to which vertices.

Example 2.19. Figure 2.1 depicts a graph $G = (V, E)$ with 10 vertices and 15 edges. We have

$$V = \{1, 2, \ldots, 10\}$$
$$E = \{\{1, 2\}, \{2, 3\}, \{3, 4\}, \{4, 5\}, \{1, 5\}, \{1, 6\}, \{2, 7\}, \{3, 8\}, \{4, 9\}, \{5, 10\},$$
$$\{6, 8\}, \{7, 9\}, \{8, 10\}, \{6, 9\}, \{7, 10\}\}.$$

To construct a subgraph of this graph, we take some of the vertices and some of the edges, but we have to make sure that we have included all the vertices of the edges that we chose. So, for example, if

$$V' = \{1, 3, 6, 7, 8\}$$
$$E' = \{\{1, 6\}, \{3, 8\}\},$$

then $H = (V', E')$ is a subgraph of G. The graph H consists of two disjoint edges, their vertices, and one other isolated vertex.

Definition 2.20 (Multigraphs, General Graphs). In the definition of a graph, if we allow E to be a multiset – i.e., allow *repeated edges* – then G is called a *multigraph* or a *graph with repeated edges*. If we also allow pairs of non-distinct elements in E – i.e., allow *loops* – then G is called a *general graph* or a *graph with repeated edges and loops*. (See Figure 2.3.)

Figure 2.3 A general graph with three vertices and four edges. Vertex 1 is incident with a loop, there are two edges whose vertices are 1 and 2, and the degrees of the three vertices are 4, 3, and 1, respectively.

Remark 2.21. In this text, most often – but not always – we are concerned with *finite simple graphs*, and sometimes we use the shorter expression *graph* for these objects. We will use the longer expression when we want to stress the lack of double edges and loops.

Remark 2.22. If $G = (V, E)$ is a general graph and $e \in E$ is an edge of G, then, unlike in the case of simple graphs, we cannot necessarily identify e by its vertices. In other words, it will be inadequate to say $e = \{v, w\}$ where $v, w \in V$. This is because there may be multiple edges connecting v and w. Hence, to be formally correct, we should say that a general graph consists of two sets V and E and an incidence map that gives two vertices for each $e \in E$. In other words, every edge will have its own name in addition to being associated with its two vertices. To avoid unnecessary clutter, in what follows we will continue to identify edges of general graphs as pairs of vertices unless there is a need to be more formal.

Definition 2.23 (Order, Adjacency, Incidence). Let $G = (V, E)$ be a general graph. Then $|V|$, the number of vertices, is called the *order* of G.

If $\alpha = \{x, y\} \in E$ – that is, α is an edge connecting the vertices x and y – then we say that x and y are *vertices* or *ends* of α, that α *joins* x and y, that x and y are *adjacent*, and that x and α are *incident*.

Definition 2.24 (The Degree of a Vertex). If x is a vertex of a graph G, then the number of edges incident with x is called the *degree* (or the *valence*) of x. In the case of general graphs, a loop at a vertex x contributes 2 to the degree of x (see Figure 2.3).

For example, in the Petersen graph of Figure 2.1, every vertex has degree 3.

Paths, Cycles, and Complete Graphs

We will be studying graph theory in chapter 10, but, for now, we want to introduce notation for some important classes of graphs.

Example 2.25 (Paths or Chains). Let n be a positive integer and let $V = [n] = \{1, 2, \ldots, n\}$, and let $E = \{\{1, 2\}, \{2, 3\}, \ldots, \{n-1, n\}\}$ be the set of consecutive pairs of elements of $[n]$. A *path* (or a *chain*) of length $n - 1$, denoted often by P_{n-1}, is the graph that has V as its vertices and E as its edges. The graph P_{n-1} has order n (i.e., it has n vertices) and $n - 1$ edges. Note that the subscript in P_k refers to the length of the path, which is the same as the number of edges in the path. This is one less than the number of vertices in the graph.[6] (See Figure 2.4 for P_9.)

Figure 2.4 A path of length 9, P_9, has 9 edges and 10 vertices.

Example 2.26 (Cycles). Let $n \geq 3$ be an integer. A *cycle of length* n is a connected simple graph with n vertices and n edges such that each vertex has degree 2. We denote a cycle of length n by C_n. (See Figure 2.5.) The graph C_n has n vertices and n edges.

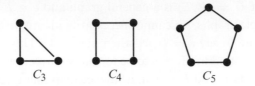

Figure 2.5 Cycles of length 3, 4, and 5.

Example 2.27 (Complete Graphs). Let K_n denote the simple graph with n vertices and all possible edges (no loops or repeated edges). Then K_n is called the *complete graph of order n*. Figure 2.6 gives drawings of K_1 through K_6.

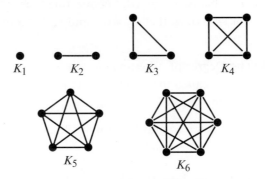

Figure 2.6 The complete graphs of orders 1 through 6.

The number of vertices of K_n is n. What is the number of edges? The first vertex is adjacent to $n-1$ other vertices, and this accounts for $n-1$ edges. The second vertex is also adjacent to $n-1$ vertices but we had already counted one of them (the edge between the first and the

---

[6] When appropriate, we make a distinction between the length and the size of an object. In graph theory, length refers to the number of edges. While this is very common, it is not universal. There are authors who use the length of a path to mean the number of vertices. In addition, many authors use the notation P_n to mean a path with n vertices (as opposed to n edges).

second vertex). So we have $n - 2$ new edges. Continuing in this way, we get that the number of edges of K_n is $(n - 1) + (n - 2) + \cdots + 1$. We had proved in Example 1.2 that this is equal to $\frac{n(n-1)}{2}$. Alternatively, we could have said that every vertex in K_n has degree $n - 1$ and there are n vertices. This sounds like $n(n - 1)$ edges, but each edge was counted twice – once for each of its two vertices – and so the correct answer is again $\frac{n(n-1)}{2}$.

Example 2.28 (Complete Bipartite Graphs). Let m and n be positive integers, and let L be a set of m vertices and R be a set of n vertices. Connect every vertex in L to every vertex in R. The resulting graph is called the *complete bipartite graph* on m and n vertices, and is denoted by $K_{m,n}$. Note that there are no edges between the vertices in R, and no edges between the vertices in L. (As examples, see $K_{1,3}$, $K_{3,3}$, and $K_{3,4}$ in Figure 2.7.) The graph $K_{m,n}$ has $m + n$ vertices, and, since each of the m vertices in L is connected to each of the n vertices in R, $K_{m,n}$ has mn edges. More generally, a *bipartite graph* is a graph whose vertices can be partitioned into two sets X and Y in such a way that all the edges have one end in X and another end in Y. In other words, a bipartite graph is a $K_{m,n}$ after you possibly delete some of the edges.

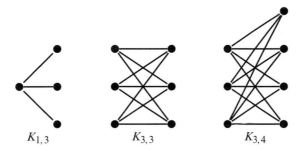

$K_{1,3}$ $\qquad$ $K_{3,3}$ $\qquad$ $K_{3,4}$

Figure 2.7 Examples of complete bipartite graphs.

In combinatorial graph theory – as opposed to topological graph theory – we are not concerned with how a graph is drawn. The drawing is just a representation that helps us "see" the vertices and the edges. All that matters is the number of vertices and their adjacencies. Hence, we consider two graphs the same if, after we appropriately relabel the vertices of one of the graphs, the set of edges of the two graphs become the same. Two such graphs are called isomorphic. We give the formal definition:

Definition 2.29 (Graph Isomorphism). Let G and H be graphs. The graph G is *isomorphic* to the graph H if there exists a 1-1, onto map $f : V(G) \rightarrow V(H)$ such that $\{x, y\}$ is an edge of G if and only if $\{f(x), f(y)\}$ is an edge of H. Such a map f is called an *isomorphism* between G and H.

Example 2.30. Consider C_4, a cycle of length 4, and label its vertices around the cycle as $1, \ldots, 4$. Now, vertices 1 and 3 are not adjacent, but each is adjacent to both 2 and 4. As such C_4 is isomorphic to the bipartite graph $K_{2,2}$. (See Figure 2.8.)

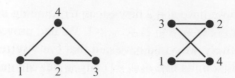

Figure 2.8 These two graphs are isomorphic graphs.

Problems

P 2.2.1. In the early 1970s, the National Football League consisted of two conferences of 13 teams each. The rules of the league specified that during the 14-week season, each team would play 11 games against teams in its own conference and three games against teams in the opposite conference. Prove that this is impossible.

P 2.2.2. Is it possible, in a finite simple graph, for all the degrees of the vertices to be distinct? Either give an example where all the degrees are distinct or prove that in a finite simple graph, we can always find at least two vertices with the same degree.

P 2.2.3. You are given a simple graph with n vertices. You also know that the graph is bipartite. What is the maximum possible number of edges in this graph?

P 2.2.4. I have a simple graph with 47 vertices. What is the maximum number of edges? What is the maximum number of edges if, in addition, I know that the graph is not connected (a graph is connected if you can go from any vertex to any other vertex by traversing the edges)?

P 2.2.5. Repeat Problem P 2.2.4 but replace 47 with another integer greater than 1. Do you see a pattern? Let the number of vertices of a simple graph be n. Let M_1 be the maximum number of edges that the graph could have, and let M_2 be the maximum possible number of edges if the graph was not connected. What is $M_1 - M_2$?

P 2.2.6. Each of nine users sends three friend requests on a social media platform. Is it possible that every user receives friend requests from precisely the three to whom she sent requests? What if the number of users was eight?

P 2.2.7. Let G be a graph whose seven vertices are Mojdeh, Mehrdokht, Māmak, Marjān, Mehrnāz, Mahshid, and Marzieh. There is an edge between two of the vertices if the two are "friends" on the social media platform that they all belong to. The list of these connections is given on the table for Warm-Up 2.12. Make a drawing of the graph G and use it to answer the following questions.

   (a) Can you find a path of length 6 in G? In other words, can you eliminate some vertices and/or some edges so that the remaining graph is a path of length 6? (If you eliminate a vertex, you have to eliminate all of the edges incident with it, and remember that the length of a path is the number of *edges* on the path.)

   (b) Can you find a cycle of length 6 in G? Again, this means that you want a cycle of length 6 after possibly eliminating some vertices and/or edges.

   (c) The *distance* between two vertices is the length of the shortest path between them. What is the distance between Māmak and Marzieh?

(d) Which two vertices are the furthest apart? In other words, the distance between which two vertices is the largest possible?

(e) What is the maximum degree of a vertex in G? What is the minimum degree?

P 2.2.8. A queen wants to build 10 castles connected by ditches. She wants exactly five ditches in the form of straight lines, and she wants four castles on every ditch. Her advisors suggested a star-shaped configuration with castles at the points of intersection of the lines (see Figure 2.9). She liked the idea of placing castles at the intersections of ditches, but decided to add a new condition. She wants one (or maybe even two) of the castles to be surrounded by ditches, and thus not subject to direct assault from the outside. Can you submit a design?[7]

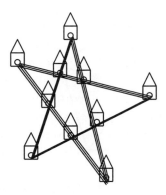

Figure 2.9 Ten castles at the pairwise intersections of five straight ditches.

P 2.2.9. By the time the queen of Problem P 2.2.8 considered all the submitted designs, an economic slowdown meant that building 10 castles was imprudent. Back to the drawing board. Since building ditches was a lot cheaper than building castles, the queen requested designs for seven castles and seven ditches. Six of the ditches were to be straight lines and one was to be circular. There would be three castles on each ditch and each castle would be at the intersection of three ditches. One of the castles would be surrounded by ditches and not in immediate danger of a direct attack. Submit a design.

P 2.2.10. A college swim conference has seven teams. During the season, each team hosts one "meet," and the meets are scheduled every other weekend throughout the season. Each meet brings together some of the teams, and those teams participate in all of the events. Is it possible to design a schedule so that each pair of teams ends up in the same meet exactly once? In other words, team A will host one meet and will go to a few others and in the course of these should be in the same meet as each of the other teams exactly once. Submit a plan. Could you make sure that each meet has the same number of teams participating?

---

[7] Mathematical folklore abounds with different versions of this puzzle.

P 2.2.11. The conference of Problem P 2.2.10 has expanded and now has 13 teams. What should they do now? Could each team still host a meet once (therefore 13 meets), and each team compete against every other team in exactly one meet?

P 2.2.12. Consider the complete graph K_4. This graph has six edges. Can you color three of the edges red and the other three blue in such a way that the graph consisting of the red edges is isomorphic to the graph consisting of the blue edges? Either show how or prove why it is not possible.

P 2.2.13. Are the two graphs in Figure 2.10 isomorphic? If so, label the vertices for the graph on the right, in order to show which vertex corresponds to which.

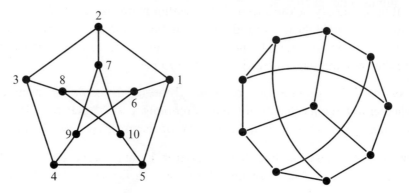

Figure 2.10 Are these two graphs isomorphic?

P 2.2.14. Repeat Problem P 2.2.12 with K_5. In other words, can you partition the edges of K_5 into two sets such that the two resulting (sub)graphs are isomorphic?

P 2.2.15. Which pairs of graphs in Figure 2.11 are isomorphic?

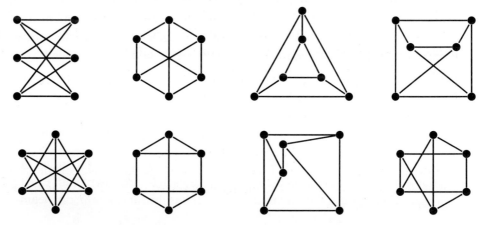

Figure 2.11 Which of these graphs are isomorphic?

2.3 Ramsey Theory

Warm-Up 2.31. *Draw a complete graph of order 5 and color each edge red or blue. (Each edge gets one color: red or blue.) Every three vertices form a triangle. We call a triangle monochromatic if all its edges have the same color. How many monochromatic triangles are there in your drawing? Now re-do the graph, but, this time, try to color the edges in a way that the number of monochromatic triangles is minimized. How many monochromatic triangles do you have now?*

Ramsey theory belongs to a thread within mathematics that says that there are patterns and order even in the most random of situations. Ramsey theory, in particular, claims that if a system is complex enough, then we will be able to find all kinds of patterns among its parts. We illustrate this vague assertion by an example.

Example 2.32. Assume that on a social media platform, two users can mutually decide whether to be connected or not. Now randomly pick six users from this platform. Regardless of whom you have picked, and regardless of the users's choices, I make a prediction.

PREDICTION: I can either find three users (from among the six) such that all three are connected to each other, or else I can find three users such that no two of them are connected to each other.

Note that, since I picked the six users randomly, it may be that none of them are connected to each other. So it would be imprudent – in fact, incorrect – to claim that there are always three users with each pair connected. What I am claiming is that *either* there are three users such that every one in that group is connected to the other two, *or* there are three users such that no pair of them are connected. In mathematics, "or" is always an "inclusive or" and so my prediction allows for the possibility that *both* events happen.

How could I possibly know such a thing? After all, the six users were picked randomly, and I have no control over who their contacts and connections are. Before giving a proof of my prediction, we reformulate the problem as one about graphs.

REFORMULATION: Construct a graph as follows. Put six vertices, one for each of the six users. Draw a blue edge between two vertices if those two are connected on the platform, and draw a red edge between the two vertices if they are not.

The resulting graph is a K_6 – it has six vertices and every pair of vertices is connected by an edge – albeit with colored edges. We have colored the edges with two colors, and we say that three vertices form a *monochromatic triangle* if the three edges of the triangle are of the same color. Our claim is that – no matter how you color the edges – there exists at least one *monochromatic triangle*.

THE ARROW NOTATION: Our claim will be written as $K_6 \rightarrow K_3, K_3$ and read "K_6 arrows K_3 or K_3." This is a mathematical statement that may or may not be true (in this case, it is true, as the next proposition shows). It claims that if we color the edges of a K_6 with red and blue (arbitrarily), then there will be at least one red K_3 or one blue K_3.[8]

Ramsey theory is interested in proving or finding counter-examples to statements of the form $K_6 \to K_3, K_3$. But it is not even clear that we can ever prove statements such as these.

Proposition 2.33.

$$K_6 \to K_3, K_3.$$

Proof. Let v be one of the vertices of K_6. Five edges are incident with v and are colored red or blue. By the pigeonhole principle, at least three of those edges will have the same color. Without loss of generality we can say that at least three of the edges are blue. Look at the other ends of these three blue edges. Either two of these are joined by a blue edge in which case those two and v together form a blue triangle, or none of them are joined by a blue edge, in which case these three vertices form a red triangle. (See Figure 2.12.) □

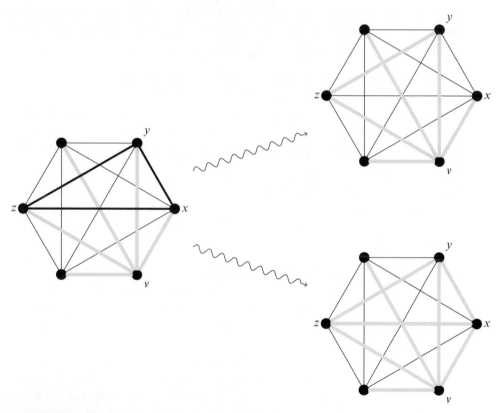

Figure 2.12 If the edges of a K_6 are colored with red and blue, then, out of the five edges incident with v, at least three will be of the same – say blue – color (figure on the left). Let x, y, and z be the other vertices of these three edges. The triangle $\triangle xyz$ either has a blue edge (top right) or all three of its edges are red (bottom right). A black and white version of this figure will appear in some formats. For the color version, please refer to the plate section.

---

[8] The arrow notation in Ramsey theory was first used by Erdős and Rado (1956).

Proposition 2.34.

$$K_5 \not\rightarrow K_3, K_3.$$

Proof. The statement $K_5 \rightarrow K_3, K_3$, if true, would have said that if you color each of the edges of a K_5 with either blue or red, then you are guaranteed a blue or a red triangle. Hence, the statement $K_5 \not\rightarrow K_3, K_3$ says that it is possible to color the edges of a K_5 with two colors and not have a monochromatic triangle. To show this, we need to come up with an example. Figure 2.13 gives an example of a coloring of the edges of K_5 without any monochromatic triangles. (You were asked to find this example in Warm-Up 2.31.) $\qquad\square$

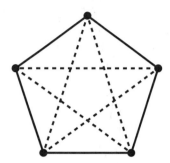

Figure 2.13 $K_5 \not\rightarrow K_3, K_3$. The solid edges are given one color and the dashed edges the second color.

Remark 2.35. We could say that Ramsey theory creates some order out of chaos. In a situation where edges are colored arbitrarily and without any planning, we are still able to guarantee the existence of certain configurations. The basic underlying principle is that if a system is complex and large enough, then we can find almost any pattern in it. In collaborative Mini-project 1, you are asked to revisit K_6 and to prove an even stronger statement than Proposition 2.33.

Propositions 2.33 and 2.34 may be unexpected and possibly interesting, but they are very specific. Using our notation, we can certainly ask many other questions. For example, does $K_6 \rightarrow K_4, K_4$? This is asking whether, if a group of six people shake hands, there will by necessity be either a group of four people who all shook hands with each other or a group of four people such that none of them shook hands. The answer is no, and you can construct an example by appropriately coloring the edges of a K_6. Then the question becomes whether there is any integer n such that $K_n \rightarrow K_4, K_4$. Our notation even suggests that maybe we can ask questions such as whether $K_7 \rightarrow K_3, K_4$. We first clarify what this notation means.

Definition 2.36 (Arrow Notation). Let $s, n,$ and m be positive integers with $n, m \le s$. Designate blue as the first color and red as the second color. Color the edges of K_s, the complete graph of order s, using blue and red. In this coloring, a *blue* K_n is a set of n vertices in the K_s such that all the edges between these vertices are blue. The statement $K_s \rightarrow K_n, K_m$ asserts – maybe truthfully, maybe not – that if we color the edges of K_s using red and blue, then we are guaranteed either a blue K_n or a red K_m.

Remark 2.37. Note that in our notation the order of the colors matters. For example, if the first color is blue and the second color is red, and you assert that $K_9 \rightarrow K_3, K_4$, then it has to be the case that in any coloring of the edges of K_9 with blue and red, there is a blue K_3 or a red K_4. In other words, finding a red K_3 will not be adequate.

Ramsey's theorem – which we shall prove after a bit of discussion – says that, given n and m, there exists a large enough integer s such that $K_s \rightarrow K_n, K_m$. The theorem does not tell us what the smallest such s is, and finding such s can actually be very difficult.

Theorem 2.38 (Ramsey 1929).[9]   *Let n and m be positive integers. Then there exists a positive integer s – which depends on n and m – such that*

$$K_s \rightarrow K_n, K_m.$$

In other words, given positive integers n and m, we can always find a large enough integer s such that no matter how we color the edges of K_s with blue and red, we can always find a blue K_n or a red K_m.

Proof. This theorem will be reworded and proved as Theorem 2.44.   □

Note that to use Ramsey's theorem, you don't start with s. You decide on the two outcomes that are acceptable to you by choosing n and m. Ramsey's theorem, then, asserts that there is some s large enough so that your wishes come true. For such an s, if you color the edges of K_s with blue and red, then you are guaranteed a blue K_n or a red K_m.

Given positive integers n and m, Ramsey's theorem does not give any indication about the values of s for which $K_s \rightarrow K_n, K_m$. If some s works, then a larger s will also work, and, likewise, if some s does not work, then neither will smaller values of s. As a result, we are interested in the smallest value of s for which the statement $K_s \rightarrow K_n, K_m$ is true. Often naming a phenomenon is the first step toward gaining some control over the situation.

Definition 2.39 (Ramsey Number). Let n and m be positive integers. Let s be the smallest positive integer such that $K_s \rightarrow K_n, K_m$. Then s is denoted by $r(n,m)$ and is called a *Ramsey number*. If $n = m$, then, sometimes, $r(n,n)$ is shortened to $r(n)$.

Remark 2.40. Let n and m be positive integers. By definition, if you claim that $N = r(n,m)$, then to prove your claim, you have to prove two things. You have to show that $K_N \rightarrow K_n, K_m$ and that $K_{N-1} \nrightarrow K_n, K_m$. The latter is done by an example. The former is done by some kind of a (possibly intricate) argument.

---

[9] Frank Ramsey (1903–1930) proved what we call Ramsey's theorem as a minor lemma in a treatise on logic. Frank Ramsey's father was also a mathematician and the president of Magdalene College, Cambridge. In his short life, Frank Ramsey wrote papers not just in mathematics, but also in philosophy – he was a close friend of Ludwig Wittgenstein – and in economics – he was a student of John Maynard Keynes. He has been described as a "militant athiest" even though his brother became the Bishop of Canterbury. Frank Ramsey was also interested in politics. His brother has been quoted as saying that "he [Frank] had got a political concern and a sort of left-wing caring-for-the-underdog kind of outlook about politics."

Example 2.41. Propositions 2.33 and 2.34 prove that $r(3,3) = 6$. Also note that $r(2,6) > 5$. To show this, designate blue as color 1 and red as color 2 and color every edge of a K_5 with the color red. Then, in this coloring, there is no blue K_2 and no red K_6, proving that $K_5 \nrightarrow K_2, K_6$. Note that it is irrelevant that there are plenty of red K_2's. By designating blue as the first color, we can satisfy the statement $K_s \rightarrow K_2, K_6$ only if we produce a blue K_2 or a red K_6.

Before we prove Ramsey's theorem, we discuss the Ramsey numbers $r(n,m)$ a bit. The next lemma gathers a few straightforward facts about these numbers.

Lemma 2.42. *Let n and m be positive integers. Then*

(a) $r(m,n) = r(n,m)$,
(b) $r(1,m) = 1$,
(c) for $m > 1$, $r(2,m) = m$, and
(d) $r(3,3) = 6$.

Proof. (a) While we do fix the colors at the beginning, there is no intrinsic difference between red and blue, and hence they could have been switched.
(b) $K_1 \rightarrow K_1, K_m$, since a K_1 has no edges and hence is already monochromatic.
(c) First we show that $r(2,m) \leq m$, by arguing that $K_m \rightarrow K_2, K_m$. The latter is true, since if you color the edges of K_m with blue and red, then either some edge is blue – a blue K_2 – or all edges are red – a red K_m. Next, we show that $r(2,m) > m - 1$, by giving an example to show $K_{m-1} \nrightarrow K_2, K_m$. If you color every edge of K_{m-1} red, then you have neither a blue K_2 nor a red K_m.
(d) This is Propositions 2.33 and 2.34. □

Does $r(n,m)$ exist? How could it not exist? The notation $r(n,m)$ denotes a specific integer. If $r(n,m) = N$, then N is an integer such that $K_N \rightarrow K_n, K_m$ while $K_{N-1} \nrightarrow K_n, K_m$. The problem is that there is no a priori reason why such a number should exist. It could be that, no matter how large a value of N you pick, you still can color the edges of a K_N with blue and red in such a way that there is no blue K_n or red K_m. To pass this hurdle and know that $r(n,m)$ exists – regardless of whether we can find it or not – we have to come up with some large integer N for which $K_N \rightarrow K_n, K_m$. This is what Ramsey's theorem promises. So, to prove Ramsey's theorem, we don't have to produce the smallest N for which $K_N \rightarrow K_n, K_m$, but we have to show that *some* N – even if it is not the smallest such N – does work.

First, let us organize in Table 2.2 the few values of $r(n,m)$ given in Lemma 2.42. We know the first two rows and the first two columns of the table, but, for the other entries, at this point we don't even know that each entry is a finite number. We prove that they are – and as a result prove Ramsey's theorem – by proving that $r(n,m) \leq r(n-1,m) + r(n,m-1)$. This means that, for example, $r(3,4) < r(2,4) + r(3,3) = 4 + 6 = 10$. Likewise, $r(4,3) \leq 10$. Knowing these, we can then say $r(4,4) \leq r(3,4) + r(4,3) \leq 10 + 10 = 20$. Continuing in this way – since we already have the first two rows and columns – we can find an upper bound for each of the entries. In particular, we will know that $r(m,n)$ is a finite number, and this is what Ramsey's theorem promises.

Table 2.2 Values of $r(n,m)$ given in Lemma 2.42.

n/m	1	2	3	4	5	$\cdots$
1	1	1	1	1	1	$\cdots$
2	1	2	3	4	5	$\cdots$
3	1	3	6			
4	1	4				
5	1	5				
$\vdots$	$\vdots$	$\vdots$				

Before proving the general statement, and to make the argument clearer, we will prove a special case. The proof of the general case will mimic this particular special case.

Example 2.43. Prove that $r(3,4) \leq r(2,4) + r(3,3) = 4 + 6 = 10$.

Proof. Showing that $r(3,4) \leq 10$ is the same as showing that $K_{10} \to K_3, K_4$. Let blue be the first color and red the second color, and arbitrarily color – using the colors blue and red – the edges of K_{10}, the complete graph of order 10. We have to demonstrate that, regardless of the coloring scheme, there is a blue K_3 or a red K_4.

Let v be one of the vertices of K_{10}. There are nine edges incident with v, and we remember $10 = r(2,4) + r(3,3) = 4 + 6$.

CLAIM: Either at least four edges incident with v are blue (the first color) or at least six edges incident with v are red (the second color).

PROOF OF CLAIM: If the claim were not true, then there would be at most three blue edges and at most five red edges incident with v. This gives a total of at most eight edges incident with v. But v is incident with nine edges. The contradiction proves the claim.

CASE 1: Assume four edges incident with v are blue. Let u_1, u_2, u_3, and u_4 be the other ends of these edges. We know $K_4 \to K_2, K_4$. Thus, among these four vertices, we will either have a blue K_2 or a red K_4. If we have a blue K_2, then this edge together with v forms a blue K_3 (because all the edges between v and the u_i's are blue), and if we have a red K_4, then we have the promised red K_4. See Figure 2.14. Thus we are done in this case.

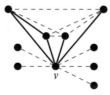

Figure 2.14 Assume at least four of the nine edges incident with v are blue (solid lines). If there is at least one blue edge among the other ends of these blue edges, then we have a blue triangle. Otherwise, the other ends of the four blue edges form a red K_4.

CASE 2: Assume six edges incident with v are red. Let $w_1, w_2, \ldots, w_6$ be the other ends of these edges. We know $K_6 \to K_3, K_3$. Thus among these six vertices will have either a blue K_3 or a red K_3. If we have a blue K_3 then we have the needed blue K_3. On the other hand, if we had a red K_3, then this red triangle together with v forms a red K_4 (because all the edges between v and the w_i's are red). So we either have a blue K_3 or a red K_4 as desired. □

We are finally ready to reword and prove Ramsey's theorem, Theorem 2.38. (For a different proof, see Problem P 2.3.20.)

Theorem 2.44. *Let n and m be integers greater than* 1. *Then*

$$r(n,m) \leq r(n-1,m) + r(n,m-1).$$

In particular, since, for all positive integers n and m, $r(n,1) = r(1,m) = 1$, $r(n,m)$ is always a finite number.

Proof. Let $s = r(n-1,m) + r(n,m-1)$. Showing that $r(n,m) \leq s$ is the same as showing $K_s \to K_n, K_m$. So, arbitrarily color the edges of K_s using blue (color 1) or red (color 2). We need to show that there is always either a blue K_n or a red K_m.

Let v be one of the s vertices of K_s. Then v is incident with $s-1$ edges, and, we claim:

CLAIM: Either at least $r(n-1,m)$ edges incident with v are blue, or at least $r(n,m-1)$ of the edges incident with v are red.

PROOF OF CLAIM: If not, the number of edges incident with v would be at most

$$r(n-1,m) - 1 + r(n,m-1) - 1 = s - 2 < s - 1.$$

This is a contradiction: the number of edges incident with v is $s-1$, which proves the claim.

CASE 1: Assume $t = r(n-1,m)$ edges incident with v are blue. Let $u_1, u_2, \ldots, u_t$ be the other ends of these edges. By the definition of $r(n-1,m)$ and since $t = r(n-1,m)$, we know $K_t \to K_{n-1}, K_m$. Thus among these t vertices we will either have a blue K_{n-1} or a red K_m. If we have a blue K_{n-1} then this edge together with v forms a blue K_n (because all the edges between v and the u_i's are blue), and if we have a red K_m, then we have the desired red K_m. Thus we are done in this case.

CASE 2: Assume $k = r(n,m-1)$ edges incident with v are red. Let $w_1, w_2, \ldots, w_k$ be the other ends of these edges, and, since $k = r(n,m-1)$, we know $K_k \to K_n, K_{m-1}$. Thus among these k vertices we will either have a blue K_n or a red K_{m-1}. Again, if we have a blue K_n then we have the needed blue K_n. On the other hand, if we have a red K_{m-1}, then this red K_{m-1} together with v forms a red K_m (because all the edges between v and the w_i's are red). So we either have a blue K_n or a red K_m as desired. □

Remark 2.45. If $s = r(n,m)$, then by definition of the Ramsey number $r(n,m)$ we know that $K_s \to K_n, K_m$. A reasonable question is whether this is all we can safely predict. For example, $K_6 \to K_3, K_3$, and so when we color the edges of a K_6 with two colors, we are guaranteed at least one monochromatic triangle. Is it possible that we are guaranteed at least two monochromatic triangles? Our proof of $K_6 \rightarrow K_3, K_3$ does not prove such a thing, but maybe a different proof can. You are asked to investigate and answer this question in Collaborative Mini-project 1.

Using Theorem 2.44, we can get an upper bound for Ramsey numbers. At least in one case, using a clever argument, we can improve the bound.

Proposition 2.46. *Let n and m be integers greater than 1. If both $r(n-1,m)$ and $r(n,m-1)$ are even integers, then*

$$r(n,m) \leq r(n-1,m) + r(n,m-1) - 1.$$

In other words, in Theorem 2.44, if both terms on the right-hand side of the inequality are even, we can never achieve equality.

Proof. Let $s = r(n-1,m) + r(n,m-1) - 1$. We have to show that $K_s \rightarrow K_n, K_m$. Color the edges of a K_s with blue and red, and let v be a vertex of K_s. Then v is incident with $s-1$ edges. If from among these edges at least $r(n-1,m)$ of them are blue or at least $r(n,m-1)$ of them are red, then we would proceed exactly as in the proof of Theorem 2.44, and we would be done. But is it possible that, from among the $s-1$ edges incident with v, exactly $r(n-1,m)-1$ of the edges are colored blue and $r(n,m-1)-1$ of the edges are colored red? The insight is to realize that v was chosen arbitrarily, and so the problem case is when, for *every* vertex of the K_s, our coloring scheme results in $r(n-1,m)-1$ blue and $r(n,m-1)-1$ red edges. If that is so, let's count the total number of blue edges in the K_s. For every vertex we have $r(n-1,m)-1$ blue edges and there are s vertices. This gives a total of $[r(n-1,m)-1]s$, but every blue edge will have been counted twice (once at each of its two vertices), and so the count of the total number of blue edges is

$$\frac{[r(n-1,m)-1]s}{2}.$$

But this is not an integer since both $r(n-1,m)-1$ and s are odd! The contradiction completes the proof.   □

Corollary 2.47 (Greenwood and Gleason 1955).

$$r(3,4) = r(4,3) = 9, \ r(3,5) = r(5,3) = 14, \ and \ r(4,4) = 18.$$

Proof. Since $r(2,4) = 4$ and $r(3,3) = 6$ are both even, Proposition 2.46 applies, and $r(4,3) = r(3,4) \leq r(2,4) + r(3,3) - 1 = 9$. Applying Theorem 2.44, we get that $r(3,5) = r(5,3) \leq r(4,3) + r(5,2) \leq 9 + 5 = 14$, and $r(4,4) \leq r(3,4) + r(4,3) \leq 18$. So far, we have proved that $r(3,4) = r(4,3) \leq 9, r(3,5) = r(5,3) \leq 14$, and $r(4,4) \leq 18$. To prove that these are equalities, we need examples to show that these numbers cannot be any smaller. In Problems P 2.3.8, *P 2.3.17*, and P 2.3.18, you are asked to construct examples to show that $r(3,4) > 8, r(3,5) > 13$, and $r(4,4) > 17$ to complete the proof.   □

Remark 2.48. We know that $r(2,2) = 2$, $r(3,3) = 6$, and $r(4,4) = 18$. Looking at this list you may guess that $r(5,5)$ is 54. However, as of this writing, finding the precise value of $r(5,5)$ remains an open problem in combinatorics. As of October 2020, it was known that $43 \leq r(5,5) \leq 48$.[10] You may want to ponder why it was hard to do a complete computer search to find the answer.

The famous mathematician Paul Erdős (1913–1996) remarked that if aliens invaded and threatened to destroy Earth unless we find $r(5,5)$ within a year, then we probably could rise to the challenge. However, if they asked for $r(6,6)$, then we should attack first.

Theorem 2.44 and Proposition 2.46 give an upper bound for Ramsey numbers. We now construct a (not so good) lower bound. (See Theorem 5.14, Corollary 5.15, and Problem P 5.1.33 for other lower bounds.)

Proposition 2.49. *Let n and m be positive integers. Then*

$$r(n,m) \geq (n-1)(m-1) + 1.$$

Proof. Let $s = (n-1)(m-1)$. We have to show that $K_s \not\rightarrow K_n, K_m$. We will color a K_s with two colors in a specific way that will contain no blue K_n and no red K_m.

Arrange the vertices of K_s in a rectangular array with $n-1$ rows and $m-1$ columns. So, in each row there are $m-1$ vertices and in each column there are $n-1$ vertices. If an edge connects two vertices in the same row, then color it red, and color all other edges blue. (See Figure 2.15, where the dashed and solid lines stand for the colors red and blue, respectively.)

Figure 2.15 An example that shows $K_6 \not\rightarrow K_3, K_4$.

Given any n vertices, at least two of them will be – by the pigeonhole principle – in the same row and hence connected by a red edge. Hence there is no blue K_n. On the other hand, given any set of m vertices, at least two of them will be in the same column. Hence the edge connecting them is blue. Thus there are also no red K_m's. The proof is complete. $\square$

Remark 2.50. Our Theorem 2.44, Proposition 2.46, and Proposition 2.49 do give lower and upper bounds for the Ramsey numbers and so they narrow the range of possible values. As we saw in Corollary 2.47, for very small parameters, our upper bounds are pretty good. However, for larger parameters our bounds are quite weak. The lower bound given in Proposition 2.49 is particularly weak, and can be strengthened considerably. (See Theorem 5.14, Corollary 5.15, and Problem P 5.1.33.) In Table 2.3, we have started with the Ramsey numbers that we know,

[10] The bound was $43 \leq r(5,5) \leq 49$ until March 2017, when Angeltveit and McKay (2017) proved that $r(5,5) \leq 48$ by computer verification, checking approximately 2 trillion separate cases.

and used them to generate upper bounds for the rest of the entries. We have also put the lower bounds that we have so far. For some entries, we used the fact that the Ramsey numbers are a non-decreasing function of the parameters, and so, for example, since $r(4,4) = 18$, we must have $r(4,5) \geq 18$. The numbers show that our lower bound increases much too slowly compared to the upper bounds. Needless to say, much more is known than what we have presented here (see Table 2.4). For example, according to what we have proved, $18 \leq r(5,5) \leq 62$ and $26 \leq r(6,6) \leq 222$, but it is known – using much more involved proofs – that $43 \leq r(5,5) \leq 48$ and $102 \leq r(6,6) \leq 161$.

Table 2.3 A summary of what we have proved for Ramsey numbers $r(n,m)$. If an entry contains a single number, then it is the precise Ramsey number. Otherwise, the two numbers in the entry give, for the Ramsey number, a lower and an upper bound that follow from what we have proved. For the best known bounds see Table 2.4.

n/m	1	2	3	4	5	6	$\cdots$
1	1	1	1	1	1	1	$\cdots$
2	1	2	3	4	5	6	$\cdots$
3	1	3	6	9	14	14-19	
4	1	4	9	18	18-31	18-50	
5	1	5	14	18-31	18-62	21-111	
6	1	6	14-19	19-50	21-111	26-222	
$\vdots$	$\vdots$	$\vdots$					

A More Colorful Ramsey's Theorem*

We can generalize what we have done by using more than two colors in coloring the edges.

Definition 2.51 (The Arrow Notation and Ramsey Numbers, Generalized). If $s, n_1, \ldots, n_k$ are positive integers, then $K_s \rightarrow K_{n_1}, K_{n_2}, \ldots, K_{n_k}$ is the statement that if we color the edges of K_s using k colors then we are guaranteed a monochromatic K_{n_1} of the first color, or a monochromatic K_{n_2} of the second color, ..., or a monochromatic K_{n_k} of the kth color. The smallest positive integer s for which the statement $K_s \rightarrow K_{n_1}, K_{n_2}, \ldots, K_{n_k}$ is true is called a *Ramsey number* and denoted by $r(n_1, n_2, \ldots, n_k)$.

Depending on the values of $s, n_1, \ldots, n_k$, the statement $K_s \rightarrow K_{n_1}, K_{n_2}, \ldots, K_{n_k}$ may be true or false. The more general version of Ramsey's theorem says that the statement is true if s is large enough, and as a result Ramsey numbers $r(n_1, \ldots, n_k)$ exist. Understanding this proof is optional, and the first-time reader may decide to skip it.

Theorem 2.52. *Let $n_1, \ldots, n_k$ be positive integers. Then there exists a positive integer s such that*

$$K_s \to K_{n_1}, K_{n_2}, \ldots, K_{n_k}.$$

In fact, for $k \geq 3$, if $m = r(n_{k-1}, n_k)$, then

$$r(n_1, n_2, \ldots, n_k) \leq r(n_1, \ldots, n_{k-2}, m).$$

Proof. The proof is by induction on k, the number of colors. The base case $k = 1$ is trivial, and we have already proved the case $k = 2$ (Theorem 2.44). So, we assume that $k \geq 3$. The positive integers $n_1, \ldots, n_k$ are given (you can think of these as our wish list for a monochromatic K_{n_1} of the first color or a monochromatic K_{n_2} of the second color, or ...), and we let $m = r(n_{k-1}, n_k)$, and

$$n = r(n_1, n_2, \ldots, n_{k-2}, m).$$

Note that m and n are both Ramsey numbers, and we know that they exist by Theorem 2.44 and the inductive hypothesis, respectively. So we know that $K_m \to K_{n_{k-1}}, K_{n_k}$, and $K_n \to K_{n_1}, \ldots, K_{n_{k-2}}, K_m$. (In the former case we are coloring the edges of K_m with two colors, while in the latter case, we are coloring each edge of K_n with one of $k - 1$ colors.) We now claim that

$$K_n \to K_{n_1}, K_{n_2}, \ldots, K_{n_{k-1}}, K_{n_k}.$$

If true, this would mean that $r(n_1, \ldots, n_k) \leq n$ and the proof would be complete. To prove the claim, color the edges of K_n using k colors. We have to show that there is either a monochromatic K_{n_1} of the first color, or a monochromatic K_{n_2} of the second color, ..., or a monochromatic K_{n_k} of the kth color. Now think of the last two colors as different hues of the same color. In other words, we are thinking that all the edges that have either of the last two colors have the same color. This means that we have colored the edges of K_n using $k - 1$ colors. By definition of n, this means that we either have a monochromatic K_{n_1} of the first color, a monochromatic K_{n_2} of the second color, ..., a monochromatic $K_{n_{k-2}}$ of the $k - 2$nd color, or a monochromatic K_m of the last color (the combined colors). We have what we want, except possibly for the last case. So assume that we have a K_m with all of its edges having the combined color. But this means that we have a K_m whose edges are colored using one of two colors (the original $(k-1)$th and kth colors). By definition of m, this means that we have a monochromatic $K_{n_{k-1}}$ of color $k - 1$ or a monochromatic K_{n_k} of color k. The proof is now complete. $\qquad\square$

Problems

P 2.3.1. Six foreign ministers meet. Among every three of them at least two have a common language. Show that we can pick three foreign ministers such that every one of them has a common language with each of the other two.

P 2.3.2. The goddess Athena comes to you in a dream and tells you that the Ramsey number $r(4, 5)$ is 25. You wake up in a calm mood and take a walk down the street. You meet a K_{25} whose edges are colored using blue and green. You introduce yourself and—without even asking—you know that at least one of the following must be true about the K_{25}:
(a) It has a blue K_5 as a subgraph, or
(b) it has a green K_5 as a subgraph, or
(c) it has a blue K_4 *and* a green K_4 as a subgraph.
Why?

P 2.3.3. Assume that you color the edges of a K_8 with red and blue. Let v be one of the vertices, and assume that among the seven edges incident with v, there are six blue edges and one red edge. Show that this coloring of K_8 has either a blue K_4 or a red K_3.

P 2.3.4. There is a group of 20 people on a social media platform that you are curious about. You look at one of their profiles, and see that this person is friends with 18 of the remaining 19 people, and not friends with just 1 of the group. Prove that, from among the 20 people, you can either find 5 such that they are all pair-wise friends or 4 such that no pair of them are friends.

P 2.3.5. You go to a play. In Act 1 Scene 3, three witches/oracles make prophecies. The first one declares that the Ramsey number $r(4, 4) = 18$. The second one pronounces that the Ramsey number $r(3, 7) = 23$, and the third one says: "Thou shalt get kings, though thou be none."

Ignoring this last statement, you color the edges of a K_{42} using red and blue. You notice that, in your coloring, one vertex v has 23 red edges, and in the whole K_{42} there is no red K_4.

Using the first two prophecies, prove that there is a blue K_5 as well as a separate blue K_7 in your coloring of K_{42}.

P 2.3.6. Let G be a simple graph. A set of vertices of G that have no edges between them is called an *independent* set of vertices. By contrast, a *clique* is a set of vertices of G such that every pair of the vertices is connected by an edge. Show that if the number of vertices of G is at least $r(n, m)$, then G either has a clique of size n or an independent set of size m.

P 2.3.7. **An Infinite Ramsey's Theorem.** Let G be an infinite simple graph. Using the definitions and the result of Problem P 2.3.6, prove that G either has an infinite clique or an infinite independent set.

P 2.3.8. Show that $r(3, 4) > 8$.

P 2.3.9. In one of your voyages, an oracle tells you that the Ramsey number $r(3, 6) = 18$. What can you say about $r(2, 3, 6)$? Give your reasoning.

P 2.3.10. What can you say about the Ramsey number $r(2, 2, 2, 3, 5)$?

P 2.3.11. Prove that

$$r(3, 3, 5, 5) \leq r(14, 14).$$

P 2.3.12. In a dream, the empress regnant Wu Zetian tells you that $r(3, 9) = 36$. In the morning, you color the edges of a K_{36} using *three* colors: red, blue, and yellow, making sure that there are no red triangles. Prove that, no matter how you colored the edges, there is either a blue or a yellow K_4, or else there is a blue *and* a yellow K_3.

P 2.3.13. Prove that $r(3,3,3) \leq 17$.

P 2.3.14. What can you say about $r(3,3,3,3)$?

P 2.3.15. Go on the Internet and search for the (5-regular) *Clebsch graph*, also known as the *Greenwood–Gleason graph*.

   (a) Find a drawing of the graph that you like and reproduce it.

   (b) How many vertices and how many edges does this graph have?

   (c) What is the degree of each vertex? Figure out what "5-regular" means and verify that this graph is 5-regular.

   (d) Does the Clebsch graph have any triangles?

   (e) You read that you can color the edges of K_{16}, the complete graph on 16 vertices, with three colors in such a way that all the vertices together with the edges for each one of the colors gives a Clebsch graph. Are the degrees of vertices and the number of edges in the Clebsch graph and in K_{16} compatible with this assertion?

   (f) Explain why the assertion in the previous part implies that $r(3,3,3) > 16$.

   (g) Give a sketch for the proof that $r(3,3,3) = 17$.

P 2.3.16. A fielding team in baseball has nine members (pitcher, catcher, first baseman, second baseman, third baseman, shortstop, left fielder, center fielder, and right fielder). The nine members of a team are standing around and throwing balls to each other. Prove that, regardless of who throws to whom, you can either find three players none of whom threw a ball to either of the other two, or you can find three players A, B, and C such that A threw a ball to B, B threw a ball to C, and A threw a ball to C. Note that if player A throws a ball to player B, we cannot assume that player B also threw a ball to player A. You may find the following steps helpful:

   STEP 1: Model the situation with a *directed graph*. A directed graph is a graph where the edges have a direction. In a directed graph, a directed path or a directed cycle is a path or a cycle that respects the directions of the edges (you are never going against the given direction of an edge). Given three vertices a, b, c of a directed K_3, if a is directed toward b, b is directed toward c, and a is directed toward c, then we say that the vertices a, b, and c are a transitive triple. Show that a directed K_3 (in other words, a triangle where each edge has a direction) is either a directed cycle or a transitive triple.

   STEP 2: Consider a directed K_4—i.e., four vertices where each pair is connected by an edge and the edge is directed one way or the other—and show that it must contain a transitive triple.

   STEP 3: Find a way to use Corollary 2.47.

   STEP 4: Combine Step 3 and Step 2 to complete the proof.

P 2.3.17. Prove that $r(3,5) > 13$. You may find the following steps useful.

   STEP 1: Put the vertices of a K_{13} around the circle and label them with $0, 1, \ldots, 12$. Let $S = \{1,5,8,12\}$. If $i < j$, the edge between vertex i and vertex j is colored red if $j - i$ is in S.[11] For example, the edge between 3 and 11 is red since $11 - 3 = 8 \in S$. Convince

---

[11] This construction may seem quite arbitrary. It seems less so if you are familiar with modular arithmetic and think of the vertices as the integers modulo 13. Then $S = \{\pm 1, \pm 5\}$ are the non-zero cubic residues modulo 13. This just means that when we do arithmetical operations, instead of the usual answers, we write the remainders when divided

yourself that this is the same as drawing a red cycle of length 13 and then putting a red edge for every two vertices that are distance 5 apart. We color all the other edges of K_{13} blue.

STEP 2: Assume that $\{i, j, k\}$ are the vertices of a red triangle. Show that this would mean that two elements of S must add up to a third element of S, and this does not happen.

STEP 3: Without drawing the blue edges, convince yourself that the graph consisting of just the blue edges is very symmetrical. More precisely, if the vertices did not have labels, would it be possible to distinguish between them?

STEP 4: By way of contradiction, assume that, in this coloring of the edges of K_{13}, there is a blue K_5 that has 0 as one of its vertices. Argue that the other vertices of the blue K_5 must come from among the elements of the following subsets: $\{2, 3, 4\}$, $\{6, 7\}$, and $\{9, 10, 11\}$. Argue that the blue K_5 must have two vertices from one of the three-element subsets. Then narrow it down to two choices, and then eliminate those possibilities as well.

STEP 5: Using Step 3, argue that the choice, in Step 4, of 0 as one of the vertices was not consequential.

STEP 6: Complete the proof and rewrite it concisely.

P 2.3.18. Show that $r(4, 4) > 17$. The following steps, in some ways similar to those for Problem $P\ 2.3.17$, may be helpful:

STEP 1: Put the vertices of a K_{17} around the circle and label them with $0, 1, \ldots, 16$. Let $S = \{1, 2, 4, 8, 9, 13, 15, 16\}$.[12] If $i < j$, assign the color red to the edge between vertex i and vertex j if $j - i$ is in S. Convince yourself that you would get the same red edges if you

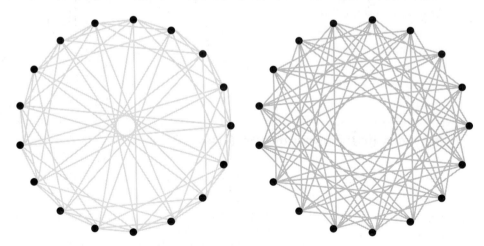

Figure 2.16 The red (left) and blue (right) edges in K_{17} without a red or a blue K_4, showing $r(4, 4) > 17$. A black and white version of this figure will appear in some formats. For the color version, please refer to the plate section.

by 13. For example, $5 + 10 = 2$, $5^2 = 12$, and $-5 = 8$ (since $5 + 8 = 0$). Modulo 13, we have $1^3 = 1 = 3^3 = 9^3$, $2^3 = 8 = 5^3 = 6^3$, $4^3 = 12 = 10^3 = 12^3$, $7^3 = 5 = 8^3 = 11^3$, and so the non-zero cubes are 1, 5, 8 = -5, and $12 = -1$.

12 The set S consists of the non-zero squares modulo 17.

start with a red cycle of length 17 and then add red edges between any two vertices whose distance on the cycle is 2, 4, or 8. Color all the other edges of K_{17} blue. See Figure 2.16.

STEP 2: Give a description of the blue edges akin to the one for the red edges in Step 1.

STEP 3: Assume that the vertices 0, i, j, and k form a red K_4. Show that i, j, k and their differences must all be in S and that this is not possible. Could there be any other red K_4's?

STEP 4: Show that there is also no blue K_4, and complete the proof.

P 2.3.19. In Problem P 2.3.18 we colored the edges of a K_{17} using red and blue in a specific way (see Figure 2.16). Show that the graph consisting of the red edges is isomorphic to the graph consisting of the blue edges.

P 2.3.20. **An Upper Bound for** $r(n,n)$**.** Give a different proof of the diagonal case of Ramsey's theorem by proving that $r(n,n) \leq 4^n$. You may find the following steps helpful:

STEP 1: Let $m_0 = 4^n = 2^{2n}$, and assume an arbitrary coloring of the edges of K_{m_0} is given. Let v_1 be one of the vertices. Show that v_1 has at least $m_1 = 2^{2n-1}$ incident edges all with the same color. Focus on the complete graph K_{m_1} whose vertices are the other vertices of these m_1 edges.

STEP 2: Let v_2 be one of the vertices of K_{m_1}. Show that at least $m_2 = 2^{2n-2}$ edges incident with v_2 (in K_{m_1}) have the same color. Focus on the complete graph K_{m_2} whose vertices are the other vertices of these m_2 edges.

STEP 3: Continue the above process and form the ordered set $S = \{v_1, v_2, \ldots, v_{2n-1}\}$. Do all the edges of the form $\{v_{47}, v_j\}$, where $48 \leq j \leq 2n - 1$, have the same color? If so, label v_{47} with that color.

STEP 4: Show that at least n elements of S are labeled with the same color and that these form a monochromatic K_n.

2.4   Schur, Van der Waerden, and Graph Ramsey Numbers*

Warm-Up 2.53. *On five pieces of paper, write the integers 1 through 5. Can you put the numbers in two piles in such a way that if you add two numbers in the same pile (you are allowed to add a number to itself), the sum is never in that same pile? What if you had only 1 through 4?*

In this optional section, we briefly and passingly mention three sets of numbers related to, and in the same spirit of, Ramsey numbers. Graph Ramsey numbers are a somewhat straightforward generalization of Ramsey numbers (and you are asked to explore them in the problems), while the proof presented here for the existence of Schur numbers is an interesting application of Ramsey's theorem. For the third sequence of numbers – the Van der Waerden numbers – we limit ourselves to the definition and a few examples (in the problems at the end of this section).

Schur's Theorem and Schur Numbers

Definition 2.54 (Partition). Let S be a set. A *partition* of S is a collection $S_1, \ldots, S_m$ of subsets of S such that $S = S_1 \cup S_2 \ldots \cup S_m$ and $S_i \cap S_j = \emptyset$ for $i \neq j$. In other words, the subsets $S_1, \ldots, S_m$ form a partition of S, if each element of S is in exactly one of these subsets. The subsets $S_1, \ldots, S_m$ are called the *parts* of the partition.

Example 2.55. Consider $[10] = \{1, 2, \ldots, 10\}$. We can partition $[10]$ into three parts as

$$\{1, 4, 10\}, \quad \{2, 3\}, \quad \{5, 6, 7, 8, 9\}.$$

This particular partition has a peculiar property: if x and y ($x = y$ is allowed) are in one of the parts then $x + y$ is *not* in that part. We say that the parts of this partition are *sum-free*. Here is another partition of $[10]$ into sum-free parts:

$$\{1, 4, 7, 10\}, \quad \{2, 3\}, \quad \{5, 6, 8, 9\}.$$

We can actually do better, and get a partition of $[13]$ with the same property:

$$\{1, 4, 10, 13\}, \quad \{2, 3, 11, 12\}, \quad \{5, 6, 7, 8, 9\}.$$

What about $[14]$? Since $4 + 10 = 3 + 11 = 7 + 7 = 14$, we cannot add 14 to any of the existing parts and keep the sum-free property, but maybe there is a different partition of $[14]$ into three parts that would have the same property. Actually there is no such partition. In other words, no matter how you partition $[14]$ into three parts, you can find an x, y, and $x + y$ in the same part.

Definition 2.56. Let A be a set consisting of integers. Then A is *sum-free* if, for all $x, y \in A$, the integer $x + y$ is not in A. (Note that x is allowed to be equal to y.)

Theorem 2.57 (Schur 1916).[13]   *Let $n \geq 1$ be an integer. Then there exists a positive integer M such that no matter how you partition $[M] = \{1, 2, \ldots, M\}$ into n parts, one of the parts will not be sum-free.*[14]

---

[13] My mathematical great grandfather (he was the thesis advisor of Richard Brauer, who was the thesis advisor of my PhD advisor Marty Isaacs) Issai Schur (1875–1941) was Russian and from a Jewish family, but studied and spent most of his professional life in Germany. He made contributions to many areas of mathematics but most notably to the representation theory of groups. He was a distinguished professor at the University of Berlin when the Nazis came to power. In April of 1933, the Law for the Restoration of the Professional Civil Service prescribed the dismissal of public servants who had unpopular political views or were of Jewish origin. Schur was suspended, then temporarily reinstated through efforts of colleagues and because of a loophole in the law, but then was dismissed in September 1935. Despite offers from universities abroad, Schur stayed in Germany until 1939, originally because he did not feel comfortable lecturing in a new language. He detested the fascists and had declared to his wife that he would rather commit suicide than appear before the Gestapo if summoned. When such a summons arrived, his wife hid it from him, had him admitted to a hospital, and went to the Gestapo herself instead. They wanted to know why Schur had not left Germany yet. The reason was actually financial. The Nazis had instituted a pre-departure flight tax which amounted to 25 percent of one's assets, and since one of their assets was foreign real estate in Lithuania, which they could neither sell nor transfer to the German government, they couldn't pay the tax. Eventually, a donor helped pay the tax, and they left for Palestine. Schur died in Tel Aviv two years later. See the edited volume *Studies in Memory of Issai Schur* (Joseph, Melnikov, and Rentschler 2003) for fascinating accounts of Schur's life and work. His collected works have also been published (Schur 1973).

[14] Schur proved this theorem as part of his work on Fermat's last theorem. In fact, he used this result to prove that if p is a large enough prime, then the equation $x^m + y^m = z^m$ has a non-zero integer solution modulo p. This means

Proof. We will prove Schur's theorem as Proposition 2.60 below.   □

Definition 2.58. Let $n \geq 1$ be an integer. Then the *Schur number* s_n is the largest integer M such that it is possible to partition $[M] = \{1, \ldots, M\}$ into n parts such that each part is sum-free.

Example 2.59. If $n = 1$, then we are partitioning our set into just one part. The set $\{1\}$ is sum-free, but the set $\{1, 2\}$ is already not sum-free since $1 + 1 = 2$. Hence, $s_1 = 1$. In Warm-Up 2.53, you were asked to show that $s_2 = 4$. In Example 2.55, we claimed that $s_3 = 13$. In Problem P 2.4.13, you are asked to write a computer program to verify this. It is also known that $s_4 = 44$ (Golomb and Baumert 1965) and $s_5 = 160$ (announced in 2018 by Heule (2018)), but no other Schur number is known exactly.

What is interesting is that we can prove the finiteness of Schur numbers (and therefore prove Schur's theorem, Theorem 2.57) by a clever use of Ramsey's theorem (Theorem 2.52).

Proposition 2.60. *Let n be a positive integer, and let s_n and $r(3, \ldots, 3)$ denote Schur and Ramsey numbers, respectively. Then*

$$s_n < r(\overbrace{3, 3, \ldots, 3}^{n}).$$

Proof. Let $N = r(\overbrace{3, 3, \ldots, 3}^{n})$.

Partition $[N] = \{1, 2, \ldots, N\}$ into n parts $S_1, S_2, \ldots, S_n$ arbitrarily. We need to show that there is a part S_i that includes integers x, y, with $x + y$ also in S_i.

Construct the complete graph K_N, and label the edge between i and j with $|i - j|$. So, for example, the edge between 4 and 9 is labeled 5 and so on. The labels are positive integers in $[N]$ (in fact, in $[N - 1]$). Thus every label is in one of the parts. Color an edge whose label is in S_i with color i. Thus we are using n colors.

By the definition of N we must have a monochromatic triangle ijk. Without loss of generality, $i > j > k$. The triangle being monochromatic means that $x = (i - j)$, $y = (j - k)$ and $z = (i - k)$ are labels in the same part. Thus x, y, z are elements that are in the same part, and, in addition, $x + y = (i - j) + (j - k) = i - k = z$. This completes the proof.   □

Graph Ramsey Numbers

In the original Ramsey's theorem, our objective was to find monochromatic complete graphs. We can replace complete graphs with other graphs and ask, for example, how large n should be such that a two-coloring of K_n will necessarily have a monochromatic cycle of length 4 of the first color or a monochromatic K_3 of the second color. Since we know that $K_9 \rightarrow K_4, K_3$, and since K_4 contains cycles of length 4, we know that any n greater than or equal to 9 will

---

that, for large enough p, we can find integers x, y, and z such that $x^m + y^m$ and z^m have the same remainder when divided by p. The proof of this result, in addition to Schur's theorem, needs a bit of elementary group theory. See Graham, Rothschild, and Spencer 2013, page 69.

work. But it is possible that even a smaller n would work. In the problems at the end of this section, you are asked to explore a few such questions. We record a definition.

Definition 2.61 (Graph Ramsey Number). Let $G_1, G_2, \ldots, G_k$ be simple graphs. Then the *graph Ramsey number* $r(G_1, \ldots, G_k)$ is the smallest integer s such that, if we color the edges of K_s with k colors, then we have either at least one G_1 of color 1, or one G_2 of color 2, or ... one G_k of color k.

Note that, after we color K_s with k colors, finding a G_1 of the first color means that you must identify a number of edges colored with the first color such that those edges and their vertices are a graph isomorphic to G_1. In other words, there may be other edges between these vertices and other edges colored with the first color that have been ignored. Also, the original Ramsey numbers $r(n_1, \ldots, n_k)$ can now be written as $r(K_{n_1}, \ldots, K_{n_k})$.

Van der Waerden Numbers

Definition 2.62 (Van der Waerden Number). Let a and s be real numbers; then a sequence of the form $a, a + s, a + 2s, a + 3s, \ldots$ is called an *arithmetic progression*. Let k and r be positive integers. Then the *Van der Waerden number* $W(k,r)$ is the smallest positive integer with the property that, if we color $\{1, 2, \ldots, W(k,r)\}$ with r colors, then we are guaranteed the existence of a monochromatic arithmetic progression of k terms.

The definition does not provide a reason for why Van der Waerden numbers exist. Maybe, no matter how large n is, I can cleverly color the integers $\{1, \ldots, n\}$ with 10 colors without having a monochromatic arithmetic progression of length 47. This is remedied by a 1927 theorem of Van der Waerden that states that $W(k,r)$ is always a finite number. Finding actual Van der Waerden numbers is another matter entirely. Some of the known ones are: $W(3,2) = 9$, $W(4,2) = 35$, $W(3,3) = 27$, $W(3,4) = 76$, and $W(5,2) = 178$.

Problems

P 2.4.1. Let $G_1, G_2, \ldots, G_k$ be graphs, and assume that the graph G_i has n_i vertices (for $i = 1, \ldots, k$). How are the two numbers $r(G_1, \ldots, G_k)$ and $r(n_1, \ldots, n_k)$ related?

P 2.4.2. Recall that P_2 is the path of length 2 and order 3. Find $r(P_2, K_3)$.

P 2.4.3. Let $G = 2K_2$ be the graph with four vertices and two disjoint edges (i.e., G is the union of two K_2's). Find $r(G, G)$.

P 2.4.4. Recall that P_3 is the path of length 3 (with 4 distinct vertices).

(a) Let $n \geq 4$ be an integer. Assume that you color the edges of K_n with red and blue, and you notice that there is a pair of disjoint red edges (i.e., two red edges that do not share a vertex). Show that this coloring of K_n has a monochromatic P_3.

(b) What is $r(P_3, P_3)$?

P 2.4.5. Recall that $K_{m,n}$ is the complete bipartite graph on m and n vertices. Find

$$r(K_{1,3}, K_{1,3}).$$

P 2.4.6. Prove that

$$r(K_{1,3}, K_3) = 7.$$

P 2.4.7. Find

$$r(K_{1,4}, K_{1,4}).$$

P 2.4.8. (Chvátal and Harary 1972) Let C_4 be the cycle of length 4. Show that

$$r(C_4, C_4) = 6.$$

You may find the following steps helpful:

STEP 1: Show $r(C_4, C_4) > 5$.

STEP 2: Assume that the edges of K_6 are colored with red and blue, and assume that this coloring has no monochromatic C_4. Can we assume, without loss of generality, that a_1, a_2, and a_3 are the vertices of a monochromatic red triangle? Let b_1, b_2, and b_3 be the other three vertices. Show that if the edges from b_1 to at least two of the a's are red, then we have a red C_4.

STEP 3: Show that if the edges from b_1 to none of the a's are red, then we must have a blue C_4. Conclude that each b has exactly one red edge to the a's.

STEP 4: Show that both of the edges $\{a_1, b_1\}$ and $\{a_1, b_2\}$ cannot be red. So far, what do the red edges look like?

STEP 5: Can any of the edges in the triangle $\{b_1, b_2, b_3\}$ be red without creating a red C_4?

STEP 6: Find a blue C_4. Complete and rewrite the proof.

P 2.4.9. A signaling network consists of six pieces of communications equipment: x_1, x_2, x_3, x_4, x_5, and x_6. Each pair of these are linked through an intermediate facility (microwave towers, trunk groups). Intermediate facilities are expensive to build and so we want to limit their number. However, if one of these facilities fails, then all the links through them become inoperative. We have set the following reliability requirement for our network: we want to be sure that if one intermediate facility fails, then there will remain at least one link between any two distinct pairs of equipment. In other words, after one facility failure, there will still be one link (through some other facility) between one of $\{x_1, x_2\}$ and one of $\{x_3, x_4\}$ as well as between one of $\{x_1, x_4\}$ and one of $\{x_2, x_6\}$, and so on. For example, if we we use a single facility to connect x_2 to x_3 and x_4, and use the same facility to connect x_6 to x_3 and x_4, then a failure of that facility will result in no connections between the pair $\{x_2, x_6\}$ and $\{x_3, x_4\}$, and this is unacceptable. What is the minimum number of facilities that we need and how should the connections be designed?

P 2.4.10. The number $r(K_n; \mathbb{Z}/2\mathbb{Z})$ is defined to be the smallest number s (if it exists), such that no matter how we label the edges of K_s with 1 or 0, we can find a copy of K_n as a subgraph of K_s such that the sum of edge labels of this copy of K_n is an even number (i.e., zero mod 2). Find $r(K_4; \mathbb{Z}/2\mathbb{Z})$.

P 2.4.11. Recall that $W(k,r)$ denotes the Van der Waerden numbers. Find $W(2,2)$.

P 2.4.12. Show that $W(3,2) = 9$.

P 2.4.13. If you know how, write a computer program to show that the Schur number s_3 is equal to 13.

2.5   Open Problems and Conjectures

Ramsey theory abounds with open problems. As has been mentioned, the value of $r(5,5)$ is still unknown. Neither is the exact value of $r(3,3,3,3)$ or the exact values of $r(K_{3,4}, K_{2,5})$, and $r(C_4, K_{11})$ known. In fact, at this point, we are incapable of calculating most Ramsey numbers. Very few Schur numbers or Van der Waerden numbers have been calculated exactly either. There are known non-trivial bounds for many of these numbers, though. For example, $43 \leq r(5,5) \leq 48$ (and it is conjectured that it is actually 43), $51 \leq r(3,3,3,3) \leq 62$, $r(K_{3,4}, K_{2,5}) \leq 21$, and $39 \leq r(C_4, K_{11}) \leq 44$. The dynamic survey of Radziszowski (2021) – which gets updated periodically – has the latest results on all Ramsey numbers and gives references to the latest techniques and advances.

In Table 2.4 – compiled from Radziszowski 2021 – much of what is known, as of this writing in January 2021, about $r(m,n)$, with $m,n \leq 10$, is summarized.

Table 2.4 Non-trivial values and bounds for the Ramsey numbers $r(n,m)$ for $n \geq m$. From Radziszowski (2021).

$n \backslash m$	3	4	5	6	7	8	9	10
3	6							
4	9	18						
5	14	25	43/48					
6	18	36/40	58/85	102/161				
7	23	49/58	80/133	115/273	205/497			
8	28	59/79	101/194	134/427	219/840	282/1532		
9	36	73/106	133/282	183/656	252/1379	329/2683	565/	
10	40/42	92/136	149/381	204/949	292/2134	343/4432	581/	798/23556
11	47/50	102/171	183/511	262/1352	405/3216	457/7647		
12	53/59	128/211	203/673	294/1865	417/			
13	60/68	138/257	233/861	347/2510	511/	817/		
14	67/77	147/307	267/1082	/3308				
15	74/87	158/364	275/1342	401/4305		873/		1313/

The slashes refer to lower and upper bounds. For example, it is known that $47 \leq r(3,11) \leq 50$ and $405 \leq r(7,11) \leq 3216$. A slash with no number on one side means that the only bounds known for that particular Ramsey number are the general bounds that apply to all the entries. The Ramsey number $r(9, 3)$ was found in 1982 by Charles Grinstead[15] and Sam Roberts, $r(8, 3)$ was found in 1992 by Brendan McKay and Ke Min Zhang, while $r(5, 4)$ was found in 1995 by Brendan McKay and Stanisław Radziszowski.

Some tantalizing and simply stated conjectures remain open. For example, the following conjecture has remained unresolved since 1976 (see Radziszowski 2021 for references and latest results).

Conjecture 2.63. *Let n and m be any pair of integers with $n \geq m \geq 3$, except $n = m = 3$. Then,*

$$r(C_n, K_m) = (n - 1)(m - 1) + 1.$$

Many special cases have been proved but, for example, it has yet to be proved that $r(C_9, K_9) = 65$. The work of finding exact values or bounds for these combinatorial problems involves a mixture of mathematical reasoning and programming savvy. Much research has focused on finding trends in the data, in the form of "asymptotic" results. Informally, this amounts to seeing how fast Ramsey numbers grow as some particular parameter increases. We mention a few problems and conjectures in this direction.

In Problem P 2.3.20, you were asked to show that $r(n, n) \leq 4^n$ (Problem P 5.1.31 improves this bound just so slightly). It is also not that difficult to show that $r(n, n) \geq 2^{n/2}$. (See Problem P 5.1.32. Corollary 5.15 gives a slightly weaker result.) This means that $\sqrt{2} \leq \sqrt[n]{r(n, n)} \leq 4$. These bounds have been improved somewhat, but an important problem in this area remains the following.

Question 2.64. Does the following limit exist? If it does, then find it.

$$\lim_{n \to \infty} \sqrt[n]{r(n, n)}.$$

Paul Erdős[16] (Erdős 1981) offered a prize of \$100 for a proof that the limit exists and \$500 for finding the limit. He remarked that the latter could be difficult, but possibly underestimated the difficulty of the former. Given the difficulty of this problem, even small improvements are significant. Conlon, Fox, and Sudakov (2015) proposed the following specific question for a small improvement on what is known.

---

[15] Charlie Grinstead graduated from Pomona College in 1974, and is currently a Professor of Mathematics at Swarthmore College.

[16] A book on combinatorics will mention the name of the Hungarian mathematician Paul Erdős (1913–1996) – or Erdős Pál in his native Hungary – often and in many different contexts. This is because Erdős – known as "Uncle Paul" to those close to him – was one of the most prolific and influential combinatorists of the twentieth century. Not only did he prove many results in combinatorics, but many areas of combinatorics – including Ramsey theory – became active areas of research because of his results, questions, and conjectures. He was a consummate traveller – he didn't have/want a permanent university job – and wherever he went he collaborated on mathematical papers and got new people interested in combinatorics. Paul Erdős published over 1400 papers and had over 500 collaborators. He was doing mathematics literally until he died at a mathematics conference in Warsaw, Poland. (He was actually working on multicolored Ramsey numbers the day before his death.) Before the age of social media, Erdős's web of collaborators piqued the interest of many, so much so that the concept of "Erdős number" was defined (many years before the similar Kevin Bacon number). You get an Erdős number of 1 if you have written a paper with Erdős, and, for $i > 1$, an Erdős number of i if you have written a paper with someone who has an Erdős number of $i - 1$. (This author's Erdős number is 2.) A large portion of research mathematicians – regardless of their specialty – have a finite, and often very small, Erdős number. Paul Hoffman's biography (Hoffman 1998) is an accessible and fascinating account of Erdős.

Question 2.65 (Conlon, Fox, and Sudakov 2015). Does there exist a positive constant ϵ such that, for all sufficiently large n, we have

$$r(n,n) \geq (1+\epsilon)\frac{\sqrt{2}n}{e}2^{n/2}?$$

In fact, for any positive constant ϵ, Question 2.65 becomes a theorem of Spencer (Spencer 1975) if $1+\epsilon$ is replaced with $1-\epsilon$.

Finally, here is a specific conjecture for graph Ramsey numbers:

Conjecture 2.66 (Conlon, Fox, and Sudakov 2015). *There is a constant c such that if H is any simple graph with n vertices and maximum degree Δ, then*

$$r(H,H) \leq 2^{c\Delta}n.$$

3   Counting, Probability, Balls and Boxes

Consider tossing a fair coin n times in a row. A **single** *is a toss of a head or a tail that is not preceded or followed by a toss with the same result. For example, HTH has three singles while HHT has only one single. If you randomly toss a fair coin n times, what is the probability that the sequence of coin tosses will have no singles?*

In this short chapter, we set the stage for the chapters on enumerative combinatorics. We will discuss a few straightforward ideas, see that most counting problems can also be thought of as questions posed in the language of probability, and, finally, we will present a framework – using the metaphor of placing balls in boxes – for formulating counting problems. This section does not contain many results, and so, if the reader is already familiar with what is presented, then she is advised to move through the chapter quickly.

3.1   The Addition, Multiplication, and Subtraction Principles

Warm-Up 3.1. *You have two grocery bags. You have* 10 *of your favorite books in the first bag and* 8 *of your favorite pencils in the second bag. No two of the books and no two of the pencils are the same. Among the books, eight are in English and two are in German. Among the pencils, five are mechanical and three are standard graphite pencils. You are going to pick a book from the first bag and a pencil from the second bag. In how many different ways can you do this if you want to make sure that you don't end up with a German book* and *a standard graphite pencil? (So, for example, it is OK if you choose a German book as long as you don't also choose a standard graphite pencil.)*

In this section, for the record, we introduce the addition, multiplication, and subtraction principles for counting. One could file all three principles under the common-sense category. As with the pigeonhole principle, none of them is that profound or really needs proof, and the real issue – which will be tackled in most of the rest of the book and not just in this chapter – is applying them. Recall – Definition 2.54 – that a set S is partitioned into subsets $S_1, \ldots, S_n$ if $S = \bigcup_{i=1}^{n} S_i$, and $S_i \cap S_j = \emptyset$ for all $1 \leq i, j \leq n$. Also recall – Definition 2.16 – that the number of elements of a finite set X is denoted by $|X|$.

The following *addition* principle does not need a proof:

Proposition 3.2 (Addition Principle). *If a finite set S is partitioned into $S_1, S_2, \ldots, S_n$ then*

$$|S| = |S_1| + \cdots + |S_n|.$$

Using the addition principle is tantamount to solving a counting problem by splitting it into cases. As such, the art of applying it depends on your ability to partition S into a reasonable number of parts. On one extreme, if $|S_i| = 1$ for all i, then you are planning to count the number of elements of S by counting each one. This is often referred to as counting by brute force! On the other extreme, if $n = 1$ – and hence $S_1 = S$ – then the addition principle is not helpful either. A more interesting principle – called the inclusion–exclusion principle and introduced in Chapter 8 – deals with the more general situation where the subsets have non-trivial intersection.

Next, we articulate the *multiplication* principle, which is basically a restatement of what multiplying two integers means.

Proposition 3.3 (Multiplication Principle). *Assume that you are to choose two items, and that you have m choices for the first item. If, no matter what item you choose as the first item, you have exactly n choices for the second item, then, in total, you have mn choices for choosing the two items.*

Proof. See Figure 3.1. Say $x_1, \ldots, x_m$ are your choices for the first item. If you pick x_1, you have n choices for the second item. Likewise, if you pick x_2 (or $x_3, \ldots,$ or x_m), you also have n choices for the second item. Hence, the total number of choices is

$$\underbrace{n + n + \cdots + n}_{m} = mn.$$

$\square$

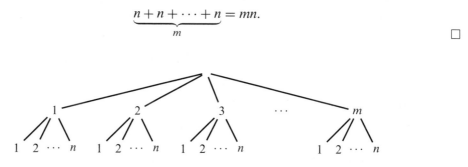

Figure 3.1 m choices followed by n choices gives a total of mn possible outcomes.

Example 3.4. How many four-digit positive integers have a 1, 2, or 3 as their ones digit? What if we also insisted on the digits being distinct?

Solution. We are looking for a four-digit number. So we have to make four choices, one for each digit. If we call the thousands digit a, the hundreds digit b, the tens digit c, and the ones digit d, then to construct a desired four-digit number, we have to make four choices.

$$\overline{}\quad \overline{}\quad \overline{}\quad \overline{}$$
$$a\qquad b\qquad c\qquad d$$

We have 9 choices for a (since 0 is not allowed), 10 choices for each of b and c, and 3 choices for d. The number of these choices is not affected by our other choices; hence the total number of such numbers is $9 \times 10 \times 10 \times 3 = 2{,}700$.

The problem becomes slightly trickier if we insist on the digits being distinct. If we proceed as before, we could say that we still have 9 choices for a, and regardless of that choice, we have 9 choices for b and then 8 choices for c. But, at this point, we wouldn't know how many choices for d we have. This is because we may or may not have used up some (or even all) of 1, 2, or 3 before. Our number of choices for d depends on our prior choices.

To remedy the problem, we change the order of choosing the digits. We have 3 choices for d, and regardless of those choices we have 8 choices for a, then 8 choices for b, and 7 choices for c. The number of choices at each stage is independent of the choices we made earlier. Thus, the total number of choices is $3 \times 8 \times 8 \times 7 = 1{,}344$. $\qquad\square$

Example 3.5. How many numbers between 1 and $100{,}000$ have exactly one digit equal to 4 and exactly one digit equal to 7?

Solution. We start with five empty slots for the digits (initial zeros are allowed and $100{,}000$ itself is not a possibility, so only need to pick five digits). A priori, these digits could be 0 through 9.

$$\underline{\quad} \quad \underline{\quad} \quad \underline{\quad} \quad \underline{\quad} \quad \underline{\quad}$$

We first place the one 4, and there are 5 choices for doing so. We then place the one 7, and there are 4 choices (since one place is already taken by the 4). For each of the remaining three spots, we have 8 choices (since we can't use 4 or 7 anymore). Hence the answer is $5 \times 4 \times 8 \times 8 \times 8 = 10{,}240$. $\qquad\square$

We could also state a rather straightforward *subtraction* principle.

Proposition 3.6 (Subtraction Principle). *Let A be a subset of a set S, and denote the complement of A in S by A^c. Then*

$$|A| = |S| - \left|A^c\right|.$$

Basically, all we are saying is that sometimes it is easier to count the objects that we *don't* want.

Example 3.7. How many integers n with $1 \leq n \leq 1500$ have at least two distinct digits?

Solution. There are a total of 1500 integers between 1 and 1500. How many of them *don't* have at least two distinct digits? The 9 one-digit numbers, 9 of the two-digit numbers (namely, $11, 22, \ldots, 99$), 9 of the three-digit numbers (namely, $111, 222, \ldots, 999$), and 1 of the four-digit numbers (namely, 1111). Hence, the total number of integers between 1 and 1500 that have at least two distinct digits is

$$1500 - (9 + 9 + 9 + 1) - 1{,}472. \qquad\square$$

Remark 3.8 (Double Counting as the Division Principle). Say you have a simple graph with five vertices and you know that the degrees of these vertices are 4, 3, 2, 2, and 1. These are the numbers of edges incident with each vertex, and if we add them we get 12. We can then argue that each edge of the graph is counted twice (once for each of its vertices), and so the number of edges of the graph is $12/2 = 6$. This is a simple example of counting something (in this case the number of edges of a graph) by first double counting (i.e., counting all the objects that we are interested in twice) and then dividing by 2. As such, you can think of the use of double counting (or triple counting, or ..., or k-fold counting) as a kind of division principle for counting. Such a labeling, however, does not help in solving problems. We will get plenty of practice in the upcoming chapters.

Remark 3.9. The addition, subtraction, and multiplication principles are pretty straightforward ideas, and, most of the time, we use them without explicitly mentioning them. Even though the principles themselves are not that profound, it is quite possible to use them incorrectly. Experience and practice are the only remedies. To get more comfortable with these principles, do more problems.

Problems

P 3.1.1. On a test there are three multiple-choice questions, and, for each question, you get to pick one of five possible choices as an answer. How many different response sheets are possible?

P 3.1.2. Consider the set of vowels {a, e, i, o, u}, and think of a "word" as any arrangement of these letters, such as aeo, ioiou, and so on. How many three-letter words are possible if no letter may be used more than once?

P 3.1.3. How many different outfits can Maria assemble if she can wear any combination of four dresses, three hats, and two pairs of shoes?

P 3.1.4. A car has five seats. In how many ways can five people be seated in the car if only three of them know how to drive?

P 3.1.5. A souvenir shop has four kinds of postcards, and I want to send a postcard to each of my seven friends. In how many ways can I do that?

P 3.1.6. How many odd five-digit integers start with an even digit?

P 3.1.7. How many five-digit numbers have at least one digit equal to 7?

P 3.1.8. Count the number of functions $f : \{1, 2, 3, 4, 5, 6, 7\} \rightarrow \{1, 2, 3, 4\}$.

P 3.1.9. Since 1980, a typical California license plate is of the form $xYYYzzz$, where x is an integer between 1 and 9 inclusive, YYY are three English letters with I, O, and Q not used in the first or the third spot, and zzz is any three-digit number (zeros allowed). How many such plates are possible?

P 3.1.10. To go to work, Mehrdad has to choose one of five shirts and one of four pairs of trousers. He also has to pick either one of three pairs of shoes or one of four pairs of sandals. If he chooses one of the shoes, then he has to pick between two pairs of socks. How many different outfits can Mehrdad assemble?

P 3.1.11. How many four-digit numbers are divisible by 5 and contain the digit 4?

P 3.1.12. The word GALOIS can be spelled by tracing paths through the array of letters given below. Steps to adjacent letters horizontally, vertically, or diagonally are allowed. What is the number of different paths which spell GALOIS?

$$
\begin{array}{cccccc}
G & G & G & S & S & S \\
G & A & A & I & I & S \\
G & A & L & O & I & S \\
G & A & L & O & I & S \\
G & A & A & I & I & S \\
G & G & G & S & S & S
\end{array}
$$

P 3.1.13. In how many ways can we spell MATHEMATICS? Only steps to adjacent letters horizontally, vertically, or diagonally are allowed.

$$
\begin{array}{ccccccccc}
 &   & T & H &   &   & T & I &   \\
 & A & T & H & E &   & A & T & I & C \\
M &   & T & H &   & M &   & T & I &   & S \\
 & A & T & H & E &   & A & T & I & C \\
 &   & T & H &   &   &   & T & I &
\end{array}
$$

P 3.1.14. A pentagon is fixed to my wall. In how many different ways can I paint the five vertices with red, blue, and green paint so that no two adjacent vertices have the same color? Justify your answer.

3.2 Probability

Warm-Up 3.10. *You have two identical copies of Murasaki Shikibu's* Tale of the Genji *and two identical copies of 'Obayd Zākāni's* Mush-o-Gorbeh. *If you close your eyes and randomly arrange the four books on a bookshelf, what is the probability that the two copies of* Tale of the Genji *will be next to each other?*

The point of this section is to point out that many counting problems can be posed in the language of discrete probability. We will give some examples, and we will discuss the definition of probability in the very basic situation when we have a finite number of equally likely cases. The following example illustrates the issue.

Example 3.11. Consider the following three questions:

(a) How many two-digit numbers have distinct and non-zero digits?
(b) What fraction of all two-digit numbers have distinct and non-zero digits?
(c) Write each two-digit number on a piece of paper, place the pieces of paper in a jar, and randomly pick one two-digit number. What is the probability that its digits will be distinct and non-zero?

Since, in the third question, each two-digit number has an equal chance of being chosen, the answers to the last two questions are the same. The first question is a counting question but very related to the other two.

To construct a two-digit number that has distinct and non-zero digits, we have to make two choices. What is the tens digit, and what is the ones digit? We have 9 choices for the tens digit, and once we have chosen the tens digit, we have 8 choices for the ones digit. Hence, the number of two-digit numbers with distinct and non-zero digits is $9 \times 8 = 72$. Alternatively, there are 90 two-digit numbers. From among these, 9 have identical digits, and 9 have a zero. Hence the number of two-digit numbers with distinct and non-zero digits is $90 - 9 - 9 = 72$.

Now the total number of two-digit numbers is $9 \times 10 = 90$, and so $\frac{72}{90}$ or 80% of all two-digit numbers have distinct and non-zero digits. The fraction $72/90 = .8$ is also the probability that a randomly chosen two-digit number will have distinct and non-zero digits.

The point is that, in order to solve the probability question, we needed to do a counting problem, and, in fact, a counting problem can often be disguised as a probability question.

Probability theory is an interesting and deep area of mathematics, and, in fact, more advanced combinatorics makes use of non-trivial inequalities that result from probability theory. The so-called "probabilistic method" is a powerful method in contemporary combinatorics. Our purpose here is much more modest. We will only introduce the most elementary vocabulary from probability theory in order to be able to state counting problems in the language of probability. As a result, we limit ourselves to situations where there are a finite number of equally likely cases. In such a situation, intuitively you can think of probability as $\frac{\text{number of favorable cases}}{\text{total number of cases}}$. To make this a tad more precise, we introduce some vocabulary.

Definition 3.12 (Experiment, Sample Space, Event). A clearly defined procedure that produces one of a given set of outcomes is called an *experiment*. The set of all of the outcomes of an experiment is called a *sample space*, and a subset of the sample space is called an *event*.

ASSUMPTION: For our purposes, we will assume *equally likely* outcomes and a *finite sample space*. In a probability course, much attention is devoted to situations where these assumptions are not valid.

Definition 3.13 (Probability of an Event). Let S be a finite sample space of equally likely outcomes, and let A be an event. Then

$$\text{Probability of } A = \Pr(A) = \frac{|A|}{|S|}.$$

Example 3.14. Three coins are flipped and the results recorded. This is an experiment, and the sample space is

$$S = \{HHH, HHT, HTH, THH, HTT, THT, TTH, TTT\}.$$

If A is the event that the number of heads be even, then

$$A = \{HHT, HTH, THH, TTT\},$$

and the probability of event A is

$$\Pr(A) = \frac{4}{8} = 0.5.$$

Problems

P 3.2.1. In a bag, you have three balls. One is red, one is blue, and one is yellow. You randomly pick a ball, look at it, put it back, shake the bag for a while, and then choose another ball. What is the probability that both balls that you looked at are yellow?

P 3.2.2. You still have a bag containing a blue, a red, and a yellow ball. You randomly pick a ball, look at it, put it back, shake the bag for a while, and then choose another ball.
   (a) What is the probability that the first ball is red?
   (b) What is the probability that exactly one of the balls is red?
   (c) What is the probability that at least one of the balls is red?
   (d) What is the probability that at most one of the balls is red?
   (e) You tell me that at least one of the balls that you picked was red. A third friend asks me "What is the probability that both picked balls are red?" What should my answer be?

P 3.2.3. You roll a regular six-sided die twice.
   (a) What is the probability that you roll five twice?
   (b) If you roll a five the first time, then what is the probability that you also roll a five on the second try?
   (c) If all we know is that at least one of your rolls was a five, then what is the probability that you rolled five twice?

P 3.2.4. You choose a positive integer less than or equal to 10,000 at random. What is the probability that your chosen integer has exactly one 4 and one 7?

P 3.2.5. You choose a six-digit integer at random. What is the probability that, somewhere in your chosen integer, you see a 47 (that is, the digit 4 followed by the digit 7)?

P 3.2.6. You choose a six-digit integer at random. What is the probability that there is at least one 4 or one 7 among the digits of the integer?

P 3.2.7. (a) It is early morning on April 10, 1912, and you are standing at the port in Southampton, England. A friend asks you: "What is the probability that the Titanic sinks on her maiden journey?" How would you answer that question, and what would be a reasonable answer?
   (b) It is mid-morning on November 3, 2020, and a friend asks you: "What is the probability that the Titanic sinks on her maiden journey?" How would you answer that question, and what would be a reasonable answer?
   (c) Comment.

P 3.2.8. Ten soccer players are standing in a circle and randomly passing a ball. When a player gets the ball, they can pass the ball to anyone except to the player who just passed them the ball.
   (a) If player A starts the exercise, what is the probability that A receives the third pass?
   (b) If player A didn't start the exercise, what is the probability that A receives the third pass?

P 3.2.9. You randomly choose two (not necessarily distinct) integers between 10 and 50 inclusive. What is the probability that their sum is even?

P 3.2.10. Take three cubes and construct three unusual dice as follows. On the faces of the first cube put 5, 6, 7, 8, 9, and 18, and call this die A. On the faces of the second cube put 2, 3, 4, 15, 16, and 17, and call it B. Finally, on the faces of the third cube put 1, 10, 11, 12, 13, and 14, and call it C.

(a) What is the probability that A beats B? Is this number greater than $1/2$?
(b) What is the probability that B beats C? Is this number greater than $1/2$?
(c) What is the probability that C beats A? Is this number greater than $1/2$?
(d) Anything strange?

P 3.2.11. The San Bernardino line of LA's Metrolink commuter rail system serves 13 stations from LA Union Station in downtown LA to San Bernardino. The Claremont stop is right in the middle (the seventh stop from either end of the line). Each day, you wake up at a random time, walk to the Claremont station, and you get on the first train that arrives, regardless of its direction. Nine days out of ten you end up in San Bernardino. If the number of trains going in each direction is the same and you arrive at Claremont station at a random time, how can this be?

P 3.2.12. A basketball player is shooting free throws.[1] She makes the first shot and misses the second one. Thereafter, the probability that she will make a shot is equal to the proportion of the shots that she has made so far. So, for example, if she makes 5 of the first 7 shots, then her chances of making the eighth shot is $5/7$. What is the probability that this basketball player will make exactly 47 of the first 100 shots? You may find the following steps helpful:

STEP 1: Let n and k be integers with $n \geq 2$ and $1 \leq k \leq n - 1$. Define $P(n,k)$ to be the probability that the basketball player makes exactly k of the first n shots. What are $P(2,1)$, $P(3,1)$, and $P(3,2)$?

STEP 2: Write a recurrence relation for $P(n,1)$.

STEP 3: Write a recurrence relation for $P(n,n-1)$.

STEP 4: For $2 \leq k \leq n - 2$, write a recurrence relation for $P(n,k)$.

STEP 5: Figure 3.2 organizes the possibilities. On the horizontal axis we have n, the number of free throws attempted, and on the vertical axis we record k, the number of free throws made. For each node (n,k), with $1 \leq n \leq 5$ and $1 \leq k \leq n-1$, record $P(n,k)$ on the graph.

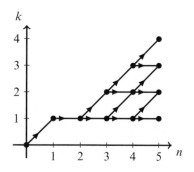

Figure 3.2 The total number of attempted free throws is n, while the number made is k. When you shoot a free throw, the number of successes either stays the same or increases by one.

[1] Adapted from Problem B-1 of the 2002 William Lowell Putnam Mathematical Competition.

STEP 6: Based on the data, make a conjecture about $P(n,k)$.

STEP 7: Prove your conjecture by induction on n. Do the inductive step in three separate cases: $k = 1$, $2 \le k \le n - 2$, and $k = n - 1$.

STEP 8: What is $P(100, 47)$?

3.2.1 Longest Run of Heads

Given a fair coin, consider a sequence of n coin tosses. The result of each toss is a head (H) or a tail (T). The *longest run of heads* is the longest sequence of consecutive heads.

Question 3.15. If you toss a fair coin eight times, what is the probability that the longest run of heads will be 3?

To answer this and related questions, we define an auxiliary function $f(n,k)$. The number $f(n,k)$ is the number of different sequences of n coin tosses whose longest run of heads is k. So $f(3,2) = 2$, since *HHT* and *THH* are the only sequences of 3 coin tosses whose longest run of heads is 2.

In the following problems, you will explore the function $f(n,k)$. In particular, you will find a recurrence relation for $f(n,k)$ (Problem P 3.2.16) and use it to answer Question 3.15 (Problem P 3.2.17).

The Number of Singles

Again consider the sequence of n tosses of a fair coin. A *single* is a toss of a head or a tail that is not preceded or followed by a toss with the same result. For example, *HTH* has three singles while *HHT* has only one single.

Question 3.16. If you toss a fair coin n times, what is the probability that you don't have any singles?

Surprisingly, the answer to Question 3.16 involves the Piṅgala–Fibonacci numbers. Let $S(n,k)$ denote the number of different sequences of n coin tosses with exactly k singles. In the problems (adapted from Bloom 1998), you are led to find a recurrence relation for $S(n,k)$ (Problem P 3.2.18) and to answer Question 3.16, which was the opening problem of the chapter (Problem *P 3.2.19*).

Problems (continued)

P 3.2.13. Let n be a positive integer, and let $h(n)$ be the number of different sequences of n tosses of a coin where every head is a single (this includes the possibility that there are no heads). Find a recurrence relation for $h(n)$ and calculate a few of its values. Have you seen these numbers before?

P 3.2.14. Let $f(n,k)$ be the number of sequences of n coin tosses whose longest run of heads is k.
    (a) Construct a table for $f(n,k)$ for small values of n and k ($0 \le k \le 5, 0 \le n \le 5$).
    (b) Find and prove an expression for each of the following:
        (i) $f(n,0) = f(n,n) =?$
        (ii) $f(n, n-1) =?$, for $n \ge 2$
        (iii) $f(n, n-2) =?$, for $n \ge 4$.

P 3.2.15. Again, let $f(n,k)$ be the number of different sequences of n coin tosses whose longest run of heads is k. Then $g(n) = f(n,1)$ is the number of different sequences of n coin tosses where there is at least one head and every head is a single. Find $g(1), g(2)$, and a recurrence relation for $g(n)$. Use these to find $g(8)$.

P 3.2.16. Can you find a recurrence relation for $f(n,k)$? Recall that we are tossing n fair coins, and $f(n,k)$ is the number of ways that we can get exactly k as the longest run of heads. You may find the following steps useful.

STEP 1: If the first toss is tails, then what is the number of ways to complete the sequence of tosses and get k as the longest run of heads?

STEP 2: Answer the same question if the first toss is heads and the second toss is tails.

STEP 3: Repeat the above if the first, the second, the third, all the way to the $(k-1)$th tosses are all heads but the kth toss is tails.

STEP 4: If the first k tosses are all heads, and if the longest run of heads is k, then what is the $(k+1)$th toss? What are the possibilities for the longest run of heads in the remaining $n-k-1$ tosses (after the $(k+1)$th toss)?

STEP 5: Using Steps 1–4, write a recurrence relation for $f(n,k)$ (in the form of a sum with possibly many terms).

P 3.2.17. We toss a fair coin eight times. For each $0 \le k \le 8$, find the probability that the longest run of heads is k.

P 3.2.18. Let $S(n,k)$ denote the number of sequences of n coin tosses with exactly k singles (heads or tails). Find a recurrence relation for $S(n,k)$.[2] You may find the following steps useful.

STEP 1. Using the function $S(n,k)$, give an expression for the number of sequences of n coin tosses with exactly k singles where the *last* toss is one of the singles.

STEP 2. Using your answer to the previous step, give an expression for the number of sequence of $n-1$ coin tosses with exactly k singles where the last toss is *not* one of the singles.

STEP 3. Let $n \ge 3$ and $1 \le k \le n$. Interpret $S(n,k)$ as zero if $k > n$, and define the following sets:

S = Sequence of n coins tosses with exactly k singles

A = Sequences in S where the last two tosses are different

B = Sequences in S where the last two tosses are the same but different from the one prior

C = Sequences in S where the last three tosses are the same.

---

[2] Problems P 3.2.18 and *P 3.2.19* are adapted from Bloom 1998.

Convince yourself that $\mathcal{A}$, $\mathcal{B}$, and $\mathcal{C}$ partition $\mathcal{S}$ (in other words, $\mathcal{A}$, $\mathcal{B}$, and $\mathcal{C}$ are pairwise disjoint and their union is all of $\mathcal{S}$).

STEP 4: How is $|\mathcal{A}|$ related to your solution for Step 1?

STEP 5: Give an expression for $|\mathcal{B}|$ (in terms of the function $S(n,k)$).

STEP 6: Argue that $|\mathcal{C}|$ is the same as the number of sequences of $n-1$ tosses where the last toss is *not* a single. Using this interpretation and your solution to Step 2, give an expression for $|\mathcal{C}|$.

STEP 7: Based on your work on the pervious steps, give a recurrence relation for $S(n,k)$.

P 3.2.19. Let n be a positive integer, and let $k(n) = S(n,0)$ be the number of sequences of n coin tosses with no singles.

(a) Find $k(1)$, $k(2)$, and a recurrence relation for $k(n)$.

(b) If you randomly toss a fair coin n times, what is the probability that the sequence of coin tosses will have no singles?

3.3   A Framework for Counting Questions: The Counting Table

Warm-Up 3.17. *Consider the equation $x_1 + x_2 = 3$. There are an infinite number of solutions to this equation. However, if we insist that x_1 and x_2 be non-negative integers, then there is only a finite number of solutions. (Why?) We are often interested in counting the number of solutions to such equations. Consider the following problem: Ananya, Anaya, Aaradhya, and Aadhya are your friends and you have five identical jars of tamarind. You want to distribute the jars somehow between your friends. You could give all the jars to one of them or distribute them more evenly. We are interested in the number of ways that this can be done. Can you translate the problem into an equivalent problem about the number of non-negative integer solutions to an equation of the form $x_1 + x_2 + \ldots + x_s = t$? (Do not solve the problem. Just translate it.)*

There are a large number of counting problems, and distinguishing them is not always straightforward. We have chosen to organize a variety of counting problems around the metaphor of balls and boxes. In this section, we present this framework. The next example illustrates our schema (and its limitations).

Example 3.18. Say that you have three identical apples and one banana. You also have two identical circular plates as well as one rectangular plate. In how many ways can you place the fruit on the plates in such a way that the number of pieces of fruit on the rectangular plate is even? In saying that the apples are identical, we mean that there is no difference between them, and if you switch the place of two of the apples, you don't get a new configuration. The same goes for the two circular plates. As long as the fruit on the circular plates is the same, we can't tell the circular plates apart. This problem suggests a whole host of problems. Let $S = \{3 \cdot A(\text{pples}), 1 \cdot B(\text{anana})\}$ be the multiset of fruit, and $T = \{2 \cdot C(\text{ircle}), 1 \cdot R(\text{ectangle})\}$ be the multiset of plates. We are asking the number of ways we can distribute a multiset of fruit among a multiset of plates when, in addition, there is some restriction on the capacity of the plates (in this case, the number of pieces of fruit on the rectangular plate must be even). We will replace the fruits and the plates with balls and boxes, and try to see the number of ways you can distribute a multiset of balls into a multiset of boxes, given some restrictions on the capacity of the boxes.

Problem 3.19. *Let*

$$S = \{m_1 \cdot 1, m_2 \cdot 2, \ldots, m_s \cdot s\}$$

be a multiset of (Soccer) Balls. Let

$$T = \{r_1 \cdot U_1, r_2 \cdot U_2, \ldots, r_t \cdot U_t\}$$

be a multiset of (Target) Boxes or Urns. In addition, a box of type U_i has capacity u_i, and there may be restrictions on these. Find the number of different ways of placing the balls into boxes.

Solution to Example 3.18. In the case of Example 3.18, to find the answer we are going to basically list all of the possibilities. We know that the rectangular plate R can have 0, 2, or 4 pieces of fruit. If R has no fruit, then either all the fruit is on one circular plate, or it is split between the two circular plates. In the latter case you can have $\{A, AAB\}, \{AA, AB\}, \{AAA, B\}$ as the possibilities. Thus, in this case, the number of possibilities is four. If there are four pieces of fruit on R, then all the fruit is on R, and there is only one possibility. Finally if there are two pieces of fruit on the rectangular plate, then the two circular plates could have $\{AA, \emptyset\}, \{AB, \emptyset\}, \{A, A\}, \{A, B\}$. Hence, in this case, there are four possibilities. We conclude that the answer is nine.   □

The general Problem 3.19 is too complicated for any one solution technique. In this text, we limit ourselves to the cases when the balls and the boxes are either all identical or all distinct. In other words, we consider the cases when

$$S = S_1 = \{1, 2, \ldots, s\}$$

is a set of distinct (distinguishable) balls, or

$$S = S_2 = \{s \cdot 1\}$$

is a multiset of identical (indistinguishable) balls. Similarly, for T, we consider the cases when

$$T = T_1 = \{U_1, U_2, \ldots, U_t\}$$

is a set of distinct (distinguishable) boxes, or

$$T = T_2 = \{t \cdot U_1\}$$

is a multiset of identical (indistinguishable) boxes. In addition, a box of type U_i has capacity u_i, and there may be restrictions on u_i for $i = 1, \ldots, t$. (See Table 3.1.)

Surprisingly, many counting problems can be cast in this framework. Sometimes the casting may seem too complicated and contrived, and often there are other easier ways of looking at a problem. A good problem solver will be able to go back and forth between the various possible metaphors to cast the problem at hand into a previously treated one. At this point we just want to see that many familiar counting problems can be translated to the above framework.

Table 3.1 Balls and boxes counting problems.

Number of Ways to Put Balls into Boxes			
$S = S_1 = \{1, 2, \ldots, s\}$, or $S = S_2 = \{s \cdot 1\}$, a multiset of balls			
$T = T_1 = \{U_1, U_2, \ldots, U_t\}$, or $T = T_2 = \{t \cdot U_1\}$, a multiset of boxes			
Box U_i contains u_i balls			
Conditions on S and $T \rightarrow$ on $u_i \downarrow$	$T = T_1$ distinct $S = S_1$ distinct	$T = T_1$ distinct $S = S_2$ identical	$T = T_2$ identical $S = S_1$ distinct	$T = T_2$ identical $S = S_2$ identical
$0 \le u_i \le 1$ Assume $t \ge s$	1	2	3	4
$u_i \ge 0$	5	6	7	8
$u_i \ge 1$	9	10	11	12
for $i = 1, \ldots, t$, $0 \le u_i \le n_i$, $n_i \in \mathbb{Z}^{>0}$	13	14		
$u_i \in N_i \subset \mathbb{Z}^{\ge 0}$ for $i = 1, \ldots, t$	15	16		

Example 3.20. We have 10 distinct books. Five of them are to be arranged on a shelf. In how many ways can this be done?

Let the five spaces on the shelf be $S = \{1, 2, 3, 4, 5\}$, the set of "soccer balls." Let the set of books be $T = \{U_1, U_2, \ldots, U_{10}\}$, the set of "target boxes." When we put a ball in a box, we have chosen the corresponding book *and* decided where on the shelf it will be. So, for example, putting the ball numbered 3 in the urn U_8 means that the eighth book goes in the third spot. When we are done, then we know which book should go in which space. Now we only have one copy of each book and so we don't want to pick a book more than once. In addition, we are only choosing 5 of the books and so some of the books are not going to be chosen. We ensure these conditions by stipulating that the capacity of each box is 0 or 1. Thus the number of ways of placing 5 distinct balls in 10 distinct boxes with no box getting more than 1 ball gives the desired answer. This is entry $\boxed{1}$ of the counting table, Table 3.1.

Note that a possible first instinct to consider the books as Balls and the spaces as Boxes would not work. This is because, in our framework, every ball is to be distributed in some box – while some boxes can be empty – and, in this problem, not all the books are put on the shelf.

Example 3.21. In how many ways can we pick 5 books out of 10 distinct books?

This time the order of the books does not matter. We just need to pick 5 of the books. So again let the set of boxes be $T = \{U_1, U_2, \ldots, U_{10}\}$ corresponding to the set of 10 distinct books. But we let the set of balls be $S = \{5 \cdot 1\}$, denoting a set of 5 identical balls. Putting a ball in the box means that we have chosen that book. Since the order of the books does not matter, we don't need to know which ball went into which box and so the balls are all identical. The capacity constraints continue to be $u_i = 0, 1$ for $i = 1, \ldots, 10$, since some books are not going to be chosen and no book can be picked more than once. Thus, the answer to the question is the number of ways that we can put 5 identical balls into 10 distinct boxes with no box getting more than 1 ball. This is entry $\boxed{2}$ of the counting table, Table 3.1.

In Table 3.1 we have listed the counting problems that fall within this framework and are treated in the coming chapters. (Take a peek at the completed counting table following Chapter 9.) For each entry of the table, we will find a solution, and, possibly various reformulations. At this point, we do want to point out the correspondence of the first and second columns to permutations and combinations, respectively.

The First Two Columns of the Counting Table

Permutations and combinations[3] of sets and multisets are very common counting problems, and we treat them first in Chapter 4. Permutations are ordered lists of some of the elements of a set or a multiset. You can think of a permutation as an arrangement of these elements in a row (as in Example 3.20). Not only do you have to choose the elements that you are going

---

[3] Replacing the expression "permutation" with *ordered list*, and the expression "combination" with *subset* or *submultiset* may be preferable. However, the terms permutations and combinations are widely used and standard.

to arrange, but you also have to know which is the first one, which is the second one, and so on. Combinations, on the other hand, are unordered collections of some of the elements of a set or a multiset. For combinations, the only thing that matters is which elements (and, in the case of multisets, how many of each element) are included (as in Example 3.21). In our counting table, Table 3.1, the first column corresponds to counting permutations while the second column corresponds to counting combinations. Recall that in our counting table we are counting the number of ways we can place balls into boxes. In both the first and the second columns, the boxes are distinguishable. That means that we have a set of boxes and each box is different than the others. If you like, you can give each box a name. For the first column, the balls are also distinguishable. Think of these balls as being numbered. When you place a numbered ball in a named box, you are choosing that box *and* deciding where in the list it goes. For example, if you put the ball numbered "47" in the box named "Pomona," then "Pomona" is going to be the 47th element in your list. In the second column, the balls are identical and you can't tell them apart. Hence, putting a ball in a named box just picks that box to be in your collection.

The rows of the counting table put restrictions on the capacity of the boxes. In the first row, each box gets at most one ball. This means that, in your permutation or combination, a specific box can appear only once. Hence, you are finding permutations and combinations of sets (as opposed to multisets). In the second row, there is no restriction on the capacities. This means that you can put as many balls as you want in any box. Hence, you could be choosing the same box several times, and that box is appearing multiple times in your permutation or combination. Hence, you are finding a permutation or a combination of a *multiset*. The third row is a variation of the second row. You can put multiple balls in a box but you have to make sure that each box gets at least one ball. Hence, every named box must appear at least once in the permutation or in the combination.

Problems

P 3.3.1. Without finding an actual numerical answer, show that the number of ways of placing seven identical balls in four distinct boxes is the same as the number of non-negative integer solutions to $x_1 + x_2 + x_3 + x_4 = 7$.

P 3.3.2. You are to organize Achara, Busarakham, Chuanchen, Duanphen, and Kanok into three teams. (A team must have at least one member.) You are interested in the number of ways that this can be done. Without finding a numerical answer, translate the problem into a problem about balls and boxes. To which entry of the counting table (Table 3.1) does this problem correspond?

P 3.3.3. Without finding an actual numerical answer, show that the number of 10-digit numbers whose digits consist of 5 fours and 5 sevens is the same as the number of ways of placing 5 identical balls in 6 distinct boxes.

P 3.3.4. You have five of each of letters A, B, C, D, E, and F. Without finding an actual number, show that the number of different 10-letter words (a word is an ordered list of letters with no regard to meaning) that you can construct is the same as the number of ways of placing 10 distinct balls into 6 distinct boxes with the added condition that no box gets more than 5 balls. Which entry of the countable table (Table 3.1) does this correspond to?

P 3.3.5. You have 47 identical cubical blocks of wood, and you are going to stack them. You will put a row of blocks down on a table, and then stack another row of contiguous blocks on top of the first row starting on the left, and continue. You don't have to use the same number of blocks on each level, but no row can have more blocks than the one below it. You want to know how many ways can this be done. For example, you could put all the 47 in one row, or put a row of 23 blocks on top of a row of 24 blocks, or make a very high tower of 47 levels with one block in each level. (See Figure 3.3.) Without finding a numerical answer, translate the problem into a problem about balls and boxes. To which entry of the counting table (Table 3.1) does this problem correspond?

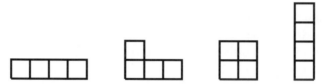

Figure 3.3 In Problem *P 3.3.5*, if instead of 47 blocks, we had 4 blocks, there would be four ways to stack them.

3.4   Bijective Maps and Counting

Warm-Up 3.22. *Let* $\mathbb{Z}$, $\mathbb{Z}^{>0}$ *denote the set of all integers and the set of all positive integers, respectively. The function* $f : \mathbb{Z} \to \mathbb{Z}$ *is given by* $f(x) = x^2$. *Is this function 1-1? What about the function* $g : \mathbb{Z}^{>0} \to \mathbb{Z}$ *given by* $g(x) = x^2$? *What is the difference – if any – between* f *and* g?

To show that two finite sets $\mathcal{M}$ and $\mathcal{N}$ have the same size, we often find a function $f : \mathcal{M} \to \mathcal{N}$ that is 1-1 and onto. The function provides a correspondence between the elements of the two sets and gives a way of pairing the elements of one set with the other. We have to make sure that f is a function – that is, any element of $\mathcal{M}$ is indeed mapped to a unique and unambiguous element of $\mathcal{N}$ – that f is onto – that is, every element of $\mathcal{N}$ is the image of some element of $\mathcal{M}$ – and that f is 1-1 – meaning that no two elements of $\mathcal{M}$ are sent to the same place. Sometimes, to show that a function f is 1-1 and onto, we show that f has an inverse. An inverse of f is a function $g : \mathcal{N} \to \mathcal{M}$ that undoes what f does. More precisely, g is the inverse of f if $g(f(x)) = x$ for all $x \in \mathcal{M}$ *and* $f(g(y)) = y$ for all $y \in \mathcal{N}$. Even though we are assuming that the reader is familiar with these concepts, to be complete we record the definitions and the related needed results:

Definition 3.23. Let $\mathcal{M}$ and $\mathcal{N}$ be two sets. A *function* (or a *map*) f from $\mathcal{M}$ to $\mathcal{N}$ is denoted by $f : \mathcal{M} \to \mathcal{N}$ and is a rule that assigns to each element of $\mathcal{M}$ exactly one element of $\mathcal{N}$.

A function $f : \mathcal{M} \to \mathcal{N}$ is *1-1* (or one-to-one or *injective*) if, for any $m_1, m_2 \in \mathcal{M}, f(m_1) = f(m_2)$ implies that $m_1 = m_2$.

A function $f : \mathcal{M} \to \mathcal{N}$ is *onto* (or *surjective*) if, for each $n \in \mathcal{N}$, there exists at least one $m \in \mathcal{M}$ with $f(m) = n$.

A function $f : \mathcal{M} \to \mathcal{N}$ is a *one-to-one correspondence* (or a *bijection*) if f is one-to-one and onto.

The identity map $1_{\mathcal{M}} : \mathcal{M} \to \mathcal{M}$ is defined by $1_{\mathcal{M}}(m) = m$, for all $m \in \mathcal{M}$.

Given three sets $\mathcal{M}$, $\mathcal{N}$, and $\mathcal{R}$ and two functions $f : \mathcal{M} \to \mathcal{N}$ and $g : \mathcal{N} \to \mathcal{R}$, the *composition* of f and g, denoted by $g \circ f$, is a function from $\mathcal{M}$ to $\mathcal{R}$ defined by $(g \circ f)(m) = g(f(m))$ for all $m \in \mathcal{M}$.

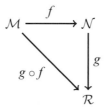

A function $g : \mathcal{N} \to \mathcal{M}$ is called the *inverse* of $f : \mathcal{M} \to \mathcal{N}$ if $f \circ g = 1_{\mathcal{N}}$ and $g \circ f = 1_{\mathcal{M}}$.

Example 3.24. Let $\mathcal{A}$ be the collection of the results of 12 consecutive coin tosses, with the condition that there are exactly 4 tails and that no two tails occur consecutively. (An example of an element of $\mathcal{A}$ is *HHTHTHTHHHHTH*.) Let $\mathcal{B}$ be the set whose elements are subsets of size 4 of $[9] = \{1, \ldots, 9\}$. Define the function $f : \mathcal{A} \to \mathcal{B}$ as follows: if x is an element of $\mathcal{A}$, then $f(x) = \{a_1, a_2, a_3, a_4\}$, where, for $1 \leq i \leq 4$, a_i is 1 plus the number of heads that occur before the ith tail. So, for example, $f(HHTHTHTHHHHTH) = \{3, 4, 5, 8\}$ (since there are, respectively, 2, 3, 4, and 7 heads before the first, second, third, and fourth tails). No two tails are assumed to be consecutive and so each tail has a different number of heads before it. As a result, this function does result in a set of four numbers between 1 and 9 inclusive, and so the function does actually map $\mathcal{A}$ to $\mathcal{B}$. The function is also 1-1, since two different lists will have different images. It is also onto, since, given a subset of size 4 of $[9]$, we can reconstruct the ordered list of heads and tails that maps onto it. This latter also tells us how to construct an inverse for the function f. Define $g : \mathcal{B} \to \mathcal{A}$ as follows: given a subset $y = \{a_1, a_2, a_3, a_4\}$ of size 4 of $[9]$, construct $g(y)$, an ordered list of heads and tails, by first writing down eight Hs, and then inserting a T after the first $a_1 - 1$, $a_2 - 1$, $a_3 - 1$, and $a_4 - 1$ heads. You should convince yourself that applying f followed by g, as well as doing g followed by f, gets you back where you started. Hence, g is the inverse function of f. The fact that we have a 1-1, onto function from $\mathcal{A}$ to $\mathcal{B}$ tells us that we can pair off the elements of $\mathcal{A}$ with elements of $\mathcal{B}$, and as a result $|\mathcal{A}| = |\mathcal{B}|$.

Proposition 3.25. *Let $\mathcal{M}$ and $\mathcal{N}$ be two finite sets. Then the following are equivalent:*

(a) $|\mathcal{M}| = |\mathcal{N}|$.
(b) *There exists a 1-1, onto function $f : \mathcal{M} \to \mathcal{N}$.*
(c) *There exists a function $f : \mathcal{M} \to \mathcal{N}$ that has an inverse $g : \mathcal{N} \to \mathcal{M}$.*

Proof. You are asked to give a proof for a (slightly stronger) version of this proposition in Problems P 3.4.14 and *P 3.4.15*. □

Consider Example 3.20. This problem asks for the number of ways of arranging 5 out of 10 books on a shelf. This is a counting problem, and the answer is a number. Hence, it is possible – and, in this case, not too hard – to come up with a formula that gives the answer.

In combinatorics, however, we often are interested in more than the answer. In that example, for instance, we were interested in showing that we can translate the problem into a different setting – that of balls and boxes. Our solution illustrated that every selection of 5 out of the 10 books and then the arrangement of these books on the shelf can be translated to a placing of 5 distinct balls into 10 distinct boxes in such a way that no box gets more than 1 ball.

In effect, we have two sets $\mathcal{M}$ and $\mathcal{N}$, where the elements of $\mathcal{M}$ are arrangements of 5 of the 10 books on a shelf and the elements of $\mathcal{N}$ are the placements of 5 distinct balls into 10 distinct boxes such that no box gets more than 1 ball. What we want is to show that $|\mathcal{M}| = |\mathcal{N}|$ by finding a 1-1, onto function from $\mathcal{M}$ to $\mathcal{N}$.

If the books are denoted by $\{B_1, \ldots, B_{10}\}$ then $B_3 B_5 B_1 B_6 B_9$ and $B_5 B_1 B_6 B_3 B_9$ are two elements of $\mathcal{M}$. In the first one, B_3 occupies the first space on the shelf, while in the second one B_3 is in the fourth space. On the other hand, if $①, ②, \ldots, ⑤$ are the 5 balls, and $U_1, \ldots, U_{10}$ are the 10 boxes, then one element of $\mathcal{N}$ is

$$① \to U_3, \quad ② \to U_5, \quad ③ \to U_1, \quad ④ \to U_6, \quad ⑤ \to U_9,$$

while a second element of $\mathcal{N}$ is

$$① \to U_5, \quad ② \to U_1, \quad ③ \to U_6, \quad ④ \to U_3, \quad ⑤ \to U_9.$$

We wanted to show in the example that $|\mathcal{M}|$ is the same as $|\mathcal{N}|$, and we did this by giving a 1-1, onto function f from $\mathcal{M}$ to $\mathcal{N}$. Which function $f : \mathcal{M} \to \mathcal{N}$ did we use? We identified the books with the boxes, and so, $B_{i_1} B_{i_2} B_{i_3} B_{i_4} B_{i_5}$, a typical element of $\mathcal{M}$, was sent to the element of $\mathcal{N}$ given by

$$① \to U_{i_1}, \quad ② \to U_{i_2}, \quad \ldots, \quad ⑤ \to U_{i_5}.$$

Often, we do not formally prove that a function is well defined, 1-1, and onto. However, the reader should convince herself that whatever rule we suggest for a correspondence between two sets is indeed a function, 1-1, and onto.

Problems

P 3.4.1. Recall that $[4] = \{1, 2, 3, 4\}$. Find the number of bijections $f : [4] \to [4]$.

P 3.4.2. Let $[47] = \{1, 2, \ldots, 47\}$ and let $2^{[47]}$ denote the set of all subsets of $[47]$ (including the empty set and $[47]$ itself). Find a bijection $f : 2^{[47]} \to 2^{[47]}$ with the property that if $A \subseteq B$, then $f(B) \subseteq f(A)$.

P 3.4.3. Let $[47] = \{1, \ldots, 47\}$, $\mathcal{A}$ be the set of subsets of size 18 of $[47]$, and $\mathcal{B}$ be the set of subsets of $[47]$ of size 29. Can you find a bijection between $\mathcal{A}$ and $\mathcal{B}$?

P 3.4.4. YYNYNN is a (meaningless) word made with only the letters Y and N. Let $\mathcal{A}$ be the set of all words that have exactly 47 letters and each of the letters is either a Y or an N. Let $2^{[47]}$ denote all subsets of $[47] = \{1, 2, \ldots, 47\}$. Can you find a bijection $f: \mathcal{A} \to 2^{[47]}$?

P 3.4.5. Let $\mathcal{A}$ be the set of all words that you can make with 20 Y's and 27 N's. Let $\mathcal{B}$ be the subsets of size 20 of $[47] = \{1, \ldots, 47\}$. Is there a bijection $f: \mathcal{A} \to \mathcal{B}$?

P 3.4.6. You have seven distinct books and one of them is Mary Astell's *Political Writings* edited by Patricia Springborg and first published by Cambridge University Press in 1996. You list all possible subsets of the books, including the empty set and the set including all seven books. Let $\mathcal{A}$ be those subsets that do not include Mary Astell's book, and $\mathcal{B}$ be those subsets that do. Give an explicit bijection $f: \mathcal{A} \to \mathcal{B}$.

P 3.4.7. You have 11 identical baskets and 47 identical balls. Let $\mathcal{A}$ be the set of possible placings of all the balls into the baskets such that each basket has at least two balls. You can't tell the balls or the baskets apart, so, for example, there is only one way to put 7 balls in one basket and 4 balls in each of the other 10 baskets. Let $\mathcal{B}$ be the set of possible placings of 36 balls in the baskets such that each basket has at least one ball. Give a bijection $f: \mathcal{A} \to \mathcal{B}$.

P 3.4.8. You are in a room with 47 other people. On a (very big) blackboard in front of the room, someone lists all the possible ways of creating a committee consisting of an odd number of people from the 48 present. Someone else, on a (also very big) blackboard in the back of the room, lists all the possible ways of creating a committee consisting of an even number of members from among the 48 present. Both the empty set and the full set of 48 people are listed as possible committees in the back of the room. Can you find a bijection between the committees listed in the back and the ones listed in the front?

P 3.4.9. Let n be a positive integer, and define $\mathcal{A}$ to be the collection of subsets of $[n]$ that have an even number of elements. Likewise, let $\mathcal{B}$ be the collection of subsets of $[n]$ that have an odd number of elements. Find a one-to-one, onto function $f: \mathcal{A} \to \mathcal{B}$.

P 3.4.10. Let $\mathcal{A}$ be the set of sequences of 47 coin tosses that have a total of 17 singles with the additional property that the last two tosses are not identical. Let $\mathcal{B}$ be the set of sequences of 45 coin tosses with 15 singles. Find a one-to-one map from $\mathcal{B}$ to $\mathcal{A}$. Is your map onto?

P 3.4.11. You go to a donut shop and they have specials on eight particular types of donuts. You want to buy a dozen. Let $\mathcal{A}$ be the set of all possible combinations of a dozen donuts – from among the eight types on sale – that you could buy. (One possibility – not recommended but understandable – would be to just buy 12 jelly-filled donuts; another would be to buy three each of glazed, chocolate-covered, Bavarian cream-filled, and apple-crumb donuts.) Let $\mathcal{B}$ be the set of all sequences of 19 coin tosses with exactly 7 heads. (One of these would be a tail, followed by a head, followed by 11 tails, followed by 6 heads.) Find a bijection between $\mathcal{A}$ and $\mathcal{B}$.

P 3.4.12. Let $n > 0$ be an integer. Let $\mathcal{A}$ be the set of divisors of n that are greater than $\sqrt{n}$. Likewise, let $\mathcal{B}$ be the set of divisors of n that are less than $\sqrt{n}$. Find a bijection between $\mathcal{A}$ and $\mathcal{B}$.

P 3.4.13. Let $\mathcal{A}$ be the set of 4×4 matrices with integer entries whose determinant has remainder 4 when divided by 7. Likewise, let $\mathcal{B}$ be the set of 4×4 matrices with integer entries whose determinant has remainder 1 when divided by 7. If A is a 4×4 matrix, define $f(A)$ to be the 4×4 matrix obtained from A by multiplying the entries of its first row by 2.

(a) Is f a function from $\mathcal{A}$ to $\mathcal{B}$? Is it 1-1? Is it onto?

(b) Can you describe a 1-1 map from $\mathcal{B}$ to $\mathcal{A}$?

(c) Without finding one, do you think that there is a 1-1, onto map from $\mathcal{A}$ to $\mathcal{B}$? Make a conjecture.

(d) Look up the Schröder–Bernstein theorem. What does it say and what does it imply here?

P 3.4.14. Let $\mathcal{A}$ and $\mathcal{B}$ be two finite sets. (Recall that, for a finite set X, the notation $|X|$ denotes the number of elements of X.)

(a) Show that there exists a 1-1 function $f \colon \mathcal{A} \to \mathcal{B}$ if and only if $|\mathcal{A}| \leq |\mathcal{B}|$.

(b) Show that there exists an onto function $f \colon \mathcal{A} \to \mathcal{B}$ if and only if $|\mathcal{A}| \geq |\mathcal{B}|$.

(c) Show that there exists a 1-1 and onto function $f \colon \mathcal{A} \to \mathcal{B}$ if and only if $|\mathcal{A}| = |\mathcal{B}|$.

(d) Show that, if there is a 1-1 function from $f \colon \mathcal{A} \to \mathcal{B}$ and a 1-1 function $g \colon \mathcal{B} \to \mathcal{A}$, then there exists a 1-1, onto function $h \colon \mathcal{A} \to \mathcal{B}$.

P 3.4.15. Let $\mathcal{A}$ and $\mathcal{B}$ be two sets and assume $f \colon \mathcal{A} \to \mathcal{B}$ is a function. Prove that f is 1-1 and onto if and only if f has an inverse.

P 3.4.16. Let X and Y be two finite sets of elements. For each of the following questions, without finding a numerical answer, translate the problem into a problem about balls and boxes. To which entry of the counting table (Table 3.1) does it correspond?

(a) What is the number of functions from X to Y?

(b) What is the number of 1-1 functions from X to Y?

(c) What is the number of onto functions from X to Y?

P 3.4.17. For real numbers a and b with $a \leq b$, the interval $[a, b]$ is defined by

$$[a, b] = \{x \in \mathbb{R} \mid a \leq x \leq b\}.$$

Is there a bijection from $[0, 1]$ to $[0, 10]$? Either give an example of such a bijection or prove that it does not exist.

P 3.4.18. Let $\mathbb{Z}^{>0}$ and $2\mathbb{Z}^{>0}$ denote the sets of positive and positive even integers, respectively. Is there a bijection from $\mathbb{Z}^{>0}$ to $2\mathbb{Z}^{>0}$? Either give an example of such a bijection or prove that it does not exist.

Collaborative Mini-project 1: Counting Monochromatic Triangles

In this mini-project, through a series of problems, you will investigate the possibilities for the number of monochromatic triangles in a two-coloring of the edges of a complete graph.

Preliminary Investigations

Please complete this section before beginning collaboration with other students.

MP1.1 Color the edges of K_6, the complete graph of order 6, using two colors (red and black). Try your best to get as few monochromatic triangles as possible. How many such triangles do you have?

MP1.2 Keep two examples of K_6's with their edges colored. One example should have as few monochromatic triangles as possible, and the other should have one or two more monochromatic triangles than the first one.

MP1.3 Color the edges of K_7 with two colors, and count the number of monochromatic triangles. (You should not spend too much time on this.)

MP1.4 Make some guesses about the minimum number of monochromatic triangles in K_6 (and maybe also in K_7).

In what follows we will try to estimate the number of monochromatic triangles in K_n by focusing on the number of the *multicolored* triangles.

MP1.5 In each of your examples for Problem MP1.2, let the vertices be 1, 2, 3, 4, 5, and 6. For vertex i, let

r_i = number of red edges that have i as a vertex,

b_i = number of black edges that have i as a vertex,

s_i = number of triangles that have i as a vertex, and for which one of the edges through i is red and the other black.

So, for example, r_2 is the number of red edges coming out of vertex 2, and s_2 is the number of triangles that have both of the following properties: (a) 2 is one of their vertices, (b) the two edges of the triangle that go through 2 are not of the same color. Such a triangle is of course multicolored. For each of your examples, complete Table MP1.1.

Table MP1.1 Red edges, black edges, and multicolored triangles.

i	r_i	b_i	s_i
1			
2			
3			
4			
5			
6			
Total			

Collaborative Investigation

Work on the rest of the questions with your team. Before getting started on the new questions, you may want to compare notes about the preliminary problems.

MP1.6 How many edges does K_6 have? What is the total number of triangles in K_6? How did you find the number of triangles?

MP1.7 Can you find b_i in terms of r_i? Can you find s_i in terms of r_i and b_i? Can you find s_i in terms of r_i? Can you prove your assertions?

MP1.8 Let $r = r_1 + r_2 + \cdots + r_6$, $b = b_1 + \cdots + b_6$, and $s = s_1 + s_2 + \cdots + s_6$. Find $r + b$. Could you have predicted the number that you got? What is the relation of s to the total number of multicolored triangles? Can you prove your assertion?

MP1.9 Write a formula that only depends on $r_1, r_2, \ldots, r_6$ for the number of monochromatic triangles in any two-coloring of K_6. Of course, if you color K_6 in two different ways using two colors, then for each of the colorings the r's will be different and thus your formula might give different answers (as it should).

MP1.10 We want to know what the smallest number of possible monochromatic triangles in a K_6 is. r_1 can be 0, 1, 2, 3, 4, or 5. Try each to see which one will result in the smallest value for your formula in Problem MP1.9. Do the same for r_2 through r_6. What is the smallest number of monochromatic triangles in K_6 according to your calculation?

MP1.11 If the number that you found in the previous part is smaller than the number of monochromatic triangles in all of your examples, then use the analysis above to help you build an example with the minimum number of monochromatic triangles.

MP1.12 Repeat the above (you don't have to gather data for parts that you have already proved) for K_7. What is the minimum number of monochromatic triangles when you color the edges of K_7 using two colors? Give an example that illustrates this.

MP1.13 We color the edges of K_6 with two colors (red and black). We notice that all the edges coming out of one of the vertices are colored red. Using an analysis similar to the above, what can you say about the minimum number of monochromatic triangles?

MP1.14 Can you give and prove a formula analogous to the one in Problem MP1.9 for the number of monochromatic triangles in any two-coloring of K_n?

Mini-project Write-Up

In your write-up of the results, include at least the following:

- Clear and precise definitions of "complete graphs," "two-coloring of a complete graph," and "monochromatic triangles"
- An example of a two-coloring of K_6 with the minimum possible number of monochromatic triangles
- A clearly stated and self-contained theorem that gives a method for finding the number of monochromatic triangles in a given two-coloring of K_n
- A proof of the your theorem from the previous part
- A clear derivation, using your theorem, for the minimum number of monochromatic triangles in an arbitrary two-coloring of K_6 and K_7
- An answer with proof to Problem MP1.13.

(Extra Credit) Try to find an actual formula for the minimum number of monochromatic triangles in a given two-coloring of K_n. You may find it helpful to consider three cases: n is even, $n = 1 \pmod 4$, and $n = 3 \pmod 4$.[1]

---

[1] Finding such a formula was the subject of Goodman 1959.

Collaborative Mini-project 2: Binomial Coefficients

In this mini-project, you explore several equivalent definitions of binomial coefficients, make a number of conjectures about these numbers, and find the number of non-negative integer solutions to an equation of the form $x_1 + x_2 + \cdots + x_n = m$.

Preliminary Investigations

Please complete this section before beginning collaboration with other students.

MP2.1 Consider Table MP2.1. The entry in row n and column k is denoted by $g(n, k)$. So what is $g(n, k)$? We are interested in paths from the box at row 0 and column 0 (i.e., the top left corner) to the box at row n and column k. We are only allowed two kinds of moves: (a) straight down to the box below $\downarrow$, (b) diagonally to the box that is one row down and one column to the right $\searrow$. So, for example, if we are at the box in row 3 and column 2 (abbreviated as the box at $(3, 2)$), then we can proceed either to the box at $(4, 2)$ or to the box at $(4, 3)$. Now, $g(n, k)$ is the number of different paths from the box at $(0, 0)$ to the box at (n, k). For example, $g(3, 1) = 3$ since there are three ways to go from $(0, 0)$ to $(3, 1)$:

$$(0, 0), \ (1, 0), \ (2, 0), \ (3, 1);$$
$$(0, 0), \ (1, 0), \ (2, 1), \ (3, 1);$$
$$(0, 0), \ (1, 1), \ (2, 1), \ (3, 1).$$

Complete a few rows of the table and look for a pattern. Express your pattern as a recurrence relation for $g(n, k)$. Check your pattern by completing some more of the table. When you are convinced that your pattern is correct, then complete the rest of the table.

MP2.2 Now consider Table MP2.2. This time the entry at row n and column k is denoted by $\binom{n}{k}$, and is the number of subsets of size k in $[n]$, a set of size n. (Recall that $[n] = \{1, 2, \ldots, n\}$ if $n \geq 1$, and $[0] = \emptyset$.) The symbol $\binom{n}{k}$ is read "n choose k" and is called a *binomial coefficient*. Figure MP2.1 gives a list of all the subsets of $\{1, 2, 3, 4, 5, 6\}$, and thus it also has within it all the subsets of $\{1, 2, 3, 4, 5\}$, $\{1, 2, 3, 4\}$, etc. Using this list, fill in some of the rows of the table and try to find a pattern. Express your pattern as an expression for $\binom{n}{k}$ (again not an explicit formula, but a recurrence relation). Complete the table.

Table MP2.1 The yet-to-be-filled table of the values of $g(n,k)$.

$n\backslash k$	0	1	2	3	4	5	6
0		XX	XX	XX	XX	XX	XX
1			XX	XX	XX	XX	XX
2				XX	XX	XX	XX
3		3			XX	XX	XX
4						XX	XX
5							XX
6							

Table MP2.2 The yet-to-be-filled table of the values of $\binom{n}{k}$.

$n\backslash k$	0	1	2	3	4	5	6
0		XX	XX	XX	XX	XX	XX
1			XX	XX	XX	XX	XX
2				XX	XX	XX	XX
3			3		XX	XX	XX
4						XX	XX
5							XX
6							

Collaborative Investigation

Work on the rest of the questions with your team. Before getting started on the new questions, you may want to compare notes about the preliminary problems.

MP2.3 Try proving your recurrence relation for $g(n,k)$.

MP2.4 Let A be the set of all subsets of size 3 in $[6] = \{1,2,\ldots,6\}$. Let B be those subsets in A that include 1, and let C be those subsets in A that do not include 1. What are $|A|$, $|B|$, and $|C|$? What is the relation among them? By definition, the size of A is $\binom{6}{3}$. Are the sizes of B and C also binomial coefficients? Explain why.

MP2.5 Prove your pattern for $\binom{n}{k}$.

MP2.6 How is $g(n,k)$ related to $\binom{n}{k}$? Give an argument for your assertion.

MP2.7 In either of the tables, find the sum of the entries in each row. Do you see a pattern? Express it as a mathematical expression involving binomial coefficients. Is there any obvious reason for your pattern?

123456

12345   12346   12356   12456   13456   23456

1234 1235 1245 1345 2345 1236 1246 1346 2346 1256 1356 2356 1456 2456 3456

123 124 134 234 125 135 235 145 245 345 126 136 236 146 246 346 156 256 356 456

12   13   23   14   24   34   15   25   35   45   16   26   36   46   56

1   2   3   4   5   6

$\emptyset$

Figure MP2.1 The subsets of $\{1,2,3,4,5,6\}$ without curly brackets or commas. Thus 1234 denotes the subset $\{1,2,3,4\}$.

MP2.8 In each row of each table there is a certain kind of symmetry. Express this as a mathematical expression using binomial coefficients. Give a proof of your expression.

NOTE: I would rather you did not use a "formula" for the binomial coefficients, and instead used their definition: $\binom{n}{k}$ is the number of subsets of size k for a set of size n. Give a similar interpretation of the other terms in your expression and give a reason why the equality holds.

MP2.9 Find the alternating sum of the entries in each row (i.e., the first entry minus the second plus the third and so on.) Make a conjecture.

MP2.10 Find the sum of the squares of the entries in each row. Do you see a pattern? Make a conjecture.

MP2.11 Starting from the top left corner, draw parallel diagonal lines through the table and add the entries on each diagonal line. Thus, for example, the first line goes only through the box at $(0,0)$. The second one goes through the boxes $(1,0)$ and $(0,1)$ (although this latter one does not have an entry). The third line goes through $(2,0)$, $(1,1)$, and $(0,2)$, and so on. Do you see a pattern for the sums? Make a conjecture.

MP2.12 We have k indistinguishable balls that have to be colored using n colors. We want to know the number of ways we can do that. The answer depends on both n and k, and so we call the answer $f(n,k)$. Using the following steps, find a formula for $f(n,k)$ that involves binomial coefficients.

Step 1: Let x_1 be the number of balls that are colored using color 1, let x_2 be the number of balls that are colored using color 2, and so on. Write an equation that $x_1, x_2, \ldots, x_n$ must satisfy. How is $f(n, k)$ related to the number of non-negative integer solutions to this equation? State this reformulation of $f(n, k)$ clearly.

Step 2: We have k indistinguishable balls that have to be colored using n colors. Answer the following questions using the function f. How many colorings are possible if we do not use the first color? How many colorings are possible if we definitely use the first color? Use your answers to these questions to write a recurrence relation for $f(n, k)$. Do the arguments that you have provided so far give a proof of the recurrence relation?

Step 3: Find $f(n, k)$ for small values of n and k. Can you find the answers in the table of binomial coefficients? Make a precise conjecture, and check your conjecture for at least one more case.

Step 4: Using the recurrence relation of Step 2, and induction (on $n + k$), prove your conjecture in Step 3.

Note: Make sure that you are clear on how you proved the recurrence relation and how you proved your conjecture of Step 3. For which one(s) did you use induction?

MP2.13   A store sells five different brands of chewing gum. In how many ways can someone choose to buy 10 packs of gum?

Mini-project Write-Up

In your write-up of the results, include at least the following:

- A definition of binomial coefficients and a table of small binomial coefficients
- A recurrence relation for binomial coefficients with proof
- A definition for the function $g(n, k)$ and its precise relation with binomial coefficients with proof
- A number of conjectures about binomial coefficients. One or two of these should be proved
- A definition for the function $f(n, k)$, and its precise relation with binomial coefficients with proof
- A solution to the chewing gum problem.

Collaborative Mini-project 3: Stirling Numbers

In this mini-project, you explore the definitions of the so-called Stirling numbers of the first and second kind, find recurrence relations for each of them, and make conjectures about these numbers.

Preliminary Investigations

Please complete this section before beginning collaboration with other students.

Stirling Numbers of the Second Kind

Let n and k be non-negative integers. We define $\left\{{n \atop k}\right\}$ to be the number of partitions of $[n]$ – a set with n elements – into exactly k non-empty subsets (blocks). We interpret (or by convention decree) that $\left\{{0 \atop 0}\right\} = 1$. The integer $\left\{{n \atop k}\right\}$ is called a *Stirling number of the second kind.*[1]

For example, $\left\{{4 \atop 3}\right\} = 6$, since the following are the only partitions of $[4]$ into 3 non-empty subsets:

$$\{\{1\},\{2\},\{3,4\}\}, \quad \{\{1\},\{3\},\{2,4\}\}, \quad \{\{1\},\{4\},\{2,3\}\},$$
$$\{\{2\},\{3\},\{1,4\}\}, \quad \{\{2\},\{4\},\{1,3\}\}, \quad \{\{3\},\{4\},\{1,2\}\}.$$

MP3.1 What is $\left\{{n \atop k}\right\}$ if $k > n$? For $n \geq 1$, what are $\left\{{n \atop 0}\right\}$, $\left\{{n \atop 1}\right\}$ and $\left\{{n \atop n}\right\}$?

MP3.2 Consider Table MP3.1. The entry in row n and column k is $\left\{{n \atop k}\right\}$. Fill in at least the first four rows of this table.

Stirling Numbers of the First Kind

We have n guests over for dinner, and we have k round tables. We would like to seat the guests around the tables in such a way that no table is empty. The tables are pretty big and we can move chairs around, and so any number of guests can sit at a given table. As usual in a circular seating arrangement, only the relative position of the guests is important. The tables are identical and their location is immaterial. (However, it does matter who is sitting to the left of whom. So, for example, a seating arrangement where A is sitting to the left of B is

---

[1] Using the notation $\left\{{n \atop k}\right\}$ for Stirling numbers of the second kind is common but not standard. Other commonly used notations are $S(n,k)$ and $s_2(n,k)$. Some people read $\left\{{n \atop k}\right\}$ as "n subset k" and they remember the meaning of $\left\{{n \atop k}\right\}$ by recalling that curly braces are also used to denote sets.

Table MP3.1 The yet-to-be-filled table of Stirling numbers of the second kind.

$n \backslash k$	0	1	2	3	4	5	6
0							
1							
2							
3			3				
4				6			
5							
6							

different from one where A is sitting to the right of B.) The symbol $\begin{bmatrix} n \\ k \end{bmatrix}$ will denote the total number of different possible seating arrangements.

More formally, the circular permutations are called cycles, and, given positive integers n and k, we define $\begin{bmatrix} n \\ k \end{bmatrix}$ as the number of permutations of n (distinct) objects into exactly k cycles. We also define $\begin{bmatrix} n \\ k \end{bmatrix} = 0$ if $n \le 0$ or $k \le 0$, except $\begin{bmatrix} 0 \\ 0 \end{bmatrix} = 1$.

The numbers $\begin{bmatrix} n \\ k \end{bmatrix}$ are known as the *Stirling numbers of the first kind,*[2] and the numbers $(-1)^{n-k} \begin{bmatrix} n \\ k \end{bmatrix}$ are called the *signed Stirling numbers of the first kind.*

MP3.3 What is $\begin{bmatrix} n \\ 1 \end{bmatrix}$?

MP3.4 Consider Table MP3.2. The entry in row n and column k is $(-1)^{n-k} \begin{bmatrix} n \\ k \end{bmatrix}$. Fill in at least the first four rows of this table.

Table MP3.2 The yet-to-be-filled table of *signed* Stirling numbers of the first kind.

$n \backslash k$	0	1	2	3	4	5	6
0							
1							
2							
3			-3				
4				-6			
5							
6							

---

[2] Other notations for Stirling numbers of first kind include $s(n,k)$, $s_1(n,k)$, and $c(n,k)$.

Collaborative Investigation

Work on the rest of the questions with your team. Before getting started on the new questions, you may want to compare notes about the preliminary problems.

MP3.5 $\begin{bmatrix} n \\ k \end{bmatrix}$ satisfies a recurrence relation of the following form:

$$\begin{bmatrix} n \\ k \end{bmatrix} = ?\begin{bmatrix} n-1 \\ k \end{bmatrix} + ??\begin{bmatrix} n-1 \\ k-1 \end{bmatrix}.$$

Make a conjecture about the coefficients, and check the recurrence relation for at least one more non-trivial case. Fill in the second table when you are confident about your conjecture.

MP3.6 Prove your conjecture about $\begin{bmatrix} n \\ k \end{bmatrix}$ by performing the following thought experiment. Guest number n arrives late. The other $n-1$ guests have already been seated. There are two possibilities: if the previous guests are using k tables then n can sit to the right of any of the guests. On the other hand if the previous guests are using only $k-1$ tables, then n has to sit by herself.

MP3.7 Find $\sum_{k=0}^{n} \begin{bmatrix} n \\ k \end{bmatrix}$ for some small values of n. Make a conjecture. Can you prove your conjecture?

MP3.8 $\begin{Bmatrix} n \\ k \end{Bmatrix}$ satisfies a recurrence relation of the following form:

$$\begin{Bmatrix} n \\ k \end{Bmatrix} = ?\begin{Bmatrix} n-1 \\ k \end{Bmatrix} + ??\begin{Bmatrix} n-1 \\ k-1 \end{Bmatrix}.$$

Make a conjecture about the coefficients, and check the recurrence relation for at least one more non-trivial case. Fill in the first table when you are confident about your conjecture.

MP3.9 Look for the Stirling numbers $\begin{Bmatrix} n \\ n-1 \end{Bmatrix}$ in a table of binomial coefficients. Make a conjecture. Can you prove it?

MP3.10 Can you make a conjecture about $\begin{Bmatrix} n \\ 2 \end{Bmatrix}$?

MP3.11 Prove your recurrence relation about $\begin{Bmatrix} n \\ k \end{Bmatrix}$ by considering the partitions of $[n-1]$ into k and $k-1$ subsets. If you have a partition of $[n-1]$ into k subsets, how can you obtain a partition of $[n]$ into k subsets? What if you had a partition of $[n-1]$ into $k-1$ subsets? Do you get all partitions of $[n]$ into k subsets from one of these two ways? Is there any double counting?

MP3.12 Are the signed Stirling numbers of the first kind related to the Stirling numbers of the second kind? Make a conjecture.

HINT: Consider Tables MP3.1 and MP3.2 as two 7×7 matrices. How are the matrices related?

MP3.13 Let n and k be integers (positive, zero, or negative), and make the following stipulations:

$$\begin{Bmatrix} 0 \\ k \end{Bmatrix} = \begin{bmatrix} 0 \\ k \end{bmatrix} = \begin{cases} 1, & \text{if } k = 0; \\ 0, & \text{otherwise.} \end{cases} \qquad \begin{Bmatrix} n \\ 0 \end{Bmatrix} = \begin{bmatrix} n \\ 0 \end{bmatrix} = \begin{cases} 1, & \text{if } n = 0; \\ 0, & \text{otherwise.} \end{cases}$$

Now use the recurrence relations that you found in Problem MP3.8 to extend the table of Stirling numbers of second kind upward and leftward (i.e., for negative n and k.) Compare with the Stirling numbers of the first kind and make a conjecture. Do the same for the Stirling numbers of the first kind.

Mini-project Write-Up

In your write-up of the results, include at least the following:

- A precise definition of Stirling numbers of the first and second kind. Illustrate your definitions with examples
- A recurrence relation with proof for each kind of Stirling number
- Completed tables of Stirling numbers of the first and second kind for small values of n and k
- Conjectures with proof for $\sum_{k=0}^{n} \begin{bmatrix} n \\ k \end{bmatrix}$ and $\begin{Bmatrix} n \\ n-1 \end{Bmatrix}$
- Conjectures for $\begin{Bmatrix} n \\ 2 \end{Bmatrix}$, for the relation of the two kinds of Stirling numbers, and for values of the tables of Stirling numbers when extended appropriately for negative values of k and n.

4 Permutations and Combinations

You are told by an oracle to distribute 47 identical coins among Cyrus, Cambyses, Bardiya, Darius, and Xerxes while making sure that each gets an odd number of coins. In how many ways can this be done?

In this chapter we count the number of permutations (e.g., ordered lists) and combinations (e.g., subsets) of sets and multisets. As discussed at the end of Section 3.3, permutations correspond to the first column of the counting table, Table 3.1, and combinations correspond to the second column of the counting table.

4.1 Permutations of a Set and Falling Factorials

Warm-Up 4.1. *Roxana, Rosalina, Gustavo and Guillermo are about to go for a ride in their car. One person is to be the driver, one will sit next to the driver, one will sit behind the driver, and the last person will sit behind the passenger. In how many ways can the four friends be placed in the car?*

In this section we will consider entry number $\boxed{1}$ of the counting table (Table 3.1).

Definition 4.2 (*s*-Permutations). Let $0 < s \leq t$ be integers. Recall that $[t] = \{1, 2, \ldots, t\}$ represents a set with t elements. An *s-permutation* of $[t]$ is an ordered arrangement of s of the t elements. A *t*-permutation of $[t]$ is just called a *permutation* of $[t]$.

Example 4.3. The 3-permutations of [4] include:

$$123, 132, 231, 213, 124, 432, \ldots$$

Definition 4.4 (The Falling Factorial Function). Let k be a non-negative integer and let x be an indeterminate (i.e., a variable). Then the *kth falling factorial function* is denoted by $(x)_k$ – read "x down k" or "x lower k" – and defined by

$$(x)_k = x(x-1)(x-2)\cdots(x-k+1), \text{ for } k > 0,$$

and $(x)_0 = 1$. As a special case, if n is a non-negative integer, then we define n factorial, written $n!$, by $n! = (n)_n$. So $0! = 1$, and, for $n \geq 1$,

$$n! = (n)_n = n(n-1)(n-2)\cdots 3 \cdot 2 \cdot 1.$$

Remark 4.5. The kth falling factorial $(x)_k$ has exactly k factors, and is a polynomial in x of degree k. So $(x)_3 = x(x-1)(x-2)$, and if we plug in 7 for x, we get $(7)_3 = 7 \times 6 \times 5 = \frac{7!}{4!}$. More generally, if $0 \leq s \leq t$ are integers, then

$$(t)_s = \frac{t!}{(t-s)!}.$$

Theorem 4.6 (Entry $\boxed{1}$ of the Counting Table). *Let $0 < s \le t$ be positive integers. Then the following integers are equal:*

(a) $\boxed{1}$ *The number of ways of placing s distinct balls into t distinct boxes such that each box gets at most one ball*

(b) *The number of s-permutations of $[t]$*

(c) $(t)_s = t(t-1)(t-2) \cdots (t-s+1) = \frac{t!}{(t-s)!}.$

In particular, the number of permutations of $[t]$ is $t!$.

Proof. (a) $=$ (b). Let $\mathcal{A}$ be the set of all possible ways of placing s distinct balls into t distinct boxes such that each box gets at most one ball. In other words, each element of $\mathcal{A}$ is one particular way of placing the balls into boxes respecting the conditions. The number of ways of placing balls in boxes with each box getting at most one ball is, then, the size of $\mathcal{A}$, denoted by $|\mathcal{A}|$. Let $\mathcal{B}$ be the set of s-permutations of $[t]$. Likewise, the number of s-permutations of $[t]$ is $|\mathcal{B}|$.

To show $|\mathcal{A}| = |\mathcal{B}|$, we give a one-to-one, onto function $f : \mathcal{A} \to \mathcal{B}$.

The function f is defined as follows. Assume you are given a placement of the balls in the boxes. Line up the balls from 1 to s and under them place the boxes that they go into. The list of boxes is then an s-permutation of t objects. If the boxes are $U_1, \ldots, U_t$, we will have

$$
\begin{array}{ccccc}
① & ② & ③ & \cdots & ⓢ \\
U_{i_1} & U_{i_2} & U_{i_3} & \cdots & U_{i_s}.
\end{array}
$$

The ordered list $U_{i_1} U_{i_2} U_{i_3} \ldots U_{i_s}$ is an s-permutation of $\{U_1, \ldots, U_t\}$. (Note that the condition that no box gets more than one ball ensures that there are no repetitions in this list.) Under the function f, this particular placement is sent to the permutation

$$i_1 i_2 \ldots i_s$$

of $[t]$. This is a well-defined function. Every placement of balls into boxes – such that no box gets more than one ball – unambiguously gives a unique s-permutation of $[t]$. The map is onto, since, given any s-permutation of $[t]$, we can reverse the above process and find a placement of balls into boxes that under f would give us the sought-after s-permutation. The map is also 1-1 since two *different* placements of balls into boxes will result in two different s-permutations.

We conclude that $|\mathcal{A}| = |\mathcal{B}|$.

(b) $=$ (c). To count the number of s-permutations of $[t]$, we line up s spaces, and ask how many choices we have for the first spot, then the second spot, and so on:

$$
\underline{\quad\quad} \quad \underline{\quad\quad} \quad \underline{\quad\quad} \quad \cdots \quad \underline{\quad\quad\quad}
$$
$$
t \qquad t-1 \qquad t-2 \qquad\quad t-s+1
$$

By the multiplication principle, the number of s-permutations of $[t]$ is then the falling factorial $(t)_s$. $\qquad\square$

Remark 4.7. The proof above was a bit too verbose. In this first example, we wanted to make the logic of the proof very clear. As we go along, we will limit ourselves to identifying the function f. Often, we leave it to the reader to convince herself that the map is well defined, 1-1, and onto. In the above proof, for example, we might have just said:

Number the balls 1 through s. If a ball numbered a goes into a box U, then the box U is chosen to be the ath in the list. Given a placement of balls into boxes (with the condition that no box gets more than one ball), we have chosen s of the boxes and decided where in the list each goes. Hence, each placement of the balls gives an s-permutation of $[t]$ and vice versa.

In such a rendering, the sets A and B and the function f are implicit. The advantage is that the reader sees the main point quickly, and does not have to wade through much verbiage. On the other hand, it should be clear that implicit in any quick description of a correspondence are two underlying sets and a function that needs to be 1-1 and onto. We invite the reader to provide the details until doing so becomes a real chore. At that point, the reader is ready to provide quick descriptions of proofs without overlooking possible shortcomings.

Example 4.8. Sophia, Maria, Anastasia, Anna, Daria, Polina, Artem, Aleksandr, Maksim, Ivan, Mikhail, and Dmitry have seats 3 through 14 of row 47 in a concert devoted to the works of Rachmaninoff. In how many ways can they be seated if no two of Anastasia, Anna, Artem, and Aleksandr are to be seated next to each other?

Solution. We decide on the order that the 12 people are going to be seated in two steps. First, we order the eight whose names don't start with an A. This amounts to permuting 8 things, and we can do this in 8! ways. Now we need to place the remaining four (the balls) in the 9 places (the boxes) before, between, and after the eight already ordered. The balls and the boxes are distinct and no box gets more than one ball. The number of ways of placing 4 distinct balls in 9 distinct boxes (with each box getting 0 or 1 ball) is $(9)_4$. (You could argue this directly as well: 9 spots to place Anastasia, then 8 spots for Anna, and so on to get $9 \times 8 \times 7 \times 6 = (9)_4$.) Hence, the total number of ways of seating the 12 friends is

$$8! \, (9)_4 = 8! \times 9 \times 8 \times 7 \times 6 = 121,927,680. \qquad \square$$

Example 4.9. A sports journal wants to write 30 articles, one per week, on the 30 basketball teams in the National Basketball Association. The 30 NBA teams are organized into 6 divisions, each with 5 teams. The journal wants to cover one division in five consecutive weeks, and then move on to a different division. It also prefers to feature the teams from the same city (the Lakers and the Clippers, both of the Pacific Division, as well as the Knicks and the Nets, both of the Atlantic Division) on consecutive weeks. How many ways can this be done?

Solution. First decide in what order the six divisions are covered. This can be done in 6! ways. Then within each division, decide the order of the teams. For four of the divisions, you just have to order 5 teams and this can be done in 5! ways. For the Pacific Division, consider Lakers-Clippers as one team. The number of ways of ordering 4 articles (one for Lakers-Clippers) is 4!. Now, the "one" team Lakers-Clippers can be covered in two ways (first the Lakers or first the Clippers), and so the total number of ways of covering the Pacific Division is 2(4!). The Atlantic Division is similar, and hence, the total number of ways of arranging the articles is

$$6! \times (5!)^4 \times (2 \times 4!)^2 = 343,985,356,800,000. \qquad \square$$

Example 4.10. Thirty people are going to run a cross country race, and, after the race, the list of first through sixth place will be announced.

(a) How many different announcements are possible?
(b) What if Emily – one of the runners – is known to be one of the six?

Solution. (a) Each possible announcements is a 6-permutation of the set [30]. Hence, the number of possible announcements is the number of 6-permutations of [30] which is $(30)_6 = \frac{30!}{24!} = 427,518,000$.

(b) Among the $(30)_6$ lists possible, $(29)_6$ do not include Emily. So, the number of possible announcements that include Emily is $(30)_6 - (29)_6 = \frac{30!}{24!} - \frac{29!}{23!} = \frac{6 \cdot 29!}{24!} = 85,503,600.$ $\square$

Example 4.11. In chess, a rook can move horizontally or vertically. Hence, two rooks cannot attack each other if they are on different rows and on different columns of the chess board.

(a) How many ways can you place 8 non-attacking rooks on an 8×8 chess board?
(b) What if the rooks were all different colors?
(c) Consider two rook placements to be the same if, by an appropriate rotation of the board, one placement would become the other. Given this definition, what is the number of ways of placing 8 non-attacking rooks on an 8×8 board?

Solution. (a) Since no two rooks can be on the same row, any placing of the rooks corresponds to a list $(1,j_1), (2,j_2), \ldots, (8,j_8)$ of coordinates for the rooks. No two rooks can be on the same column either. Hence $j_1, j_2, \ldots, j_8$ are 8 distinct integers between 1 and 8. In other words, $j_1 j_2 \ldots j_8$ is a permutation of [8]. This map from the rook placements to permutations of [8] is clearly 1-1 since two rook placements give two different permutations. It is also onto, since every permutation of [8] can be translated back to a placement for the rooks. (If $r_1 r_2 \ldots r_8$ is a permutation of [8], then $(1,r_1), (2,r_2), \ldots, (8,r_8)$ are the coordinates for placing 8 non-attacking rooks on an 8×8 chess board.)

So the number of ways of placing 8 non-attacking rooks on an 8×8 board is the same as the number of permutations of [8], which is 8!.

(b) Place the rooks in two steps: first find the positions, and then decide on the color arrangement. From the previous part, the number of positions is 8!. Given any set of coordinates for the rooks, the number of permutations of the 8 different colors is 8!. Hence the total number of ways of placing 8 rooks each differently colored, is $(8!)^2$.

(c) If we want to do this problem without resorting to many cases, then we should first get an understanding of the symmetries of an 8×8 board. This is done in group theory courses and is beyond the scope of the present volume (see Shahriari 2017, Section 8.2). $\square$

Remark 4.12. The number of ways of placing 8 non-attacking rooks on an 8×8 chess board is the same as the number of 8×8 matrices with exactly one 1 in every row and every column, and zeros in all the other entries. This is because, in order to have the rooks not be able to attack each other, you have to place each in a different row and a different column than all the others.

Also recall that an elementary product in an 8×8 matrix is a product of eight entries of the matrix such that each comes from a different row and a different column than all the others. (The determinant of a square matrix is the sum of all the signed elementary products of a matrix.) Hence, the number of placements of 8 non-attacking rooks on an 8×8 board is the same as the number of elementary products in an 8×8 matrix.

Circular Permutations

Definition 4.13 (Circular Permutations). Let n be a positive integer. An ordering of the elements of $[n]$ around a circle is called a *circular permutation* of $[n]$. In a circular permutation, any of the elements can be considered the starting point.

Remark 4.14. In a circular permutation, what matters is which element is to the right of which other element, and there is no "first" element. So, for example, 123456 and 234561 are the same circular permutation of $[6]$ since, if you write them around a circle, you can't tell them apart.

Example 4.15. There are only two different circular permutations of $[3]$. These are 123 and 132.

Theorem 4.16 (Circular Permutations). *Let $0 < r \leq n$ be integers. The number of circular r-permutations of $[n]$ is $\frac{(n)_r}{r} = \frac{n!}{r(n-r)!}$. In particular, the number of circular permutations of $[n]$ is $(n-1)!$.*

Proof. There are $(n)_r$ r-permutations of $[n]$. Each r-permutation gives a circular r-permutation but r of them give the same circular r-permutation. The result follows. $\square$

Example 4.17. In how many ways can you seat Weza, Lukeny, Nataniela, Raila, Kendra, Denzel, Winny, Elsabe, Nuria, and Nyura around a round table in such a way that the twins Nuria and Nyura are not seated next to each other? In seating the guests around the table, the actual seats do not matter. What matters is who is sitting on which side of whom.

Solution. If we did not have the restriction that Nuria and Nyura are not to be seated next to each other, we could seat the 10 people in 9! ways around a round table. We will subtract from this the number of ways of seating the 10 people where the twins are seated next to each other. Thinking of the twins as one person, we have 9 people to seat around a table. This can be done in 8! ways, and then we have to decide the order in which Nuria and Nyura are sitting. Hence, the total number of ways of seating the 10 guests in such a way that Nuria and Nyura are next to each other is $2 \times 8!$. We conclude that the total number of ways of seating the 10 guests without seating the twins next to each other is

$$9! - 2 \times 8! = 282,240. \qquad \square$$

Problems

P 4.1.1. Let the function $f\colon \mathbb{R} \to \mathbb{R}$ be defined by $f(x) = (x)_2$. Graph the function f. For which values of x is $f(x) < 0$? Answer the same questions for the function $g\colon \mathbb{R} \to \mathbb{R}$, defined by $g(x) = (x)_3$.

P 4.1.2. For which real numbers x is $(x)_2 = (x)_3$? For which real numbers x is $(x)_t = (x)_{t-1}$?

P 4.1.3. A standard deck of playing cards consists of 52 cards in each of the four suits of Spades, Hearts, Diamonds, and Clubs. Each suit contains 13 cards: Ace, 2, 3, 4, 5, 6, 7, 8, 9, 10, Jack, Queen, and King. In how many ways can you order a standard deck of cards so that all the cards of the same suit are next to each other (the cards in each suit do not have to be in order)?

P 4.1.4. How many integers with no repeating digits and greater than 6300 can you write down without using the digits 4 or 7?

P 4.1.5. Let m and n be fixed positive integers. A sports federation has decreed that the flag of each team must consists of m vertical bands, each colored with one of n colors in such a way that adjacent bands are colored differently. How many flags can be constructed in this way? It may be assumed that one edge of each flag is distinguished by being attached to a flagpole, so that ABC and CBA represent different flags.

P 4.1.6. Lì, Wěi, Fāng, Xiùyīng, Nà, Mǐn, Jìng, and Qiáng are to be seated in a row. In how many ways can this be done if

(a) there are no restrictions on the seating order
(b) Lì and Wěi must sit next to each other
(c) Fāng, Xiùyīng, Nà, Mǐn, and Jìng must sit next to each other (in some order)
(d) the eight have paired up so that there are four couples and each couple must sit together?

P 4.1.7. You write letters to Cesar, Yenny, Paula, Tingyu, and Hannah, and then address five different envelopes, one to each of them. Later on, you ask a work-study student to put the letters into the envelopes. The student assumes that all the letters are the same and puts them into the envelopes at random, one letter per envelope. What is the probability that only Yenny and Tingyu get the correct letters but that nobody else does?

P 4.1.8. Let $\mathcal{A}$ be the set of all permutations of [47], and let $\mathcal{B}$ be the set of all bijective maps from [47] to [47]. Describe a bijection $f\colon \mathcal{A} \to \mathcal{B}$.

P 4.1.9. Let $A = \{1, 4, 7, 11, 13\}$ and $B = \{2, 5, 8, 12, 15\}$. Find the number of 1-1, onto maps $f\colon [47] \to [47]$ that map A onto B. In other words, we want to count maps that not only are 1-1 and onto and send elements of [47] to elements of [47], but also send every element of A to an element of B.

P 4.1.10. You order ten different books online. From among these, three are for your sister. The books are ordered from different vendors and arrive randomly and one by one. What is the probability that the three books for your sister arrive consecutively?

P 4.1.11. You are to seat eight people at one side of a long dinner table. From among the eight, only Neema and Leonardo are left-handed. In how many ways can you seat the guests so that neither Neema nor Leonardo are sitting to the right of a right-handed person?

P 4.1.12. To make an activity mobile for your friend's baby, you want to attach four pairs of colorful socks to a (small) circular hula hoop, and it happens that the right and the left socks are different in the socks that you are going to use. In how many ways can you organize the socks around the circle if you want to make sure that the two socks of each pair are placed next to each other?

P 4.1.13. You are to assign seats to seven Americans and seven Russians around a circular table. You do not want two Americans or two Russians to sit next to each other. In how many ways can you do this? (The actual seats do not matter. We are only interested in the seating arrangements, that is, who is sitting next to whom and how many seats away from whom.)

P 4.1.14. You have seven of your friends, including Elaine, over for dinner, and the eight of you are to seat around a circular dinner table. In how many ways can you arrange the seating so that you are not sitting exactly across the table from Elaine?

P 4.1.15. Ms. Ashton is convening a meeting. Afsaneh, Ahang, Anahita, and Arezoo are representing the city of Shimashki, while Nikolaos, Melissa, Lycus, and Kassandra are representing the region of Thrace. The eight representatives and Ms. Ashton are to be seated around a circular table. However, we do not want any two representative of Shimashki to sit next to each other. Likewise, no two representative of Thrace are to be seated next to each other. In how many ways can this be done?

P 4.1.16. Assume that a year has 365 days and you want to write down a list of 30 dates of the year (you are writing only the day and the month, such as June 2nd or June 23rd).

(a) If you allow repetition, then how many such lists are possible?

(b) If you don't allow repetition—i.e., all the dates are distinct—then how many lists are possible?

(c) If, from among all the possible lists (including the ones with repetition), you randomly choose one, then what is the probability that all the dates on the list will be distinct? (Use a simple computer program or a spreadsheet to calculate an actual number.)

(d) Answer the previous question but replace 30 with a positive integer r.

P 4.1.17. **The Birthday Problem.**

(a) In a room of 30 people, what is the probability that at least 2 people will have the same birthday?

(b) What if the room had 20 people? Or 40 people?

4.2 Combinations of Sets and Binomial Coefficients

Warm-Up 4.18. *Aditsan, Bly, Chayton, and Diyani are your friends. You need to ask two of them for a favor, but you can't decide which two. Assume that, on separate pieces of paper, you write down pairs of their names (so on one piece you have Aditsan and Diyani, while on another one you have Bly and Diyani, and so on). You put these pieces of paper in a bag, and randomly choose one. What is the probability that you pick Aditsan and Chayton? Now assume that, again* *on four separate pieces of paper, you write the name of each friend (so on one piece you have Aditsan, while on another you have Bly, and so on) and fold the pieces of paper and randomly put them in a row. Now you choose the first two for your chore. What is the probability that you pick Aditsan and Chayton (in any order)?*

In this section, we consider entry $\boxed{2}$ of the counting table, Table 3.1 (as well as the trivial entries $\boxed{3}$ and $\boxed{4}$).[1] For n a positive integer, we continue to use $[n] = \{1, \ldots, n\}$ as the prototype of a set with n elements, and remind the reader that $[0] = \emptyset$.

Definition 4.19. Let $0 \le s \le t$ be integers. A subset of size s of $[t]$ is called an *s-subset* or an *s-combination* of $[t]$. You can think of an s-combination as an unordered selection of s of the t elements of $[t]$.

The number of s-subsets of $[t]$ is denoted by $\binom{t}{s}$. The number $\binom{t}{s}$ – for reasons to become clear later – is called a *binomial coefficient* (and is read "t choose s").

The actual collection of s-subsets of $[t]$ – as opposed to the *number* of such subsets – is denoted by $\binom{[t]}{s}$.

Example 4.20. The binomial coefficient $\binom{0}{0} = 1$ since the number of subsets of size 0 in $[0] = \emptyset$ is 1.

The collection of subsets of size 2 of $[3]$ is

$$\binom{[3]}{2} = \{\{1,2\}, \{1,3\}, \{2,3\}\},$$

and so $\binom{3}{2} = 3$.

Theorem 4.21 (Entry $\boxed{2}$ of the Counting Table). *Let $0 \le s \le t$ be integers. Then the following integers are equal:*

(a) $\boxed{2}$ *The number of ways of placing s identical balls into t distinct boxes such that each box gets at most one ball*

(b) *The number of s-subsets (or s-combinations) of $[t]$, that is,* $\binom{t}{s}$

(c)

$$\frac{t!}{s!\,(t-s)!}.$$

Proof. (a) = (b). Given s identical balls, and given the restriction that each box gets at most one ball, each placement of the balls in t boxes amounts to choosing a subset of size s of the t boxes. Hence, the number of ways of placing the s balls into the t boxes with each box getting at most one ball is the same as the number of s-subsets of $[t]$. This number has been denoted $\binom{t}{s}$.

---

[1] While not necessary, the reader may benefit from working through Collaborative Mini-project 2 on binomial coefficients before reading through the next several sections.

(b) = (c). Consider the s-permutations of $[t]$. In Theorem 4.6, we saw that the number of s-permutations of $[t]$ is $(t)_s = \frac{t!}{(t-s)!}$.

On the other hand, we can construct an s-permutation of $[t]$ by first picking s elements of $[t]$ and then finding an s-permutation of the chosen s elements. There are $\binom{t}{s}$ ways of picking s elements of $[t]$, and, given any choice of s elements, there are $s!$ ways of permuting them. So, the number of s-permutations of $[t]$ is also $\binom{t}{s}s!$.

Thus

$$(t)_s = \binom{t}{s}s! \qquad \Rightarrow \qquad \binom{t}{s} = \frac{(t)_s}{s!} = \frac{t!}{s!\,(t-s)!}. \qquad \square$$

We will come back to binomial coefficients in Chapter 5. At that point, we will learn a number of different ways of manipulating binomial coefficients, obtaining binomial identities, and using binomial coefficients. For now, we just record one straightforward identity:

Lemma 4.22. *Let n and k be non-negative integers with $k \le n$. Then*

$$\binom{n}{k} = \binom{n}{n-k}.$$

Proof. By Theorem 4.21c, *both* $\binom{n}{k}$ and $\binom{n}{n-k}$ are equal to $\frac{n!}{k!(n-k)!}$. Hence the two binomial coefficients are equal. $\square$

Entries $\boxed{3}$ and $\boxed{4}$ of the Counting Table

Before proceeding, we quickly dispose of two of the entries in the counting table (Table 3.1).

Theorem 4.23 (Entries $\boxed{3}$ and $\boxed{4}$ of the Counting Table). *Let $0 < s \le t$ be integers. Then*

(a) $\boxed{3}$ *There is exactly one way to place s distinct balls into t identical boxes such that each box gets at most one ball.*

(b) $\boxed{4}$ *There is exactly one way to place s identical balls into t identical boxes such that each box gets at most one ball.*

Proof. In both cases, since each box can contain a maximum of one ball, the required placement amounts to placing each ball in a separate box. But since the boxes are identical, it doesn't matter which box any of the balls goes in. Hence there is only one way to accomplish either of these tasks. $\square$

Problems

P 4.2.1. How many of subsets of size 17 of $[47]$ contain 2, 4, and 7, and do not include 1, 8, and 17?

P 4.2.2. Let $n \ge 3$ be an integer. How many triangles can you find in K_n, the complete graph of order n?

P 4.2.3. A four-digit ternary sequence is a sequence a_1, a_2, a_3, a_4 where each a_i is one of 0, 1, or 2.
   (a) How many such four-digit ternary sequences have exactly two ones?
   (b) How many such four-digit ternary sequences have at least two ones?

P 4.2.4. One hundred (distinct) students are to be assigned to one of three sections of combinatorics. The three sections have capacities of 25, 35, and 40, respectively.
   (a) In how many ways can you assign the students to the sections?
   (b) Suppose that 50 of the students are sophomores while the other 50 are juniors. We want the small section to be reserved for sophomores, and the medium sized section to only have juniors. The big section can have a mixture of the two. How many ways are there to assign the students?

P 4.2.5. In the English Premier League for football (same as soccer in the United States) there are 20 teams. After every season, the three lowest-ranked teams are relegated to the Championship League (which is the second level in the English Football League – there are five levels of national leagues with many more regional level leagues below them) and three teams from the 24-team Championship League are promoted to the Premier League (the two top-ranked teams automatically qualify, the next four teams compete in the playoffs, and the winner is also promoted). In a certain betting house, before the season begins, you fill up a card with your predictions for the Premier League. You have to decide which teams are going to be first, second, and third, as well as which three teams are going to be relegated to the Championship League. (For the three last-place teams, the order doesn't matter.) You don't know anything about English football and want to cover all bases. How many cards do you need to buy in order to cover all the possibilities? (Assume that there are enough tie-breaking rules that, at the end of the season, there are no ties in the rankings.)

P 4.2.6. I have four identical oranges and six identical apples. I want to arrange these in a row on my window sill. In how many ways can this be done?

P 4.2.7. You want to write down a 3×4 matrix whose entries are either 0 or 1. You want the matrix to have 2 or 3 ones (and the rest zeros). How many such matrices are there?

P 4.2.8. You put every four-digit number in a bag and randomly pick one. What is the probability that the digits of the number that you picked form an increasing sequence? (For example, 2468 has increasing digits while 8642 and 1354 don't.)

P 4.2.9. Find the number of ways to pair off 10 police officers into partners for a patrol.

P 4.2.10. A class of 32 students is asked to split into lab groups. If there are 10 groups of 3 students and 1 group of 2 students, in how many ways is this possible?

P 4.2.11. You want to schedule a one-hour webinar three times. The available time slots start at 8, 9, 10, and 11 a.m., and, 12, 1, 2, 3, and 4 p.m. In how many ways can this be done if no two webinars are to be scheduled one after another in consecutive time slots?

P 4.2.12. How many sequences $a_1, a_2, \ldots, a_{12}$ are there consisting of four 0's and eight 1's, if no two consecutive terms are both 0's?

P 4.2.13. Toss a fair coin ten times in a row. What is the probability of 8 or more heads?

P 4.2.14. Jorge, Nahid, Mary, Crystal, Akira, Vivian, Ellen, and Rafael are eight students. We want to organize them into teams in such a way that one team has three members, two teams have two members, and one team has one member. In how many ways can this be done?

P 4.2.15. The local chapter of a political party has 25 members and 2 factions. Ten of the members belong to faction B and 15 belong to faction H. The chapter gets five tickets to a national meeting, and the members decide to distribute the tickets randomly. The president of the chapter, claiming to have used a random draw, announces the five winners. None of the five is from faction B.

    (a) Assuming that the drawing was truly random, what is the probability that none of the tickets goes to faction B?

    (b) Do you think the draw is suspect if none of the five tickets goes to faction B?

P 4.2.16. I am interested in getting fragrant plants for my garden. The nursery has seven kinds of shrubs: Azara, Buddleia, Clethra, Daphne, Gardenia, Lonicera, and Plumeria. It also has four kinds of vines: Hoya carnosa, Ipomoea alba, Jasminum, and Wisteria. In how many ways can I choose four kinds of plants for my garden if I want two shrubs and two vines, but I do not want my selection to include both Lonicera and Jasminum at the same time?

P 4.2.17. In a standard deck of 52 cards (see Problem P 4.1.3), the Jacks, Queens, and Kings are called picture cards. There are 12 of these. If the 52 cards are distributed in a random manner to 4 players in such a way that each player receives 13 cards, what is the probability that each player will receive 3 picture cards?

P 4.2.18. A standard deck of 52 cards (see Problem P 4.1.3) contains four Queens. Randomly distribute the cards to four players so that each player gets 13 cards. What is the probability that all four Queens are in one player's hand?

P 4.2.19. You shuffle a standard deck of cards (see Problem P 4.1.3) repeatedly so that the cards are ordered randomly. What is the probability that the top two cards are a pair? (A pair is two cards with the same value, such as two 4's or two Queens.)

P 4.2.20. From a standard deck of cards (see Problem P 4.1.3), you are given a set of five cards randomly. For each of the following, find the probability that you will have that configuration among your five cards:

    (a) four Aces

    (b) four of a kind (i.e., four Jacks, or four 5's, or ...)

    (c) a full house (three of a kind and a pair, such as three 4's and two 7's)

    (d) a straight (a set of five consecutive cards regardless of the suit, such as 3, 4, 5, 6, and 7). Note that an Ace can be part of two types of straights: Ace, 2, 3, 4, 5 and 10, Jack, Queen, King, Ace.

P 4.2.21. Consider a standard deck of cards (see Problem P 4.1.3). You are dealt a random set of five cards. Let r be the number of pairs that you can identify among your cards. (For example, $r = 4$ if your hand consists of 4, 4, 4, 7, and 7, while $r = 6$ if your hand is 4, 4, 4, 4, and 7.) What are the possible values of r? For each value of r, find the corresponding probability (that your hand has that value of r).

P 4.2.22. In a letter dated November 22, 1693, Samuel Pepys (1633–1703) asked Isaac Newton (1643–1727) a probability question. He posed three scenarios and wanted to know which one had the greatest chance of success. The scenarios were: throwing six dice in hopes of obtaining at least one 6; throwing twelve dice in hopes of obtaining at least two 6's; and throwing eighteen dice in hopes of obtaining at least three 6's. Calculate the probability of success in each of the three cases.[2]

P 4.2.23. We want to estimate the number of fish in a big pond. Assume that the number of fish is k. We catch 10 fish from the pond, mark them and return them to the pond. Two days later (after the marked fish have mixed themselves amongst the unmarked ones), we catch 20 fish and observe that 2 out of the 20 fish have been marked.

(a) Assuming that the fish are caught at random, and that the pond has k fish, what is the probability that 2 out of the 20 fish have been marked?

(b) What value of k maximizes the probability? This value of k is the *maximum likelihood estimate* of the number of fish in the pond.

4.2.1 Lattice Paths

Consider the 9×7 grid of Figure 4.1 (you can think of the grid lines as north-south and east-west streets in a particularly orderly town). We are interested in paths – called *lattice paths* – that go from one point to another through the grid lines. Many combinatorial counting

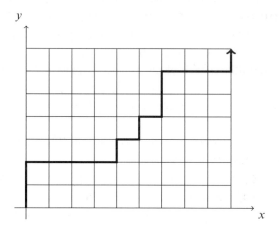

Figure 4.1 Lattice paths go from $(0,0)$ to $(9,7)$ along the grid lines. For a NE path, the only permissible moves are to go one unit up or one unit to the right.

---

[2] Samuel Pepys was the President of the Royal Society of London from 1684 to 1686, and Newton's *Principia* was presented to the society in this period. Pepys is known today mainly as a "diarist," since he kept a detailed and intimate personal diary that was published after his death. Newton went back and forth with Pepys to clarify the question, give an answer, and then explain his solution to Pepys, who confessed that "...I must not pretend to soe much Conversation wth Numbers, as presently to comprehend as I ought to doe, all ye force of that wch you are pleas'd to assigne for ye Reason of it ..." Pepys's original question read as follows: "A – has 6 dice in a Box, wth wch he is to fling a 6. B – has in another Box 12 Dice, wth wch he is to fling 2 Sixes. C – has in another Box 18 Dice, wth wch he is to fling 3 Sixes. Q. whether B & C have not as easy a Taske as A, at even luck?" See Stigler 2006 for a fascinating discussion of the mathematics, including errors, in Newton's logical argument (Newton had a correct numerical calculation of the probabilities but offered a faulty logical argument for his final conclusion). You can find the complete "The Diary of Samuel Pepys" online at www.pepysdiary.com/. You can even sign up to receive a daily diary entry!

problems can be translated to counting the number of such paths, and this interpretation can even be used to prove a number of identities. Depending on our goal, we may impose different rules on the lattice paths. For now, consider paths that go from $(0,0)$ to $(9,7)$ by only going to the right (east) and up (north). In other words, from a given node, you can only go to the right or go up. These are called north-east, or NE, paths (In Problem MP2.1 of Collaborative Mini-project 2, for example, we used a different kind of a path.) One such NE lattice path is shown in Figure 4.1.

Problems (continued)

P 4.2.24. Draw a 4×4 grid (akin to the 9×7 grid of Figure 4.1) with corners at $(0,0)$, $(4,0)$, $(0,4)$, and $(4,4)$. For integers i and j with $0 \le i, j \le 4$, at every node at (i,j) write the number of NE lattice paths from $(0,0)$ to (i,j). Comment.

P 4.2.25. What is the number of different NE lattice paths from $(0,0)$ to $(9,7)$ (see Figure 4.1)? If n and k are positive integers, what is the number of different NE lattice paths from $(0,0)$ to (n,k)?

P 4.2.26. If n_1, n_2, k_1, and k_2 are positive integers with $n_2 \ge n_1$ and $k_2 \ge k_1$, then how many different NE lattice paths are there from (n_1, k_1) to (n_2, k_2)?

P 4.2.27. What is the number of different NE lattice paths from $(0,0)$ to $(9,7)$ that pass through $(6,3)$?

P 4.2.28. What is the number of different NE lattice paths from $(0,0)$ to $(9,7)$ that pass through either $(6,3)$ or $(5,4)$?

P 4.2.29. What is the number of different NE lattice paths from $(0,0)$ to $(9,7)$ that pass through $(2,3)$ but avoid $(4,6)$?

P 4.2.30. The grid segment from $(4,3)$ to $(5,3)$ is blocked. What is the number of different NE lattice paths from $(0,0)$ to $(9,7)$ that avoid this segment?

P 4.2.31. What is the number of different NE lattice paths from $(0,0)$ to $(9,7)$ that pass through $(6,3)$ or $(6,4)$ (or both)?

4.3   Permutations of Multisets and Multinomial Coefficients

Warm-Up 4.24. *For our purposes, a* word *is just an ordered list of letters. How many four-letter words are possible if we only use letters A, J, and Z? (Examples of such a words would be* ZAJA *and* AAAZ*.) From among these words, how many are just a reordering (i.e., an anagram) of the word* JAZZ*? (An example would be* AZJZ*.)*

Definition 4.25 (Permutations of Multisets). Let s be a non-negative integer, and let T be a multiset. An *s-permutation* (or an *s-list*) of T is an ordered arrangement of s of the objects of T.

Example 4.26. If $T = \{3 \cdot U_1, 5 \cdot U_2, U_3\}$, then $U_2 U_2 U_1 U_2$ as well as $U_3 U_1 U_2 U_1$ are examples of 4-permutations of T.

Theorem 4.27 (Entry $\boxed{5}$ of the Counting Table). *Let s and t be non-negative integers. Let $R = \{\infty \cdot U_1, \infty \cdot U_2, \ldots, \infty \cdot U_t\}$ be a multiset with t different types of elements and with all repetition numbers equal to ∞. Then the following integers are equal:*

(a) $\boxed{5}$ *The number of ways of placing s distinct balls into t distinct boxes (when there is no restriction on the capacities of the boxes)*

(b) *The number of s-permutations of the multiset R*

(c) t^s.

Proof. (a) = (b) We want a 1-1, onto map from the set of placements of s distinct balls into t distinct boxes to the set of s-permutations of R. Let the set of boxes be $\{U_1, U_2, \ldots, U_t\}$, and the set of balls be $\{1, 2, \ldots, s\}$. When a ball is placed in a box, that box is chosen and the number of the ball tells us where in the list the box goes. We can put more than one ball in any box, and so a box can appear multiple times in a list. Any given placement of balls in the boxes gives an s-permutation of the multiset R.

For example, if $s = 4$ and $t = 5$, then one possible placement of balls in boxes is

$$\textcircled{1} \to U_5, \quad \textcircled{2} \to U_2, \quad \textcircled{3} \to U_2, \quad \textcircled{4} \to U_1.$$

From this placement, we get $U_5 U_2 U_2 U_1$, which is indeed a 4-permutation of the multiset $R = \{\infty \cdot U_1, \ldots, \infty \cdot U_5\}$.

The map is well defined (there is no ambiguity about the image of any placement of balls into boxes) and 1-1 (two different placements will differ in at least one place and hence are mapped to two different s-permutations). The map is also onto, since we can reverse the process. Given any s-permutation of R, we know exactly where each ball should go. The ball numbered j should go to the jth listed U in our permutation.

Table 4.1 Balls and boxes counting problems – so far.

Number of Ways to Put Balls into Boxes			
Balls (S) and Boxes (T)			
$\|S\| = s, \|T\| = t$			
The ith box contains u_i balls.			
Conditions on S & $T \to$ on $u_i \downarrow$	T distinct S distinct	T distinct S identical	T identical S distinct	T identical S identical
$0 \leq u_i \leq 1$ Assume $t \geq s$	$\boxed{1}$ $(t)_s$	$\boxed{2}$ $\binom{t}{s}$	$\boxed{3}$ 1	$\boxed{4}$ 1
$u_i \geq 0$	$\boxed{5}$ t^s			

(b) = (c) To construct an s-permutation of R, we have to fill s spots in a row with elements of R. We have t choices for the first spot. Regardless of what we choose for the first spot, we have another t choices for the second spot, and another t choices for the third spot and so on. Hence, the total number of s-permutations is $t \times t \cdots \times t = t^s$. □

The Multinomial Coefficient

Theorem 4.27 gave the number of s-permutations of a multiset with infinite repetition numbers. What if the repetition numbers are not infinite? One such case, when the sum of repetition numbers is n and we want the number of n-permutations of the multiset, will be tackled now. We first need a definition.

Definition 4.28 (The Multinomial Coefficient). Let n be a non-negative integer, and let $n_1, n_2, \ldots, n_k$ be non-negative integers with $n_1 + n_2 + \cdots + n_k = n$. Then define

$$\binom{n}{n_1 \quad n_2 \quad \cdots \quad n_k} = \frac{n!}{n_1! \, n_2! \cdots n_k!}.$$

$\binom{n}{n_1 \ n_2 \ \cdots \ n_k}$ is called a *multinomial coefficient*.

Remark 4.29. In the case $k = 2$, the multinomial coefficient becomes the binomial coefficient. This is because if $n_1 + n_2 = n$, then

$$\binom{n}{n_1 \quad n_2} = \frac{n!}{n_1! \, n_2!} = \binom{n}{n_1} = \binom{n}{n_2}.$$

Theorem 4.30. *Let $n_1, \ldots, n_t$ be non-negative integers with $n = n_1 + \cdots + n_t$. Also, let R denote the multiset $\{n_1 \cdot V_1, \ldots, n_t \cdot V_t\}$. Then the following numbers are equal:*

(a) *The number of n-permutations of R*
(b) *The number of ways of placing n distinct balls into t distinct boxes in such a way that the ith box gets exactly n_i balls*
(c)

$$\binom{n}{n_1 \quad n_2 \quad \cdots \quad n_t} = \frac{n!}{n_1! \, n_2! \cdots n_t!}.$$

Proof. (a) = (c) We want to make a list of all of the n elements in R, and hence every object has to go somewhere in the list. Construct a list in steps. First, decide where to place the n_1 copies of V_1. We have $\binom{n}{n_1}$ choices. Now, we have $n - n_1$ spots left. How many choices for placing $n_2 \cdot V_2$? The answer is $\binom{n-n_1}{n_2}$. Continue, and get that the final answer is

$$\binom{n}{n_1}\binom{n - n_1}{n_2}\binom{n - n_1 - n_2}{n_3} \cdots \binom{n - n_1 - \cdots - n_{t-1}}{n_t} = \frac{n!}{n_1! \, n_2! \cdots n_t!}.$$

(b) = (c) Choose which balls go in the first box, then choose, from among the remaining, how many go into the second box, and so on. The count again will be

$$\binom{n}{n_1}\binom{n-n_1}{n_2}\binom{n-n_1-n_2}{n_3}\cdots\binom{n-n_1-\cdots-n_{t-1}}{n_t} = \frac{n!}{n_1!\,n_2!\cdots n_t!}.$$

□

Example 4.31. The number of permutations of the letters in SHAHRIARI is the same as the number of 9-permutations of the multiset $M = \{1 \cdot S, 2 \cdot H, 2 \cdot A, 2 \cdot R, 2 \cdot I\}$. By Theorem 4.30, the answer is

$$\binom{9}{1\,2\,2\,2\,2} = \frac{9!}{(2!)^4} = 22{,}680.$$

Remark 4.32. As we have often seen, any result about counting can be recast in several ways. For example, let $R = \{\infty \cdot V_1, \ldots, \infty \cdot V_t\}$ be a multiset with infinite repetition numbers. Let $p(n_1, \ldots, n_t)$ be the number of n-permutations of R with the condition that, for $1 \leq i \leq t$, the number of V_is in the permutation is exactly n_i. Then, by Theorem 4.30, $p(n_1, \ldots, n_t)$ is equal to the multinomial coefficient $\binom{n}{n_1\,n_2\,\ldots\,n_t}$.

Entries 13 and 15 : The Preliminary Version

We have found the number of s-permutations of a multiset in two cases – when the repetition numbers are infinite (Theorem 4.27) and when the sum of the repetition numbers is exactly s (Theorem 4.30). More general cases (of permutations) are treated by entries 13 and 15 of the counting table (Table 3.1). We will now get a preliminary answer for these entries. These will be revisited later (see Theorem 9.47 and Corollary 9.48).

Theorem 4.33 (Entry 13 of the Counting Table). *Let t be a positive integer, and let $n_1, n_2, \ldots, n_t$ be non-negative integers. Let $n = n_1 + n_2 + \cdots + n_t$ and let s be a non-negative integer with $s \leq n$. Let $R = \{n_1 \cdot U_1, n_2 \cdot U_2, \ldots, n_t \cdot U_t\}$. Then the following numbers are equal:*

(a) 13 *The number of ways of placing s distinct balls into t distinct boxes such that, for $i = 1, \ldots t$, the ith box gets no more than n_i balls*
(b) *The number of s-permutations of the multiset R*
(c)

$$\sum_{\substack{m_1+\cdots+m_t=s \\ 0 \leq m_i \leq n_i}} \binom{s}{m_1\,m_2\,\cdots\,m_t}.$$

Proof. (a) = (b) We define a map from the set of all legitimate – meaning those that satisfy the conditions on capacities – placements of balls into boxes to the set of all s-permutations of R. Given a particular placement of the balls into boxes, list the balls in a row, and, under each ball, place a copy of the box that they are placed in. If the set of boxes is $T = \{U_1, \ldots, U_t\}$, then this procedure gives a list of s of the U's, with the provisions that repetition is allowed and, for $1 \leq i \leq t$, each of the U_i may occur no more than n_i times. This is, indeed, an s-permutation of R, and we have a well-defined mapping. Two different placings of the balls give two different permutations and hence the map is 1-1. It is also onto, since, given an s-permutation of R, we can reverse the process, and arrive at a placement of balls into boxes, with the condition that the ith box gets no more than n_i balls.

(b) = (c) To construct s-permutations of R, first pick an s-submultiset of R – that is, an s-combination – and then find all the permutations of that submultiset. Picking an s-submultiset amounts to finding $m_1, m_2, \ldots, m_t$ such that $m_1 + \cdots + m_t = s$ and $0 \leq m_i \leq n_i$. The number of s-permutations of such a multiset of size s is $\binom{s}{m_1 \, \cdots \, m_t}$. Hence, for each set of eligible integers $m_1, m_2, \ldots, m_t$ – that is, the ones for which $m_1 + \cdots + m_t = s$ and $0 \leq m_i \leq n_i$ – we get a multinomial coefficient, and the sum of these multinomials is the desired answer. The result follows. $\qquad\square$

Entry $\boxed{15}$ (of the counting table, Table 3.1) is basically the same as entry $\boxed{13}$. The only difference is in the kind of restriction put on u_i. For $\boxed{13}$, a legitimate constraint is $0 \leq u_5 \leq 47$, while for $\boxed{15}$, a constraint could be that u_5 is a multiple of 47. In other words, for $\boxed{15}$ we can limit the u's to any subset of the integers – like multiples of 47. Hence, N_i consist of all the integers that we want to allow for box i. If, for example, N_1 is the set of all positive integers, then, as in entry $\boxed{5}$, we are allowing $u_i \geq 0$. On the other hand, if $N_1 = \{0, 1\}$, then, as in entry $\boxed{1}$, box 1 can have no more than 1 ball in it. But we could also let $N_1 = \{0, 5, 47\}$, meaning that the only choices we are allowing are that we put 0, 5, or 47 balls in box 1! In other words, $\boxed{1}$, $\boxed{5}$, $\boxed{9}$, and $\boxed{13}$ are just special cases of $\boxed{15}$. In those earlier entries, a very special kind of restriction was allowed, while in $\boxed{15}$, the restrictions can be just about anything. The proof for $\boxed{15}$ is basically just a repeat of the argument for $\boxed{13}$.

Theorem 4.34 (Entry $\boxed{15}$ of the Counting Table). *Let t and s be positive integers. Let $R = \{\infty \cdot U_1, \infty \cdot U_2, \ldots, \infty \cdot U_t\}$. Let $N_1, N_2, \ldots, N_t$ be sets of non-negative integers. Then the following numbers are equal:*

(a) $\boxed{15}$ *The number of ways of placing s distinct balls into t distinct boxes such that, for $i = 1, \ldots t$, the number of balls in the ith box is an integer in N_i*

(b) *The number of s-permutations of the multiset R such that, for $i = 1, \ldots, t$, the number of U_i's used is an integer in N_i*

(c)

$$\sum_{\substack{m_1 + \cdots + m_t = s \\ m_i \in N_i}} \binom{s}{m_1 \, m_2 \, \cdots \, m_t}.$$

Proof. (a) = (b) As usual, let the set of boxes be $T = \{U_1, \ldots, U_t\}$. Given a particular placement of the balls into boxes, list the balls in a row, and, under each ball, place a copy of the box that it is placed in. This is an s-permutation of R satisfying the given conditions. We can reverse the process and go from an s-permutation of R with the given conditions to a placement of balls into boxes satisfying the same conditions. Hence, our correspondence is bijective, and the two numbers are equal.

(b) = (c) To construct a desired s-permutation of R, first pick an s-submultiset of R – in such a way that the number of U_i's chosen is an integer in N_i – and then find all the possible permutations of that submultiset. Picking an s-submultiset amounts to finding $m_1, m_2, \ldots, m_t$ such that $m_1 + \cdots + m_t = s$ and $m_1 \in N_1, m_2 \in N_2, \ldots$ The number of s-permutations of such a multiset of size s is $\binom{s}{m_1 \ \cdots \ m_t}$. The result follows.   □

Example 4.35. How many four-letter words can we make from the letters A, B, and C in such a way that we use two B's and an even number of A's? Put another way, how many 4-permutations of $\{\infty \cdot A, \infty \cdot B, \infty \cdot C\}$ have an even number of A's and two B's?

Solution. First we find all the 4-combinations with the given properties; then for each we get a multinomial that gives us the number of its permutations. Adding these gives us the final answer.

If we are going to have two B's, then the only question is what the other two elements are going to be. They can either both be A or neither be A. In the latter case, they both have to be C. So the possible submultisets are $\{2 \cdot A, 2 \cdot B\}$ and $\{2 \cdot B, 2 \cdot C\}$. Each of these has $\binom{4}{2\ 2} = \binom{4}{2} = 6$ permutations. Hence the total number of 4-permutations satisfying the conditions is $6 + 6 = 12$.

Note that, in the first part of the solution – when we were finding the desired submultisets – we were really finding all integer solutions to $x_1 + x_2 + x_3 = 4$ where x_1 is even, $x_2 = 2$, and $x_3 \geq 0$.   □

Problems

P 4.3.1. You are organizing 14 (distinct) people into four working groups of sizes 2, 3, 4, and 5. In how many ways can you do this?

P 4.3.2. From among 14 (distinct) people, you are to organize three teams of sizes 2, 3, and 4. In how many ways can you do this?

P 4.3.3. In the game of chess, rooks can move either horizontally or vertically. Thus if you want non-attacking rooks (rooks that do not attack each other), you should not place any two rooks in the same row or in the same column. For this problem, we are given 5 red and 3 blue rooks.

(a) In how many ways can the 8 rooks be placed on an 8×8 chess board so that no two of the rooks can attack each other?

(b) In how many ways can the 8 rooks be placed on a 12×12 chess board so that no two rooks can attack each other?

P 4.3.4. In how many ways can 2 red and 4 blue rooks be placed on an 8×8 chess board so that no two rooks can attack one another?

P 4.3.5. In how many ways can you place six non-attacking life forms on an 8×8 chess board if two are Humans, two are Klingons, and two are Romulans?

NOTE: Humans, Klingons, and Romulans move as rooks do. All humans are alike, as are all Klingons and all Romulans.

P 4.3.6. How many ways are there to rearrange the letters in the word INEDIBILITY so that D does not immediately follow E?

P 4.3.7. In how many different ways can the letters of the word MISSISSIPPI be arranged if all four S's cannot appear consecutively?

P 4.3.8. How many of the rearrangements of the letters in the word SESQUIPEDALIAN do not contain either QUIP or ED as a contiguous subword?

P 4.3.9. You randomly rearrange the letters in the word WISCONSIN. What is the probability that (just like the word WISCONSIN) there are no pairs of consecutive vowels and the W is adjacent to an I?

P 4.3.10. For us a word is just a sequence of letters, regardless of meaning. Let $\mathcal{S}$ be the set of four letter words made with the letters C, O, O, and L.

    (a) What is $|\mathcal{S}|$?

    (b) How many elements of $\mathcal{S}$ contain a CO?

    (c) How many elements of $\mathcal{S}$ contain a CL?

    (d) How many elements of $\mathcal{S}$ contain an OO?

P 4.3.11. n pairs of identical twins are to stand in a row. In how many ways can you arrange the $2n$ people, assuming that each person is indistinguishable from their twin?

P 4.3.12. Anushka, Kashvi, Mishka, Sahana, Tanvi, and Zara are to be organized into three teams, each of size two. In how many ways can this be done?

P 4.3.13. A math class has 18 (distinct) students, and the instructor wants to organize them into collaborative teams. The teams are to have either three or four members. In how many ways can this be done?

P 4.3.14. At some point, you chose a five-digit pin but you don't quite remember what it was. You do remember that 0, 4, and 7 were the only digits that you used and you did not use any digit three times. How many different pins satisfy these conditions?

P 4.3.15. A certain site requires that you choose a safe key consisting of three letters followed by three digits. The only other requirement is that you cannot use both the letter O and the digit 0. How many different safe keys are possible?

P 4.3.16. Let M be a multiset with repetition numbers $n_1 = 1, n_2, \ldots, n_k$, and let $n = n_1 + n_2 + \cdots + n_k$. Find the number of circular permutations of M.

P 4.3.17. The NBA consists of two conferences (Western and Eastern), each with 15 teams organized into 3 divisions. Every team in the NBA plays 82 games in the regular season (generally teams play 3.5 games a week). During the regular season, each team has to play each of the other 4 teams in its own division 4 times, and each of the 15 teams in the other conference 2 times. As for the remaining 10 teams in its own conference (but not in its own division), 4 of the teams are played 3 times each, while the remaining 6 are played 4 times each. A rotating five-year schedule determines which 4 teams are to be played only

3 times. The Los Angeles Clippers are in the Pacific Division of the Western conference. How many different schedules (by a schedule, we just mean the order of teams played— the actual assigning of the dates and times is a more complicated matter) are possible for the LA Clippers,

(a) if we already know which four teams the Clippers are going to play 3 times during the season, or

(b) if we don't know which four teams are slated for being played 3 times?[3]

P 4.3.18. You are flipping a fair coin, and are interested in getting a head.

(a) What is the probability that you get the first head on the fourth toss?

(b) What is the probability that you get the second head on the seventh toss?

P 4.3.19. Let i and k be positive integers with $i \leq k$. If you keep tossing a fair coin, what is the probability that the ith head comes exactly on the kth toss?

P 4.3.20. A car dealer has a large number of three models of a particular car, and wants to arrange six of them in a row in the showroom. Assume that all cars of a particular model are identical. Given each of the following restrictions, in how many ways can she arrange the cars?

(a) There are no other restrictions.

(b) The dealer wants to exhibit two of each model.

(c) The dealer wants to exhibit at least two of the sports model and at least one of each of the other two models.

P 4.3.21. A manufacturer has shipped 100 of their product to a customer. After the fact, they realized that 15 of the items are missing a part (but they don't have any way of identifying the defective items).

(a) If the customer randomly picks 10 of the items and tests them, then what is the probability that at least one of the tested items is defective?

(b) If the customer randomly picks 20 items to test, then what is the probability of finding at least one defective item? What is the probability of finding two defective items?

(c) If the customer tests the items one by one in a random sequence, then what is the probability that the first defective machine will be the kth one examined? What is the probability that the last defective machine will be the kth one examined?

4.4   Combinations of Multisets and Counting Integer Solutions

Warm-Up 4.36. *You are a consultant to a company and want to interview some of the stakeholders. You have a list of* 36 *names and have been assured that* 12 *are managers,* 12 *are union workers, and* 12 *are customers (you don't know who has what role). You pick three people to interview. How many different mixes of interviewees are possible? (One possibility would*

---

[3] In selecting its actual schedule, in addition to the above scheduling formula, the NBA considers "strength of schedule," travel costs, court availability, conflicts with NHL games on the same courts, and the desires of NBA's TV partners.

be that you are interviewing three customers; another would be that you are interviewing two managers and one union worker.)

We have counted permutations of sets and multisets and combinations of sets. It is natural to consider counting combinations of multisets.

Definition 4.37 (*s*-Combinations of Multisets). Let M be a multiset. A submultiset of M of size s is called an *s-combination* (or an *s*-submultiset) of M.

Remark 4.38. Note that we have now used *s*-combination for both *s*-subsets of a set and *s*-submultisets of a multiset. If you hear "the *s*-combinations" then it may not be clear whether we are looking for *s*-subsets or *s*-submultisets. To make the context clear, and to have a complete mathematical sentence, you have to say "the *s*-combinations of M." Then, depending on whether M is a set or a multiset, you will be looking for subsets or submultisets. Sometimes you may hear "the *s*-combinations of the set M repetition allowed." Allowing repetitions is the same as giving an infinite repetition number to elements of M and looking for submultisets of size s.

Definition 4.39. Let t and s be non-negative integers. Define the symbol $\left(\!\!\binom{t}{s}\!\!\right)$ (read "t multichoose s") to denote the number of s-submultisets of a multiset M with t types of elements and with infinite repetition numbers. In other words, $\left(\!\!\binom{t}{s}\!\!\right)$ is the number of ways to select s objects from a group of t objects, when order is irrelevant and repetition is allowed. Note that, either by an appropriate interpretation of the definition or by fiat, we have $\left(\!\!\binom{0}{0}\!\!\right) = 1$.

Theorem 4.40 (Entry $\boxed{6}$ of the Counting Table). *Let s be a non-negative integer, and t be a positive integer. Let*

$$R = \{\infty \cdot U_1, \infty \cdot U_2, \cdots, \infty \cdot U_t\}, \text{ and } W = \{s \cdot \mid, (t-1) \cdot \star\}$$

be multisets. Then the following numbers are equal:

(a) $\boxed{6}$ *The number of ways of placing s identical balls into t distinct boxes (with no capacity restrictions)*

(b)

$$\left(\!\!\binom{t}{s}\!\!\right),$$

  the number of s-combinations (i.e., submultisets of size s) of R

(c) *The number of non-negative integer solutions to*

$$x_1 + x_2 + \cdots + x_t = s$$

(d) *The number of permutations of W*

(e)

$$\binom{s+t-1}{s} = \binom{s+t-1}{t-1}.$$

Proof. (a) = (b) If you want to place s identical balls into t distinct boxes, then – since the balls are indistinguishable – all you have to do is to choose s boxes from among the t boxes, but with the provision that you are allowed repeated choice of the same boxes. This is evidently the same as choosing a submultiset of size s from R.

(b)= (c) Given a submultiset of size s from R, let c_i be the number of U_i's in the submultiset. Since the total size is s, we have

$$c_1 + \cdots + c_t = s.$$

Hence, every s-combination of R gives one non-negative integer solution – namely, $x_1 = c_1$, ..., $x_t = c_t$ – to the equation $x_1 + \cdots + x_t = s$. This map is well defined, 1-1, and onto, and hence the number of solutions is the same as the number of s-combinations of R.

(c)= (d) We want to provide a map from the set of permutations of W to the set of non-negative integer solutions to $x_1 + \cdots + x_t = s$. Any permutation of W is a list consisting of $t - 1$ stars and s |'s. Given such a permutation, let c_1 be the number of |'s before the first star, c_2 the number of |'s after the first star and before the second star, and so on until, finally, c_t is the number of |'s after the last star. Then $c_1 + \cdots + c_t = s$ and hence $x_1 = c_1, \ldots, x_t = c_t$ is a solution in non-negative integers for $x_1 + \cdots + x_t = s$. This map is well defined, 1-1, and onto. The latter follows since we can reverse our steps and, from any non-negative solution to the equation $x_1 + \cdots + x_t = s$, construct a permutation of W.

(d)= (e) By Theorem 4.30, the number of permutations of W is given by the multinomial coefficient

$$\binom{s+t-1}{s \quad t-1} = \binom{s+t-1}{s} = \binom{s+t-1}{t-1}.$$

The direct argument is not that difficult either. The multiset W has $s+t-1$ elements. Hence, to find a permutation of W, all you have to do is to place the $t-1$ stars into a list of $s+t-1$ open slots. All the other slots will be taken by |'s. To do this, choose $t-1$ spots from among the $s+t-1$ spots. Hence the number of permutations of W is $\binom{s+t-1}{t-1}$.   □

Remark 4.41. The various parts of the above theorem can come in handy in specific counting problems. However, maybe the take-home message is

$$\left(\!\!\binom{n}{k}\!\!\right) = \binom{n+k-1}{k}.$$

In other words, if you have an inexhaustible supply of n objects, then the number of ways of choosing k of them, with repetition allowed but with no regard to the order, is $\binom{n+k-1}{k}$.

Remark 4.42. We have claimed (at the end of Section 3.3) that the first column of the counting table, Table 3.1, corresponds to permutations, while the second column corresponds to combinations. While this is generally a useful heuristic, Theorem 4.40d shows that entry $\boxed{6}$, in addition to giving the number of s-combinations of a multiset, also counts the number of permutations of a multiset with a limited number of two types of elements. By now, it should come as no surprise that a counting problem can often be solved in a multitude of ways, and that the most effective counting techniques will have broad applications.

Remark 4.43. The integer solutions to an equation are also referred to as the *integral* solutions to the equation.

Example 4.44. A bakery bakes eight types of donuts, and a box contains one dozen donuts. How many different types of boxes are possible? What if each box had to contain at least one of each kind?[4]

Solution. We want the number of 12-combinations of $D = \{\infty \cdot d_1, \ldots, \infty \cdot d_8\}$. By Theorem 4.40, the answer is $\left(\!\binom{8}{12}\!\right) = \binom{19}{12} = \binom{19}{7} = 50{,}388$.

If we want each box to contain at least one of each kind of donut, then we first put one of each in the box, and then decide how to fill the remaining four spots. So, we need to find the number of 4-combinations of the multiset D. Again by Theorem 4.40, the answer is $\left(\!\binom{8}{4}\!\right) = \binom{11}{4} = 330$.

The two questions could also be cast in terms of the number of integer solutions to the equation $x_1 + \cdots + x_8 = 12$. In this equation, x_i represents the number of donuts of type i that we want in the box. There are eight type of donuts – hence eight variables x_1 through x_8 – and they have to add up to 12, since each box contains 12 donuts. For the first question, we need each $x_i \geq 0$ since we allow the possibility of not including every kind of donut. For the second question, we need each $x_i \geq 1$. Now, to solve the latter, we let $y_i = x_i - 1$. From $x_i \geq 1$ we get that $y_i \geq 0$. Plugging $x_i = y_i + 1$ into the original equation, we get the equivalent equation $y_1 + \cdots + y_8 = 4$. The difference is that we now want the number of *non-negative integer* solutions to $y_1 + \cdots + y_8 = 4$. This allows us to use Theorem 4.40 again, and get $\left(\!\binom{8}{4}\!\right)$ for an answer. $\qquad\square$

Using the little trick of the last example, we can now use our solution to entry $\boxed{6}$ to also solve entry $\boxed{10}$ of the counting table (Table 3.1).

Theorem 4.45 (Entry $\boxed{10}$ of the Counting Table). *Let $1 \leq t \leq s$ be integers. Then the following numbers are equal:*

(a) $\boxed{10}$ *The number of ways of placing s identical balls into t distinct boxes in such a way that each box gets at least one ball*

(b) *The number of positive integer solutions – that is, $x_i \geq 1$ – to*

$$x_1 + x_2 + \cdots + x_t = s$$

(c)

$$\left(\!\binom{t}{s-t}\!\right) = \binom{s-1}{s-t} = \binom{s-1}{t-1}.$$

Proof. (a)$=$ (c) Since the balls are identical and each box needs at least one ball, first place one ball in each box. At this point we have $s - t$ identical balls left and we need to distribute them among t distinguishable boxes. We are back to entry $\boxed{6}$ of the counting table and hence, by Theorem 4.40, the answer is

$$\left(\!\!\binom{t}{s-t}\!\!\right) = \binom{s-1}{s-t} = \binom{s-1}{t-1}.$$

(b)= (c) Let $y_i = x_i - 1$, for $i = 1, \ldots, t$. Then $x_i \geq 1$ implies that, for $1 \leq i \leq t$, $y_i \geq 0$. Plug $x_i = y_i + 1$ into the original equation and get

$$y_1 + \cdots + y_t = s - t.$$

Now the number of non-negative integer solutions to this equation is given by Theorem 4.40 and is $\left(\!\!\binom{t}{s-t}\!\!\right) = \binom{s-1}{t-1}$. $\qquad\square$

Remark 4.46. How should you remember Theorems 4.40 and 4.45? It is helpful to remember that $\left(\!\!\binom{n}{k}\!\!\right) = \binom{n+k-1}{k}$. But it is also useful to learn how to re-prove this identity when you need to. In other words, often it may be easier to reenact the proof process than to remember precisely what the theorem said.

For example, many times we can translate a counting problem into a question about the *number* of solutions to

$$x_1 + x_2 + \cdots + x_n = k, \quad \text{where} \quad x_i \geq 0 \quad \text{are integers.} \tag{4.1}$$

The way to remember the answer to this question is to say that we have k ones that we want to distribute among $x_1, \ldots, x_n$. If we place $(n-1)$ stars, $\star$, in a row, then there are n spaces before, between, and after them. The placing of ones in these spaces tells us what the values of $x_1, \ldots, x_n$ need to be to get a solution to the above equation:

$$\underbrace{\qquad}_{x_1} \star \underbrace{\qquad}_{x_2} \star \quad \cdots \quad \star \underbrace{\qquad}_{x_n}.$$

Hence, the answer to our original question – the number of non-negative integer solutions to $x_1 + \cdots + x_n = k$ – is the same as the number of permutations of the following multiset:

$$\{(n-1) \cdot \star, k \cdot |\}.$$

The total number of objects is $n + k - 1$. Imagine $n + k - 1$ spaces in a row. As soon as you decide where the stars go (or where the vertical bars go), you are done. Hence the final answer is

$$\binom{n+k-1}{n-1} = \binom{n+k-1}{k} = \left(\!\!\binom{n}{k}\!\!\right).$$

Now if the original problem translated into an equation like Equation 4.1, with the exception that, instead of $x_i \geq 0$, we need other conditions of the form $x_i \geq k_i$, then a simple change of variable – $y_i = x_i - k_i$ – will return us to Equation 4.1. If you wish, you could avoid changing variables, and instead just argue that you need to pre-distribute some of the ones according to the constraints, and then distribute the remaining ones using the same argument that we used before.

Example 4.47. I have 13 plants that I want to arrange on a shelf. (They will be arranged one next to another in a row). I have one each of Alstroemeria, Aquilegia, Arabis, Arctotis, and Aster as well as eight identical Violas. In how many ways can I do this if no two of the plants whose name starts with an A are next to each other?

Solution. We are going to order the plants on the shelf. We can start by placing the A's first and then deciding where the Violas go. Alternatively, we can first place the Violas, and then worry about the A's. This choice gives us two ways of solving this problem.

You can order the A's on the shelf in 5! ways (permutations of 5 distinct objects). Given an order for A's, in how many ways can you place the Violas? Let x_1 be the number of Violas that you put before the first A, and x_6 the number of Violas that are going to be placed after the last A. For $2 \leq i \leq 5$, let x_i be the number of Violas between the $(i-1)$th and ith A's:

$$\underset{x_1}{\underline{\quad}} \; A \; \underset{x_2}{\underline{\quad}} \; A \; \underset{x_3}{\underline{\quad}} \; A \; \underset{x_4}{\underline{\quad}} \; A \; \underset{x_5}{\underline{\quad}} \; A \; \underset{x_6}{\underline{\quad}}$$

What are the restrictions on the x_i? They are all non-negative integers and they have to add up to 8. Moreover, since we don't want two A's next to each other, x_2 through x_5 have to be at least 1. Thus, we want the number of non-negative integer solutions to

$$x_1 + x_2 + x_3 + x_4 + x_5 + x_6 = 8,$$

with the added condition that, for $2 \leq i \leq 5$, we have $x_i \geq 1$. This latter condition can be taken care of by a change of variables: for $2 \leq i \leq 5$, let $y_i = x_i - 1$. (Alternatively, you could just put one Viola between each of those A's and then only worry about how to distribute the remaining four Violas.) We now want the number of non-negative integer solutions to

$$x_1 + y_2 + \cdots + y_5 + x_6 = 4,$$

and the number of such solutions is $\left(\!\!\binom{6}{4}\!\!\right) = \binom{6+4-1}{4} = \binom{9}{4}$. Thus the total number of ways to arrange the plants is

$$5! \binom{9}{4} = \frac{9!}{4!} = 15,120.$$

For an alternative method, we place the Violas first. Since the Violas are all identical, there is only one way to put them on the shelf. We then have to decide where the A's go. There are 9 spots before, between, and after the Violas, and we can put at most one A in each of those spots. Hence, we need to pick 5 out of the 9 spots, and then decide which A goes in which spot. We have $\binom{9}{5}$ choices for the former and 5! choices for the latter. This also gives $\binom{9}{5} \times 5! = 15,120$ for the total count. □

We now solve the opening problem of the chapter.

Example 4.48. Cyrus, Cambyses, Bardiya, Darius, and Xerxes are gathered around a campfire. You are told by an oracle to distribute 47 identical coins among them while making sure that each gets an odd number of coins. In how many ways can this be done?

Table 4.2 Balls and boxes counting problems – so far.

Number of Ways to Put Balls into Boxes			
Balls (S) and Boxes (T)			
$\|S\| = s,\ \|T\| = t$			
The ith box contains u_i balls.			
Conditions on S & $T \rightarrow$ on $u_i\ \downarrow$	T distinct S distinct	T distinct S identical	T identical S distinct	T identical S identical
$0 \leq u_i \leq 1$ Assume $t \geq s$	$\boxed{1}\ \ (t)_s$	$\boxed{2}\ \ \binom{t}{s}$	$\boxed{3}\ \ 1$	$\boxed{4}\ \ 1$
$u_i \geq 0$	$\boxed{5}\ \ t^s$	$\boxed{6}\ \ \left(\!\!\binom{t}{s}\!\!\right)$		
$u_i \geq 1$		$\boxed{10}\ \ \left(\!\!\binom{t}{s-t}\!\!\right)$		

Solution. Since the number of coins each gets has to be an odd number, everyone has to get at least one coin. So, first distribute one coin to each of them. This leaves 42 coins, and each of the five has to get an even number of further coins. Pair the coins up so that you have 21 identical pairs of coins. You have to decide how many of these pairs to give to each of the five (distinguishable) individuals. Let x_1 be the number of pairs given to Cyrus. Likewise, x_2, x_3, x_4, and x_5 are, respectively, the number of pairs of coins given to Cambyses, Bardiya, Darius, and Xerxes. We want the number of non-negative integer solutions to

$$x_1 + x_2 + x_3 + x_4 + x_5 = 21.$$

The answer is $\left(\!\!\binom{5}{21}\!\!\right) = \binom{25}{21} = 12{,}650$.   □

Problems

P 4.4.1. Ananya, Anaya, Aaradhya, and Aadhya are your friends and you have five identical jars of tamarind. You want to distribute the jars somehow between your friends. You could give all the jars to one of them or distribute them more evenly. In how many ways can this be done?[5]

---

[5] Warm-Up 3.17 was about this problem.

P 4.4.2. I want to buy exactly 10 jars of various herbs and spices, and I am only interested in Cinnamon, Curry, Cumin, Caraway, Coriander, and Chervil. The supermarket has plenty of each. How many different combinations are possible?

P 4.4.3. (a) You are proofreading a textbook and you are looking for six kinds of grammatical errors. If a book has 13 such errors, in how many ways could these errors be distributed among the error types?

(b) In (a), suppose that we do not distinguish the types of errors, but we do keep a record of the page on which an error occurred. In how many ways can we find 20 errors in 70 pages?

P 4.4.4. A Geiger counter records the impact of six different kinds of radioactive particles over a period of time. How many ways are there to obtain a count of 30?

P 4.4.5. In Problem *P 1.3.17*, assume that the people are distinguishable. Solve this variation directly.

P 4.4.6. Noosha, Neema, Borna, Kiavash, Shooka, and Tascha are cousins. I have 30 identical copies of Jalal al-din Rumi's *Masnavi* to distribute among them. I want to distribute all the books and I want each of them to get at least one copy. In how many ways can I do this?

P 4.4.7. I have 34 identical donuts and I want to distribute them among Farzin, Farshad, Faranak, Fariborz, and Farzaneh. My only restrictions are that each person gets a positive even number of donuts and that all the donuts are distributed. In how many ways can I do this?

P 4.4.8. How many integral solutions of

$$x_1 + x_2 + x_3 + x_4 = 30$$

satisfy $x_1 \geq 2$, $x_2 \geq 0$, $x_3 \geq -5$, and $x_4 \geq 8$?

P 4.4.9. How many integer solutions are there to

$$x_1 + x_2 + x_3 + x_4 + x_5 = 47,$$

if, for $1 \leq i \leq 5$, $5 \leq x_i \leq 30$?

P 4.4.10. How many non-negative integer solutions are there to $x_1 + x_2 + \cdots + x_8 = 47$, where exactly three of the variables are equal to zero? What if we wanted at least three variables equal to zero?

P 4.4.11. Find the number of non-negative integer solutions to

$$x_1 + \cdots + x_7 \leq 47.$$

P 4.4.12. Let t be a positive integer, and s be a non-negative integer. You are interested in the number of non-negative integer solutions to $x_1 + \cdots + x_t \leq s$, and you ask for guidance. The following two vague ideas are suggested:

(a) Add a slack variable x_{t+1}, and consider the equation $x_1 + \cdots + x_t + x_{t+1} = s$.

(b) Consider a sequence of equations of the form $x_1 + \cdots + x_t = i$, where i ranges from 0 to s.

Can you make either work? If both ideas give an answer, then, by equating the two answers, get an identity involving multichoose numbers.

P 4.4.13. Find the number of non-negative integer solutions to the equation

$$4x_1 + 4x_2 + x_3 + x_4 = 24.$$

P 4.4.14. I have four identical jars of Saffron, and eight identical jars of Sage. I want to arrange them in the top row of my spice rack. In how many ways can I do this, if I do not want any two jars of Saffron to be next to each other?

P 4.4.15. In your restaurant, you have a dinner table which is a long 1×13 board, and life forms that come for dinner sit on one side of it, next to one another. You want to seat five Romulans at this table in such a way that no two of them sit next to each other. In how many ways can you accomplish this? Note that all Romulans look alike, and after you are done seating the Romulans, some chairs will be empty.

P 4.4.16. Let n and k be positive integers with $k \leq n$. In the market, n identical cans of tuna are lined up in a row on a shelf. You want to pick k of them. Of course, regardless of which k cans of tuna you pick, you will end up with k identical cans of tuna. So in that respect, you only have one choice. What intrigues you, however, is the pattern left on the shelf. Are the first, second, third, fifth, and eighth spots empty, for example?

    (a) How many ways can you pick k cans from the row of n cans?

    (b) How many ways can you do this if you don't want to pick any two consecutive cans?

    (c) Let ℓ be an appropriately small non-negative integer. In how many ways can you pick the cans if you want at least ℓ cans of tuna left between each pair that you choose?

P 4.4.17. A publisher has published Ferdawsī's *Shāhnāmeh*, Ibn Khaldūn's *Muqaddimah*, Ovid's *Metamorphoses*, and Sartre's *Being and Nothingness*, and has many copies of each.

    An employee of this firm randomly arranges on a shelf eight copies of the *Shāhnāmeh* and six copies of the *Muqaddimah*. What is the probability that no two copies of the *Muqaddimah* are next to each other?

P 4.4.18. The same publisher as in Problem P 4.4.17 still has many copies of each of the four books. In one room we have a large number of boxes. Each of these boxes has 12 of the books. No two boxes have an identical collection and all possible collections are represented. An employee randomly picks a box. What is the probability that there are at least three copies of the *Metamorphoses* and at least two copies of *Being and Nothingness* in the box?

P 4.4.19. The same publisher as in Problem P 4.4.17 still has many copies of each of the four books. Yet another employee randomly arranges 12 books on a shelf. What is the probability that on the shelf there are exactly three of each of the four books and, in addition, all books of the same kind are adjacent to one another?

P 4.4.20. How many ways are there to distribute 11 identical apples and 1 banana among Tammy, Ellen, and Elaine, if everybody is to get at least 1 piece of fruit?

P 4.4.21. Let S be a multiset with elements of k different types and with all repetition numbers equal to 2. Let $T(k, m)$ denote the number of m-combinations of S. Find and prove a recurrence relation of the form

$$T(k, m) = T(k - 1, ?) + T(k - 1, ?) + T(k - 1, ?).$$

4.5   More Problems

It is a good exercise to do counting problems without knowing whether that problem falls within a specific rubric. To consolidate your dexterity for the various counting techniques of this chapter, you may want to try some of the problems in this section.

Problems

P 4.5.1. Tiridates, Mithridates, Artabanus, Gotarzes, Orodes, Vologases, and Phraates are seven Parthian rulers, and each has minted coins with their own image on it. While not actually true, you should assume that the coins of a particular ruler are all identical to one another.

  (a) If a dealer has 1 coin from each ruler, in how many ways can I choose 4 of them?
  (b) If a dealer has 10 coins from each ruler, in how many ways can I choose 4 of them?
  (c) If a dealer has 1 coin from each ruler, in how many ways can she arrange 4 of them in a row?
  (d) If a dealer has 10 coins from each ruler, in how many ways can she arrange 4 of them in a row?
  (e) If a dealer has 2 coins from Tiridates, 1 coin from Mithridates, and 4 coins from Orodes, in how many ways can she arrange all of these 7 coins in a row?

P 4.5.2. You have 47 distinct socks, as well as 13 different batches of commemorative coins, each from a different country. Each batch has 20 identical coins in it.

  (a) You take one coin from each batch, and for 13 days in a row, each day, you randomly choose a sock and put a coin in it. What is the probability that after the 13 days, no sock will contain more than one coin?
  (b) You have chosen 13 identical silver coins commemorating the 10th anniversary of the Slovak National Uprising in 1944. In how many different ways can you distribute these coins among the socks (no restrictions on how many coins in each sock)? What percentage of the possibilities have no more than one coin in any one sock?
  (c) You have chosen seven identical $1\frac{1}{2}$-rupee coins dated 1964 and commemorating Jawaharlal Nehru, and six identical 50-litas coins from 2011 commemorating the 150th birthday of Gabrielė Petkevičaitė-Bitė. In how many ways can you distribute these coins among the socks? What percentage of the possibilities have no more than one coin in any one sock?
  (d) In how many ways can you choose 47 coins from your collection if you are to make sure that you pick at least 3 coins from each batch?
  (e) You have chosen 20 identical 2-Deutsche Mark copper-nickel coins commemorating Willy Brandt, 14 identical 1-rand silver coins commemorating Desmond Tutu, and 13 identical bronze medals commemorating Jimmy Carter. In how ways can you distribute them among the socks so that each sock gets one coin?

P 4.5.3. If you roll six dice, what is the probability of each of the following scenarios?

    (a) Each of the six different numbers will appear exactly once.

    (b) You get three 6's, two 5's, and one 3.

    (c) You get exactly three 6's.

    (d) Three of the dice are 6's and the other three also have identical values, but not 6.

    (e) At least three dice give the same value.

P 4.5.4. A donut shop carries 19 different types of donuts, and donuts of the same type are identical.

    (a) In how many ways can you pick a dozen donuts?

    (b) You have seven distinct small containers, each labeled for one day of the week and each meant for one donut. In how many ways can you put one donut in each of these containers? (The donuts in the different containers need not be different.)

    (c) Using the same containers as in the previous part, in how many ways can you put one donut in each of these containers, but making sure that no two of the donuts are of the same type?

    (d) Using the same containers as in the previous two parts, in how many ways can you put three jelly-filled donuts, three chocolate glazed donuts, and one cream-filled donut in the containers?

    (e) In how many ways can you pick 12 different donuts (no two of the same type), and then choose 2 of the 12 to eat on the spot?

    (f) In how many ways can you pick 2 distinct donuts to eat, and then pick another 10 distinct donuts, different (in type) from the 2 that you ate, to take with?

    (g) In how many ways can you buy a dozen donuts, making sure that there are at least three different types of donuts in the mix?

    (h) In how many ways can you place three jelly-filled donuts, three chocolate glazed donuts, and one cream-filled donut on a 10×10 chess board, making sure that no two donuts can attack each other? (Donuts move as rooks do—each can attack other donuts in the same row or in the same column.)

P 4.5.5. In a particular ice cream shop, you are given a bowl to go down a row of 47 flavors of ice cream. As you pass a flavor, you decide whether to take it or not, and you must choose exactly 7 of the flavors by the end. Before you start, your friend alerts you that the 23rd flavor is coffee, your favorite. For each of the following, decide in how many ways you can pick 7 of the flavors, and observe the given restriction.

    (a) No other restriction.

    (b) After you pick a flavor, you want to move a little so as not to cause a traffic jam, and so you don't want to pick any two flavors that are next to each other.

    (c) After you have picked your fourth flavor, you want to skip at least three flavors before picking the fifth one.

    (d) One of the chosen flavors must be coffee.

    (e) You want coffee to be right in the middle of your bowl, and so you want coffee to be the fourth flavor chosen.

(f) You want coffee to be the fourth flavor chosen, and you don't want to pick any two flavors that are next to each other.

P 4.5.6. Seven distinct people walk through the door. Three of them are wearing red hats, while the other four are wearing blue hats. You randomly seat all of them in one row. Find the probability of each of the following events:

(a) Cecelia and Seena—two of the seven—are seated next to each other.

(b) There are exactly two people sitting between Cecilia and Seena.

(c) All those with red hats are sitting next to each other.

(d) All those with the same colored hat are sitting next to each other.

(e) No two with a red hat sit next to each other.

(f) No two with a blue hat sit next to each other.

P 4.5.7. I have potted (separately, each in a different pot) four basil, five chives, and six oregano plants. Assuming that all plants of the same kind are identical, in how many ways can I accomplish each of the following tasks?

(a) Choose seven plants for my niece but make sure that she gets at least two oreganos and one chives.

(b) Choose five plants to give to my nephew.

(c) Arrange all 15 plants on a window sill.

(d) Arrange only four of the plants on a smaller window sill but make sure that both the number of basil and the number of chives plants are odd.

(e) Arrange all 15 plants on a window sill but make sure that no 2 oregano plants are next to each other.

P 4.5.8. Yuqing, Yuehan, Danai, Sophia, and Sanami are waiting for an elevator at the ground floor of a hotel. Each of them knows where they are going, but you don't, and so you assume that each is getting off at random, and independent of the others, at one of seven floors. Find the probability of each of the following scenarios:

(a) All five get off at the same floor.

(b) No two get off at the same floor.

(c) Two of the five get off at the same floor but the other three each get off alone.

(d) At least two of the five get off at the same floor.

(e) Yuqing and Yuehan get off at the same floor.

(f) Yuqing and Yuehan get off at the same floor, and, at a different floor, Sophia and Sanami get off together. Danai gets off alone at yet a different floor.

(g) They get off in different floors and in alphabetical order.

(h) They get off in alphabetical order but some (or all) of them could get off at the same floor.

P 4.5.9. In the previous problem, to find the number of ways that at least two of the five passengers get off at the same floor from among seven floors, a student argues as follows. "Choose which two passengers in $\binom{5}{2} = 10$ ways, then pick which floor in 7 ways, then decide where the other three are going to get off in $7 \times 7 \times 7$ ways. Hence, the number of ways that two of the five get off at the same floor is $10 \times 7^4 = 24,010$ ways." Is this correct? If not, point out the problem, and give a correct count.

P 4.5.10. You have n distinct pairs of shoes, and you put all of the shoes in a large duffle bag. After the shoes get thoroughly mixed up in the bag, you randomly pick two individual shoes from the bag and declare them a pair. You continue until you have paired all shoes, and so again you have n "pairs" of shoes. Find the probability for each of the following scenarios:
   (a) You have paired each shoe with its original pair.
   (b) Each left shoe is paired with a right shoe.
   (c) Your favorite three pairs of shoes have been paired properly. (Assume $n \geq 3$.)

P 4.5.11. **Dimension of Vector Spaces of Polynomials.**[6] Polynomials with real coefficients of degree less than or equal to n in the variable x form a finite-dimensional real vector space. (A *real* vector space is one where the scalars are real numbers.) Recall that the dimension of a finite-dimensional vector space is the number of elements in its basis. The set $\{1, x, x^2, \ldots, x^n\}$ forms a basis for this vector space, and so its dimension is $n + 1$. Now consider polynomials in more than one variable. An example of a polynomial in three variables would be $x^2yz^3 - 4x^3yz + 10x^3$. This polynomial is the linear combination of three monomials x^2yz^3, x^3yz, and x^3, and the degrees of these monomials are, respectively, 6, 5, and 3. In general, a polynomial in m variables of the form $x_1^{\alpha_1} x_2^{\alpha_2} \cdots x_m^{\alpha_m}$, with $\alpha_1, \ldots, \alpha_m$ non-negative integers, is called a *monomial* and its *degree* is defined to be $\alpha_1 + \cdots + \alpha_m$. A general polynomial in m variables is a linear combination of monomials, and the degree of such a polynomial is equal to the largest degree among its monomials. So, for example, the degree of $x^2yz + y^4z$, a polynomial in three variables, is 5. Let $\mathbb{R}_n[x_1, \ldots, x_m]$ denote the real vector space of polynomials of degree less than or equal to n in m variables. The set of monomials of degree less than or equal to n forms a basis for $\mathbb{R}_n[x_1, \ldots, x_m]$.
   (a) List all monomials of degree less than or equal to 3 in two variables.
   (b) What is the dimension of $\mathbb{R}_3[x, y]$?
   (c) What is the dimension of $\mathbb{R}_n[x_1, \ldots, x_m]$, the real vector space of polynomials in m variables of degree less than or equal to n?

---

[6] For students who have had linear algebra.

5   Binomial and Multinomial Coefficients

*Let n be a non-negative integer. The $n + 1$ binomial coefficients $\binom{n}{0}$, $\binom{n}{1}$, ..., $\binom{n}{n}$, respectively, count the number of subsets of size $0, 1, \ldots, n$ of a set with n elements. As a result, the average of these numbers is $2^n/(n+1)$ and often not an integer. But the average of the **squares** of these numbers is always an integer. Why?*

5.1   Binomial Coefficients

Let n be a non-negative integer. Recall that, for $n > 0$, $[n] = \{1, 2, \ldots, n\}$, $[0] = \emptyset$, and for $0 \le k \le n$, the binomial coefficient $\binom{n}{k}$ is defined to be the number of subsets of size k of $[n]$. We showed in Theorem 4.21c that

$$\binom{n}{k} = \frac{n!}{k! \, (n-k)!}.$$

The binomial coefficients are a set of numbers that come up often. They appear in many counting problems and they satisfy many identities. In this chapter, we explore some of their properties. Since we have a formula for these numbers, we can certainly use the formula to prove identities involving the binomial coefficients. However, we can also use induction—often using the identity of Theorem 5.2c below—as well as "combinatorial arguments" and the "binomial theorem." The latter will be treated in Section 5.2.

5.1.1   Combinatorial Proofs and Binomial Coefficient Identities

Warm-Up 5.1. *Forty-seven individuals are gathered in a room. You first choose 7 from the group of 47, and then 4 from the 7, and give each of the 4 a blue scarf. In how many ways can you do that? What if you pick 4 from the group of 47, give each a blue scarf, and then choose 3 more people from the remaining 43? Are the two answers the same? Why or why not? If possible, generalize and give an identity.*

The next theorem gives examples of combinatorial proofs of counting identities. In such proofs we either show that we can count the same set of objects in two different ways and, as a result, get an identity, or we find a bijection between two sets of objects showing that the two sets have equal numbers of elements.

Theorem 5.2. *Let $0 \leq k \leq n$ be integers; then*

(a)
$$\binom{n}{k} = \binom{n}{n-k}$$

(b)
$$\binom{n}{0} + \binom{n}{1} + \cdots + \binom{n}{n} = 2^n$$

(c) *For $1 \leq k \leq n$, we have*
$$\binom{n}{k} = \binom{n-1}{k} + \binom{n-1}{k-1}$$

(d)
$$\binom{n}{0}^2 + \binom{n}{1}^2 + \cdots + \binom{n}{n}^2 = \binom{2n}{n}.$$

Proof. (a) We already proved this identity in Lemma 4.22. But here we give a combinatorial proof. Pair each k-subset of $[n]$ with its complement. The complement is an $(n-k)$-subset of $[n]$, and the pairing shows that $\binom{n}{k}$, the number of k-subsets of $[n]$, is equal to $\binom{n}{n-k}$, the number of $(n-k)$-subsets of $[n]$.

(b) We know that the total number of subsets of $[n]$ is 2^n (for each element we have two choices: put it in the subset or not). However, the total number of subsets of $[n]$ can also be found by adding the number of subsets of size 0, the number of subsets of size 1, and so on up to the number of subsets of size n. This is, by definition, $\binom{n}{0} + \cdots + \binom{n}{n}$.

(c) Consider the k-subsets of $[n]$. These subsets of size k come in two varieties: those that include n and those that do not. To find a k-subset of $[n]$ that does not include n is exactly the same as finding a k-subset of $[n-1]$, and there are $\binom{n-1}{k}$ of these. On the other hand, if you want a k-subset of $[n]$ that includes n, then you add n to a subset of size $k-1$ of $[n-1]$. There are precisely $\binom{n-1}{k-1}$ of these. Since we have counted each k-subset of $[n]$ exactly once, we must have
$$\binom{n}{k} = \binom{n-1}{k} + \binom{n-1}{k-1}.$$

(d) Let S be a set with $2n$ elements. Color n of the elements red and n of the elements blue. Now count the total number of subsets of size n in S. On the one hand, by the definition of binomial coefficients, the answer is $\binom{2n}{n}$.

But we can count the number of n-subsets of S differently. Each n-subset of S has $0, 1, \ldots,$ or n red elements. If it has k red elements then it must have $n-k$ blue elements, and the number of ways of choosing k red elements and $n - k$ blue elements from S is $\binom{n}{k}\binom{n}{n-k}$. Hence

$$\binom{2n}{n} = \binom{n}{0}\binom{n}{n} + \binom{n}{1}\binom{n}{n-1} + \cdots + \binom{n}{n}\binom{n}{0}.$$

Using the identity $\binom{n}{n-k} = \binom{n}{k}$ gives the desired result. $\qquad\square$

5.1.2 The Karaji–Jia Triangle

Warm-Up 5.3. *Use the Google search engine to search for each of the three terms "Tartaglia's triangle," "Yang Hui's triangle," and "Khayyam's triangle." Anything remarkable?*

The binomial coefficients depend on two variables n and k, and to tabulate the values of $\binom{n}{k}$ we need a two-dimensional table. Theorem 5.2c gives a recurrence relation for these numbers, and we can use it – and appropriate initial conditions – to build the table of values for the binomial coefficients. In fact, we will use this method – finding a recurrence relation and then using it to create a table of values – over and over again in the coming chapters.

For the initial conditions, we use the fact that, by definition,

$$\binom{0}{k} = \begin{cases} 1 & \text{if } k = 0 \\ 0 & \text{if } k > 0 \end{cases}, \qquad \binom{n}{0} = 1, \text{ for all } n \geq 0.$$

These initial conditions give the first row and the first column of the table. We then use the recurrence relation from Theorem 5.2c to fill in the rest. In other words, every entry – other than the entries of the first row and the first column – is the sum of two entries in the previous row, one directly above and one diagonally to the left; see Table 5.1. Note that the second column is just the counting numbers: $1, 2, \ldots$, while the third column is the sequence of triangular numbers (see Remark 1.3): $1, 3, 6, 10, \ldots$

The table of binomial coefficients—given in Table 5.1—is called *Pascal's triangle* by most mathematicians, but we will call it the *Karaji–Jia triangle* in honor of Abū Bakr Karaji (circa 953 Karaj–1029 Baghdad) and Jiǎ Xiàn (circa 1010–1070).

History of the Triangle

Blaise Pascal (1623–1662) did, in fact, publish in 1665 (posthumously), a treatise, *Traité du Triangle Arithmétique*, devoted to a systematic study of the triangle of binomial coefficients. He was able to present with proof the different aspects of the theory, and his point of view and presentation, for the most part, would satisfy us, the modern reader. This justifies calling the triangle "Pascal's triangle" as French mathematicians of the eighteenth century began to do. In fact, A. W. F. Edwards asserts "that the Arithmetical Triangle should bear Pascal's name cannot be disputed" (Edwards 1987, p. ix). However, the general study of binomial coefficients—sometimes as numbers that count subsets of certain sizes and sometimes as the numbers that come up in the binomial theorem, Theorem 5.23—and the specific organization

Table 5.1 The Karaji–Jia triangle – most commonly known as Pascal's triangle, but sometimes as Khayyam's, Yang Hui's, or Tartaglia's triangle – organizes the binomial coefficients and highlights the recurrence relation for generating them. For $k > n$, the entries in each row are zero.

$n\backslash k$	0	1	2	3	4	5	6	7	8	9	10	...
0	1											
1	1	1										
2	1	2	1									
3	1	3	3	1								
4	1	4	6	4	1							
5	1	5	10	10	5	1						
6	1	6	15	20	15	6	1					
7	1	7	21	35	35	21	7	1				
8	1	8	28	56	70	56	28	8	1			
9	1	9	36	84	126	126	84	36	9	1		
10	1	10	45	120	210	252	210	120	45	10	1	
⋮												

of the number into a triangle predates Pascal by many centuries. In fact, for some time—we are talking at least 600 years—scholars in India, China, Iran, the wider Islamic world, North Africa, southern Europe, and the Hebrew tradition had been studying binomial coefficients, discovering their properties, and putting them to use. Pinpointing a clear starting point and mapping the routes of transmission are difficult for a number of reasons. Many crucial texts have been lost, and, in fact, not all extant manuscripts have been studied—or, for that matter, catalogued or translated—properly. We know of many works only through the commentary of later authors. The definite history of the triangle of binomial coefficients cannot yet be written and what we know is provisional, at best. A second issue is that the scholars of the past did not necessarily share our interests, and their motives and their purposes could be very different from us. For example, we can glean combinatorial arguments in a work on divination by the sixth-century Indian scholar Varāhamihira. In a chapter on perfumes, the discussion is about the various fragrances that can be created by combining four of a collection of sixteen standard ingredients (see Kusuba and Plofker 2013, p. 51). In India, there was less interest in the binomial theorem, but a long list of scholars and commentators studied binomial coefficients in the context of counting arguments. According to Edwards 1987, p. 27–33, by 1150 CE, the mathematician Bhāskara II (circa 1114 Bijapur–1185 Ujjain), in his book *Līlāvatī*, not only had the formula for binomial coefficients but also the formula for multinomial coefficients. It could easily be argued that the name of at least one of the scholars in the Indian tradition should also be attached to the triangle. Figure 5.1 and Figure 5.2 both explicitly give the triangle of binomial coefficients. Figure 5.1 is from manuscript numbered 2718 of Aya Sofia, housed in Suleymaniye library in Istanbul, Turkey,

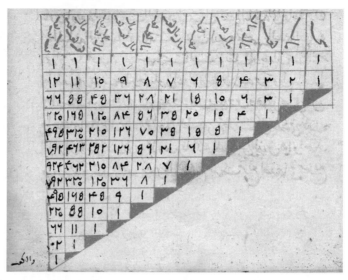

Figure 5.1 Karajī's triangle (circa 1000 CE), as described by al-Samaw'al in mid twelfth century. Courtesy of Directorate of the Turkish Institute of Manuscripts, Süleymaniye Manuscript Library Ayasofya, Collection 02718, folio 50. A black and white version of this figure will appear in some formats. For the color version, please refer to the plate section.

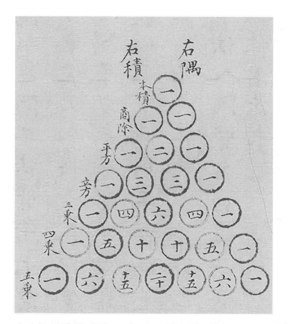

Figure 5.2 Jiǎ Xiàn's triangle (circa 1050 CE), as depicted by Zhu Shijie, around 1303 CE. Reproduced by kind permission of the Syndics of Cambridge University Library. A black and white version of this figure will appear in some formats. For the color version, please refer to the plate section.

and is the only extant copy—copied in 1324 CE—of al-Bāhir fī al-ḥisāb (the splendid book on arithmetic) with diagrams. Al-Bāhir was written by al-Samaw'al al-Maghribī (circa 1130 Baghdad–1180 Marāghe) when he was 19 years old, contains much innovative mathematics, gives an inductive argument for the binomial theorem, and explicitly credits Abū Bakr Karajī (c. 953 Karaj–1029 Baghdad) for the material on the triangle. (See Bajri, Hannah, and Montelle 2015.) Abū Bakr Karajī (when written in Arabic—as opposed to Persian—one would add the definite article "al" to his name making it al-Karajī) most likely hails from the town of Karaj (just north-west of modern-day Tehran), was a scholar with interests in arithmetic, algebra, astronomy, and surveying, and worked in Baghdad around the year 1000 CE. His work on the binomial theorem is not itself extant but was well known in the Islamic world. In particular, the mathematician and poet Omar Khayyam (1048–1151) used the binomial theorem, and wrote a now-lost treatise on extraction of nth roots. As a result, in Iran, the triangle of binomial coefficients is often referred to as *Khayyam's triangle*. See Berggren 2016, especially p. 66), where the author states that "[The triangle] might with more justice be called al-Karajī's triangle." Figure 5.2 is from the treatise Siyuan yujian, written in 1303 CE, by Zhu Shijie (1260–1320)—the figure uses rod numerals, and, in fact, has one typo—but there are earlier versions of the triangle in the 1261 treatise Xiangjie Jiuzhang Suanfa by Yáng Huī (circa 1238–1298). Yáng Huī's was using the triangle to present algorithms for root extraction, and credits Jiǎ Xiàn (circa 1010–1070)—some 50 years after Karajī —with coming up with the triangle. Jiǎ Xiàn was from Kaifeng and lived during the Song dynasty, and his writing is now lost. In China, the triangle is often referred to as the *Yang Hui triangle* (see Lam 1980 and Bréard 2013). The triangle and its properties were also known in the Hebrew tradition and in North Africa (see Katz 1992 and Djebbar 2013). In Europe, the Jewish mathematician Levi ben Gershon (circa 1288–1344), the German mathematicians Petrus Apianus (1495–1552) and Michael Stifel (1487–1567), and Italian mathematicians Niccoló Tartaglia (1500–1577), and Gerolamo Cardano (1501–1576) all published versions of the triangle long before Pascal, and in Italy, in fact, the triangle is sometimes referred to as *Tartaglia's triangle*. For much on the triangle and its history see Edwards 1987 and Wilson and Watkins 2013. The Karaji–Jia triangle is a wonderful testament to the international nature of mathematics, and our peculiar naming of the triangle is as much an ode to the scribes and librarians that, over the centuries, preserved for us the magnificent manuscripts that contain Figure 5.1 and Figure 5.2. The "Karaji–Jia" triangle is also a more convenient name than the "Karaji–Jia–Bhaskara–Yang Hui–Khayyam–ben Gershon–Tartaglia–Pascal" triangle.

Remark 5.4. Every identity about binomial coefficients can be interpreted as a statement about the Karaji–Jia triangle of Table 5.1. For example, Theorem 5.2a says that each row of the triangle is a palindrome. By Theorem 5.2b the sum of each row is 2^n and according to Theorem 5.2d the sum of the squares of every row is $\binom{2n}{n}$. You may wonder about the sum of the entries in a *column* of the Karaji–Jia triangle. Each column has potentially infinite non-zero entries but one could ask what happens if you sum the numbers in a column of the Karaji–Jia triangle up to some specific row. For column k, the first non-zero entry is $\binom{k}{k}$, and so we are asking for an identity for

$$\binom{k}{k} + \binom{k+1}{k} + \cdots + \binom{n}{k},$$

for some positive integers k and n, with $k \leq n$. Try finding a pattern. You are asked to prove the pattern in Problem P 5.1.6. It is a useful identity, called the *hockey-stick* identity.

Another curious thing happens if you add the binomial coefficients along the diagonals of the Karaji–Jia triangle running upward from the left. The first few are: 1, 1, 1 + 1, 1 + 2, 1 + 3 + 1, 1 + 4 + 3. Again, try finding a pattern. You are asked about it in Problem P 5.1.14. Likewise, in Problem *P 5.1.7*, you are asked to add the numbers along the diagonals going downward from the left.

5.1.3 Unimodality of Binomial Coefficients

Warm-Up 5.5. *Consider the real-valued function*

$$f(x) = \frac{2^{10}}{\sqrt{5\pi}} e^{\frac{-(x-5)^2}{5}}.$$

For integers k with $0 \leq k \leq 10$, use a computer (or a calculator) to compare the values of $\binom{10}{k}$ with approximate values of $f(k)$. Are they close? What does the graph of $y = f(x)$ look like?

When we look across any row of the Karaji–Jia triangle, we see that the row starts with a 1, then increases to a maximum, and then decreases. Any sequence of numbers with this property is called unimodal.

Definition 5.6. A sequence of real numbers $s_0, s_1, \ldots, s_n$ is *unimodal* if there is an integer t such that $0 \leq t \leq n$ and

$$s_0 \leq s_1 \leq \ldots \leq s_t, \quad \text{and} \quad s_t \geq s_{t+1} \geq \ldots \geq s_n.$$

Lemma 5.7. *Let n and k be positive integers with $k \leq n$. Then*

(a)

$$\binom{n}{k} = \frac{n}{k}\binom{n-1}{k-1}$$

(b)

$$\binom{n}{k} = \frac{n-k+1}{k}\binom{n}{k-1}.$$

Proof. Both of these follow directly from the formula $\binom{n}{k} = \frac{n!}{k!(n-k)!}$. $\square$

Theorem 5.8. *Let n be a positive integer. Then the sequence*

$$\binom{n}{0}, \binom{n}{1}, \binom{n}{2}, \ldots, \binom{n}{n}$$

is unimodal.

In fact, if n is even, then

$$\binom{n}{0} < \binom{n}{1} < \binom{n}{2} < \cdots < \binom{n}{n/2},$$

$$and \ \binom{n}{n/2} > \binom{n}{n/2+1} > \cdots > \binom{n}{n-1} > \binom{n}{n},$$

and, if n is odd, then

$$\binom{n}{0} < \binom{n}{1} < \binom{n}{2} < \cdots < \binom{n}{\frac{n-1}{2}}, \quad \binom{n}{\frac{n-1}{2}} = \binom{n}{\frac{n+1}{2}},$$

$$and \ \binom{n}{\frac{n+1}{2}} > \binom{n}{\frac{n+3}{2}} > \cdots > \binom{n}{n-1} > \binom{n}{n}.$$

Proof. It follows from Lemma 5.7.b that

$$\frac{\binom{n}{k}}{\binom{n}{k-1}} = \frac{n-k+1}{k}.$$

Hence, $\binom{n}{k} > \binom{n}{k-1}$ if and only if $n - k + 1 > k$. The latter is true if and only if $k < \frac{n+1}{2}$. Also $\binom{n}{k} = \binom{n}{k-1}$ if and only if $n - k + 1 = k$, and this is true if and only if $k = \frac{n+1}{2}$. All the claims of the theorem now follow. □

Definition 5.9. Let x be a real number. Then the *floor* of x, denoted by $\lfloor x \rfloor$, is the greatest integer which is less than or equal to x. Likewise, the *ceiling* of x, denoted by $\lceil x \rceil$, is the smallest integer that is greater than or equal to x.

Remark 5.10. The floor is also called the greatest integer function and is denoted by $[x]$.
    Note that if n is an even integer, then $\lfloor n/2 \rfloor = n/2 = \lceil n/2 \rceil$, while if n is odd, $\lfloor n/2 \rfloor = \frac{n-1}{2}$, while $\lceil n/2 \rceil = \frac{n+1}{2}$.

Corollary 5.11. *Let n be a positive integer. Among the binomial coefficients*

$$\binom{n}{0}, \binom{n}{1}, \ldots, \binom{n}{n},$$

the largest one is

$$\binom{n}{\lfloor n/2 \rfloor} = \binom{n}{\lceil n/2 \rceil}.$$

5.1.4 Ramsey Numbers: An Upper and a Lower Bound

Warm-Up 5.12. *The 25 prizes for a game show are hidden as entries of a 5 × 5 matrix. Each entry when unveiled will reveal one of the prizes. Each entry is referred to by a pair of integers* (i,j)*, where i is the row number and j is the column number of the entry. So, for example, the* $(3,4)$ *entry is the entry in row 3 and column 4 of the matrix. You have decided to examine the entries one by one according to the following rule: you examine the entry* (i,j) *before the entry*

(n,m), if $i+j$ is smaller than $n+m$, or if $i+j = n+m$ and $i < n$. Which entry do you examine first? Which entry do you examine in the 10th step?

Recall (Definition 2.39) that the Ramsey number $r(n,m)$ is the smallest positive integer s such that an arbitrary two-coloring of the edges of the complete graph K_s results in a K_n of the first color or a K_m of the second color (depicted as $K_s \to K_n, K_m$). We proved that the number $r(n,m)$ exists, and, in fact, by Theorem 2.44, for $n,m > 1$, $r(n,m) \leq r(n-1,m) + r(n,m-1)$. Our current purpose is to give, using binomial coefficients, explicit upper and lower bounds for $r(n,m)$. For the upper bound, if you look at Table 2.3, you will see that it is basically the Karaji–Jia triangle on its side (except that we modified some entries using Proposition 2.46). The proof of the next theorem illustrates a common practice: we use induction and the defining identity, Theorem 5.2c, of the Karaji–Jia triangle, as a one-two punch.

Theorem 5.13. *Let n and m be positive integers, and let $r(n,m)$ denote the two-color Ramsey number. Then*

$$r(n,m) \leq \binom{n+m-2}{n-1} = \frac{(n-1+m-1)!}{(n-1)! \, (m-1)!}.$$

Proof. The proof is by induction on $n + m$. When we say that we are going to do this by induction, it means that we go through all the possible choices for n and m, in some order, and in each case use the fact that we have proved the theorem for the "previous" cases. Since there are two variables n and m, a priori, there are multiple ways of going through all the cases. We have decided to use induction on $n + m$. This means that we first prove the result for the entries for which $n + m = 2$ (this is only possible if $n = m = 1$), then for entries for which $n + m = 3$ (i.e., $n = 1$, $m = 2$ or $n = 2$, $m = 1$), then for entries for which $n + m = 4$, and so on. For the base cases, we note that if $n = 1$ (or, similarly, if $m = 1$), we have

$$r(1,m) = 1 = \binom{m-1}{0}, \qquad r(n,1) = 1 = \binom{n-1}{n-1}.$$

Now, assume both n and m are greater than 1. Using Theorem 2.44, the inductive hypothesis (since both $n + (m-1)$ and $(n-1) + m$ are less than $n + m$), and the recurrence relation for binomial coefficients, we have

$$r(n,m) \leq r(n-1,m) + r(n,m-1) \leq \binom{(n-1)+m-2}{(n-1)-1} + \binom{n+(m-1)-2}{n-1}$$
$$= \binom{n+m-3}{n-2} + \binom{n+m-3}{n-1} = \binom{n+m-2}{n-1}. \qquad \square$$

Proposition 2.49 provided a weak lower bound for Ramsey numbers. We will now provide a better one. (See also Problems P 5.1.33–P 5.1.35.) The proof given here goes back to Erdős (1947). It is almost universally presented using the language of probabilities, and heralded as the prototypical example of "probabilistic methods." To prove that the Ramsey number $r(n,n)$ is greater than m, the most obvious track – the one we took in the proof of Proposition 2.49 – would be to color, using two colors, the edges of the complete graph K_m in such a way that no monochromatic K_n is created. In contrast to such a "constructive" proof, the next argument is just an existence proof.

Theorem 5.14. *Let n and m be integers greater than 2 such that* $\binom{m}{n} < 2^{\binom{n}{2}-1}$. *Let $r(n,n)$ denote the two-color (diagonal) Ramsey number. Then*

$$r(n,n) > m.$$

Proof. We need to show that there exists a two-coloring of the edges of the complete graph K_m with no monochromatic K_n. How many different ways can we color the edges of K_m using two colors? The complete graph K_m has $\binom{m}{2}$ edges, and we have two choices of colors for each edge. Hence, the total number of ways of coloring the edges of K_m using two colors is

$$A = 2^{\binom{m}{2}}.$$

From among these, how many of the colorings create a monochromatic K_n? A precise answer is difficult, but we can easily give an upper bound. To have a monochromatic K_n, first pick n of the vertices, then pick the color, and then color all the edges between those n vertices using that color. Finally, color every other edge using one of the two colors. You can do this in

$$B = \binom{m}{n} \times 2 \times 2^{\binom{m}{2}-\binom{n}{2}}$$

ways. Of course, any coloring that has more than one monochromatic K_n has been counted more than once, but the number of offending colorings – those with monochromatic K_n's – is at most B. Now if $B < A$, then there must be two-colorings of K_m without a monochromatic K_n, and the proof will be complete. To prove this, we use the – admittedly strange – condition on m and n:

$$B = \binom{m}{n} 2^{\binom{m}{2}-\binom{n}{2}+1} < 2^{\binom{n}{2}-1} 2^{\binom{m}{2}-\binom{n}{2}+1} = 2^{\binom{m}{2}} = A. \qquad \square$$

By estimating $\binom{m}{n}$, we can have a more straightforward upper bound than Theorem 5.14. A bit of care in such an estimation will yield, for $n \geq 3$, $r(n,n) > 2^{n/2}$. (See Problem P 5.1.32.) A sloppier estimate is to say that $\binom{m}{n} = \frac{m(m-1)\cdots(m-n+1)}{n!} < \frac{m \cdot m \cdots m}{n!} < m^n$. As a result, if $m \leq 2^{\frac{1}{n}\binom{n}{2}-\frac{1}{n}}$, then $m^n \leq 2^{\binom{n}{2}-1}$, and the condition of the theorem is satisfied. Since $\frac{1}{n}\binom{n}{2} = \frac{n-1}{2}$, we have the following:

Corollary 5.15. *Let $n \geq 3$ be an integer, then*

$$r(n,n) > 2^{\frac{n}{2}-\frac{1}{2}-\frac{1}{n}}.$$

5.1.5 Ming–Catalan Numbers*

Warm-Up 5.16. *For each of the rows 0 through 4 of the Karaji–Jia triangle, find the average of the squares of the entries in that row.*

Consider the sequence

$$\frac{2 \times 1}{2 \times 1} = 1,$$

$$\frac{4 \times 3 \times 2 \times 1}{3 \times 2 \times 2 \times 1} = 2,$$

$$\frac{6 \times 5 \times 4 \times 3 \times 2 \times 1}{4 \times 3 \times 3 \times 2 \times 2 \times 1} = 5,$$

$$\frac{8 \times 7 \times 6 \times 5 \times 4 \times 3 \times 2 \times 1}{5 \times 4 \times 4 \times 3 \times 3 \times 2 \times 2 \times 1} = 14,$$

$$\vdots$$

whose nth term is

$$c_n = \frac{(2n)!}{(n+1)!\, n!} = \frac{1}{n+1}\binom{2n}{n}.$$

Starting with $n = 0$ in this formula, this sequence (of integers) starts as

$$1,\ 1,\ 2,\ 5,\ 14,\ 42,\ 132,\ 429,\ 1430,\ 4862,\ \ldots$$

We first name this sequence.

Definition 5.17. Let n be a non-negative integer; then the sequence $c_0, c_1, \ldots, c_n, \ldots$ defined by $c_n = \dfrac{1}{n+1}\binom{2n}{n}$ is called the sequence of *Ming–Catalan* numbers. (This sequence is almost universally known as just the *Catalan* numbers.[1])

You are asked, in the problems, to verify the following proposition giving equivalent definitions of the Ming–Catalan numbers.

Proposition 5.18. *Let n be a non-negative integer. Then the following numbers are equal:*

(a) $c_n = \frac{1}{n+1}\binom{2n}{n}$, *the nth Ming–Catalan number*
(b) $\binom{2n}{n} - \binom{2n}{n-1}$
(c) $\binom{2n-1}{n-1} - \binom{2n-1}{n+1}$.

In addition, the Ming–Catalan numbers satisfy the following recurrence relation:

$$c_n = \frac{2(2n-1)}{n+1} c_{n-1}.$$

Proof. You are asked to provide the details in Problem P 5.1.28.   □

---

[1] The Catalan numbers are named after Eugène Catalan (1814–1894), a Belgian mathematician who studied them in connection with triangulations of polygons. However, much earlier, around 1730, these numbers make an appearance in the writings of the Mongolian scholar Ming Antu (or Ming'antu), who worked in the Qing court, in connections with series expansions of $\sin m\theta$ (see Larcombe 2000). The systematic combinatorial study of these numbers started with the work of Leonhard Euler (1707–1783) and Johann Segner (1704–1777), and, before being known as Catalan numbers, these numbers were referred to as the *Euler–Segner* numbers. See Pak 2015 for a history of these numbers in different traditions.

Remark 5.19. Our definition of the Ming–Catalan numbers does not make it clear that these numbers are *integers*. One can prove this directly from the definition, or note that, by Proposition 5.18, $c_n = \binom{2n}{n} - \binom{2n}{n-1}$ and hence must be an integer.

Remark 5.20. The Ming–Catalan numbers appear surprisingly often and in seemingly unrelated combinatorial places.[2] A few of these are discussed in Collaborative Mini-project 7, which is focused on the Ming–Catalan numbers. The intimate connection with NE lattice paths is explored in Section 5.4.3 and the ensuing problems.

We are now ready to discuss the opening problem of the chapter.

Corollary 5.21. *The average of the squares of a row of the Karaji–Jia triangle is an integer. In fact, the sequence of these averages is exactly the sequence of Ming–Catalan numbers.*

Proof. Let n be a positive integer. The entries in the nth row of the Karaji–Jia triangle are $\binom{n}{0}, \binom{n}{1}, \ldots, \binom{n}{n}$. There are $n + 1$ entries and the average of their squares is

$$\frac{1}{n+1}\left(\binom{n}{0}^2 + \binom{n}{1}^2 + \cdots + \binom{n}{n}^2\right).$$

By Theorem 5.2d, this average is equal to $\frac{1}{n+1}\binom{2n}{n}$ and hence it is the nth Ming–Catalan number. By Proposition 5.18, this is also equal to $\binom{2n}{n} - \binom{2n}{n-1}$ and an integer.

As an aside, note that the sum $\binom{n}{0} + \binom{n}{1} + \cdots + \binom{n}{n}$ of the entries of a row of the Karaji–Jia triangle is 2^n (Theorem 5.2b) and so the average of the entries of a row of the triangle is $2^n/(n + 1)$, and this is an integer only if n is one less than a power of 2. □

Problems

P 5.1.1. (a) Which one is bigger: $\binom{47}{22}$ or $\binom{47}{21}$?

(b) Which one is bigger: $\binom{47}{20}$ or $\binom{47}{27}$?

(c) Is the following number an integer?

$$\frac{100 \times 99 \times 98 \times \cdots \times 54}{47!}.$$

P 5.1.2. I add up the entries in the fifth row of the Karaji–Jia triangle and square the result. I get $(1+5+10+10+5+1)^2 = 1024$. I then add up the entries in the tenth row of the triangle and I get $1 + 10 + 45 + \cdots + 10 + 1 = 1024$. Is this a coincidence? Prove your assertion.

P 5.1.3. Let $0 \le k \le n$ be integers. We have already proved, in two different ways, that $\binom{n}{k} = \binom{n}{n-k}$ (Lemma 4.22 and Theorem 5.2a). Give a third proof by counting, in two different ways, the number of NE lattice paths (see Section 4.2.1) from $(0,0)$ to $(k, n - k)$.

---

[2] See Stanley 2015 for over 200 such occurrences.

P 5.1.4. You have 47 friends, 7 identical copies of Spinoza's *Ethics*, 4 identical copies of Epicurus's *Letter to Herodotus*, and 2 identical copies of Lucretius's *On the Nature of Things*.
  (a) You first choose 13 of your friends and invite them to a party. You then decide who from among them is going to get a copy of *Ethics*. You then choose 2 of the remaining 6 invitees to get a copy of the *On the Nature of Things*. The final remaining 4 invitees get a copy of *Letter to Herodotus*. In how many different ways can this be done?
  (b) This time you choose 7 of your friends and give them a copy of Spinoza's book, then from the remaining friends choose 4 to give Epicurus's book, and then from those remaining choose 2 to give Lucretius's book. In how many ways can this be done?
  (c) Are the two answers the same?
  (d) Can you reformulate either of the questions as a balls-and-boxes problem and give the answer as a multinomial coefficient?

P 5.1.5. Let $c \leq b \leq a$ be non-negative integers. Give two proofs, one combinatorial, for

$$\binom{a}{b}\binom{b}{c} = \binom{a}{c}\binom{a-c}{b-c}.$$

P 5.1.6. **The Hockey-Stick Identity.** What is the sum of the entries in a column of the Karaji–Jia triangle? Let k and n be positive integers with $k \leq n$. Prove that

$$\binom{k}{k} + \binom{k+1}{k} + \cdots + \binom{n}{k} = \binom{n+1}{k+1}.$$

Why would this identity be called the "hockey-stick" identity?

P 5.1.7. **Another Hockey Stick.** Start with your favorite 1 in the first column of the Karaji–Jia triangle, Table 5.1, and add a few of the terms running diagonally down and to the right. For example,

$$1 + 4 + 10 = 15$$
$$1 + 5 + 15 = 21$$
$$1 + 4 + 10 + 20 + 35 = 70$$
$$1 + 5 + 15 + 35 + 70 = 126.$$

Notice that the resulting sums are also binomial coefficients. Do you see a pattern? State it as a mathematical statement and prove it. Is this the same as the hockey-stick identity of Problem P 5.1.6 or a new one?

P 5.1.8. Use the hockey-stick theorem of Problem *P 5.1.7* (or the one from Problem P 5.1.6) to prove that[3]

$$\left(\!\binom{t+1}{s}\!\right) = \left(\!\binom{t}{s}\!\right) + \left(\!\binom{t}{s-1}\!\right) + \cdots + \left(\!\binom{t}{1}\!\right) + \left(\!\binom{t}{0}\!\right).$$

---

[3] You were asked to give a combinatorial proof of this identity in Problem P 4.4.12.

P 5.1.9. Let n and k be non-negative integers, and let A be the number of non-negative integer solutions to

$$x_1 + x_2 + \cdots + x_{n+1} \le k.$$

(a) By considering equations of the form $x_1 + x_2 + \cdots + x_{n+1} = j$ for $0 \le j \le k$, and by using the hockey-stick identity of Problem *P 5.1.7*, find A.

(b) By using a slack variable as suggested in Problem P 4.4.12, find A in a different way, and, as a result, give a combinatorial proof of the hockey-stick identity of Problem *P 5.1.7*.

P 5.1.10. Let n and m be positive integers with $m \le n$. Start with $\binom{n}{m}$ and, using Theorem 5.2c, write it as $\binom{n-1}{m} + \binom{n-1}{m-1}$. Now apply the same identity to $\binom{n-1}{m-1}$ (the term further left in the triangle) and write it as a sum of two binomial coefficients. So $\binom{n}{m} = \binom{n-1}{m} + \binom{n-2}{m-1} + \binom{n-2}{m-2}$. Continue this process for a while, and get an identity for $\binom{n}{m}$. Explain the identity as a statement about the triangle. Is this related to Problem *P 5.1.7*?

P 5.1.11. Let n and m be positive integers, with $m \le n$. Start with $\binom{n}{m}$ and repeat Problem *P 5.1.10*, but instead of unwinding $\binom{n-1}{m-1}$, apply Theorem 5.2c to $\binom{n-1}{m}$ (the term further right in the triangle). In other words, you would start by writing

$$\binom{n}{m} = \binom{n-1}{m-1} + \binom{n-1}{m} = \binom{n-1}{m-1} + \binom{n-2}{m-1} + \binom{n-2}{m} = \cdots$$

How far can you go? Have you seen this identity before? What if you stop earlier?

P 5.1.12. We choose our favorite row of the Karaji–Jia triangle, start at the beginning, find the alternating sum of the entries, and then, before we get to the end of the row, we suddenly stop. As examples, if we had picked row 7, we may be looking at any of the following:

$$1 = 1$$
$$1 - 7 = -6$$
$$1 - 7 + 21 = 15$$
$$1 - 7 + 21 - 35 = -20$$
$$1 - 7 + 21 - 35 + 35 = 15$$
$$1 - 7 + 21 - 35 + 35 - 21 = -6.$$

Can you find the resulting integers, give or take a negative sign, in the triangle? Generalize and, for integers $n > 0$ and $t \ge 0$ with $t < n$, find an identity for

$$\binom{n}{0} - \binom{n}{1} + \binom{n}{2} - \cdots + (-1)^t \binom{n}{t}.$$

Does your formula work for $t = n$?

P 5.1.13. Let n and k be positive integers. Then show that

(a)

$$\left(\!\!\binom{n}{k}\!\!\right) = \left(\!\!\binom{??}{n-1}\!\!\right)$$

(b)

$$\left(\binom{n}{k}\right) = \left(\binom{?}{k-1}\right) + \left(\binom{n-1}{?}\right).$$

P 5.1.14. Consider the sum of the binomial coefficients along the diagonals of the Karaji–Jia triangle running upward from the left. The first few are: $1, 1, 1 + 1, 1 + 2, 1 + 3 + 1, 1 + 4 + 3$. Make a conjecture and prove it.

P 5.1.15. Exactly 2^{1000} balls are fed into the top of a triangular shaped sorter. Figure 5.3 shows the beginning of the machine. The sorter actually contains 1001 rows of hexagons, with each row having one more hexagon than the previous one. If we assume a perfect sort, so that whenever two balls reach a junction one goes left and the other right, how many balls wind up in the left-most compartment at the bottom? How many wind up in compartment k (from the left)? Prove your assertion.[4]

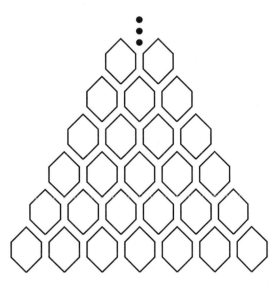

Figure 5.3 In this sorter, when two balls reach a junction, one goes left and the other right.

P 5.1.16. Let n and k be fixed positive integers with $k \leq n$. Consider the following thought experiment. Use a set of index cards, and, on each card, on one side write an element of $[n] = \{1, \ldots, n\}$ and on the other side write a subset of size k of $[n]$ that includes the element on the flip side. How many different such index cards are there? Do the count in two different ways, and, as a result, give a combinatorial proof of Lemma 5.7a.

P 5.1.17. Find one binomial coefficient equal to the following expression:

$$\binom{n}{k} + 3\binom{n}{k-1} + 3\binom{n}{k-2} + \binom{n}{k-3}.$$

---

[4] From Bose and Manvel 1984.

P 5.1.18. Let $n \le m$ be positive integers. Give a simple expression for

$$\binom{n}{0}\binom{m}{0} + \binom{n}{1}\binom{m}{1} + \cdots + \binom{n}{n}\binom{m}{n}.$$

P 5.1.19. Use a combinatorial argument to prove Chu's identity.[5] Let m_1, m_2, and n be positive integers, with n no larger than m_1 or m_2. Prove

$$\sum_{k=0}^{n} \binom{m_1}{k}\binom{m_2}{n-k} = \binom{m_1 + m_2}{n}.$$

Was the condition that n is no larger than m_1 or m_2 necessary?

P 5.1.20. Let n and m be positive integers with $m \le n$. Start with $\binom{n}{m}$ and repeat Problems P 5.1.10 and *P 5.1.11*, but this time use the identity $\binom{n}{k} = \binom{n-1}{k} + \binom{n-1}{k-1}$ on every term. So you start as

$$\binom{n}{m} = \binom{n-1}{m} + \binom{n-1}{m-1} = \binom{n-2}{m} + 2\binom{n-2}{m-1} + \binom{n-2}{m-2}.$$

What identity do you get? Have you seen it before?

P 5.1.21. (a) We have 10 sophomores, 17 juniors, and 20 seniors gathered in a hall. What is a question whose answer would be

$$\binom{10}{2}\binom{17}{0}\binom{20}{5} + \binom{10}{1}\binom{17}{1}\binom{20}{5} + \binom{10}{0}\binom{17}{2}\binom{20}{5}?$$

Does your question, or Chu's identity of Problem *P 5.1.19*, lead to a simpler expression for the above?

(b) Find and prove a formula for

$$\sum \binom{m_1}{r}\binom{m_2}{s}\binom{m_3}{t},$$

where the summation extends over all non-negative integers r, s, and t with sum $r + s + t = n$.

P 5.1.22. We have a group of math majors consisting of n sophomores and n juniors. We want to form a smaller group that has a total of n students in it, but from among that group we want to designate one of the students as a departmental liaison. The liaison needs to be a junior, but there is no other restriction on the students chosen for the smaller group.

(a) In how many different ways can we choose the smaller group, and designate a liaison? (Two groups are considered different if their members are different or if their members are the same but the student chosen to be a liaison is different.)

(b) Let k be a positive integer. In how many ways can you pick k juniors, a liaison from among the k juniors, and $n - k$ sophomores?

---

[5] This identity appears in the Chinese mathematician Chu Chi-kie's manuscript of 1303 (see Needham 1959). It was rediscovered in the eighteenth century by Alexandre-Théophile Vandermonde, and it is commonly called the *Vandermonde convolution* for binomial coefficients.

(c) By answering the first question in two different ways, give a simple expression for

$$\binom{n}{1}^2 + 2\binom{n}{2}^2 + 3\binom{n}{3}^2 + \cdots + n\binom{n}{n}^2.$$

P 5.1.23. Vary the scenario in Problem P 5.1.22 somewhat, and use it and a combinatorial argument to find a different identity for binomial coefficients.

P 5.1.24. For us, a "word" is just an ordered sequence of letters.

(a) How many words of length 47 can you make from 1 O, 7 A's, 9 B's, and 30 X's if all the A's appear before the O, all the B's appear after the O, and the O is the 17th letter in the word?

(b) Answer the previous question without the assumption that O is in the 17th position. Do this problem once directly, and once using your answer to the previous part. Equating the two answers, you should have an identity for binomial coefficients.

(c) Let r, s, and n be positive integers with $n \geq r+s+1$. How many words of length n can you make from r A's, s B's, one O, and $n-r-s-1$ X's if all the A's come before the O and all the B's come after the O? Answer this question in two different ways, and, as a result, give a combinatorial proof of a binomial identity.[6]

P 5.1.25. Vary the scenario in Problem P 5.1.24 somewhat, and use it and a combinatorial argument to find a different identity involving binomial coefficients.

P 5.1.26. Find and prove an identity of the form

$$\sum_{k=0}^{n} \frac{(2n)!}{k!^2 (n-k)!^2} = \binom{?}{n}^2.$$

P 5.1.27. Let $m \leq n$ be positive integers. Find and prove an identity of the form

$$\sum_{k=0}^{m} \frac{m! (n-k)!}{n! (m-k)!} = \frac{?+1}{?-?+1}.$$

P 5.1.28. **Proof of Proposition 5.18.** Let n be a non-negative integer and let

$$c_n = \frac{(2n)!}{(n+1)! \, n!} = \frac{1}{n+1}\binom{2n}{n}$$

be the nth Ming–Catalan number. Show that

$$c_n = \binom{2n-1}{n-1} - \binom{2n-1}{n+1}$$
$$= \binom{2n}{n} - \binom{2n}{n-1}$$
$$= \frac{2(2n-1)}{n+1} c_{n-1}.$$

P 5.1.29. Prove, by induction, that $\binom{2n}{n} < 4^{n-1}$ for all $n \geq 5$.

6 Adapted from DeTemple 2004.

P 5.1.30. You read in a calculus book that, for positive integers k, we always have $k! \geq \left(\dfrac{k}{e}\right)^k$. Use this fact and show that, for a positive integer n and $1 \leq k \leq n$, we have

$$\left(\frac{n}{k}\right)^k \leq \binom{n}{k} \leq \left(\frac{ne}{k}\right)^k.$$

P 5.1.31. Let $n \geq 5$ be an integer. Recall—Definition 2.39—that $r(n,n)$ denotes a Ramsey number. Prove that $r(n,n) < 4^{n-2}$.

P 5.1.32. Let n be an integer greater than 2. Show that the Ramsey number $r(n,n)$ is greater than $\lceil 2^{n/2} \rceil$. You may find the following steps useful.

STEP 1: Check that $r(n,n) > \lceil 2^{n/2} \rceil$ for $n = 3, 4, 5$. Hence, can assume $n \geq 6$.

STEP 2: Let $\alpha = \dfrac{n2^{\frac{n-1}{2} - \frac{1}{n}}}{e}$, and show that $\left(\dfrac{\alpha e}{n}\right)^n = 2^{\binom{n}{2}-1}$.

STEP 3: Let $m = \lceil 2^{n/2} \rceil$, and show that, for $n \geq 6$, $\alpha > (1.1)2^{n/2} > m$.

STEP 4: Use Problem P 5.1.30 and Theorem 5.14 to show that $r(n,n) > m$.

P 5.1.33. Let n_1 and n_2 be integers greater than 2. Mimic the proof of Theorem 5.14 to prove that, for an integer m, if

$$\binom{m}{n_1} 2^{-\binom{n_1}{2}} + \binom{m}{n_2} 2^{-\binom{n_2}{2}} < 1,$$

then the Ramsey number $r(n_1, n_2)$ (Definition 2.39) is bigger than m.

P 5.1.34. Note that Proposition 2.49, Theorem 5.14, and Corollary 5.15 all give a lower bound for the Ramsey number $r(n,n)$.

(a) Using computer software when necessary, use each of these three results to give a lower bound for $r(6,6)$. Which one gives the best bound and which ones are easy to use?

(b) Repeat the previous question for $n = 100$. You don't have to find precise answers; finding orders of magnitude suffices.

(c) Comment.

P 5.1.35. Use the bound given in Problem P 5.1.33 and some computer software to find a rough lower bound for $r(20, 47)$. Compare your bound to the one given by Proposition 2.49.

5.2 The Binomial Theorem

Warm-Up 5.22. *You may remember, from high school algebra, how to expand $(x + y)^3$. Don't rely on your memory; instead, find this expansion by writing $(x + y)^3 = (x + y)(x + y)^2$ and multiplying it out. (To be pedantic, you should first find an expansion of $(x+y)^2$ by writing it as $(x+y)(x+y)$ and multiplying out using the distributive law.) Now use your formula for $(x+y)^3$ to find $(10.1)^3$ in your head (no calculators and no paper or pencils).*

In this section, we prove the binomial theorem and use it to get more identities for binomial coefficients.

Theorem 5.23 (The Binomial Theorem). *Let n be a positive integer. Then for all x and y we have*

$$(x + y)^n = x^n + \binom{n}{1}x^{n-1}y + \binom{n}{2}x^{n-2}y^2 + \cdots + \binom{n}{n-1}xy^{n-1} + y^n.$$

In other words,

$$(x + y)^n = \sum_{k=0}^{n} \binom{n}{k}x^{n-k}y^k.$$

Proof. To expand the expression $(x+y)^n = (x+y)(x+y)\cdots(x+y)$, we have to multiply out the n factors. Each resulting term will be the product of exactly one of x or y from each factor. Hence the typical term will be $x^{n-k}y^k$. This term comes about when we choose y from k of the factors and x from the remaining $n - k$ factors. The coefficient of $x^{n-k}y^k$ in the result will be the same as the number of times we produced $x^{n-k}y^k$. In how many ways can we choose k y's and $(n - k)$ x's? The number of such choices is $\binom{n}{k}\binom{n-k}{n-k} = \binom{n}{k}$. The binomial identity now follows.[7]   □

The theorem says that, for example, if you want to expand $(x + y)^4$, then the terms will be x^4, x^3y, x^2y^2, xy^3, and y^4, and the coefficients will be given by the fourth row – the top-most row is the zeroth row – of the Karaji–Jia triangle. Hence

$$(x + y)^4 = x^4 + 4x^3y + 6x^2y^2 + 4xy^3 + y^4.$$

Remark 5.24. The binomial theorem has many uses, and it can be thought of as an "application" of binomial coefficients. You want to expand $(x+y)^n$, and the coefficients that show up are binomial coefficients. In fact, some of the earliest systematic studies of binomial coefficients and their triangle (see Section 5.1.2) were for the purpose of expanding $(x + y)^n$. The other way to think of the binomial theorem – our approach in this section – is to point out that $(x + y)^n$ compresses a whole row of the Karaji–Jia triangle into a simple expression, and we can use this to investigate binomial coefficients. This – using algebraic identities to encapsulate interesting sequences – is a preview of "generating functions" which will be more fully explored in Chapter 9.

A special case of the binomial theorem – replacing x by 1, and y by x – will turn out to be very useful.

Corollary 5.25. *Let n be a positive integer. Then*

$$(1 + x)^n = \sum_{k=0}^{n} \binom{n}{k}x^k.$$

---

[7] Problem P 5.2.11, asks you to give a different proof using induction. Problem P 5.4.1 guides you through a combinatorial proof.

The identity given in Corollary 5.25 is an algebraic identity – the two sides are identical expressions – and hence it is valid for all values of x. We can also manipulate the identity – multiply both sides by like terms, take derivatives, etc. – and get new identities. In the next proposition, we see a sample of identities for the binomial coefficients that can be proved in this manner.

Proposition 5.26. *Let n be a positive integer, then*

(a)
$$\binom{n}{0} + \binom{n}{1} + \binom{n}{2} + \cdots + \binom{n}{n} = 2^n$$

(b)
$$\binom{n}{0} - \binom{n}{1} + \binom{n}{2} - \cdots + (-1)^n\binom{n}{n} = 0$$

(c) *For $n \geq 1$,*
$$\binom{n}{1} + 2\binom{n}{2} + 3\binom{n}{3} + \cdots + n\binom{n}{n} = n2^{n-1}$$

(d) *For $n \geq 1$,*
$$\binom{n}{1} + 2^2\binom{n}{2} + 3^2\binom{n}{3} + \cdots + n^2\binom{n}{n} = n(n+1)2^{n-2}.$$

Proof. (a) We have already given a combinatorial proof of this identity in Theorem 5.2b, where we argued that both sides count the number of subsets of $[n]$. Here, we notice that this identity follows trivially from Corollary 5.25, if we plug in 1 for x.

(b) Plugging in $x = -1$, in the equation of Corollary 5.25, gives this identity. (Problem P 5.1.12 asked you to prove this using induction.)

(c) Take the derivative of both sides of the equation in Corollary 5.25. We get

$$n(1+x)^{n-1} = \binom{n}{1} + 2\binom{n}{2}x + 3\binom{n}{3}x^2 + \cdots + n\binom{n}{n}x^{n-1}. \tag{5.1}$$

Now, plug in $x = 1$ to get the desired identity.

(d) Multiply both sides of Equation 5.1 of the previous part by x, take another derivative, and plug in $x = 1$. We get, consecutively,

$$nx(1+x)^{n-1} = \binom{n}{1}x + 2\binom{n}{2}x^2 + 3\binom{n}{3}x^3 + \cdots + n\binom{n}{n}x^n$$

$$n(1+x)^{n-1} + n(n-1)x(1+x)^{n-2} = \binom{n}{1} + 2^2\binom{n}{2}x + 3^2\binom{n}{3}x^2 + \cdots + n^2\binom{n}{n}x^{n-1}$$

$$n2^{n-1} + n(n-1)2^{n-2} = \binom{n}{1} + 2^2\binom{n}{2} + 3^2\binom{n}{3} + \cdots + n^2\binom{n}{n}.$$

Now, $n2^{n-1} + n(n-1)2^{n-2} = n2^{n-2}[2+n-1] = n(n+1)2^{n-2}$, giving the desired identity.
$\square$

Problems

P 5.2.1. p is a prime number and n is some positive integer. We know that p divides the following sum:

$$\sum_{k=0}^{n} \binom{n}{k} 6^k.$$

What can you say about p? What can you say about n?

P 5.2.2. Evaluate the sum

$$\sum_{k=0}^{n} (-1)^k \binom{n}{k} 10^k.$$

P 5.2.3. Find a simple expression that equals

$$\binom{47}{0} + 46\binom{47}{1} + 46^2\binom{47}{2} + \cdots + 46^{47}\binom{47}{47}.$$

P 5.2.4. Find a simple expression that equals

$$\binom{n}{0} + 3\binom{n}{1} + 5\binom{n}{2} + \cdots + (2n+1)\binom{n}{n}.$$

P 5.2.5. We have seen (Proposition 5.26) that the alternating sum of the binomial coefficients in one row of the Karaji–Jia triangle is zero. What if we find the alternating sum, but also multiply each term by the column number? In other words, what is the sum

$$0\binom{n}{0} - 1\binom{n}{1} + 2\binom{n}{2} - 3\binom{n}{3} + 4\binom{n}{4} - \cdots + (-1)^n n\binom{n}{n}?$$

P 5.2.6. Find a simple expression that equals each of the following two binomial sums:

(a)

$$7\binom{n}{1} + 2 \times 7^2\binom{n}{2} + 3 \times 7^3\binom{n}{3} + \cdots + n \times 7^n\binom{n}{n}$$

(b)

$$-7\binom{n}{1} + 2 \times 7^2\binom{n}{2} - 3 \times 7^3\binom{n}{3} + \cdots + (-1)^n \times n \times 7^n\binom{n}{n}$$

(c) Find a third identity by averaging the previous two.

P 5.2.7. (a) For a positive integer n, find a simple expression that equals

$$1 + \frac{1}{2}\binom{n}{1} + \frac{1}{3}\binom{n}{2} + \cdots + \frac{1}{n+1}\binom{n}{n}.$$

(b) Evaluate the sum

$$1 - \frac{1}{2}\binom{n}{1} + \frac{1}{3}\binom{n}{2} - \cdots + (-1)^n \frac{1}{n+1}\binom{n}{n}.$$

P 5.2.8. Let $n \geq 1$ be an integer. You have n friends, and you want to invite an even number of them over for dinner (in spite of conventional definitions of a dinner party, 0 is a perfectly fine even number). In how many ways can you do that? In how many ways could you invite an odd number of friends over? Which one—inviting an even number or an odd number of friends—gives you more options?

P 5.2.9. We notice the following:

$$\binom{2}{1} = 2 = 2\binom{2}{2}$$

$$\binom{3}{1} + 3\binom{3}{3} = 6 = 2\binom{3}{2}$$

$$\binom{4}{1} + 3\binom{4}{3} = 16 = 2\binom{4}{2} + 4\binom{4}{4}$$

$$\binom{5}{1} + 3\binom{5}{3} + 5\binom{5}{5} = 40 = 2\binom{5}{2} + 4\binom{5}{4}$$

$$\binom{6}{1} + 3\binom{6}{3} + 5\binom{6}{5} = 96 = 2\binom{6}{2} + 4\binom{6}{4} + 6\binom{6}{6}.$$

Is there a pattern? Does it continue? If so, express it as a mathematical statement and prove it. Can you give a formula for the sequence 2, 6, 16, 40, 96, ... of the sums?

P 5.2.10. We notice

$$\binom{1}{0} + 2\binom{1}{1} = 3$$

$$\binom{2}{0} + 2\binom{2}{1} + \binom{2}{2} = 3 \times 2$$

$$\binom{3}{0} + 2\binom{3}{1} + \binom{3}{2} + 2\binom{3}{3} = 3 \times 2^2$$

$$\binom{4}{0} + 2\binom{4}{1} + \binom{4}{2} + 2\binom{4}{3} + \binom{4}{4} = 3 \times 2^3.$$

Is there a continuing pattern? If so express it in mathematical terms and prove it.

P 5.2.11. **Binomial Theorem via Induction.** Give a different proof of the binomial theorem, Theorem 5.23, using induction and Theorem 5.2c.

P 5.2.12. Give a combinatorial proof of Proposition 5.26c. In other words, come up with a counting problem that can be solved in two different ways, with one method giving $n2^{n-1}$ and the other $\binom{n}{1} + 2\binom{n}{2} + \cdots + n\binom{n}{n}$.

P 5.2.13. Give a combinatorial proof of Proposition 5.26d. In other words, come up with a counting problem that can be solved in two different ways, with one method giving $\binom{n}{1} + 2^2\binom{n}{2} + \cdots + n^2\binom{n}{n}$ and the other $n2^{n-1} + n(n-1)2^{n-2} = n(n+1)2^{n-2}$.

P 5.2.14. Let $n \geq 3$. Assume that you have a room of n people. You want to choose a non-empty committee of unknown size and designate a chair, a vice-chair, and a treasurer. We are interested in the number of ways this can be done. For each of the following scenarios, do the count in two different ways, and write your responses as identities for binomial coefficients.

(a) The three positions are to be filled by different people.

(b) One person on the committee can be given multiple or even triple positions.

Do your identities work for $n = 1$ and $n = 2$?

P 5.2.15. Use the binomial theorem, to find a closed form for the following two sums:

(a)

$$\sum_{k=0}^{n} k(k-1)(k-2)\binom{n}{k}$$

(b)

$$\sum_{k=0}^{n} k^3 \binom{n}{k}.$$

Compare your solutions to your combinatorial arguments in Problem P 5.2.14. Which approach do you prefer?

P 5.2.16. (a) Find

$$\binom{2n}{1} + \binom{2n}{3} + \cdots + \binom{2n}{2n-1}.$$

(b) Find the greatest common divisor of

$$\binom{2n}{1}, \binom{2n}{3}, \ldots, \binom{2n}{2n-1}.$$

P 5.2.17. In our uses of Corollary 5.25, we have been thinking of x as a number (real or complex). Could x be an $m \times m$ matrix? In other words, if n is a non-negative integer, A an arbitrary $m \times m$ matrix, and I_m the $m \times m$ identity matrix, then is

$$(I_m + A)^n = \sum_{k=0}^{n} \binom{n}{k} A^k?$$

Would our original proof of Corollary 5.25 still work? What about expanding Theorem 5.23 to matrices? In other words, if A and B are $m \times m$ matrices, then is

$$(A + B)^n = \sum_{k=0}^{n} \binom{n}{k} A^{n-k} B^k?$$

If true, give a proof; if not, give a counter-example.

5.3   Multinomials and the Multinomial Theorem

Warm-Up 5.27. *You have* 10 *different books and* 4 *friends. You want to give three books to each of Vanessa and David and two books to each of Alan and Luis. In how many ways can this be done? Can you give your answer as one multinomial coefficient?*

We now turn our attention to the multinomial coefficients and their properties. Recall that if $n_1, \ldots, n_t$ are non-negative integers with $n = n_1 + \cdots + n_t$, then the multinomial coefficient $\binom{n}{n_1 \; n_2 \; \cdots \; n_t}$ is defined by

$$\binom{n}{n_1 \quad n_2 \quad \cdots \quad n_t} = \frac{n!}{n_1! \; n_2! \; \cdots \; n_t!}. \tag{5.2}$$

Note that in the multinomial coefficient (5.2), n is always equal to the sum $n_1 + \cdots + n_t$, and Theorem 4.30 has given two uses for this multinomial coefficient. It counts the number of ways of placing n distinct balls into t distinct boxes as long as we want to put exactly n_1 balls into the first box, n_2 balls into the second box, …and n_t balls in the tth box. It also counts the number of permutations of n objects, if the objects are of t types with repetition numbers n_1, $n_2, \ldots, n_t$.

Finally, binomial coefficients are a special case of multinomial coefficients. In fact,

$$\binom{n}{k} = \binom{n}{k \quad n-k}.$$

Because of this, we can ask whether any results about binomial coefficients can be generalized to the more general setting of multinomial coefficients. Here, we will generalize two things: the recurrence relation for binomial coefficients and the binomial theorem.

If we write binomial coefficients as multinomial coefficients, then the recurrence relation for binomial coefficients – Theorem 5.2c – becomes

$$\binom{n}{k \quad n-k} = \binom{n-1}{k \quad n-k-1} + \binom{n-1}{k-1 \quad n-k}.$$

In direct analogy, we now prove the following:

Theorem 5.28. *Let* $n_1, \ldots, n_t$ *be non-negative integers with* $n = n_1 + \cdots n_t$. *Then*

$$\binom{n}{n_1 \; n_2 \; \cdots \; n_t} = \binom{n-1}{n_1 - 1 \; n_2 \; \cdots \; n_t} + \binom{n-1}{n_1 \; n_2 - 1 \; \cdots \; n_t} + \cdots + \binom{n-1}{n_1 \; n_2 \; \cdots \; n_t - 1}.$$

Proof. Let $R = \{n_1 \cdot U_1, n_2 \cdot U_2, \ldots, n_t \cdot U_t\}$ be a multiset with t different types of elements and with repetition numbers $n_1, \ldots, n_t$. We know (Theorem 4.30) that the number of n-permutations of R is given by the multinomial coefficient (5.2). We will now count these permutations in a different way. Note that each permutation of R has to start with one of

$U_1, \ldots, U_t$. If an n-permutation starts with U_i, then it can be followed by any $(n-1)$-permutation of $R_i = \{n_1 \cdot U_1, \ldots, (n_i - 1) \cdot U_i, \ldots, n_t \cdot U_t\}$. Hence, the number of permutations of R that start with U_i is the same as the number of permutations of R_i. The latter (again by Theorem 4.30) is given by

$$\binom{n-1}{n_1 \ \cdots \ n_i - 1 \ \cdots n_t}.$$

So, the sum – over i – of these multinomial coefficients must equal the multinomial coefficient (5.2), and the proof is complete. $\qquad\square$

We now generalize the binomial theorem (Theorem 5.23) and prove the multinomial theorem.

Theorem 5.29 (Multinomial Theorem). *Let n be a positive integer. For all $x_1, x_2, \ldots, x_t$, we have*

$$(x_1 + x_2 + \cdots + x_t)^n = \sum \binom{n}{n_1 \ n_2 \ \ldots \ n_t} x_1^{n_1} x_2^{n_2} \cdots x_t^{n_t},$$

where the summation extends over all non-negative integer solutions $n_1, n_2, \ldots, n_t$ of $n_1 + n_2 + \cdots + n_t = n$.

Remark 5.30. In the summation, we have one term for each solution (in non-negative integers) to $y_1 + y_2 + \cdots + y_t = n$. We find a solution $y_1 = n_1, \ldots, y_t = n_t$, and this gives us a term – $\binom{n}{n_1 \ n_2 \ \cdots \ n_t} x_1^{n_1} x_2^{n_2} \ldots x_t^{n_t}$ – for the summation. Hence, the number of terms in the summation is the same as the *number* of non-negative integer solutions to $y_1 + y_2 + \cdots + y_t = n$, which is $\left(\binom{t}{n}\right) = \binom{n+t-1}{n}$ (see Theorem 4.40).

Proof. In the product

$$(x_1 + x_2 + \cdots + x_t)(x_1 + x_2 + \cdots + x_t) \cdots (x_1 + x_2 + \cdots + x_t),$$

the typical term is

$$x_1^{n_1} x_2^{n_2} \cdots x_t^{n_t} \quad \text{with} \quad n_1 + n_2 + \cdots + n_t = n \quad \text{and, for } 1 \le i \le n, \quad n_i \ge 0.$$

The coefficient of this term is equal to the number of times we get this term in the product. To get this term, we first pick n_1 factors (out of a total of n factors) from which we choose x_1. Then, from the remaining $n - n_1$ factors, we choose x_2 from another n_2 factors, and so on. These choices are independent of each other, and hence the total number of ways of getting this term is

$$\binom{n}{n_1}\binom{n-n_1}{n_2}\binom{n-n_1-n_2}{n_3}\cdots\binom{n-n_1-\cdots-n_{t-1}}{n_t} = \frac{n!}{n_1! \, n_2! \cdots n_t!}.$$

$\qquad\square$

Problems

P 5.3.1. Find

$$\sum \binom{7}{k_1 \ \ k_2 \ \ k_3 \ \ k_4},$$

where k_1, k_2, k_3, and k_4 are non-negative integers ranging over all possible choices such that $k_1 + k_2 + k_3 + k_4 = 7$.

P 5.3.2. Consider

$$A = \sum \binom{6}{a \ \ b \ \ c},$$

where the sum is over all non-negative integers a, b, and c with $a + b + c = 6$.

(a) How many terms does the sum defining A have?

(b) Find A.

(c) Find $\sum(-1)^a \binom{6}{a \ \ b \ \ c}$, where the sum is over all non-negative integers a, b, and c with $a + b + c = 6$.

P 5.3.3. Find $\sum(-1)^{a+c} \binom{47}{a \ \ b \ \ c \ \ d}$, where the sum is over all non-negative integers a, b, c, and d with $a + b + c + d = 47$.

P 5.3.4. What is the coefficient of x^4 in $(1 + x + x^2)^7$? What is the sum of *all* of the coefficients?

P 5.3.5. Find the constant term in $(x + x^{-1} + 2x^{-4})^{17}$.

P 5.3.6. What is the coefficient of $x^9 y^3$ in $(x + 3x^2 + 2y)^{10}$?

P 5.3.7. What is the coefficient of x^{10} in the expansion of $(3 + 2x^2 + x^3)^9$?

P 5.3.8. Fermat's Little Theorem from the Multinomial Theorem. Let n be a positive integer, and let p be a prime number. Then Fermat's little theorem says that, no matter what your choice of n or p, $n^p - n$ is divisible by p. For example, $6^{13} - 6 = 13,060,694,010 = 13 \times 1,004,668,770$. Use the multinomial theorem, Theorem 5.29, to prove Fermat's little theorem.[8] You may find the following steps useful.

STEP 1: Without any calculation, could you say whether $11! = 39,916,800$ is divisible by 13 or not? Is it divisible by 14?

STEP 2: Assume that $k_1, \ldots, k_n$ are non-negative integers *less* then p, with $k_1 + \ldots + k_n = p$. Is $\binom{p}{k_1 \ \ k_2 \ \cdots \ k_n}$ divisible by p?

STEP 3: Write the expansion of $(x_1 + \cdots + x_n)^p$ given by the multinomial theorem, and let $x_1 = x_2 = \cdots = x_n = 1$. How many of the terms in the expansion equal 1? Which ones are divisible by p?

STEP 4: Show that $n^p = n + pm$ and complete the proof.

---

[8] Adapted from Osler 2002.

5.4   Binomial Inversion, Sums of Powers, Lattice Paths, Ming–Catalan Numbers, and More*

In this optional section, we invite the reader to explore additional topics by working on sets of problems. The problems are organized into six projects: (1) a combinatorial proof of the binomial theorem, (2) log concavity of sequences, (3) the inverse of the Karaji–Jia triangle and sums of powers, (4) the connection between NE lattice paths and Ming–Catalan numbers, (5) a certain duality in sampling without replacement, and (6) some counting problems related to Knight–Knave puzzles. Each topic is split into smaller manageable problems, and each set of problems is independent of the others.

Problems

P 5.4.1. **Combinatorial Proof of the Binomial Theorem.** The binomial theorem, Theorem 5.23, is not just about integers. It works even if x and y are arbitrary real (or complex) numbers. As a result, a purely combinatorial proof is not quite possible. But with a little help from algebra, we can bootstrap a combinatorial argument to a proof.[9] The fact from algebra that we need is that a polynomial of degree n with real or complex coefficients has at most n (real or complex) roots.[10] Use this fact and give a combinatorial proof of the binomial theorem. You may find the following steps useful.

STEP 1: Let r, n, and i be non-negative integers with $i \leq n$, and consider $[r + 1] = \{1, 2, \ldots, r, r + 1\}$. You want to make an ordered list of size n, with repetition allowed in such a way that the element $r + 1$ appears exactly i times in your list. How many different such lists are possible?

STEP 2: In the previous part, i could be 0 through n. What are you counting, if you sum, from $i = 0$ to $i = n$, your answers to the previous part? Do this same count in a different way, and get an identity for $(1 + r)^n$.

STEP 3: If r and n are non-negative integers, then, using your answer to the previous part, prove that

$$(1 + r)^n = \sum_{k=0}^{n} \binom{n}{k} r^k.$$

STEP 4: Define the polynomial $p(x)$ by $p(x) = (1+x)^n - \sum_{k=0}^{n} \binom{n}{k} x^k$. What are the possible degrees of $p(x)$? What is the maximum number of roots (a root is a number α such that $p(\alpha) = 0$) for this polynomial? In the previous part, did we already find some roots for this

---

[9] Adapted from Ramras 2003.
[10] This fact will be familiar from high school algebra, but, for a rigorous understanding of it, consult any text in abstract algebra (for example, see Corollary 19.27 of my *Algebra in Action* (Shahriari 2017)).

polynomial? Give a proof that $p(x)$ must be the zero polynomial, and hence, for any real or complex number x, we have

$$(1 + x)^n = \sum_{k=0}^{n} \binom{n}{k} x^k.$$

STEP 5: In the above, replace x with y/x, do some algebra, and prove the binomial theorem, Theorem 5.23.

5.4.1   Log Concavity and Unimodal Sequences

We saw in Theorem 5.8 that the sequence of integers in every row of the Karaji–Jia triangle is unimodal (Definition 5.6). Many combinatorial sequences end up being unimodal, and, in fact, many interesting sequences have a stronger property that we shall define now.

Definition 5.31. Let $a_0, a_1, \ldots, a_n, \ldots$ be a sequence of positive real numbers. We say that this sequence is *log concave* if, for all $k \geq 1$, we have

$$\frac{a_k}{a_{k-1}} \geq \frac{a_{k+1}}{a_k},$$

or, equivalently, $a_k^2 \geq a_{k-1} a_{k+1}$.[11]

In Problems P 5.4.2 and P 5.4.3, you are asked to show that a log concave sequence of positive real numbers is always unimodal, and that the rows of the Karaji–Jia triangle are indeed log concave. An interesting theorem that goes back to Newton relates roots of a polynomial with the log concavity of the sequence of its coefficients.

Theorem 5.32. *Let $a_0, a_1, \ldots, a_n$ be a sequence of positive real numbers. If the polynomial $a_n x^n + a_{n-1} x^{n-1} + \cdots + a_1 x + a_0$ has only real roots, then the sequence $a_0, a_1, \ldots, a_{n-1}, a_n$ is log concave, and, hence, unimodal.*

So, for example, the sequence of coefficients of $3x^4 + x^3 + 2x^2 + x + 1$ is not unimodal, and so this polynomial must have some complex roots. In Problems P 5.4.6 and *P 5.4.7* (using Problem P 5.4.5), you are asked to prove this theorem for $n = 2$, 3, and 4. Similar techniques can lead to the general proof.

Problems (continued)

P 5.4.2. (a) Prove that a log concave sequence of positive numbers is unimodal.
(b) Does a unimodal sequence of positive numbers have to be log concave? Either prove that it is, or give a counter-example.

---

[11] The condition is equivalent to $\log(a_k) \geq \frac{\log(a_{k-1}) + \log(a_{k+1})}{2}$ which accounts for calling the sequence *log concave*.

P 5.4.3. Directly (and without recourse to Theorem 5.32) prove that each row of the Karaji–Jia triangle is log concave.

P 5.4.4. Let n be a positive integer. Show that the polynomial $\binom{n}{0} + \binom{n}{1}x + \cdots + \binom{n}{n}x^n$ has only real roots, and, as a result, the log concavity of a row of the Karaji–Jia triangle also follows from Theorem 5.32.

P 5.4.5. A polynomial of degree n with real coefficients has at most n (real or complex) roots, and it has exactly n (real or complex) roots if you count the multiplicity of the roots.[12] Accepting these facts,[13] and as a prelude to the proof (for $n = 3$ and 4) of Theorem 5.32, prove the following facts:
   (a) If a polynomial has only real roots, then so does its derivative.
   (b) If $p(x)$, a polynomial of degree n, has only real roots, then so does the polynomial $x^n p(1/x)$.

P 5.4.6. Prove Theorem 5.32 for $n = 2$ and $n = 3$ by showing the following stronger results:
   (a) If $a_2x^2 + a_1x + a_0$ has only real roots, then $a_1^2 \geq 4a_0a_2$.
   (b) If $a_3x^3 + a_2x^2 + a_1x + a_0$ has only real roots, then $a_1^2 \geq 3a_0a_2$ and $a_2^2 \geq 3a_1a_3$.

P 5.4.7. Prove Theorem 5.32 for $n = 4$. In other words, assume that $p(x) = a_4x^4 + \cdots + a_1x + a_0$ has positive real coefficients and real roots, and prove that the sequence a_4, a_3, a_2, a_1, a_0 is log concave.

5.4.2   The Inverse of the Karaji–Jia Triangle, Binomial Inversion, and Sums of Powers

If we write the Karaji–Jia triangle as a matrix, then, sometimes, we are able to use a matrix equation to express certain combinatorial identities (or questions). In such situations, to rearrange the matrix equation, it is very helpful to have the inverse of the Karaji–Jia triangle. We will give examples of the use of this "binomial inversion" in the problems. First, let's define the matrix.

Definition 5.33 (The Matrix of Binomial Coefficients C_n). The matrix C_n is an $(n+1) \times (n+1)$ matrix whose rows and columns are numbered $0, 1, \ldots, n$, and whose (i,j) entry (for $0 \leq i$, $j \leq n$) is $\binom{i}{j}$. (As usual, $\binom{0}{0} = 1$ and $\binom{i}{j} = 0$ if $j > i$). The matrix C_n is the first $n + 1$ rows and columns of the Karaji–Jia triangle written as a matrix. So, for example,

$$C_6 = \begin{bmatrix} 1 & 0 & 0 & 0 & 0 & 0 & 0 \\ 1 & 1 & 0 & 0 & 0 & 0 & 0 \\ 1 & 2 & 1 & 0 & 0 & 0 & 0 \\ 1 & 3 & 3 & 1 & 0 & 0 & 0 \\ 1 & 4 & 6 & 4 & 1 & 0 & 0 \\ 1 & 5 & 10 & 10 & 5 & 1 & 0 \\ 1 & 6 & 15 & 20 & 15 & 6 & 1 \end{bmatrix}.$$

---

[12] The real or complex number α is a root of multiplicity m of the polynomial $p(x)$ if $p(x) = (x - \alpha)^m q(x)$ for some polynomial $q(x)$, with $q(\alpha) \neq 0$.
[13] For a proof see Corollary 19.27 of my *Algebra in Action* (Shahriari 2017).

The rows of this matrix are linearly independent, and so the rank of the matrix is $n + 1$ and it is invertible. In fact, the determinant of C_n is 1, which – if you remember your matrix algebra – means that the inverse of C_n has integer entries as well. (This is because, if the square matrix A is invertible, then $A^{-1} = \frac{1}{\det(A)}\mathrm{adj}(A)$, where $\mathrm{adj}(A)$ is the adjoint of A, and if A is integer valued, then so is $\mathrm{adj}(A)$.) As an example,

$$
C_6^{-1} = \begin{bmatrix}
1 & 0 & 0 & 0 & 0 & 0 & 0 \\
-1 & 1 & 0 & 0 & 0 & 0 & 0 \\
1 & -2 & 1 & 0 & 0 & 0 & 0 \\
-1 & 3 & -3 & 1 & 0 & 0 & 0 \\
1 & -4 & 6 & -4 & 1 & 0 & 0 \\
-1 & 5 & -10 & 10 & -5 & 1 & 0 \\
1 & -6 & 15 & -20 & 15 & -6 & 1
\end{bmatrix},
$$

which looks awfully similar to the original matrix C_6. We call the above matrix – whose (i, j) entry (for $0 \le i, j \le n$) is $(-1)^{i+j}\binom{i}{j}$ – the matrix of "signed" binomial coefficients. In Problem *P 5.4.11* (which is based on Problems P 5.4.9 and P 5.4.10), you are asked to prove that the inverse of the matrix of binomial coefficients is the matrix of the signed binomial coefficients.

As an application, in the problems, we will first use the inverse of C_n to write powers of k – e.g., k, k^2, k^3, ... – as linear combinations of binomial coefficients. Then, we use these linear combinations to find closed formulas for sums of powers of k.[14] In particular, we find formulas for

$$\sum_{k=1}^{n} k^2 = 1^2 + 2^2 + \cdots + n^2 \qquad\qquad \text{Problem P 5.4.14}$$

$$\sum_{k=1}^{n} k^3 = 1^3 + 2^3 + \cdots + n^3 \qquad\qquad \text{Problem P 5.4.15}$$

$$\sum_{k=1}^{n} k^4 = 1^4 + 2^4 + \cdots + n^4 \qquad\qquad \text{Problem P 5.4.18.}$$

The early problems show how to find an expression for the sums of powers as long as we know how to write powers of k as linear combinations of binomial coefficients (Problems P 5.4.14 and P 5.4.15 which depend on Problem P 5.1.6). We then turn to the problem of finding these linear combinations (Problems P 5.4.16 and *P 5.4.17*), and this is where we will use the inverse of the matrix of binomial coefficients.

---

[14] This set of problems has been adapted from Merris 1996, pp. 31–35. We will come back to sums of powers in Theorem 9.55 and Problem P 9.4.36 and develop a different method using Bernoulli numbers.

P 5.4.8. For a positive integer n, the matrix C_n (Definition 5.33) denotes the $(n+1) \times (n+1)$ matrix of binomial coefficients. We denote by C_n^t the transpose of C_n.

(a) Explicitly find $C_3 C_3^t$. What does the resulting matrix look like? Are you surprised?

(b) If n is a positive integer, and i and j are non-negative integers no more than n, what is the (i,j) entry of the matrix $C_n C_n^t$? Make a conjecture.

(c) Prove your conjecture.[15]

P 5.4.9. Let j be a non-negative integer. In Corollary 5.25, we showed that

$$(x + 1)^j = \sum_{k=0}^{j} \binom{j}{k} x^k.$$

You can think of this as writing the polynomial $(x + 1)^j$ as a linear combination of the polynomials $1, x, x^2, \ldots, x^j$. Write x^j as a linear combination of $1, x + 1, (x + 1)^2, \ldots, (x + 1)^j$.

P 5.4.10. The set of all polynomials with real coefficients of degree less than or equal to 5 is a vector space. The dimension of this vector space is 6 and so every basis for this vector space has six elements. Two such bases are

$$S = \{1, x, x^2, x^3, x^4, x^5\}, \quad B = \{1, x + 1, (x + 1)^2, (x + 1)^3, (x + 1)^4, (x + 1)^5\}.$$

In linear algebra, given two bases S and B for the same vector space, we can find the *change of basis matrix* from S to B. This square matrix has as many columns as the dimension of the vector space, and its ith column is the coordinate vector of the ith element of S written in terms of B.[16] You can find the change of basis matrix from S to B as well as the change of basis matrix from B to S. These two matrices will always be inverses of each other. For the two bases listed above for polynomials of degree less than or equal to 5, find both of the change of basis matrices, and verify that they are inverses of each other. Have you seen these matrices before?

P 5.4.11. **The Inverse of the Karaji–Jia Triangle.** Let n be a positive integer. Prove that the inverse of the $(n + 1) \times (n + 1)$ matrix C_n – the matrix of binomial coefficients – is the matrix of signed binomial coefficients (see the example in Section 5.4.2).

P 5.4.12. **Binomial Inversion.** We have a mystery sequence of integers $a_0, a_1, \ldots, a_n$. We are given the sequence of integers $b_0, b_1, \ldots, b_n$, defined by

$$b_0 = \binom{0}{0} a_0$$

$$b_1 = \binom{1}{0} a_0 + \binom{1}{1} a_1$$

[15] Adapted from Edelman and Strang 2004.

[16] The reader is urged to consult a linear algebra text for review.

$$b_2 = \binom{2}{0}a_0 + \binom{2}{1}a_1 + \binom{2}{2}a_2$$

$$\vdots$$

$$b_n = \binom{n}{0}a_0 + \binom{n}{1}a_1 + \cdots + \binom{n}{n}a_n.$$

Find each of $a_0, a_1, \ldots, a_n$ in terms of the b's.

P 5.4.13. Let C_4 be the 5×5 matrix of binomial coefficients, as defined in Definition 5.33. We define two other 5×5 matrices A_4 and P_4 (whose rows and columns are also indexed $0, \ldots, 4$). The matrix A_4 is upper-triangular (i.e., all entries below the main diagonal are zero) and, for $0 \le i \le j \le 4$, the (i,j) entry of A_4 is an unknown real number $a_{i,j}$. For $0 \le i,j \le 4$, the (i,j) entry of P_4 is i^j (for the purposes of this definition, we define $0^0 = 1$).

(a) What are the entries in the last row of $C_4 A_4$?

(b) If $P_4 = C_4 A_4$, then each of the 25 entries of P_4 is equal to the corresponding entry of $C_4 A_4$. Write down the five equalities for the entries in the last row.

(c) If $P_4 = C_4 A_4$, then find A_4.

(d) Using the values that you found for A_4, now rewrite the last equality that you found in part (b).

P 5.4.14. We note that, for a non-negative integer k,

$$k^2 = \binom{k}{1} + 2\binom{k}{2}.$$

Let n be a positive integer. Use the above identity to find a closed expression for

$$\sum_{k=1}^{n} k^2 = 1^2 + 2^2 + \cdots + n^2.$$

P 5.4.15. We note that, for a non-negative integer k,

$$k^3 = \binom{k}{1} + 6\binom{k}{2} + 6\binom{k}{3}.$$

Let n be a positive integer, and use the above identity to find a closed expression for

$$\sum_{k=1}^{n} k^3 = 1^3 + 2^3 + \cdots + n^3.$$

P 5.4.16. We want to generate identities such as those given in Problems P 5.4.14 and P 5.4.15. Thus, for a positive integer k and a non-negative integer n, write

$$k^0 = a_{0,0}\binom{k}{0}$$

$$k^1 = a_{0,1}\binom{k}{0} + a_{1,1}\binom{k}{1}$$

$$k^2 = a_{0,2}\binom{k}{0} + a_{1,2}\binom{k}{1} + a_{2,2}\binom{k}{2}$$

$$\vdots$$

$$k^m = a_{0,m}\binom{k}{0} + a_{1,m}\binom{k}{1} + a_{2,m}\binom{k}{2} + \cdots + a_{m,m}\binom{k}{m}$$

$$\vdots$$

$$k^n = a_{0,n}\binom{k}{0} + a_{1,n}\binom{k}{1} + a_{2,n}\binom{k}{2} + \cdots + a_{n,n}\binom{k}{n}.$$

Our eventual intent (see Problem $P\ 5.4.17$) is to find $a_{r,m}$ for $r = 0,\ldots,m$, and $m = 0,\ldots,n$ so that all the above identities are correct for all values of k. (Actually, from Problems P 5.4.14 and P 5.4.15, we already know $a_{r,m}$ for $r \leq 3$.) For convenience we define $a_{r,m} = 0$ for $m < r \leq n$.

In addition to C_n, the matrix of binomial coefficients, we define two other matrices A_n and P_n. All of these matrices are $(n+1) \times (n+1)$ and their rows and columns are numbered $0, 1, \ldots, n$. The (i,j) entry of A_n is $a_{i,j}$, and the (i,j) entry of P_n is i^j. For the purpose of defining P_n, we define $0^0 = 1$. Find a relationship among the matrices P_n, C_n, and A_n.

$P\ 5.4.17.$ Continuing Problem P 5.4.16, use the matrix equation that was found in that problem, as well as the inverse of C_n (Problem $P\ 5.4.11$) to find a formula for $a_{r,m}$ for $r = 0,\ldots,m$.

$P\ 5.4.18.$ Use the result of Problem $P\ 5.4.17$ to write k^4 as a linear combination of binomial coefficients. Then use this and the hockey-stick identity of Problem P 5.1.6 to find a closed formula for

$$\sum_{k=1}^{n} k^4.$$

5.4.3   Dyck Paths, Ming–Catalan Numbers, and the Chung–Feller Theorem

We introduced north-east (abbreviated NE) lattice paths in Section 4.2.1, but here is a recap. Fix the positive integers n and k, and assume that, on the plane, you want to go from the point $(0,0)$ to the point (n,k) using only unit steps to the right or straight up. In other words, if you are at the point with coordinates (a,b), then you have the choice of going (east) to $(a+1,b)$ or (north) to $(a,b+1)$. These are called NE lattice paths, and any NE lattice path from $(0,0)$ to (n,k) will need n steps to the right (E steps) and k steps straight up (N steps). Since the total number of steps to be taken is $n+k$, to choose a particular path, you just need to decide which of the $n+k$ steps are E steps (or, equivalently, how many are N steps). Hence, there are $\binom{n+k}{n} = \binom{n+k}{k}$ NE lattice paths from $(0,0)$ to (n,k) (this was Problem $P\ 4.2.25$). Going from (n_1,k_1) to (n_2,k_2) using NE lattice paths is equivalent to going from $(0,0)$ to $(n_2 - n_1, k_2 - k_1)$, and so finding the total number of NE paths for this more general setting is just as straightforward (this was Problem P 4.2.26). Here, we introduce a variation.

Definition 5.34. Let n be a positive integer. A NE lattice path from $(0,0)$ to (n,n) that never goes below the diagonal line $y = x$ is called a *Dyck path*.[17] In other words, a Dyck path can never visit a point with coordinates (a,b) where $b < a$.

Example 5.35. All the Dyck paths for $n = 3$ are given in Figure 5.4.

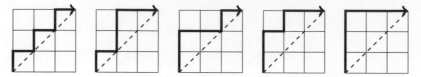

Figure 5.4 All the Dyck paths from $(0,0)$ to $(3,3)$.

We introduced the Ming–Catalan numbers in Section 5.1.5 (see Definition 5.17). Maybe surprisingly, the number of Dyck paths from $(0,0)$ to (n,n) is the nth Ming–Catlan number c_n. Much more surprising and unexpected is the Chung–Feller theorem. Any NE lattice path from $(0,0)$ to (n,n) has n N steps (going one unit up). For a Dyck path all such steps are above the line $y = x$. Let k be an integer with $0 \leq k \leq n$, and let S_k denote all NE lattice paths from $(0,0)$ to (n,n) with the property that exactly k of the n N steps are above the line $y = x$. The set S_n is then the collection of all Dyck paths. The Chung–Feller theorem (Chung and Feller 1949) says that, regardless of the value of k, the number of NE lattice paths in S_k is the nth Ming–Catlan number! The problems guide you through a proof of these and related facts. In counting special classes of lattice paths it is helpful to "translate" and "reflect" paths.

Definition 5.36. To specify a NE lattice path, you need to know the ordered sequence of N and E steps as well as either the starting point or the ending point of the path. Let P and Q be two NE lattice paths. If the ordered sequence of N and E steps is identical for these two paths, then we say that Q is a *translate* of P. If the ordered sequence of N and E steps for these two paths are exactly opposite to each other, then we say that Q is a *reflection* of P. (See Figure 5.5.)

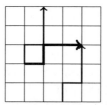

Figure 5.5 If P is a path from $(1,2)$ to $(4,3)$ with the sequence *ENEE*, then reflection of P ending at $(4,3)$ is a path from $(3,0)$ to $(4,3)$, while the reflection of P starting at $(1,2)$ is a path from $(1,2)$ to $(2,5)$. The two reflections are translates of each other, and the sequence of steps for both is *NENN*.

---

[17] Named after the German mathematician Walther von Dyck (1856–1934).

P 5.4.19. Consider the NE lattice path P that starts at $(0,0)$ and whose (ordered) sequence of N and E steps is given by

NENENNEEEENENNNENN.

(a) Where does the path P end?
(b) Does the path ever go below the line $y = x$? If so, what is the first point on the path whose coordinate is of the form $(h + 1, h)$, where h is a non-negative integer?
(c) Where is the starting point for a reflection of P that ends at $(10, 11)$?
(d) Split the path P into two equal parts, reflect the first half (starting at $(0,0)$), and translate the second half so that it starts where the reflection of the first half ends. Where does the resulting path end?

P 5.4.20. Let a_1, a_2, b_1, and b_2 be non-negative integers with $a_1 \leq a_2$ and $b_1 \leq b_2$, and let P be a NE path from (a_1, b_1) to (a_2, b_2). Let Q be the path that is a reflection of P starting at (a_1, b_1), and let R be a reflection of P that ends at (a_2, b_2). For each of the paths Q and R, give their starting and end points.

P 5.4.21. Let n be a positive integer. Define $\mathcal{A}$ to be the set of NE paths from $(0,0)$ to (n,n) that visit some point below the line $y = x$. In other words, every NE path in $\mathcal{A}$ goes through a point (x, y) with $y < x$. For each $P \in \mathcal{A}$ we want to define a related path $P^\star$. First, find the first point p on P of the form $(h + 1, h)$ for some $0 \leq h \leq n - 1$. Then, think of the path P as consisting of two shorter paths, one following the other. The path P_1 is the portion of P from $(0,0)$ to $(h+1, h)$, and the path P_2 is the portion of P from $(h+1, h)$ to (n, n). Now $P^\star$ is defined as the reflection of P_1 starting at $(0,0)$ followed by a translate of P_2 that starts where this reflection of P_1 ends. Let $\mathcal{B} = \{P^\star \mid P \in \mathcal{A}\}$.

(a) Show that $\mathcal{B}$ is exactly the collection of *all* NE lattice paths from $(0,0)$ to $(n-1, n+1)$.
(b) What is $|\mathcal{A}|$?

P 5.4.22. **Counting Dyck Paths.** Prove that the number of Dyck paths from $(0,0)$ to (n,n) is $c_n = \frac{1}{n+1}\binom{2n}{n}$, the nth Ming–Catalan number.

P 5.4.23. Let n be a positive integer. If you arrange the integers $1, 2, \ldots, 2n$ in a $2 \times n$ array in such a way that the integers in each row and each column are an increasing sequence, then you have what is called a $2 \times n$ *standard Young tableau*. Figure 5.6 shows all the 2×3 standard Young tableaux. Give a bijection between the set of all $2 \times n$ standard Young tableaux and all Dyck paths from $(0,0)$ to (n,n).

1	2	3
4	5	6

1	2	4
3	5	6

1	2	5
3	4	6

1	3	4
2	5	6

1	3	5
2	4	6

Figure 5.6 In a standard Young tableau, the integers in every row and column form an increasing sequence. All 2×3 standard Young tableaux are listed.

P 5.4.24. Consider the NE lattice path P that starts at $(0,0)$, ends at $(7,7)$, and whose (ordered) sequence of N and E steps is given by

$$NENEEENENNNEEN.$$

Define the integers h and k as follows: h is the smallest non-negative integer such that the point $(h+1,h)$ is on P, and k is the smallest integer greater than h such that (k,k) is on P. Let P_1 be the portion of P from $(0,0)$ to $(h+1,h)$, let P_2 be the portion of P from $(h+1,h)$ to (k,k), and, finally, let P_3 be the portion of P from (k,k) to (n,n). We define the NE lattice path $P^\star$ as a translation of P_2 starting at $(0,0)$ followed by a translation of P_1 followed by P_3.

(a) For the specific path P defined above, what are h and k, and what is the sequence of N and E steps for the path $P^\star$?

(b) Can we carry out the above procedure for all NE lattice paths from $(0,0)$ to (n,n)? If not, give a description of paths Q for which we can find the integers h and k, and as a result define the path $Q^\star$.

(c) In our description of $P^\star$, we translated P_2 to start at $(0,0)$, and then translated P_1 so that the translation of P_1 would start where the translation of P_2 ended. However, we didn't do anything to P_3. Verify that the reflection of P_2 followed by a translation of P_1 ends exactly where P_3 starts.

(d) For each of P_1, P_2, and P_3, count the number of N steps that are above the line $y = x$. Do the same for the three parts of the path $P^\star$.

(e) For i a non-negative integer, we define S_i to be the collection of NE lattice paths that have i N steps above the line $y = x$. To which S_i does our path P belong? What about $P^\star$?

P 5.4.25. **Chung–Feller Theorem.** Let n be a positive integer. For i an integer with $0 \le i \le n$, define S_i to be the collection of all NE lattice paths from $(0,0)$ to (n,n) that have exactly i N steps above the line $y = x$. The Chung–Feller theorem (Chung and Feller 1949) says that, for $0 \le i \le n$,

$$|S_i| = \frac{1}{n+1}\binom{2n}{n}.$$

It is remarkable that the size of S_i does not depend on i and that it is a Ming–Catalan number. In fact, the total number of NE lattice paths from $(0,0)$ to (n,n) is $\binom{2n}{n}$, and there are $n + 1$ sets $S_0, S_1, \ldots, S_n$, and so the Chung–Feller theorem says that the NE lattice paths are uniformly distributed among these sets.

Prove the Chung–Feller theorem by giving, for $0 \le i \le n - 1$, a bijection between S_i and S_{i+1}, and thereby proving that

$$|S_i| = |S_{i+1}|.$$

You may find the following steps helpful:[18]

---

[18] This proof is from Jewett and Ross 1988. See Callen 1995 for a different bijective proof.

STEP 1: As a warm-up, do Problem P 5.4.24.

STEP 2: Let i be a fixed non-negative integer with $0 \leq i \leq n-1$, and let $P \in S_i$. Proceeding as in Problem P 5.4.24, define h to be the smallest non-negative integer such that $(h+1, h)$ is on P, and k is the smallest positive integer greater than h such that (k, k) is on P.

STEP 3: Split P into three parts. P_1 is the part of P starting at $(0,0)$ and ending at $(h+1, h)$. P_2 is the part of P starting at $(h+1, h)$ and ending at (k, k). Finally P_3 is the part of P starting at (k, k) and ending at (n, n).

STEP 4: Define a NE path $P^\star$ from $(0,0)$ to (n, n) by switching the place of P_1 and P_2 and appropriately translating them.

STEP 5: Show that $P^\star \in S_{i+1}$.

STEP 6: Show that the map sending P to $P^\star$ is a bijection from S_i to S_{i+1}.

STEP 7: Rewrite what you have, to give a complete and concise proof of the Chung–Feller theorem.

P 5.4.26. Does our proof of the Chung–Feller theorem in Problem *P 5.4.25* give a different proof (from your proof in Problem P 5.4.22) that the number of Dyck paths is a Ming–Catalan number?

P 5.4.27. Count the number of NE lattice paths from $(0,0)$ to (n, n) that, except for $(0,0)$ or (n, n), do not touch or go below the line $y = x$.

P 5.4.28. For each of the following scenarios, find the number of ways that can you flip a coin 14 times and get 7 heads and 7 tails:

(a) There are no other restrictions.

(b) At no time in the process should the number of heads be less than the number of tails.

(c) The only times in the process when the number of heads equals the number of tails is before the tosses begin and after all the 14 tosses are completed.

P 5.4.29. You have n identical chocolate bars, and every day you eat half of a chocolate bar. Some days, you only have full bars left and on those days you have to split one bar into two and eat one-half of it. Other days, you have a choice of either eating an existing half bar (one that you split some days ago) or, if you still have whole ones left, splitting a whole one and eating half. You keep track of what you have done by recording a sequence of w's and h's. As an example, if $n = 3$, your final sequence may be *whwhwh* or *wwhwhh*. What is the number of possible sequences?[19]

P 5.4.30. **The Ballot Problem.** Candidates A and B stand for an election. Candidate A receives a votes and candidate B receives b votes, with $a \geq b$. The votes are counted one by one. In how many ways could the votes be ordered so that, starting with the first vote and throughout the count, candidate A maintains a lead?[20] You may find the following steps helpful:

STEP 1: Show that the problem is equivalent to the following problem on NE lattice paths: how many NE lattice paths starting at $(0,0)$ and ending at (b, a) never cross below or touch the line $y = x$ except at $(0,0)$?

[19] Adapted from Bayer and Brandt 2014.
[20] For more on the ballot problem, see Renault 2007.

STEP 2: The NE lattice paths from $(0,0)$ to (b,a) that cross or touch the line $y = x$ (at some point other than $(0,0)$) start with either an E step or a N step. Count the former by comparing them to all NE lattice paths from $(1,0)$ to (b,a).

STEP 3: To count the "bad" NE lattice paths that start with a N step, consider the first point on the path of the form (k,k), with $k > 0$. Reflect the first portion of the path between $(0,0)$ and (k,k). Argue that this gives a bijection between these paths and all NE paths from $(1,0)$ to (b,a).

STEP 4: By subtracting the number of NE lattice paths violating our condition from the total, complete the problem. Simplify your answer so that it is of the form $\frac{?}{??}\binom{a+b}{a}$.

P 5.4.31. **The Ballot Problem Generalized.** Let k be a positive integer. Generalizing Problem P 5.4.30, assume that candidate A receives a votes and candidate B receives b votes, and $a \geq kb$. Prove that the number of ways that the ballots can be ordered so that, throughout the vote count, the number of votes of candidate A is more than k times the number of votes of candidate B is

$$\frac{a - kb}{a + b}\binom{a+b}{a}.$$

P 5.4.32. Let a and b be positive integers with $a > b$. Let $Q = (a-1,b)$ and $R = (a,b-1)$ be two points in the plane, and let D be the "diagonal" line segment from $(a-b,0)$ to (a,b).

(a) Assume a NE lattice path starts on (a point with integer coordinates on) D, stays below D, and ends at R. If you reflect this path (and keep its starting point unchanged), then where will the path end? Will the reflected path intersect D (at a point other than its starting point)?

(b) Find a bijection between NE lattice paths from $(0,0)$ to R and those NE lattice paths from $(0,0)$ to Q that touch or cross D.

(c) Prove that the number of lattice paths from $(0,0)$ to Q is more than the number of lattice paths from $(0,0)$ to R. Conclude that

$$\binom{a+b-1}{b} > \binom{a+b-1}{b-1}.$$

(d) Assume that $k < \frac{n+1}{2}$. By appropriately choosing a and b, use the previous part to give a proof (a different proof was given in Theorem 5.8) that

$$\binom{n}{k} > \binom{n}{k-1}.$$

5.4.4 Duality and Sampling without Replacement

Assume that you want to examine 7 elements of a set of size 47. A probabilist may speak of sampling with or without replacement. To *sample with replacement* would mean that you randomly pick one element, examine it, and then return it to the batch before picking a second element. This would mean that after seven attempts, you may have actually examined some elements more than once, and the total number of distinct elements chosen may be less than seven. The advantage of sampling with replacement is that every time you choose an element, the probability of picking a particular element is constant – in our case $1/47$ – making certain probability calculations simpler. On the other hand, *sampling without replacement* would mean to genuinely choose 7 distinct elements from among the 47. You could number the elements 1 through 47 and randomly choose one, then another, and so on, until you have 7 elements, or you could write down all subsets of size 7 of $[47] = \{1, \ldots, 47\}$ and then pick one of them randomly. In either method, the probability that a given element will be chosen is the same. (See Problems P 5.4.33 and P 5.4.36.)

In the problems – adapted from Barnier and Jantosciak (2002) and Davidson and Johnson (1993) – you are asked to explore a certain duality among the parameters in sampling without replacement, as well as to answer questions such as the following:

Problem 5.37. *You have a group of* 12 *people in a room. You also have seven red hats and five blue hats, as well as eight black belts and four brown belts. Each person is randomly given a hat and a belt. What is the probability that there are exactly five people with red hats and black belts?*

In the above problem, an interesting duality emerges with the insight that the sought-after probability does not change if you distribute only the belts randomly. In other words, you could decide a priori who gets the blue hats and who gets the red hats, and, as long as you distribute the belts randomly, there would be no effect on the final probability.

Problems (continued)

P 5.4.33. Let $[47] = \{1, \ldots, 47\}$. We use two methods for picking a random sample of size 7 from [47]. In the first method, we pick one element randomly, then pick a second one from the remaining 46, and so on until we have 7 elements. In the second method, we first write down all subsets of size 7 of [47], and then randomly pick one of the subsets. In the first method, we are picking a random 7-permutation, while in the second method, we are picking a random 7-combination.

(a) What is the number of 7-permutations of [47]? How many of these contain 23? If we use the first method, what is the probability that 23 is in our sample of 7?

(b) What is the number of 7-combinations of [47]? How many of these contain 23? If we use the second method, what is the probability that 23 is in our sample of 7?

(c) What is the number of 7-permutations of [47] that contain both 4 and 7? What is the probability that a randomly chosen 7-permutation of [47] would contain both 4 and 7?

(d) What is the number of 7-combinations of [47] that contain both 4 and 7? What is the probability that a randomly chosen 7-combination of [47] would contain both 4 and 7?

(e) What is the probability that a randomly chosen 7-permutation of [47] contains both 4 and 7, and 4 comes before 7?

P 5.4.34. Let n, k, and r be positive integers with $r \le k \le n$. Let S be a fixed subset of size r of [n]. Compare the probabilities that a random k-permutation or k-combination of [n] will contain the elements of S.

P 5.4.35. In Problem P 5.4.33, you compared two supposedly different – but ultimately the same – notions of "what is the probability that 23 is in a random sample of size 7 from [47]?" In this problem, we try a "third" method. First, make all possible notecards that have a subset of size 7 of [47] on one side, and one element of [47] on the other side. So, for example, one notecard would have $\{3, 5, 7, 25, 43, 44, 45\}$ on one side and 36 on the other. Mix all the notecards and randomly choose one.

(a) What is the probability that the single number on one side of the notecard is an element of the subset of size 7 on the other side?

(b) What is the probability that the single number is 23 *and* that 23 is also among the elements of the subset of size 7 on the other side?

(c) Keep the cards that have 23 somewhere on them and throw the rest away. Now pick a random card. What is the probability that 23 is on both sides of the card?

(d) How does your answer in the previous parts differ from your answers to Problem P 5.4.33? Are you surprised?

P 5.4.36. You are contemplating buying a box of 47 used light bulbs, and you assume that k – for some $0 \le k \le 47$ – of the 47 light bulbs are defective. To estimate k, you want to test four of the light bulbs. Here are three possible sampling methods for picking four light bulbs (the first two methods are examples of sampling without replacement, while the third is sampling with replacement):

(a) Number the light bulbs from 1 to 47, and randomly pick a subset of size 4 of [47] = $\{1, \ldots, 47\}$.

(b) Choose one light bulb randomly from among the 47 light bulbs, then randomly choose a second from among the remaining 46, and so on until you have four light bulbs.

(c) Choose one light bulb randomly from among the 47 light bulbs, test it and then return it to the batch. Now choose a second one, and test it, and return it, and so on until you have tested four light bulbs.

Assume that – regardless of the sampling method used – you find 2 light bulbs defective.

(a) For each sampling method, what is the probability – in terms of k – of having two defective light bulbs in a sample of size four?

(b) Does any pair of sampling methods give the same probabilities?

(c) You argue that since 50% of the random sample was defective, then it is likely that about 50% of the original batch is defective as well. Make a table that shows, for each sampling method, for $22 \le k \le 25$, the probability that, in a random sample of four, two items will be defective. For comparative purposes, also include the probabilities for $k = 2, 5, 10, 20, 27, 37, 42$, and 45.

(d) Which value of k maximizes the probability? Does the value of k depend on the sampling method?

(e) Are the differences in the probabilities stark enough to give you confidence in predicting a single value for k, or would you rather give a range of possible values for k? Comment.

P 5.4.37. Act E is to pick a random element of [47]. Act E' is to pick the element 23. Act S is to pick a random subset of size 7 of [47]. Act S' is to pick the subset $\{1, 2, 3, 4, 5, 6, 23\}$ of [47].

(a) Do E and S. What is the probability that the element that you picked is in the subset that you picked? Would it have made a difference if you had performed E and S in some order versus doing them simultaneously?

(b) Do E' and then S. What is the probability that the element that you picked (i.e., 23) is in the subset that you picked?

(c) Do S' and then E. What is the probability that the element that you picked is in the subset that you picked?

(d) Do E' and then S'. What is the probability that the element that you picked is in the subset that you picked?

(e) Let p be the probability that a randomly chosen subset of size 7 of [47] will contain 23. Let q be the probability that a randomly chosen element of [47] will be a member of $\{1, 2, 3, 4, 5, 6, 23\}$. Is $p = q$? Why?

P 5.4.38. Let i, b, r, and n be positive integers with $i \leq r \leq b \leq n$. Consider four acts. Act R is to randomly choose r elements of the set $[n]$ and put a red dot on them. Act B is to randomly choose b elements of the set $[n]$ and put a black dot on them. Act R' is to put a red dot on $1, 2, \ldots, r$. Act B' is to put a black dot on $1, \ldots, b$. In each of the following scenarios, find the probability that exactly i elements have both dots. Are any of them equal?

(a) Do R' and B.

(b) Do B' and R.

(c) Do R and B.

P 5.4.39. Let i, r, k, and n be positive integers with $i \leq r, k \leq n$. A combinatorics class has n students, and r of the students have already taken a probability class. In an exam, you are asked the following: "If you randomly choose a sample of k students from the class, what is the probability that exactly i students in your sample will already have had probability?" You argue as follows: there are $\binom{n}{k}$ total ways of choosing a sample of size k. From among these choices, $\binom{r}{i}\binom{n-r}{k-i}$ will have exactly i probability students. (First choose i from among the r, and then choose the rest of the sample from among the students that have not had probability.) Hence the desired probability is

$$\frac{\binom{r}{i}\binom{n-r}{k-i}}{\binom{n}{k}}.$$

When you want to transcribe your answer, you accidentally switch the places of the parameters r and k (number of designated elements and sample size), and you report

$$\frac{\binom{k}{i}\binom{n-k}{r-i}}{\binom{n}{r}}$$

as the desired probability. When you get home, you realize your mistake and are very worried. But when you find that you get your exam back, you got full credit on this problem. Why? Comment.

P 5.4.40.   You have a group of 12 people in a room. You also have seven red hats and five blue hats, as well as eight black belts and four brown belts. Each person is randomly given a hat and a belt. What is the probability that there are exactly five people with red hats and black belts?

5.4.5   Island of Logica and Knight–Knave Puzzles

On the island of Logica, there are two types of people, Knights and Knaves. The Knights always tell the truth and the Knaves always lie. (See Smullyan 1978 for fascinating Knight–Knave puzzles.) In the ensuing set of problems, adapted from and closely following Levin and Roberts (2013), we consider only one type of Knight–Knave puzzle: n is a fixed positive integer, and you ask a group of n people on the island of Logica how many from among the group are Knights, and each of them responds with an integer between 0 and n. You record the responses as a multiset of integers. We think of this multiset as a "puzzle" and a "solution to the puzzle" would be a determination of the number of Knights in the group consistent with the responses. (If you figure out the number of the Knights in the group, you also will know who, among the respondents, is a Knight and who is a Knave.)

Example 5.38. Let $n = 5$. The puzzle we are presented with is $\{0, 1, 1, 2, 2\}$. This means that in a group of 5 people on the island of Logica, one person said that there are no Knights among the group, two people said that there is 1 Knight among them, and another two people said that there are 2 Knights among them. We immediately know that the people giving the answer 0 or the answer 1 are lying. (If you were telling the truth, then you yourself would be a Knight and so there couldn't be 0 Knights. Likewise, if there was only 1 Knight, then only that person would be telling the truth and so there would have only been one response of 1.) If among this group there were 3, 4, or 5 Knights, then we would have seen those numbers in the puzzle since Knights tell the truth. We conclude that there is only one possible solution: the two that responded 2 are Knights, and every one else is a Knave. This puzzle has exactly one solution.

In general, if there are k Knights in the group, then we would get the response k exactly k times. So, as another example, the puzzle $\{2, 2, 3, 3, 3\}$ has three different possible solutions. It could be that there are 0, 2 or 3 Knights. In the first case, everyone is lying and everyone is a Knave. In the second case, the two people that said 2 are Knights and the others are Knaves, and, finally, it could be that the three that said 3 are Knights and the other two are Knaves. By contrast, the puzzle $\{0, 1, 1, 3\}$ does not have any solutions at all.

Definition 5.39. Let n be a fixed positive integer. By a *Knight–Knave puzzle* we mean a multiset P that records the responses of a group of n inhabitants of the island of Logica to the question: "How many Knights are among you?" In other words, let $a_0, a_1, \ldots, a_n$ be non-negative integers with $a_0 + a_1 + \cdots + a_n = n$; then any multiset of the form

$P = \{\underbrace{0, \ldots, 0}_{a_0}, \underbrace{1, \ldots, 1}_{a_1}, \ldots, \underbrace{n, \ldots, n}_{a_n}\}$ – that contains a_i copies of the integer i – is a Knight–Knave puzzle.

Given a Knight–Knave puzzle P, a *solution* to the puzzle – if it exists – is a non-negative integer k such that, given the responses in P, it is possible – in other words, logically consistent – that the number of Knights in the group is k.

In the problems, you are asked to *count* the number (of this kind) of Knight–Knave puzzles, to count the number of "puzzle–solution" pairs, and to find a combinatorial bijection between puzzles and "puzzle–solution" pairs.

Problems (continued)

P 5.4.41. A group of 10 people were standing on a street corner on the island of Logica. You asked "how many among you are Knights?" and the responses were $\{1, 2, 2, 3, 4, 4, 4, 4, 5, 6\}$. Given this set of responses, how many of the 10 could be Knights? What are the possibilities?

P 5.4.42. For a fixed positive integer n, how many different Knight–Knave puzzles are there? For a fixed non-negative integer k with $k \le n$, how many different Knight–Knave puzzles have k as a solution?

P 5.4.43. Let n be a fixed positive integer. As we have seen in Example 5.38, some Knight–Knave puzzles have multiple solutions, some have unique solutions, and some have no solutions. How many puzzle–solution pairs are there? In other words, find the size of the set

$$\mathcal{S} = \{(P, k) \mid P \text{ a Knight–Knave puzzle, the non-negative integer } k \text{ a solution to } P\}.$$

Furthermore, verify that the number of puzzle–solution pairs is the same as the number of Knight–Knave puzzles found in Problem P 5.4.42.

P 5.4.44. Let n be a fixed positive integer. If you randomly pick an integer k with $0 \le k \le n$, then what is the expected number of Knight–Knave puzzles with k as a possible solution? (In other words, if, for $0 \le k \le n$, s_k is the number of puzzles with k as a possible solution, then what is the average of $s_0, \ldots, s_n$?) Is this average always an integer? Have you seen this number before?

P 5.4.45. Let n be a fixed positive integer. Show that the number of Knight–Knave puzzles that have no zeros is the same as the number of puzzle–solution pairs (see Problem *P 5.4.43*) where the solution is not zero.

P 5.4.46. Let n be a fixed positive integer. In Problem *P 5.4.43*, you showed that the number of Knight–Knave puzzles is the same as the number of puzzle–solution pairs. This is unexpected since some of the puzzles have no solutions and others have multiple solutions. It is remarkable that there are just the right number of puzzles with multiple solutions to make up for the ones with no solutions. Give an alternative combinatorial proof of this fact by exhibiting a bijection between Knight–Knave puzzles and puzzle–solution pairs. You may find the following steps helpful:

STEP 1: If a puzzle has a 0 (i.e., the multiset defining the puzzle has 0 as a member), then could everyone in the group be a Knave? What if a puzzle does not have a 0?

STEP 2: Consider the puzzle $\{0, 0, 1, 1, 2, 2, 2, 3, 4\}$ for $n = 9$. Does this puzzle have a solution? There are *two* zeros among the responses. What if we make the 0's into 2's and the 2's into 0's? Does the resulting puzzle $\{0, 0, 0, 1, 1, 2, 2, 3, 4\}$ have a solution? What?

STEP 3: Assume that a puzzle has 0 as a member and that the repetition number of 0 is k. Show that if we switch the 0's and the ks in the puzzle (i.e., turn the 0's into k's and turn the k's into 0's), then the resulting puzzle will have at least one solution. What would that be? Could there be other solutions?

STEP 4: Define a map from puzzles to puzzle–solution pairs by using ideas from the previous steps. Deal with puzzles that don't have zeros separately from the ones that do.

STEP 5: Is your map from Step 4 a bijection? Complete the proof.

P 5.4.47. **Mann–Shanks Primality Criterion.**[21] We reorganize the table of binomial coefficients in a peculiar way, by shifting each row appropriately to the right. We will start row n in column $2n$. The first five rows of the resulting table are depicted in Table 5.2.

Table 5.2 Start row n of the Karaji–Jia triangle in column $2n$. What can you say about columns where all their entries are divisible by the row number?

	0	1	2	3	4	5	6	7	8	9	10	11	...
0	1												
1			1	1									
2					1	2	1						
3							1	3	3	1			
4									1	4	6	4	
5											1	5	
⋮													

Some numbers in the table are divisible by their row number and some are not. (The row numbers start with row $n = 0$.) For example, in the row for $n = 3$, the ones are not divisible by 3 while the threes are. Look for a *column* such that *every* integer in that column is divisible by its row number. Make a conjecture about these columns. Express your conjecture as a statement about binomial coefficients. You don't have to prove your conjecture.

---

[21] From Mann and Shanks 1972.

5.5 Open Problems and Conjectures

We highlight two open problems about binomial coefficients.

Singmaster's Conjecture

The integer 1 appears an infinite number of times in the Karaji–Jia triangle. No other positive integer a appears infinitely many times, since from the $(a+1)$th row on, every entry in the triangle – except the first and the last non-zero entry in each row – is bigger than a. But is there a positive integer a that appears more than eight times in the Karaji–Jia triangle? The answer is not known! The current record holder for the integer that appears the greatest number of times in the Karaji–Jia triangle is 3003, which appears exactly eight times in the triangle:

$$3003 = \binom{3003}{1} = \binom{78}{2} = \binom{15}{5} = \binom{14}{6} = \binom{14}{8} = \binom{15}{10} = \binom{78}{76} = \binom{3003}{3002}.$$

Conjecture 5.40 (Singmaster's Conjecture (Singmaster 1971)). *There is a constant c, such that no integer greater than 1 appears in the Karaji–Jia triangle more than c times.*

The constant c could be 8 or it could be higher. Singmaster himself (Singmaster 1971) proved in 1971 that a positive integer n can appear in the triangle at most $2 + 2\log_2(n)$ times. Over the years, this bound has been improved, but no constant bound – even a very large one – has been proved.

The Binomial Coefficients modulo an Integer

Take your favorite integer k, and replace every entry of the Karaji–Jia triangle by its remainder when you divide by k. Figure 5.7 shows what you get in the two cases $k = 2$ and $k = 6$. On the table on the left – that is, when $k = 2$ – count the number of ones in each row. You can prove that this number – which is the same as the number of odd entries in every row of the triangle – is always a power of 2. Are there any other patterns? Much is known

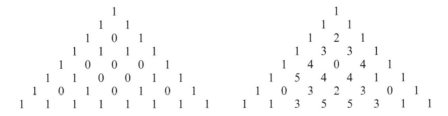

Figure 5.7 The Karaji–Jia triangle modulo 2 and modulo 6.

when $k = p$ is a prime or $k = p^m$ is a prime power (see Granville 1992, 1997), but very little is known when k is divisible by two or more primes. For example, for such a k, the number of non-zero entries in a row of the triangle is not known. Here is an example of a very particular conjecture for $k = 16$.

Conjecture 5.41 (Rowland 2011). *Let n be a positive integer, and let r be an odd positive integer less than 16. Let a be the number of integers, in the nth row of the triangle, that have remainder r when divided by 16. Then, if a is divisible by 3, it is also divisible by 2.*

6 Stirling Numbers

Let n and k be positive integers with n ≥ k. Each time you open your favorite language app, an advertisement randomly chosen from among k possible choices pops up. Far from being annoyed, in fact, you would like to see each of the k ads. Assuming that the algorithm is equally likely to choose any of the ads each time, what is the probability that it takes you exactly n tries to see all the k ads?

6.1 Stirling Numbers of the Second Kind

Warm-Up 6.1. *Kaitlyn and Megan together take a course in Latin American literature in translation. They buy one copy each of* The President, The Labyrinth of Solitude, The Death of Artemio Cruz, The Time of the Hero, One Hundred Years of Solitude, The House of the Spirits, *and* The Motorcycle Diaries. *At the end of the course, they want to split the books between themselves. In how many ways can this be done if the only condition is that each gets at least one book?*

   While the methods of the previous chapters – namely the use of permutations and combinations – can be used to solve many counting problems, other straightforward sounding problems will benefit from new tools, and cannot directly be reduced to permutations and combinations. Consider the portion of our counting table, Table 3.1, given in Table 6.1.[1]

   Entry $\boxed{5}$ was straightforward. We have s distinct balls and t distinct boxes, and no restriction on the box capacities. There are t distinct choices for where to put the first ball, the same t choices for where the second ball should go, and so on. Hence the total number of choices is t^s.

   Entries $\boxed{7}$, $\boxed{9}$, and $\boxed{11}$ are not as obvious. We will begin with entry $\boxed{11}$.

Example 6.2. We have four distinct balls – ①, ②, ③, ④ – and three identical boxes. In how many ways can we place the balls into boxes such that no box is empty? This is the same as asking: In how many ways can we group the balls into three non-empty parts?

   Here they are:

   $\{①, ②, ③ \; ④\}$, $\{1, 24, 3\}$, $\{1, 23, 4\}$, $\{12, 3, 4\}$, $\{13, 2, 4\}$, $\{14, 2, 3\}$.

   Hence the answer is 6.

---

[1] While not necessary, the reader may benefit from working through Collaborative Mini-project 3 on Stirling numbers before tackling this chapter.

Table 6.1 Entries $\boxed{5}$, $\boxed{7}$, $\boxed{9}$, and $\boxed{11}$ of the counting table.

Number of Ways to Put Balls into Boxes		
Balls (S) and Boxes (T) $\|S\| = s, \|T\| = t$ The ith box contains u_i balls.		
Conditions on S & $T \rightarrow$ on $u_i \downarrow$	S distinct T distinct	S distinct T identical
$u_i \geq 0$	$\boxed{5}$ t^s	$\boxed{7}$?
$u_i \geq 1$	$\boxed{9}$?	$\boxed{11}$?

Remark 6.3. The answer in Example 6.2 was 6. We found it by brute force and by enumerating all the possibilities. This method is clearly not going to work with much larger examples, and even if it did work, it does not give us any insight into the problem. Where did the answer 6 come from? The expressions $\binom{4}{2}$, 3!, and $1 + 2 + 3$ all give 6. Are any of them relevant to this problem? For the particular case of Example 6.2 – grouping 4 balls into 3 non-empty parts – we could argue that exactly one of the parts has to have two balls and that is the only choice we have to make. Hence, the answer is $\binom{4}{2} = 6$. But this argument cannot be generalized to other cases. For example, what if we have 47 balls and 10 boxes? In enumerative combinatorics, we often encounter new counting problems. One place to start is to "name" the solution to the problem. In other words, even if we don't know the answer, we give a name – often a function of one or more variables – to the answer to the problem.[2] We then proceed to find properties of the function. If we can find a recurrence relation for the function, then we can calculate many of the values. The recurrence relation will also allow us to prove further properties of the function – possibly even a formula – using induction.

Definition 6.4 (Stirling Numbers of the Second Kind: Entry $\boxed{11}$ of the Counting Table). Let s and t be non-negative integers. We define

$$\begin{Bmatrix} s \\ t \end{Bmatrix}$$

---

[2] "Sometimes naming a thing—giving it a name or discovering its name—helps one to begin to understand it. Knowing the name of a thing *and* knowing what that thing is for gives me even more of a handle on it." Octavio Butler, *Parable of the Sower*, Chapter 7.

to be the number of ways of putting s distinct balls into t identical boxes with at least one ball per box. Equivalently, $\left\{{s \atop t}\right\}$ is the number of ways of partitioning a set of s elements (i.e., $[s] = \{1, 2, \ldots, s\}$ if $s > 0$, and $[0] = \varnothing$ if $s = 0$) into t non-empty parts. $\left\{{s \atop t}\right\}$ is called a *Stirling Number of the second kind*.[3]

Remark 6.5. We read $\left\{{s \atop t}\right\}$ as "s subset t" or "s squig t." Also, note that $\left\{{0 \atop 0}\right\}$ is 1. This could be mandated separately as a part of the definition, but that is really not necessary. If we have no balls and no boxes, can I place each of the balls in some box in a way that none of the boxes is empty? The answer is yes. I am done before I start, since I don't have any balls to worry about, and there are no boxes to remain empty. Hence, the conditions – distributing all the balls and making sure that all the boxes are non-empty – are satisfied trivially. In fact, if you had to do the same task, you would carry it out exactly the same way that I did – i.e., do nothing – and hence there is exactly one way of placing 0 distinct balls in 0 identical boxes with no empty boxes.

We now know that the answer to entry $\boxed{11}$ is $\left\{{s \atop t}\right\}$. But the problem is that we don't know how to calculate this $\left\{{s \atop t}\right\}$. A few cases are easy to calculate.

Lemma 6.6. *Let $s \geq 1$ and $t \geq 0$ be integers. Then*

(a) $\left\{{s \atop t}\right\} = 0$ *if* $t > s$
(b) $\left\{{s \atop s}\right\} = 1$
(c) $\left\{{s \atop 1}\right\} = 1$
(d) $\left\{{s \atop s-1}\right\} = \binom{s}{2}$
(e) $\left\{{s \atop 2}\right\} = 2^{s-1} - 1$.

Proof. Recall that $\left\{{s \atop t}\right\}$ is the number of ways of putting s distinct balls into t identical boxes while making sure that no box goes empty.

(a) There are too many boxes, and hence there are 0 ways of making sure no box goes empty.
(b) There are as many boxes as balls, and so we must put each ball in a separate box.
(c) There is only one box, and so all the balls go into that one box.
(d) We have one fewer box than the number of balls, and so, just as in Example 6.2, we only have to decide which two balls go into the same box. There are exactly $\binom{s}{2}$ ways of choosing two balls to share a box.
(e) You are going to put the s balls into two boxes. All you have to do is to decide which balls go in the first box (since the rest go into the second box), and, other than choosing the empty set or the full set of balls, you can choose any subset of the balls for the first box. Hence, there are $2^s - 2$ ways of deciding which balls go in the first box. But this is assuming that there is a "first" box. Since the boxes are identical, we have double counted. In our count, every possibility got counted twice, once when you pick a set A of the balls for the first box, and a second time when you pick A^c, the complement of A, for the first box. In other words, splitting the balls as (A, A^c) is the same as splitting them as (A^c, A), and, in

---

[3] Some denote the Stirling numbers of the second kind by $S(s, t)$ or $s_2(s, t)$.

our original count, we counted these as two different possibilities. Hence, accounting for the double count, the final answer is $\frac{1}{2}(2^s - 2) = 2^{s-1} - 1$.　　　　　□

Table 6.2 gives some of the values of $\left\{{s \atop t}\right\}$. To find the rest, we prove a recurrence relation for the Stirling numbers of the second kind.

Table 6.2 A preliminary table of small Stirling numbers of the second kind.

The small values of $\left\{{s \atop t}\right\}$									
$s \backslash t$	0	1	2	3	4	5	6	7	8
0	1	0	0	0	0	0	0	0	0
1	0	1	0	0	0	0	0	0	0
2	0	1	1	0	0	0	0	0	0
3	0	1	3	1	0	0	0	0	0
4	0	1	7	6	1	0	0	0	0
5	0	1	15		10	1	0	0	0
6	0	1	31			15	1	0	0
7	0	1	63				21	1	0
8	0	1	127					28	1

Theorem 6.7. *Let s and t be positive integers; then*[4]

$$\left\{{s \atop t}\right\} = t\left\{{s-1 \atop t}\right\} + \left\{{s-1 \atop t-1}\right\}.$$

Proof. Thought experiment: consider s distinct balls. We want to partition the balls into t non-empty parts. Single out one of the balls, say ⓢ. Now partition the rest and then add ⓢ. Either you add ⓢ to an existing part or you make ⓢ a part all by itself. Now, in order to have t non-empty parts at the end, in the former case you must have partitioned the first $s - 1$ balls into t non-empty parts, and in the latter case you must have partitioned them into $t - 1$ non-empty parts. We write

$$\{\text{partitions of } [s] \text{ into } t \text{ non-empty parts}\} = A \cup B,$$

---

[4] Discovered by Japanese mathematician Masanobu Saka, who published it in his 1782 book *Sanpō-Gakkai* (The Sea of Learning on Mathematics). See Knuth 2013.

where

$$A = \{\text{those partitions where } \text{\textcircled{s}} \text{ is in a set by itself}\}$$
$$B = \{\text{those partitions where } \text{\textcircled{s}} \text{ is grouped with } \geq 1 \text{ other elements}\}.$$

Now, $|A| = \left\{{s-1 \atop t-1}\right\}$ since no choices have to be made for $\text{\textcircled{s}}$, and the number of ways of partitioning the other $s-1$ balls into $t-1$ non-empty parts is $\left\{{s-1 \atop t-1}\right\}$. On the other hand, to construct a partition in B, we first partition $s-1$ distinct balls into t non-empty parts – we can do this in $\left\{{s-1 \atop t}\right\}$ ways – and then we place $\text{\textcircled{s}}$ into one of these existing parts. We have t choices for where to put $\text{\textcircled{s}}$ and so $|B| = t\left\{{s-1 \atop t}\right\}$. The proof is now complete. $\square$

In Table 6.3, using the recurrence relation, we have calculated more values of $\left\{{s \atop t}\right\}$. While at this point we don't have a "formula" for the Stirling numbers of the second kind, using the recurrence relation we can calculate any desired value.

Table 6.3 A table of values for small Stirling numbers of the second kind.

					The small values of $\left\{{s \atop t}\right\}$						
$s\backslash t$	0	1	2	3	4	5	6	7	8	9	10
0	1	0	0	0	0	0	0	0	0	0	0
1	0	1	0	0	0	0	0	0	0	0	0
2	0	1	1	0	0	0	0	0	0	0	0
3	0	1	3	1	0	0	0	0	0	0	0
4	0	1	7	6	1	0	0	0	0	0	0
5	0	1	15	25	10	1	0	0	0	0	0
6	0	1	31	90	65	15	1	0	0	0	0
7	0	1	63	301	350	140	21	1	0	0	0
8	0	1	127	966	1701	1050	266	28	1	0	0
9	0	1	255	3025	7770	6951	2646	462	36	1	0
10	0	1	511	9330	34105	42525	22827	5880	750	45	1

Entries ⑨ and ⑦

We now turn to entries ⑨ and ⑦ of the counting table, Table 3.1.

Example 6.8. Assume that we want to partition $[4] = \{1, 2, 3, 4\}$ into three non-empty *distinct* parts. In other words, we have 4 distinct balls and 3 distinct boxes, and we want to distribute the balls into the boxes while making sure that none of the boxes are empty (entry $\boxed{9}$ of the counting table). If we had allowed empty boxes, the count would be easy. There are three choices for each ball, and the choices are independent of each other. So the total number of choices would be 3^4. However, some of these choices require putting all the balls into just one or two non-empty parts. We could try to adjust the count by subtracting the choices that lead to empty parts – and we will do that in Section 8.1 – but here we will use a different strategy. The boxes are distinct, and so assume each has an identifying label on it. First take the labels off so that the boxes become identical, distribute the balls, and then put the labels back on. The total number of ways of partitioning [4] into three non-empty parts is $\{^4_3\} = 6$ (entry $\boxed{11}$). Putting the labels back on, you have three choices for where the first label goes, two choices for the second label, and one choice for the last one. Hence, the total number of ways of placing 4 distinct balls into 3 distinct non-empty boxes is $3! \{^4_3\} = 36$.

Theorem 6.9 (Entry $\boxed{9}$ of the Counting Table). *Let s and t be non-negative integers, and let $R = \{\infty \cdot U_1, \infty \cdot U_2, \dots, \infty \cdot U_t\}$ be a multiset with t types of elements, each with an infinite repetition number. Then the following numbers are equal:*

(a) $\boxed{9}$ *The number of ways of placing s distinct balls into t non-empty distinct boxes*
(b) *The number of s-permutations of R that contain each of $U_1, \dots, U_t$ at least once*
(c)

$$ t! \left\{ {s \atop t} \right\}. $$

Proof. (a) $=$ (c) First partition the s balls into t non-empty parts and then label the parts with the name of each box. There are $\{^s_t\}$ ways of doing the former, and $t!$ ways of doing the latter.

(a) $=$ (b) Think of the U's as t distinct boxes, and have s balls numbered 1 through s. Putting a ball in a box chooses the box *and* determines the place of the box in a list (if balls numbered 23 and 47 go into the red box, then the red box is the 23rd and the 47th element in our ordered list). Hence, every placing of s distinct balls into t non-empty distinct boxes corresponds to an ordered list of s of the elements of R in such a way that each U appears at least once. □

Theorem 6.10 (Entry $\boxed{7}$ of the Counting Table). *Let s and t be non-negative integers. The number of ways of placing s distinct balls into t identical boxes (with empty boxes allowed), or equivalently the number of partitions of [s] into t or fewer parts, is*

$$ \sum_{k=0}^{t} \left\{ {s \atop k} \right\}. $$

Proof. You could put the balls into $0, 1, 2, \dots, t$ non-empty boxes. The result follows. □

We now answer the opening problem of the chapter.

Problem 6.11. *Let n and k be positive integers with $n \geq k$. Each time you open your favorite language app, an advertisement randomly chosen from among k possible choices pops up. Far from being annoyed, in fact, you would like to see each of the k ads. Assuming that the* *algorithm is equally likely to choose any of the ads each time, what is the probability that it takes you exactly n tries to see all the k ads?*

Solution. For each try, there are k choices, and so the total number of sequences of ads that you could possibly see is k^n. From among these, how many include all the k ads in such a way that all n tries were necessary to see all k? It must have been that one of the ads is shown exactly once on the nth try. Choose which one. There are k choices. Now we are left with $k-1$ ads and $n-1$ tries. Think of the $n-1$ tries as distinct balls (the first try is different from the 30th try by virtue of being the first) and the $k-1$ ads as distinct boxes. We have to put the balls into the boxes in such a way that no box remains empty. The number of ways of doing this is $(k-1)! \left\{ {n-1 \atop k-1} \right\}$. Hence, the total number of ways of getting all the k ads in n tries (and not earlier) is $k(k-1)! \left\{ {n-1 \atop k-1} \right\} = k! \left\{ {n-1 \atop k-1} \right\}$. The sought-after probability is then

$$\frac{k!}{k^n} \left\{ {n-1 \atop k-1} \right\}.$$

□

The Bell Numbers

Consider the table of Stirling numbers of the second kind, Table 6.3. Just as we did with the table of binomial coefficients, we can ask many questions regarding these numbers. We have already noted (Lemma 6.6) that the diagonal entries are always 1, and the numbers right below the main diagonal are the triangular numbers $\binom{s}{2}$. What can we say about the *sum* of the entries in each row? The (s, t) entry of the table is $\left\{ {s \atop t} \right\}$, the number of partitions of $[s]$ into t non-empty parts. So the sum of every row is the *total* number of partitions of $[s]$. This is an interesting sequence of numbers. Before proceeding, we name this sequence.

Definition 6.12 (Bell Number). Let s be a non-negative integer. Then the *Bell number* $B(s)$ is defined by

$$B(s) = \sum_{k=0}^{s} \left\{ {s \atop k} \right\}.$$

In other words, $B(s)$ is the total number of ways of partitioning the set $[s]$.[5]

---

[5] Bell numbers are named after E. T. Bell, who wrote a paper about them in 1938. Bell himself wrote that these numbers had been rediscovered many times, and had been studied by many mathematicians. One of the earliest appearances of Bell numbers is in Japan around the year 1500. In an incense-comparing game called *genji-kou*, a host would burn five different packets of incense. The guests would have to decide which, if any, of the incenses were the same. So a solution may be {{1, 3}, {2, 4, 5}}, indicating that the first and the third were the same incense, as were the other three. So, the set of all possible solutions is all the partitions of [5], and the number of such solutions is $B(5) = 52$. For ease of remembering, each of the 52 partitions was named after one chapter of the classic eleventh-century *Tale of Genji*. Following this tradition, Japanese mathematicians such as Seki Takakazu (circa 1642–1708) and his student Yoshisuke Matsunaga (1690–1744) studied partitions of finite sets systematically. Matsunaga, for example, gave a recurrence relation for the Bell numbers (Problem P 6.1.23) and posed and gave a solution to the problem of finding n if we know that $B(n) = 678, 570$. For more on the history of set partitions, see Knuth 2013.

Example 6.13. Just multiply the matrix of the Stirling numbers by a column vector consisting of all ones, to find the first few Bell numbers:

$$1, 1, 2, 5, 15, 52, 203, 877, 4140, 21147, 115975, \ldots$$

We now turn to a specific process of generating a sequence of functions $f_0, f_1, \ldots$ Begin with the function $f_0(x) = e^x$, and, for $n > 0$, define $f_n(x)$ to be x times the derivative of $f_{n-1}(x)$. In other words, keep taking derivatives and multiplying by x. Reading the definition, the reader will naturally wonder about the connection with the Stirling numbers, the Bell numbers, or even combinatorics. The sequence of functions may seem quite arbitrary. But, as soon as we compute a few of the functions, the connection will become clear. In fact, we will get quite a remarkable result at the end of our investigation. We have:

$$f_0(x) = e^x \qquad\qquad\qquad f_0'(x) = e^x$$
$$f_1(x) = xe^x \qquad\qquad\qquad f_1'(x) = e^x + xe^x$$
$$f_2(x) = xe^x + x^2e^x \qquad\qquad f_2'(x) = e^x + 3xe^x + x^2e^x$$
$$f_3(x) = xe^x + 3x^2e^x + x^3e^x \qquad f_3'(x) = e^x + 7xe^x + 6x^2e^x + x^3e^x$$
$$f_4(x) = xe^x + 7x^2e^x + 6x^3e^x + x^4e^x \qquad f_4'(x) = e^x + 15xe^x + 25x^2e^x + 10x^3e^x + x^4e^x$$
$$f_5(x) = xe^x + 15x^2e^x + 25x^3e^x + 10x^4e^x + x^5e^x$$

If you compare the results with Table 6.3 of the Stirling numbers of the second kind, the connection is clear. The coefficient of $x^k e^x$ in $f_n(x)$ is $\left\{{n \atop k}\right\}$.[6]

Proposition 6.14. *Let $f_0(x) = e^x$, and, for $n > 0$, define $f_n(x) = x\frac{d}{dx}(f_{n-1}(x))$. Then*

$$f_n(x) = \left[\left\{{n \atop 0}\right\} + \left\{{n \atop 1}\right\}x + \cdots + \left\{{n \atop n}\right\}x^n\right]e^x = e^x \sum_{k=0}^{n} \left\{{n \atop k}\right\}x^k.$$

Proof. We have defined the function $f_n(x)$ recursively with $f_0(x) = e^x$ and, for $n > 0$, $f_n(x)$ is $xf_{n-1}'(x)$. This we know. What we want to prove is that this definition will give $f_n(x) = e^x \sum_{k=0}^{n} \left\{{n \atop k}\right\}x^k$. We will prove this by induction on n.

We have seen that the result holds for small values of n. For the inductive step, we assume the formula to be correct for n, and we will prove the claim for $n+1$. So, we are assuming that we know that $f_n(x) = e^x \sum_{k=0}^{n} \left\{{n \atop k}\right\}x^k$. We can now find $f_{n+1}(x)$ using the recursive definition of the function, but then we have to show that what we get is equal to $e^x \sum_{k=0}^{n+1} \left\{{n+1 \atop k}\right\}x^k$.

Using the product rule for derivatives, we have

$$f_n'(x) = e^x\left[\left\{{n \atop 0}\right\} + \left\{{n \atop 1}\right\}x + \left\{{n \atop 2}\right\}x^2 + \cdots + \left\{{n \atop k}\right\}x^k + \cdots + \left\{{n \atop n}\right\}x^n\right]$$
$$+ e^x\left[\left\{{n \atop 1}\right\} + 2\left\{{n \atop 2}\right\}x + \cdots + (k+1)\left\{{n \atop k+1}\right\}x^k + \cdots + n\left\{{n \atop n}\right\}x^{n-1}\right]$$

---

[6] The German mathematician Johann August Grünert (1797–1872) studied the functions $f_n(x)$ and proved the formula in Theorem 6.23b for the coefficients that appear in these functions. See Boyadzhiev 2012.

$$= e^x \left[\left\{ {n \atop 0} \right\} + \left\{ {n \atop 1} \right\} + \left(\left\{ {n \atop 1} \right\} + 2 \left\{ {n \atop 2} \right\} \right) x + \cdots + \left(\left\{ {n \atop k} \right\} + (k+1) \left\{ {n \atop k+1} \right\} \right) x^k \right.$$
$$\left. + \cdots + \left(\left\{ {n \atop n-1} \right\} + n \left\{ {n \atop n} \right\} \right) x^{n-1} + \left\{ {n \atop n} \right\} x^n \right].$$

Now, except for $k = n$, the coefficient of x^k is $\left\{ {n \atop k} \right\} + (k+1) \left\{ {n \atop k+1} \right\}$, which by the recurrence relation for Stirling numbers of the second kind (Theorem 6.7) is equal to $\left\{ {n+1 \atop k+1} \right\}$. Also, $\left\{ {n \atop n} \right\} = 1 = \left\{ {n+1 \atop n+1} \right\}$. Thus

$$f_n'(x) = e^x \left[\left\{ {n+1 \atop 1} \right\} + \left\{ {n+1 \atop 2} \right\} x + \cdots + \left\{ {n+1 \atop k+1} \right\} x^k + \cdots + \left\{ {n+1 \atop n} \right\} x^{n-1} + \left\{ {n+1 \atop n+1} \right\} x^n \right].$$

Multiplying by x, and adding a superfluous $e^x \left\{ {n+1 \atop 0} \right\} x^0 = 0$ term, we get the expression that we want for $f_{n+1}(x)$. □

The unexpected fact that the Stirling numbers appear in these functions is not the end of the story.[7] We continue by recalling the power series expansion of e^x:

$$e^x = 1 + x + \frac{x^2}{2!} + \frac{x^3}{3!} + \cdots + \frac{x^k}{k!} + \cdots$$

By repeatedly taking derivatives and multiplying by x, we get power series expansions for $f_1, \ldots, f_n, \ldots$

$$f_1(x) = x + \frac{2x^2}{2!} + \frac{3x^3}{3!} + \cdots + \frac{kx^k}{k!} + \cdots$$

$$f_2(x) = x + \frac{2^2 x^2}{2!} + \frac{3^2 x^3}{3!} + \cdots + \frac{k^2 x^k}{k!} + \cdots$$

$$f_3(x) = x + \frac{2^3 x^2}{2!} + \frac{3^3 x^3}{3!} + \cdots + \frac{k^3 x^k}{k!} + \cdots$$

$$\vdots$$

$$f_n(x) = x + \frac{2^n x^2}{2!} + \frac{3^n x^3}{3!} + \cdots + \frac{k^n x^k}{k!} + \cdots \tag{6.1}$$

From Proposition 6.14, we know that $f_n(1) = e \sum_{k=0}^{n} \left\{ {n \atop k} \right\} = B(n)e$. But we can plug $x = 1$ into Equation 6.1, and get a remarkable formula for $B(n)$. We have now proved:

Proposition 6.15. *Let n be a non-negative integer. Then*

$$B(n) = \frac{1}{e} \sum_{k=0}^{\infty} \frac{k^n}{k!}.$$

(In the case of $n = 0$ and $k = 0$, we make the stipulation that $0^0 = 1$ in this formula.)

---

[7] Also, see Problem P 6.1.26 for a generalization.

Remark 6.16. You can read Proposition 6.15 as a statement that is giving you a formula for the Bell numbers. As such, the formula is unexpected. While we have an infinite series on the right-hand side, we can approximate it by finding a partial sum, and use the partial sum and some estimate of the remaining terms to find $B(n)$. For example,

$$B(6) \approx \frac{1}{e} \sum_{r=0}^{10} \frac{r^6}{r!} \approx 202.98.$$

Thus, it is reasonable to expect that $B(6) = 203$, and this can be made precise with an analytical estimate of the ignored terms.

On the other hand, we can read Proposition 6.15 as telling us about the sum of certain series and a remarkable sequence of integers that appears in those sums. Namely, by Proposition 6.15, we have

$$\frac{1}{0!} + \frac{1}{1!} + \frac{1}{2!} + \frac{1}{3!} + \cdots + \frac{1}{n!} + \cdots = e$$

$$\frac{0}{0!} + \frac{1}{1!} + \frac{2}{2!} + \frac{3}{3!} + \cdots + \frac{n}{n!} + \cdots = e$$

$$\frac{0^2}{0!} + \frac{1^2}{1!} + \frac{2^2}{2!} + \frac{3^2}{3!} + \cdots + \frac{n^2}{n!} + \cdots = 2e$$

$$\frac{0^3}{0!} + \frac{1^3}{1!} + \frac{2^3}{2!} + \frac{3^3}{3!} + \cdots + \frac{n^3}{n!} + \cdots = 5e$$

$$\frac{0^4}{0!} + \frac{1^4}{1!} + \frac{2^4}{2!} + \frac{3^4}{3!} + \cdots + \frac{n^4}{n!} + \cdots = 15e,$$

$$\vdots$$

and the sequence of integers, $1, 1, 2, 5, 15, \ldots$ appearing on the right-hand side is precisely the Bell numbers!

Remark 6.17. To approximate $B(n)$, you can use any mathematical software to find the partial sums of the series in Proposition 6.15. For small values of n, a symbolic algebra software such as Maple can actually give you exact answers. In Maple, you could try the following to get $B(7)$ (in Maple, $\exp(1)$ stands for e, and in the following code, we first define Bn as a function of n):

```
> Bn:=n -> sum(k^{n}/k!, k=0..infinity)/exp(1);
> Bn(7);
```

Entries $\boxed{5}$, $\boxed{7}$, $\boxed{9}$, $\boxed{11}$ in Perspective

We have now completed our analysis of entries $\boxed{5}$, $\boxed{7}$, $\boxed{9}$, and $\boxed{11}$ of the counting table. See the partial Table 6.4. These four entries are each a bit different from each other, but they are related to one another. In particular, we got entry $\boxed{9}$ directly from entry $\boxed{11}$. Recall the reason that $\boxed{9} = t! \boxed{11}$:

Table 6.4 Entries $\boxed{5}$, $\boxed{7}$, $\boxed{9}$, **and** $\boxed{11}$ **of the counting table reconsidered.**

	Number of Ways to Put Balls into Boxes	
	Balls (S) and Boxes (T)	
	$\|S\| = s, \|T\| = t$	
	The ith box contains u_i balls.	
Conditions on S & $T \rightarrow$ on $u_i \downarrow$	S distinct T distinct	S distinct T identical
$u_i \geq 0$	$\boxed{5}$ t^s	$\boxed{7}$ $\sum_{k=0}^{t} \left\{ {s \atop k} \right\}$
$u_i \geq 1$	$\boxed{9}$ $t! \left\{ {s \atop t} \right\}$	$\boxed{11}$ $\left\{ {s \atop t} \right\}$

- we first partition s distinct objects into t non-empty bundles ($\left\{ {s \atop t} \right\}$ ways of doing this as in $\boxed{11}$),
- then we label the bundles with t labels ($t!$ ways of doing this).

We may wonder why we can't we get entry $\boxed{5}$ from entry $\boxed{7}$ in the same way (See Table 6.4). In other words, if we want to find the number of ways of placing s distinct balls into t distinct boxes (and allow empty boxes), then mimicking what we did before,

- we may first partition the s distinct balls into t (possibly empty) parts (and we know – entry $\boxed{7}$ – there are $\left\{ {s \atop 0} \right\} + \left\{ {s \atop 1} \right\} + \cdots + \left\{ {s \atop t} \right\}$ ways of doing this),
- then we label the parts with t labels (?? ways of doing this).

The complication is that the answer to the second step – the number of labelings – depends on the number of non-empty parts.

For example, for $s = 4$, $t = 3$, if the parts are $\{1\}, \{3\}, \{2,4\}$, then there are six ways of labeling them. On the other hand, if the parts are $\{1,2,3,4\}, \{\}, \{\}$, then we only have to decide which one of the three boxes gets to be non-empty, and so there are only three possible labelings. As a result, we cannot multiply entry $\boxed{7}$ by one number to find the answer to entry $\boxed{5}$. But all is not lost. We can do the – admittedly more complicated – count, if we know the number of non-empty parts. First, we refine our count for entry $\boxed{9}$ of the counting table. Recall that, for $k > 0$, the falling factorial $(t)_k$ is defined – Definition 4.4 – to be $t(t-1)(t-2)\cdots(t-k+1)$, and $(t)_0 = 1$.

Lemma 6.18. *Let s and t be non-negative integers, and let k be an integer with $0 \le k \le t$. The number of ways of putting s distinct balls into t distinct boxes such that exactly k boxes are non-empty is*

$$(t)_k \left\{ {s \atop k} \right\}.$$

Proof. First put the s balls into k non-empty parts – you can do this in $\left\{ {s \atop k} \right\}$ ways – and then label the k boxes with labels from 1 to t – you can do this in $t(t-1)\cdots(t-k+1)$ ways. (Note that the labels on the empty boxes do not matter.) $\qquad\square$

We can now carry out the project of relating entry $\boxed{5}$ to entry $\boxed{7}$ of the counting table. If we put s distinct balls into t possibly empty boxes, then the number of non-empty boxes could be anywhere from 0 to t. Lemma 6.18 gives a count for each of these, and so we can write down a new expression for entry $\boxed{5}$. But we already had a very nice expression – namely t^s – for entry $\boxed{5}$. We have thus proved:

Theorem 6.19. *Let s and t be positive integers; then*

$$t^s = (t)_0 \left\{ {s \atop 0} \right\} + (t)_1 \left\{ {s \atop 1} \right\} + (t)_2 \left\{ {s \atop 2} \right\} + \cdots + (t)_t \left\{ {s \atop t} \right\}.$$

Example 6.20. Let $s = 5$ and $t = 3$. Then, we have

$$(3)_0 \left\{ {5 \atop 0} \right\} + (3)_1 \left\{ {5 \atop 1} \right\} + (3)_2 \left\{ {5 \atop 2} \right\} + (3)_3 \left\{ {5 \atop 3} \right\}$$

$$= 1 \cdot 0 + 3 \cdot 1 + 6 \cdot 15 + 6 \cdot 25 = 3^5.$$

For the case $t = 5$ and $s = 3$, we get

$$(5)_0 \left\{ {3 \atop 0} \right\} + (5)_1 \left\{ {3 \atop 1} \right\} + (5)_2 \left\{ {3 \atop 2} \right\} + (5)_3 \left\{ {3 \atop 3} \right\} + (5)_4 \left\{ {3 \atop 4} \right\} + (5)_5 \left\{ {3 \atop 5} \right\}$$

$$= 1 \cdot 0 + 5 \cdot 1 + 20 \cdot 3 + 60 \cdot 1 + 120 \cdot 0 + 120 \cdot 0 = 5^3.$$

Remark 6.21. The original answer for entry $\boxed{5}$ of the counting table, Table 3.1, was a pretty straightforward t^s (Theorem 4.27). What was the point of looking for a much more complicated expression? In combinatorics, we really like counting the same configuration in multiple ways. By so doing, we often get additional insight into the problem. In the case here, what we obtained was *not* a useful way of approaching $\boxed{5}$. Instead, the double count allowed us to prove the remarkable identity of Theorem 6.19 that relates Stirling numbers of the second kind to falling factorials and exponents. We will use this identity for several purposes.

In Section 6.3, it is used to find the relation between the two kinds of Stirling numbers (the Stirling numbers of the first kind will be defined in Section 6.2). In Problem P 6.1.30, you are asked to use this identity to give an actual formula for $\left\{{s \atop t}\right\}$ (see Theorem 6.23). Here, we rewrite Theorem 6.19 to give a new recurrence relation for the Stirling numbers.

Corollary 6.22. *Let s and t be positive integers; then*

$$\left\{{s \atop t}\right\} = \frac{1}{t!}\left[t^s - (t)_0\left\{{s \atop 0}\right\} - (t)_1\left\{{s \atop 1}\right\} - \cdots - (t)_{t-1}\left\{{s \atop t-1}\right\}\right]$$

$$= \frac{t^s}{t!} - \frac{1}{t!}\left\{{s \atop 0}\right\} - \frac{1}{(t-1)!}\left\{{s \atop 1}\right\} - \cdots - \frac{1}{1!}\left\{{s \atop t-1}\right\}.$$

Formulas for Stirling Numbers of the Second Kind

To calculate the Stirling number $\left\{{s \atop t}\right\}$, we can use the recurrence relation of Theorem 6.7 that relies on $\left\{{s-1 \atop t}\right\}$ and $\left\{{s-1 \atop t-1}\right\}$ (the numbers in the previous row of the table of Stirling numbers of the second kind), or the recurrence relation of Corollary 6.22 that uses the values of $\left\{{s \atop i}\right\}$ for $0 \leq i \leq t-1$ (the earlier Stirling numbers in the same row). In fact, using the latter, we can find not too complicated formulas for $\left\{{s \atop t}\right\}$ for small values of t (see Problems P 6.1.14 and P 6.1.16). It is also possible to write down actual formulas – albeit quite messy ones – for $\left\{{s \atop t}\right\}$. We will record the results here, but you are asked to provide the proofs in the problems.

Theorem 6.23. *Let s and t be positive integers; then*

(a) *(Problem P 6.1.29)*

$$\left\{{s \atop t}\right\} = \frac{1}{t!}\sum_{\substack{m_1+\cdots+m_t=s \\ m_i>0}}\binom{s}{m_1 \; m_2 \; \cdots \; m_t}$$

(b) *(Problem P 6.1.30 or P 6.1.31)*

$$\left\{{s \atop t}\right\} = \sum_{r=0}^{t}(-1)^{t-r}\binom{t}{r}\frac{r^s}{t!} = \sum_{r=0}^{t}\frac{(-1)^{t-r}r^s}{r!\,(t-r)!}.$$

Proof. You are asked to prove the first identity in Problem P 6.1.29, and the second identity in Problem P 6.1.30 or *P 6.1.31*. We will give yet another proof of the second identity in Proposition 8.14, using the inclusion–exclusion principle. □

  Problem *P 6.1.33* asks you to reduce the task of finding Stirling numbers of the second kind to matrix multiplication. Another approach to Stirling numbers of the second kind – an approach based on generating functions – will be explored in Section 9.4.3 and the ensuing problems. As an example, in Problem P 9.4.29, we prove that the sequence of numbers in any row of the table of Stirling numbers of the second kind is unimodal.

Problems

P 6.1.1. I have six distinguishable students and three distinct tasks.
  (a) In how many ways can I assign each of the six students to one of the three tasks?
  (b) In how many ways can I assign each of the six students to one of the three tasks so that exactly two of the tasks are assigned?

P 6.1.2. I have four identical servings of potatoes. I also have one serving of each of bee balm, basil, bay leaves, borage, and burnet (i.e., five kinds of herbs). For each serving of potatoes I can use any number (zero to five) of the herbs, and I do want to use all my herbs. In how many different ways could I do this?

  Note: For example, one way would be to put all the five herbs in one of the servings and serve the other three bland. Another would be to use bee balm and basil on one of the dishes, bay leaves on the second, borage and burnet on the third, and leave the fourth one bland.

P 6.1.3. Jorge, Nahid, Mary, Crystal, Akira, Vivian, Ellen, and Rafael are waiting to be organized into teams. The only condition is that the number of teams be no more than six and be an even number. In how many ways can this be done?

P 6.1.4. I have one apple, one orange, one banana, one pear, and one melon. I also have three friends: Noosha, Shooka, and Borna. I distribute the fruit among my friends randomly.
  (a) What is the probability (to three decimal places) that each of my friends gets at least one piece of fruit?
  (b) What is the probability (to three decimal places) that exactly two of my friends get some piece of fruit (and one goes empty-handed)?

P 6.1.5. Find the number of ways to assign six different jobs to four (distinct) workers so that each job gets a worker and each worker gets at least one job.

P 6.1.6. Hercules, Ophiuchus, and Daedalus are students in a class. The following works of Aristotle are to be distributed among them in such a way that each gets at least one: *Organon*, *Physica*, *Metaphysica*, *Politica*, and *Rhetorica*. In how many ways can this be done?

  Note: There is exactly one copy of each work and they all have to be distributed.

P 6.1.7. You are a school teacher in the year 4747. Your students are Theodoric, Athalaric, Theodahad, and Witiges. However, each of these comes to class with 7 of their clones, and so the total number of students in your class is 32. (This makes the classroom experience more bearable since at any given time only one of the 8 clones has to be paying attention, while the other 7 can check notifications.) Every day, you create an ordered list of 8 names of students, who then take turns to take a yoga break during the class. Your only restriction is that each list of 8 should have at least one of each of the 4 different students on it. How many such lists are possible? (Each student is identical to their clones, and you can't tell the clones apart. Thus your lists will include a number of repeats.)

P 6.1.8. Consider the letters $\{A, H, M, T\}$. Make seven-letter words from just these letters, making sure that each of the letters is used at least once.

(a) How many such words are there?

(b) If you choose one of these words at random, what is the probability that the word MATH appears consecutively somewhere in the word?[8]

P 6.1.9. Let $S = \{5 \cdot a, 7 \cdot b, 11 \cdot c, 13 \cdot d\}$ be a multiset of four elements with repetition numbers 5, 7, 11, and 13. How many ordered lists of size eight of these elements are there that contain exactly two a's and at least one each of b, c, and d?

P 6.1.10. Akira, Crystal, Ellen, Jorge, Nahid, Rafael, Oscar, Kanishka, and Vivian enter an elevator. At each of the first four stops, at least one person leaves the elevator. After the fourth stop, there are two people left in the elevator and they leave at the fifth stop. In how many ways could this have happened?

P 6.1.11. How many onto functions are there from $\{1, 2, 3, 4, 5, 6\}$ onto $\{1, 2, 3, 4\}$? Can you generalize?

P 6.1.12. Tascha, Kiavash, Neema, Noosha, Borna, and Shooka are to be organized into two, three, or four (non-empty) teams. In how many ways can this be done? (Other than being non-empty, there is no other restriction on the sizes of the teams.) If we choose one of these possibilities at random, what is the probability that Kiavash and Neema end up on the same team?

P 6.1.13. Give a combinatorial argument to find and prove a formula of the form

$$\left\{{n \atop n-2}\right\} = \binom{n}{?} + ? \binom{n}{?}.$$

P 6.1.14. Let n be a positive integer. Use Corollary 6.22 to find and prove a formula of the form

$$\left\{{n \atop 3}\right\} = \frac{1}{?}(?^{n-1} - ?^n + 1).$$

P 6.1.15. Re-do Problem P 6.1.14, but this time do not use Corollary 6.22. Give either a combinatorial proof – by mimicking the argument for Corollary 6.22 – or an inductive proof.

P 6.1.16. Let n be a positive integer. Find a formula for $\left\{{n \atop 4}\right\}$.

P 6.1.17. Let n be a positive integer. Prove that

$$\left\{{n \atop 1}\right\} + 4\left\{{n \atop 2}\right\} + 12\left\{{n \atop 3}\right\} + 24\left\{{n \atop 4}\right\} + 24\left\{{n \atop 5}\right\} = 5^{n-1}.$$

P 6.1.18. In Example 6.8, we claimed that there are 36 ways to distribute 4 distinct balls into 3 distinct boxes with the condition that none of the boxes ends up empty. A student argues as follows: we have three choices for each ball and that gives a total of 3^4 possibilities, but some of these possibilities involve empty boxes. There are three ways of putting all the balls in one box (just choose one of the three boxes), and there are 2^4 ways of putting the balls into two boxes (two choices for each ball). So the total number of ways of distributing 4 distinct balls into 3 distinct boxes should be $3^4 - 3 - 2^4 = 62$. What went wrong? Can you fix this method?

---

[8] Problem P 8.1.18 gives a variation of this problem.

P 6.1.19. Let s and t be positive integers. In Theorem 6.19, we showed that

$$t^s = \sum_{i=0}^{t} (t)_i \left\{{s \atop i}\right\}.$$

Show that, if we wish, we can change the range of the index i and write

$$t^s = \sum_{i=0}^{s} (t)_i \left\{{s \atop i}\right\} = \sum_{i=0}^{s} \left\{{s \atop i}\right\} i! \binom{t}{i}.$$

P 6.1.20. Let k be a positive integer and m a non-negative integer. In Problem P 5.4.16, we wrote

$$k^m = a_{0,m}\binom{k}{0} + a_{1,m}\binom{k}{1} + \cdots + a_{r,m}\binom{k}{r} + \cdots + a_{m,m}\binom{k}{m},$$

where, for $0 \le r \le m$, $a_{r,m}$ was an as-yet-to-be-determined constant. In Problem *P 5.4.17*, you were asked to show that $a_{r,m} = \sum_{k=0}^{r}(-1)^{r+k}\binom{r}{k}k^m$. Use Theorem 6.19 (or Problem *P 6.1.19*) to find a different expression for $a_{r,m}$ in terms of the Stirling numbers of the second kind.

P 6.1.21. Using the recurrence relation (Theorem 6.7) for the Stirling numbers of the second kind as a model, for n and k non-negative integers, we define an integer $P(n,k)$ as follows: for all $n \ge 0$, $P(n,0) = 1$, and, for all $k > 0$, $P(0,k) = 0$. In addition, for n and k positive integers, we define

$$P(n,k) = P(n-1,k) + kP(n-1,k-1).$$

Make a table of values for $P(n,k)$ for small values of n and k. Have you seen these numbers before? Prove your assertion.

P 6.1.22. Let n be a positive integer, and assume that we know the Bell numbers $B_0, B_1, \ldots, B_n$. Let i be an integer with $0 \le i \le n-1$. You have n objects and one of them is your favorite teddy bear. In how many ways can you partition the n objects into non-empty parts such that the teddy bear is in a part with exactly i other objects?

P 6.1.23. Let n be a positive integer, and let $B(n)$ denote the nth Bell number. Find and prove a formula of the form[9]

$$B(n) = ?B(0) + ?B(1) + \cdots + ?B(n-1).$$

P 6.1.24. Let n be a non-negative integer. In analogy with the Bell numbers, define $E_n = \sum_{k=0}^{n} k! \left\{{n \atop k}\right\}$. Prove[10] that, for $n > 0$,

$$E_n = \binom{n}{0}E_0 + \binom{n}{1}E_1 + \cdots + \binom{n}{n-1}E_{n-1}.$$

---

[9] This formula was among the discoveries of the Japanese mathematician Yoshisuke Matsunaga (1690–1744) in the eighteenth century.

[10] Adapted from Dasef and Kautz 1997.

P 6.1.25. Define $f_0(x) = x^5$, and, for $n > 0$, define $f_n(x) = xf'_{n-1}(x)$.
  (a) Find a formula for $f_n(x)$.
  (b) In analogy with Proposition 6.14, verify that

$$f_n(x) = \begin{Bmatrix} n \\ 0 \end{Bmatrix} f_0(x) + \begin{Bmatrix} n \\ 1 \end{Bmatrix} xf'_0(x) + \cdots + \begin{Bmatrix} n \\ 5 \end{Bmatrix} x^5 f_0^{(5)}(x),$$

  where $f'_0, \ldots, f_0^{(5)}$ are the successive derivatives of f_0.

P 6.1.26. Let $f(x)$ be an infinitely differentiable function. Define $f_0(x) = f(x)$, and, for $n > 0$, define $f_n(x) = x\frac{d}{dx}(f_{n-1}(x))$. Prove that

$$f_n(x) = \begin{Bmatrix} n \\ 0 \end{Bmatrix} f(x) + \cdots + \begin{Bmatrix} n \\ k \end{Bmatrix} x^k \frac{d^k}{dx^k}(f(x)) + \cdots + \begin{Bmatrix} n \\ n \end{Bmatrix} x^n \frac{d^n}{dx^n}(f(x)).$$

P 6.1.27. **Alternative Proof of Theorem 6.19.** Use the result of Problem P 6.1.26 to give a different proof of Theorem 6.19.

P 6.1.28. Accept the identities of Theorem 6.23, and use each of them to calculate $\begin{Bmatrix} 6 \\ 3 \end{Bmatrix}$.

P 6.1.29. **Proof of Theorem 6.23a.** Let s and t be positive integers. Using Theorem 6.9 and Theorem 4.34, prove that

$$\begin{Bmatrix} s \\ t \end{Bmatrix} = \frac{1}{t!} \sum_{\substack{m_1+\cdots+m_t=s \\ m_i>0}} \binom{s}{m_1\ m_2\ \ldots\ m_t}.$$

P 6.1.30. **Proof of Theorem 6.23b.** Let s and t be positive integers. Use Problems *P 5.4.16*, *P 5.4.17* and *P 6.1.19* to prove that

$$\begin{Bmatrix} s \\ t \end{Bmatrix} = \sum_{r=0}^{t} \frac{(-1)^{t-r} r^s}{r!\ (t-r)!}.$$

P 6.1.31. **Another Proof of Theorem 6.23b.** Start with Equation 6.1 and multiply it by the series expansion for e^{-x}. What is the coefficient of x^k in the product? Compare with Proposition 6.14 to get a different proof of Theorem 6.23b.

P 6.1.32. **Alternative Proof of Proposition 6.15.** Let n be a non-negative integer. Use Theorem 6.23b to give a different proof of

$$B(n) = \frac{1}{e} \sum_{r=0}^{\infty} \frac{r^n}{r!}.$$

(In the case of $n = 0$ and $r = 0$, we make the stipulation that $0^0 = 1$ in this formula.) You may find the following steps helpful:

STEP 1: If you switch the order of summation in $\sum_{k=0}^{M} \sum_{r=0}^{k} f(k,r)$, what will the range values for the new sums be? In other words,

$$\sum_{k=0}^{M} \sum_{r=0}^{k} f(k,r) = \sum_{r=?}^{?} \sum_{k=?}^{?} f(k,r).$$

STEP 2: Show $\sum_{k=r}^{M} \frac{(-1)^{k-r}}{(k-r)!} = \sum_{s=0}^{M-r} \frac{(-1)^s}{s!}$ and find the limit of the latter as $M \to \infty$.

STEP 3: Let M be any integer greater than or equal to n, write

$$B(n) = \sum_{k=0}^{M} \left\{ {n \atop k} \right\},$$

replace $\left\{ {n \atop k} \right\}$ with the expression from Theorem 6.23b, switch the order of summation, simplify, and take the limit as $M \to \infty$.

STEP 4: Rewrite the proof completely and concisely.

P 6.1.33. I am sitting at a computer that has some software for doing matrix algebra (e.g., it can do matrix addition and multiplication and find inverses of matrices). The positive integer n is fixed (and not too large). I want to input a few straightforward matrices, manipulate them using the software, and find a table of Stirling numbers of the second kind that contains the numbers $\left\{ {s \atop t} \right\}$ for $s, t = 0, \ldots, n$. What should I do?

6.2   Stirling Numbers of the First Kind

Warm-Up 6.24. *Annisa, Baskoro, Fadhlan, Intan, Nurul, and Putri go to a restaurant together. There are no tables for six and so they are offered two identical circular tables. In how many ways can they be seated if no one is going to sit alone? We are not concerned about who gets what seat but rather the relative seating position of each friend. So it matters who is sitting at your table and who is sitting to the right of whom.*

We now introduce a new set of numbers, the so-called Stirling numbers of the first kind. Here, we also partition $[n]$ into k non-empty parts, but we add a twist by ordering each part as a circular permutation. Alternatively, we can think of these new sets of numbers as counting certain classes of permutations of n objects. Recall that the number of permutations of $[n] = \{1, 2, \ldots, n\}$ is $n!$, and that the number of circular permutations of $[n]$ is $(n-1)!$. We think of a circular permutation as an arrangement of the elements around a circle where the actual place of an object is irrelevant but what matters is the relative place of each element relative to the others (Definition 4.13). In other words, we care about which element is to the right of which other element. Here, we are interested in the number of arrangements of elements of $[n]$ around k circles, as opposed to just one circle. The direct connection of these numbers with Stirling numbers of the second kind is *not* clear from the definition, but the relation is fascinating, and will be discussed in Section 6.3.

Definition 6.25 (Stirling Numbers of the First Kind). Let n and k be non-negative integers. Define $\left[{n \atop k} \right]$ to be the number of permutations of n objects into exactly k cycles, or, equivalently, the number of ways of seating n people at k (non-empty) identical circular tables. $\left[{n \atop k} \right]$ is called a *Stirling number of the first kind*.[11] We interpret $\begin{bmatrix} 0 \\ 0 \end{bmatrix}$ to be equal to 1, and we also define the *signed Stirling numbers of the first kind*, denoted by $s(n,k)$, by

$$s(n,k) = (-1)^{n+k}\begin{bmatrix} n \\ k \end{bmatrix}.$$

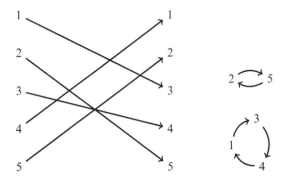

Figure 6.1 The permutation 35412 can be represented as a 1-1, onto function from [5] to [5], but also as $\begin{pmatrix} 1\,2\,3\,4\,5 \\ 3\,5\,4\,1\,2 \end{pmatrix}$ and (1 3 4)(2 5).

Remark 6.26 (Cycle Notation for Permutations). For us, in this text, 35412 is a permutation of $[5] = \{1,2,3,4,5\}$. It is an arrangement of elements of [5] into a particular order. In addition to 35412, there are other ways of representing this same permutation. One way is to write a two-line representation, such as $\begin{pmatrix} 1\,2\,3\,4\,5 \\ 3\,5\,4\,1\,2 \end{pmatrix}$. This notation indicates that there was an "original" order, namely 12345, and then shows the new positions. The two-line representation also reminds us that this permutation is really a 1-1, onto function from [5] to [5], and we could depict it as in the figure in the left of Figure 6.1. In this way of thinking, it is not that 3 is moved to the first position, but that 1 is mapped to 3. If you have studied group theory, then you are familiar with the *cycle notation*, yet another notation for permutations. We first think of a permutation as a 1-1, onto function from [5] to [5], as in Figure 6.1, and then we ask a series of questions, each based on the answer to the previous one. Where is 1 mapped to? Answer: 3. Where then is 3 mapped to? Answer: 4. We record the answers, so far, as "(1 3 4", 1 goes to 3 goes to 4. Where is 4 mapped to? Answer: back to 1. We think of this as a cycle (see the figure on the right of Figure 6.1), and denote it by closing the parenthesis: (1 3 4). We read this as "1 goes to 3 goes to 4 goes to 1." Now 2 goes to 5 and 5 goes back to 2, also. This is denoted by a cycle as well: (2 5). Hence, we can write the permutation 35412 as (1 3 4)(2 5). We think of this as a "product" of a 3-cycle (a cycle of size 3) and a 2-cycle. Likewise, (1 3 4 5)(2) is the product of a 4-cycle and a 1-cycle, and denotes the permutation 32451. Both of these two permutations were products of two disjoint

---

[11] Other notations used for Stirling numbers of the first kind include $c(n,k)$ and $|s(n,k)|$. Note that the notation $\begin{bmatrix} n \\ k \end{bmatrix}$ is also commonly used for the so-called Gaussian coefficients, an unrelated set of numbers.

cycles. On the other hand, the permutation (1 2 3 4 5) is made up of just one cycle. Writing permutations as a product of disjoint cycles ends up being very useful in group theory[12]. Now if k and n are positive integers, then the Stirling number of first kind, $\begin{bmatrix} n \\ k \end{bmatrix}$, counts the number of permutations of $[n]$ (elements of the symmetric group S_n, in the parlance of group theory) that are the product of exactly k disjoint cycles.

Example 6.27. What is $\begin{bmatrix} 3 \\ 2 \end{bmatrix}$? In other words, in how many ways can we seat three people around two (non-empty) tables? Since the two tables are to be non-empty, two people have to sit at one table and the third person has to sit alone. There are three ways to decide who is sitting alone (or $\binom{3}{2} = 3$ ways to choose which two are going to sit together), and there are no other choices to be made. Hence $\begin{bmatrix} 3 \\ 2 \end{bmatrix} = 3$. In fact, if we have n people to seat around $n - 1$ non-empty tables, then all we have to do is to decide which two are sitting together, and this can be done in $\binom{n}{2}$ ways. Hence, for all positive integers n, $\begin{bmatrix} n \\ n-1 \end{bmatrix} = \binom{n}{2}$ are the triangular numbers.

Example 6.28. If $k > n$ are integers, there is no way to seat n people around k non-empty tables, and so, for $k > n$, $\begin{bmatrix} n \\ k \end{bmatrix} = 0$. There is only one way to seat n people around n (non-empty) tables, and so $\begin{bmatrix} n \\ n \end{bmatrix} = 1$, for $n \geq 0$.

$\begin{bmatrix} n \\ 1 \end{bmatrix}$ is the number of ways of seating n people around a single table. This is the number of circular permutations of $[n]$ and is equal to $(n - 1)!$, for $n \geq 1$ (Theorem 4.16).

Remark 6.29. Already it is unclear how to give a nice closed formula for $\begin{bmatrix} n \\ 2 \end{bmatrix}$. We give a preliminary table of small Stirling numbers of the first kind in Table 6.5. As has been the case with many tables of this kind, we will be able to complete it – at least for small values of n and k – after we find a recurrence relation for $\begin{bmatrix} n \\ k \end{bmatrix}$.

Table 6.5 A preliminary table for small Stirling numbers of the first kind.

The small values of $\begin{bmatrix} n \\ k \end{bmatrix}$								
$n \backslash k$	0	1	2	3	4	5	6	7	8
0	1	0	0	0	0	0	0	0	0
1	0	1	0	0	0	0	0	0	0
2	0	1	1	0	0	0	0	0	0
3	0	2	3	1	0	0	0	0	0
4	0	6		6	1	0	0	0	0
5	0	24			10	1	0	0	0
6	0	120				15	1	0	0
7	0	720					21	1	0
8	0	5040						28	1

<hr>

[12] See any undergraduate abstract algebra book such as my *Algebra in Action* (Shahriari 2017).

Theorem 6.30. *Let n and k be positive integers. We then have*

$$\begin{bmatrix} n \\ k \end{bmatrix} = (n-1)\begin{bmatrix} n-1 \\ k \end{bmatrix} + \begin{bmatrix} n-1 \\ k-1 \end{bmatrix}.$$

Proof. Thought experiment: n people are invited to dinner and you are one of them. You arrive late and everyone else is already seated at identical circular tables. There are k tables and, after everyone is seated, none should be empty. Where do you sit?

You either join an already existing table, in which case all the k tables have someone at them; or you are seated all by yourself.

Hence,

{ways of seating n people at k (non-empty) identical circular tables} $= A \cup B$,

where

$\qquad A = \{$seating positions in which n seats alone$\}$

$\qquad B = \{$seating positions in which n picks a spot in one of the existing tables$\}$.

Now $|A| = \begin{bmatrix} n-1 \\ k-1 \end{bmatrix}$, since, if you ignore n, then you have a seating configuration for $n-1$ people around $k-1$ non-empty tables. Finally, $|B| = (n-1)\begin{bmatrix} n-1 \\ k \end{bmatrix}$, since you first seat everyone other than n around k tables (in $\begin{bmatrix} n-1 \\ k \end{bmatrix}$ ways), and then n decides to sit to the right of a particular person (and there are $n-1$ choices for that). $\square$

Table 6.6, using the recurrence relation, completes Table 6.5. As you may expect, there are many patterns in this table of numbers. Problems $P\ 6.2.7$, $P\ 6.2.8$, $P\ 6.2.10$, $P\ 6.2.11$, and $P\ 6.2.17$ give some examples. As another example, in Problem $P\ 6.2.27$ you are asked to prove that the sequence of numbers in any row of this table is unimodal.

Table 6.6 A table of values for small Stirling numbers of the first kind.

	The small values of $\begin{bmatrix} n \\ k \end{bmatrix}$								
$n \backslash k$	0	1	2	3	4	5	6	7	8
0	1	0	0	0	0	0	0	0	0
1	0	1	0	0	0	0	0	0	0
2	0	1	1	0	0	0	0	0	0
3	0	2	3	1	0	0	0	0	0
4	0	6	11	6	1	0	0	0	0
5	0	24	50	35	10	1	0	0	0
6	0	120	274	225	85	15	1	0	0
7	0	720	1,764	1,624	735	175	21	1	0
8	0	5,040	13,068	13,132	6,769	1,960	322	28	1

Stirling Numbers and Rising and Falling Factorials

For k a non-negative integer, recall that in Definition 4.4 we defined $(x)_k$, the kth falling factorial function, by

$$(x)_k = x(x-1)(x-2)\cdots(x-k+1), \quad \text{for } k > 0,$$

and $(x)_0 = 1$. So far, we have substituted for x a positive integer n, but what if we keep x an indeterminate and multiply $(x)_k$ out? Since there are k terms, we will get a polynomial of degree k, but what are the coefficients? Let's look at some examples:

$$(x)_0 = 1$$
$$(x)_1 = x$$
$$(x)_2 = -x + x^2$$
$$(x)_3 = 2x - 3x^2 + x^3$$
$$(x)_4 = -6x + 11x^2 - 6x^3 + x^4$$
$$(x)_5 = 24x - 50x^2 + 35x^3 - 10x^4 + x^5.$$

Comparing these polynomials to the entries in each row of the table of Stirling numbers of the first kind, Table 6.6, the similarity is striking, and we suspect that the correspondence is not a coincidence. The coefficients seem to be the signed Stirling numbers of the first kind (which we denoted by $s(n, k)$ – Definition 6.25). But why? To prove this, it is actually helpful to also consider *rising factorials*. Let k be a non-negative integer and x an indeterminate. We define $x^{(0)} = 1$ and, for $k \geq 1$,

$$x^{(k)} = x(x+1)(x+2)\cdots(x+k-1).^{13}$$

The rising factorial – just like the falling factorial – is the product of k terms, and hence if we multiply it out we get a polynomial of degree k as well. We look at some examples:

$$x^{(0)} = 1$$
$$x^{(1)} = x$$
$$x^{(2)} = x + x^2$$
$$x^{(3)} = 2x + 3x^2 + x^3$$
$$x^{(4)} = 6x + 11x^2 + 6x^3 + x^4$$
$$x^{(5)} = 24x + 50x^2 + 35x^3 + 10x^4 + x^5.$$

This time – having seen the examples of falling factorials, this is less surprising – the coefficients are the (unsigned) Stirling numbers of the first kind. We now prove both of these patterns. In fact, the pattern for the falling factorials will follow from the one for rising factorials, and we will prove the latter in an interesting – you could say roundabout – way.

---

[13] This notation could be confusing, and we will only use it sporadically. An alternative (and popular) notation for rising and falling factorials is $x^{\overline{k}}$ and $x^{\underline{k}}$, respectively.

We will prove that the coefficients in the expansion of rising factorials have the same initial values *and* they satisfy the same recurrence relation as the Stirling numbers of the first kind. Hence, they must be the same.

Theorem 6.31. *Let n be a non-negative integer. Then*

(a) $x^{(n)} = \begin{bmatrix} n \\ 0 \end{bmatrix} + \begin{bmatrix} n \\ 1 \end{bmatrix}x + \begin{bmatrix} n \\ 2 \end{bmatrix}x^2 + \cdots + \begin{bmatrix} n \\ n \end{bmatrix}x^n$, *and*

(b) $(x)_n = s(n,0) + s(n,1)x + \cdots + s(n,n)x^n$.

Proof. (a) By definition, $x^{(0)} = 1$, and, for $n > 0$, $x^{(n)} = x(x+1)\cdots(x+n-1)$, and so, for all $n \geq 0$, $x^{(n)}$ is a polynomial of degree n. Let $b(n,k)$ denote the coefficient of x^k in the expansion of $x^{(n)}$. Hence, by definition,

$$x^{(n)} = b(n,0) + b(n,1)x + \cdots + b(n,n)x^n.$$

We need to show that, for all non-negative values of n and k, $b(n,k) = \begin{bmatrix} n \\ k \end{bmatrix}$. First note that if $n = 0$, then $x^{(0)} = 1$, and so $b(0,k) = \begin{cases} 1 & \text{if } k = 0 \\ 0 & \text{otherwise} \end{cases} = \begin{bmatrix} 0 \\ k \end{bmatrix}$. Likewise, for $n > 0$, $b(n,0) = 0 = \begin{bmatrix} n \\ 0 \end{bmatrix}$. Hence, $b(n,k) = \begin{bmatrix} n \\ k \end{bmatrix}$ if either n or k is equal to zero. For n and k both positive integers, we have

$$x^{(n)} = (x+n-1)x^{(n-1)}$$
$$= (x+n-1)\left[b(n-1,0) + b(n-1,1)x + \cdots + b(n-1,n-1)x^{n-1}\right]$$
$$= b(n-1,0)x + \cdots + b(n-1,n-2)x^{n-1} + b(n-1,n-1)x^n$$
$$+ (n-1)\left[b(n-1,0) + b(n-1,1)x + \cdots + b(n-1,n-1)x^{n-1}\right].$$

If we combine like terms in the above expression for $x^{(n)}$, we see that the coefficient of x^k is $b(n-1,k-1)+(n-1)b(n-1,k)$. On the other hand, originally, the coefficient of x^k in $x^{(n)}$ was defined to be $b(n,k)$. So we conclude that $b(n,k) = b(n-1,k-1)+(n-1)b(n-1,k)$. This is the same recurrence relation as the one satisfied by $\begin{bmatrix} n \\ k \end{bmatrix}$. Thus $b(n,k)$ and $\begin{bmatrix} n \\ k \end{bmatrix}$ satisfy the same initial conditions (when $n = 0$ or $k = 0$) and the same recurrence relation. Hence, they must be equal, and the identity is proved.

(b) The identity is valid for $n = 0$ by inspection. For $n > 0$, substitute $-x$ for x in the identity from part (a). Get

$$\begin{bmatrix} n \\ 0 \end{bmatrix} - \begin{bmatrix} n \\ 1 \end{bmatrix}x + \begin{bmatrix} n \\ 2 \end{bmatrix}x^2 + \cdots + (-1)^n\begin{bmatrix} n \\ n \end{bmatrix}x^n = -x(-x+1)\cdots(-x+n-1).$$

Factoring the -1's from the right-hand side gives

$$\sum_{k=0}^{n}(-1)^k\begin{bmatrix} n \\ k \end{bmatrix}x^k = (-1)^n x(x-1)\cdots(x-n+1)$$

$$\Rightarrow \quad (x)_n = \sum_{k=0}^{n}(-1)^{k+n}\begin{bmatrix} n \\ k \end{bmatrix}x^k,$$

and the proof is complete, since $(-1)^{k+n} \begin{bmatrix} n \\ k \end{bmatrix} = s(n, k)$ are exactly the signed Stirling numbers of the first kind. □

Example 6.32. Consider the polynomial $p(x) = x(x + 1)(x + 2)(x + 3)$. You can think of this polynomial as either a rising factorial or a falling factorial. In fact, as a rising factorial, $p(x) = x^{(4)}$, and, as a falling factorial, $p(x) = (x + 3)_4$. Thus, using Table 6.6, the table of Stirling numbers of the first kind, Theorem 6.38 gives two expressions for $p(x)$:

$$p(x) = x^{(4)} = 6x + 11x^2 + 6x^3 + x^4, \text{ and}$$

$$p(x) = (x + 3)_4 = -6(x + 3) + 11(x + 3)^2 - 6(x + 3)^3 + (x + 4)^4.$$

Looking at these two expressions, it is not obvious that they are the same polynomial. However, Theorem 6.38 assures us that they are, and the skeptical (and patient) reader can check it out for herself.

Stirling Numbers and the Harmonic Series

Consider the partial sums of the harmonic series:

$$1 \qquad\qquad = \tfrac{1}{1!}$$

$$1 + \tfrac{1}{2} \qquad\qquad = \tfrac{3}{2!}$$

$$1 + \tfrac{1}{2} + \tfrac{1}{3} \qquad = \tfrac{11}{3!}$$

$$1 + \tfrac{1}{2} + \tfrac{1}{3} + \tfrac{1}{4} = \tfrac{50}{4!}.$$

What is the sequence $1, 3, 11, 50, \ldots$ of the numerators of these partial sums? Surprisingly, this sequence of numbers is exactly one of the columns of Table 6.6 of Stirling numbers of the first kind!

Theorem 6.33. *Let n be a positive integer. Then*

$$1 + \frac{1}{2} + \frac{1}{3} + \cdots + \frac{1}{n} = \frac{\begin{bmatrix} n+1 \\ 2 \end{bmatrix}}{n!}.$$

Proof. It is straightforward to give an inductive proof using the recurrence relation of Theorem 6.30. Here, we give a combinatorial argument.[14] By definition, $\begin{bmatrix} n+1 \\ 2 \end{bmatrix}$ is the number of ways of seating $n + 1$ people at 2 non-empty identical circular tables. Let Fred be one of the $n + 1$ people. How many ways can you seat $n + 1$ people at 2 non-empty tables, if there are exactly t people at the table that Fred is *not* sitting at? First, choose the t people (from among everyone except Fred), seat them at a round table, and then seat the remaining $n - t + 1$ people

---

[14] Adapted from Benjamin, Preston, and Quinn 2002.

(including Fred) at a different round table. Since there are $(m-1)!$ circular permutations of m objects, the total number of such choices is

$$\binom{n}{t}(t-1)!\,(n-t)! = \frac{n!}{t}.$$

Now t can be any integer between 1 and n, and so the total number of ways of seating $n+1$ people at two identical circular tables is

$$\frac{n!}{1} + \frac{n!}{2} + \cdots + \frac{n!}{n} = n!\left[1 + \frac{1}{2} + \cdots + \frac{1}{n}\right].$$

Putting this number equal to $\left[{n+1 \atop 2}\right]$ gives the result. $\qquad\square$

Problems

P 6.2.1. Let n be a positive integer, and x an indeterminate. Then what is the identity of the form

$$x^{(n)} = (?)_n,$$

relating rising and falling factorials?

P 6.2.2. How many permutations of $[n]$ have one cycle of size 3, two cycles of size 2, and $n-7$ cycles of size 1?

P 6.2.3. How many permutations of $[n]$ have exactly $n-5$ cycles of size 1?

P 6.2.4. You list a sequence of seven distinct emojis in order. You then give your favorite bot the task of making a (possibly meaningless) "sentence" from a combination of (possibly meaningless) words and some of your emojis, by the following method. The bot is to make k words, where $1 \le k \le 7$, using exactly once each of the 47 primary characters in the Devanagari script. The bot then replaces k of the emojis with the words it has made. The resulting sentence (made of k strings of letters and $7-k$ emojis, still in the original order except with some or maybe all emojis having being replaced) is sent to every one of your contacts. You request that the bot never send the same sentence twice. How many sentences are possible? Can you write your answer as a rising factorial? If your bot sends a message every second, about how long before it stops?

P 6.2.5. You want to put s distinct balls into t distinct boxes, where the boxes have no capacity restrictions (so, for example, some boxes could go empty). However, in each box, you also want to (linearly) order the balls. In how many ways can you do that? Can you express the answer as a rising factorial?

P 6.2.6. You have invited 13 guests (other than yourself) to a dinner party and have 5 circular tables. One of the tables is special but the other four are identical. Two of the guests are the guests of honor, and you want to sit at the special table, flanked at either side by these two. You also want three other people at the special table, but you don't really care who those people might be. Everyone else is to be seated at the other tables, with the condition that no table goes empty. In how many ways can the guests be seated? (As in all circular permutations, for any of the tables, we are only interested in the relative position of each guest. The actual seats are identical.)

P 6.2.7. Find and prove an identity of the form

$$\begin{bmatrix} n \\ n-2 \end{bmatrix} = ? \binom{n}{?} + ? \binom{n}{?}.$$

P 6.2.8. Find and prove an identity of the form

$$\begin{bmatrix} n \\ n-3 \end{bmatrix} = \binom{n}{4} \binom{n}{?}.$$

P 6.2.9. In Theorem 6.31, we showed that $\begin{bmatrix} n \\ n-2 \end{bmatrix}$ is the coefficient of x^{n-2} in the expansion of the rising factorial $x^{(n)}$. In Problem $P\,6.2.7$, you found a formula for $\begin{bmatrix} n \\ n-2 \end{bmatrix}$. Can you combine these two facts to get some kind of an identity?

P 6.2.10. In Table 6.6 of Stirling numbers of the first kind, what is the sum of each row? In other words, for $n \geq 0$, find $\sum_{k=0}^{n} \begin{bmatrix} n \\ k \end{bmatrix}$.

P 6.2.11. Let n be a non-negative integer. What can you say about

(a)

$$\begin{bmatrix} n \\ 0 \end{bmatrix} + 2\begin{bmatrix} n \\ 1 \end{bmatrix} + 2^2\begin{bmatrix} n \\ 2 \end{bmatrix} + \cdots + 2^n\begin{bmatrix} n \\ n \end{bmatrix}?$$

(b)

$$\begin{bmatrix} n \\ 0 \end{bmatrix} + 47\begin{bmatrix} n \\ 1 \end{bmatrix} + 47^2\begin{bmatrix} n \\ 2 \end{bmatrix} + \cdots + 47^n\begin{bmatrix} n \\ n \end{bmatrix}?$$

P 6.2.12. If we make a table of *signed* Stirling numbers of the first kind, $s(n, k)$, what is the sum of each row? In other words, let $n \geq 0$ be an integer, and find $\sum_{k=0}^{n} s(n, k)$.

P 6.2.13. Let $n \geq 0$ be an integer, and let $s(n, k)$, as usual, denote a signed Stirling number of the first kind. What can you say about

$$s(n, 0) + 47s(n, 1) + 47^2 s(n, 2) + \cdots + 47^n s(n, n)?$$

Prove your assertions.

P 6.2.14. Give a proof of Theorem 6.33 by induction.

P 6.2.15. Let $n \geq 0$ and $k > 0$ be integers. Extend the combinatorial argument in the proof of Theorem 6.33 to k tables and prove

$$\frac{1}{n!}\begin{bmatrix} n+1 \\ k \end{bmatrix} = \frac{1}{0!}\begin{bmatrix} 0 \\ k-1 \end{bmatrix} + \frac{1}{1!}\begin{bmatrix} 1 \\ k-1 \end{bmatrix} + \frac{1}{2!}\begin{bmatrix} 2 \\ k-1 \end{bmatrix} + \frac{1}{3!}\begin{bmatrix} 3 \\ k-1 \end{bmatrix} + \cdots + \frac{1}{n!}\begin{bmatrix} n \\ k-1 \end{bmatrix}.$$

P 6.2.16. Give a different derivation of the identity of Problem P 6.2.15 by unwinding the recurrence relation of Theorem 6.30.[15]

P 6.2.17. Let n be a positive integer. Prove

$$\begin{bmatrix} n \\ 1 \end{bmatrix} + 2\begin{bmatrix} n \\ 2 \end{bmatrix} + \cdots + n\begin{bmatrix} n \\ n \end{bmatrix} = \begin{bmatrix} n+1 \\ 2 \end{bmatrix}.$$

[15] Yet a third proof using generating functions will be given in Probelm $P\,9.4.27$.

You may find the following steps helpful:

STEP 1: Let $f(x) = x^{(n)}$, the rising factorial function. Find $f'(x)$ using the product rule.

STEP 2: Find $\frac{f'(1)}{n!}$ and $f'(1)$.

STEP 3: Take a derivative of both sides of the equation in Theorem 6.31a to get a different expression for $f'(1)$.

STEP 4: Use Theorem 6.33 to finish the proof.

P 6.2.18. Re-prove the result of Problem *P 6.2.17* by induction.

P 6.2.19. What can you say about

$$\lim_{n\to\infty} \frac{\left[{n \atop 2}\right]}{\left[{n \atop 1}\right]}?$$

P 6.2.20. **Asymptotic Behavior of** $\left[{n \atop 2}\right]$**.** Let n be a positive integer.
   (a) Prove that

$$\ln(n) \leq 1 + \frac{1}{2} + \cdots + \frac{1}{n} \leq 1 + \ln(n).$$

   (b) Prove

$$(n-1)! \, \ln(n-1) \leq \left[{n \atop 2}\right] \leq (n-1)! \, [1 + \ln(n-1)].$$

   (c) For the case $n = 8$, compare the lower and upper bounds (from the previous part) to the actual value of $\left[{8 \atop 2}\right]$.
   (d) Show that

$$\lim_{n\to\infty} \frac{\left[{n \atop 2}\right]}{(n-1)! \, \ln(n-1)} = 1.$$

P 6.2.21. Remark 6.26 explained that each permutation of $[n]$ is the "product" of disjoint cycles. What is the *average* number of cycles in a permutation of $[n]$? Is this number related to a partial sum of the harmonic series?[16]

P 6.2.22. How many ways are there to seat the elements of $[10]$ around three identical tables in such a way that 1, 4, and 7 are the smallest numbers at their table? (So 1, 4, and 7 are sitting at different tables, and, for example, at the table with 4, all other guests are numbers 5 or more.)

P 6.2.23. Let $n > 1$ be an integer. Find the number of permutations of $[n]$ that have two cycles and 1 and 2 are in different cycles.

P 6.2.24. Let $n > 1$ be an integer, and let k be an integer with $0 \leq k \leq n - 3$. Find the number of permutations of $[n]$ into two cycles such that 1 and 2 are in the same cycle and there are exactly k elements to the left of 1 and to the right of 2. (The k elements are between 1 and 2.) Your answer can be in terms of Stirling numbers of the first kind.

---

[16] Problems *P 6.2.21*–P 6.2.25 are adapted from Benjamin, Preston, and Quinn 2002.

P 6.2.25. Give a combinatorial proof that

$$\begin{bmatrix} n+1 \\ 2 \end{bmatrix} = n! + (n-1)! \begin{bmatrix} 2 \\ 2 \end{bmatrix} + \frac{(n-1)!}{2!} \begin{bmatrix} 3 \\ 2 \end{bmatrix}$$
$$+ \cdots + \frac{(n-1)!}{(n-2)!} \begin{bmatrix} n-1 \\ 2 \end{bmatrix} + \frac{(n-1)!}{(n-1)!} \begin{bmatrix} n \\ 2 \end{bmatrix}.$$

Use this to prove that, if $H_n = 1 + \frac{1}{2} + \cdots + \frac{1}{n}$, then

$$\sum_{i=1}^{n} H_i = nH_n - n.$$

P 6.2.26. (Re)prove Proposition 6.14, using a method similar to the way we proved Theorem 6.31a.

P 6.2.27. Use Theorem 5.32 to prove that the sequence of numbers in a row of the table for Stirling numbers of the first kind is log concave (Definition 5.31) and hence unimodal (Definition 5.6).

P 6.2.28. Let n and k be positive integers. Using the recurrence relation (Theorem 6.30) for the Stirling numbers of the first kind as a model, define an integer $Y(n,k)$ as follows: for all $n \geq 1$, $Y(n,1) = 1$, and, for all $k > 1$, $Y(1,k) = 0$. In addition, for all integers n and k larger than 1, we define

$$Y(n,k) = Y(n-1,k) + (n-1)Y(n-1,k-1).$$

Make a table of values for $Y(n,k)$ for small values of n and k. Have you seen these numbers before? Prove your assertion.[17]

P 6.2.29. Let n be a positive integer, and let $d(n)$ denote the number of integer divisors of n. In other words, $d(n)$ counts the number of integers m with $1 \leq m \leq n$ and m a factor of n. So, for example, $d(6) = 4$ since the divisors of 6 are 1, 2, 3, and 6. Let

$$A = \frac{1}{n} \sum_{k=1}^{n} d(k)$$

be the average of $d(1), d(2), \ldots, d(n)$. Estimate A by finding a number a (the expression for a will be in terms of the Stirling numbers of the first kind) such that $a \leq A \leq a + 1$. You may find the following steps helpful:

STEP 1: Let $S = \{(i,j) \mid i \text{ divides } j, 1 \leq i \leq j \leq n\}$. To familiarize yourself with S, write down its elements for $n = 4$.

STEP 2: Show that $|S| = \sum_{k=1}^{n} d(k)$.

STEP 3: Fix k with $1 \leq k \leq n$, and let $S_k = \{(k,j) \mid k \text{ divides } j, 1 \leq j \leq n\}$. Show that $S_k = \left\lfloor \frac{n}{k} \right\rfloor$.

[17] From Merris 1996, Problem 14, page 111.

STEP 4: What is the relationship between $\mathcal{S}$ and $\{\mathcal{S}_k \mid 1 \le k \le n\}$?

STEP 5: Show that $n \sum_{k=1}^{n} \frac{1}{k} - n \le |S| \le n \sum_{k=1}^{n} \frac{1}{k}$.

STEP 6: Use Theorem 6.33 to complete your solution.

P 6.2.30. **Classifying Permutations According to the Number of Increasing Segments.** Consider a listing of elements of $[7] = \{1, 2, \ldots, 7\}$ where order matters. This is the same as a permutation of $[7]$ in one line format. For each such permutation, we can count the number of increasing segments. For example, the permutation 2351647 of $[7]$ consists of three increasing segments, namely 235, 16, and 47. On the other hand, the permutation 7123456 has two increasing segments (7 and 123456). Let n be a positive integer, and let k be an integer with $1 \le k \le n$. Denote by $f(n, k)$ the *number* of permutations of $[n]$ with k increasing segments. The purpose of this problem is to explore the function $f(n, k)$ and to find a recurrence relation for it.

(a) What are $f(3, 1), f(3, 2)$, and $f(3, 3)$?

(b) Fix a positive integer n. Then what can you say about $f(n, 1), f(n, n)$, and $\sum_{k=1}^{n} f(n, k)$?

(c) Consider the permutation 2351647 of $[7]$. I want to insert the symbol 8 somewhere in this list to create a permutation of $[8]$. In how many places can I insert 8 so that the resulting permutation continues to have three increasing segments? In how many ways can I insert 8 if I want the resulting permutation to have four increasing segments? Is there a way to insert 8 so that the resulting permutation will have five increasing segments?

(d) Find a recurrence relation for $f(n, k)$. It may be helpful to think of every permutation of $[n]$ with k increasing segments as being constructed from some permutation of $[n - 1]$ with either k or $k - 1$ increasing segments. Given a permutation of $[n - 1]$ with k increasing segments, in how many ways can you get a permutation of $[n]$ also with k increasing segments? Given a permutation of $[n - 1]$ with $k - 1$ increasing segments, in how many ways can you get a permutation of $[n]$ with k increasing segments?

(e) Make a table, akin to the Karaji–Jia triangle, that shows the values of $f(n, k)$ up to $n = 5$. What is $f(5, 3)$?

(f) Is the sequence $f(n, 1), f(n, 2), \ldots, f(n, n)$ unimodal? Is it symmetric? Make a conjecture.

P 6.2.31. $\binom{p}{k}$ **modulo p.** Let p be a prime number. Prove that, for $2 \le k \le p - 1$, $\binom{p}{k}$ is divisible by p.[18] You may find the following ideas and approach helpful:

THE SETUP: Let $\mathbb{F}_p$ denote the field of integers modulo p. This means that $\mathbb{F}_p$ has p elements: $0, 1, \ldots, p-1$, and we add and multiply these elements modulo p. In other words, the "sum" or "product" of two numbers is equal to the remainder of their usual sum or product when divided by p. As an example, for $p = 7$, in $\mathbb{F}_7$, $4 + 5 = 2$ and $4 \times 5 = 6$. The fact that $\mathbb{F}_p$ is a *field* basically means that, in this world, just like the world of real numbers, we can add, subtract, multiply, and divide by non-zero elements (and these operations follow

---

[18] This problem uses some ideas from abstract algebra. For background material, see Shahriari 2017, Section 1.3 and Chapter 19, for example.

reasonable rules). Let $q(x)$ be a polynomial with coefficients in $\mathbb{F}_p$; then if $\alpha \in \mathbb{F}_p$, so is $q(\alpha)$ (we perform all operations modulo p). A fact from field theory is that a polynomial of degree k with coefficients in a field (such as $\mathbb{F}_p$ or $\mathbb{R}$) has at most k roots, and if α is a root, then $(x - \alpha)$ is a factor of the polynomial.

THE RELEVANT POLYNOMIAL AND ITS ROOTS: Let $q(x) = x^p - x$. Using Fermat's little theorem (see Problem P 5.3.8), find p different roots for $q(x)$ in $\mathbb{F}_p$. Conclude that, in $\mathbb{F}_p$, $q(x) = (x)_p$, the falling factorial.

THE FINAL ARGUMENT: Compare the representation of $(x)_p$ in Theorem 6.31b with the one in the previous step, to conclude that p is a factor of $s(p,k)$, and hence $\left[{p \atop k}\right]$, for $2 \le k \le p-1$.

6.3 How Are the Stirling Numbers Related?

Warm-Up 6.34. *Assume that A, B, C, and D are all $n \times n$ matrices, and that we know $AB + C = D$. If A is an invertible matrix, find an expression for B in terms of A, C, and D.*

So far, the relation between the Stirling numbers of the first and the second kind is unclear. Why are they are both called Stirling numbers? For both, we are counting the number of certain kinds of partitions of $[n]$ into k non-empty parts. For Stirling numbers of the second kind, the parts are unordered subsets of $[n]$, while for Stirling numbers of the first kind, the parts are cyclically ordered. Even so, the relationship between the two sets of numbers is subtle. We can't just relate $\left\{{n \atop k}\right\}$ to $\left[{n \atop k}\right]$, but, given an appropriate *collection* of Stirling numbers of one kind, we can find a corresponding collection of Stirling numbers of the other kind. As an example, let S_5 denote the 6×6 partial matrix of Stirling numbers of the second kind (for $0 \le k,n \le 5$, the (n,k) entry is $\left\{{n \atop k}\right\}$ – compare with Table 6.3):

$$S_5 = \begin{bmatrix} 1 & 0 & 0 & 0 & 0 & 0 \\ 0 & 1 & 0 & 0 & 0 & 0 \\ 0 & 1 & 1 & 0 & 0 & 0 \\ 0 & 1 & 3 & 1 & 0 & 0 \\ 0 & 1 & 7 & 6 & 1 & 0 \\ 0 & 1 & 15 & 25 & 10 & 1 \end{bmatrix}.$$

If you just compare entries, the entries of S_5 do not seem related to Stirling numbers of the first kind. But if we find the *inverse* of this matrix, we get the matrix of signed Stirling numbers of the first kind! This is an amazing relationship, and we will give an interesting proof of it. So, for example, let s_5 denote the 6×6 partial matrix of signed Stirling numbers of the first kind (for $0 \le k,n \le 5$, the (n,k) entry is $s(n,k) = (-1)^{n+k}\left[{n \atop k}\right]$ – compare with Table 6.6):

$$s_5 = \begin{bmatrix} 1 & 0 & 0 & 0 & 0 & 0 \\ 0 & 1 & 0 & 0 & 0 & 0 \\ 0 & -1 & 1 & 0 & 0 & 0 \\ 0 & 2 & -3 & 1 & 0 & 0 \\ 0 & -6 & 11 & -6 & 1 & 0 \\ 0 & 24 & -50 & 35 & -10 & 1 \end{bmatrix}.$$

Multiply the two matrices, and you will get the 6×6 identity matrix! But why is that, and how are we going to prove it? The proof – maybe unsurprisingly – uses some linear algebra.

Consider the vector space of polynomials of degree less than or equal to n. This is a vector space of dimension $n + 1$. Let

$$\mathcal{B}_1 = \{1, x, x^2, \dots, x^n\}$$

$$\mathcal{B}_2 = \{(x)_0, (x)_1, \dots, (x)_n\} = \{1, x, x(x-1), \dots, x(x-1)\dots(x-n+1)\}.$$

$\mathcal{B}_1$ and $\mathcal{B}_2$ are both bases for the vector space.

What is the change of basis matrix (sometimes called the transition matrix) from $\mathcal{B}_2$ to $\mathcal{B}_1$?[19] From linear algebra, we know that to find the change of basis matrix from a basis $\mathcal{B}_2$ to another basis $\mathcal{B}_1$, we write each element of $\mathcal{B}_2$ as a linear combination of elements of $\mathcal{B}_1$. The scalars in each linear combination form a column of the change of basis matrix. We can find the change of basis matrix from $\mathcal{B}_1$ to $\mathcal{B}_2$ similarly, and linear algebra tells us that these two matrices are inverses of each other. To write the change of basis matrix from $\mathcal{B}_2$ to $\mathcal{B}_1$, we need to write each falling factorial in terms of powers of x. We have already done this in Theorem 6.31b. To refresh your memory, here are the first few falling factorials:

$$(x)_0 = 1$$
$$(x)_1 = x$$
$$(x)_2 = -x + x^2$$
$$(x)_3 = 2x - 3x^2 + x^3$$
$$(x)_4 = -6x + 11x^2 - 6x^3 + x^4$$
$$(x)_5 = 24x - 50x^2 + 35x^3 - 10x^4 + x^5.$$

As we proved in Theorem 6.31b, the coefficients are the signed Stirling numbers of the first kind. Each set of coefficients becomes a column, and so the change of basis matrix from $\mathcal{B}_2$ to $\mathcal{B}_1$ is the transpose of the matrix of the signed Stirling numbers of the first kind. We have proved:

Corollary 6.35. *Let n be a non-negative integer and let s_n be the matrix of signed Stirling numbers of the first kind, i.e., s_n is the $(n+1) \times (n+1)$ matrix whose (i,j) entry, for $0 \le i, j \le n$, is $s(i,j) = (-1)^{i+j}\begin{bmatrix} i \\ j \end{bmatrix}$. Then s_n^t is the change of basis matrix from $\mathcal{B}_2$ to $\mathcal{B}_1$.*

---

[19] See Problem P 5.4.10 for a quick description of a change of basis matrix.

What is the transition matrix from $\mathcal{B}_1$ to $\mathcal{B}_2$? To calculate examples, we have to write powers of x as a linear combination of falling factorials:

$$1 = 1$$
$$x = x$$
$$x^2 = x + x(x-1)$$
$$x^3 = x + 3x(x-1) + x(x-1)(x-2)$$
$$x^4 = x + 7x(x-1) + 6x(x-1)(x-2) + x(x-1)(x-2)(x-3).$$

The coefficients are the Stirling numbers of the second kind, and we will prove this pattern next. (We actually have already seen something similar in Theorem 6.19.)

Theorem 6.36. *Let n be a non-negative integer and x an indeterminate; then*[20]

$$x^n = \left\{ {n \atop 0} \right\}(x)_0 + \left\{ {n \atop 1} \right\}(x)_1 + \left\{ {n \atop 2} \right\}(x)_2 + \cdots + \left\{ {n \atop n} \right\}(x)_n.$$

Proof. Consider the polynomial

$$f(x) = x^n - \left\{ {n \atop 0} \right\}(x)_0 - \left\{ {n \atop 1} \right\}(x)_1 - \left\{ {n \atop 2} \right\}(x)_2 - \cdots - \left\{ {n \atop n} \right\}(x)_n.$$

This is a polynomial in x of degree n or less, and we want to prove that it is actually the zero polynomial. We already have seen – Theorem 6.19 – that, for any positive integers s and t, we have

$$t^s = (t)_0 \left\{ {s \atop 0} \right\} + (t)_1 \left\{ {s \atop 1} \right\} + (t)_2 \left\{ {s \atop 2} \right\} + \cdots + (t)_t \left\{ {s \atop t} \right\}.$$

Note that if $t > s$, then $\left\{ {s \atop t} \right\} = 0$, and so the final terms of the above sum – the terms after $(s)_s \left\{ {s \atop s} \right\}$ – will be zero. Hence, replacing s with n, and, if $t > n$, we have

$$t^n = (t)_0 \left\{ {n \atop 0} \right\} + (t)_1 \left\{ {n \atop 1} \right\} + (t)_2 \left\{ {n \atop 2} \right\} + \cdots + (n)_n \left\{ {n \atop n} \right\}.$$

This means that $f(t) = 0$ for every positive integer $t > n$. So the polynomial f has an infinite number of roots,[21] but a non-zero polynomial with real coefficients of degree less than or equal to n has at most n roots.[22] Thus, $f(x) = 0$, which completes the proof. □

Corollary 6.37. *Let n be a non-negative integer, and let S_n be the matrix of Stirling numbers of the second kind, i.e., S_n is the $(n+1) \times (n+1)$ matrix whose (i,j) entry is $S(i,j) = \left\{ {i \atop j} \right\}$. Then S_n^t is the change of basis matrix from $\mathcal{B}_1$ to $\mathcal{B}_2$.*

---

[20] James Stirling (1692–1770), a Scottish mathematician, after whom the Stirling numbers are named, studied the expansion of x^n as linear combinations of falling factorials, and made tables of the coefficients.
[21] The real or complex number α is a root of a polynomial f if $f(\alpha) = 0$.
[22] For a proof, see Shahriari 2017, Corollary 19.27.

So the crux of the matter is that, if we write falling factorials as a linear combination of powers of x, the scalars are signed Stirling numbers of the first kind, and vice versa: if we write powers of x as a linear combination of falling factorials, the scalars are Stirling numbers of the second kind. As a result, the change of basis matrix from $\mathcal{B}_2$ to $\mathcal{B}_1$ is s_n^t, and the change of basis matrix from $\mathcal{B}_1$ to $\mathcal{B}_2$ is S_n^t. Hence, $s_n^t S_n^t = I_n$ (for consistency, we are denoting the $(n+1) \times (n+1)$ identity matrix by I_n), which means that $S_n s_n = I_n$, and so we have:

Theorem 6.38. *Let S_n be the matrix of Stirling numbers of the second kind, and let s_n be the matrix of signed Stirling numbers of the first kind. Then*

$$s_n = (S_n)^{-1}.$$

Remark 6.39. We organized the proof of Theorem 6.38 using change of basis matrices. We didn't have to do that. To show that S_n is the inverse of s_n, we could have just multiplied the two matrices and shown, entry by entry, that the resulting matrix is the identity matrix.

More Matrices

The statement that S_n, the matrix of Stirling numbers of the second kind, is the inverse of s_n, the matrix of signed Stirling numbers of the first kind, is clear, succinct, and easy to remember. To prove this theorem, we used Theorem 6.19 and Theorem 6.31b, and the statements of these are more complicated. But we can use matrices to organize these results as well. Define two

column vectors $U_n = \begin{bmatrix} (n)_0 \\ (n)_1 \\ \vdots \\ (n)_n \end{bmatrix}$ and $H_n = \begin{bmatrix} n^0 \\ n^1 \\ \vdots \\ n^n \end{bmatrix}$ of, respectively, falling factorials and powers

of n. Then, we can summarize Theorem 6.19 and Theorem 6.31b as

$$S_n U_n = H_n, \quad \text{and} \quad s_n H_n = U_n. \tag{6.2}$$

You are asked to provide the details in Problem $P\,6.3.5$. This interpretation allows us to see that, if we accept that s_n is the inverse of S_n, then the two matrix equations are equivalent:

$$S_n U_n = H_n \;\Leftrightarrow\; s_n S_n U_n = s_n H_n \;\Leftrightarrow\; I_n U_n = s_n H_n \;\Leftrightarrow\; U_n = s_n H_n.$$

Now let C_n be the matrix of binomial coefficients (Definition 5.33), and define two other matrices P_n and F_n (see Problem P 5.4.13). All of these matrices are $(n+1) \times (n+1)$ and their rows and columns are numbered $0, 1, \ldots, n$. The (i,j) entry of P_n is i^j (we stipulate that $0^0 = 1$) and F_n is a diagonal matrix whose ith diagonal entry is $1/i!$. We have seen that the inverse of C_n is the matrix of signed binomial coefficients (Problem P 5.4.11), and, in Problem P 6.1.33, you were asked to show that

$$(S_n)^t = F_n C_n^{-1} P_n, \tag{6.3}$$

where $(S_n)^t$ is the transpose of S_n. The matrix equations of Theorem 6.38, Equation 6.2, and Equation 6.3 pack quite a bit of combinatorial information in a compact matrix form.

Problems

P 6.3.1. Let n and m be non-negative integers and let $\left[\begin{smallmatrix} n \\ m \end{smallmatrix}\right]$ and $\left\{\begin{smallmatrix} n \\ m \end{smallmatrix}\right\}$ denote, as usual, the (unsigned) Stirling numbers of the first and second kind.

(a) Calculate

$$\left\{\begin{matrix} 5 \\ 3 \end{matrix}\right\}\left[\begin{matrix} 3 \\ 3 \end{matrix}\right] - \left\{\begin{matrix} 5 \\ 4 \end{matrix}\right\}\left[\begin{matrix} 4 \\ 3 \end{matrix}\right] + \left\{\begin{matrix} 5 \\ 5 \end{matrix}\right\}\left[\begin{matrix} 5 \\ 3 \end{matrix}\right].$$

(b) Let k and j be positive integers with $k > j$. What can you say about

$$\left\{\begin{matrix} k \\ j \end{matrix}\right\}\left[\begin{matrix} j \\ j \end{matrix}\right] - \left\{\begin{matrix} k \\ j+1 \end{matrix}\right\}\left[\begin{matrix} j+1 \\ j \end{matrix}\right] + \left\{\begin{matrix} k \\ j+2 \end{matrix}\right\}\left[\begin{matrix} j+2 \\ j \end{matrix}\right] + \cdots + (-1)^{k+j}\left\{\begin{matrix} k \\ k \end{matrix}\right\}\left[\begin{matrix} k \\ j \end{matrix}\right]?$$

Prove your assertion.

P 6.3.2. Given Theorem 6.38, after defining the Stirling numbers of the second kind, we could have defined the Stirling numbers of the first kind as the absolute value of the entries of the inverse of the matrix of Stirling numbers of the second kind. Could we get a new set of interesting integers if we start with the matrix of binomial coefficients? In other words, why don't we have "binomial coefficients of the second kind"?

P 6.3.3. Let s be a positive integer and let $B(s) = \sum_{k=0}^{s} \left\{\begin{smallmatrix} s \\ k \end{smallmatrix}\right\}$ denote the sth Bell number (see Definition 6.12). Calculate

$$\left[\begin{matrix} 1 \\ 1 \end{matrix}\right] B(1)$$

$$-\left[\begin{matrix} 2 \\ 1 \end{matrix}\right] B(1) + \left[\begin{matrix} 2 \\ 2 \end{matrix}\right] B(2)$$

$$\left[\begin{matrix} 3 \\ 1 \end{matrix}\right] B(1) - \left[\begin{matrix} 3 \\ 2 \end{matrix}\right] B(2) + \left[\begin{matrix} 3 \\ 3 \end{matrix}\right] B(3)$$

$$\vdots$$

Do you see a pattern? Generalize and prove.

P 6.3.4. Let D_n be the diagonal $(n+1) \times (n+1)$ matrix whose rows and columns are numbered $0, 1, \ldots, n$, and, for $0 \leq i \leq n$, whose (i,i) entry is $(-1)^i$. So, for example, $D_2 = \begin{bmatrix} 1 & 0 & 0 \\ 0 & -1 & 0 \\ 0 & 0 & 1 \end{bmatrix}$.

(a) What is D_n^{-1}?

(b) If A is any $(n+1) \times (n+1)$ matrix with rows and columns numbered $0, \ldots, n$, then describe the matrix $D_n A D_n$.

(c) What is the inverse of the matrix of the (unsigned) Stirling numbers of the first kind? First make a conjecture, and then prove it.

P 6.3.5. Use Theorem 6.19 and Theorem 6.31b to prove the matrix equations of Equation 6.2.

P 6.3.6. Re-do Problem P 6.1.20, but this time use only the matrix equations of Equation 6.3 and the one found in Problem P 5.4.16.

P 6.3.7. Let n and k be non-negative integers, and make the following stipulations:

$$\left\{\begin{matrix} 0 \\ -k \end{matrix}\right\} = \left[\begin{matrix} 0 \\ -k \end{matrix}\right] = \begin{cases} 1, & \text{if } k = 0; \\ 0, & \text{otherwise.} \end{cases} \qquad \left\{\begin{matrix} -n \\ 0 \end{matrix}\right\} = \left[\begin{matrix} -n \\ 0 \end{matrix}\right] = \begin{cases} 1, & \text{if } n = 0; \\ 0, & \text{otherwise.} \end{cases}$$

Now assume that the recurrence relations of Theorem 6.7 and Theorem 6.30 work for *all* non-zero integers (positive or negative). Prove that, for non-negative integers n and k,

$$\left\{\begin{matrix} -n \\ -k \end{matrix}\right\} = \left[\begin{matrix} k \\ n \end{matrix}\right] \quad \text{and} \quad \left[\begin{matrix} -n \\ -k \end{matrix}\right] = \left\{\begin{matrix} k \\ n \end{matrix}\right\}.$$

6.3.1   Lah Numbers or the Stirling Numbers of the Third Kind*

Stirling numbers of the first kind count the partitions of $[n]$ into k circularly ordered non-empty parts. Stirling numbers of the second kind count the partitions of $[n]$ into k (unordered) non-empty parts. Lah numbers[23] also count certain partitions of $[n]$.

Definition 6.40. Let n and k be non-negative integers. Define the *Lah number* (or a *Stirling number of the third kind*) $L(n,k)$ to be the number of partitions of $[n]$ into k non-empty ordered lists. Either by proper interpretation of this definition or by convention, $L(0,0) = 1$.

Example 6.41. If $n = 3$ and $k = 2$, then partitions of $[3]$ into two linearly ordered lists are

$$\{12, 3\}, \{21, 3\}, \{13, 2\}, \{31, 2\}, \{23, 1\}, \{32, 1\}.$$

Hence $L(3, 2) = 6$.

In the problems, you are asked to prove the following properties of the Lah numbers.[24]

Proposition 6.42. *Let n and k be positive integers with $k \le n$, and let $L(n, k)$ denote the Lah numbers. Then*

(a) *(Problem P 6.3.9)* $L(n, k) = L(n - 1, k - 1) + (n + k - 1)L(n - 1, k).$

(b) *(Problem P 6.3.10)* $L(n, k) = \dbinom{n}{k} \dfrac{(n - 1)!}{(k - 1)!}.$

(c) *(Problem P 6.3.12)*

$$x^{(n)} = L(n, 0)(x)_0 + L(n, 1)(x)_1 + \cdots + L(n, k)(x)_k + \cdots + L(n, n)(x)_n.$$

(d) *(Problem P 6.3.14) Let S_n and $|s|_n$ denote, respectively, the $(n + 1) \times (n + 1)$ matrices of the (unsigned) Stirling numbers of the second kind and the (unsigned) Stirling numbers of the first kind. Likewise let L_n denote the $(n + 1) \times (n + 1)$ matrix of Lah numbers. Then $L_n = |s|_n S_n.$*

---

[23] Named after Ivo Lah (1896–1979), a Slovenian mathematician and actuary, who introduced them in 1955 (Lah 1955).

[24] See Petkovšek and Pisanski 2007 and Daboul et al. 2013 for these, and more on Lah numbers.

(e) *(Problem P 6.3.16)* $L(n,k) = \sum_{j=0}^{n} \begin{bmatrix} n \\ j \end{bmatrix} \begin{Bmatrix} j \\ k \end{Bmatrix}$.

(f) *(Problem P 6.3.17) Let $f(x) = e^{1/x}$. Then $f^{(n)}(x)$, the nth derivative of f, is given by*

$$f^{(n)}(x) = (-1)^n e^{1/x} \left[\frac{L(n,1)}{x^{n+1}} + \frac{L(n,2)}{x^{n+2}} + \cdots + \frac{L(n,n)}{x^{2n}} \right].$$

Problems (continued)

P 6.3.8. Let n be a non-negative integer. Find straightforward formulas for Lah numbers $L(n,0)$, $L(n,1)$, $L(n,n-1)$, and $L(n,n)$. If k is a non-negative integer with $k > n$, then what is $L(n,k)$?

P 6.3.9. **Proof of Proposition 6.42a.** Let n and k be positive integers with $k \leq n$, and let $L(n,k)$ denote the Lah numbers. Prove that

$$L(n,k) = L(n-1,k-1) + (n+k-1)L(n-1,k).$$

P 6.3.10. **Proof of Proposition 6.42b.** Let n and k be positive integers with $k \leq n$. By directly counting the number of partitions of $[n]$ into k non-empty ordered lists, show that the Lah number $L(n,k)$ is equal to $\binom{n}{k} \frac{(n-1)!}{(k-1)!}$.

P 6.3.11. **Another Proof of Proposition 6.42b.** Let n and k be positive integers with $k \leq n$. Use induction and the recurrence relation of Proposition 6.42a to give a second proof of Proposition 6.42b.

P 6.3.12. **Proof of Proposition 6.42c.** Let n be a non-negative integer. Prove that

$$x^{(n)} = L(n,0)(x)_0 + L(n,1)(x)_1 + \cdots + L(n,k)(x)_k + \cdots + L(n,n)(x)_n,$$

where, for a non-negative integer $k \leq n$, $L(n,k)$ denotes a Lah number. You may find the following setup and the subsequent steps helpful:

SETUP: Let n and m be positive integers. Assume that you have n distinct balls and m distinct boxes. The boxes have no capacity restrictions, and, in particular, can be empty. You want to place all the balls in the boxes but you also want to assign a specific order to the balls in each box.

STEP 1: Place the balls in the boxes one at a time. How many choices do you have for the first ball? How many choices do you have for the second ball? In how many ways can you complete the task?

STEP 2: Let $0 \leq i \leq m$ be an integer. In how many ways can you place the balls into boxes with the additional condition that exactly i of the boxes are not empty?

STEP 3: Using Step 1 and Step 2, show that the identity we want to prove is true if $x = m$, a positive integer.

STEP 4: Use an argument similar to the one in the proof of Theorem 6.36 to complete the proof.

P 6.3.13. Let $V = \mathcal{P}_n(\mathbb{R})$ be the vector space of polynomials of degree less than or equal to n. Let $\mathcal{B}_1 = \{1, x, \ldots, x^n\}$, $\mathcal{B}_2 = \{(x)_0, (x)_1, \ldots, (x)_n\}$, and $\mathcal{B}_3 = \{x^{(0)}, x^{(1)}, \ldots, x^{(n)}\}$.

(a) Prove that $\mathcal{B}_1$, $\mathcal{B}_2$, and $\mathcal{B}_3$ are bases for V.

(b) What is the change of basis matrix (see Problem P 5.4.10) from $\mathcal{B}_3$ to $\mathcal{B}_2$?

(c) What is the change of basis matrix from $\mathcal{B}_3$ to $\mathcal{B}_1$?

P 6.3.14. **Proof of Proposition 6.42d.** Let n be a non-negative integer. Let S_n and $|s|_n$ denote, respectively, the $(n+1) \times (n+1)$ matrices of the (unsigned) Stirling numbers of the second kind and the (unsigned) Stirling numbers of the first kind.[25] Likewise, let L_n denote the $(n+1) \times (n+1)$ matrix whose (i, j) entry is the Lah number $L(i, j)$. Prove that $L_n = |s|_n S_n$.

P 6.3.15. Let L_n be the $(n+1) \times (n+1)$ matrix of Lah numbers. What is L_n^{-1}? Prove your assertion.

P 6.3.16. **Proof of Proposition 6.42e.** Let n and k be non-negative integers with $k \leq n$. As usual, $L(n, k)$ denotes a Lah number. Prove that

$$L(n, k) = \sum_{j=0}^{n} \begin{bmatrix} n \\ j \end{bmatrix} \begin{Bmatrix} j \\ k \end{Bmatrix}.$$

P 6.3.17. **Proof of Proposition 6.42f.** Let n and k be positive integers with $k \leq n$. As usual, $L(n, k)$ denotes a Lah number. Let $f(x) = e^{1/x}$. Then prove that $f^{(n)}(x)$, the n-derivative of f, is given by

$$f^{(n)}(x) = (-1)^n e^{1/x} \left[\frac{L(n, 1)}{x^{n+1}} + \frac{L(n, 2)}{x^{n+2}} + \cdots + \frac{L(n, n)}{x^{2n}} \right].$$

---

[25] We are using the admittedly strange symbol $|s|_n$ for the unsigned Stirling numbers of the first kind since we already used s_n for the signed Stirling numbers of the first kind.

7 Integer Partitions

In my flower shop, every morning, I take 30 identical Azaleas and plant them in some number of identical pots. While I like having a lot of options, I am considering adopting one of two guidelines. Should I decree that I will always use 7 (non-empty) pots? Or should I instead decide that I can use as many pots as I wish but that the greatest number of Azaleas in any one pot will be exactly 7? Which one of these requirements is more flexible and gives me more options?

7.1 Partitions of an Integer

Warm-Up 7.1. *I have five identical pencils. In how many ways can I organize them into three (non-empty) piles?*

In this chapter, we will turn our attention to entries $\boxed{8}$ and $\boxed{12}$ of the counting table, Table 3.1. In Chapter 6, we studied partitions of a finite set. Here, we consider the partitions of an integer.

Definition 7.2. Let n and k be non-negative integers. A *partition* of n with k parts is a sequence $\lambda_1, \lambda_2, \ldots, \lambda_k$ of k positive integers such that

$$\sum_{i=1}^{k} \lambda_i = n, \quad \text{and} \quad \lambda_1 \geq \lambda_2 \geq \cdots \geq \lambda_k > 0.$$

The integers $\lambda_1, \ldots, \lambda_k$ are the *parts* of the partition.

If $\lambda = (\lambda_1, \lambda_2, \ldots, \lambda_k)$ is a partition of n, then we write $\lambda \vdash n$. If $\lambda \vdash n$, then, sometimes, we augment λ with a finite number of zeros. While the resulting sequence is not technically a partition of n, for convenience we continue to think of it as another representation of the same partition λ (with the same number of parts).

For $n > 0$, we denote by $p_k(n)$ the number of partitions of n with exactly k parts, and the total number of partitions of n is denoted by $p(n)$. We define $p_0(0) = 1$, and, for $k > 0$, $p_k(0) = 0$. As a result, $p(0) = 1$.

Example 7.3. The only partition of 4 into 3 parts is $(2, 1, 1)$. Hence, $p_3(4) = 1$.

Recall that $\left\{ {4 \atop 3} \right\} = 6$, since we had six different partitions of the set $[4] = \{1, 2, 3, 4\}$ into 3 non-empty parts. Each of these partitions consisted of one subset with 2 elements and two subsets with 1 element each. While $p_3(4)$ only counts the various possible subset sizes, $\left\{ {4 \atop 3} \right\}$ counts the actual ways you can partition the set.

All the possible partitions of 4 (into any number of parts) are:

$$(4), (3,1), (2,2), (2,1,1), (1,1,1,1).$$

Hence, $p(4) = 5$.

Example 7.4. If n is any positive integer, then there is only one way to partition the integer n into n parts – namely, $(1, 1, \ldots, 1)$ – or into 1 part – namely, (n) – and, hence, both $p_n(n)$ and $p_1(n)$ are equal to 1.

Also $p_{n-1}(n) = 1$, since the only partition of n into $n-1$ parts is

$$(2, \underbrace{1, 1, \ldots, 1}_{n-2}).$$

In Table 7.1, we have begun constructing a table of values for $p_k(n)$ and $p(n)$.

Table 7.1 A preliminary table for small partition numbers.

	The small values of $p_k(n)$ and $p(n)$									
$n \backslash k$	0	1	2	3	4	5	6	7	8	p(n)
0	1	0	0	0	0	0	0	0	0	1
1	0	1	0	0	0	0	0	0	0	1
2	0	1	1	0	0	0	0	0	0	2
3	0	1	1	1	0	0	0	0	0	3
4	0	1	2	1	1	0	0	0	0	5
5	0	1			1	1	0	0	0	
6	0	1				1	1	0	0	
7	0	1					1	1	0	
8	0	1						1	1	

Theorem 7.5 (Entries $\boxed{8}$ and $\boxed{12}$ of the Counting Table). *Let s and t be non-negative integers.*

(a) $\boxed{12}$ *The number of ways of placing s identical balls into t identical boxes such that no box remains empty is $p_t(s)$.*

(b) $\boxed{8}$ *The number of ways of placing s identical balls into t identical boxes (allowing empty boxes) is*

$$p_0(s) + p_1(s) + p_2(s) + \cdots + p_t(s),$$

Proof. (a) Since the balls and the boxes are indistinguishable, the only thing that matters is the number of balls in each box. Hence, each distribution of the balls in the boxes corresponds to finding positive integers $\lambda_1, \lambda_2, \ldots, \lambda_t$ such that $\lambda_1 + \cdots \lambda_t = s$. Since two permutations of the same integers are considered the same, we may as well order them in non-increasing order and also assume that $\lambda_1 \geq \lambda_2 \geq \cdots \geq \lambda_t \geq 1$. By definition of $p_t(s)$, the number of ways that we can do this is equal to $p_t(s)$.

(b) To distribute the balls we will use i non-empty boxes with i ranging from 0 to t (note that the number of ways of placing s balls into 0 non-empty boxes is zero unless $s = 0$, in which case it is 1). By the previous part, the number of ways of placing the balls into boxes using i non-empty boxes is $p_i(s)$. The result now follows.   □

To be able to compute values of $p_k(n)$, as in the past, we discover a recurrence relation.

Theorem 7.6. *Let n and k be positive integers with $n \geq k$. Then*

$$p_k(n) = p_{k-1}(n-1) + p_k(n-k).$$

Proof. A partition of n into k parts consists of a sequence $\lambda_1, \lambda_2, \ldots, \lambda_k$ of positive integers such that $\sum_i \lambda_i = n$, and $\lambda_1 \geq \ldots, \lambda_k > 0$. From among these, we let $\mathcal{A}$ be those partitions for which $\lambda_k = 1$, and we define $\mathcal{B}$ to be those partitions for which $\lambda_k \geq 2$. Clearly,

$$p_k(n) = |\mathcal{A}| + |\mathcal{B}|.$$

CLAIM 1: $|\mathcal{A}| = p_{k-1}(n-1)$.

PROOF OF CLAIM 1: Given any k-part partition of n with $\lambda_k = 1$, we can get a $(k-1)$-part partition of $n-1$ by removing λ_k. Conversely, given any $(k-1)$-part partition of $n-1$, we can get an element of $\mathcal{A}$ by adding a $\lambda_k = 1$ to the end. These two maps are inverses of each other, and so both are 1-1 and onto, proving the claim.

CLAIM 2: $|\mathcal{B}| = p_k(n-k)$.

PROOF OF CLAIM 2: Given a k-part partition of $n-k$, add 1 to each part to get a k-part partition of n where all parts are at least 2. Conversely, given a k-part partition of n with every part greater than or equal to 2, subtract 1 from each part to get a k-part partition of $n-k$. The two maps are evidently inverses of each other. Hence, they are both 1-1 and onto, and the claim is proved.   □

Using the recurrence relation, we can now complete the table of small values for $p_k(n)$. See Table 7.2.

Remark 7.7. Many subtle relations can be found in the table of partition numbers. For example, the great Indian mathematician Srinivasa Ramanujan (1887–1920) discovered that, for all non-negative integers n, $p(5n+4)$ is divisible by 5. The first two instances of this – namely $p(4) = 5$ and $p(9) = 30$ – can be seen in Table 7.2.

Example 7.8. I have 10 identical Azaleas. In how many ways can I organize them into three, four, or five groups for planting? Other than being non-empty, there is no restriction on how many flowers I put into each group.

Solution. The Azaleas are identical, and so are the groups (there is no difference between groups of sizes 5, 3, and 2 and groups of sizes 3, 2, and 5.) Hence, we are looking for ways to

Table 7.2 A table of values for small partition numbers.

	The small values of $p_k(n)$ and $p(n)$											
$n \backslash k$	0	1	2	3	4	5	6	7	8	9	10	p(n)
0	1	0	0	0	0	0	0	0	0	0	0	1
1	0	1	0	0	0	0	0	0	0	0	0	1
2	0	1	1	0	0	0	0	0	0	0	0	2
3	0	1	1	1	0	0	0	0	0	0	0	3
4	0	1	2	1	1	0	0	0	0	0	0	5
5	0	1	2	2	1	1	0	0	0	0	0	7
6	0	1	3	3	2	1	1	0	0	0	0	11
7	0	1	3	4	3	2	1	1	0	0	0	15
8	0	1	4	5	5	3	2	1	1	0	0	22
9	0	1	4	7	6	5	3	2	1	1	0	30
10	0	1	5	8	9	7	5	3	2	1	1	42

partition the integer 10 into three, four, or five parts. Using Table 7.2, the answer is $p_3(10) + p_4(10) + p_5(10) = 8 + 9 + 7 = 24$. $\qquad \square$

Ferrers Diagrams

A graphical device allows us to visualize the partitions of n.

Definition 7.9 (Ferrers Diagram). Let $\lambda = (\lambda_1, \lambda_2, \ldots, \lambda_k)$ be a partition of n into k parts. Place n dots in k rows such that the ith row from the top has λ_i dots, and the dots are left-aligned. This diagram – see Figure 7.1 – is called the *Ferrers diagram* of λ.[1]

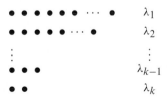

Figure 7.1 The Ferrers diagram of $(\lambda_1, \ldots, \lambda_k)$.

---

[1] Named after Norman Macleod Ferrers (1829–1903), a British mathematician who used such a diagram to give a proof of Theorem 7.15.

Remark 7.10. Many arguments about partitions become more visual when combined with Ferrers diagrams. For example, Theorem 7.6 amounts to the following: given a Ferrers diagram of a partition of n into k parts, either the last row has one dot or every row has at least two dots. In the former case, eliminate the last row and get a partition of $n - 1$ into $k - 1$ parts. In the latter case, eliminate a dot from every row and get a partition of $n - k$ into k parts. Conversely, given a partition of $n - 1$ into $k - 1$ parts, just add one row with one dot to get a partition of n into k parts, and given a partition of $n - k$ into k parts, add a dot to each row to get a partition of n into k parts. Thus, there is a 1-1 correspondence between partitions of n into k parts and the union of the partitions of $n - 1$ into $k - 1$ parts and the partitions of $n - k$ into k parts.

Definition 7.11 (Conjugate Partition). Let λ be a partition of n, and let F be the Ferrers diagram of λ. The transpose of F, F^t, is also the Ferrers diagram of a partition of n. This partition – whose Ferrers diagram is F^t – is called the *conjugate* of λ and denoted by λ'.

Example 7.12. Consider $\lambda = (4, 2, 1)$, a partition of 7. The conjugate partition of λ is $\lambda' = (3, 2, 1, 1)$, as can be seen in Figure 7.2 showing their Ferrers diagrams.

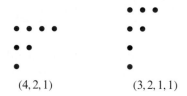

Figure 7.2 $\lambda = (4, 2, 1)$ and $\lambda' = (3, 2, 1, 1)$ are conjugate partitions.

Sometimes, by just playing with the (Ferrers) diagrams, we can get new theorems! For example, what happens if we start with all the partitions of a positive integer n into exactly k parts and find their transposes? We will first look at two examples, and using these as prototypes, we will state a theorem.

Example 7.13. If $n = 4$ and $k = 2$, then $p_2(4) = 2$, and the two partitions of 4 into two parts are $(3, 1)$ and $(2, 2)$. The transpose of these two are, respectively, the partitions $(2, 1, 1)$ and $(2, 2)$. See Figure 7.3.

Example 7.14. If $n = 6$ and $k = 3$, then $p_3(6) = 3$. The three partitions of 6 into three parts are $2, 2, 2$, $3, 2, 1$, and $4, 1, 1$. Their conjugates, respectively, are $3, 3$, $3, 2, 1$, and $3, 1, 1, 1$. See Figure 7.4.

We are now ready to resolve the opening problem of the chapter. The problem wanted us to compare the number of ways of partitioning 30 into seven parts – that is, $p_7(30)$ – with the number of ways of partitioning 30 into any number of parts, in such a way that the largest part is exactly 7. The next theorem shows that the two counting problems have the same answer.

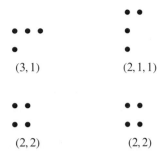

Figure 7.3 Partitions of 4 into two parts (on the left) and their conjugates (on the right).

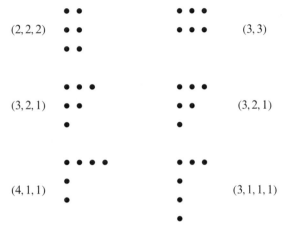

Figure 7.4 Partitions of 6 into three parts (on the left) and their conjugates (on the right).

Theorem 7.15. *Let n and k be positive integers with $k \le n$. Then the number of partitions of n into k parts, $p_k(n)$, is equal to the number of those partitions of n (into any number of parts) whose largest part is exactly k.*

Proof. Given any partition λ of n into k parts, λ', the conjugate partition, will continue to be a partition of n, and its largest part will be exactly k. On the other hand, given any partition μ of n whose largest part is exactly k, the conjugate partition μ' will have exactly k parts. Since the conjugate of a conjugate is the original partition, these two maps are inverses of each other, and hence both are 1-1 and onto. $\qquad\square$

We prove one final theorem on partitions using Ferrers diagrams.

Definition 7.16 (Self-Conjugate Partitions). Let n be a positive integer, λ a partition of n, and λ' the conjugate of λ. If $\lambda = \lambda'$, then λ is called *self-conjugate*.

Theorem 7.17. *Let n be a positive integer. Let A be the set of self-conjugate partitions of n, and let the set B consist of those partitions of n that have distinct odd parts. Then*

$$|A| = |B|.$$

Proof. First, we give a map $f : B \to A$. In other words, given a partition $\lambda \in B$, we construct a partition $f(\lambda) \in A$.

Let λ be the partition $(2k_1 + 1, 2k_2 + 1, \ldots, 2k_m + 1)$ with $k_1 > k_2 > \cdots > k_m$. Now construct the Ferrers diagram of $f(\lambda)$ as follows. Place $k_1 + 1$ dots in the first row *and* in the first column of $f(\lambda)$. Note that, since the first row and the first column share one dot, the total number of dots used so far is $2k_1 + 1$. Now, place an *additional* $k_2 + 1$ dots in the second row and in the second column. Again, we have added $2k_2 + 1$ dots to the Ferrers diagram. Note that the second row (and the second column) now have $k_2 + 2$ dots. In the next step, add an additional $k_3 + 1$ dots to the third row and to the third column of the Ferrers diagram of $f(\lambda)$. Continue in this fashion, and, in the last step, put an additional $k_m + 1$ dots in the mth row and in the mth column. See Figure 7.5.

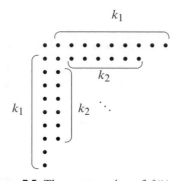

Figure 7.5 The construction of $f(\lambda) \in A$.

Note that when we got to row ℓ, we already had $\ell - 1$ dots from the previous steps – one for each prior row – and we then added an additional $k_\ell + 1$ dots. Hence, row ℓ has $k_\ell + \ell = (k_\ell + 1) + (\ell - 1)$ dots. Since $k_{\ell+1} + (\ell + 1) \le k_\ell + \ell$ – recall that $k_{\ell+1} < k_\ell$ – we conclude that the construction indeed gives the Ferrers diagram of a partition. Now, in the construction of the Ferrers diagram of $f(\lambda)$, we placed exactly the same number of dots as in the Ferrers diagram of λ, and so $f(\lambda)$ is a partition of n. Moreover, the construction clearly creates a symmetric – with respect to the main diagonal – Ferrers diagram. This means that $f(\lambda)$ is self-conjugate, and so it is in A.

Next, we reverse the process and get a map $g : A \to B$. Given a self-conjugate partition $\mu \in A$, we know that the first row and in the first column of μ have the same number of dots. Call this number $k_1 + 1$. Put all of these dots – the ones in the first row and the first column of μ – as the dots in the first row of $g(\mu)$. Hence, the first row of $g(\mu)$ has $2k_1 + 1$ dots. Put the remaining dots in the second row and the second column of μ in the second row of $g(\mu)$, and so on. Clearly, we get a partition of n with distinct odd parts – the parts are distinct since, for each row, we are using at least two fewer dots – and, hence, $g(\mu) \in B$.

The two maps f and g are inverses of each other. Hence, they are both 1-1 and onto, and so $|A| = |B|$. $\square$

Table 7.3 Balls and boxes counting problems – the first three rows.

	Number of Ways to Put Balls into Boxes			
	Balls (S) and Boxes (T)			
	$\lvert S \rvert = s,\ \lvert T \rvert = t$			
	The ith box contains u_i balls.			
$S\ \&\ T \rightarrow$	T distinct	T distinct	T identical	T identical
$u_i \downarrow$	S distinct	S identical	S distinct	S identical
$0 \le u_i \le 1, t \ge s$	$\boxed{1}\quad (t)_s$	$\boxed{2}\quad \binom{t}{s}$	$\boxed{3}\quad 1$	$\boxed{4}\quad 1$
	Theorem 4.6	Theorem 4.21	Theorem 4.23	Theorem 4.23
$u_i \ge 0$	$\boxed{5}\quad t^s$	$\boxed{6}\quad \left(\!\binom{t}{s}\!\right)$	$\boxed{7}\quad \sum_{k=0}^{t} \left\{ {s \atop k} \right\}$	$\boxed{8}\quad \sum_{k=0}^{t} p_k(s)$
	Theorem 4.27	Theorem 4.40	Theorem 6.10	Theorem 7.5
$u_i \ge 1$	$\boxed{9}\quad t!\left\{ {s \atop t} \right\}$	$\boxed{10}\quad \left(\!\binom{t}{s-t}\!\right)$	$\boxed{11}\quad \left\{ {s \atop t} \right\}$	$\boxed{12}\quad p_t(s)$
	Theorem 6.9	Theorem 4.45	Definition 6.4	Theorem 7.5

Partitions of integers have many, often subtle, properties. We have seen two methods for approaching these partitions: recurrence relations and Ferrers diagrams. Yet another – possibly the most powerful – approach is that of generating functions, which will be explored in Chapter 9. (See Section 9.4.2 and the ensuing problems.)

Remark 7.18. Having introduced partition numbers, we have finally finished the first three rows – referred to as the *twelvefold way*[2] – of the counting table, Table 3.1. See Table 7.3.

Problems

P 7.1.1. I have eight identical coins. In how many ways can I distribute them between five identical Romulans without giving all the coins to a single Romulan (i.e., $2 \le$ (number of Romulans that get a coin) ≤ 5)?

P 7.1.2. Re-do Problem *P 1.3.17* by first translating it to the language of partitions.

P 7.1.3. (a) Chaos, Ge, Tartarus, Eros, Erebus, and Night are the names of six Romulan spaceships, each colored differently. Uranus, Mountains, Pontus, Cyclpes, and Hecatonchires are five Romulans. The spaceships are going to take the five Romulans and lots of cargo to the planet Oceanus. Four of the spaceships will be filled with

---

[2] I learned of the twelvefold way and this approach to organizing counting problems from Stanley 2012. Richard Stanley attributes the idea of the twelvefold way to Gian Carlo Rota, and the terminology "twelvefold way" to Joel Spencer.

cargo (and carry no passengers), and two will be taking passengers. In how many ways can this be done, assuming that no spaceship flies empty?

(b) Do the same problem assuming that the spaceships are indistinguishable and all the Romulans are clones of each other.

P 7.1.4. Let $S = \{x_1, x_2, \ldots, x_5\}$, and let $T = \{y_1, y_2, \ldots, y_6\}$. Consider $\mathcal{F} = \{f\colon S \to T \mid |\mathrm{Im}(f)| \le 2\}$, the collection of functions from S to T whose images have at most two elements.

(a) Find $|\mathcal{F}|$. Note that, for example, a function that sends all the x's to y_1 is different from a function that sends all the x's to y_2.

(b) For a function $f \in \mathcal{F}$, define a multiset A_f consisting of the numbers of preimages of elements in T. For example, if $f(x_1) = f(x_2) = y_1$ and $f(x_3) = f(x_4) = f(x_5) = y_4$, then y_1 has two preimages, y_4 has three preimages, and all other ys have no preimages. So $A_f = \{4 \cdot 0, 2, 3\}$. Say functions f and g are in the same class if $A_f = A_g$. Find the number of classes of functions in $\mathcal{F}$.

P 7.1.5. Let n be a positive integer. How is $p(n)$, the total number of partitions of the integer n, related to the number of non-negative integer solutions to

$$x_1 + 2x_2 + 3x_3 + \cdots + nx_n = n?$$

P 7.1.6. Let n be a positive integer. Find a simple expression for each of the following:

(a) $p_{n-2}(n)$ for $n \ge 4$
(b) $p_{n-3}(n)$ for $n \ge 6$
(c) $p_2(n)$.

P 7.1.7. Let n and j be positive integers with $j \le n$. Let $\mathcal{A}$ be the set of partitions of n that have at least j parts equal to 1. What is $|\mathcal{A}|$ in terms of the partition function p?

P 7.1.8. Let n and k be positive integers with $k \le n$. Let $\mathcal{A}$ be the set of partitions of n that have at least one part equal to k. Find $|\mathcal{A}|$ in terms of the partition function p.

P 7.1.9. Let k be a fixed positive integer. Prove that, for integers $n \ge 2k$,

$$p_{n-k}(n) = p(k).$$

P 7.1.10. Let r and k be non-negative integers. Show that the number of partitions of $2r + k$ into $r + k$ parts is equal to $p_r(2r)$, regardless of the value of k.

P 7.1.11. Let k and n be positive integers with $k \le n$. Prove that

$$p_k(n) = p_k(n - k) + p_{k-1}(n - k) + \cdots + p_1(n - k) + p_0(n - k).$$

P 7.1.12. Let n and k be positive integers with $k \le n$. Let $A(n,k)$ denote the number of partitions of n with k or fewer parts. In other words, $A(n, k) = p_k(n) + p_{k-1}(n) + \cdots + p_0(n)$. Using a combinatorial argument, give a recurrence relation for $A(n, k)$.

P 7.1.13. Let n and m be positive integers. Let $P(n, m)$ be the number of partitions of n with no part bigger than m. Use a combinatorial argument to give a recurrence relation for $P(n, m)$.

P 7.1.14. Let n and k be positive integers with $k \le n$. Are the numbers $A(n, k)$ and $P(n, k)$, defined in Problems P 7.1.12 and P 7.1.13, related? If so, why would that be?

P 7.1.15. Let $\mathcal{A}$ be the set of all partitions of n with no part equal to 1. Show that $|\mathcal{A}| = p(n) - p(n - 1)$.

P 7.1.16. Let k and n be positive integers with $k > n/2$. Prove that $p_k(n) = p_{k-1}(n-1)$.

P 7.1.17. Let $\mathcal{A}$ be the collection of all partitions of 8 into four parts. For each $\mu \in \mathcal{A}$, find the partition of 4 that you get by subtracting one from each part of μ. (See Figure 7.6.) Which partitions of 4 can you get in this way? Generalize, prove your assertion, and write your conclusion as an identity about partition numbers.

Figure 7.6 For Problem P 7.1.17, start with a partition of 8 into four parts and take one off each part.

P 7.1.18. Let $n = 6$. Find all partitions of n (into any number of parts) where the first two parts are equal. Also find all partitions of n where each part is at least 2. Is there a relationship? Generalize, make a conjecture, and prove it.

P 7.1.19. Let $n = 6$. Find the number of partitions of n into distinct parts (i.e., no two parts should be the same). Also, find the number of partitions of n into parts $\lambda_1 \geq \lambda_2 \geq \ldots \geq \lambda_k > 0$ such that the difference between two consecutive parts is 0 or 1 and $\lambda_k = 1$. Consider a couple of other values for n and make a conjecture. Prove your conjecture.

P 7.1.20. Let $n = 6$. Find the number of partitions of n into parts of even size. Also, find the number of partitions of n into parts such that each part occurs an even number of times. Try other values for n and make a conjecture. Can you prove your conjecture?

P 7.1.21. What is the number of partitions of 6 into three parts? What is the number of partitions of $12 = 2 \times 6$ into three parts of size < 6? By looking at some examples, make a conjecture. Prove your conjecture.

P 7.1.22. Let n be a positive integer.[3]

(a) Prove that

$$\left\lfloor \frac{n}{2} \right\rfloor + \left\lfloor \frac{n+3}{2} \right\rfloor = n + 1.$$

(b) Let A be the set of partitions of $n + 6$ into three parts where the smallest part is 2 or more. Give a combinatorial proof that $|A| = p_3(n) + p_2(n) + p_1(n)$.

(c) Prove that

$$p_3(n+6) = p_3(n) + p_2(n) + p_1(n) + p_2(n+3) + p_1(n+3).$$

Is this formula valid for $n = 0$?

(d) Prove that

$$p_3(n+6) = p_3(n) + n + 3.$$

Is this formula valid for $n = 0$?

[3] Problems P 7.1.22, P 7.1.23, and P 7.1.24 are adapted from Bindner and Erickson 2012.

P 7.1.23. Let n be a non-negative integer, and define the function f by

$$f(n) = \text{ closest integer to } \frac{n^2}{12}.$$

(a) Prove that $f(n+6) = f(n) + n + 3$.
(b) Prove that $p_3(n) = f(n)$.

P 7.1.24. **The Number of Triangles with Integer Sides and Fixed Perimeter.**[4] Let a, b, and c be positive integers with $a \leq b \leq c$. We say that the triple (a, b, c) satisfies the *triangle inequality* if $c < a + b$. Let $\mathbb{Z}^{>0}$ denote the positive integers, and, for an integer $n \geq 3$, let

$$T(n) = \{(a, b, c) \in (\mathbb{Z}^{>0})^3 \mid a \leq b \leq c < a + b, \text{and } a + b + c = n\}.$$

(a) Give an argument that a triangle with sides a, b, and c exists if and only if (a, b, c) satisfies the triangle inequality.
(b) Show that $(a, b, c) \in T(2n)$ if and only if $(n - a, n - b, n - c)$ is a partition of n into three non-empty parts.
(c) Show that if n is even, then $|T(n)| = p_3(n/2)$.
(d) Let n be odd. Show that $(a, b, c) \in T(n)$ if and only if $(a + 1, b + 1, c + 1) \in T(n + 3)$.
(e) Show that if n is odd, then $|T(n)| = p_3(\frac{n+3}{2})$.
(f) Prove that

$$|T(n)| = \begin{cases} \text{the integer closest to } \frac{n^2}{48} & \text{if } n \text{ is even} \\ \text{the integer closest to } \frac{(n+3)^2}{48} & \text{if } n \text{ is odd.} \end{cases}$$

7.2   Asymptotics, Pentagonal Number Theorem, and More*

There is much more to be said about partitions, and we will pick up the topic again in Chapter 9 (see Section 9.4.2, and especially Theorem 9.53 and Theorem 9.54). In this optional section, problems and projects on (1) compositions of an integer, (2) an asymptotic bound for partition numbers $p_k(n)$, and (3) a version of Euler's pentagonal number theorem are introduced through problems.

7.2.1   Compositions of an Integer

For $\lambda_1, \lambda_2, \ldots, \lambda_k$ a partition of n into k parts, in order to pick only one of the possible orderings of the parts, we insisted that $\lambda_1 \geq \lambda_2 \geq \cdots \geq \lambda_k$. We didn't want reordering the parts to result in a new partition. If we relax this condition, and let the order of parts matter, then we get a composition of an integer.

---

[4] Triangles with integer sides will be revisited in Problems P 9.4.21 and P 9.4.22.

Definition 7.19. Let n be a positive integer. An ordered list of positive integers $\mu_1, \ldots, \mu_k$ such that $\mu_1 + \cdots + \mu_k = n$ is a *composition* of n into k parts.

In other words, every composition of n into k parts corresponds to a solution in positive integers of $x_1 + x_2 + \cdots + x_k = n$.

Example 7.20. The number of partitions of 6 into three parts, $p_3(6)$, is 3, and these partitions are $\{4, 1, 1\}$, $\{3, 2, 1\}$, and $\{2, 2, 2\}$. However, there are 10 compositions of 6 into three parts: $4 + 1 + 1$, $1 + 4 + 1$, $1 + 1 + 4$, $3 + 2 + 1$, $3 + 1 + 2$, $2 + 3 + 1$, $2 + 1 + 3$, $1 + 3 + 2$, $1 + 2 + 3$, and $2 + 2 + 2$.

Problems

P 7.2.1. Let n and k be positive integers with $k \leq n$. What is the number of compositions of n into k parts? What is the total number of compositions of n (into any number of parts)?

P 7.2.2. Let n and k be positive integers with $k \leq n$. Arrange n sticks in a row.
  (a) In how many ways can you place $k - 1$ stars in between the n sticks so that there are no two stars next to each other (and none before or after the sticks)?
  (b) In how many ways can you place any number of stars (including zero stars) in between the n sticks so that there are no two stars next to each other (and none before or after the sticks)?
  (c) Is there any relationship between these questions and those of Problem P 7.2.1?

P 7.2.3. Let n be a positive integer. What is the number of compositions of n (into any number of parts) such that every part is greater than 1? Are you surprised?

7.2.2 An Asymptotic Bound for $p_k(n)$.

Since straightforward formulas are not available for partition numbers, we may want to find good approximations. To give you one such approximation, in the problems you are asked to prove the following[5]:

Proposition 7.21. *Let n and k be positive integers with $k \leq n$. If, as usual, $p_k(n)$ is the number of partitions of n into k parts, then*

$$\frac{1}{k!}\binom{n-1}{k-1} \leq p_k(n) \leq \frac{1}{k!}\binom{n + \frac{(k+1)(k-2)}{2}}{k-1},$$

and, for a fixed k,

$$\lim_{n \to \infty} \frac{p_k(n)}{\frac{n^{k-1}}{k!\,(k-1)!}} = 1.$$

[5] Adapted from Mazur 2010.

P 7.2.4. You hear the following claim: calculating $p_3(8)$ amounts to finding all the ways that you can write 8 as a sum of three positive integers, and so the answer should be the same as the number of positive integer solutions to $x_1 + x_2 + x_3 = 8$.

(a) Explain why this is not correct.

(b) Let s be the number of positive integer solutions to $x_1 + x_2 + x_3 = 8$. Find s. Does s give an over-count or an under-count for $p_4(8)$?

(c) One partition of 8 into 3 parts is $(4, 3, 1)$. This one partition gives rise to how many different solutions to $x_1 + x_2 + x_3 = 8$? Answer the same question for each of the partitions of 8 into three parts.

(d) Which is bigger, $p_4(8)$ or $s/6$? Give an argument that doesn't rely on finding the actual numbers.

P 7.2.5. Let n and k be positive integers with $k \leq n$. Let s be the number of positive integer solutions to $x_1 + x_2 + \cdots + x_k = n$. Prove that $k! \, p_k(n) \geq s$. Conclude that

$$p_k(n) \geq \frac{1}{k!}\left(\binom{k}{n-k}\right).$$

P 7.2.6. (a) Write down – as in Problem P 7.2.4 – all five partitions of 8 into three parts. For each partition $(\lambda_1, \lambda_2, \lambda_3)$ of 8, consider $(\lambda_1+2, \lambda_2+1, \lambda_3+0)$. These are partitions of which number?

(b) Each of the (new) partitions constructed in the previous part gives rise to how many positive integer solutions to $x_1 + x_2 + x_3 = 11$? Do we get all the positive integer solutions of this equation in this way?

(c) Which is bigger, $6p_3(8)$ or the number of positive integer solutions to $x_1 + x_2 + x_3 = 11$? Answer the question without finding the answers, but then do check the actual numbers.

P 7.2.7. Let n and k be positive integers with $k \leq n$. Let $\lambda = (\lambda_1, \ldots, \lambda_k)$ be a partition of n into k parts. Let $\mu = (\lambda_1 + (k-1), \ldots, \lambda_{k-2} + 2, \lambda_{k-1} + 1, \lambda_k + 0) \vdash m$.

(a) Find m.

(b) Does μ always have distinct parts?

(c) By permuting the parts of μ, how many positive integer solutions to $x_1 + x_2 + \cdots + x_k = m$ can we find?

(d) Which is bigger, $k! \, p_k(n)$ or the number of positive integer solutions to $x_1 + x_2 + \cdots + x_k = m$?

(e) Prove that

$$p_k(n) \leq \frac{1}{k!}\left(\binom{k}{m-k}\right).$$

(f) Conclude that

$$p_k(n) \le \frac{1}{k!} \binom{m-1}{m-k} = \frac{1}{k!} \binom{n + \frac{(k+1)(k-2)}{2}}{k-1}.$$

P 7.2.8. Let k be a fixed positive integer, and, for positive integers n, let $A(n) = \frac{1}{k!} \binom{n-1}{k-1}$ and $B(n) = \frac{1}{k!} \binom{n + \frac{(k+1)(k-2)}{2}}{k-1}$ be the lower and upper bounds for $p_k(n)$ in Proposition 7.21. Finally, let $C(n) = \frac{n^{k-1}}{k!\,(k-1)!}$.

(a) For a fixed integer a, what is

$$\lim_{n \to \infty} \frac{(n+a)_k}{n^k},$$

where $(n+a)_k$ is, as usual, a falling factorial?

(b) For a fixed integer a, what is

$$\lim_{n \to \infty} \frac{\binom{n+a}{k-1}}{\frac{n^{k-1}}{(k-1)!}} ?$$

(c) Show that

$$\lim_{n \to \infty} \frac{A(n)}{C(n)} = 1 = \lim_{n \to \infty} \frac{B(n)}{C(n)}.$$

(d) Assuming that $A(n) \le p_k(n) \le B(n)$, prove that

$$\lim_{n \to \infty} \frac{p_k(n)}{C(n)} = 1.$$

P 7.2.9. After doing Problems P 7.2.4–P 7.2.8, write a concise outline for the proof of Proposition 7.21.

7.2.3 Euler's Pentagonal Number Theorem

Let n be a positive integer. Let $\mathcal{O}$ be the set of partitions of n into an odd number of distinct parts. Likewise, let $\mathcal{E}$ be the set of partitions of n into an even number of distinct parts. We want to compare $|\mathcal{E}|$ with $|\mathcal{O}|$. Admittedly, this is a strange question, but the lure is both the proof of the result in this section, and its application for finding an amazing recurrence relation for partition numbers, Theorem 9.54 of Chapter 9.

Example 7.22. CASE $n = 7$. The partitions (7) and $(4, 2, 1)$ are the only partitions of 7 into an odd number of distinct parts, while $(6, 1)$, $(5, 2)$, and $(4, 3)$ are the only partitions of 7 into an even number of distinct parts. Thus, in this case $|\mathcal{E}| = |\mathcal{O}| + 1$.

CASE $n = 9$. Partitions (9), $(6, 2, 1)$, $(5, 3, 1)$, are $(4, 3, 2)$ the partitions of 9 into an odd number of distinct parts, while $(8, 1)$, $(7, 2)$, $(6, 3)$, and $(5, 4)$ are the partitions of 9 into an even number of distinct parts. In this case $|\mathcal{E}| = |\mathcal{O}|$.

C_{ASE} $n = 12$. Partitions (12), $(9,2,1)$, $(8,3,1)$, $(7,4,1)$, $(7,3,2)$, $(6,5,1)$, $(6,4,2)$, and $(5,4,3)$ are the partitions of 12 into an odd number of distinct parts, and $(11,1)$, $(10,2)$, $(9,3)$, $(8,4)$, $(7,5)$, $(6,3,2,1)$, and $(5,4,2,1)$ are the only partitions of 12 into an even number of distinct parts. In this case $|\mathcal{E}| = |\mathcal{O}| - 1$.

The examples only suggest that the two counts – namely $|\mathcal{E}|$ and $|\mathcal{O}|$ – are going to be close to each other. It is remarkable that we can pinpoint the relationship precisely. We start with a definition.

Definition 7.23 (Generalized Pentagonal Numbers). The sequence of integers 1, 5, 12, 22, 35, ..., $\frac{k(3k-1)}{2}$, ... are called the *pentagonal numbers*. (See Figure 7.7.) If we allow zero and negative integer values for k, then we get the *generalized pentagonal numbers*. In other words, a generalized pentagonal number is any integer of the form $\frac{k(3k-1)}{2}$ for $k = 0, \pm 1, \pm 2, \ldots$ The sequence of the generalized pentagonal numbers begins as

$$0, 1, 2, 5, 7, 12, 15, 22, 26, 35, 40, 51, 57, 70, 77, 92, 100, \ldots$$

Figure 7.7 The pentagonal numbers 1, 5, 12, and 22.

The relationship between $|\mathcal{E}|$ and $|\mathcal{O}|$ depends on whether n is a generalized pentagonal number or not.

Proposition 7.24 (Euler's Pentagonal Number Theorem[6]). *Let n be a positive integer. Let $\mathcal{O}$ be the set of partitions of n into an odd number of distinct parts. Likewise, let $\mathcal{E}$ be the set of partitions of n into an even number of distinct parts. Then*

$$|\mathcal{E}| = |\mathcal{O}| + (-1)^k \quad \text{if } n = \frac{k(3k-1)}{2} \text{ for some } k = \pm 1, \pm 2, \ldots$$

$$|\mathcal{E}| = |\mathcal{O}| \quad \text{otherwise.}$$

---

[6] The Swiss mathematician Leonhard Euler (1707–1783) first conjectured and proved an equivalent version of this theorem (see Theorem 9.54). The connection with partition numbers was noticed by Adrien-Marie Legendre (1752–1833). Euler's father studied theology, was friends with the Bernoulli family, and wanted Euler to also become a priest. With the intervention of Johann Bernoulli (1667–1748), a prominent mathematician and Euler's tutor, he switched to mathematics. Euler lived most of his adult life in Saint Petersburg and Berlin, and is considered among the greatest mathematicians of all time. Pierre-Simon Laplace (1749–1827) is purported to have said "Read Euler, read Euler, he is the master of us all." For a wonderful selection of Euler's mathematics, see Dunham 1999.

The problems guide you through a proof of this result.[7] It is helpful to settle on the following notation.

Definition 7.25. Let n be a positive integer and let λ be a partition of n. In the Ferrers diagram of λ, we let $s(\lambda)$ to be the set of dots in the bottom row of the diagram. Furthermore, starting with the top row of the Ferrers diagram, consider the set of rows whose sizes are consecutive integers, and such that the last row in the set is not equal in size to the next row in the diagram. The set of dots at the ends of these rows will be denoted by $\ell(\lambda)$.

Example 7.26. Let $n = 21$ and λ be the partition $(6, 5, 4, 2, 2, 2)$. Then $|s(\lambda)| = 2$ and $|\ell(\lambda)| = 3$. The sets $s(\lambda)$ and $\ell(\lambda)$ are illustrated in Figure 7.8.

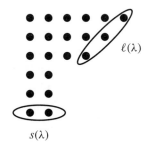

Figure 7.8 $s(\lambda)$ and $\ell(\lambda)$ for $\lambda = (6, 5, 4, 2, 2, 2)$.

Problems (continued)

P 7.2.10. (a) Consider the pentagonal figures in Figure 7.7. Show that, if we continue this sequence, the kth figure will have exactly $\frac{k(3k-1)}{2}$ dots.

(b) Multiply the sequence of the pentagonal numbers by 3. Are these numbers related to the triangular numbers (see Remark 1.3)? How?

(c) Consider the integers that are of the form $\frac{k(3k\pm1)}{2}$ for some non-negative integer k. How are these numbers related to the generalized pentagonal numbers?

P 7.2.11. Let n be a positive integer and $\lambda \vdash n$. In the Ferrers diagram for λ, take the dots in $s(\lambda)$ (Definition 7.25) and, starting from the top row, add one of these dots to each row until you run out. Call the resulting partition $U(\lambda)$. For each of the following partitions decide whether λ and/or $U(\lambda)$ is in $\mathcal{E}$, $\mathcal{O}$, or neither.

(a) $\lambda = (9, 8, 7, 5, 4, 3)$

(b) $\lambda = (11, 10, 8, 6)$

---

[7] This proof of the theorem is due to Fabian Franklin (1853–1939) (Franklin 1881). Both of Franklin's parents were born in Poland, but he was born in Hungary, and his family moved to the US when he was four. Johns Hopkins University opened in 1876, modeled after the German research universities. Franklin, who had been working as an engineer and surveyor for a railroad company, entered Johns Hopkins in 1877 and got his PhD in mathematics in 1880. He eventually became a professor at John Hopkins but left in 1895 to pursue a successful third career as a journalist and writer.

(c) $\lambda = (11, 10, 9, 8, 7, 6)$

(d) $\lambda = (9, 8, 7, 5, 3)$

(e) $\lambda = (9, 8, 7, 6, 5)$.

Let λ be a partition with $|\ell(\lambda)| \geq |s(\lambda)|$. Is it true that if $\lambda \in \mathcal{E}$, then $U(\lambda) \in \mathcal{O}$, and if $\lambda \in \mathcal{O}$, then $U(\lambda) \in \mathcal{E}$? If not, give an additional condition on λ to make this true.

P 7.2.12. Let n be a positive integer and $\lambda \vdash n$. In the Ferrers diagram for λ, take the dots in $\ell(\lambda)$ (Definition 7.25) and create a new row (appropriately placed) with them. Call the resulting partition $D(\lambda)$. For each of the following partitions, decide whether λ and/or $D(\lambda)$ is in $\mathcal{E}$, $\mathcal{O}$, or neither.

(a) $\lambda = (9, 8, 7, 5, 4)$

(b) $\lambda = (11, 10, 8, 2)$

(c) $\lambda = (10, 9, 8, 7, 6)$

(d) $\lambda = (9, 8, 7, 6, 4, 3)$

(e) $\lambda = (8, 7, 6, 5)$.

Let λ be a partition with $|\ell(\lambda)| < |s(\lambda)|$. Is it true that if $\lambda \in \mathcal{E}$, then $D(\lambda) \in \mathcal{O}$, and if $\lambda \in \mathcal{O}$, then $D(\lambda) \in \mathcal{E}$? If not, give an additional condition on λ to make this true.

P 7.2.13. The maps U and D are defined as in Problems P 7.2.11 and P 7.2.12. Let $k > 1$ be an integer, and consider the partition $\lambda = (2k - 1, 2k - 2, \ldots, k)$ of $n = (2k - 1) + \cdots + k = \frac{k(3k-1)}{2}$. Does there exist a partition $\mu \vdash n$ with distinct parts such that $U(\mu) = \lambda$? What about $D(\mu) = \lambda$?

P 7.2.14. Let n be a positive integer, and let the map U be defined as in Problem P 7.2.11. We call a partition $\lambda \vdash n$ *problematic* if $|s(\lambda)| \leq |\ell(\lambda)|$, and either $\lambda \in \mathcal{E}$ and $U(\lambda) \notin \mathcal{O}$ or $\lambda \in \mathcal{O}$ and $U(\lambda) \notin \mathcal{E}$.

(a) Show that a problematic $\lambda \in \mathcal{E}$ exists if and only if n is a pentagonal number $\frac{k(3k-1)}{2}$ with k even.

(b) Show that a problematic $\lambda \in \mathcal{O}$ exists if and only if n is a pentagonal number $\frac{k(3k-1)}{2}$ with k odd.

Moreover, in either case, the problematic λ is unique.

P 7.2.15. Find results similar to those in Problem P 7.2.14, for the operator D of Problem P 7.2.12.

P 7.2.16. Let n be a positive integer and assume that n is not a generalized pentagonal number. Define the function f on partitions $\lambda \vdash n$ as follows:

$$f(\lambda) = \begin{cases} U(\lambda) & \text{if } |s(\lambda)| \leq |\ell(\lambda)| \\ D(\lambda) & \text{if } |s(\lambda)| > |\ell(\lambda)|. \end{cases}$$

Show that $f : \mathcal{E} \to \mathcal{O}$ as well as $f : \mathcal{O} \to \mathcal{E}$. Also show that applying f twice in succession gives the identity map on $\mathcal{E}$ and on $\mathcal{O}$. Conclude that $f : \mathcal{E} \to \mathcal{O}$ is a bijection.

P 7.2.17. Let n be a non-zero generalized pentagonal number. Assume $n = \frac{k(3k\pm1)}{2}$.

(a) If k is even, show that $|\mathcal{E}| = |\mathcal{O}| + 1$.

(b) If k is odd, show that $|\mathcal{O}| = |\mathcal{E}| + 1$.

P 7.2.18. After doing Problems P 7.2.10–P 7.2.17, write a concise outline of the proof of Proposition 7.24.

7.3 Open Problems and Conjectures

The theory of integer partitions is vast and has deep connections to other areas of mathematics including analysis, number theory, and representation theory. Pursuing any of these connections is beyond the scope of this book, and so we will limit ourselves to a few open problems that can be stated without too much added jargon.[8]

Recall that $p(n) = p_1(n) + \cdots + p_n(n)$ is the total number of partitions of a positive integer n. Let x be a large positive integer. There are heuristic arguments that suggest that about half of the values of $p(n)$, for $1 \leq n \leq x$, should be even and about half should be odd. While billions and billions of values of $p(n)$ have been calculated, the following remain open problems:

Problem 7.27. (a) *When is $p(n)$ even and when is $p(n)$ odd? Give a reasonably straightforward condition on n that determines the parity of $p(n)$.*

(b) *Is* $\displaystyle \lim_{x \to \infty} \frac{|\{n \mid 1 \leq n \leq x \text{ and } p(n) \text{ even}\}|}{x} = \frac{1}{2}$?

(c) *When is $p(n)$ a prime number? Is it a prime infinitely often?*

An exciting result is a theorem of Ken Ono from 1995 (Oro 1995). He proved, using the advanced machinery of modular forms, that in every arithmetic progression there are an infinite number of terms n with $p(n)$ even, and, in any arithmetic progression, if there is at least one n with $p(n)$ odd, then there are infinitely many such n.

As we have seen, many combinatorial results about partitions compare one set of partitions with another. There are many open problems and conjectures of this type. We mention a recent such conjecture.

Conjecture 7.28 (Berkovich and Uncu 2019). *Let n, s, and L be positive integers with $L \geq 3$. Let $\mathcal{A}$ be the set of those partitions of n with smallest part greater than s and largest part no more than $s + L$. Let $\mathcal{B}$ be the set of those partitions of n with smallest part equal to s, largest part no more than $s + L$, and no part equal to $s + L - 1$. Then, for large enough n, $|\mathcal{A}| \leq |\mathcal{B}|$.*

Berkovich and Uncu (2019) prove the cases when $s = 1$ and $s = 2$, but pose the more general situation as a conjecture.

---

[8] For a very readable foray into the subject at the undergraduate level, see Andrews and Eriksson 2004. For a more advanced treatment, see Andrews 1998.

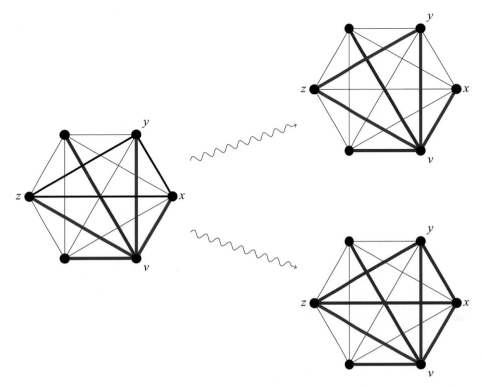

Figure 2.12 If the edges of a K_6 are colored with red and blue, then, out of the five edges incident with v, at least three will be of the same – say blue – color (figure on the left). Let x, y, and z be the other vertices of these three edges. The triangle $\triangle xyz$ either has a blue edge (top right) or all three of its edges are red (bottom right).

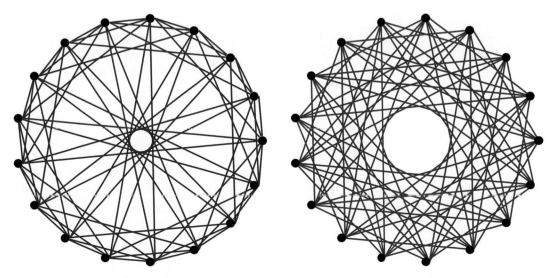

Figure 2.16 The red (left) and blue (right) edges in K_{17} without a red or a blue K_4, showing $r(4,4) > 17$.

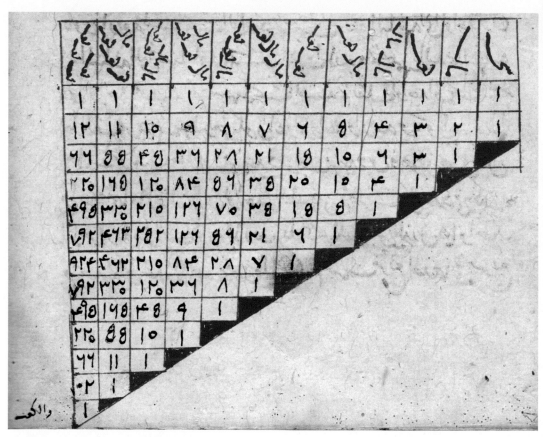

Figure 5.1 Karajī's triangle (circa 1000 CE), as described by al-Samaw'al in in mid twelfth century. Courtesy of Directorate of the Turkish Institute of Manuscripts, Süleymaniye Manuscript Library Ayasofya, Collection 02718, folio 50.

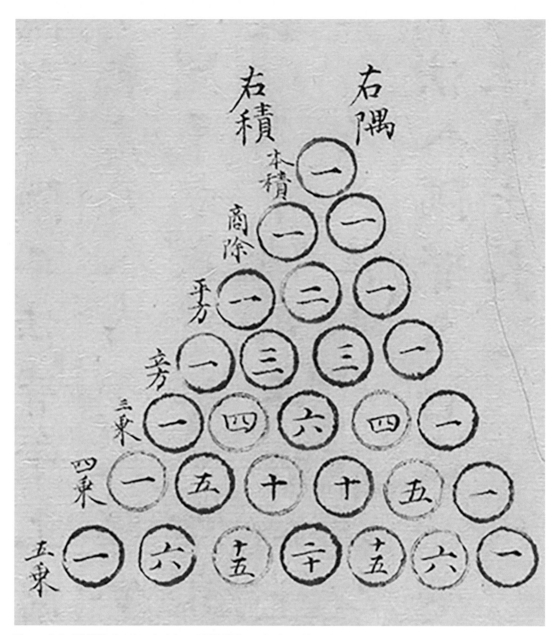

Figure 5.2 Jiǎ Xiàn's triangle (circa 1050 CE), as depicted by Zhu Shijie, around 1303 CE. Reproduced by kind permission of the Syndics of Cambridge University Library.

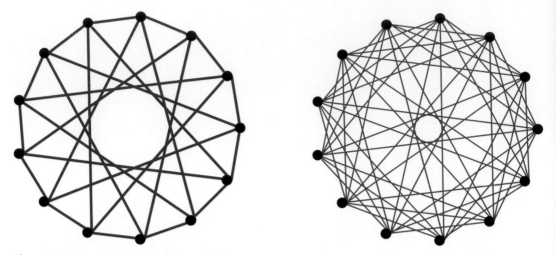

Figure D.5 The red (left) and blue (right) edges in K_{13} without a red triangle or a blue K_5 showing $r(3, 5) > 13$.

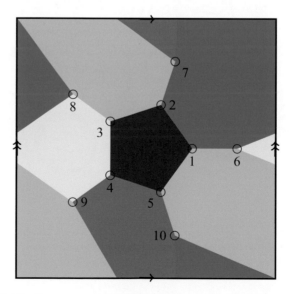

Figure D.17 The Petersen graph embedded on a torus, and its regions colored.

Collaborative Mini-project 4: Generating Functions

In this mini-project, we explore the use of power series to answer combinatorial questions. A *power series* is an expression of the form $a_0 + a_1 x + \cdots + a_n x^n + \cdots$, and we use it as a tool for studying the sequence of its coefficients $a_0, a_1, \ldots, a_n, \ldots$ We have already seen several examples of such a technique. In Example (d) in the Introduction, we used power series to solve a recurrence relation, and in the proof of Proposition 6.15, the power series expansion of e^x and related series was used to prove a formula for Bell numbers.

Preliminary Investigations

Please complete this section before beginning collaboration with other students.

MP4.1 Let $a(n)$ denote the number of integer solutions to

$$y_1 + y_2 + y_3 = n,$$

where $0 \le y_1 \le 1$, $0 \le y_2 \le 2$, and $1 \le y_3 \le 3$. Find the sequence

$$a(0), a(1), a(2), \ldots$$

Do this directly by finding the actual number of solutions for each n.

Generating Functions

Let

$$h_0, h_1, h_2, \ldots, h_n, \ldots$$

be a sequence of numbers. The *(ordinary) generating function* of this sequence is

$$g(x) = h_0 + h_1 x + h_2 x^2 + \cdots + h_n x^n + \cdots.$$

Usually in combinatorics we are not interested in plugging in values for x in a generating function. Rather, we are interested in the coefficients. Thus, most of the time, we do not have to be concerned with convergence questions and can manipulate the series as "formal power series." We are, however, interested in finding closed forms for our generating functions if possible. In this way, a function can carry all the information in the sequence, and manipulations of the function can provide insight into the behavior of the sequence.

EXAMPLE. The generating function for the sequence

$$1, 1, 1, \ldots$$

is

$$g(x) = 1 + x + x^2 + \cdots = \frac{1}{1-x}. \tag{7.1}$$

The nth term of the sequence is, of course, 1, but it is also the coefficient of x^n in the Taylor series[1] expansion of $\frac{1}{1-x}$. Equation 7.1 is very useful, since by manipulating it – that is, by substituting more complicated expressions for x, by differentiating and integrating, and by multiplying it by other expressions – we can get other series expansions.

MP4.2 Can you find a closed form for the generating function of

$$1, 2, 3, \ldots?$$

HINT: First write down the power series, and then use Equation 7.1 to find a closed form. When in doubt, differentiate.

Collaborative Investigation

Work on the rest of the questions with your team. Before getting started on the new questions, you may want to compare notes about the preliminary problems.

MP4.3 Consider the polynomial

$$f(x) = (1 + x)(1 + x + x^2)(x + x^2 + x^3).$$

Let $b(n)$ be the coefficient of x^n in $f(x)$. Find the sequence

$$b(0), b(1), b(2), \ldots$$

Compare this sequence with the one in Problem MP4.1. Explain what you found. Then generalize your observation and give an explanation for its validity.

MP4.4 What is the closed form for the generating function of

$$1, 2, 4, 8, 16, 32, \ldots?$$

HINT: Again start with Equation 7.1 and replace x with . . .

MP4.5 Can you give a closed form for the generating function of

$$\binom{m}{0}, \binom{m}{1}, \binom{m}{2}, \ldots, \binom{m}{m}?$$

---

[1] If you do not remember that the Taylor series of $\frac{1}{1-x}$ is $1 + x + x^2 + \cdots$, please consult any calculus book such as my *Approximately Calculus* (Shahriari 2006). You do *not* need a deep understanding of Taylor series to understand generating functions.

MP4.6 Let m be a positive integer and consider

$$g_m(x) = \frac{1}{(1-x)^m}.$$

Find the sequence for which $g_m(x)$ is the generating function.

The following steps may be helpful:

STEP 1: Find the Taylor series of $g_m(x)$ for small values of m. To do this, you could start with Equation 7.1 and take its derivative.

STEP 2: The coefficient of x^n in $g_m(x)$ is a binomial coefficient. By comparing your examples with a table of binomial coefficients, make a conjecture.

STEP 3: Use induction and derivatives to prove your conjecture.

MP4.7 We can use generating functions to prove general identities. The following steps leads you to one such identity for binomial coefficients:

STEP 1: Expand $(1+x)^5$ using the binomial theorem. Leave the binomial coefficients as they are (i.e., do not find the actual numbers). Do the same for $(1+x)^7$.

STEP 2: We know

$$(1+x)^5(1+x)^7 = (1+x)^{12}.$$

What is the coefficient of x^4 on the right-hand side? What is the coefficient of x^4 on the left-hand side? Can you write what you found as an identity for binomial coefficients? Did this process just suggest an identity or did it actually *prove* the identity?

STEP 3: Generalize your answer to the previous steps and prove a general identity for binomial coefficients.

MP4.8 In how many ways can you distribute 30 (identical) slices of pizza to five (distinct) students so that none of them gets more than seven slices?

Answer this question using generating functions by following the steps below:

STEP 1: First generalize the problem. We have n slices of pizza, and five students. No student is allowed to eat more than seven slices. The students are distinguishable, but the slices of pizza are not. Let $c(n)$ be the number of ways to distribute the slices of pizza.

STEP 2: Let $g(x)$ be the generating function for the sequence

$$c(0), c(1), c(2), \ldots, c(n), \ldots$$

Use an argument similar to Problem MP4.3 to find an expression for $g(x)$ as a product of polynomials.

STEP 3: Write $g(x)$ as

$$g(x) = \left(\frac{?-?}{1-?}\right)^? = \frac{1}{(1-?)^?}(?-?)^?$$

STEP 4: Since we want $c(30)$, we need to find the coefficient of x^{30} in the Taylor series expansion of $g(x)$. Hence, we do not need to find all the terms of the Taylor expansion.

Use Problem MP4.6 to expand – as far as necessary – one of the factors of $g(x)$. Expand the other one – as far as necessary – by multiplying it out. Use these to find $c(30)$.

Mini-project Write-Up

In your write-up of the results, include at least the following:

- A careful definition of an ordinary generating function of a sequence
- Examples of sequences and their generating functions
- In your examples, illustrations of how you manipulate known generating functions to get new ones
- Proof of a binomial identity using generating functions (as in Problem MP4.7)
- An example of finding the number of integral solutions to a linear equation where the variables have both lower and upper bounds (as in Problem MP4.3)
- A complete solution to Problem MP4.8 regarding the distribution of pizza slices.

Collaborative Mini-project 5: Graphic Sequences and Planar Graphs

In this mini-project, we investigate some properties of graphs. In Section 2.2, we saw some basic definitions of graphs. Here, we consider degree sequences, trees, and planar graphs (all to be defined below).

Preliminary Investigations

Please complete this section before beginning collaboration with other students.

Preliminaries on Graphs: A Review

Recall that a (simple) *graph* G consists of two sets V and E (or, to be more precise, $V(G)$ and $E(G)$). The set V is the set of *vertices* and E is a (possibly empty) set of distinct pairs of elements of V. Elements of E are called the *edges* of the graph. If V and E are, respectively, the sets of vertices and of edges of G, then we write $G = (V, E)$. If v_1 and v_2 are two vertices of a graph G – in other words, v_1, $v_2 \in V$ – then an element of E could possibly be $\{v_1, v_2\}$. If so, then we say that the edge $e = \{v_1, v_2\}$ *joins* the vertices v_1 and v_2, or v_1 (or v_2) and e are *incident*. In this mini-project, we only consider finite (simple) graphs and not multigraphs (where E can be a multiset and multiple edges are allowed), or general graphs (where E is allowed to include pairs of identical vertices and loops are allowed). In other words, we do not allow double edges or loops.

Chains and Cycles

Let $G = (V, E)$ be a graph. A sequence of m edges of the form

$$\{x_0, x_1\}, \{x_1, x_2\}, \ldots, \{x_{m-1}, x_m\},$$

where the vertices $x_0, \ldots, x_m$ are distinct, is called a *chain* or a *path* of length m. The same sequence of edges where all the vertices are distinct except $x_0 = x_m$ is called a *cycle*.

Degrees, and Degree Sequences

Let G be a graph, and let v be a vertex of G. The *degree* of v is the number of edges incident with v. The *degree sequence* of G is the (finite) sequence of all the degrees of the vertices of G arranged in non-increasing order:

$$d_1, d_2, \cdots, d_p, \text{ or } (d_1, d_2, \ldots, d_p).$$

A non-increasing sequence of non-negative integers is called *graphic* if there exists a graph whose degree sequence is precisely that sequence.

MP5.1 What is the degree sequence of K_5, the complete graph with five vertices? What is the degree sequence of a chain of length 4?

MP5.2 Is the sequence $(5, 5, 4, 4, 0)$ graphic? What about $(1, 1, 1, 1, 1, 1)$? What about $(3, 3, 2, 2, 2, 2)$?

Cycles and Trees

A graph G is *connected* if, for any two distinct vertices a and b, there is a chain from a to b. A *tree* is a connected graph with no cycles. A *forest* is a graph with no cycles. In other words, a forest is a disjoint union of trees.

MP5.3 Let G be a finite tree with p vertices and q edges. Look at some examples and make a conjecture about the relationship of p and q.

Planar Graphs

A *planar graph* is a graph that can be drawn in the plane with no edges crossing.

When you draw a planar graph in the plane with no edges crossing, the plane is divided up into *regions*. One of these regions, the outside region, has infinite area. We let r denote the number of regions in a plane drawing of a planar graph.

In what follows you can assume that a plane drawing of a cycle divides the plane into two regions, one inside and one outside. A more general version of this fact, known in topology as the *Jordan curve theorem*, is in fact difficult to prove rigorously.

MP5.4 Show that K_4 is planar by exhibiting a plane drawing of K_4.

MP5.5 We have three houses and three wells. Can we join each of the houses to each of the wells with non-crossing paths? Make a conjecture.

Collaborative Investigation

Work on the rest of the questions with your team. Before getting started on the new questions, you may want to compare notes about the preliminary problems.

MP5.6 Our goal is to find an algorithm that easily identifies graphic sequences. We are looking for an algorithm that starts with a non-increasing sequence of non-negative integers and somehow creates a simpler sequence. We want the algorithm to be such that if, after the repeated application of the algorithm, we arrive at a degree sequence that is clearly graphic, then the original sequence is graphic also. In addition, we want to investigate whether our algorithm can discern non-graphic sequences as well. In other words, if we start with a non-increasing sequence of non-negative integers and, after repeatedly applying our algorithm, we arrive at a non-graphic sequence, then will the original sequence be necessarily non-graphic as well?

Investigate degree sequences and try to achieve the above goal by completing and answering the questions in the following steps:

STEP 1: Is the following sequence graphic:

$$(3, 2, 2, 2, 1, 1, 1)?$$

Could we have answered this question without drawing a graph, and by using the fact that the sequence $(1, 1, 1, 1, 1, 1)$ is graphic? In other words, the latter sequence is graphic, and therefore there is a graph whose degree sequence is $(1, 1, 1, 1, 1, 1)$. Can we modify this graph and get a graph whose degree sequence is $(3, 2, 2, 2, 1, 1, 1)$?

STEP 2: Use the ideas of the last step to decide whether the following sequence is graphic:

$$(6, 5, 5, 4, 3, 3, 2, 2, 2).$$

You should have an algorithm that starts with an alleged degree sequence and gives a simpler sequence. If the repeated application of the algorithm results in a graphic sequence, then the original sequence should be graphic also.

STEP 3: Does the converse of your algorithm work? In other words, assume that you started with a sequence and, using your procedure of the previous problem, you arrived at a *non-graphic* sequence. Can you conclude that your original sequence is non-graphic? Answer this question by looking at some examples and making a conjecture. The proof of your conjecture is not very straightforward. Do you agree? (You do not have to find a proof.)

MP5.7 Prove your conjecture in Problem MP5.3.

HINT: Use induction, and use a vertex at the end of a longest path in G.

MP5.8 Are all trees planar? Convince yourself.

MP5.9 Let G be a connected planar graph. Let v, e, and r be the number of vertices, edges, and regions, respectively. By looking at some examples, make a conjecture about

$$v - e + r.$$

Are both the two conditions (connectedness and planarity) necessary for your conjecture?

MP5.10 Prove your conjecture of Problem MP5.9 by induction on the number of cycles.

HINT: If you remove one edge of one cycle, you will have a graph with fewer cycles. What happens to v, e, and r?

MP5.11 Let $n \geq 3$. Let G be a planar graph with n vertices. Let M be the maximum possible number of edges. By looking at some examples, make a conjecture about M. Use Problem MP5.9 to prove your conjecture.

MP5.12 Let $n \geq 3$. Let G be a triangle-free planar graph with n vertices (i.e., there are no three vertices in G such that each pair of these vertices is connected by an edge.) Let M be the maximum possible number of edges. By looking at some examples, make a conjecture about M.

MP5.13 Using your conjectures (in Problem MP5.11 or MP5.12), can you prove your conjecture of Problem MP5.5?

MP5.14 Is K_5 planar? Answer using your conjectures (in Problem MP5.11 or MP5.12).

Mini-project Write-Up

In your write-up of the results, include at least the following:

- A definition of graphic sequence and an algorithm for identifying graphic sequences Examples should be given and, when appropriate, a proof or a conjecture noted
- A proof of the relation between the number of vertices and edges of a tree
- A conjecture with proof about $v - e + r$ in planar connected graphs
- Applications to maximum number of edges in planar graph and in triangle-free planar graphs
- A proof of your answers to Problems MP5.5 and MP5.14.

Collaborative Mini-project 6: Connectivity of Graphs

In this mini-project, we investigate connectivity properties of graphs.[1] We assume the reader is, by now, familiar with the basic vocabulary of graphs, introduced both in Section 2.2 and in Collaborative Mini-project 5.

Preliminary Investigations

Please complete this section before beginning collaboration with other students.

Connectivity of Graphs

A graph is *connected* if any two distinct vertices can be joined by a path; otherwise it is called *disconnected*. A maximal connected subgraph of a graph G is a *component* of G.

The *vertex-connectivity* or simply *connectivity* of a graph G is the minimum number of vertices whose removal from G results in a disconnected graph or the graph with just one vertex (and no edges). We denote this number by $\kappa(G)$. (When we remove a vertex, we also remove all edges incident with it.)

The *edge-connectivity* of a graph G is the minimum number of edges whose removal from G results in a disconnected graph or the graph with just one vertex. We denote this number by $\lambda(G)$. (When we remove edges, we do not remove any vertices.)

Among the vertices of G, one (or more) has the least degree, and one (or more) has the largest degree. The minimum degree of a graph G is denoted by $\delta(G)$, and the maximum degree of a graph G is denoted by $\Delta(G)$.

MP6.1 Find $\kappa(G)$, $\lambda(G)$, $\delta(G)$, and $\Delta(G)$ for each of the following graphs:
  (i) G is a disconnected graph consisting of two (disjoint) copies of K_4.
  (ii) G consists of just one vertex, and no edges.
  (iii) G is a path of length ℓ (Example 2.25). This means that the vertices of G are $\{1, 2, \ldots, \ell, \ell + 1\}$, and the edges are $\{1, 2\}$, $\{2, 3\}$, ..., $\{\ell - 1, l\}$, and $\{\ell, \ell + 1\}$.
  (iv) G is a cycle of length ℓ (Example 2.26). This means that the vertices of G are $\{1, 2, 3, \ldots, \ell\}$, and the edges are $\{1, 2\}$, $\{2, 3\}$, ..., $\{\ell - 1, \ell\}$, and $\{\ell, 1\}$.
  (v) $G = K_n$, the complete graph with n vertices.

---

[1] The topics in this mini-project are not discussed later in the text. Hence, this material can be explored before or after working through chapter 10.

MP6.2 Find an example of a simple graph G with 7 vertices such that $\kappa(G) = 1$ and $\lambda(G) = 3$. In your example, what are $\delta(G)$ and $\Delta(G)$?

Collaborative Investigation

Work on the rest of the questions with your team. Before getting started on the new questions, you may want to compare notes about the preliminary problems.

MP6.3 Assume that G has n vertices and $\lambda(G) = n - 1$. What can you say about G?

MP6.4 Can you generalize your construction of the example in Problem MP6.2 to get a graph with vertex-connectivity equal to one but arbitrarily large edge-connectivity?

MP6.5 Find an example of a graph G with $\kappa(G) = 2$, $\lambda(G) = 3$, and $\delta(G) = 4$.

HINT: Start with two K_5's.

MP6.6 Is there any relationship between $\lambda(G)$ and $\delta(G)$? Prove it.

MP6.7 A friend argues as follows. If $\lambda(G) = k$, then there are k edges whose removal results in a disconnected graph. Instead remove just one vertex from each of those edges. When you remove a vertex, you also remove the edges incident with it. As a result, removing these vertices will result in removing the original k edges as well. Hence, the resulting graph will be disconnected. Hence, $\kappa(G) \leq \lambda(G)$. By considering the graph in Figure MP6.1, show that this argument, in its present form, is incorrect. What can go wrong? Do you still think that $\kappa(G) \leq \lambda(G)$? Make a conjecture.

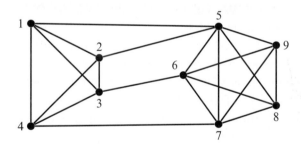

Figure MP6.1 A graph with nine vertices.

MP6.8 Is $\kappa(G) \leq \lambda(G)$, for complete graphs and for graphs with $\lambda(G) = 0$?

MP6.9 You have n vertices. You split them into two non-empty subsets, X and Y. You create an edge $\{x, y\}$ for every $x \in X$ and $y \in Y$. Could you have $n - 1$ edges? Could you have less than $n - 1$ edges?

MP6.10 Let $n \geq m$ be positive integers, and $G = K_{n,m}$ be the complete bipartite graph with $n + m$ vertices. Let S be a set consisting of a vertices of G of degree m and b vertices of G of degree n. Let T be the remaining set of vertices of G. Let $f(a, b)$ denote the number of edges with one end in S and the other end in T. Assuming that neither S nor T is an empty set, what is the minimum value of $f(a, b)$?

MP6.11   Let $n \geq m$ be positive integers, and $G = K_{n,m}$. Find $\kappa(G)$, $\lambda(G)$, $\delta(G)$, and $\Delta(G)$.

   HINT: Problem MP6.10 may be relevant.

MP6.12 Consider the graph in Figure MP6.1 with nine vertices.

   (i) Check that removing the edges $F = \{\{1,5\}, \{2,5\}, \{3,6\}, \{4,7\}\}$ will result in a disconnected graph, and in fact $\lambda(G) = 4$.

   (ii) Removing F will result in two components, X and Y. Find a vertex $x \in X$ and a vertex $y \in Y$ such that $\{x, y\}$ was not an edge of G.

   (iii) For each edge in F, pick one of its end points, making sure that you never pick x or y. Call this collection of vertices K.

   (iv) Check that removing the vertices in K and the edges incident with them causes the graph to be disconnected. After removing K, is x connected to y?

   (v) What can you conclude about $\kappa(G)$?

MP6.13 Let G be a simple graph. Prove that

$$\kappa(G) \leq \lambda(G) \leq \delta(G) \leq \Delta(G).$$

   HINT: Because of Problem MP6.8 you can assume that the graph G is neither disconnected nor complete. Mimic the steps of Problem MP6.12. In generalizing Problem MP6.12ii, use Problems MP6.9 and MP6.3 to show that you can find the desired vertices x and y.

Separating Sets and Independent Paths

Let G be a simple connected graph and assume that s and t are two vertices of G. A set S of vertices (or edges) of G is said to *separate* s and t if the removal of the elements of S from G produces a disconnected graph in which s and t lie in different components. In other words, if you remove S then there is no path from s to t.

If s and t are two vertices in a graph G, then two paths in G from s to t are called *independent* if they have only the vertices s and t in common.

MP6.14 Consider the graph in Figure MP6.2. Find the following numbers: the minimum number of vertices that separate u and v, the minimum number of edges that separate u and v, the

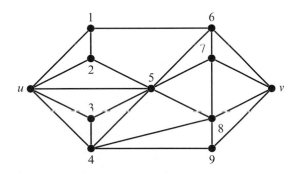

Figure MP6.2 A graph with 11 vertices.

maximum number of pairwise independent paths from u to v, and the maximum number of pairwise edge-disjoint paths from u to v.

MP6.15 Let G be a simple graph, and let u and v be two vertices of G. Assume that $\{u, v\}$ is not an edge of G. Define:

$$a = \text{the minimum number of vertices of } G \text{ that separate } u \text{ and } v$$

$$b = \text{the minimum number of edges of } G \text{ that separate } u \text{ and } v$$

$$c = \text{the maximum number of pairwise independent paths from } u \text{ to } v$$

$$d = \text{The maximum number of pairwise edge-disjoint paths from } u \text{ to } v.$$

It is a theorem (that we are not proving) that $a = c$ and $b = d$. To show $a = c$, you have to show $a \le c$ and $c \le a$. One of these is much easier to prove than the other. Which one? Prove it. Prove the corresponding statement for b and d.

k-Connected Graphs

If k is an integer, we say that the simple graph G is k-connected if $\kappa(G) \ge k$.

MP6.16 Let G be a simple graph with n vertices, and let k be a positive integer with $k \le n - 1$. Assume the theorem mentioned in Problem MP6.15, and prove a relationship between the k-connectivity of the graph G and the number of pairwise independent paths between any two non-adjacent vertices.

Mini-project Write-Up

In your write-up of the results, include at least the following:

- For a graph G, clear definitions and examples of $\kappa(G)$, $\lambda(G)$, $\delta(G)$, and $\Delta(G)$
- The examples requested in Problems MP6.4 and MP6.5
- A complete proof of the statement in Problem MP6.13
- Definitions, with examples, of separating sets and independent paths. How is finding the minimum size of a separating set different from finding $\kappa(G)$?
- Restatements of the problems and proofs requested in Problems MP6.15 and MP6.16.

8   The Inclusion–Exclusion Principle

You compose 10 different emails, each to a different friend. The prank-ware on your email server permutes the recipients' email addresses and randomly sends each of the messages to one of the individuals on your list. What is the probability that none of your friends receives the email intended for them? Would this probability go up or down if instead of 10 emails, you had written 10,000 emails?

8.1   The Inclusion–Exclusion Principle

Warm-Up 8.1. *A coffee company spent ten cents per person to question patrons about their coffee preferences. Of the persons interviewed, 250 liked latte, 200 liked cappuccino, 80 liked both, and 100 did not like either choice. What was the total amount of money the company had to pay?*

The inclusion–exclusion principle is a counting method that generalizes familiar ideas. The addition principle told us that, if a set was partitioned into subsets, we could count the number of elements in the set by adding the number of elements in each part. In a partition, the subsets cover the original set – that is, the union of the subsets is the original set – and the pairwise intersection of the subsets is the empty set. The inclusion–exclusion principle extends the addition principle to the case when we have a collection of subsets that cover the original set but – unlike a partition – may have non-trivial intersections.

Remark 8.2. The inclusion–exclusion principle is also an example of a *sieve* method. In a sieve method you start with a set larger than what you are interested in and systematically throw out extraneous elements. Historically, an early use of a sieve method was by Eratosthenes.[1] For the sieve of Eratosthenes, you start with the integers from 2 to n and, one by one, cross out the multiples of the first integer that has not yet been crossed out (you don't cross out that integer itself and you stop when you get to $\sqrt{n}$). At the end of the process, you are left with all the primes up to n.

Definition 8.3. Let S be a set and $A \subseteq S$. Then the *complement* of A (in S), written A^c, is defined by

$$A^c = S - A = \{x \in S \mid x \notin A\}.$$

Remark 8.4. Note that, given a set A, to be able to talk about the complement of A, we first have to have an ambient set S. The notation A^c does not mention S and, as such, it could be ambiguous. However, most often it is clear from the context what the bigger set S is.

---

[1] Born in Cyrene – which is in today's Libya – in 276 BCE, and died in Alexandria, Egypt in 194 BCE.

Before the main theorem, in the next example we do a special case of the inclusion–exclusion principle.

Example 8.5. Let S be a finite set, and let A_1 and A_2 be two subsets of S. These two subsets may or may not overlap (see Figure 8.1). We want to find an expression for the number of elements of S that are in at least one of A_1 or A_2. In other words, we want to find $|A_1 \cup A_2|$. In the case of just two sets, the Venn diagram of Figure 8.1 can be helpful. We want to count the elements that are in $A_1 \cup A_2$. We start by finding $|A_1| + |A_2|$. However, by doing so, we have double counted those elements that are in both A_1 and A_2, namely the elements of $A_1 \cap A_2$. To compensate, we subtract one of those counts to get

$$|A_1 \cup A_2| = |A_1| + |A_2| - |A_1 \cap A_2|.$$

Instead of wanting to count $|A_1 \cup A_2|$, sometimes we want to count the elements that are in neither A_1 nor A_2. This amounts to finding $\left|A_1^c \cap A_2^c\right|$. We don't have to make a new argument, since the elements that we want to count are exactly the elements of S that we did *not* count in our earlier count. Hence,

$$\left|A_1^c \cap A_2^c\right| = |S| - |A_1 \cup A_2| = |S| - |A_1| - |A_2| + |A_1 \cap A_2|.$$

When the number of sets increases, it is much harder to draw Venn diagrams. As a prelude to the argument in the general case, we give a different argument for the case of two sets as well.

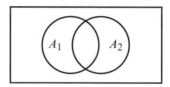

Figure 8.1 $|A_1 \cup A_2| = |A_1| + |A_2| - |A_1 \cap A_2|$, and $\left|A_1^c \cap A_2^c\right| = |S| - |A_1 \cup A_2|$.

CLAIM: If A_1 and A_2 are subsets of a finite set S, then

$$|A_1 \cup A_2| = |A_1| + |A_2| - |A_1 \cap A_2|. \tag{8.1}$$

PROOF OF CLAIM: Every term in each side of Equation 8.1 counts certain elements of S. We prove the claim by showing that every element of S – regardless of where it is – makes the same contribution to both sides of Equation 8.1.

Let $x \in S$. Either x is a member of $A_1 \cup A_2$ or it is not.

CASE 1: If x is an element of S that is not a member of $A_1 \cup A_2$, then it makes a contribution of 0 to the left-hand side of Equation 8.1. The element x is also not an element of A_1, A_2, or $A_1 \cap A_2$. Hence it makes a contribution of $0+0-0 = 0$ to the right-hand side of Equation 8.1. Hence, in this case, the contribution of x to both sides is the same.

CASE 2: If x is a member of $A_1 \cup A_2$, then x makes a contribution of 1 to the left-hand side of Equation 8.1. We consider three subcases depending on whether x is just in A_1, just in A_2, or in both.

Subcase 1: x is in A_1 but not in A_2. In this case, the contribution of x to the right-hand side of Equation 8.1 is $1 + 0 - 0 = 1$.

Subcase 2: x is in A_2 but not in A_1. Similarly, in this case, the contribution of x to the right-hand side of Equation 8.1 is $0 + 1 - 0 = 1$.

Subcase 3: x is in A_1 and in A_2. In this case, the contribution of x to the right-hand side of Equation 8.1 is $1 + 1 - 1 = 1$.

Thus, in all cases, an element of S makes the same contribution to both sides of Equation 8.1, and the proof is complete. □

We now turn to the general situation, and we will follow the method of proof of Example 8.5. Let S be a finite set, and let $A_1, \ldots, A_m$ be m subsets of S. We want to find an expression for the number of elements of S that are in at least one of these subsets – namely, $|A_1 \cup A_2 \cup \ldots \cup A_m|$ – or an expression for the number of elements of S that are in *none* of the subsets, that is, $\left| A_1^c \cap A_2^c \cap \ldots \cap A_m^c \right|$. These two questions are intimately related since every element of S has one of two options: it can be either (1) in at least one the subsets or (2) in none of the subsets. So

$$|A_1 \cup A_2 \cup \ldots \cup A_m| + \left| A_1^c \cap A_2^c \cap \ldots \cap A_m^c \right| = |S| .$$

Thus, we just need to find one of these.

Theorem 8.6 (Inclusion–Exclusion Principle).   *Let S be a set, m a positive integer, and, for $1 \leq i \leq m$, A_i is a subset of S. Then*

$$
\left| A_1^c \cap A_2^c \cap \cdots \cap A_m^c \right| = |S| - \sum_{i=1}^m |A_i| + \sum_{1 \leq i < j \leq m} |A_i \cap A_j|
$$
$$
- \sum_{1 \leq i < j < k \leq m} |A_i \cap A_j \cap A_k| + \cdots + (-1)^m |A_1 \cap \cdots \cap A_m| .
$$

(8.2)

Remark 8.7.   The indices in the sums of Equation 8.2 should not confuse you. In the first sum – $\sum_{i=1}^m |A_i|$ – we have one term for each subset $\{i\}$ of size 1 of $[m] = \{1, \ldots, m\}$. For the second sum – $\sum_{1 \leq i < j \leq m} |A_i \cap A_j|$ – we have one term for each subset $\{i, j\}$ of size 2 of $[m]$. For the third sum, we have one term for each subsets $\{i, j, k\}$ of size 3 of $[m]$, and so on.

Example 8.8.   If $m = 3$, and A_1, A_2, and A_3 are subsets of a set S, then, by Theorem 8.6, we have

$$\left| A_1^c \cap A_2^c \cap A_3^c \right| = |S| - |A_1| - |A_2| - |A_3| + |A_1 \cap A_2| + |A_2 \cap A_3| + |A_1 \cap A_3|$$
$$|A_1 \cap A_2 \cap A_3| .$$

Note that the number of terms on the right-hand side of Equation 8.2 is

$$\binom{m}{0} + \binom{m}{1} + \binom{m}{2} + \cdots + \binom{m}{m} = 2^m .$$

Proof of Theorem 8.6. The left-hand side of Equation 8.2 counts the number of objects of S that are in none of the sets $A_1, \ldots, A_m$. We will show that an object in none of these sets will make a net contribution of 1 to the right-hand side of the equation, *and* an object that lives in at least one of these sets will make a net contribution of 0 to the right-hand side of the equation. This will prove the needed equality.

Let $x \in S$ be a member of none of the sets $A_1, \ldots, A_m$. Then the contribution of x to the right-hand side is $1 - 0 + 0 - \cdots + (-1)^m 0 = 1$, as promised.

Let $y \in S$ be a member of exactly n of the sets $A_1, \ldots, A_m$, for some fixed n with $1 \leq n \leq m$. Then, what is the contribution of y to the right-hand side of Equation 8.2? As an example, consider the term $\sum |A_i \cap A_j \cap A_k|$. The element y will be counted once each time it is in three of the As. Since y is assumed to be in n of the subsets, there are $\binom{n}{3}$ ways to pick three sets A_i, A_j, and A_k so that y is in their intersection. Thus, y contributes $\binom{n}{3}$ to $\sum |A_i \cap A_j \cap A_k|$. A similar argument works for the other intersections, and hence, the total contribution of y to the right-hand side will be

$$1 - n + \binom{n}{2} - \binom{n}{3} + \cdots + (-1)^m \binom{n}{m}.$$

Since $n \leq m$, for any $n < k \leq m$ we have $\binom{n}{k} = 0$. Eliminating the zeros at the end of the above sum and using Proposition 5.26b, we get that the contribution of y to the right-hand side is

$$\binom{n}{0} - \binom{n}{1} + \binom{n}{2} - \cdots + (-1)^n \binom{n}{n} = 0. \qquad \square$$

Just for the record, we will also give the expression for the number of elements that are in at least one of m subsets.

Corollary 8.9. *Let S be a set, and, for $1 \leq i \leq m$, A_i is a subset of S. Then*

$$|A_1 \cup A_2 \cup \cdots \cup A_m| = \sum_{i=1}^m |A_i| - \sum_{1 \leq i < j \leq m} |A_i \cap A_j| + \sum_{1 \leq i < j < k \leq m} |A_i \cap A_j \cap A_k|$$
$$+ \cdots + (-1)^{m+1} |A_1 \cap \cdots \cap A_m|.$$

Proof. The proof immediately follows from Theorem 8.6, since

$$|A_1 \cup A_2 \cup \cdots \cup A_m| = |S| - |A_1^c \cap A_2^c \cap \cdots \cap A_m^c|. \qquad \square$$

Example 8.10. You pick a random integer between 1 and 5000 inclusive. What is the probability that the integer is not divisible by 4, 6, or 7?

Solution. Let $S = [5000] = \{n \in \mathbb{Z} \mid 1 \leq n \leq 5000\}$. Let A_1, A_2, and A_3 denote the set of elements of S that are divisible, respectively, by 4, 6, and 7. We need $|A_1^c \cap A_2^c \cap A_3^c|$. Starting with 1, every fourth integer is divisible by 4, and so the number of integers in S divisible by 4 is $5000/4 = 1250$. Every sixth integer is divisible by 6, but $5000/6 \approx 833.34$. This means that if we count off the numbers that are divisible by 6, the 833rd one will still be less than 5000 but the 834th one will be larger. So $|A_2| = \left\lfloor \frac{5000}{6} \right\rfloor = 833$. Note that an integer is divisible by 4 *and* 6 if and only if it is divisible by 12, their least common multiple, and so on. Thus, we have

$$|A_1| = \left\lfloor \frac{5000}{4} \right\rfloor = 1250 \qquad\qquad |A_1 \cap A_2| = \left\lfloor \frac{5000}{12} \right\rfloor = 416$$

$$|A_2| = \left\lfloor \frac{5000}{6} \right\rfloor = 833 \qquad\qquad |A_1 \cap A_3| = \left\lfloor \frac{5000}{28} \right\rfloor = 178$$

$$|A_3| = \left\lfloor \frac{5000}{7} \right\rfloor = 714 \qquad\qquad |A_2 \cap A_3| = \left\lfloor \frac{5000}{42} \right\rfloor = 119$$

$$|A_1 \cap A_2 \cap A_3| = \left\lfloor \frac{5000}{84} \right\rfloor = 59.$$

By the inclusion–exclusion principle, we have

$$\left| A_1^c \cap A_2^c \cap A_3^c \right| = 5000 - 1250 - 833 - 714 + 416 + 178 + 119 - 59 = 2857,$$

and so the sought-after probability is $\frac{2857}{5000} \approx 0.57$. □

Example 8.11. You have one each of the letters A, B, D, E, F, I, L, N, O, T, and you randomly arrange them in a row. (For example, you may have ABEDONTFIL, ALIFTNOBED, BATONFIDEL or ABLETOFIND.) What is the probability that you will have at least one of the words ABLE, TO, or FIND as consecutive letters (and in the right order)?

Solution. We let S be the set of all possible permutations of the 10 letters. We let A_1 denote the set of permutations in S in which the word ABLE occurs as consecutive letters. Likewise, A_2 and A_3 denote those permutations of S in which, respectively, the words TO and FIND occur as consecutive letters. We need $|A_1 \cup A_2 \cup A_3|$. The size of S is just the total number of permutations of 10 distinct objects. In other words, $|S| = 10! = 3{,}628{,}800$. To calculate the size of A_1, we think of ABLE as one letter, and, hence, we need to find the total number of permutations of seven objects. Thus $|A_1| = 7!$. To count the number of elements in $A_1 \cap A_2$, we think of each of ABLE and TO as one letter. This leaves us with a total of six letters (two of which are "long" letters) and so $|A_1 \cap A_2| = 6!$. Continuing in this manner, we get

$$|A_1| = 7! = 5040 \qquad |A_1 \cap A_2| = 6! = 720 \qquad |A_1 \cap A_2 \cap A_3| = 3! = 6.$$
$$|A_2| = 9! = 362{,}880 \qquad |A_1 \cap A_3| = 4! = 24$$
$$|A_3| = 7! = 5040 \qquad |A_2 \cap A_3| = 6! = 720$$

By the inclusion–exclusion principle (version of Corollary 8.9), we have

$$|A_1 \cup A_2 \cup A_3| = 7! + 9! + 7! - 6! - 4! - 6! + 3! = 371{,}502.$$

Thus, the probability that, in a random permutation of these 10 letters, we can find at least one of the words ABLE, TO, or FIND is $\frac{371{,}502}{3{,}628{,}800} \approx 0.1024$. In other words, while only one of the permutations is ABLETOFIND, about 10.24% of the permutations of the ten letters contain one or more of the three words. □

We record one special case of the inclusion–exclusion principle that comes up often. It follows immediately from Theorem 8.6.

Corollary 8.12. *Let S be a set, and A_1, A_2, ..., A_m be subsets of S. Assume that $\left| A_{i_1} \cap A_{i_2} \cap \cdots \cap A_{i_k} \right|$ depends only on k and not on which k subsets are used. Let*

$$\alpha_0 = |S|$$
$$\alpha_1 = |A_1| = |A_2| = \cdots = |A_m|$$
$$\alpha_2 = |A_1 \cap A_2| = |A_1 \cap A_3| = \cdots = |A_{m-1} \cap A_m|$$
$$\alpha_3 = |A_1 \cap A_2 \cap A_3| = |A_1 \cap A_2 \cap A_4| = \cdots = |A_{m-2} \cap A_{m-1} \cap A_m|$$
$$\vdots$$
$$\alpha_m = |A_1 \cap A_2 \cap \cdots \cap A_m|.$$

Then

$$\left| A_1^c \cap A_2^c \cap \cdots \cap A_m^c \right| = \alpha_0 - \binom{m}{1}\alpha_1 + \cdots + (-1)^k \binom{m}{k}\alpha_k + \cdots + (-1)^m \alpha_m.$$

Example 8.13. In an old library, you find many copies of each of 10 different books. Upon examination, you find that each is one of the *Ten Books on Architecture* written circa 30 BCE by the Roman architect and military engineer Marcus Vitruvius Pollio. You want to arrange five (not necessarily distinct) books on a bookshelf. How many of such arrangements will include at least one copy of Books 8 (on aquaducts), 9 (on the sciences), and 10 (on machines)?

An Incorrect Solution. Could we argue as follows? First choose the three spots where Books 8, 9, and 10 are going, then find all permutations of these books in those chosen three spots, and finally, pick a book for the remaining two spots? If we carry this plan, we would get

$$\binom{5}{3} \, 3! \, 10^2 = 6000.$$

But this would be *incorrect*, since a permutation such as $89X89$ (where X denotes Book 10) would have been counted several times. (Do you see why?) □

Solution. Let S be the set of 5-permutations of 10 objects (with infinite repetition numbers). The subset A_1 is the subset of S consisting of permutations *without* the eighth object. Likewise, A_2, and A_3 are the subsets of S consisting of permutations *without* the ninth and tenth objects, respectively. We want to find $\left| A_1^c \cap A_2^c \cap A_3^c \right|$, and we will use the inclusion–exclusion principle. Using the notation of Corollary 8.12, we have

$$\alpha_0 = |S| = 10^5 \qquad\qquad \alpha_1 = |A_1| = |A_2| = |A_3| = 9^5$$
$$\alpha_2 = |A_1 \cap A_2| = |A_1 \cap A_3| = |A_2 \cap A_3| = 8^5, \qquad \alpha_3 = |A_1 \cap A_2 \cap A_3| = 7^5.$$

As an example, $\alpha_2 = 8^5$, since α_2 counts the number of 5-permutations that do not include two of the objects, and, hence, we have 8 remaining choices for each of the 5 spots.

Now, by Corollary 8.12,

$$|A_1^c \cap A_2^c \cap A_3^c| = 10^5 - \binom{3}{1}9^5 + \binom{3}{2}8^5 - 7^5 = 4350.$$

$\square$

A Second Solution. Like most counting problems, there are several ways of doing this problem correctly (and a plethora of incorrect solutions). Here is another count not using the inclusion–exclusion principle. You can think of this as a modification of the incorrect solution.

First, count the number of 5-permutations of 10 objects (with infinite repetition numbers) that have exactly one occurrence of each of objects numbered 8, 9, and 10. Choose the positions that contain these three objects, then find all permutations of these in those three positions, and then choose the remaining two objects. We get a count of

$$\binom{5}{3} 3! \, 7^2 = 2940.$$

Second, count the number of 5-permutations of 10 objects (with infinite repetition numbers) that have exactly two occurrences of one of objects numbered 8, 9, and 10, and one occurrence of the other two. There are $\binom{3}{1}$ choices for an object – from among 8, 9, 10 – to be repeated twice, $\binom{5}{4}$ places to place these four objects, $\binom{4}{2\,1\,1}$ ways of permuting the four objects, and finally seven choices for the last object to complete the permutation. This gives a count of

$$\binom{3}{1}\binom{5}{4}\binom{4}{2\,1\,1} 7 = 1260.$$

Third, count the number of permutations with exactly three occurrences of one of 8, 9, 10. There are $\binom{3}{1}$ ways of choosing which one occurs three times, and $\binom{5}{3\,1\,1}$ ways of ordering them, giving a count of

$$\binom{3}{1}\binom{5}{3\,1\,1} = 60.$$

Fourth, count the number of permutations with exactly two occurrences of *two* of 8, 9, and 10. There are $\binom{3}{2}$ ways of picking the two, and $\binom{5}{2\,2\,1}$ ways of ordering the five. This amounts to

$$\binom{3}{2}\binom{5}{2\,2\,1} = 90.$$

Adding the various possibilities, we get $2940 + 1260 + 60 + 90 = 4350$.

$\square$

Stirling Numbers of the Second Kind, Revisited

Let s and t be non-negative integers with $t \le s$. Recall that the Stirling number of the second kind $\{{}^s_t\}$ is the number of partitions of $[s]$ into t non-empty parts (Definition 6.4). Using the inclusion–exclusion principle, we give another proof for the remarkable formula of Theorem 6.23b. You were already asked to give two proofs in Problems P 6.1.30 and

P 6.1.31, but this new proof is the only one that gives some insight into the actual terms in the formula, and explains why some terms are added while others are subtracted.

Proposition 8.14. *Let s and t be non-negative integers. Then*

$$\begin{Bmatrix} s \\ t \end{Bmatrix} = \sum_{r=0}^{t} \frac{(-1)^{t-r} r^s}{r! \ (t-r)!}.$$

Proof. Let s and t be non-negative integers with $t \le s$, and let $\mathcal{N}$ denote the set consisting of possible ways of placing s distinct balls into t distinct boxes in such a way that each box gets at least one ball. Finding the size of $\mathcal{N}$ was entry $\boxed{9}$ of the counting table, Table 3.1 – see Table 7.3 for the first 12 entries – and, as explained in Theorem 6.9, we have

$$|\mathcal{N}| = t! \ \begin{Bmatrix} s \\ t \end{Bmatrix}.$$

Now let $\mathcal{M}$ denote the set consisting of the possible ways of placing s distinct balls into t distinct boxes with no capacity restrictions. Entry $\boxed{5}$ of the counting table – the argument is straightforward: we have t choices for each ball – gave us that

$$|\mathcal{M}| = t^s.$$

Now $\mathcal{N}$ is a subset of $\mathcal{M}$ and we can look at their relationship once more. Let $T = \{U_1, \ldots, U_t\}$ be the set of boxes (or urns), and, for $1 \le i \le t$, let A_i denote the subset of $\mathcal{M}$ consisting of the possible ways of placing s distinct balls in t distinct boxes in such a way that U_i is empty. Then, clearly,

$$\mathcal{N} = A_1^c \cap A_2^c \cap \cdots \cap A_t^c.$$

Note that, for any choice of $i_1, \ldots, i_k$, $A_{i_1} \cap A_{i_2} \cap \ldots \cap A_{i_k}$ denotes the possible ways of placing s distinct balls into $t - k$ distinct boxes – since we are insisting that boxes $i_1, \ldots, i_k$ remain empty – with no capacity restrictions, and hence its size is $(t - k)^s$. Using the notation of Corollary 8.12, we have

$$\alpha_0 = |\mathcal{M}| = t^s$$

$$\alpha_1 = |A_i| = (t-1)^s$$

$$\alpha_2 = |A_i \cap A_j| = (t-2)^s$$

$$\vdots$$

$$\alpha_t = |A_1 \cap \cdots \cap A_t| = (t-t)^s.$$

Using the inclusion–exclusion principle – in the form of Corollary 8.12 – we now have

$$t! \left\{ {s \atop t} \right\} = |\mathcal{N}| = |A_1^c \cap A_2^c \cap \cdots \cap A_t^c|$$

$$= t^s - \binom{t}{1}(t-1)^s + \binom{t}{2}(t-2)^s + \cdots + (-1)^t \binom{t}{t}(t-t)^s$$

$$= \sum_{k=0}^{t} (-1)^k \binom{t}{k}(t-k)^s.$$

We had started by assuming that $t \leq s$. However, the above argument makes sense even if $s < t$. In such a case, $\mathcal{N}$ will be the empty set and the scary-looking sum must add up to zero. Now, regardless of the values of s and t, and solving for $\left\{ {s \atop t} \right\}$, we have proved that

$$\left\{ {s \atop t} \right\} = \sum_{k=0}^{t} \frac{(-1)^k (t-k)^s}{k! \, (t-k)!}.$$

If we let $t - k = r$, then $k = t - r$, and as k varies from 0 to t, r varies from t to zero. Making this change of variable, we get the desired formula. $\square$

Problems

P 8.1.1. In a student government election, 450 students who voted for the Dunia-Badiambila ticket were polled. Out of these, 310 people liked Dunia and 420 people liked Badiambila. Assuming that the people who voted for this ticket liked at least one person on the ticket, how many of those polled liked both candidates on the ticket?

P 8.1.2. An advertising agency finds that, of its 470 clients, 150 listen to National Public Radio (NPR), 100 watch cable news, and 130 regularly listen to news podcasts. Out of these, 75 listen to NPR and watch cable news, 95 watch cable news and listen to podcasts, 85 listen to both NPR and podcasts, and 70 do all three. How many of the clients use none of these media? How many of these clients use *only* podcasts?

P 8.1.3. How many permutations of the letters in SCRIPPS have no two consecutive letters the same?

P 8.1.4. In the fall of 2047, the number of students at Pomona College who sign up for combinatorics is 225. All these students participate in the first session of the class's mentor session and order pizza. Each student eats at most one slice from each kind of pizza. A total of 100 slices of the mushroom pizza, 75 slices of the pepperoni pizza, and 50 slices of the green pepper pizza are eaten. From among the students, seven students have eaten both a slice of the mushroom and a slice of the pepperoni pizza, four have eaten a slice of the mushroom and a slice of the green pepper pizza, and 31 have eaten a slice of the pepperoni and a slice of the green pepper pizza. Only one student ate a slice of each three kinds of pizza. How many of the students ate at least one slice of pizza?

P 8.1.5. Find the number of positive integers n such that $1 \leq n \leq 10{,}000$ and such that n has no common prime divisors with $10{,}000$.

P 8.1.6. Find the number of integers between 1 and 10^{12} that are neither perfect squares nor perfect cubes nor perfect fourth powers.

P 8.1.7. Let S be the set of integers between 10,000 and 20,000 inclusive.

    (a) How many elements of S are a multiple of 7?

    (b) How many elements of S are not a multiple of 7, 11, or 13?

P 8.1.8. Here are some excerpts from "Knot 10," titled "Chelsea Buns," in *A Tangled Tale* by Lewis Carroll (Carroll 1885).

> "How very, very sad!" exclaimed Clara; and the eyes of the gentle girl filled with tears as she spoke, "Sad—but very curious when you come to look at it arithmetically," was her aunt's less romantic reply. "Some of them have lost an arm in their country's service, some a leg, some an ear, some an eye—"
> "And some, perhaps, *all*!" Clara murmured dreamily, as they passed the long rows of weather-beaten heroes basking in the sun.
>
> ...
>
> "To return to the arithmetic," Mad Mathesis resumed—the eccentric old lady never let slip an opportunity of driving her niece into a calculation—"what percentage do you suppose must have lost all four—a leg, an arm, an eye, and an ear?"
> "How *can* I tell?" gasped the terrified girl. She knew well what was coming.
> "You can't, of course, without *data*," her aunt replied: "but I'm just going to give you"
>
> ...
>
> "Say that 70 per cent have lost an eye—75 per cent an ear—80 per cent an arm—85 per cent a leg—that'll do it beautifully. Now, my dear, what percentage, *at least*, must have lost all four?"

Let p be the percentage of fighters that have lost all four. Find a lower and an upper bound for p. Show how these bounds can be achieved.

P 8.1.9. You have three copies of Fyodor Dostoyevsky's *Crime and Punishment*, four copies of Leo Tolstoy's *Anna Karenina*, and two copies of Nikolai Gogol's *Dead Souls*. In how many ways can you arrange the nine books on a shelf in such a way that not all the books of any one author are next to each other? (So ACCCDADAA is not allowed while ACAAADCCD is.)

P 8.1.10. If you are randomly dealt five cards from a standard deck of cards (see Problem P 4.1.3), then what is the probability that among your five cards there will be at least one Jack, one Queen, and one King?

P 8.1.11. An experimenter has access to three prisms and a number of magnets. She has used each prism in an experiment 12 times, each pair of prisms in the same experiment together 6 times, and all 3 prisms in the same experiment together 4 times. In eight experiments, none of the prisms were used. How many experiments were performed altogether?

P 8.1.12. Assume that you write down every integer from 0 to 9,999,999 as a seven-digit number $\overline{abcdefg}$ (so, for example, you write 0000015 for 15). For how many of these numbers are the first three digits, $\overline{abc}$, exactly the same, and in the same order, as $\overline{def}$ (the three digits starting with the fourth one) or $\overline{efg}$ (the last three digits) or both? For example, 5348534 and 4444444 would be such numbers.

P 8.1.13. If you are randomly dealt 10 cards from a standard deck of cards (see Problem P 4.1.3), then what is the probability that you have at least 1 four of a kind?

P 8.1.14. In Problem P 4.2.20, you were asked to find the probability of getting four of a kind if you were dealt a set of five cards from a standard deck. In Problem *P 8.1.13*, you were asked the same question except that you were dealt a 10-card hand. What makes the two problems different, and why would the latter problem be in this chapter while the former problem was in Chapter 4?

P 8.1.15. The northern hemisphere's winter solstice is called Yaldā night in Iran, and, on this night, in addition to eating pomegranates (and other fruits and nuts), it is customary to take turns reading the poems (*ghazals*) of the fourteenth-century poet Hafez (1315–1390), as a starting point for thinking about what the future may bring. An extended family of 17 puts seven of Hafez's *ghazals* in a bag. Each takes turn randomly picking one *ghazal*, reciting it, and then putting it back in the bag. We are interested in the probability that at the end, each of the seven *ghazals* has been picked at least once.

   (a) Find the probability in terms of Stirling numbers (and without recourse to the inclusion–exclusion principle).
   (b) Find the probability using the inclusion–exclusion principle but without directly invoking Proposition 8.14. You can leave your answer as a sum involving binomial coefficients and fractions.

P 8.1.16. How many permutations of the 26 letters of the English alphabet contain at least one of the sequences YOU, HAD, or TIME (as consecutive letters in the right order)?

P 8.1.17. How many permutations of the letters M, A, T, H, I, S, C, O, O, L do not include any of the sequences MATH, IS, or COOL (as consecutive letters in the right order)?

P 8.1.18. Consider the letters $\{A, H, M, T\}$. Make 10-letter words from just these letters, making sure that each of the letters is used at least once.

   (a) How many such words are there?
   (b) If you choose one of these words at random, what is the probability that the word MATH appears consecutively at least once somewhere in the word?
   (c) How is this problem different than Problem P 6.1.8? Why is this one in this chapter?

P 8.1.19. Find the number of primes less than 100 without actually finding all the primes.

P 8.1.20. Zakiya, Cristal, Sylvia, Ezequiel, and Jia are five of my students, and, in the next 15 days, every day, I am going to meet with one of them. For each of the following scenarios, determine the number of different possible schedules.

   (a) There are no other restrictions.
   (b) I want to meet each of them three times.
   (c) I want to meet each of them three times, but I don't want to meet Zakiya three times in a row.
   (d) I want to meet each of them three times, but I don't want to meet any of them three times in a row.
   (e) I want to meet each of them three times, but I want to meet each of them once in the first five days, once in the second five days, and once in the final five days.

(f) I want to meet each of them three times, I want to meet each of them once in the first five days, once in the second five days, and once in the final five days, and I don't want to see any of them two days in a row.

P 8.1.21. k is a positive integer. The four walls and the ceiling of a room are to be painted with k colors available. How many ways can this be done if bordering sides of the room must have different colors? (The four walls are distinct walls, and the ceiling shares common borders with all of the four walls.)

P 8.1.22. You are a fan of Eric Arthur Blair (1903–1950) – much better known by his pen name, George Orwell – and have three identical copies of each of *Burmese Days*, *The Road to Wigan Pier*, and *Homage to Catalonia*. You want to arrange these nine books on one shelf of your library but with the proviso that no two copies of the same book should be next to each other. In how many ways can you do this? You may find the following steps helpful:

STEP 1: Convince yourself that a straightforward application of the inclusion–exclusion principle – as could be used for Problem P 8.1.9 – will not work.

STEP 2: Let X be the set of all permutations of the multiset $M = \{3 \cdot B, 3 \cdot W, 3 \cdot C\}$, and, for $1 \leq i \leq 8$, let A_i be the set of elements of M where two identical elements are in positions i and $i + 1$. In terms of these sets, what are we looking for?

STEP 3: Consider $WBBBWCCWC$, a permutation of M and an element of X. This permutation belongs to which A_i?

STEP 4: Convince yourself that $\sum_{i=1}^{8} |A_i|$ is three times the number of permutations of the multiset $\{z, B, 3 \cdot W, 3 \cdot C\}$. (We are thinking of z as a stand-in for the symbol BB.)

STEP 5: There are two ways for a permutation to be in the intersection of two A_is. Either there is an occurrence of three consecutive copies of a book starting at position i, or there are two occurrences of two consecutive copies of a book starting at positions i and j with $j > i + 1$. For the first type, give an argument for why $\sum_{1 \leq i \leq 7} |A_i \cap A_{i+1}|$ is the same as three times the number of permutations of the multiset $\{z, 3 \cdot W, 3 \cdot C\}$. For the second type, give an argument for why $\sum_{\substack{1 \leq i < j \leq 8 \\ j \neq i+1}} |A_i \cap A_j|$ is the same as three times the number of permutations of the multiset $\{y, z, B, W, 3 \cdot C\}$.

STEP 6: In how many ways can an element of X be in three of the A_is? What about four, five, or six of the A_is?

STEP 7: Proceed carefully, as in Step 5, to find all the necessary ingredients for the use of the inclusion–exclusion principle (in the form of Equation 8.3). Complete the solution.

8.1.1 Bonferroni's Inequalities

Let S be a set, and let $A_1, A_2, \ldots, A_m$ be subsets of S. Introduce the following notation:

$$E = \left| A_1^{\,c} \cap A_2^{\,c} \cap \cdots \cap A_m^{\,c} \right|$$

$$N_0 = |S|$$

$$N_1 = \sum_{i=1}^{m} |A_i|$$

$$N_2 = \sum_{1 \le i < j \le m} |A_i \cap A_j|$$

$$\vdots$$

$$N_{m-1} = \sum_{1 \le i_1 < \cdots < i_{m-1} \le m} |A_{i_1} \cap \cdots \cap A_{i_{m-1}}|$$

$$N_m = |A_1 \cap A_2 \cap \cdots \cap A_m|.$$

With this notation, the inclusion–exclusion principle (Theorem 8.6) becomes

$$E = N_0 - N_1 + N_2 + \cdots + (-1)^m N_m = \sum_{r=0}^{m} (-1)^r N_r. \qquad (8.3)$$

This expression for E has $m + 1$ terms ($N_0, -N_1, \ldots, (-1)^m N_m$). Usually, we would like to find E, but, if m is large, the exact computation may involve too many calculations. What if we approximate E by truncating Equation 8.3 and only using t of the terms? That is, let t be an integer less than or equal to m, and use the approximation

$$E \approx \sum_{r=0}^{t-1} (-1)^r N_r.$$

To estimate the error of this approximation, we let e_t denote the difference between the true value of E and the approximate value of E. In other words,

$$e_t = E - \sum_{r=0}^{t-1} (-1)^r N_r.$$

Bonferroni's inequalities assert that truncating the sum gives an error which is no larger than the absolute value of the first term that was neglected. Furthermore, the sum is either an overestimate or an underestimate, according to whether t is odd or even, respectively. We state this result here. You are asked to prove it in Problem *P 8.1.24.*

Proposition 8.15 (Bonferroni's Inequalities). *Using the notation above, let $e_t = E - \sum_{r=0}^{t-1} (-1)^r N_r$ denote the difference between the true value of E and the approximate value of E obtained by using the first t terms from the inclusion–exclusion principle of Equation 8.3. Then*

(a) $0 \le e_t \le N_t$ *if t is even, and*
(b) $-N_t \le e_t \le 0$ *if t is odd.*

P 8.1.23. Let S be a set, and let $A_1, A_2, \ldots, A_m$ be subsets of S. Using the notation of Bonferroni's inequality, let

$$e_t = E - \sum_{r=0}^{t-1}(-1)^r N_r = E - N_0 + N_1 - N_2 + \cdots - (-1)^{t-1}N_{t-1}.$$

Assume that $s \in S$, and that s belongs to n of the sets $A_1, A_2, \ldots, A_m$. If $n > 0$, what is the net contribution of s to e_t? In other words, each term on the right-hand side of the equation for e_t counts certain elements of S (sometimes multiple times), and we want to know the net number of times the element s is counted. What if $n = 0$?

P 8.1.24. **Proof of Bonferroni's Inequalities.** Prove Proposition 8.15. You may find the following steps relevant.

STEP 1: Do Problem P 8.1.23.

STEP 2: If I know the contribution of each $s \in S$ to the right-hand side of the formula for e_t, then how do I find e_t?

STEP 3: If t is even, then show that $e_t \geq 0$, and, if t is odd, then show that $e_t \leq 0$.

STEP 4: Assume t is even, and, as a result, $t + 1$ is odd. How is e_{t+1} related to e_t? Use the previous step to show that $e_t \leq N_t$.

STEP 5: Complete the proof and rewrite it concisely.

P 8.1.25. In Problem *P 8.1.19*, you were asked to find the number of primes up to 100 without actually finding the primes. This can be done using the inclusion–exclusion principle and using four subsets $A_1, \ldots, A_4$ of $S = \{2, \ldots, 100\}$. These, respectively, consist of multiples of 2, 3, 5, and 7. Approximate the number of primes up to 100 by using the first three terms (instead of five) of Equation 8.3. What is the error, and how does the actual error compare with the estimate of the error given by Bonferroni's inequalities of Proposition 8.15? What if you had used the first four terms?

P 8.1.26. Continuing with the ethos of Problems *P 8.1.19* and P 8.1.25, and the notation of Equation 8.3 and Proposition 8.15, we want to *approximate* the number of primes up to 1000 without finding them. Let $S = \{2, \ldots, 1000\}$, and let $P = \{2, 3, 5, 7, 11, 13, 17, 19, 23, 29, 31\}$.

(a) Assume that we start with S and eliminate all integers in S that are multiples of an integer in P except the elements of P themselves. Give an argument for why the remaining integers are exactly all the primes up to 1000.

(b) Explain (but do not carry out) how we could use the inclusion–exclusion principle using 11 subsets of S to find the number of primes up to 1000.

(c) For this problem, in the inclusion–exclusion principle of Equation 8.3, we have $m = 11$. Use $t = 3$ terms to approximate the number of primes up to 1000. In other words, use the approximation $E \approx N_0 - N_1 + N_2$.

(d) Use a computer program or search the Internet to find the exact number of primes up to 1000. How far off was your approximation? Why didn't I ask for a better approximation by insisting that you use $t = 4$?

8.2   Combinations of a Multiset

Warm-Up 8.16. *What is the number of non-negative integer solutions to $x_1 + x_2 + \cdots + x_7 = 20$ if $x_1 \geq 3$ and $x_2 \geq 5$?*

Let a multiset $R = \{n_1 \cdot V_1, n_2 \cdot V_2, \ldots, n_t \cdot V_t\}$ be given. Recall that a submultiset of size s of R is called an s-combination of R. We have already seen a count of the number of s-combinations of R in two (extreme) cases:

- If all the repetition numbers $n_1, n_2, \ldots, n_t$ are equal to 1 – in other words, R is an ordinary set – then the number of s-combinations of R is just the number of subsets of size s of a set with t elements, and is equal to $\binom{t}{s}$. (Entry $\boxed{2}$ of the counting table, Theorem 4.21.)
- If all the repetition numbers $n_1, n_2, \ldots, n_t$ are greater than or equal to s – in other words, for the purposes of finding submultisets of size s, there is an inexhaustible supply of each type of elements – then the number of s-combinations is the same as the number of non-negative integral solutions to $x_1 + x_2 + \cdots + x_t = s$, and is equal to $\left(\binom{t}{s}\right) = \binom{s+t-1}{s}$. (Entry $\boxed{6}$ of the counting table, Theorem 4.40.)

The general case – entry $\boxed{14}$ of the counting table, Table 3.1 – can be done using the inclusion–exclusion principle.

Theorem 8.17 (Entry $\boxed{14}$ of the Counting Table). *Let s and t be non-negative integers, and $n_1, \ldots, n_t$ be positive integers. Let $R = \{n_1 \cdot V_1, n_2 \cdot V_2, \cdots, n_t \cdot V_t\}$ be a multiset with t types of elements with repetition numbers $n_1, \ldots, n_t$. Finally, let $R^* = \{\infty \cdot V_1, \infty \cdot V_2, \cdots, \infty \cdot V_t\}$ be a multiset with t types of elements with infinite repetition numbers. Then the following integers are equal:*

(a) $\boxed{14}$ *The number of ways of placing s identical balls in t distinct boxes in such a way that, for $1 \leq i \leq t$, the number of balls in box i is less than or equal to n_i*

(b) *The number of integral solutions to*

$$x_1 + x_2 + \cdots + x_t = s$$

that, for $1 \leq i \leq t$, satisfy $0 \leq x_i \leq n_i$

(c) *The number of s-combinations of R*

(d)

$$\left| A_1^c \cap A_2^c \cap \ldots \cap A_t^c \right|,$$

where A is the set of s-combinations of R^* and, for $1 \le i \le t$, A_i is those elements of A with more than n_i of the element V_i. (A_i^c is $A - A_i$, the complement of A_i in A.)

Proof. The reasoning for the equality of the first three numbers is identical to that of Theorem 4.40, and numbers (c) and (d) are evidently equal. □

Using formulation 8.17d, we translate the problem of finding the number of combinations of a multiset – formulation 8.17c – to one that can be solved using the inclusion–exclusion principle. We illustrate the method with one example.

Example 8.18. Assume that you have 4 clones of the robot R2-D2, 7 clones of the robot C-3PO, 10 clones of a Death trooper, and 26 clones of a Stormtrooper. In how many ways can we choose 17 of these? Some examples would be {4 · R2-D2, 3 · C-3PO, 10 · Death trooper}, {17 · Stormtrooper}, or {7 · Death trooper, 10 · Stormtrooper}.

Let $T = \{4 \cdot \text{R2-D2}, 7 \cdot \text{C-3PO}, 10 \cdot \text{Death trooper}, 26 \cdot \text{Stormtrooper}\}$ be a multiset with four types of elements and with repetition numbers 4, 7, 10, and 26. We want the total number of submultisets of size 17 (or, what is the same thing, the number of 17-combinations of T). We begin with the admittedly much bigger multiset

$$T^* = \{\infty \cdot \text{R2-D2}, \infty \cdot \text{C-3PO}, \infty \cdot \text{Death trooper}, \infty \cdot \text{Stormtrooper}\}.$$

Let S be the set of submultisets of size 17 of T^*, and define four subsets (of S) A_1, A_2, A_3, and A_4 as follows:

$$A_1 = \{\text{submultisets of size 17 of } T^* \text{ with more than 4 R2-D2s}\}$$

$$A_2 = \{\text{submultisets of size 17 of } T^* \text{ with more than 7 C-3POs}\}$$

$$A_3 = \{\text{submultisets of size 17 of } T^* \text{ with more than 10 Death troopers}\}$$

$$A_4 = \{\text{submultisets of size 17 of } T^* \text{ with more than 26 Stormtroopers}\}.$$

The sets A_1, A_2, A_3, and A_4 are subsets of S, and our original question amounts to finding $|A_1^c \cap A_2^c \cap A_3^c \cap A_4^c|$. ($A_i^c = S - A_i$ is the complement of A_i in S.) To find this, we can use the inclusion–exclusion principle. But we first notice that A_4 is the empty set, since you can't have more than 26 Stormtroopers in any 17-combination of T^*. Hence, $A_1^c \cap A_2^c \cap A_3^c \cap A_4^c = A_1^c \cap A_2^c \cap A_3^c$, and we are looking for $|A_1^c \cap A_2^c \cap A_3^c|$.

The size of S, $|S|$, is found using Theorem 4.40 (entry $\boxed{6}$ of the counting table). Or more directly, we say that the number of 17-combinations of T^* is the same as the number of integral solutions to $x_1 + x_2 + x_3 + x_4 = 17$ where, for $1 \le i \le 4$, $x_i \ge 0$. We know this to be $\left(\!\binom{4}{17}\!\right) = \binom{20}{17} = 1140$.

The size of A_1, $|A_1|$, is the number of 12-combinations of T^*. This is because a submultiset of size 17 of T^* with more than 4 R2-D2s has at least 5 R2-D2s, and we need to pick another 12 elements from T^*. This is the same as the number of non-negative integer solutions to $x_1 + x_2 + x_3 + x_4 = 12$, which is $\left(\!\binom{4}{12}\!\right) = \binom{15}{12} = 455$. Similarly, $|A_2|$ is the number of 9-combinations of T^*, or the number of non-negative integer solutions to $x_1 + x_2 + x_3 + x_4 = 9$, which is $\left(\binom{4}{9}\right) = \binom{12}{9} = 220$. Also, $|A_3|$ is the number of 6-combinations of T^*, which is $\left(\binom{4}{6}\right) = \binom{9}{6} = 84$.

To use the inclusion–exclusion principle, we also have to find $|A_i \cap A_j|$ for $1 \le i < j \le 3$. The size of $A_1 \cap A_2$ is the number of 4-combinations of T^*, which is $\left(\binom{4}{4}\right) = 35$; $|A_1 \cap A_3|$ is the number of 1-combinations of T^*, which is 4. Finally, both $|A_2 \cap A_3|$ and $|A_1 \cap A_2 \cap A_3|$ are 0. Thus,

$$|A_1^c \cap A_2^c \cap A_3^c| = |S| - \sum_{i=1}^{3} |A_i| + \sum_{1 \le i < j \le 3} |A_i \cap A_j| - |A_1 \cap A_2 \cap A_3|$$

$$= 1140 - 455 - 220 - 84 + 35 + 4 + 0 - 0 = 420.$$

The reader may want to see if it is possible to solve this problem without the inclusion–exclusion principle, and by directly counting the 420 possibilities.

Remark 8.19. To find the number of integer solutions to

$$x_1 + x_2 + \cdots + x_t = s$$

such that, for $0 \le i \le t$, we have $0 \le x_i \le n_i$, we can proceed just as in Example 8.18. After all, in that example, we found the number of integer solutions to $x_1 + x_2 + x_3 + x_4 = 17$, with the added conditions that $0 \le x_1 \le 4, 0 \le x_2 \le 7, 0 \le x_3 \le 10$, and $0 \le x_4 \le 26$. In general, this is practical as long as t, the number of variables, is not too large.

Problems

P 8.2.1. Determine the number of 10-combinations of the multiset

$$S = \{\infty \cdot a, 3 \cdot b, 5 \cdot c, 7 \cdot d\}.$$

P 8.2.2. Determine the number of integral solutions of the equation

$$x_1 + x_2 + x_3 + x_4 = 20,$$

where $1 \le x_1 \le 6, 0 \le x_2 \le 7, 4 \le x_3 \le 8$, and $2 \le x_4 \le 6$.

P 8.2.3. We have m identical slices of pizza, and n students. We want to distribute the slices to the students in such a way that each student gets 2, 3, or 4 slices. We let $F(m,n)$ denote the number of ways that this can be done. Note that students are individuals and hence are distinguishable.

(a) Find $F(15,4)$.

(b) Find a recurrence relation for $F(m,n)$.

P 8.2.4. A bakery sells seven kinds of donuts. How many ways are there to choose one dozen donuts if no more than three donuts of any kind are used?

P 8.2.5. How many ways are there to distribute 24 identical balls into 5 distinct boxes with at most 5 balls in any of the first 3 boxes?

P 8.2.6. Hercules, Ophiuchus, Daedalus, Icarus, Iapyx, and Perdix are soccer players. In how many ways can you distribute 47 identical soccer balls among them such that Hercules, Ophiuchus, and Daedalus get at most 7 balls each, and Icarus, Iapyx, and Perdix get at least 3 balls each?

P 8.2.7. Throw six dice. What is the probability of getting a sum equal to 15? What about 19?

P 8.2.8. Let r and n be positive integers with $n \leq r \leq 6n$. You have to put a present in each of r identical stockings. You have succumbed to the latest advertising blitz and bought six of each of n of the latest gizmos. In how many ways can you put one gizmo in each of the stockings, making sure that each type of gizmo has been used?

8.3   Permutations with Forbidden Positions

8.3.1   The Hat-Check Problem and Derangements

Warm-Up 8.20. *Find the number of* 4×4 *matrices whose entries are* 0 *or* 1, *there is only one non-zero entry in each row and each column, and there are zeros on the diagonal.*

We begin with the opening problem of the chapter, which can be rephrased as what is known as the hat-check problem.

Question 8.21 (The Hat-Check Problem). At a party, n patrons check their hats. At the end of the night, the clerk returns the hats randomly. What is the probability that no one receives their own hat? In particular, is the probability lower or higher if $n = 10$ versus $n = 10,000$?

There are $n!$ ways to permute the hats. We need to know how many among these ways result in no patron receiving their own hat. This is an example of permutations with forbidden positions. We are given a set X with $|X| = n$ with the additional stipulation that each element has a specified position. (That is, we know which one is the *first* element, which one is the *second* element, and so on.) We want the number of permutations of X in which no element is in its specified position. Other examples of forbidden positions will be treated later in the section.

Definition 8.22. Let n be a positive integer, and $[n] = \{1, 2, \dots, n\}$. A *derangement* of $[n]$ is a permutation $i_1 i_2 \dots i_n$ of $[n]$ such that $i_1 \neq 1$, $i_2 \neq 2$, $\dots$, $i_n \neq n$. Let D_n denote the number of derangements of $[n]$.

We have named our problem. Finding the number of possibilities for returning the hats such that no one gets their own hat is the same as finding D_n. After finding D_n for very small values of n, we will use the inclusion–exclusion principle to find a formula for D_n.

Example 8.23. We have

n	Derangements	D_n
1	$\emptyset$	0
2	21	1
3	231, 312	2
4	2143, 2341, 2413, 3142, 3412, 3421, 4123, 4312, 4321	9

Theorem 8.24. *Let n be a positive integer, and let D_n denote the number of derangements of $[n]$. Then*

$$D_n = n!\left(1 - \frac{1}{1!} + \frac{1}{2!} - \frac{1}{3!} + \cdots + (-1)^n\frac{1}{n!}\right) = n!\sum_{k=0}^{n}(-1)^k\frac{1}{k!}.$$

Proof. Let S be the set of all permutations of $[n] = \{1, 2, \ldots, n\}$, and, for $1 \le j \le n$, let A_j be those elements of S for which j is in its natural position. Evidently, D_n is $\left|A_1^c \cap A_2^c \cap \ldots \cap A_n^c\right|$. The number of permutations of $[n]$ for which k of the elements of $[n]$ are presupposed to be in their natural positions is $(n-k)!$. Hence, we have

$$\alpha_0 = |S| = n!$$
$$\alpha_1 = \left|A_j\right| = (n-1)! \qquad \text{for } 1 \le j \le n$$
$$\alpha_2 = \left|A_i \cap A_j\right| = (n-2)! \qquad \text{for } 1 \le i < j \le n$$
$$\vdots$$
$$\alpha_k = \left|A_{i_1} \cap A_{i_2} \cap \ldots \cap A_{i_k}\right| = (n-k)! \qquad \text{for } 1 \le i_1 < \cdots < i_k \le n.$$

Thus, by the inclusion–exclusion principle (version of Corollary 8.12), we have

$$D_n = n! - \binom{n}{1}(n-1)! + \binom{n}{2}(n-2)! - \binom{n}{3}(n-3)! + \cdots + (-1)^n\binom{n}{n}0!$$
$$= n! - \frac{n!}{1!} + \frac{n!}{2!} - \frac{n!}{3!} + \cdots + (-1)^n\frac{n!}{n!}$$
$$= n!\left(1 - \frac{1}{1!} + \frac{1}{2!} - \frac{1}{3!} + \cdots + (-1)^n\frac{1}{n!}\right).$$

$\square$

Example 8.25. Continuing Example 8.23, we now have

$$D_5 = 5!\left(1 - \frac{1}{1} + \frac{1}{2} - \frac{1}{6} + \frac{1}{24} - \frac{1}{120}\right) = 44.$$

Remark 8.26. In Theorem 6.33, we saw that the Stirling numbers of the first kind appeared in the partial sums of the harmonic series. Likewise, Theorem 8.24 – in addition to being a formula for the number of derangements – shows the appearance of the integers D_n in the partial sums of the series expansion of $1/e$. In other words, we have

$$1 - \frac{1}{1!} = \frac{0}{1!} = \frac{D_1}{1!}$$

$$1 - \frac{1}{1!} + \frac{1}{2!} = \frac{1}{2!} = \frac{D_2}{2!}$$

$$1 - \frac{1}{1!} + \frac{1}{2!} - \frac{1}{3!} = \frac{2}{3!} = \frac{D_3}{3!}$$

$$1 - \frac{1}{1!} + \frac{1}{2!} - \frac{1}{3!} + \frac{1}{4!} = \frac{9}{4!} = \frac{D_4}{4!}$$

$$1 - \frac{1}{1!} + \frac{1}{2!} - \frac{1}{3!} + \frac{1}{4!} - \frac{1}{5!} = \frac{44}{5!} = \frac{D_5}{5!}$$

$$\vdots$$

We will come back to the connection between D_n and the exponential function in Problem P 9.4.11 of Chapter 9, but we are now ready to answer the opening problem of this chapter.

Corollary 8.27. *Let D_n denote the number of derangements of $[n]$. Then*

$$\lim_{n \to \infty} \frac{D_n}{n!} = \frac{1}{e}.$$

Moreover, regardless of the value of n, as long as $n \geq 6$, in the hat-check Question 8.21, the probability that no one receives their own hat, rounded to the thousandths place, is 0.368.

Proof. Since $e^x = 1 + x + \frac{x^2}{2!} + \cdots + \frac{x^n}{n!} + \cdots$, by plugging in $x = -1$ and using Theorem 8.24, we get that $\lim_{n \to \infty} \frac{D_n}{n!} = \frac{1}{e}$. Moreover, since $1/e = 1 - 1 + 1/2! - 1/3! + \cdots$ is an alternating series and the absolute values of the terms are decreasing, if we truncate the series and use it to approximate $1/e$, the error will be no more than the absolute value of the next term. Hence, $1/e$ and $D_n/n!$ differ by less than $1/(n+1)!$. For $n \geq 7$, this is less than 0.0000249. Now $1/e \approx 0.367879$ and so, for $n \geq 7$, rounded to the thousandths place we have 0.368. For $n = 6$, the value of $D_n/n!$ is approximately 0.36806, and so we can use the same three decimal approximation for $n \geq 6$. $\qquad\square$

Our most common – and practical in terms of calculating a sequence of numbers – strategy has been to find a recurrence relation. We will now do so for derangements.

Proposition 8.28. *Let n be a positive integer, and let D_n denote the number of derangements of $[n]$. Then $D_1 = 0$, $D_2 = 1$,*

$$D_n = (n-1)(D_{n-2} + D_{n-1}) \text{ for } n \geq 3,$$

and

$$D_n = nD_{n-1} + (-1)^n \text{ for } n \geq 2.$$

Proof. Let $n \geq 3$, and let d_n be the number of derangements of $[n]$ that have a 2 in the first position. To construct a derangement of $[n]$, you cannot have a 1 in the first position. If 2 is in the first position, then, from among the remaining elements $1, 3, 4, \ldots, n$, the latter $n - 2$ elements – that is, 3 through n – can be placed anywhere except for their natural positions, and 1 can be placed anywhere. The same situation would be true if 3 were in the first position. Namely, there is no restriction on where 1 is placed, and all the remaining $n - 2$ elements – that is, $2, 4, \ldots, n$ – can be placed anywhere except for their natural positions. We conclude that d_n, the number of derangements of $[n]$ that have 2 in the first position, is the same as the number of derangements that have $3, 4, \ldots$, or n in the first position. Since, any derangement starts with one of these $n - 1$ elements, we conclude that $D_n = (n - 1)d_n$.

Now if a derangement starts with a 2, then it is one of two types. Either we have a 1 in the second position or we do not have a 1 in the second position. In the former case, the remaining part of the derangement is exactly a derangement of $\{3, \ldots, n\}$, and there are D_{n-2} of these. In the latter case, following the initial 2, we have a derangement of $\{1, 3, 4, \ldots, n\}$, and there are D_{n-1} of these. Thus $d_n = D_{n-2} + D_{n-1}$, and the first recurrence relation follows.

Rewriting $D_n = (n - 1)(D_{n-2} + D_{n-1})$, we have, for $n \geq 3$,

$$D_n - nD_{n-1} = -\left[D_{n-1} - (n - 1)D_{n-2}\right].$$

Since this relation is true for all $n \geq 3$, we can keep applying it (or, in the language of Section 1.4, unwind this recurrence relation), to get

$$
\begin{aligned}
D_n - nD_{n-1} &= -\left[D_{n-1} - (n - 1)D_{n-2}\right] \\
&= (-1)^2\left[D_{n-2} - (n - 2)D_{n-3}\right] \\
&= (-1)^3\left[D_{n-3} - (n - 3)D_{n-4}\right] \\
&\vdots \\
&= (-1)^{n-2}[D_2 - 2D_1] \\
&= (-1)^{n-2}[1 - 0] = (-1)^n.
\end{aligned}
$$
$\square$

Using the recurrence relation, it is now easy to generate as much of the sequence D_n as we like:

n	1	2	3	4	5	6	7	8	9	10
D_n	0	1	2	9	44	265	1,854	14,833	133,496	1,334,961

8.3.2   Permutations with Forbidden Positions

Warm-Up 8.29. *Find the number of 4×4 matrices whose entries are 0 or 1, there is only one non-zero entry in each row and each column, and there are zeros in the entries of the 2×2 square on the bottom left of the matrix.*

Let $[n] = \{1, 2, \ldots, n\}$. The set n has $n!$ permutations and each of these corresponds to one possible way of placing n non-attacking rooks on an $n \times n$ board. As an alternative – if you don't like non-attacking rooks – each permutation corresponds to an $n \times n$ matrix of 0's and 1's with the condition that each row and each column has exactly one 1. The correspondence is straightforward. The permutation $i_1 i_2 \ldots i_n$ corresponds to the placing of the rooks – or the 1's – in positions $(1, i_1)$, $(2, i_2)$, $\ldots$ (n, i_n). (While the choice is inconsequential, we use "matrix coordinates." Hence, position $(3, 5)$ refers to the square or matrix entry in row three – from the top – and column five – from the left.)

Example 8.30. For $n = 5$, the permutations 34125 and 13254 respectively correspond to

$$\begin{bmatrix} 0 & 0 & 1 & 0 & 0 \\ 0 & 0 & 0 & 1 & 0 \\ 1 & 0 & 0 & 0 & 0 \\ 0 & 1 & 0 & 0 & 0 \\ 0 & 0 & 0 & 0 & 1 \end{bmatrix} \text{ and } \begin{bmatrix} 1 & 0 & 0 & 0 & 0 \\ 0 & 0 & 1 & 0 & 0 \\ 0 & 1 & 0 & 0 & 0 \\ 0 & 0 & 0 & 0 & 1 \\ 0 & 0 & 0 & 1 & 0 \end{bmatrix}.$$

With this correspondence, a derangement is a placing of n non-attacking rooks on an $n \times n$ board, with the condition that we are forbidden to place any rooks on the diagonal. Hence, D_n, the number of derangements, is the total number of ways of placing n non-attacking rooks on an $n \times n$ board with forbidden positions down the diagonal. See Figure 8.2.

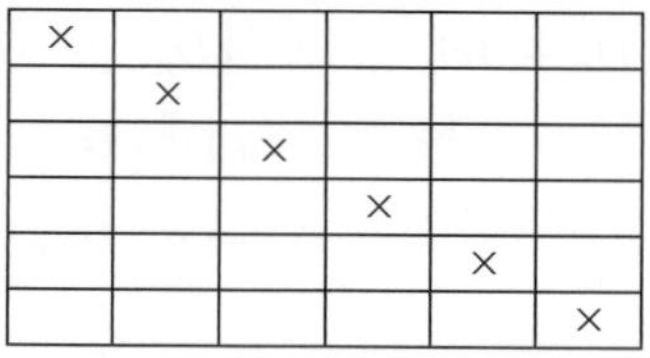

Figure 8.2 The crossed-out positions are forbidden. The number of ways of placing six non-attacking rooks on this board with forbidden positions is the number of derangements of [6], D_6.

Given this interpretation, we can generalize the problem of counting derangements. We are given an $n \times n$ board with forbidden positions and we want the number of ways of placing n non-attacking rooks on the board without using any of the forbidden positions.

The solution depends on the particular configuration of the forbidden positions, but, using the inclusion–exclusion principle, we can switch the problem around, and instead of counting the number of ways of placing non-attacking rooks in the permitted spaces, count the possibilities for placing non-attacking rooks in the forbidden squares. This flipping of perspectives can sometimes be helpful.

Let n be a positive integer and $[n] = \{1, \ldots, n\}$. For $1 \le i \le n$, let X_i be a subset of $[n]$ denoting the column numbers of the forbidden squares in row i. Let S_n denote the set of all permutations of $[n]$ (or the set of all non-attacking rook placements on an $n \times n$ board), and, for $1 \le j \le n$, let A_j denote the subset of S_n consisting of the set of non-attacking rook placements on an $n \times n$ board such that the rook in row j is in X_j. We want to find $|A_1^c \cap A_2^c \cap \ldots \cap A_n^c|$.

For $1 \leq i \leq n$, we have $|A_i| = |X_i| (n-1)!$, since we have $|X_i|$ choices for rook placement in row i and $(n-1)!$ choices for placing the rest of the rooks. Thus, if r_1 denotes the number of forbidden squares, then

$$\sum_{i=1}^{n} |A_i| = (|X_1| + \cdots + |X_n|)(n-1)! = r_1(n-1)!.$$

For $1 \leq i < j \leq n$, $|A_i \cap A_j|$ is the number of ways of placing two non-attacking rooks in rows i and j and in forbidden positions, times $(n-2)!$, the number of ways of placing $n-2$ non-attacking rooks anywhere in rows other than i and j. Again, let r_2 be the number of ways of placing two non-attacking rooks *in* the forbidden positions, and we have

$$\sum_{1 \leq i < j \leq n} |A_i \cap A_j| = r_2(n-2)!.$$

We can continue in the same way. If r_k is the number of ways of placing k non-attacking rooks *in* the forbidden positions, then

$$\sum_{1 \leq i_1 < i_2 < \cdots < i_k \leq n} |A_{i_1} \cap A_{i_2} \cap A_{i_3} \cap \ldots \cap A_{i_k}| = r_k(n-k)!.$$

We can now apply the inclusion–exclusion principle to find $|A_1^c \cap A_2^c \cap \ldots \cap A_n^c|$. We record the result as a theorem.

Theorem 8.31. *The number of ways of placing n identical non-attacking rooks on an $n \times n$ board with forbidden positions is equal to*

$$n! - r_1(n-1)! + r_2(n-2)! - \cdots + (-1)^k r_k(n-k)! + \cdots + (-1)^n r_n,$$

where r_k is the number of ways of placing k non-attacking rooks on the $n \times n$ board in such a way that each of the k rooks is in a forbidden position.

Remark 8.32. Theorem 8.31 is useful when the number of forbidden squares is manageable, and, hence, it is easier to find the number of non-attacking rook placements in the forbidden positions rather than in the non-forbidden positions. For derangements (see Figure 8.2), $r_k = \binom{n}{k}$ and Theorem 8.31 – as expected – becomes identical to Theorem 8.24.

Example 8.33. We want to find the number of possible ways of placing eight non-attacking rooks on the 8×8 board with forbidden positions of Figure 8.3, without placing any rooks on the forbidden positions. (The forbidden positions are marked by $\times$'s.) Using the notation of Theorem 8.31, for $1 < k \leq 8$, we let r_k be the number of ways of placing k non-attacking rooks on the forbidden positions.

Before we calculate the r_i, consider the configuration of forbidden positions. Let F_1 be the forbidden positions in the first three rows, and F_2 be the forbidden positions in the last three columns. Given the particular configuration of the board in Figure 8.3, no rook placed in F_1

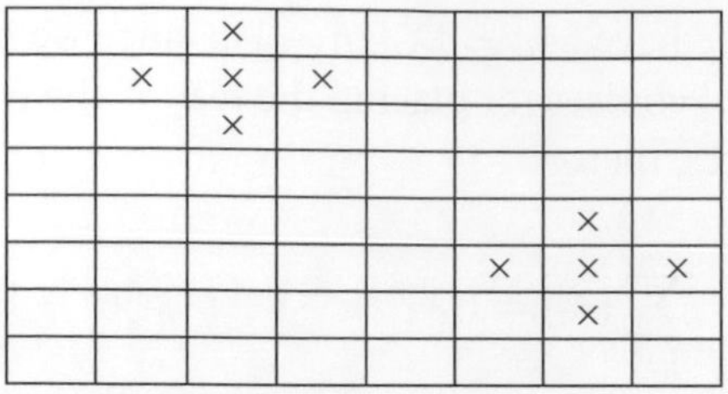

Figure 8.3 The board with forbidden configuration of Example 8.33.

can attack a rook in F_2 and vice versa. For $1 \leq i \leq 3$, let s_i denote the number of ways of placing i non-attacking rooks in F_1. The number of ways of placing one non-attacking rook in F_1 is the same as the total number of forbidden positions in F_1, and so $s_1 = 5$. We can also check directly that s_2, the number of ways of placing two non-attacking rooks in F_1, is 4. Furthermore, there is no way to put three non-attacking rooks on F_1 and so $s_3 = 0$. The configuration F_2 is identical to F_1 and so our calculations, so far, apply to it as well.

Turning to the r_i's, to place one rook on the forbidden position, you place it on either F_1 or F_2, and so $r_1 = 5 + 5 = 10$.

We can place two non-attacking rooks in the forbidden positions by either putting the two rooks in F_1, one in F_1 and one in F_2, or both in F_2. Respectively, there are 4, 5×5, and 4 ways of doing this. Hence, $r_2 = 4 + 25 + 4 = 33$.

Similarly, $r_3 = 0 + 4 \times 5 + 5 \times 4 + 0 = 40$. In this count, the first 0 refers to the number of ways of placing three rooks in F_1. The number of ways of placing two non-attacking rooks in F_1 and one rook in F_2 is 4×5, and likewise 5×4 is the number of possibilities for placing one rook in F_1 and two non-attacking rooks in F_2. Finally, you cannot put three non-attacking rooks in F_2.

The only way to place four non-attacking rooks in the forbidden positions of the board in Figure 8.3 is to place two of the rooks in F_1 and two of the rooks in F_2. This can be done in $4 \times 4 = 16$ ways.

Clearly, $r_5 = \cdots = r_8 = 0$. Hence, by Theorem 8.31, the total number of ways of placing eight non-attacking rooks in the non-forbidden positions of the board of Figure 8.3 is

$$8! - 10 \times 7! + 33 \times 6! - 40 \times 5! + 16 \times 4! = 9264.$$

Problems

P 8.3.1. When they arrive at a party, nine people leave their umbrellas at the entrance. When they leave, the umbrellas are distributed randomly. What is the probability that exactly four people get their own umbrellas back?

P 8.3.2. Seven people check their hats when entering a museum. When the museum closes, the hats are returned to the same seven people, but randomly. What is the probability that at least two people will get their own hats back?

P 8.3.3. Consider the permutations of the letters in the word BLUEJACK.

    (a) In how many permutations are exactly four of the letters in their original spots?

    (b) In how many permutations are the letters B, L, U, and E not in their original spots?

P 8.3.4. Eight negotiators sit around a round table with eight seats. For their next session, you want to seat the negotiators in such a way that every one has a new person to their left. In how many ways can you do this?

P 8.3.5. In how many ways can you seat the negotiators in Problem P 8.3.4 so that each negotiator faces a different person across the table at the next session?

P 8.3.6. Keicha and Amadou have identical decks of cards. Each deck has n cards, with $n \geq 10$. The players shuffle the decks, and, at each turn of a game, each reveals the card at the top of their deck. If, at any turn, the two cards are exactly the same, Keicha wins. If, throughout the game, they draw different cards, then Amadou wins. What is the probability of Keicha winning? What about Amadou's chances? Would the probabilities be substantially different if $n = 10$ versus $n = 100$?

P 8.3.7. For each of the 6×6 boards with forbidden positions in Figure 8.4, find the number of ways to place six non-attacking rooks in the non-forbidden positions.

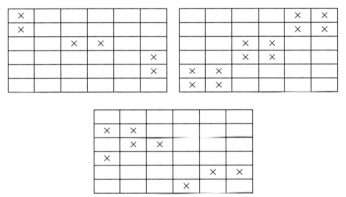

Figure 8.4 Boards with forbidden configurations for Problem P 8.3.7.

P 8.3.8. A corporation has 7 available positions $y_1, \ldots, y_7$ and 10 applicants $x_1, \ldots, x_{10}$. The set of positions each applicant is qualified for is given, respectively, by $\{y_1, y_2, y_6\}$, $\{y_2, y_6, y_7\}, \{y_3, y_4\}, \{y_1, y_5\}, \{y_6, y_7\}, \{y_3\}, \{y_2, y_3\}, \{y_1, y_3\}, \{y_1\}, \{y_5\}$.

    (a) Draw a 10×7 board with forbidden positions such that each placing of m non-attacking rooks on the board (in the allowed positions) corresponds to one way of choosing m qualified applicants for m of the positions.

    (b) What is the largest m possible?

P 8.3.9. Count the permutations $i_1 i_2 i_3 i_4 i_5 i_6$ of $\{1, 2, 3, 4, 5, 6\}$, where $i_1 \neq 1, 5$, $i_3 \neq 2, 3, 5$, $i_4 \neq 4$, and $i_6 \neq 5, 6$.

P 8.3.10. In how many ways can you place six non-attacking slices of pizza on the 6×6 board with forbidden positions of Figure 8.5, if three of the slices are vegetable supreme, two are sausage, and one is anchovy?

Note: pizza slices move as rooks do, and slices with the same topping are identical.

Figure 8.5 Board with forbidden configurations for Problem P 8.3.10.

P 8.3.11. Fereydoon, Faranak, Roudabeh, Tahmineh, and Rostam are students in a class. Each is to be assigned a different dialogue from the following list: Crito, Euthyphro, Protagoras, Meno, and Parmenides. Fereydoon is not interested in the Crito or the Protagoras; Faranak is not interested in the Meno; Roudabeh is not interested in the Euthyphro or the Parmenides; Tahmineh is not interested in the Euthyphro; and Rostam is not interested in the Crito or the Protagoras. In how many ways can we assign the five dialogues to the five students, given the above constraints?

P 8.3.12. Find the number of 5×5 matrices of the form

$$
\begin{bmatrix}
0 & 0 & \star & \star & \star \\
\star & 0 & 0 & \star & \star \\
\star & \star & \star & 0 & \star \\
\star & \star & \star & 0 & 0 \\
\star & \star & \star & \star & 0
\end{bmatrix}
$$

such that each star is either a 0, 1, 2, or 3, *and* each row and each column has exactly one non-zero entry.

P 8.3.13. You are in charge of scheduling a student-run coffee shop. There are six two-hour shifts every day starting at 10 a.m., and you have six students (Ārash, Bābak, Cyrus, Dārāb, Esfandiār, and Fereydoon) to schedule. Each student is to be assigned exactly one of the shifts. The only restrictions are that Bābak and Esfandiār cannot do the second or the fourth shift and Cyrus and Dārāb cannot do the third shift.

(a) How many valid schedules are possible?

(b) You are still in charge of scheduling the same coffee shop and you have the same student workers with the same shift restrictions as before. However, since the coffee shop has been losing money, the management wants a manager present at each shift, in order to compel the student workers to follow the new rules and to charge extra for adding sprouts and cucumbers to the sandwiches. There are three managers, named Giv, Hooshang, and Īraj, and they don't have any particular time constraints, but you have to assign exactly two shifts to each of them. How many valid schedules are possible now?

NOTE: A schedule consists of an assignment of a student and a manager to each shift without violating any of the restrictions.

P 8.3.14. I know how to make seven dishes, and I prepare one per day for this week. Next week, I want to prepare the same dishes but in a different order so that no dish is preceded by the same dish as this week. In how many possible ways can I plan next week's menu?

P 8.3.15. You are stranded in an island, and you desperately want to know the value of D_{11}, the number of derangements of 11 objects. You remember from your calculus days that $e \approx 2.718281828$,[2] and, for some reason, you know that $11! = 39,916,800$. You have a calculator that has enough battery for a single arithmetic operation (an addition, subtraction, multiplication, or division). What is your best guess for the value of D_{11}? When you are off the island, you find the true value of D_{11}. How far off were you? You also wonder what would have happened if you had used the approximation $e \approx 2.71$. Would it have made a difference?

P 8.3.16. In analogy with derangement numbers, define a sequence of numbers by $F_1 = 1$, $F_2 = 2$, and

$$F_n = (n-1)(F_{n-2} + F_{n-1}), \text{ for } n \geq 3.$$

Have you seen this sequence? Prove your assertion.

P 8.3.17. Let n be a positive integer, and k a non-negative integer with $k \leq n$. A derangement is a permutation of $[n]$ where none of the elements of $[n]$ is in its natural place. Consider permutations of $[n]$ where the first k elements of $[n]$ are not in their natural places. Let $D_{n,k}$ denote the number of such permutations, so that $D_{n,n} = D_n$ is the number of derangements.[3]

(a) Translate the problem of finding $D_{n,k}$ to a problem about placing rooks on a board with forbidden positions.

(b) Find a formula for $D_{n,k}$.

(c) Let $0 \leq k \leq n-1$, and let $\mathcal{B}$ be those permutations of $[n]$ where the first k elements are not in their natural positions, while the $(k+1)$th element *is* in its natural position. Can we write $|\mathcal{B}| = D_{m,\ell}$ for an appropriate choice of m and ℓ?

(d) Let $0 \leq k \leq n-1$. Use a combinatorial argument to prove that

$$D_{n,k+1} = D_{n,k} - D_{n-1,k}.$$

P 8.3.18. Let D_n be the number of derangements of n objects. When is D_n an even integer and when is it odd? Make a conjecture and prove it.

P 8.3.19. Let n be a positive integer, and, for a positive integer i, as usual, let D_i denote the number of derangements of i objects. For convenience, also define $D_0 = 1$. Using a combinatorial argument, find a formula for

$$\binom{n}{0}D_0 + \binom{n}{1}D_1 + \cdots + \binom{n}{n}D_n.$$

[2] A mnemonic for remembering this approximation to e is to remember that the 7th president of the US was Andrew Jackson and he was first elected in 1828.

[3] Adapted from Spivey 2010.

P 8.3.20. Let n be a positive integer, and define $P_n = \frac{D_n}{n!}$ to be the probability that, in the hat-check problem of Question 8.21, no hat is returned to the right person.[4]

(a) Using Proposition 8.28, find a recurrence relation for P_n of the form

$$P_n = ?P_{n-2} + ??P_{n-1}.$$

(b) Rewrite the recurrence relation of the previous part to get a relation between $P_n - P_{n-1}$ and $P_{n-1} - P_{n-2}$.

(c) Starting with $P_n - P_{n-1}$, and repeatedly using the formula from the previous part, unwind the formula to get

$$P_n - P_{n-1} = \frac{(-1)^n}{n!}.$$

P 8.3.21. **Another Proof of Theorem 8.24.** The proof of Theorem 8.24 in the text used the inclusion–exclusion principle. The proof of Proposition 8.28 did not. You used the latter, in Problem P 8.3.20, to prove that $P_n - P_{n-1} = (-1)^n/n!$, where $P_n = D_n/n!$. Use this formula to give a second proof of Theorem 8.24. You may find the following steps helpful:

STEP 1: In finding $\sum_{i=2}^{n}(P_i - P_{i-1})$, do any of the terms cancel?

STEP 2: Use Step 1 and the formula from Problem P 8.3.20 to give a formula for P_n.

STEP 3: Give a formula for D_n, completing the proof.

P 8.3.22. **Binomial Inversion Proof of Theorem 8.24.** Let n be a positive integer, and let D_n, as usual, denote the nth derangement number. Define $D_0 = 1$, and define the $(n + 1) \times 1$ column vector $E = \begin{bmatrix} D_0 \\ D_1 \\ \vdots \\ D_n \end{bmatrix}$. Let C_n be the $(n + 1) \times (n + 1)$ matrix of binomial coefficients (Definition 5.33).

(a) Do Problem P 8.3.19, and rewrite the result as giving the result of the matrix product $C_n E$.

(b) Use the inverse of C_n (Problem P 5.4.11 or, what amounts to the same thing, the binomial inversion of Problem P 5.4.12) to solve the equation from the previous part for E.

(c) Do you have a new proof of Theorem 8.24?

P 8.3.23. How many rooks can you place on a three-dimensional $10 \times 10 \times 10$ chess board if no two may attack each other?

8.3.3   Derangements and Bell Numbers

Let n be a positive integer and let s be a fixed non-negative integer with $s \leq n$. As usual, let $B(j)$ denote the jth Bell number (Definition 6.12) and D_m be the number of

---

[4] Problems P 8.3.20–P 8.3.22 were adapted from Hathout 2004.

derangements of $[m]$. In the problems, you are asked to prove, for positive integers n and s with $s \le n$, that[5]

$$\sum_{m=0}^{n} m^s \binom{n}{m} D_m = n! \sum_{j=0}^{s} (-1)^j \binom{s}{j} n^{s-j} B(j). \tag{8.4}$$

To give a combinatorial proof, we propose a thought experiment:

Thought Experiment 8.34. Let n and s be fixed integers with $0 \le s \le n$. Let $T = \{T_1, T_2, \ldots, T_s\}$ be a set of s distinct tiles that need to be painted, and let $[n] = \{1, \ldots, n\}$ be a set of colors. There are $n!$ artisans ready to assign colors – one per tile – to the tiles. Let S_n denote the set of permutations of $[n]$. Now each of the artisans is associated with a different permutation of $[n]$. In fact, since there are exactly $n!$ permutations of $[n]$, and there are $n!$ artisans, there is a bijection between the permutations and the artisans, and we take the liberty of referring to each artisan by the permutation that is associated with her. In coloring the tiles, the artisans must follow one cardinal rule: if an artisan is associated with the permutation $\mu \in S_n$, then she can only use those colors – remember that $1, 2, \ldots, n$ are the list of the colors – that are not in their natural positions in μ (in the language of group theory, artisan μ can only use the colors that are not fixed by μ). For example, if $n = 4$, and $\mu = 1432$, then the artisan μ is allowed to use colors 2 and 4 (but not 1 or 3).[6] So, if $s = 3$, then this artisan could suggest any of the following eight schemes for coloring T_1, T_2, and T_3: 222, 224, 242, 244, 422, 424, 442, 444. (The scheme 242 means that the artisan is assigning color 2 to T_1 and T_3 while assigning color 4 to T_2.) For $\mu \in S_n$, we let $N(\mu)$ – which depends on n, s, and μ – to be the number of ways that the artisan associated with μ could assign colors to elements of T. So, for $n = 4$, $s = 3$, and $\mu = 1432$, we have $N(\mu) = 8$. Each artisan signs her work and so two sets of identical colorings but by two different artisans can be distinguished.

In Problems P 8.3.24–P 8.3.32, you will use the metaphor, the notation, and the definitions of Thought Experiment 8.34 to prove and explore the identity of Equation 8.4. The bird's-eye view is to find $\sum_{\mu \in S_n} N(\mu)$ in two different ways.

Problems (continued)

P 8.3.24. Adapting the notation of Thought Experiment 8.34, if $n = 8$, $s = 7$, and $\mu = 12436578$, then what is $N(\mu)$? More generally, for m a non-negative integer, let B_m denote those permutations of $[n]$ where exactly m elements of $[n]$ are not in their natural positions (and $n - m$ elements of $[n]$ are fixed and in their natural positions). What is $|B_m|$, and what is $N(\mu)$ for $\mu \in B_m$?

---

[5] Adapted from Clarke and Sved 1993.
[6] If we use the cycle notation – preferred by group theorists – for permutations (see Remark 6.26), then the permutation $\mu = 1432$ would be denoted by $(1)(2\,4)(3)$ or just $(2\,4)$, making it clear that 1 and 3 are fixed (in our parlance, they are in their natural position) while 2 and 4 are swapped.

P 8.3.25. Adapt the notation of Thought Experiment 8.34, and let B_m be defined as in Problem P 8.3.24. What is

$$\sum_{\mu \in B_m} N(\mu)?$$

Show that

$$\sum_{\mu \in S_n} N(\mu) = \sum_{m=0}^{n} m^s \binom{n}{m} D_m.$$

P 8.3.26. Let $n = 9$, $s = 7$, and $\mu = 124365789 \in S_9$. Continuing with Thought Experiment 8.34, consider the artisan associated with μ. If this artisan rebels and ignores the cardinal rule, then in how many ways can she color the tiles in T in such a way that the color of the tiles T_3, T_5, and T_6 violates the cardinal rule?

P 8.3.27. Let $n = 9$, $s = 7$. Continuing with Thought Experiment 8.34, recall that there are nine colors, $1, \ldots, 9$, and seven tiles, $T_1, \ldots, T_7$. Consider the coloring 2851364, which means that T_1 has color 2, T_2 has color 8, $\ldots$, and T_7 has color 4. For how many of the 9! artisans would this coloring constitute a violation of the cardinal rule for each of the tiles T_3, T_4, and T_6?

P 8.3.28. With the notation of Thought Experiment 8.34, assume that all the $n!$ artisans ignored the cardinal rule. Let X be the set of all colorings of tiles by all the artisans (and all ignoring the cardinal rule). Each artisan has signed her work, and so the same assignments of colors to tiles may appear multiple times in X (and these are considered different elements of X).
   (a) What is $|X|$?
   (b) For how many elements of X is the tile T_1 colored with color numbered 1 *and* this coloring of T_1 is a violation of the cardinal rule?
   (c) For how many elements of X does the color of the tile T_1 violate the cardinal rule?
   (d) For how many elements of X do the colors of the tiles T_1 and T_2 both violate the cardinal rule?
   (e) Let $1 \le j \le s$. Show that the number of elements of X for which each of the colors of the tiles $T_1, \ldots, T_j$ constitute a violation of the cardinal rule is

$$n! \, n^{s-j} B(j).$$

P 8.3.29. Continuing with Thought Experiment 8.34 and Problem P 8.3.28, we make the following definitions:

   $X =$ the set of all colorings of the tiles by all the artisans and with no regard to the cardinal rule

   $A_1 =$ elements of X where the color of the tile T_1 is in violation of the cardinal rule

   $A_2 =$ elements of X where the color of the tile T_2 is in violation of the cardinal rule

   $\vdots$

   $A_s =$ elements of X where the color of the tile T_s is in violation of the cardinal rule.

   (a) How are $\left| A_1^c \cap A_2^c \cap \cdots \cap A_s^c \right|$ and $\sum_{\mu \in S_n} N(\mu)$ related?

(b) Prove that

$$\sum_{\mu \in S_n} N(\mu) = n! \sum_{j=0}^{s} (-1)^j \binom{s}{j} n^{s-j} B(j).$$

(c) Use the previous part and Problem P 8.3.25 to give a complete proof of Equation 8.4.

P 8.3.30. What does the identity of Equation 8.4 turn into if $s = 0$? Have you seen this before?

P 8.3.31. Let n be a positive integer, and, as usual, let D_n denote the number of derangements of n objects. Consider

$$\binom{1}{1} D_1 = 0$$

$$\binom{2}{1} D_1 + 2\binom{2}{2} D_2 = 2$$

$$\binom{3}{1} D_1 + 2\binom{3}{2} D_2 + 3\binom{3}{3} D_3 = 12$$

$$\binom{4}{1} D_1 + 2\binom{4}{2} D_2 + 3\binom{4}{3} D_3 + 4\binom{4}{4} D_4 = 72.$$

Use the identity of Equation 8.4 to find the sum

$$\binom{n}{1} D_1 + 2\binom{n}{2} D_2 + 3\binom{n}{3} D_3 + \cdots + n\binom{n}{n} D_n.$$

P 8.3.32. Let $n \geq 2$ be an integer. Use the identity of Equation 8.4, in the case when $s = 2$, to prove

$$\sum_{m=0}^{n} \sum_{k=0}^{m} \frac{(-1)^k m^2}{(n-m)! \, k!} = n^2 \quad 2n + 2.$$

9 Generating Functions

You want to arrange **17** *books on the shelf of a bookstore. The shelf is dedicated to the three Toni Morrison novels published between 1977 and 1987:* **Song of Solomon, Tar Baby,** *and* **Beloved.** *You have many copies of each, but on the shelf you want an even number of* **Song of Solomon,** *at least three copies of* **Tar Baby,** *and at most four copies of* **Beloved.** *How many different arrangements are possible? (Copies of the same book do not have to be next to each other.)*

In this chapter, we turn to the powerful technique of generating functions for solving combinatorial problems. In combinatorics, we are often investigating a sequence of numbers $h_0, h_1, \ldots, h_n, \ldots$ We may be looking for a formula for h_n, and sometimes we may have a recurrence relation or other partial information about h_n. To the untrained eye, it may seem unlikely that by considering the power series $h_0 + h_1 x + \cdots + h_n x^n + \cdots$ – that is, constructed using every number in that sequence – we can glean any new information about our sequence of numbers. After all, if we already know what h_n is, then what use is this power series, and if we don't know what h_n is, then how are we going to deal with the more complicated power series?

Actually, we have already seen a number of examples of the use of generating functions in the previous chapters. Power series were used in Example (d) in the Introduction to solve a recurrence relation; in Proposition 5.26 we manipulated the expansion of $(1 + x)^n$ to get binomial coefficient identities; in the proof of Proposition 6.15, we used the power series expansion of e^x and related series to prove a formula for Bell numbers; and we used the algebraic identities of Theorem 6.31 to prove the relation between the two kinds of Stirling numbers. The basic idea is that we can use power series to encode a sequence of integers. Manipulating the underlying function may then provide an avenue for extracting additional information about the sequence. We begin with ordinary generating functions and then turn to exponential generating functions.[1]

9.1 Ordinary Generating Functions

Warm-Up 9.1. *Consider the function* $f(x) = \frac{1}{1+x^2}$ *and the polynomial* $g(x) = a_0 + a_1 x + a_2 x^2 + a_3 x^3$. *Determine values for* a_0, a_1, a_2, *and* a_3 *such that* $f(0) = g(0), f'(0) = g'(0), f''(0) = g''(0)$, *and* $f'''(0) = g'''(0)$. *Do you remember a name associated with such a polynomial* $g(x)$?

---

[1] While not necessary, the reader may benefit from working through Collaborative Mini-project 4 on generating functions before tackling this chapter.

Definition 9.2. Let $h_0, h_1, \ldots, h_n, \ldots$ be an infinite sequence of numbers. The *(ordinary)* *generating function* (or *ordinary power series*) of this sequence is

$$g(x) = h_0 + h_1 x + h_2 x^2 + \cdots + h_n x^n + \cdots,$$

and we write

$$g \xleftrightarrow{\text{ops}} \{h_n\}_{n=0}^{\infty}$$

("ops" stands for ordinary power series). Given a finite sequence $a_0, \ldots, a_k$ of numbers, we think of it as an infinite sequence by appending an infinite string of 0's to the end of it. The generating function of this finite sequence will then be a polynomial $f(x) = a_0 + a_1 x + \cdots + a_k x^k$, and we write $f \xleftrightarrow{\text{ops}} \{a_n\}_{n=0}^{k}$.

Remark 9.3. Recall from calculus that a *power series* is an infinite series of the form

$$\sum_{k=0}^{\infty} a_k x^k,$$

where, for all k, a_k is a real number.

In combinatorics we consider $\sum_{k=0}^{\infty} a_k x^k$ mostly as a *formal* power series. This means that, most often and especially at the more elementary level, we do not think of the power series as a function and we do not plug in values for x. Rather we manipulate the series formally and are mostly interested in what happens to the coefficients. As such, at the elementary level, we are interested in algebraic properties of power series more than analytic properties of such series. If we were to plug in values for x, and sometimes we may do that, we should note that – depending on the power series – one of three things could happen: the series could converge only for $x = 0$; it could converge for all values of x (this would happen, for example, in the case of polynomials where only a finite number of the coefficients are non-zero); or it could converge for all x with $|x| < R$, where R is a positive real number called the *radius of convergence* (in this last case, the series may or may not converge for $x = \pm R$).

Some power series have a closed form. In other words, the Taylor series (at $x = 0$)[2] of a known function generates the same power series. If f is a real-valued function that is defined and is infinitely differentiable at and around $x = 0$ (more precisely, $f : A \to \mathbb{R}$ is infinitely differentiable, where A is an open set containing zero), then its Taylor series at $x = 0$ is the power series

$$f(0) + f'(0)x + \frac{f''(0)}{2!} x^2 + \cdots + \frac{f^{(k)}(0)}{k!} x^k + \cdots \qquad (9.1)$$

(See Problem *P 9.1.23*.) If we plug in a real number r for x in Equation 9.1, the resulting series may or may not converge. Even if the series converges, what it converges to may not be equal to $f(r)$. On the other hand, *if* the function f is equal (in an open interval that includes 0) as a function to some power series, then that power series must be the Taylor series given in

---

[2] This may be a good time to review the basic idea of a Taylor series from a calculus book such as my *Approximately Calculus* (Shahriari 2006) Chapter 15.

Equation 9.1. These are issues that are addressed in detail in courses in analysis. But, for us, the power series of Equation 9.1 *is* the power series for $f(x)$ regardless of convergence issues. The function $f(x)$ carries – in a compact form – all the coefficients of the power series of Equation 9.1, and that is all we want. If you want the coefficient of x^k, you just find $f^{(k)}(0)/k!$, and you can do this even if the Taylor series converges just for $x = 0$ (which it always does). With considerable abuse of notation, we will sometimes write $f(x) = f(0) + f'(0)x + \cdots + \frac{f^{(k)}(0)}{k!}x^k + \cdots$, without specifying the radius of convergence. The advantage is that we can find the Taylor series (at $x = 0$) of other functions by manipulating this "equation." You can multiply both sides by a power of x, take derivatives and antiderivatives, and even replace x by a power of x. This formal manipulation of Taylor series allows us to efficiently store many infinite sequences of numbers using a function. Instead of writing that $f(x)$ is equal to $f(0) + \cdots + \frac{f^{(k)}(0)}{k!}x^k + \cdots$ – which, if we are thinking of these as functions, could be incorrect – the more precise notation would be $f(x) \xleftrightarrow{\text{ops}} \left\{\frac{f^{(k)}(0)}{k!}\right\}_0^{\infty}$.

Example 9.4. We remind the reader of a few important power series from calculus. Remember that the equal sign here does *not* mean that the function and the power series are the same function. It only signifies that this power series is the Taylor series at $x = 0$ for this function.

$$\frac{1}{1-x} = 1 + x + x^2 + x^3 + \cdots = \sum_{k=0}^{\infty} x^k$$

$$e^x = 1 + x + \frac{x^2}{2!} + \frac{x^3}{3!} + \cdots = \sum_{k=0}^{\infty} \frac{1}{k!} x^k$$

$$\ln(1+x) = x - \frac{x^2}{2} + \frac{x^3}{3} - \cdots = \sum_{k=1}^{\infty} \frac{(-1)^{k+1} x^k}{k}.$$

In analysis, we often start with a function, ask for its power series, and investigate the convergence properties of the series. In combinatorics, we often start with a sequence of numbers, create the generating function for the sequence, and ask for a closed form.

Example 9.5. The ordinary generating function for the sequence

$$1, \ 1, \ 1, \ \ldots$$

is $1 + x + x^2 + \cdots$ and has the closed form $\frac{1}{1-x}$. Hence, we can write

$$\frac{1}{1-x} \xleftrightarrow{\text{ops}} \{1\}_{n=0}^{\infty}.$$

Note that, as functions, $\frac{1}{1-x} = \sum_{n=0}^{\infty} x^n$ only for $|x| < 1$.

Example 9.6. The ordinary generating function for the sequence

$$\underbrace{1, \ 1, \ \ldots, \ 1}_{k}, \ 0, \ 0, \ \ldots$$

is $s(x) = 1 + x + x^2 + \cdots + x^{k-1}$. This is a partial sum of a geometric series, and its closed form is found by noticing that $s(x) - xs(x) = 1 - x^k$, and hence, for $x \neq 1$, $s(x) = \frac{1-x^k}{1-x}$. Hence, we can write

$$\frac{1-x^k}{1-x} \xleftrightarrow{\text{ops}} \{1\}_{n=0}^{k-1}.$$

Example 9.7. The ordinary generating function for the sequence

$$\frac{1}{0!}, \frac{1}{1!}, \frac{1}{2!}, \frac{1}{3!}, \ldots$$

is $1 + x + \frac{x^2}{2!} + \frac{x^3}{3!} + \cdots$ and has the closed form e^x. So, we write

$$e^x \xleftrightarrow{\text{ops}} \left\{\frac{1}{n!}\right\}_{n=0}^{\infty}.$$

Example 9.8. Let n be a positive integer. Consider the (finite) sequence consisting of the binomial coefficients in one row of the Karaji–Jia triangle:

$$\binom{n}{0}, \binom{n}{1}, \binom{n}{2}, \ldots, \binom{n}{n}.$$

In the case of a finite sequence, if you so desire, while not necessary, you can imagine a string of trailing zeros to make the sequence into an infinite sequence. The ordinary generating function for this sequence is

$$\binom{n}{0} + \binom{n}{1}x + \binom{n}{2}x^2 + \cdots + \binom{n}{n}x^n \left(+ 0x^{n+1} + \cdots\right).$$

By the binomial theorem (Corollary 5.25), we know that the closed form for this generating function is $(1+x)^n$. Hence,

$$(1+x)^n \xleftrightarrow{\text{ops}} \left\{\binom{n}{k}\right\}_{k=0}^{n} \quad \text{or} \quad \left\{\binom{n}{k}\right\}_{k=0}^{\infty}.$$

Example 9.9. Consider the sequence of squares of non-negative integers:

$$0^2, 1^2, 2^2, 3^2, \ldots, n^2, \ldots$$

The ordinary generating function of this sequence is

$$g(x) = x + 4x^2 + 9x^3 + \cdots + n^2 x^n + \cdots.$$

Can we find a closed form for this power series? We start with the power series for $1/(1-x)$ and take derivatives. We get

$$\frac{1}{(1-x)^2} = 1 + 2x + 3x^2 + \cdots + nx^{n-1} + \quad.$$

The coefficient of x^n in this power series is $n+1$, and we have shown that

$$\frac{1}{(1-x)^2} \xleftrightarrow{\text{ops}} \{n+1\}_{n=0}^{\infty}.$$

But this is not what we were looking for. We want to square each of the coefficients. Taking another derivative would result in a coefficient of $(n + 1)(n + 2)$ for x^n, and this is not a square. The trick – or technique – for resolving the issue is to first multiply the power series of $1/(1 - x)^2$ by an x, and *then* take the derivative. Since the derivative of $x/(1 - x)^2$ is $\frac{(1-x)^2+2x(1-x)}{(1-x)^4} = \frac{1+x}{(1-x)^3}$, we have

$$\frac{x}{(1 - x)^2} = x + 2x^2 + 3x^3 + \cdots + nx^n + \cdots$$

$$\frac{1 + x}{(1 - x)^3} = 1 + 2^2x + 3^2x^2 + \cdots + n^2x^{n-1} + \cdots .$$

To get $g(x)$, we have to multiply this power series by x. Hence, $g(x) = \frac{x(1+x)}{(1-x)^3}$, and

$$\frac{x(1 + x)}{(1 - x)^3} \overset{\text{ops}}{\longleftrightarrow} \{n^2\}_{n=0}^{\infty} .$$

Remark 9.10. In Example 9.9, we used the fact that if you have a power series for the function $f(x)$, then to find the power series for $xf(x)$ and $f'(x)$, you, respectively, just multiply by x or take the derivative of the power series for $f(x)$. The power series for an antiderivative of $f(x)$ is also an antiderivative of the power series for $f(x)$, and the power series for $f(x^2)$ is what you get when you plug in x^2 for x in the power series of $f(x)$. These and similar facts do need a proof – many follow from the fact that the power series for a function $f(x)$ is uniquely given by Equation 9.1 – but most of the proofs are not particularly complicated or illuminating (see Problem P 9.1.24, for example). Hence, we are skipping the proofs here.

Multiplying Power Series

As already noted, given the generating function for one sequence of numbers, we want to be able to generate generating functions for other related sequences. Here, we ask: what happens when you multiply two power series?

Example 9.11. In the calculations in Example 9.9, we found, using derivatives, that

$$\frac{1}{(1 - x)^2} = 1 + 2x + 3x^2 + \cdots + (n + 1)x^n + \cdots .$$

Could we have found this without taking derivatives and by using the fact that $\frac{1}{(1-x)^2} = \frac{1}{1-x}\frac{1}{1-x}$?

We know that $\frac{1}{1-x} = 1 + x + x^2 + \cdots$, so do we get

$$(1 + x + x^2 + \cdots)(1 + x + x^2 + \cdots) = 1 + 2x + 3x^2 + \cdots + (n + 1)x^n + \cdots?$$

As an example, consider the x^2 term. Why did we get $3x^2$ as a term in this power series? Every term of the resulting series is the product of one term from the first parenthesis and one term from the second parenthesis, and the products $1 \cdot x^2$, $x \cdot x$, and $x^2 \cdot 1$ are the ones that give us an x^2 term. Since there are three such terms (and the coefficient of each is 1), we end up with $3x^2$. Likewise, to get an x^n term, we need a term of the form x^r multiplied by a term of the form x^{n-r}. This can happen for $r = 0, 1, \ldots, n$, and so the coefficient of x^n is $n + 1$. We conclude – using a different argument from Example 9.9 – that the ordinary generating function for the sequence

$$1, 2, 3, \ldots$$

is $\frac{1}{(1-x)^2}$, and we can write

$$\frac{1}{(1-x)^2} \xleftrightarrow{\text{ops}} \{n + 1\}_{n=0}^{\infty}.$$

Proposition 9.12. *Assume that*

$$f \xleftrightarrow{\text{ops}} \{a_n\}_{n=0}^{\infty}, \text{ and } g \xleftrightarrow{\text{ops}} \{b_n\}_{n=0}^{\infty}.$$

Then

$$fg \xleftrightarrow{\text{ops}} \{c_n\}_{n=0}^{\infty},$$

where $c_n = a_0 b_n + a_1 b_{n-1} + \cdots + a_n b_0 = \sum_{r=0}^{n} a_r b_{n-r}$.

Proof. We have

$$f(x) = a_0 + a_1 x + \cdots + a_n x^n + \cdots$$
$$g(x) = b_0 + b_1 x + \cdots + b_n x^n + \cdots.$$

When you multiply f and g, for every r with $0 \le r \le n$, the product of the $a_r x^r$ term of f with the $b_{n-r} x^{n-r}$ term of g gives an x^n term. Hence, in fg, the coefficient of x^n will be $a_0 b_n + a_1 b_{n-1} + \cdots + a_n b_0$, and

$$f(x)g(x) = a_0 b_0 + (a_1 b_0 + a_0 b_1)x + \cdots + \left(\sum_{r=0}^{n} a_r b_{n-r} \right) x^n + \cdots. \qquad \square$$

Remark 9.13. We can apply Proposition 9.12 to finite sequences as well. If $a_0, a_1, \ldots, a_k$ is a finite sequence, we append 0's at the end, turn it into the infinite sequence $a_0, a_1, \ldots, a_k, 0, 0, \ldots$, and then apply the proposition.

Definition 9.14. Given two sequences $a_0, a_1, \ldots$ and $b_0, b_1, \ldots$, construct a third sequence $c_0, c_1, \ldots$ by the rule

$$c_n = \sum_{i=0}^{n} a_r b_{n-r}.$$

The sequence $\{c_n\}_0^{\infty}$ is called the *convolution* of $\{a_n\}_0^{\infty}$ and $\{b_n\}_0^{\infty}$.

Given Definition 9.14, Proposition 9.12 says that the product of the generating functions for two sequences is the generating function of the convolution of the two sequences.

Example 9.15. The convolution of 1, 2, 1 and 1, 3, 3, 1 is

$$1 \times 1, 1 \times 3 + 2 \times 1, 1 \times 3 + 2 \times 3 + 1 \times 1, 1 \times 1 + 2 \times 3 + 1 \times 3,$$
$$1 \times 0 + 2 \times 1 + 1 \times 3, 1 \times 0 + 2 \times 0 + 1 \times 1 = 1, 5, 10, 10, 5, 1.$$

Often, to find the convolution of two sequences, it is less confusing to just use Proposition 9.12. In our example, the generating function for 1, 2, 1 is $1 + 2x + x^2$, and the generating function for 1, 3, 3, 1 is $1 + 3x + 3x^2 + x^3$, and

$$(1 + 2x + x^2)(1 + 3x + 3x^2 + x^3) = 1 + 5x + 10x^2 + 10x^3 + 5x^4 + x^5.$$

In fact, we could have noticed that $1 + 2x + x^2 = (1 + x)^2$ and $1 + 3x + 3x^2 + x^3 = (1 + x)^3$. Then it would be clear that

$$(1 + x)^2(1 + x)^3 = (1 + x)^5 = 1 + 5x + 10x^2 + 10x^3 + 5x^4 + x^5.$$

To use the notation of Proposition 9.12, we would write $(1 + x)^2 \overset{\text{ops}}{\longleftrightarrow} \{1, 2, 1\}$, $(1 + x)^3 \overset{\text{ops}}{\longleftrightarrow}$ $\{1, 3, 3, 1\}$, and $(1+x)^5 \overset{\text{ops}}{\longleftrightarrow} \{1, 5, 10, 10, 5, 1\}$. Since $(1+x)^2(1+x)^3 = (1+x)^5$, the convolution of $1, 2, 1$ and $1, 3, 3, 1$ is $1, 5, 10, 10, 5, 1$.

Remark 9.16. In manipulating sequences, sometimes you have to be careful about the range of the indices of the various sums. For example, in a convolution of two series, we have $c_n = \sum_{r=0}^{n} a_r b_{n-r}$, and the sum has to start at 0 and stop at n. To streamline the work, we can adopt the convention that, given a sequence $\{a_\ell, a_{\ell+1}, \ldots, a_k\}$, where ℓ and k are two non-negative integers (or even if one or both are negative), we define $a_n = 0$ for $n < \ell$ or $n > k$. In other words, we extend every sequence to a doubly infinite sequence by appending extra zeros when necessary. Given this convention, we can write $c_n = \sum_{r=0}^{\infty} a_r b_{n-r}$ since $b_{-1} = b_{-2} = \cdots = 0$. We could even write $c_n = \sum_{r=-\infty}^{\infty} a_r b_{n-r}$, and it would be correct.

We can extend Proposition 9.12 to the products of more than two sequences:

Proposition 9.17. *Let t be a positive integer, and, for $1 \le i \le t$, let f_i be the ordinary generating function for the sequence*

$$a_0^{(i)}, a_1^{(i)}, \ldots, a_n^{(i)}, \ldots$$

Then the product of these generating functions, $f_1 f_2, \ldots, f_t$, is itself the generating function for the sequence $c_0, c_1, \ldots, c_n, \ldots$, where c_n is the sum of all terms of the form $a_{n_1}^{(1)} a_{n_2}^{(2)} \cdots a_{n_t}^{(t)}$ – a product of t terms, one from each of the original sequences – with $n_1, \ldots, n_t$ non-negative integers and $n_1 + \cdots + n_t = n$.

In other words, if

$$f_1 \overset{\text{ops}}{\longleftrightarrow} \left\{ a_n^{(1)} \right\}_{n=0}^{\infty}, f_2 \overset{\text{ops}}{\longleftrightarrow} \left\{ a_n^{(2)} \right\}_{n=0}^{\infty}, \ldots, f_t \overset{\text{ops}}{\longleftrightarrow} \left\{ a_n^{(t)} \right\}_{n=0}^{\infty},$$

then

$$f_1 \ldots f_t \overset{\text{ops}}{\longleftrightarrow} \{c_n\}_{n=0}^{\infty}, \text{ where } c_n = \sum_{\substack{n_1 + n_2 + \cdots + n_t = n \\ n_i \ge 0}} a_{n_1}^{(1)} a_{n_2}^{(2)} \cdots a_{n_t}^{(t)}.$$

Example 9.18. Assume

$$f(x) = a_0 + a_1 x + a_2 x^2 + \cdots + a_n x^n + \cdots$$

$$g(x) = b_0 + b_1 x + b_2 x^2 + \cdots + b_n x^n + \cdots$$

$$h(x) = c_0 + c_1 x + c_2 x^2 + \cdots + c_n x^n + \cdots .$$

In other words, f, g, and h are the ordinary generating functions for the sequences $\{a_n\}_{n=0}^{\infty}$, $\{b_n\}_{n=0}^{\infty}$, and $\{c_n\}_{n=0}^{\infty}$, respectively. Then the product fgh is the ordinary generating function of some sequence $\{h_n\}_{n=0}^{\infty}$. What is h_4? By definition, we have

$$fgh = h_0 + h_1 x + h_2 x^2 + h_3 x^3 + h_4 x^4 + \cdots .$$

Applying Proposition 9.17, with $t = 3$ and $n = 4$, we have

$$h_4 = \sum_{\substack{n_1 + n_2 + n_3 = 4 \\ n_i \geq 0}} a_{n_1} b_{n_2} c_{n_3}.$$

This means that, for each non-negative solution to $n_1 + n_2 + n_3 = 4$, we have a term for h_4. There are $\left(\binom{3}{4} \right) = \binom{6}{4} = 15$ such solutions, and so there are 15 terms in the sum defining h_4. One of the solutions is $n_1 = 2$, $n_2 = 0$, and $n_3 = 4$, and so one of the terms is $a_2 b_0 c_2$. This makes sense, since in finding the product fgh, one of the terms that we get is $(a_2 x^2)(b_0)(c_2 x^2) = a_2 b_0 c_2 x^4$. Writing out all the terms – each corresponding to one non-negative integer solution to $n_1 + n_2 + n_3 = 4$ – we get

$$h_4 = a_4 b_0 c_0 + a_3 b_1 c_0 + a_3 b_0 c_1 + a_2 b_2 c_0 + a_2 b_1 c_1 + a_2 b_0 c_2 + a_1 b_3 c_0 + a_1 b_2 c_1$$

$$+ a_1 b_1 c_2 + a_1 b_0 c_4 + a_0 b_4 c_0 + a_0 b_3 c_1 + a_0 b_2 c_2 + a_0 b_1 c_3 + a_0 b_0 c_4.$$

This example makes the proof of Proposition 9.17 – at least in this case – clear. To get any term of fgh, you have to multiply one term of f with one term of g and with one term of h. So any such term has the form

$$(a_{n_1} x^{n_1})(b_{n_2} x^{n_2})(c_{n_3} x^{n_3}) = a_{n_1} b_{n_2} c_{n_3} x^{n_1 + n_2 + n_3},$$

where n_1, n_2, and n_3 are non-negative integers. To get x^4, you need $n_1 + n_2 + n_3 = 4$, as claimed before.

Proof of Proposition 9.17. To get any term of $f_1 f_2 \cdots f_t$, you have to multiply one term from each $f_1, f_2, \ldots, f_t$. So any such term has the form

$$(a_{n_1}^{(1)} x^{n_1})(a_{n_2}^{(2)} x^{n_2}) \cdots (a_{n_t}^{(t)} x^{n_t}) = a_{n_1}^{(1)} a_{n_2}^{(2)} \cdots a_{n_t}^{(t)} x^{n_1 + n_2 + \cdots + n_t},$$

where $n_1, n_2, \ldots, n_t$ are non-negative integers. To get x^n, we need $n_1 + n_2 + \cdots + n_t = n$, and when we collect all the x^n terms, the coefficient of x^n will be the sum of all the terms of the form $a_{n_1}^{(1)} a_{n_2}^{(2)} \cdots a_{n_t}^{(t)}$ where $n_1, \ldots, n_t$ are non-negative integers that satisfy $n_1 + \cdots + n_t = n$. This is exactly what the proposition claims. $\square$

Remark 9.19. In the sum defining c_n in Proposition 9.17, we get one term for each non-negative integer solution to $x_1 + x_2 + \cdots + x_t = n$. Hence, by Theorem 4.40, the number of terms in this sum is $\left(\!\binom{t}{n}\!\right) = \binom{n+t-1}{t-1}$.

One special case of Proposition 9.17, when each $f_i = \frac{1}{1-x} \overset{\text{ops}}{\longleftrightarrow} \{1\}_{n=0}^{\infty}$, generalizes Example 9.11, and will be particularly useful.

Corollary 9.20.

$$\frac{1}{(1-x)^t} \overset{\text{ops}}{\longleftrightarrow} \left\{\left(\!\binom{t}{n}\!\right)\right\}_{n=0}^{\infty}.$$

In other words,

$$\frac{1}{(1-x)^t} = \sum_{n=0}^{\infty} \left(\!\binom{t}{n}\!\right) x^n = \sum_{n=0}^{\infty} \binom{n+t-1}{n} x^n.$$

Proof. For $1 \leq i \leq t$, let $f_i = \frac{1}{1-x} \overset{\text{ops}}{\longleftrightarrow} \{1\}_{n=0}^{\infty}$. In other words, each f_i is the generating function for the sequence of all 1's. Now find the product of these generating functions using Proposition 9.17. Since all the terms of all of the sequences are 1, the $a_{n_1}^{(1)} a_{n_2}^{(2)} \cdots a_{n_t}^{(t)}$ term in the definition of c_n will be a product of all 1's, and equal to 1. So

$$f_1 f_2 \cdots f_t = \frac{1}{(1-x)^t} \overset{\text{ops}}{\longleftrightarrow} \left\{ \sum_{\substack{x_1+x_2+\cdots+x_t=n \\ x_i \geq 0}} 1 \right\}_{n=0}^{\infty} = \left\{\left(\!\binom{t}{n}\!\right)\right\}_{n=0}^{\infty}.$$

The final equality follows since $\sum_{\substack{x_1+x_2+\cdots+x_t=n \\ x_i \geq 0}} 1$ just counts the number of non-negative integer solutions to $x_1 + x_2 + \cdots + x_t = n$. This is $\left(\!\binom{t}{n}\!\right) = \binom{n+t-1}{n}$ (by Theorem 4.40). $\qquad\square$

Power Series Expansions of Functions

When you want information about – or possibly even a formula for – a sequence $h_0, h_1, \ldots,$ $h_n, \ldots,$ the generating function technique, presented in this section, suggests that you consider the ordinary power series $g(x) = h_0 + h_1 x + \cdots + h_n x^n + \cdots$. It may be that finding a closed form for $g(x)$ is easier than finding a formula for h_n directly. After finding the closed form, you already know that h_n is the coefficient of x^n in the power series expansion of $g(x)$ (at $x = 0$). Finding an actual formula for h_n from $g(x)$ could be tricky, but there are several possible options. We begin by illustrating one approach.

Example 9.21. Let $g(x) = \frac{13x-2}{(1-4x)^2}$. How can we find the coefficient of x^{47} in the power series expansion of $g(x)$? We begin with related power series that we already know. From Corollary 9.20, we know that

$$\frac{1}{(1-x)^2} = \sum_{n=0}^{\infty} \left(\!\binom{2}{n}\!\right) x^n = \sum_{n=0}^{\infty} \binom{n+1}{n} x^n = \sum_{n=0}^{\infty} (n+1) x^n.$$

Plugging in $4x$ for x, we get

$$\frac{1}{(1-4x)^2} = \sum_{n=0}^{\infty}(n+1)(4x)^n = \sum_{n=0}^{\infty}(n+1)4^n x^n.$$

Now $\frac{13x-2}{(1-4x)^2} = 13x\frac{1}{(1-4x)^2} - 2\frac{1}{(1-4x)^2}$, and so

$$\frac{13x-2}{(1-4x)^2} = 13x\sum_{n=0}^{\infty}(n+1)4^n x^n - 2\sum_{n=0}^{\infty}(n+1)4^n x^n$$

$$= \sum_{n=0}^{\infty}13(n+1)4^n x^{n+1} - \sum_{n=0}^{\infty}2(n+1)4^n x^n.$$

From here on, we have to do a little bit of algebra to combine the two infinite sums into one sum. In the first sum, we replace $n+1$ with n to get that it is equal to $\sum_{n=1}^{\infty}13n4^{n-1}x^n$. Adding an additional 0th term to this sum does no harm, but allows us to combine the two series:

$$\frac{13x-2}{(1-4x)^2} = \sum_{n=0}^{\infty}\left[13n4^{n-1} - 2(n+1)4^n\right]x^n$$

$$= \sum_{n=0}^{\infty}(5n-8)4^{n-1}x^n.$$

Now h_n is the coefficient of x^n, and so

$$h_n = (5n-8)4^{n-1}.$$

We conclude that the coefficient of x^{47} in the power series expansion of $g(x)$ is $h_{47} = 227 \times 4^{46}$.

If you are looking for one specific term of the sequence – in Example 9.21, we were looking for h_{47}, the coefficient of x^{47} – then you can use a symbolic algebra software such as SageMath, Maple, or Mathematica. In Maple, after defining the generating function g, use the command `coeftayl(g, x=0, 47)` to get the coefficient of x^{47}.

The coefficient of x^n in the power series expansion of $g(x)$ is equal to $g^{(n)}(0)/n!$, so a second approach would be to find a pattern by taking successive derivatives of g. If you suspect a pattern, then you can make a conjecture for the nth derivative of g, and prove it by induction.

Depending on the closed form of g, a third approach is to try to find a power series expansion of g based on power series expansions that we already know. This is what we did in Example 9.21, and just as in that example, a power series expansion that comes in very handy is that of Corollary 9.20:

$$\frac{1}{(1-x)^t} = \sum_{n=0}^{\infty}\left(\binom{t}{n}\right)x^n = \sum_{n=0}^{\infty}\binom{n+t-1}{n}x^n.$$

Often, if g is a rational function – that is, the ratio of two polynomials – then we can use the method of partial fractions from calculus to split g into more manageable pieces, and then expand each fraction using some variation of the power series that we already know. Example 9.22 illustrates this technique further. However, stepping back a bit, it is quite remarkable that calculus concepts and techniques can be so helpful in solving discrete counting problems.

A fourth technique, if g is a rational function, is to use the generating function to find a recurrence relation for the set of coefficients. We will illustrate how in Example 9.23.

Example 9.22. Let

$$g(x) = \frac{11x^2 - 6x + 1}{1 - 5x + 3x^2 + 9x^3}.$$

What is the coefficient of x^n in the power series expansion of $g(x)$? Put differently, $g(x)$ is the ordinary power series for which sequence?

The basic technique – if $g(x)$ is a rational function – is the method of partial fractions! We use partial fractions to write the rational function as a sum of simpler expressions. We first have to factor the denominator. You could notice that -1 is a root of $9x^3 + 3x^2 - 5x + 1$, and so $(x+1)$ is a factor. You would then divide by $x+1$ and then factor the remaining quadratic polynomial. However, in general, the better way to do algebraic manipulations is to use a symbolic algebra software such as SageMath,[3] Maple, or Mathematica. In SageMath, you would run `factor(9*x^3+3*x^2-5*x+1)` to get $(x+1)(3x-1)^2$ as the factorization of the denominator. We then write

$$g(x) = \frac{11x^2 - 6x + 1}{(x+1)(3x-1)^2} = \frac{A}{x+1} + \frac{B}{3x-1} + \frac{C}{(3x-1)^2},$$

where A, B, and C are yet-to-be-determined constants. There are several ways of finding A, B, and C. One is to combine the fractions on the right-hand side, and then equate the two numerators, remembering that the identity has to be valid for all values of x. You can also just use your symbolic algebra software to find the partial fraction decomposition of g. In SageMath, you would first define g by `g=(11*x^2-6*x+1)/(1-5*x+3*x^2+9*x^3)` and then run `g.partial_fraction(x)`. Either way, you get

$$g(x) = \frac{9/8}{1+x} + \frac{7/24}{3x-1} + \frac{1/6}{(3x-1)^2}.$$

Now, we can find the coefficient of x^n in each of the fractions and add them up to find the coefficient of x^n in g. For the first two fractions, we start with $\frac{1}{1-x} = 1 + x + x^2 + \cdots$. Substituting $-x$ for x, we have $\frac{1}{1+x} = 1 - x + x^2 + \cdots + (-1)^n x^n + \cdots$. Likewise, replacing x with $3x$ and then multiplying by a minus sign, we get $\frac{1}{3x-1} = \frac{-1}{1-3x} = -1 - 3x - 3^2 x^2 - \cdots - 3^n x^n - \cdots$. For the third fraction, we replace x with $3x$ in Corollary 9.20 for $t = 2$, and get that $\frac{1}{(1-3x)^2} = \sum_{n=0}^{\infty} (n+1)(3x)^n$. Putting these all together, we get

[3] SageMath (www.sagemath.org/) and its cloud version CoCalc (https://cocalc.com/app) are alternatives to Maple and Mathematica.

$$g(x) = \frac{9/8}{1+x} + \frac{7/24}{3x-1} + \frac{1/6}{(3x-1)^2}$$

$$= \frac{9}{8} \sum_{n=0}^{\infty} (-1)^n x^n - \frac{7}{24} \sum_{n=0}^{\infty} 3^n x^n + \frac{1}{6} \sum_{n=0}^{\infty} (n+1)3^n x^n$$

$$= \sum_{n=0}^{\infty} \left[\frac{9}{8}(-1)^n - \frac{7}{24}3^n + \frac{1}{6}(n+1)3^n \right] x^n$$

$$= \sum_{n=0}^{\infty} \left[\frac{9}{8}(-1)^n - \frac{1}{8}3^n + \frac{n}{6}3^n \right] x^n.$$

So the coefficient of x^n in the power series expansion of $g(x)$ is

$$\frac{9}{8}(-1)^n - \frac{1}{8}3^n + \frac{n}{6}3^n.$$

Example 9.23. Let n be a non-negative integer, and let a_n be the coefficient of x^n in the power series expansion of

$$g(x) = \frac{1}{7x^2 - 2x + 1}.$$

If we are interested in a_n, we may proceed as in the previous examples and do a partial fraction decomposition of $g(x)$, and then proceed to find a somewhat messy formula for a_n. However, as long as $g(x)$ is a rational function, we can use the generating function to just find a recurrence relation for a_n. We have seen that recurrence relations are ideal for calculating actual values of a_n for small values of n. We write

$$\frac{1}{7x^2 - 2x + 1} = a_0 + a_1 x + a_2 x^2 + \cdots + a_n x^n + \cdots$$

$$\Rightarrow (7x^2 - 2x + 1)(a_0 + a_1 x + a_2 x^2 + \cdots + a_n x^n + \cdots) = 1.$$

We know how to find the coefficients of the product of two power series using Proposition 9.12, and so we have

$$a_0 = 1$$
$$-2a_0 + a_1 = 0$$
$$7a_0 - 2a_1 + a_2 = 0$$
$$7a_1 - 2a_2 + a_3 = 0$$
$$\vdots$$
$$7a_{n-2} - 2a_{n-1} + a_n = 0$$
$$\vdots$$

So, $a_0 = 1$, $a_1 = 2$, and for $n > 1$, we have

$$a_n = 2a_{n-1} - 7a_{n-2}.$$

Using the recurrence relation, we know that, for example, $a_2 = -3$, $a_3 = -20$, $a_4 = -19$, and $a_5 = 102$.

Problems

P 9.1.1. Assume $f(x)$ is the ordinary generating function for the sequence $a_0, a_1, a_2, \ldots$ What is the generating function for the sequence $0, 0, a_0, a_1, a_2, \ldots$ in terms of $f(x)$?

P 9.1.2. Find the first few terms and the ordinary generating function of the following sequence:

$$\left(\!\binom{5}{0}\!\right), \left(\!\binom{5}{1}\!\right), \left(\!\binom{5}{2}\!\right), \left(\!\binom{5}{3}\!\right), \ldots$$

Is this a finite or an infinite sequence?

P 9.1.3. Let n be a non-negative integer. Find the ordinary generating function of the (finite) sequence consisting of one row of the table of the Stirling numbers of the first kind (Definition 6.25):

$$\begin{bmatrix} n \\ 0 \end{bmatrix}, \begin{bmatrix} n \\ 1 \end{bmatrix}, \begin{bmatrix} n \\ 2 \end{bmatrix}, \cdots \begin{bmatrix} n \\ n \end{bmatrix}.$$

Express your answer is terms of a familiar function of x. Do the same for the signed Stirling numbers of the first kind.

P 9.1.4. Find the coefficient of x^n in the power series expansion of

$$\frac{1}{(1-7x)^4}.$$

P 9.1.5. Find the coefficient of x^{47} in the power series expansion of

$$\frac{1}{(x-5)^3}$$

and multiply it by 5^{50}. What do you get?

P 9.1.6. Find (in closed form) the ordinary generating function of the sequence whose nth term, for $n \geq 0$, is $n+3$. In other words, find the generating function for

$$3, 4, 5, \ldots, n+3, \ldots$$

P 9.1.7. Determine (in closed form) the ordinary generating function for the sequence of cubes:

$$0, 1, 8, \ldots, n^3, \ldots$$

P 9.1.8. Find (in closed form) the ordinary generating function for the sequence of reciprocals of integers:

$$1, \frac{1}{2}, \frac{1}{3}, \ldots, \frac{1}{n}, \ldots$$

P 9.1.9. In Example 9.9, we showed that

$$\sum_{n=0}^{\infty} n^2 x^n = \frac{x(1+x)}{(1-x)^3}.$$

By plugging in $x = 1$, can we use this result to find $\sum_{n=0}^{\infty} n^2$? How, or why not?

P 9.1.10. Find a closed form for

$$g(x) = \sum_{n=2}^{\infty} (2^{n-1} - 1)x^n.$$

P 9.1.11. Let n be a non-negative integer. What is the coefficient of x^n in the power series expansion of

$$\frac{8x - 1}{(1 - 4x)^2}?$$

P 9.1.12. Let n be a non-negative integer. What is the coefficient of x^n in the power series expansion of

$$\frac{x^2}{(1 - x^3)^6}?$$

P 9.1.13. Let

$$g(x) = \frac{1}{1 - 3x}\left[\frac{x}{1 - 5x} + 1\right].$$

Assume that $g(x)$ is the ordinary power series for the sequence $\{h_n\}_{n=0}^{\infty}$. Find a closed formula for h_n.

P 9.1.14. Let

$$g(x) = \frac{4x}{(1 - 4x)(1 - 3x)}.$$

Assume that $g(x)$ is the ordinary power series for the sequence $\{b_k\}_{k=0}^{\infty}$. Find a closed formula for b_k.

P 9.1.15. Let

$$g(x) = \frac{4x}{(1 - 4x)(1 - 3x)}.$$

Assume that $g(x)$ is the ordinary power series for the sequence $\{b_k\}_{k=0}^{\infty}$. Without finding a formula for b_k, find a recurrence relation for the sequence b_k, and give the needed initial conditions.

P 9.1.16. Let

$$g(x) = \frac{x}{1 + x - 16x^2 + 20x^3}.$$

What is the coefficient of x^n in the power series expansion of $g(x)$?

P 9.1.17. Let

$$g(x) = \frac{4 - x + 2x^2}{1 - 3x + 7x^2 - 5x^3}.$$

Assume that $g(x) \overset{\text{ops}}{\longleftrightarrow} \{a_n\}_{n=0}^{\infty}$. Find a recurrence relation for a_n, and use it to find a_6. What is $g^{(6)}(0)$, the value of the sixth derivative of $g(x)$ at $x = 0$?

P 9.1.18. Let $g(x) = \frac{1}{1-x+x^2}$. Let h_n be the coefficient of x^n in the power series expansion of g at $x = 0$. Find a recurrence relation for h_n and use it to find $h_0, \ldots, h_7$. What is the Taylor polynomial of degree 7 for $g(x)$ at $x = 0$?

P 9.1.19. Let $f(x) = (\ln(1+x))^2$. Without taking any derivatives, and starting with the power series expansion of $\ln(1 + x)$, find the coefficient of x^5 in the power series expansion of $f(x)$. What is $f^{(5)}(0)$? ($f^{(5)}(x)$ is the fifth derivative of $f(x)$.)

P 9.1.20. Let $f(x) = \frac{e^x}{1-x}$. Without taking any derivatives, find the coefficient of x^5 in the power series expansion of $f(x)$.

P 9.1.21. $f(x)$ is the ordinary generating function for the sequence $a_0, a_1, \ldots, a_n, \ldots$ Express, in terms of $f(x)$, the ordinary generating function of the following sequence:

$$a_2 + 3a_1 - a_0, a_3 + 3a_2 - a_1, \ldots, a_{n+2} + 3a_{n+1} - a_n, \ldots$$

P 9.1.22. Let $g(x) = \frac{1}{1+x+x^2}$. Assume that $g(x)$ is the ordinary power series for a sequence $\{h_n\}_{n=0}^{\infty}$.
(a) Find a recurrence relation for h_n.
(b) Find $h_0, \ldots, h_5$.
(c) Let α and β be the roots of $x^2 + x + 1$. Even though α and β are complex numbers, write

$$g(x) = \frac{1}{(x - \alpha)(x - \beta)} = \frac{A}{x - \alpha} + \frac{B}{x - \beta},$$

where A and B are constants. Find A and B.
(d) Find a closed formula for h_n.
(e) Any surprises? Comment.

P 9.1.23. Assume that a function $f(x)$ is infinitely differentiable for every real number (a much weaker condition would have sufficed for this problem). Further assume that, as a function, and for all $x \in \mathbb{R}$,

$$f(x) = a_0 + a_1 x + a_2 x^2 + \cdots + a_n x^n + \cdots,$$

where $a_0, a_1, \ldots, a_n, \ldots$ are real numbers. In other words, for every $x \in \mathbb{R}$, the series $\sum_{n=0}^{\infty} a_n x^n$ converges to $f(x)$. Show that

$$a_n = \frac{f^{(n)}(0)}{n!}.$$

P 9.1.24. Given a function $f(x)$, we defined its power series expansion (at zero) by Equation 9.1. Show that, given this definition, the power series expansion of $xf(x)$ is what you get by multiplying the power series for $f(x)$ with x.

9.2 Combinations of Multisets and Solving Recurrence Relations

Having developed some tools, we are finally ready to apply them to counting problems. In this section, we use ordinary generating functions to count combinations of multisets and to "solve" recurrence relations.

9.2.1 Combinations of Multisets

Warm-Up 9.24. *You are at the "Arabic literature in translation" aisle of your favorite bookstore, and want to buy* 47 *books to give as presents. Every one of your selections is going to come from among Tayeb Salih's* Season of Migration to the North, *Naguib Mahfouz's* Miramar, *Mahmoud Darwish's* Memory for Forgetfulness, *and Elias Khoury's* Gate of the Sun. *You have decided to buy at least one copy of each, but you want to make sure that you have an even number of* Miramar *while an odd number of each of the other three books. Consider the question: "In how many ways can you choose the* 47 *books and still satisfy your conditions?" Translate this question to a different one about the number of certain integer solutions to a linear equation.*

The second column of Table 3.1 – when the s balls are identical but the t boxes are distinct – is all about counting "combinations." (Also see the discussion at the end of Section 3.3.) We are choosing "boxes," and the "balls" tell us how. The t distinct empty boxes are set on the floor and you have to decide where to put the s identical balls. If the capacity of a box is more than one, then we can choose that box more than once, and so we are counting combinations of multisets. We now want to, finally, turn to entry $\boxed{16}$, where again we want the number of s-combinations of a multiset, but this time the restrictions on each item can be quite arbitrary. It could be that some box accepts only an even number of balls while another one takes one, two, or three balls. We can attack this kind of counting problem using generating functions. Entry $\boxed{16}$ of the counting table is the most general of its column, and the method used for attacking it can also be used for the other entries in this column as well. But as you may expect, carrying out this method of generating functions is not always very straightforward. We begin with a definition.

Definition 9.25. Let $T = \{n_1, n_2, \ldots\}$ be a (finite or infinite) set of non-negative integers, define the power series (or possibly the polynomial) χ_T by

$$\chi_T = x^{n_1} + x^{n_2} + x^{n_3} + \cdots,$$

and call it the *(ordinary) indicator power series* of T.

Example 9.26. Let $T = \{0, 1, 2, \ldots, 10\}$. Then the indicator power series of T is

$$\chi_T = 1 + x + x^2 + \cdots + x^{10}.$$

In this particular case, χ_T is the partial sum of a geometric series, and we can find a closed form for it by noticing – as can be done with all partial sums of geometric series – that $\chi_T - x\chi_T = 1 - x^{11}$. Hence,

$$\chi_T = \frac{1 - x^{11}}{1 - x}.$$

Example 9.27. Let $T = \{2, 4, 6, \cdots\}$ be the set of positive even integers. Then the indicator power series of T is

$$\chi_T = x^2 + x^4 + x^6 + \cdots.$$

What is a closed form for χ_T? We know that

$$\frac{1}{1 - x} = 1 + x + x^2 + x^3 + \cdots.$$

Replacing x with x^2, we get

$$\frac{1}{1 - x^2} = 1 + x^2 + x^4 + x^6 + \cdots.$$

This is very close to χ_T but not yet equal to it. We can get χ_T either by subtracting 1 from both sides, or by multiplying both sides by x^2:

$$\chi_T = x^2 + x^4 + x^6 + \cdots = \frac{1}{1 - x^2} - 1 = \frac{x^2}{1 - x^2}.$$

We will now show how to use ordinary generating functions to find the number of integer solutions to $x_1 + x_2 + \cdots + x_t = s$, where the restrictions on each x_i are quite general.

Theorem 9.28. *Let t be a positive integer and s a non-negative integer. For $1 \le i \le t$, let N_i be a set of non-negative integers, and let χ_{N_i} be the indicator power series for N_i. Define a sequence $\{c_s\}_{s=0}^{\infty}$ by*

$$c_s = \textit{number of those solutions to } x_1 + \cdots + x_t = s \textit{ for which } x_i \in N_i.$$

Let

$$f \xleftrightarrow{\text{ops}} \{c_s\}_{s=0}^{\infty}.$$

Then

$$f = \chi_{N_1} \chi_{N_2} \cdots \chi_{N_t}.$$

Proof. For $1 \le i \le t$, each N_i is a set of non-negative integers, and

$$\chi_{N_i} = a_0^{(i)} + a_1^{(i)} x + \cdots + a_n^{(i)} x^n + \cdots,$$

where

$$a_k^{(i)} = \begin{cases} 1 & \text{if } k \in N_i \\ 0 & \text{if } k \notin N_i. \end{cases}$$

In other words, all the coefficients for χ_{N_i} are 0's and 1's, and we have a coefficient of one precisely for the powers of x where the exponent is a member of N_i. As a result, χ_{N_i} is the generating function of the sequence $a_0^{(i)}, a_1^{(i)}, \ldots$, and we can apply Proposition 9.17 to find the product of t generating functions:

$$\chi_{N_1} \chi_{N_2} \cdots \chi_{N_t} \overset{\text{ops}}{\longleftrightarrow} \{d_s\}_{s=0}^{\infty},$$

where

$$d_s = \sum_{\substack{x_1+x_2+\cdots+x_t=s \\ x_i \geq 0}} a_{x_1}^{(1)} a_{x_2}^{(2)} \cdots a_{x_t}^{(t)} = \sum_{\substack{x_1+x_2+\cdots+x_t=s \\ x_i \in N_i}} 1 = c_s.$$

Basically, we get a contribution to d_s exactly if $x_1 + \cdots + x_t = s$ and each a_{x_i} is equal to 1. This happens exactly each time all the restrictions (that is, $x_i \in N_i$, for $1 \leq i \leq t$) are met. $\square$

Combining Theorem 9.28 with the usual arguments (as those for Entry $\boxed{6}$ in Theorem 4.40) gives us a technique for addressing entry $\boxed{16}$ of the counting table.

Theorem 9.29 (Entry $\boxed{16}$ of the Counting Table). *Let t be a positive integer and s a non-negative integer. For $1 \leq i \leq t$, let N_i be a set of non-negative integers, and let χ_{N_i} be the indicator power series for N_i. Let $M = \{\infty \cdot U_1, \infty \cdot U_2, \ldots, \infty \cdot U_t\}$ be a multiset with t types of elements and with infinite repetition numbers. Then the following integers are equal:*

(a) *The number of those solutions to $x_1 + \cdots + x_t = s$ where, for $1 \leq i \leq t$, $x_i \in N_i$*
(b) *The number of those s-combinations (i.e., submultisets of size s) of M in which, for $1 \leq i \leq t$, the number of U_i is an element of N_i*
(c) $\boxed{16}$ *The number of ways of placing s identical balls in t distinct boxes such that the number of balls in the ith box is an element of N_i*
(d) *The coefficient of x^s in the function $\chi_{N_1} \chi_{N_2} \cdots \chi_{N_t}$.*

Example 9.30. It is Tuesday, and you want to order 30 tacos for a group of friends. There are five types of taco available: Al Pastor (A), Carnitas (C), Fish (F), Hongos (H), and Potato (P). You want the number of A to be odd, the number of C to be a multiple of 6, the number of H to be at most 5, and the number of P to be 0 or 1. There is no restriction on the number of F. In how many different ways can you place your order?

Solution. Instead of an order of size 30, we consider an order of size n, and we let h_n be the number of different possible orders of size n with the same conditions as in the problem. This may seem as complicating the problem, and it is, but it allows for the use of generating functions. For $n = 1$, the only possibility is $\{A\}$. For $n = 2$, the possible orders are $\{A, F\}, \{A, H\}$, and $\{A, P\}$. Finally, for $n = 3$, we have $\{A, A, A\}, \{A, F, F\}, \{A, F, H\}, \{A, F, P\}, \{A, H, H\}$, and $\{A, H, P\}$, for a total of six possibilities.

We have a multiset $\{\infty \cdot A, \infty \cdot C, \infty \cdot F, \infty \cdot H, \infty \cdot P\}$, and we want to count the number of n-combinations that satisfy certain conditions. This is the same as finding the number of non-negative integer solutions to $x_1 + \cdots + x_5 = n$, where x_1 is odd, x_2 is a multiple of six, there are no extra restrictions on x_3, $0 \le x_4 \le 5$, and $0 \le x_5 \le 1$. We let $g(x) \overset{\text{ops}}{\longleftrightarrow} \{h_n\}_{n=0}^{\infty}$. By Theorem 9.29, this generating function is the product of five different indicator power series. The indicator power series for odd numbers is $x + x^3 + x^5 + \cdots$, which, using Example 9.27, equals $x(1 + x^2 + x^4 + \cdots) = \frac{x}{1-x^2}$. The indicator power series for multiples of six is $1 + x^6 + x^{12} + \cdots = \frac{1}{1-x^6}$ (just replace x with x^6 in the power series expansion of $1/(1-x)$). We also have $1 + x + \cdots = \frac{1}{1-x}$ and $1 + x + \cdots + x^5 = \frac{1-x^6}{1-x}$ (see Example 9.26). Thus

$$g(x) = (x + x^3 + x^5 + \cdots)(1 + x^6 + x^{12} + \cdots)(1 + x + x^2 + \cdots)$$
$$\times (1 + x + \cdots + x^5)(1 + x)$$
$$= \frac{x}{1-x^2} \frac{1}{1-x^6} \frac{1}{1-x} \frac{1-x^6}{1-x}(1+x) = \frac{x}{(1-x)^3}.$$

Now, recall from Corollary 9.20 that

$$\frac{1}{(1-x)^k} = \sum_{n=0}^{\infty} \left(\binom{k}{n} \right) x^n = \sum_{n=0}^{\infty} \binom{n+k-1}{n} x^n.$$

Applying this in the case $k = 3$, we get

$$g(x) = \frac{x}{(1-x)^3} = x \sum_{n=0}^{\infty} \binom{n+2}{n} x^n = \sum_{n=0}^{\infty} \binom{n+2}{2} x^{n+1}.$$

So, the coefficient of x^{n+1} in this power series is $\binom{n+2}{2}$, and so the coefficient of x^n is $\binom{n+1}{2} = \frac{n(n+1)}{2}$. But, by definition of $g(x)$, this coefficient was h_n. So, $h_n = \frac{n(n+1)}{2}$, and, rather unbelievably, we have found a formula for h_n. We conclude that there are $\binom{31}{2} = 465$ ways to order 30 tacos with the given restrictions. $\square$

While it is pretty amazing that manipulating power series allowed us to find the answer to the specific counting problem of Example 9.30, you may argue that the restrictions that we had put on the different types of tacos was artificial. Restrictions of this type (e.g., the number of some item be a multiple of five) is more common than you may first think. The following corollary of Theorem 9.29 will connect these kinds of conditions to finding the number of non-negative integer solutions to certain types of equations.

Corollary 9.31. *Let s be a non-negative integer and let t, $a_1, \ldots, a_t$ be positive integers. For $1 \le i \le t$, let N_i be the set of non-negative integer multiples of a_i, and let*

$$\chi_{N_i} = 1 + x^{a_i} + x^{2a_i} + \cdots = \frac{1}{1-x^{a_i}}$$

be the indicator power series for N_i. Then the following integers are equal:

(a) The number of non-negative integer solutions to $a_1 x_1 + a_2 x_2 + \cdots + a_t x_t = s$

(b) *The number of those non-negative integer solutions to $y_1 + \cdots + y_t = s$ where, for $1 \leq i \leq t$,*
    y_i is a multiple of a_i
(c) *The coefficient of x^s in the function $\chi_{N_1} \chi_{N_2} \cdots \chi_{N_t}$.*

Proof. For $1 \leq i \leq t$, let $y_i = a_i x_i$. Then $x_1, \ldots, x_t$ is a non-negative integer solution to $a_1 x_1 + a_2 x_2 + \cdots + a_t x_t = s$ if and only if $y_1, \ldots, y_t$ is a non-negative integer solution to $y_1 + \cdots + y_t = s$ where each y_i is a multiple of a_i. The rest of the assertions follow directly from Theorem 9.29.
$\square$

Even more specifically, if $x_1, x_2, \ldots, x_n$ are non-negative integers satisfying $x_1 + 2x_2 + \cdots + nx_n = n$, then

$$\underbrace{1 + 1 + \cdots + 1}_{x_1} + \underbrace{2 + 2 + \cdots + 2}_{x_2} + \cdots + \underbrace{n + n + \cdots + n}_{x_n} = n,$$

and this is a partition of the integer n (Definition 7.2). So, the number of non-negative integer solutions to $x_1 + 2x_2 + \cdots + nx_n = n$ is equal to $p(n)$, the total number of partitions of n. (This was Problem P 7.1.5.) We will come back to partitions and their generating functions in subsection 9.4.2, but, for now, we record the combination of our observation with Corollary 9.31:

Corollary 9.32. *Let n be a positive integer. Then the following integers are equal:*

(a) *$p(n)$, the number of partitions of the integer n*
(b) *The number of non-negative integer solutions to $x_1 + 2x_2 + \cdots + nx_n = n$*
(c) *The number of non-negative integer solutions to $y_1 + y_2 + \cdots + y_n = n$ where, for $1 \leq i \leq n$,*
    y_i is a multiple of i
(d) *The coefficient of x^n in the power series expansion of*

$$\frac{1}{(1 - x)(1 - x^2) \cdots (1 - x^n)}.$$

9.2.2 Solving Recurrence Relations

Warm-Up 9.33. *If $f(x) \stackrel{\text{ops}}{\longleftrightarrow} \{a_n\}_{n=0}^{\infty}$, then what is the coefficient of x^n in the power series expansion of $5x^2 f'(x)$?*

Example 9.23 showed how we can get a recurrence relation from a generating function. We now want to illustrate how to get a generating function from a recurrence relation! We have seen many examples of recurrence relations for sequences. In Sections 1.3 and 1.4 (see also Remark 1.11) we saw three methods for working with a recurrence relation in order to possibly get a formula for the underlying sequence. Our fourth technique for solving recurrence relations is to use generating functions. If the underlying recurrence relation happens to be a *linear* recurrence relation, then the technique for solving such recurrences presented in Section 1.4 works best. The generating function technique provides an alternative approach and it could, in theory, be applicable to non-linear recurrence relations as well. We will demonstrate through examples.

Example 9.34 (The Piṅgala–Fibonacci Sequence). The Piṅgala–Fibonacci sequence is defined recursively by $F_0 = 0$, $F_1 = 1$, and, for $n \geq 2$, $F_n = F_{n-2} + F_{n-1}$ (see subsection 1.2.1). In Example 1.18, we already solved this recurrence relation and found a formula for F_n. Here, we do the same, using generating functions. Let

$$g(x) \xleftrightarrow{\text{ops}} \{F_n\}_{n=0}^{\infty}.$$

In other words (remember, in what follows, that equality does not indicate equality of functions; see Remark 9.3),

$$g(x) = \sum_{n=0}^{\infty} F_n x^n = x + x^2 + 2x^3 + 3x^4 + \cdots + F_n x^n + \cdots.$$

We now try to take advantage of the recurrence relation. Hence, we try to construct from $g(x)$ two other power series that respectively have F_{n-1} and F_{n-2} as the coefficients of x^n. We can achieve this by multiplying g by x and x^2 respectively:

$$xg(x) = \sum_{n=0}^{\infty} F_n x^{n+1} = x^2 + x^3 + 2x^4 + 3x^5 + \cdots + F_{n-1} x^n + \cdots$$

$$x^2 g(x) = \sum_{n=0}^{\infty} F_n x^{n+2} = 0x^2 + x^3 + x^4 + 2x^5 + \cdots + F_{n-2} x^n + \cdots.$$

If we add these two power series and collect terms, then – for most terms – we will have a coefficient of the form $F_{n-1} + F_{n-2}$ which by the recurrence relation is equal to F_n. This means that the resulting power series will be very close to the original $g(x)$:

$$xg(x) + x^2 g(x) = x^2 + 2x^3 + 3x^4 + \cdots + \underbrace{(F_{n-1} + F_{n-2})}_{F_n} x^n + \cdots$$

$$= g(x) - x.$$

Note that the expression on the right was the same as $g(x)$ except that it was missing one term. We can now solve the above for $g(x)$, and get $g(x) \left[x + x^2 - 1 \right] = -x$, and so

$$g(x) = \frac{x}{1 - x - x^2}. \tag{9.2}$$

We have found the generating function for the Piṅgala–Fibonacci sequence. But we can go further, and use this generating function to find a formula for F_n. Let $\alpha = \frac{1+\sqrt{5}}{2}$ and $\beta = \frac{1-\sqrt{5}}{2}$, and note that

$$1 - x - x^2 = (1 - \alpha x)(1 - \beta x).$$

Using the method of partial fractions, we write $\frac{x}{1-x-x^2} = \frac{A}{1-\alpha x} + \frac{B}{1-\beta x}$ and solve for the constants A and B. We get that $A = \frac{1}{\alpha - \beta}$ and $B = -\frac{1}{\alpha - \beta}$. So

$$g(x) = \frac{x}{1-x-x^2} = \frac{1}{\alpha - \beta}\left[\frac{1}{1-\alpha x} - \frac{1}{1-\beta x}\right].$$

Now, we know $\frac{1}{1-x} \overset{\text{ops}}{\longleftrightarrow} \{1\}_{n=0}^{\infty}$. Replacing x with αx, we have that $\frac{1}{1-\alpha x}$ is equal to $1 + \alpha x + \cdots + \alpha^n x^n + \cdots$, and the expression for $\frac{1}{1-\beta x}$ is similar. Thus, the coefficient of x^n in $g(x)$ is $\frac{1}{\alpha - \beta}\left[\alpha^n - \beta^n\right]$. Plugging in the values for α and β, we get

$$F_n = \frac{1}{\sqrt{5}}\left[\left(\frac{1+\sqrt{5}}{2}\right)^n - \left(\frac{1-\sqrt{5}}{2}\right)^n\right].$$

The skeptical reader should feel free to plug in a few small values of n to see that this – admittedly strange – formula does agree with the corresponding values of the Piṅgala– Fibonacci sequence.

Remark 9.35. Example 9.34 demonstrates how we use generating functions to analyze recurrence relations. We try to combine various kinds of power series in order to take advantage of the recurrence relation. In the case of Example 9.34, we added two power series in order to get $F_{n-2} + F_{n-1}$ as the coefficient of x^n. We then used the fact that $F_{n-2} + F_{n-1} = F_n$ to get back a power series very close to the original $g(x)$. Often, we have to be careful with the early terms and treat them separately. In Example 9.34, it was clear that we were getting the terms of $g(x)$ starting with the x^2 term, but we had to manually adjust for the x term.

Example 9.36. A sequence h_n is defined as follows:

$$h_0 = 1, \; h_1 = -1, \; h_2 = 3, \text{ and, for } n \geq 3, \; h_n = 5h_{n-1} - 3h_{n-2} - 9h_{n-3}.$$

Given the recurrence relation, we can find further values of h_n. For example, let $n = 3$, and get $h_3 = 5(3) - 3(-1) - 9(1) = 9$. Likewise, we get $h_4 = 5(9) - 3(3) - 9(-1) = 45$. We want to find the generating function for the sequence h_n, and use it to possibly find a closed formula for h_n. Let

$$g(x) \overset{\text{ops}}{\longleftrightarrow} \{h_n\}_{n=0}^{\infty},$$

and we write

$$g(x) = h_0 + h_1 x + h_2 x^2 + h_3 x^3 + \cdots + h_n x^n + \cdots$$
$$5xg(x) = 5h_0 x + 5h_1 x^2 + 5h_2 x^3 + \cdots + 5h_{n-1}x^n + \cdots$$
$$-3x^2 g(x) = -3h_0 x^2 - 3h_1 x^3 - \cdots - 3h_{n-2}x^n - \cdots$$
$$-9x^3 g(x) = -9h_0 x^3 - \cdots - 9h_{n-3}x^n + \cdots.$$

Adding the last three expressions, we see that, starting with x^3, the recurrence relation comes into play. For the earlier powers, we use the initial values to get

$$5xg(x) - 3x^2g(x) - 9x^3g(x) = \overbrace{5h_0}^{5} x + \overbrace{(5h_1 - 3h_0)}^{-8} x^2$$
$$+ \overbrace{(5h_2 - 3h_1 - 9h_0)}^{h_3} x^3 + \cdots$$
$$+ \overbrace{(5h_{n-1} - 3h_{n-2} - 9h_{n-3})}^{h_n} x^n + \cdots$$
$$= 5x - 8x^2 + (g(x) - h_0 - h_1x - h_2x^2)$$
$$= g(x) - 11x^2 + 6x - 1.$$

Solving for $g(x)$, we get

$$g(x) = \frac{11x^2 - 6x + 1}{1 - 5x + 3x^2 + 9x^3}.$$

In Example 9.22, we found the coefficient of x^n in $g(x)$. As a result, we have

$$h_n = \frac{9}{8}(-1)^n - \frac{1}{8}3^n + \frac{n}{6}3^n.$$

You should check that indeed $h_3 = 9$ and $h_4 = 45$. Even the fact that h_n is always an integer is not obvious from this formula!

Example 9.37 (Stirling Numbers of the Second Kind). We have lauded the efficacy of recurrence relations often. Since we have a recurrence relation for the Stirling numbers of the second kind, can we find a generating function for them? The Stirling number $\left\{{n \atop k}\right\}$ is a function of two variables, and so we fix k, and focus on the Stirling numbers in one *column* of the table of Stirling numbers of the second kind (Table 6.3). Let

$$f_k(x) \stackrel{\text{ops}}{\longleftrightarrow} \left\{ \left\{{n \atop k}\right\} \right\}_{n=0}^{\infty}.$$

Since $\left\{{n \atop k}\right\}$ is zero when $n < k$, we have

$$f_k(x) = \sum_{n=k}^{\infty} \left\{{n \atop k}\right\} x^n = \left\{{k \atop k}\right\} x^k + \left\{{k+1 \atop k}\right\} x^{k+1} + \cdots + \left\{{n \atop k}\right\} x^n + \cdots.$$

To take advantage of the recurrence relation (Theorem 6.7),

$$\left\{{n \atop k}\right\} = k\left\{{n-1 \atop k}\right\} + \left\{{n-1 \atop k-1}\right\},$$

we have

$$kxf_k(x) = k\begin{Bmatrix} k \\ k \end{Bmatrix} x^{k+1} + k\begin{Bmatrix} k+1 \\ k \end{Bmatrix} x^{k+2} + \cdots + k\begin{Bmatrix} n-1 \\ k \end{Bmatrix} x^n + \cdots$$

$$xf_{k-1}(x) = \begin{Bmatrix} k-1 \\ k-1 \end{Bmatrix} x^k + \begin{Bmatrix} k \\ k-1 \end{Bmatrix} x^{k+1} + \cdots + \begin{Bmatrix} n-1 \\ k-1 \end{Bmatrix} x^n + \cdots.$$

Add these two expressions, note that $\begin{Bmatrix} k-1 \\ k-1 \end{Bmatrix} x^k = \begin{Bmatrix} k \\ k \end{Bmatrix} x^k$, and use the recurrence relation to get

$$kxf_k(x) + xf_{k-1}(x) = f_k(x).$$

Solving for $f_k(x)$, and then unwinding the resulting recurrence relation, we get

$$f_k(x) = \frac{xf_{k-1}(x)}{1-kx} = \frac{x^2 f_{k-2}(x)}{(1-kx)[1-(k-1)x]} = \cdots = \frac{x^k f_0(x)}{\prod_{r=1}^{k}(1-rx)} = x^k \prod_{r=1}^{k} \frac{1}{(1-rx)}.$$

Problems

P 9.2.1. You want to invite n life forms for dinner. You want to make sure that
    (a) the number of Humans is even
    (b) the number of Romulans is 0 or 1
    (c) the number of Klingons can be anything
    (d) the number of Borgs can be anything.
    Note that the number of Humans (or Romulans, or Klingons, or Borgs) available to be invited is infinite, and that all Humans are alike, as are all Romulans, all Klingons, and all Borgs.
      In how many ways can you do this?

P 9.2.2. I have an unlimited number of apples, bananas, oranges, pears, and pomegranates. In how many ways can I make a basket of 47 fruits if I want all of the following conditions to be satisfied:
    (a) No restriction on the number of apples.
    (b) The number of bananas should be even.
    (c) The number of oranges should be odd.
    (d) The number of pears should be 0 or 1.
    (e) The number of pomegranates should be 0 or 1.

P 9.2.3. Let h_n denote the number of bags consisting of n apples, oranges, bananas, pears, chocolate bars, and carrots such that there are an even number of apples, at most two oranges, a multiple of three number of bananas, at most one pear, and no restriction on the number of chocolate bars or carrots. Find an explicit formula for h_n.

P 9.2.4. How many different bags of fruit are possible if each bag has n fruits chosen from apples, bananas, oranges, pears, pomegranates, and dates, subject to the following conditions:
    (a) The number of apples is a multiple of six.

(b) The number of bananas is at most five.

(c) Any number of oranges and/or pears is possible.

(d) The number of pomegranates is at most one.

(e) The number of dates is at most one.

P 9.2.5. We have defined the initial terms of the Piṅgala–Fibonacci sequence as $F_0 = 0$ and $F_1 = 1$. What would happen to the generating function (see Example 9.34) if you had started with 1 instead of 0? In other words, define a sequence $\{f_n\}_{n=0}^\infty$ by $f_0 = f_1 = 1$, and $f_n = f_{n-1} + f_{n-2}$ for $n \geq 2$. What is the ordinary generating function for the sequence $\{f_n\}_{n=0}^\infty$ and what is a closed formula for f_n?

P 9.2.6. Let T be a set of non-negative integers, and define a sequence of 0's and 1's, called the *indicator sequence* of T, as follows:

$$a_i = \begin{cases} 1 & \text{if } i \in T \\ 0 & \text{if } i \notin T \end{cases} \quad \text{for } i = 0, 1, 2, \cdots.$$

In other words, the placing of the 1's coincides with the elements of T. Show that χ_T, the ordinary indicator power series of T (Definition 9.25), is the ordinary generating function of the sequence a_i, i.e.,

$$\chi_T \xleftrightarrow{\text{ops}} \{a_i\}_{i=0}^\infty.$$

P 9.2.7. Let h_n be the number of non-negative integer solutions to

$$x_1 + x_2 + \cdots + x_7 = n,$$

with the conditions that x_2 is even, x_3 is a multiple of 5, and $0 \leq x_4 \leq 4$. (There are no extra restrictions on the rest of the variables.) What is the generating function of h_n? Give an expression for h_{47}.

P 9.2.8. Let n be a non-negative integer. You have an unlimited supply of oranges, apples, bananas, and pears, and you are going to make bags consisting of n of these fruits. However, you insist that the numbers of apples and oranges must be odd. Let h_n be the number of possible such bags.

(a) What is the ordinary generating function for the sequence $\{h_n\}_{n=0}^\infty$?

(b) Find a partial fraction decomposition for this generating function.

(c) Find a formula for h_n.

(d) What is h_{13}?

P 9.2.9. Let h_n be the number of non-negative integer solutions to

$$x_1 + 3x_2 + 5x_3 = n.$$

Find the ordinary generating function for h_n. Use a symbolic algebra software to find h_{47}.

P 9.2.10. You are at the grocery store, want to buy a pack of gum, and have a seemingly endless supply of pennies, nickels, dimes, and quarters in your pocket. The gum costs 47 cents. You are wondering in how many ways you can make this amount using your coins. Let h_n be the number of ways to have a total of n cents using pennies, nickels, dimes, and quarters.

Find the ordinary generating function for h_n. Finally, use a computer algebra system to answer the original question.

P 9.2.11. You have an inexhaustible supply of pomegranates, dates, pistachios, and figs. Let h_n be the number of ways of choosing n of these, with the condition that the number of pomegranates be even, the number of pistachios be a multiple of 7, the number of figs be a multiple of 4, and with no limit on the number of dates. Let $f(x)$ be the ordinary generating function for h_n.

(a) Give a closed-form expression for $f(x)$.

(b) Reinterpret h_n as the number of certain kinds of partitions of the integer n. (Be specific about what kind of partitions are being counted.) Using this interpretation, write down all such partitions for $n = 10$. What is h_{10}? How do these partitions of 10 relate to the original question about pomegranates, dates, pistachios, and figs?

(c) Using a symbolic algebra software, find h_{47}.

P 9.2.12. Corollary 9.32 argued that $p(n)$, the number of partitions of a positive integer n, is the coefficient of x^n in the power series expansion of

$$f(x) = \frac{1}{(1 - x)(1 - x^2) \cdots (1 - x^n)}.$$

Let $1 \leq m \leq n$, and give an interpretation, in terms of partitions of m, of the coefficient of x^m in the power series expansion of $f(x)$. What if $m > n$?

P 9.2.13. With the aid of generating functions, solve Warm-up 9.24. Can you come up with a way to get the answer without generating functions?

P 9.2.14. Define a sequence of numbers $a_0, a_1, \ldots, a_n$ by $a_0 = 0$, $a_1 = 1$, and, for $n \geq 2$, $a_n = 3a_{n-2}$.

(a) Find a few terms of the sequence, guess a pattern, and prove your pattern using induction.

(b) Re-prove the formula for a_n using ordinary generating functions.

P 9.2.15. Let $r(n)$ be defined as follows: $r(0) = 0$, and, for positive integers n, we have $r(n) = 2r(n - 1) + 2^n$. Find the ordinary generating function for $\{r(n)\}_{n=0}^{\infty}$, and use it to find a closed formula for $r(n)$.

P 9.2.16. Find the generating function for the sequence

$$a_0, a_1, a_2, \ldots, a_n, \ldots,$$

where $a_0 = 3$ and

$$a_n + 2a_{n-1} = n + 3.$$

Write the generating function in as closed a form as possible.

P 9.2.17. Consider the sequence

$$-2, \ -1, \ 17, \ -79, \ 293, \ \ldots$$

defined by $h_0 = -2$, $h_1 = -1$, and, for $n \geq 2$, $h_n = -5h_{n-1} - 6h_{n-2}$.

(a) Use the technique for solving homogeneous linear recurrence relations from Section 1.4 to find a closed formula for h_n.

(b) Re-do the previous part using generating functions.

P 9.2.18. Consider the sequence

$$-2, \ -3, \ 8, \ 112, \ \ldots$$

defined by $h_0 = -2$, $h_1 = -3$, and, for $n \geq 2$,

$$h_n = 8h_{n-1} - 16h_{n-2}.$$

Find an explicit formula for h_n.

P 9.2.19. A sequence $\{h_n\}_{n=0}^{\infty}$ is defined by $h_0 = h_1 = h_2 = 1$, and, for $n \geq 3$, $h_n = 10h_{n-1} - 31h_{n-2} + 30h_{n-3}$. Find the ordinary generating function for $\{h_n\}_{n=0}^{\infty}$, and use it to find a closed formula for h_n.

P 9.2.20. You work at a car dealership that sells three models: a pickup truck, an SUV, and a compact hybrid. Your job is to park the vehicles in a row. Each pickup truck and each SUV takes up two spaces while each hybrid takes up one space. Let n be a positive integer and let a_n be the number of ways of arranging vehicles in n spaces. In Problem P 1.3.8, you were asked to find a recurrence relation for a_n.
 (a) Using that recurrence relation, find a closed form for the ordinary power series for the sequence $\{a_n\}_{n=0}^{\infty}$.
 (b) Use partial fractions and write the generating function as the sum of two fractions.
 (c) Find a formula for a_n.

P 9.2.21. Let the sequence $\{h_n\}_{n=0}^{\infty}$ be defined by $h_0 = 3$, and, for $n \geq 1$, $h_n = h_{n-1} + 3n - 2$.
 (a) Find the ordinary generating function for $\{h_n\}_{n=0}^{\infty}$, and use it to find a closed formula for h_n.
 (b) Find a closed formula for h_n without recourse to generating functions, and by using one of the techniques of Chapter 1.

P 9.2.22. Let the sequence $\{h_n\}_{n=0}^{\infty}$ be defined by $h_0 = 1$, and, for $n \geq 1$, $h_n = h_{n-1} + 2n - 3$.
 (a) Find the first few terms of the sequence, guess a pattern, and prove it by induction on n.
 (b) Ignoring your solution to the previous part, find the ordinary generating function for $\{h_n\}_{n=0}^{\infty}$, and use it to find a closed formula for h_n.

P 9.2.23. Let the sequence $\{h_n\}_{n=0}^{\infty}$ be defined by $h_0 = 2$, and, for $n \geq 1$, $h_n = h_{n-1} + n^2 - 1$.
 (a) Find the ordinary generating function of the sequence $\{h_n\}_{n=0}^{\infty}$.
 (b) Find a closed formula for h_n.

P 9.2.24. We are interested in sequences made up of 0's and 1's such that there is at least one instance of consecutive 0's. For $n \geq 0$, let h_n be the number of such sequences of length n.
 (a) Find $h_0, \ldots, h_3$.
 (b) Find a recurrence relation for h_n. Use it to find h_6.
 (c) Find the ordinary generating function for the sequence $\{h_n\}_{n=0}^{\infty}$.

P 9.2.25. Let n be a non-negative integer, and $E(n)$ the number of sequences $a_1, a_2, \ldots, a_n$ such that each a_i is 0, 1, 2, 3, or 4 *and* the number of 3's in the sequence is even. For example, $E(0) = 1$, and $E(1) = 4$. In Problem P 1.3.16, you were asked to show that, for $n \geq 1$,

$$E(n) = 3E(n-1) + 5^{n-1}.$$

Let $g(x) \overset{\text{ops}}{\longleftrightarrow} \{E(n)\}_{n=0}^{\infty}$. Find a closed form for $g(x)$, and use it to find a formula for $E(n)$.

P 9.2.26. Let $a_0 = 0$ and, for $k \geq 1$, assume

$$a_k = 3a_{k-1} + 4^k.$$

Let $f(x)$ be the ordinary power series for the sequence $\{a_k\}_{k=0}^{\infty}$. Find a closed formula for $f(x)$, and use it to find a closed formula for a_k.

P 9.2.27. Let n be a non-negative integer. You have an inexhaustible supply of each of three James Baldwin books: *Go to Tell It on the Mountain*, *Notes of a Native Son*, and *If Beale Street Could Talk*. You are going to make a box consisting of n of these books. Consider the following three scenarios:

(a) The box has at least two of each title.

(b) The box has at most two copies of *Notes of a Native Son*.

(c) The box has an even number of *Go to Tell It on the Mountain*.

For each scenario, let h_n denote the number of different boxes that you could assemble, and find the ordinary generating function of h_n. Use this generating function to determine a closed formula for h_n. Could you find—in each scenario—h_n without using generating functions?

P 9.2.28. Let $f(n)$ be the number of sequences

$$a_1, a_2, a_3, \ldots, a_n$$

such that a_i is 0, 1, or 2 for $i = 1, \ldots, n$, and such that there are no adjacent 1's and no adjacent 2's. For convenience we define $f(0) = 1$.

(a) By considering the position of the last 0 in the sequence, prove that

$$f(n) = 2 + 2f(0) + 2f(1) + \cdots + 2f(n-2) + f(n-1).$$

(b) Let $A(x) = f(0) + f(1)x + f(2)x^2 + \cdots$ be the ordinary generating function for $f(n)$. Show that

$$A(x) = (1 + 2x + 2x^2 + \cdots)(1 + f(0)x + f(1)x^2 + \cdots)$$

$$= (1 + \frac{2x}{1-x})(1 + xA(x)).$$

(c) Find a closed formula for $A(x)$.

(d) What is $f(10)$? (You can use a computer if you wish.)

P 9.2.29. Let n be a non-negative integer. A hoard of n identical silver coins is to be distributed among 6 distinct pirates in such a way that no one receives more than 12 coins. Let $d(n)$ be the number of ways this can be done.

(a) Translate the problem into one about the number of integer solutions to an equation.

(b) Write an expression for the ordinary generating function of $\{d(n)\}_{n=0}^{\infty}$. Use it to find $d(47)$.

(c) Find an expression for $d(47)$ using the inclusion–exclusion principle.

P 9.2.30. Let n be a non-negative integer no greater than 58, and let h_n be the number of integer solutions to $x_1 + \cdots + x_4 + y_1 + \cdots + y_6 = n$, with $1 \le x_i \le 4$ for $1 \le i \le 4$, and $1 \le y_j \le 7$ for $1 \le j \le 6$.

(a) Find the ordinary generating function for $\{h_n\}_{n=0}^{58}$.

(b) Use the generating function to find h_{24}.

(c) Using the inclusion–exclusion principle, and without recourse to generating functions, find h_{24}.

P 9.2.31. Consider Theon's ladder[4] of whole numbers:

n	a_n	b_n
0	0	1
1	1	1
2	2	3
3	5	7
4	12	17
5	29	41
$\vdots$	$\vdots$	$\vdots$

The ladder defines two sequences of numbers, $\{a_n\}_{n=0}^{\infty}$ and $\{b_n\}_{n=0}^{\infty}$, by $a_0 = 0$, $b_0 = 1$, and, for $n \ge 1$,

$$a_n = a_{n-1} + b_{n-1}$$
$$b_n = a_n + a_{n-1}.$$

Let $A(x) \xleftrightarrow{\text{ops}} \{a_n\}_{n=0}^{\infty}$ and $B(x) \xleftrightarrow{\text{ops}} \{b_n\}_{n=0}^{\infty}$.

(a) Find a closed form for both $A(x)$ and $B(x)$.

(b) Using a partial fraction decomposition, find the coefficient of x^n in $B(x)$ and give a formula for b_n.

(c) Do the same for a_n.

(d) Without using generating functions, prove that $b_n^2 - 2a_n^2 = (-1)^n$. Conclude that

$$\lim_{n \to \infty} \frac{b_n}{a_n} = \sqrt{2}.$$

P 9.2.32. In analogy with Theon's ladder of Problem *P 9.2.31*, consider two (interrelated) sequences $\{a_n\}_{n=0}^{\infty}$ and $\{b_n\}_{n=0}^{\infty}$ defined by $a_0 = 0$, $b_0 = 1$, and, for $n \ge 1$,

$$a_n = a_{n-1} + b_{n-1}$$
$$b_n = a_n + b_{n-1}.$$

---

[4] Attributed to Theon of Smyrna (circa 140 CE).

The first few terms of these sequences are given in the following table:

n	a_n	b_n
0	0	1
1	1	2
2	3	5
3	8	13
4	21	34
5	55	89
$\vdots$	$\vdots$	$\vdots$

Let $A(x) \overset{\text{ops}}{\longleftrightarrow} \{a_n\}_{n=0}^{\infty}$ and $B(x) \overset{\text{ops}}{\longleftrightarrow} \{b_n\}_{n=0}^{\infty}$, and find a closed form for both $A(x)$ and $B(x)$.

9.3 Exponential Generating Functions

Warm-Up 9.38. *You have many copies of each of three books of Nikos Kazantzakis:* Zorba the Greek, Christ Recrucified, *and* The Last Temptation. *You want to arrange seven of these books on a bookshelf, with the restriction that each of the three novels is in the line-up at least once and at most three times. In how many ways can this be done? Leave your answer as a sum of multinomial coefficients.*

The ordinary generating function is not the only way to create a power series from a sequence of numbers. For some purposes, an *exponential* generating function will be more useful.

Definition 9.39. Let $h_0, h_1, \ldots, h_n, \ldots$ be a sequence of numbers. The *exponential generating function* of this sequence is

$$G(x) = h_0 + h_1 x + \frac{h_2}{2!}x^2 + \cdots + \frac{h_n}{n!}x^n + \cdots .$$

We write

$$G \overset{\text{egf}}{\longleftrightarrow} \{h_n\}_{n=0}^{\infty}$$

("egf" stands for exponential generating function). Some authors use $g^{(e)}(x)$ to denote an exponential generating function.

Remark 9.40. Note that if $G \overset{\text{egf}}{\longleftrightarrow} \{h_n\}_{n=0}^{\infty}$, then the coefficient of x^n in $G(x)$ is $\frac{h_n}{n!}$. Hence, h_n is $n! \, b_n$, where b_n is the coefficient of x^n in the power series expansion of $G(x)$ at $x = 0$.

Example 9.41. The exponential generating function for the sequence $1, 1, \ldots$ of all 1's is

$$1 + x + \frac{x^2}{2!} + \frac{x^3}{3!} + \cdots = e^x.$$

In other words,

$$e^x \xleftrightarrow{\text{egf}} \{1\}_{n=0}^{\infty},$$

and, plugging in ax for x, we have

$$e^{ax} \xleftrightarrow{\text{egf}} \{a^n\}_{n=0}^{\infty}.$$

Example 9.42. Let n be a fixed positive integer, and consider the sequence of falling factorials

$$(n)_0, \ (n)_1, \ \ldots, \ (n)_n,$$

where $(n)_k = \frac{n!}{(n-k)!}$. Let

$$G(x) \xleftrightarrow{\text{egf}} \{(n)_k\}_{k=0}^{n}.$$

Then

$$G(x) = (n)_0 + (n)_1 x + \frac{(n)_2}{2!}x^2 + \cdots + \frac{(n)_n}{n!}x^n$$

$$= \frac{n!}{n! \ 0!} + \frac{n!}{(n-1)! \ 1!}x + \frac{n!}{(n-2)! \ 2!}x^2 + \cdots + \frac{n!}{(n-n)! \ n!}x^n$$

$$= \binom{n}{0} + \binom{n}{1}x + \binom{n}{2}x^2 + \cdots + \binom{n}{n}x^n$$

$$= (1+x)^n.$$

Remark 9.43. Ordinary generating functions were a perfectly good way of using a power series to encode a sequence of numbers. Why bother with the seemingly more complicated *exponential* generating functions? For some sequences of numbers, an exponential generating function just works better in the sense that you may find a more manageable closed form for the function. However, the efficacy of exponential generating functions becomes clear when we multiply them. Multiplying ordinary power series was naturally connected to the number of non-negative integer solutions to $x_1 + \cdots + x_s = t$ (Proposition 9.17), and as such ended up providing a useful vehicle for counting combinations (the entries in the second column of the counting table, Table 3.1 – see the comments at the beginning of subsection 9.2.1). As we shall momentarily see, multiplying exponential generating functions gives a natural connection to *permutations* (the entries of the first column of the counting table, Table 3.1 – see the comments before Definition 9.45).

Proposition 9.44. *Let t be a positive integer, and, for $1 \le i \le t$, let f_i be the exponential generating function for the sequence*

$$a_0^{(i)}, a_1^{(i)}, \ldots, a_n^{(i)}, \ldots$$

Then, the product of these exponential generating functions, $f_1 f_2, \ldots, f_t$, is itself the exponential generating function for the sequence $c_0, c_1, \ldots, c_n, \ldots$, where c_n is the sum of all terms of the form $\binom{n}{n_1 \; n_2 \; \ldots \; n_t} a_{n_1}^{(1)} a_{n_2}^{(2)} \cdots a_{n_t}^{(t)}$ – a product of $\binom{n}{n_1 \; n_2 \; \ldots \; n_t}$ with t terms, one from each of the original sequences – with $n_1, \ldots, n_t$ non-negative integers and $n_1 + \cdots + n_t = n$.

In other words, if

$$f_1 \overset{\text{egf}}{\longleftrightarrow} \left\{a_n^{(1)}\right\}_{n=0}^{\infty}, f_2 \overset{\text{egf}}{\longleftrightarrow} \left\{a_n^{(2)}\right\}_{n=0}^{\infty}, \ldots, f_t \overset{\text{egf}}{\longleftrightarrow} \left\{a_n^{(t)}\right\}_{n=0}^{\infty},$$

then

$$f_1 \ldots f_t \overset{\text{egf}}{\longleftrightarrow} \{c_n\}_{n=0}^{\infty}, \text{ where } c_n = \sum_{\substack{n_1+n_2+\cdots+n_t=n \\ n_i \geq 0}} \binom{n}{n_1 \; n_2 \; \ldots \; n_t} a_{n_1}^{(1)} a_{n_2}^{(2)} \cdots a_{n_t}^{(t)}.$$

Proof. By definition, we have

$$f_1(x) = \frac{a_0^{(1)}}{0!} + \frac{a_1^{(1)}}{1!}x + \cdots + \frac{a_n^{(1)}}{n!}x^n + \cdots$$

$$f_2(x) = \frac{a_0^{(2)}}{0!} + \frac{a_1^{(2)}}{1!}x + \cdots + \frac{a_n^{(2)}}{n!}x^n + \cdots$$

$$\vdots$$

$$f_t(x) = \frac{a_0^{(t)}}{0!} + \frac{a_1^{(t)}}{1!}x + \cdots + \frac{a_n^{(t)}}{n!}x^n + \cdots.$$

Let $f = f_1 f_2 \cdots f_t$ be the product of $f_1, \ldots, f_t$. When we multiply these power series, how do we get an x^n term? Each power series contributes a power of x, and the exponents have to add to n. You may use the term $\frac{a_{n_1}^{(1)}}{n_1!}x^{n_1}$ from the first power series, the term $\frac{a_{n_2}^{(1)}}{n_2!}x^{n_2}$ from the second power series, and so on. The product of these will give an x^n term if and only if $n_1 + n_2 + \cdots + n_t = n$. Hence, the coefficient of x^n in f is

$$\sum_{\substack{n_1+n_2+\cdots+n_t=n \\ n_i \geq 0}} \frac{a_{n_1}^{(1)}}{n_1!} \frac{a_{n_2}^{(2)}}{n_2!} \cdots \frac{a_{n_t}^{(t)}}{n_t!}.$$

Now, c_n is the coefficient of $x^n/n!$ in f, and hence

$$c_n = n! \sum_{\substack{n_1+n_2+\cdots+n_t=n \\ n_i \geq 0}} \frac{a_{n_1}^{(1)}}{n_1!} \frac{a_{n_2}^{(2)}}{n_2!} \cdots \frac{a_{n_t}^{(t)}}{n_t!}$$

$$= \sum_{\substack{n_1+n_2+\cdots+n_t=n \\ n_i \geq 0}} \binom{n}{n_1 \; n_2 \; \ldots \; n_t} a_{n_1}^{(1)} a_{n_2}^{(2)} \cdots a_{n_t}^{(t)}.$$

$\square$

Permutations of Multisets

The first column of Table 3.1 – when the s balls and the t boxes are both distinct – counts "permutations." (Also see the discussion at the end of Section 3.3.) We are permuting the "boxes," and the "balls" tell us how. The balls are distinguishable and so we can number them: ball 1, ball 2, and so on. The boxes are also distinguishable and they have names: U_1, ..., U_t. When you place a ball in a box, you have chosen that box to be in your permutation, and the number of the ball tells you where in the permutation the box is. Each placing of the balls is some kind of a permutation of the boxes. If you are allowed to put more than one ball in a box, then that box is appearing more than once in your permutation, and hence we have a permutation of a multiset. Here, we are coming back to entries $\boxed{13}$ and $\boxed{15}$. These are also counting permutations, but the restriction on the capacity of the boxes (or equivalently on the number of elements of each type that appear in a permutation) could be a bit more complicated. Using multinomial coefficients, we already gave a preliminary solution for each of these entries in Theorems 4.33 and 4.34. Here we continue that discussion and provide an (exponential) generating function approach to these entries.

Definition 9.45. Let $T = \{n_1, n_2, \ldots\}$ be a (finite or infinite) set of non-negative integers, define (in analogy with Definition 9.25) $\mathfrak{X}_T$ by

$$\mathfrak{X}_T = \frac{x^{n_1}}{n_1!} + \frac{x^{n_2}}{n_2!} + \frac{x^{n_3}}{n_3!} + \cdots,$$

and call it the *exponential indicator generating function* of T.

Example 9.46. Let T be the set of non-negative even integers. Then the exponential indicator generating function of T is

$$\mathfrak{X}_T = 1 + \frac{x^2}{2!} + \frac{x^4}{4!} + \cdots + \frac{x^{2k}}{(2k)!} + \cdots.$$

Starting with the power series for e^x, we can find a closed form for $\mathfrak{X}_T$:

$$e^x = 1 + \frac{x}{1!} + \frac{x^2}{2!} + \cdots + \frac{x^m}{m!} + \cdots$$

$$e^{-x} = 1 - \frac{x}{1!} + \frac{x^2}{2!} + \cdots + (-1)^m \frac{x^m}{m!} + \cdots$$

$$\frac{1}{2}\left(e^x + e^{-x}\right) = 1 + \frac{x^2}{2!} + \frac{x^4}{4!} + \cdots + \frac{x^{2k}}{(2k)!} + \cdots.$$

Hence,

$$\mathfrak{X}_T = \frac{\left(e^x + e^{-x}\right)}{2}.$$

Theorem 4.34 provided a preliminary version for entry $\boxed{15}$ of the counting table. In fact, entry $\boxed{13}$ was a special case and was treated in Theorem 4.33. We now revise these theorems using exponential generating functions.

Theorem 9.47 (Entry $\boxed{15}$ of the Counting Table). *Let t be a positive integer, and let $R =$* $\{\infty \cdot V_1, \infty \cdot V_2, \ldots, \infty \cdot V_t\}$ *be a multiset with t types of elements with infinite repetition numbers.* *Let $N_1, N_2, \ldots, N_t$ be sets of non-negative integers, and, for $1 \le i \le t$, let $\mathfrak{X}_i$ be the exponential* *indicator generating function of N_i.*

Then the following three numbers are equal:

(a) $\boxed{15}$ *The number of ways of placing s distinct balls into t distinct boxes such that, for $i =$* $1, \ldots t$, *the number of balls in the ith box is an integer in N_i*

(b) *The number of s-permutations of the multiset R such that, for $i = 1, \ldots, t$, the number of* V_i's *used is an integer in N_i*

(c)

$$h_s = \sum_{\substack{m_1 + \cdots + m_t = s \\ m_i \in N_i}} \binom{s}{m_1 \ m_2 \ \cdots \ m_t}.$$

Moreover, for the sequence $\{h_s\}_{s=0}^{\infty}$, we have

(d)

$$\mathfrak{X}_1 \cdots \mathfrak{X}_t \overset{\text{egf}}{\longrightarrow} \{h_s\}_{s=0}^{\infty}.$$

Proof. Except for the statement about the exponential generating function of $\{h_s\}_{s=0}^{\infty}$, everything was proved in Theorem 4.34. By Proposition 9.44 and the definition of the exponential indicator generating functions, $\mathfrak{X}_1 \cdots \mathfrak{X}_t$ is the exponential generating function for the sequence

$$\left\{ \sum_{\substack{m_1 + m_2 + \cdots + m_t = s \\ m_i \ge 0}} \binom{s}{m_1 \ m_2 \ \cdots \ m_t} a_{m_1}^{(1)} a_{m_2}^{(2)} \cdots a_{m_t}^{(t)} \right\}_{s=0}^{\infty},$$

where $a_{m_i}^{(i)}$ is zero unless $m_i \in N_i$, in which case $a_{m_i}^{(i)} = 1$. Now, $m_1 \in N_1, m_2 \in N_2, \ldots, m_t \in N_t$, and $m_1 + m_2 + \cdots + m_t = s$, if and only if $a_{m_1}^{(1)} a_{m_2}^{(2)} \cdots a_{m_t}^{(t)} = 1$. The result follows. □

If $N_i = \{0, 1, \ldots, n_i\}$, then $\mathfrak{X}_{N_i} = 1 + x + \frac{x^2}{2!} + \cdots + \frac{x^{n_i}}{n_i!}$ is a partial sum of the power series for e^x. By specializing Theorem 9.47 to this case, we get the generating function technique for confronting entry $\boxed{13}$ of the counting table.

Corollary 9.48 (Entry $\boxed{13}$ of the Counting Table). *Let $R = \{n_1 \cdot V_1, n_2 \cdot V_2, \ldots, n_t \cdot V_t\}$ be* *a multiset with t different types of elements and with repetition numbers $n_1, \ldots, n_t$ (where n_i is* *allowed to be ∞). For non-negative integers s, define*

$$h_s = \text{ the number of s-permutations of R.}$$

Let $G(x) \overset{\text{egf}}{\longleftrightarrow} \{h_s\}_{s=0}^{\infty}$. Then

$$G(x) = f_{n_1}(x) f_{n_2}(x) \ldots f_{n_t}(x),$$

where, for $1 \leq i \leq t$,

$$f_{n_i}(x) = \begin{cases} 1 + x + \frac{x^2}{2!} + \cdots + \frac{x^{n_i}}{n_i!} & \text{if } n_i < \infty \\ e^x & \text{if } n_i = \infty. \end{cases}$$

Example 9.49. Assume that you are seating your guests on one side of a $1 \times n$ dinner table. Your guests could be Humans, Klingons, or Romulans. All Humans are considered alike, as are all Klingons, and all Romulans, and there is an inexhaustible supply of each life form. Your only condition is that the number of seated Humans must be even. How many seating arrangements are possible?

Solution. For n a non-negative integer, let h_n denote the number of possible seating arrangements (in which the number of Humans is even). By inspection, we have $h_0 = 1$, $h_1 = 2$, $h_2 = 5$, and $h_3 = 14$. The fourteen possibilities for sitting three guests are:

$$HHK, \ HHR, \ HKH, \ HRH, \ KHH, \ RHH,$$
$$KKK, \ KKR, \ KRK, \ RKK, \ KRR, \ RKR, \ RRK, \ RRR.$$

If you have experience with combinatorial sequences and see the sequence $1, 2, 5, 14$, you may suspect a connection to the Ming–Catalan numbers (see Definition 5.17 and Collaborative Mini-project 7 on Ming–Catalan numbers). But you would be wrong! Define

$$G(x) \xleftrightarrow{\text{egf}} \{h_n\}_{n=0}^{\infty}.$$

Then by Theorem 9.47, and using Example 9.46, we have

$$G(x) = \underbrace{\left(1 + \frac{x^2}{2!} + \frac{x^4}{4!} + \cdots\right)}_{\text{Humans}} \underbrace{\left(1 + x + \frac{x^2}{2!} + \cdots\right)}_{\text{Klingons}} \underbrace{\left(1 + x + \frac{x^2}{2!} + \cdots\right)}_{\text{Romulans}}$$

$$= \frac{e^x + e^{-x}}{2} e^x e^x = \frac{1}{2}\left(e^{3x} + e^x\right)$$

$$= \frac{1}{2}\left(\sum_{n=0}^{\infty} 3^n \frac{x^n}{n!} + \sum_{n=0}^{\infty} \frac{x^n}{n!}\right) = \frac{1}{2}\left(\sum_{n=0}^{\infty} (3^n + 1)\frac{x^n}{n!}\right).$$

Note that h_n is $n!$ times the coefficient of x^n in $G(x)$, and so

$$h_n = \frac{3^n + 1}{2}. \qquad \square$$

Finally, we address the opening problem of the chapter.

Example 9.50. You want to arrange 17 books on the shelf of a bookstore. The shelf is dedicated to the three Toni Morrison novels published between 1977 and 1987: *Song of Solomon*, *Tar Baby*, and *Beloved*. You have many copies of each, but on the shelf you want an even number of *Song of Solomon*, at least three copies of *Tar Baby*, and at most four copies of *Beloved*. How many different arrangements are possible?

Solution. For n a non-negative integer, let h_n be the number of ways of arranging n books on the shelf, given the constraints of the problem. We then have $h_0 = h_1 = h_2 = 0$, $h_3 = 1$ (TTT is the only possibility), and $h_4 = 5$ (the possibilities are $TTTB, TTBT, TBTT, BTTT$, and $TTTT$). Let

$$G(x) \overset{\text{egf}}{\longleftrightarrow} \{h_n\}_{n=0}^{\infty}.$$

Then by Theorem 9.47, we have

$$G(x) = \underbrace{\left(1 + \frac{x^2}{2!} + \frac{x^4}{4!} + \cdots\right)}_{\text{Song of Solomon}} \underbrace{\left(\frac{x^3}{3!} + \frac{x^4}{4!} + \cdots\right)}_{\text{Tar Baby}} \underbrace{\left(1 + x + \frac{x^2}{2!} + \frac{x^3}{3!} + \frac{x^4}{4!}\right)}_{\text{Beloved}}$$

$$= \frac{e^x + e^{-x}}{2}\left(e^x - 1 - x - \frac{x^2}{2}\right)\left(1 + x + \frac{x^2}{2!} + \frac{x^3}{3!} + \frac{x^4}{4!}\right).$$

While it is possible to multiply out the expression for $G(x)$ and find the coefficient of x^n in each term, it is somewhat tedious to do so. Instead we use a computer algebra system to find the coefficient of x^{17} in the Taylor series expansion of $G(x)$ at $x = 0$. (In Maple, for example, after defining the expression for G, we use `coeftayl(G,x=0,17)`.) Remembering that the coefficient of x^{17} is $\frac{h_{17}}{17!}$, we have that h_{17} is 17! times this coefficient, and we get that

$$h_{17} = 18{,}072{,}242.$$ $\qquad\square$

Problems

P 9.3.1. What is the exponential generating function for

$$1, \ -1, \ 1, \ \ldots, \ (-1)^n, \ \cdots.$$

P 9.3.2. The function xe^{3x} is the exponential generating function for which sequence?

P 9.3.3. If $G(x) \overset{\text{egf}}{\longleftrightarrow} \{a_n\}_{n=0}^{\infty}$, then what is the exponential generating function for the sequence $\{3a_n - 2\}_{n=0}^{\infty}$?

P 9.3.4. The function $\frac{1}{(1-x)^2}$ is the exponential generating function for which sequence?

P 9.3.5. What is the exponential generating function for the following sequence?

$$0!, \ -1! \times 5, \ 2! \times 5^2, \ -3! \times 5^3, \ \ldots, (-1)^k k! \times 5^k, \ \ldots$$

P 9.3.6. Find the exponential generating function for each of the following sequences:
(a) $0!, \ -1!, 2!, \ -3!, \ldots, (-1)^k k!, \ldots$
(b) $0, 0!, \ -1!, 2!, \ -3!, \ldots, (-1)^{k-1}(k-1)!, \ldots$

P 9.3.7. Let h_n denote the number of ways to color the squares of a $1 \times n$ board with one of the colors red, white, blue, and green, in such a way that the number of squares colored red is even and the number of squares colored white is odd. Find an explicit formula for h_n.

P 9.3.8. You have an unlimited supply of the first four of the seven Maya Angelou autobiographies: *I Know Why the Caged Bird Sings*, *Gather Together in My Name*, *Singin' and Swingin' and Gettin' Merry Like Christmas*, and *The Heart of a Woman*. Let n be a non-negative integer. In how many ways can you arrange n of the books on a shelf, with the condition that the first two autobiographies appear an odd number of times?

P 9.3.9. Determine the number of n-digit numbers with all digits odd, such that 1 and 3 each occur a *positive* even number of times.

P 9.3.10. You want to invite n life forms to sit together on one side of a long $1 \times n$ dinner table. You have exactly the same restrictions as in Problem P 9.2.1. In how many ways can you do this?

P 9.3.11. At a dinner party on the Starship Enterprise, three life forms are present: Humans, Klingons, and Romulans. The dinner table is a long 1×13 board, and the life forms sit on one side of it, next to one another. There are 56 Humans, 32 Klingons, and 35 Romulans present, and so only a total of 13 sit at the table. The only condition is that the number of Klingons and also the number of Romulans at the table should be odd. In how many different ways can this be done? Assume that all Humans look alike, as do all Klingons, and all Romulans.

P 9.3.12. Find the number of sequences $a_1 a_2 \cdots a_{47}$ of length 47, such that both of the following hold:
(a) $a_i = 0, 1, 2$, or 3, for $1 \le i \le 47$.
(b) The number of zeros in the sequence is even.

P 9.3.13. Find the number of sequences $a_1, a_2, \ldots, a_{47}$ of length 47 that satisfy all of the following:
(a) $a_i \in \{0, 1, 2, 3\}$ for $1 \le i \le 47$.
(b) The number of 0's is even.
(c) There are either one or two 3's.
(d) There is no restriction on the number of 1's or 2's.

P 9.3.14. Continuing with Warm-Up 9.38, assume that you want to arrange n copies of the three books on a bookshelf, keeping the restriction that each book appears at least once and at most three times. Let $h(n)$ be the number of possibilities, and let $G(x)$ be the exponential generating function for $h(n)$.
(a) Give an expression for $G(x)$.
(b) Use a symbolic algebra software to expand $G(x)$.
(c) Give the values of $h(n)$ for $3 \le n \le 9$.

P 9.3.15. Let T be a set of non-negative integers, and let $\{a_i\}_{i=0}^{\infty}$ be the indicator sequence of T, as defined in Problem P 9.2.6. Show that $\mathcal{X}_T$, the exponential generating function of T, is the exponential generating function of the sequence $\{a_i\}_{i=0}^{\infty}$. In other words, show that

$$\mathcal{X}_T \overset{\text{egf}}{\longleftrightarrow} \{a_i\}_{i=0}^{\infty}.$$

P 9.3.16. Assume that $G(x)$ is the exponential generating function for

$$a_0, a_1, \ldots, a_n, \ldots,$$

and $H(x)$ is the exponential generating function for

$$c, a_0, a_1, \ldots, a_{n-1}, \ldots,$$

where c is a constant. How are G and H related?

P 9.3.17. Let $G(x)$ and $H(x)$ be exponential generating functions for the sequences $\{a_n\}$ and $\{b_n\}$, respectively. The product $G(x)H(x)$ will be the exponential generating function for some sequence $\{c_n\}$. Find a formula of the following form for c_n in terms of the original two sequences:

$$c_n = \sum_{?}^{?} ??a_k b_{n-k}.$$

P 9.3.18. Consider the functions e^x, e^{2x}, and e^{3x}, and write $e^x \xleftrightarrow{\text{egf}} \{a_n\}_{n=0}^{\infty}$, $e^{2x} \xleftrightarrow{\text{egf}} \{b_n\}_{n=0}^{\infty}$, and $e^{3x} \xleftrightarrow{\text{egf}} \{c_n\}_{n=0}^{\infty}$.

(a) What are closed formulas for a_n, b_n, and c_n?

(b) We know $e^x e^{2x} = e^{3x}$, and so a_n, b_n, and c_n are related as in Problem P 9.3.17. Write the relationship as an identity. Can you give a more direct proof of this identity?

P 9.3.19. Let $\{F_n\}_{n=0}^{\infty}$ denote the Piṅgala–Fibonacci sequence. Recall that $F_0 = 0$, $F_1 = 1$, and, for $n \geq 2$, $F_n = F_{n-2} + F_{n-1}$. Let $F(x) \xleftrightarrow{\text{egf}} \{F_n\}_{n=0}^{\infty}$. Find a differential equation satisfied by $F(x)$.

P 9.3.20. Consider a "combination lock"[5] that has five buttons labeled $1, \ldots, 5$. You can set the combination to be either the empty set (interpreted as a broken lock) or any ordered sequence of disjoint non-empty subsets of $[5] = \{1, 2, 3, 4, 5\}$. For example, $\emptyset$, $\{3\} - \{2, 5\} - \{1\}$, and $\{2, 3, 4\}$ are three different valid combinations, while $\{3, 4\} - \{1, 2\} - \{2\}$ and $\{1, 2\} - \emptyset - \{3, 4\}$ are not. The combinations $\{1, 2\} - \{4\}$ and $\{4\} - \{1, 2\}$ are different combinations (since the order of the sets matters), but $\{1, 2, 3\} - \{4, 5\}$ is the same combination as $\{2, 1, 3\} - \{5, 4\}$, since $\{1, 2, 3\}$ and $\{2, 1, 3\}$, as well as $\{4, 5\}$ and $\{5, 4\}$, are the same sets. To open the lock, you sequentially go through the sets in the combination and, for each set, simultaneously push all the buttons in that set. Let n be a non-negative integer, and let a_n be the number of such combination locks with n buttons. So, for example, $a_0 = 1$, since $\emptyset$ is the only possibility, while $a_2 = 6$, since all the possibilities are

$$\emptyset, \ \{1\}, \ \{2\}, \ \{1, 2\}, \ \{1\} - \{2\}, \ \{2\} - \{1\}.$$

(a) Let n and k be positive integers with $k \leq n$. Let A be a fixed subset of $[n] = \{1, 2, \ldots, n\}$ of size k. Is it true that there are a_{n-k} combinations that start with A?

(b) Based on your answer to the previous part, give a recurrence relation for a_n. (Be careful with the case when $k = 0$ for the starting set.)

(c) Find $a_0, a_1, \ldots, a_7$.

---

[5] Problem adapted from Benjamin 1996.

(d) Let $G(x) \overset{\text{egf}}{\longleftrightarrow} \{a_n\}_{n=0}^{\infty}$. Using the result of Problem P 9.3.17, $e^x G(x)$ is the exponential generating function for which sequence? How is this sequence related to $\{a_n\}_{n=0}^{\infty}$?

(e) Find a closed form for $G(x)$.

(f) Use a computer algebra system and your expression for $G(x)$ to compute a_{15}.

9.4 Generating Functions for Partitions, Stirling Numbers, Bernoulli Numbers, and More*

We have barely scratched the surface with what can be done with generating functions. For a wonderful treatment of generating functions, see Wilf (2006). In the problems in this optional section, we ask you to explore a few more topics. These are: (1) an example of a generating function in two variables; (2) additional techniques for getting new generating functions from known ones; (3) generating functions of partitions and their applications; (4) generating functions for Stirling numbers of the second kind and Bell numbers; (5) Bernoulli numbers; (6) summing the reciprocals of Catalan numbers; and (7) the connections between ordinary generating functions, exponential generating functions, and Laplace transforms. Each topic is broken down into smaller problems, and the problems for each topic are independent of the other topics.

Problems

P 9.4.1. **A Generating Function in Two Variables.** To commemorate the 50th anniversary of the first Moon landing in 1969, the US Postal Service issued two stamps. One stamp is the iconic photograph of Aldrin on the surface of the Moon, taken by Armstrong. The second stamp is a photograph of the Moon with a dot showing the landing site of the lunar module, Eagle, in the Sea of Tranquility.

(a) You have 27 (identical) copies of the first stamp, and 21 of the second stamp. In how many ways can you distribute these stamps among seven (distinct) friends in such a way that every friend gets at least one of each stamp, and no friend gets more than four copies of the second stamp? If possible, do not use generating functions.

(b) You still have the same seven friends, but r copies of the first stamp and s copies of the second stamp. You still want each friend to get at least one of each stamp and at most four of the second stamp. Given these restrictions, for non-negative integers r and s, let $f(r, s)$ be the number of ways that you can distribute all the stamps among your friends. Let

$$g(x, y) = \sum_{s=0}^{\infty} \sum_{r=0}^{\infty} f(r, s) x^r y^s$$

be the generating function, in two variables, for the integers $f(r, s)$. Let $h(x)$ be the ordinary generating function for $\{a_r\}_{r=0}^{\infty}$, where a_r is the number of ways of distributing

r of the first type of stamp among the seven friends such that each friend gets at least one stamp. Likewise, let $k(y)$ be the ordinary generating function for $\{b_s\}_{s=0}^{\infty}$, where b_s is the number of ways of distributing s of the second type of stamp among the seven friends such that each friend gets at least one and no more than four stamps. Argue that

$$g(x, y) = h(x)k(y).$$

(c) Continuing with the notation of the previous part, find a closed form for $g(x, y)$.

(d) Using a computer algebra system, find the coefficient of $x^{27}y^{21}$ in the Taylor series expansion of the function $g(x, y)$ from the previous part, and confirm that it matches your answer from part (a). In Maple, after defining $g(x, y)$, you can use the command

```
coeftayl(g(x,y),[x,y]=[0,0],[27,21]);
```

P 9.4.2. Let r and s be non-negative integers, and consider the following system of two equations and eight unknowns:

$$\begin{cases} x_1 + x_2 + x_3 + x_4 & = r \\ y_1 + 2y_2 + 3y_3 + 4y_4 & = s. \end{cases}$$

Let $f(r, s)$ be the number of non-negative integer solutions to this system, where x_1 and x_2 are less than or equal to 3. Find a closed form for

$$\sum_{r=0}^{\infty} \sum_{s=0}^{\infty} f(r, s) x^r y^s.$$

9.4.1 New Generating Functions from Old

We have seen many examples where we start with a known generating function, manipulate it, and get a generating function for another sequence. Let $f(x)$ be the generating function of a sequence. In the problems, you are asked to explore $\frac{f(x)+f(-x)}{2}$ (Problem P 9.4.3) and $\frac{f(x)}{1-x}$ (Problem P 9.4.8). These functions are generating functions of which sequences? As applications, you are asked to find the sum of every other term in one row of the Karaji–Jia triangle (Problem P 9.4.5), as well as the exponential generating function for the sequence of derangments (Problem P 9.4.11).

Problems (continued)

P 9.4.3. Assume that $f(x)$ is the ordinary generating function of the sequence $\{a_n\}_{n=0}^{\infty}$. Let

$$g(x) = \frac{f(x) + f(-x)}{2}.$$

Then $g(x)$ is the ordinary generating function for which sequence? Does a similar claim work for exponential generating functions?

P 9.4.4. Assume that $f(x)$ is the ordinary generating function of the sequence $\{a_n\}_{n=0}^{\infty}$. Let

$$g(x) = f(x^2).$$

Then $g(x)$ is the ordinary generating function for which sequence? Does a similar claim work for exponential generating functions? What about $h(x) = (f(x))^2$?

P 9.4.5. (a) Let n be a positive integer. Find the ordinary generating function for the following sequence:

$$\binom{n}{0}, 0, \binom{n}{2}, 0, \binom{n}{4}, 0, \ldots$$

(The sequence ends with $\binom{n}{n}$ or 0, depending on whether n is even or odd, respectively.)

(b) What is

$$\binom{11}{0} + \binom{11}{2} + \binom{11}{4} + \binom{11}{6} + \binom{11}{8} + \binom{11}{10}?$$

More generally, find

$$\sum_{k=0}^{\lfloor n/2 \rfloor} \binom{n}{2k},$$

the sum of every other term in one row of the Karaji–Jia triangle.

P 9.4.6. Assume that f is the ordinary generating function for the sequence $\{a_n\}_{n=0}^{\infty}$. What is the ordinary generating function for the sequence

$$0, a_1, 0, a_3, 0, a_5, \cdots?$$

P 9.4.7. Let n be a positive integer. Find the exponential generating function for the following sequence:

$$0, (n)_1, 0, (n)_3, 0, (n)_5, 0, \ldots$$

(The sequence ends with 0 or $(n)_n$, depending on whether n is even or odd, respectively.)

P 9.4.8. If f is the ordinary generating function for the sequence $\{a_n\}_{n=0}^{\infty}$, then $\dfrac{f}{1-x}$ is the ordinary generating function for which sequence?

P 9.4.9. Consider the sequence of partial sums of the harmonic series:

$$s_1 = 1, s_2 = 1 + \frac{1}{2}, s_3 = 1 + \frac{1}{2} + \frac{1}{3}, \ldots, s_n = 1 + \frac{1}{2} + \cdots + \frac{1}{n}.$$

Let $s_0 = 0$ and find the ordinary power series (in closed form) for $\{s_n\}_{n=0}^{\infty}$.

P 9.4.10. Let $\{F_n\}_{n=0}^{\infty}$ denote the Pingala–Fibonacci sequence. Recall that $F_0 = 0$, $F_1 = 1$, and, for $n \geq 2$, $F_n = F_{n-2} + F_{n-1}$. Let $g(x) \overset{\text{ops}}{\longleftrightarrow} \{F_n\}_{n=0}^{\infty}$. In Example 9.34, we showed (see Equation 9.2) that $g(x) = \frac{x}{1-x-x^2}$.

(a) What is the ordinary generating function for the sequence

$$F_0, F_0 + F_1, F_0 + F_1 + F_2, \ldots, F_0 + F_1 + \cdots + F_n, \ldots?$$

(b) What is the ordinary generating function for the sequence $\{F_{n+2} - 1\}_{n=0}^{\infty}$?

(c) Are the two generating functions from the two previous parts related? How?

(d) Is there any relation to Problem *P 1.2.7*?

P 9.4.11. **Exponential Generating Function for the Number of Derangements.** Let n be a non-negative integer, and let D_n denote the number of derangements of $[n]$ (Definition 8.22).

(a) Let $a_n = (-1)^n \frac{1}{n!}$. What is the ordinary power series for $\{a_n\}_{n=0}^\infty$?

(b) Using Theorem 8.24 and Problem P 9.4.8, find the ordinary power series of $\left\{ \frac{D_n}{n!} \right\}_{n=0}^\infty$.

(c) What is a closed form for the exponential generating function for $\{D_n\}_{n=0}^\infty$?

9.4.2   A Generating Function Approach to Partitions

Let n be a non-negative integer. In Chapter 7, we defined a partition of n with k parts to be a sequence $\lambda_1, \lambda_2, \ldots, \lambda_k$ of k positive integers such that $\sum_{i=1}^k \lambda_i = n$, and $\lambda_1 \geq \lambda_2 \geq \cdots \geq \lambda_k > 0$ (Definition 7.2). We used the notation $p_k(n)$ to denote the number of partitions of n into k parts, and $p(n) = \sum_{k=0}^n p_k(n)$ denoted the total number of partitions of n (into any number of parts). For example, 6 has 11 partitions, as follows:

$$(6), \ (5,1), \ (4,2), \ (3,3), \ (4,1,1), \ (3,2,1), \ (2,2,2), \ (3,1,1,1),$$
$$(2,2,1,1), \ (2,1,1,1,1), \ (1,1,1,1,1,1).$$

So $p(6) = 11$. In Chapter 7, we found a recurrence relation for $p_k(n)$ (Theorem 7.6), and used Ferrers diagrams (Definition 7.9) to find interesting connections between different types of integer partitions. Generating functions provide a powerful third method for gleaning information about partitions. We have already seen in Corollary 9.32 that, for a fixed n, $p(n)$ is the coefficient of x^n in the power series expansion of

$$\frac{1}{(1-x)(1-x^2) \cdots (1-x^n)}.$$

This is *not quite* the generating function for $p(0), p(1), p(2), \ldots, p(n), \ldots$, since, for $m > n$, we need to have more factors in the denominator. Note that $\dfrac{1}{1-x^m} = 1 + x^m + x^{2m} + \cdots$. As a result, if $m > n$, and if we include additional terms of the form $(1 - x^m)$ in the denominator, then there is no effect on the coefficient of x^n in the power series expansion, and that coefficient remains $p(n)$. For the same reason, the coefficient of x^k for $k \leq n$ in the above product is also $p(k)$. (Also see Problem P 9.2.12.) Thus, to actually get the whole sequence $p(0), p(1), \ldots, p(n), \ldots$ as the coefficients of a power series, we need an infinite product. We record the result here. (In Problem P 9.4.16, you are asked to reassemble the various pieces of the proof.)

Proposition 9.51. *Let*

$$f(x) = p(0) + p(1)x + p(2)x^2 + \cdots + p(n)x^n + \cdots$$

be the ordinary generating function for the sequence $\{p(n)\}_{n=0}^{\infty}$ *of partition numbers. Then*

$$f(x) = (1 + x + x^2 + \cdots)(1 + x^2 + x^4 + \cdots)(1 + x^3 + x^6 + \cdots)\cdots$$
$$= \frac{1}{(1-x)(1-x^2)(1-x^3)\cdots}.$$

Since we have not considered infinite products before, Problem P 9.4.14 asks you to contemplate the nature of these. We then turn to applications. Consider two types of partitions: partitions of n into *distinct* parts, and partitions of n into *odd* parts.

Example 9.52. Let $n = 6$. We have seen that 6 has 11 partitions altogether. From among these, only the following do not have a repeating part:

$$(6),\ (5,1),\ (4,2),\ (3,2,1).$$

Hence, the number of partitions of 6 into distinct parts is 4. On the other hand, only the following partitions of 6 have only odd numbers as the parts:

$$(5,1),\ (3,3),\ (3,1,1,1),\ (1,1,1,1,1,1).$$

So, the number of partitions of 6 into odd parts is also 4.

Is it a coincidence that the number of partitions of 6 into distinct parts is the same as the number of partitions of 6 into odd parts? The answer is not at all obvious, and the power of generating functions is at full display as you prove the following in Problem P 9.4.17:

Theorem 9.53. *Let n be a positive integer; then the number of partitions of n into odd parts is the same as the number of partitions of n into distinct parts.*

Finally, we return to the truly amazing Euler's pentagonal number theorem. Recall Definition 7.23, of a generalized pentagonal number. You will combine the previous version, Proposition 7.24, and generating functions to prove the following:

Theorem 9.54 (Euler's Pentagonal Number Theorem, Revisited). *Let $n > 1$ be an integer. Then*

$$\prod_{k=1}^{\infty}(1 - x^k) = 1 - x - x^2 + x^5 + x^7 - x^{12} - x^{15} + \cdots,\ and$$
$$p(n) = p(n-1) + p(n-2) - p(n-5) - p(n-7) + p(n-12) + p(n-15) + \cdots,$$

where, in both identities, the sequence $1, 2, 5, 7, 12, 15, \ldots$ is the sequence of non-zero generalized pentagonal numbers.

The unusual recurrence relation for $p(n)$ does provide a reasonably efficient method for calculating $p(n)$ for small values of n.

P 9.4.12. Let n be a non-negative integer, and let a_n denote the number of partitions of n where the parts are 1, 2, or 5. (By this definition or by convention, $a_0 = 1$.) Is a_n the same as the number of ways of paying n dollars using one-, two-, and five-dollar bills? Find the ordinary generating function for $\{a_n\}_{n=0}^{\infty}$.

P 9.4.13. Let m be a fixed non-negative integer. For a non-negative integer n, let $s(n)$ be the number of partitions of n into parts, where all the parts are less than or equal to m. What is the ordinary generating function for $s(0), s(1), \ldots, s(n), \ldots$?

P 9.4.14. In this problem, we are concerned with the following question: "Is an infinite product of power series well defined? What about an infinite product of polynomials?" In other words, if we write down an infinite product of power series, will the result always be a power series with coefficients that are finite numbers? (As usual, we are not concerned with issues of convergence.)

   (a) Is the product $(1 + x)(1 + x) \cdots$, where the term $(1 + x)$ appears an infinite number of times, well defined? In particular, if this product makes sense, then what would the resulting constant term and the coefficient of x be?

   (b) Is the product $(1 + x)(1 + x^2) \cdots (1 + x^n) \cdots$ well defined? For finding the coefficient of x^{47}, which terms are relevant?

   (c) Let $P_1, \ldots, P_n, \ldots$ be an infinite sequence of power series. Assume that, for all positive integers i, the constant term of P_i is 1, and that, for each positive integer k, there is an x^k term in only a finite number of these power series. Prove that the infinite product $\prod_{i=1}^{\infty} P_i$ is well defined.

P 9.4.15. Show that the following integers are equal:

   (a) The number of integer solutions to

$$x_1 + 2x_2 + \cdots + 47x_{47} = 47,$$

   where, for $1 \le i \le 47, 0 \le x_i \le 1$

   (b) The number of partitions of the integer 47 into distinct parts

   (c) The coefficient of x^{47} in the finite product

$$(1 + x)(1 + x^2) \cdots (1 + x^{47})$$

   (d) The coefficient of x^{47} in the infinite product

$$\prod_{n=1}^{\infty} (1 + x^n).$$

P 9.4.16. Generating Function for Partitions. Let n be a non-negative integer, and let $p(n)$ denote the total number of partitions of n. Further, let $f(x)$ be defined by

$$f(x) = (1 + x + x^2 + \cdots)(1 + x^2 + x^4 + \cdots)(1 + x^3 + x^6 + \cdots) \cdots$$

$$= \frac{1}{(1-x)(1-x^2)(1-x^3) \cdots}.$$

Proposition 9.51 states that $f(x)$ is the ordinary generating function of the sequence

$$p(0),\ p(1),\ p(2),\ \ldots,\ p(n),\ \ldots$$

Write a self-contained proof of this result.

P 9.4.17. Distinct Parts versus Odd Parts: Proof of Theorem 9.53.

(a) Let $q(n)$ denote the number of partitions of n into *distinct* parts, and let $Q(x) \overset{\text{ops}}{\longleftrightarrow} \{q(n)\}_{n=0}^{\infty}$. Show

$$Q(x) = (1 + x)(1 + x^?)(1 + x^{??})(1 + x^{???}) \cdots$$

(b) Let $r(n)$ denote the number of partitions of n into *odd* parts, and let $R(x) \overset{\text{ops}}{\longleftrightarrow} \{r(n)\}_{n=0}^{\infty}$. Show

$$R(x) = \frac{1}{1-x} \frac{1}{1-x^?} \frac{1}{1-x^{??}} \cdots$$

(c) In $Q(x)$, write $1 + x^r$ as $\frac{1-x^{2r}}{1-x^r}$ and simplify. Compare with $R(x)$ and prove Theorem 9.53.

P 9.4.18. Proof of Euler's Pentagonal Number Theorem. Consider the infinite product

$$\prod_{n=1}^{\infty}(1 - x^n) = (1 - x)(1 - x^2) \cdots (1 - x^n) \cdots .$$

(a) Show that the coefficient of x^m in this product is $|\mathcal{E}| - |\mathcal{O}|$, where $\mathcal{E}$ is the set of partitions of m into an even number of distinct parts, and $\mathcal{O}$ is the set of partitions of m into an odd number of distinct parts.

(b) Using Proposition 7.24, prove the first identity of Theorem 9.54:[6]

$$\prod_{n=1}^{\infty}(1 - x^n) = 1 - x - x^2 + x^5 + x^7 - x^{12} - x^{15} + x^{22} + x^{26} - \cdots .$$

Are the coefficients always 0, 1, or -1? What integers appear as exponents of terms with a non-zero coefficient? Which ones of these have a -1 as a coefficient? What are the next few terms in the expansion?

(c) Would you agree with the following?

$$\prod_{n=1}^{\infty}(1 - x^n) = \sum_{k=-\infty}^{\infty} (-1)^k x^{\frac{k(3k-1)}{2}} .$$

---

[6] This proof of the theorem is due to Franklin (1881). For an exposition of Euler's original two proofs, see Andrews 1983 and Andrews and Bell 2012.

P 9.4.19. **Proof of Euler's Pentagonal Number Theorem, Continued.** Let n be a positive integer, and consider the trivial identity

$$\prod_{n=0}^{\infty} \frac{1}{1-x^n} \prod_{n=0}^{\infty}(1-x^n) = 1.$$

As usual, let $p(n)$ denote the number of partitions of the integer n. Using Problems P 9.4.16 and P 9.4.18, rewrite the identity as

$$\left(\sum_{n=0}^{\infty} p(n)x^n\right)\left(1 - x - x^2 + x^5 + x^7 - x^{12} - x^{15} + x^{22} + x^{26} - \cdots\right) = 1.$$

(a) The coefficient of x^{11} on the right-hand side is zero. What is the coefficient of x^{11} on the left-hand side? What do you conclude?

(b) Prove the second identity of Theorem 9.54. In other words, prove the recurrence relation

$$p(n) = p(n-1) + p(n-2) - p(n-5) - p(n-7)$$
$$+ p(n-12) + p(n-15) + \cdots,$$

where $p(m) = 0$ if $m < 0$, and the sequence $1, 2, 5, 7, 12, \ldots$ is the sequence of non-zero generalized pentagonal numbers.

(c) Starting with the data in Table 7.2, use the recurrence relation from the previous part to compute $p(13)$.

P 9.4.20. We want to find $p(17)$, the total number of partitions of the integer 17.

(a) Using just the recurrence relation of Theorem 7.6, how would we find $p(17)$? Describe a method for doing so, but do not carry it out.

(b) Using the recurrence relation for $p(n)$ given in Euler's pentagonal theorem, find $p(17)$.

(c) Use a symbolic algebra software (such as SageMath, Maple, or Mathematica) to find $p(17)$ using a generating function.

P 9.4.21. **On Triangles with Integer Sides and a Given Perimeter.**[7] Let $n \geq 3$. Let $\mathcal{T}(n)$ be the set of triples (a, b, c) where a, b, and c are positive integers with $a + b + c = n$, and $a \leq b \leq c < a + b$. $((a, b, c) \in \mathcal{T}(n)$ if and only if there is a triangle in the plane with sides equal to a, b, and c. See Problem P 7.1.24.) For $n \geq 3$, let $t(n) = |\mathcal{T}(n)|$, and, for $0 \leq n \leq 2$, define $t(n) = 0$. Let $f(x) \overset{\text{ops}}{\longleftrightarrow} \{t(n)\}_{n=0}^{\infty}$.

(a) Define a function $g \colon \mathbb{R}^3 \to \mathbb{R}^3$ by $g\begin{bmatrix} a \\ b \\ c \end{bmatrix} = \begin{bmatrix} b - a \\ a + b - c - 1 \\ c - b \end{bmatrix}$. Find an inverse for this function, and conclude that g is 1-1 and onto.

(b) Let $\mathcal{R}(n)$ be the set of triples (α, β, γ), where α, β, and γ are non-negative integers with $2\alpha + 3\beta + 4\gamma = n - 3$. Show that $\mathcal{R}(n)$ is the image of $\mathcal{T}(n)$ under the map g. (In other words, if you apply g to elements of $\mathcal{T}(n)$, you will get precisely the elements of $\mathcal{R}(n)$.)

---

[7] Problems P 9.4.21 and P 9.4.22 are adapted from Bindner and Erickson 2012.

(c) Prove that, for $n \geq 3$, $t(n)$ is equal to the number of those partitions of $n - 3$ where the parts are 2, 3, or 4.

(d) Find a closed form for $f(x)$, the ordinary function for $t(0), t(1), \ldots, t(n), \ldots$

P 9.4.22. Continuing with the definitions and notation of Problem P 9.4.21, let $t(n)$ be the number of positive integer triples (a, b, c) where $a \leq b \leq c < a + b$ and $a + b + c = n$. Use the generating function for $\{t(n)\}_{n=0}^{\infty}$ to find a recurrence relation for $t(n)$.

9.4.3 Generating Functions for Stirling and Bell Numbers

In Problem P 9.1.3 and Example 9.37, we considered ordinary generating functions for Stirling numbers of the first and second kind. Here, we mostly focus on exponential generating functions. Recall, from Section 6.2, that $\begin{bmatrix} n \\ k \end{bmatrix}$ – a Stirling number of the first kind – is the number of permutations of n with exactly k cycles (or, equivalently, the number of ways of seating n people around k non-empty identical circular tables). Recall also, from Section 6.1, that $\begin{Bmatrix} n \\ k \end{Bmatrix}$ – a Stirling number of the second kind – denotes the number of ways that we can partition the set $[n] = \{1, \ldots, n\}$ into k non-empty parts. In addition, $B(n) = \sum_{k=0}^{n} \begin{Bmatrix} n \\ k \end{Bmatrix}$ is called a Bell number. Let $S_k(x)$ to be the exponential generating function for the kth *column* of the table of Stirling numbers of the second kind, Table 6.3. The first non-zero entry in each column is the diagonal entry, and so, for $k \geq 0$,

$$S_k(x) = \sum_{n=k}^{\infty} \frac{\begin{Bmatrix} n \\ k \end{Bmatrix} x^n}{n!}.$$

Thus, $S_0(x) = 1$; as an example, $S_4(x)$ begins with $\frac{1}{4!} x^4 + \frac{10}{5!} x^5 + \frac{65}{6!} x^6 + \frac{350}{7!} x^7 + \cdots$. In Problem P 9.4.23, you are asked to prove that, for $k \geq 0$,

$$S_k(x) = \frac{1}{k!} \left(e^x - 1 \right)^k.$$

The proof is interesting because it uses a theorem from differential equations that certain initial-value differential equations have unique solutions. If two sequences of numbers have the same initial value and satisfy the same recurrence relation, then they are automatically equal to each other. The corresponding situation for continuous functions is not as straightforward. There certainly are differential equations (the continuous version of recurrence relations) with initial values that have multiple solutions. Hence, just showing that two functions satisfy the same differential equation and the same initial value does not necessarily mean that they are the same function.

Next, we let $G(x)$ be the exponential generating function for the Bell numbers. In other words,

$$G(x) = B(0) + B(1)x + \frac{B(2)x^2}{2!} + \cdots + \frac{B(n)x^n}{n!} + \cdots.$$

We will use a simple differential equation and get assistance from Problems P 6.1.23 and P 9.3.17 to show, in Problem P 9.4.25, that

$$G(x) = e^{e^x - 1}.$$

It is possible – although not particularly efficient – to use these generating functions and a symbolic algebra software to find actual values of Stirling numbers or Bell numbers. We can also use these functions to prove further properties of these numbers. I mention one such application that we will not pursue here. In analysis one can prove that the radius of convergence of a power series $a_0 + a_1 x + a_2 x^2 + \cdots + a_n x^n + \cdots$ is given by $\lim \sup_{n \to \infty} \dfrac{1}{\sqrt[n]{|a_n|}}$.
(Do not worry if you don't what lim sup is.) We can also prove that the radius of convergence for $G(x)$ is ∞. Using these two facts – neither of which we have proved – and a bit more of standard analysis, we get that

$$\lim_{n \to \infty} \sqrt[n]{\frac{B(n)}{n!}} = 0,$$

which can be reinterpreted as saying that $\frac{B(n)}{n!}$ goes to zero faster than any function of the form c^n, for $0 < c < 1$ a constant. The point is that, using generating functions, one can learn about asymptotic behavior – i.e., the behavior as n gets larger and larger – of sequences of numbers.

In Problem P 9.4.26, we turn to Stirling numbers of the first kind, and find – again using derivatives – a closed form for their exponential generating functions. We use it (Problem P 9.4.27) to give an alternative proof of the identity of Problem P 6.2.15.

Finally, in Problem P 9.4.28, we find a relation for the ordinary generating function of the sequence of numbers in one *row* of the Stirling numbers of the second kind. We use this relation, in Problem P 9.4.29, to prove that this sequence of numbers is unimodal.

Problems (continued)

P 9.4.23. Let k be a non-negative integer, and let $S_k(x)$ be the exponential generating function for the kth column of the table of Stirling numbers of the second kind (the initial column in the table is numbered 0):

$$S_k(x) = \sum_{n=k}^{\infty} \frac{\left\{ {n \atop k} \right\} x^n}{n!}.$$

(a) Show that, for $k \geq 1$,

$$S_k'(x) = k S_k(x) + S_{k-1}(x). \tag{9.3}$$

(b) For $k \geq 0$, define

$$g_k(x) = \frac{1}{k!} (e^x - 1)^k.$$

Show that, for $k \geq 1$, $g_k(x)$ satisfies the same differential equation as Equation 9.3. In other words, for $k \geq 1$, $g'_k(x) = kg_k(x) + g_{k-1}(x)$.

(c) Let $k \geq 1$, and consider the differential equation $y' = ky + f(x)$, where $f(x)$ is a continuous function of x. Then a theorem from differential equations says that there is a *unique* function, defined on a closed interval containing zero, that satisfies this differential equation *and* the initial condition $y(0) = 0$. Use this theorem to argue that, for $k \geq 0$,

$$S_k(x) = \frac{1}{k!}(e^x - 1)^k.$$

P 9.4.24. Let n and m be non-negative integers with $m \leq n$. We have an alphabet consisting of m distinct letters. Let $h(n,m)$ be the number of those n-letter words from this alphabet that use each of the m letters at least once.

(a) What is $h(n,m)$? (Answer with no recourse to generating functions.)

(b) Fix m, and let $G(x) = \sum_{n=0}^{\infty} h(n,m)\frac{x^n}{n!}$ be the exponential generating function for $\{h(n,m)\}_{n=0}^{\infty}$. Find a closed form for $G(x)$.

(c) By comparing $G(x)$ with the result of Problem P 9.4.23, give a different derivation of $h(n,m)$.

(d) Explain how we can retrieve the result for entry 9 of the counting table using exponential generating functions.

P 9.4.25. For n a non-negative integer, let the Bell number $B(n)$ denote the total number of partitions of the set $[n]$, and let $G(x)$ be the exponential generating function of $\{B(n)\}_{n=0}^{\infty}$. In other words,

$$G(x) = B(0) + B(1)x + \frac{B(2)}{2!}x^2 + \cdots + \frac{B(n)}{n!}x^n + \cdots.$$

Prove that

$$G(x) = e^{e^x - 1}.$$

You may find the following steps helpful:

STEP 1: The function $e^x G(x)$ is the exponential generating function for which sequence? Problem *P 9.3.17* can be helpful in writing the answer as a finite sum.

STEP 2: Use Problem P 6.1.23 to simplify your answer to Step 1.

STEP 3: Find $G'(x)$ and use Step 2 to prove that $G'(x) = e^x G(x)$.

STEP 4: Solve the differential equation $G'(x) = e^x G(x)$ with the initial condition $G(0) = 1$, to complete the proof.

P 9.4.26. Let k be a non-negative integer, and let $F_k(x)$ be the exponential generating function for the kth column of the table of Stirling numbers of the first kind (the initial column in the table is numbered 0):

$$F_k(x) = \frac{\begin{bmatrix} k \\ k \end{bmatrix}}{k!}x^k + \frac{\begin{bmatrix} k+1 \\ k \end{bmatrix}}{(k+1)!}x^{k+1} + \cdots + \frac{\begin{bmatrix} n \\ k \end{bmatrix}}{n!}x^n + \cdots.$$

(a) What is $F_0(x)$? Find a closed form for $F_1(x)$.

(b) Using the recurrence relation for the Stirling numbers of the first kind, prove that

$$F'_k(x) = \frac{F_{k-1}(x)}{1-x}.$$

(c) Use the previous part to find a closed form for $F_2(x)$.

(d) Repeat and find a closed form for $F_3(x)$.

(e) Make a conjecture for the closed form of $F_k(x)$.

(f) Use induction to prove your conjecture.

P 9.4.27. Let k be a non-negative integer, and, as in Problem P 9.4.26, let $F_k(x) = \sum_{n=k}^{\infty} \frac{\left[{n \atop k}\right]}{n!} x^n$. In Problem P 9.4.26, you proved that $F'_k(x) = \frac{F_{k-1}(x)}{1-x}$. Use this fact, and the result of Problem P 9.4.8, to find an identity involving the Stirling numbers of the first kind. Is your identity related to the identity of Problem P 6.2.15?

P 9.4.28. For the most part, we have considered generating functions for each *column* of the table of Stirling numbers. In Problem P 9.1.3, the ordinary generating function for one row of the Stirling numbers of the first kind was investigated. Here, we consider the ordinary generating function for one row of the Stirling numbers of the second kind. Define $g_0(x) = 1$, and for $n \geq 1$, let

$$g_n(x) = \left\{{n \atop 0}\right\} + \left\{{n \atop 1}\right\} x + \cdots + \left\{{n \atop k}\right\} x^k + \cdots + \left\{{n \atop n}\right\} x^n.$$

Prove that, for $n \geq 1$,

$$g_n(x) = x\left(g_{n-1}(x) + g'_{n-1}(x)\right).$$

P 9.4.29. **Unimodality of Stirling Numbers of the Second Kind.**[8] Let n be a non-negative integer. As in Problem P 9.4.28, let $g_n(x) = \sum_{k=0}^{n} \left\{{n \atop k}\right\} x^k$ be the ordinary generating function for the nth row of the table of Stirling numbers of the second kind.

(a) Let $H_n(x) = g_n(x)e^x$. Prove that

$$H_{n+1}(x) = xH'_n(x).$$

(b) For $n = 0, 1, 2$, show that $g_n(x)$ has n real, distinct, non-positive roots.

(c) What is $\lim_{x \to -\infty} H_n(x)$?

(d) Let k be a positive integer. Use Rolle's theorem and the previous part to argue that if $H_n(x)$ has k real, distinct, non-positive roots, then $H'_n(x)$ will have at least k real, distinct, negative roots.

(e) Let k be a non-negative integer. Show that if $g_n(x)$ has k real, distinct, non-positive roots, then $H_{n+1}(x)$ has $k+1$ such roots.

(f) Using induction on n, prove that, for n a non-negative integer, $g_n(x)$ has n real, distinct, non-positive roots.

---

[8] Adapted from Harper 1967.

(g) For n a non-negative integer, are all the roots of $g_n(x)$ real, distinct, and non-positive?

(h) Use Theorem 5.32 to prove that the numbers in any row of the table of Stirling numbers of the second kind are log concave and hence unimodal.

9.4.4 Bernoulli Numbers

Consider the function

$$F(x) = \frac{x}{e^x - 1}.$$

The function $F(x)$ is the exponential generating function for a sequence of numbers B_0, B_1, ..., B_n, ... called the *Bernoulli numbers*.[9] In other words,

$$\frac{x}{e^x - 1} = B_0 + B_1 x + \frac{B_2}{2!}x^2 + \cdots + \frac{B_n}{n!}x^n + \cdots .$$

The Bernoulli numbers can be calculated using a recurrence relation (see Problem P 9.4.32):

$$B_n = -\frac{1}{n+1}\left[\binom{n+1}{n-1}B_{n-1} + \binom{n+1}{n-2}B_{n-2} + \cdots + \binom{n+1}{0}B_0\right].$$

These numbers are called Bernoulli numbers since Jacob Bernoulli (1654–1705) found – without proof – a nice formula for summing the kth powers of the first n integers using these numbers.[10]

Theorem 9.55. *Let k and n be positive integers. Then*

$$1^k + 2^k + \cdots + (n-1)^k = \sum_{i=0}^{k}\binom{k}{i}B_i\frac{n^{k+1-i}}{k+1-i}.$$

In Problem P 9.4.35, you are asked to give a proof – following Nunemacher and Young (1987) – for Theorem 9.55. The Bernoulli numbers come up in many other situations as well. Another famous such occurrence – which we will not pursue here – is due to Euler. For an integer $s > 1$, define the infinite series $\zeta(s) = \sum_{n=1}^{\infty}\frac{1}{n^s} = \frac{1}{1^s} + \frac{1}{2^s} + \frac{1}{3^s} + \cdots$. ($\zeta$ is called the Riemann zeta function.) Then, if s is an *even* integer, we have

$$\zeta(s) = (2\pi)^s\frac{|B_s|}{2(s!)}.$$

As an example, since $B_2 = 1/6$, we have that

$$\frac{1}{1^2} + \frac{1}{2^2} + \cdots + \frac{1}{n^2} + \cdots = \frac{\pi^2}{6}.$$

---

[9] Named after Jacob Bernoulli (1655–1705), but also studied independently and at about the same time by the Japanese mathematician Seki Takakazu (1642–1708).

[10] An alternative approach – using the inverse of the Karaji–Jia triangle – for summing kth powers of the first n integers was given in subsection 5.4.2.

Problems (continued)

P 9.4.30. A function $f(x)$ is called an *even* function if $f(-x) = -f(x)$ for all real numbers x. Show that if f is an even function, then in any power series expansion of f, the coefficients of the odd powers of x will be zero.

P 9.4.31. Let $F(x) = \frac{x}{e^x-1}$, and let $B_0, B_1, \ldots, B_n, \ldots$ be the sequence of Bernoulli numbers. Prove that $F(x) + \frac{x}{2}$ is an even function (see Problem P 9.4.30). Conclude that $B_1 = -\frac{1}{2}$ and $B_n = 0$ if n is an odd integer greater than 1.

P 9.4.32. Let $B_0, B_1, \ldots, B_n, \ldots$ be the sequence of Bernoulli numbers. Prove that $B_0 = 1$ and, for $n \geq 1$,

$$B_n = -\frac{1}{n+1} \sum_{k=0}^{n-1} \binom{n+1}{k} B_k.$$

You may find the following steps helpful:

STEP 1: Rewrite $\frac{x}{e^x-1} = a_0 + a_1 x + \cdots + a_n x^n + \cdots$ as $(e^x - 1)(a_0 + a_1 x + \cdots + a_n x^n + \cdots) = x$, and replace e^x with its power series expansion.

STEP 2: Multiply out the product of the two series from the previous step. What should the constant term, the x term, and more generally the x^n term be so that the product is equal to x?

STEP 3: Prove that $a_0 = 1$ and, for $n > 1$,

$$a_{n-1} = -\frac{1}{2!} a_{n-2} - \frac{1}{3!} a_{n-3} - \cdots - \frac{1}{n!} a_0.$$

STEP 4: How is B_n related to a_n? You can now complete the proof.

P 9.4.33. Use the recurrence relation of Problem P 9.4.32 to find $B_0, \ldots, B_{10}$.

P 9.4.34. Let $B_0 = 1, B_1, \ldots, B_n, \ldots$ be the sequence of Bernoulli numbers. Prove that, for $n \geq 2$,

$$\sum_{k=0}^{n-1} \binom{n}{k} B_k = 0, \quad \text{and} \quad B_n = \sum_{k=0}^{n} \binom{n}{k} B_k.$$

P 9.4.35. **Proof of Theorem 9.55.** Let k and n be positive integers, and let $B_0 = 1, B_1, \ldots, B_n, \ldots$ be the sequence of Bernoulli numbers. Prove that

$$1^k + 2^k + \cdots + (n-1)^k = \binom{k}{0} B_0 \frac{n^{k+1}}{k+1} + \binom{k}{1} B_1 \frac{n^k}{k} + \cdots + \binom{k}{k} B_k \frac{n^1}{1}.$$

You may find the following steps useful.[11]

STEP 1: What is the coefficient of x^k in the power series expansion of $\frac{e^{nx}-1}{x}$?

STEP 2: Recall that $\frac{x}{e^x-1}$ is the exponential generating function of Bernoulli numbers. Use Step 1 to find the coefficient of x^k in the power series expansion of

---

[11] Adapted from Nunemacher and Young 1987.

$$\frac{e^{nx} - 1}{x} \frac{x}{e^x - 1}.$$

STEP 3: Note that

$$1 + e^x + e^{2x} + \cdots + e^{(n-1)x} = \frac{e^{nx} - 1}{x} \frac{x}{e^x - 1},$$

and use it to give a different expression for the coefficient of x^k in the power series expansion of $\frac{e^{nx}-1}{x} \frac{x}{e^x-1}$.

STEP 4: Compare your answers in Steps 2 and 3, and complete the proof.

P 9.4.36. Use Theorem 9.55 and Problem P 9.4.33 to find formulas for

$$1^4 + 2^4 + \cdots + n^4$$

and

$$1^5 + 2^5 + \cdots + n^5.$$

9.4.5   Summing the Reciprocals of the Ming–Catalan Numbers

We first defined the ubiquitous Ming–Catalan numbers in subsection 5.1.5, and subsequently, in subsection 5.4.3, you explored the connection of these numbers with NE lattice paths. Recall that, for n a non-negative integer, we defined (Definition 5.17) the nth Ming–Catalan number c_n by

$$c_n = \frac{1}{n+1}\binom{2n}{n}.$$

The first few terms of the sequence are 1, 1, 2, 5, 14, 42, 132, 429, 1430, 4862, ..., and these numbers appear in a multitude of combinatorial situations. In the upcoming Mini-project 7, you will explore a few of these, and you will also find and use the ordinary generating function for these numbers. Here, following Koshy and Gao (2012), we consider the ordinary generating function for the sequence of the reciprocals of the Ming–Catalan numbers:

$$1, 1, \frac{1}{2}, \frac{1}{5}, \frac{1}{14}, \frac{1}{42}, \cdots$$

We will see a plausible closed formula for this generating function – the interested reader is urged to consult Koshy and Gao (2012) for the details of the derivation – and use it to show the following:

Proposition 9.56. *If $\{c_n\}_{n=0}^{\infty}$ denotes the sequence of the Ming–Catalan numbers, then*

$$1 + 1 + \frac{1}{2} + \frac{1}{5} + \frac{1}{14} + \frac{1}{42} + \cdots + \frac{1}{c_n} + \cdots = 2 + \frac{4\sqrt{3}}{27}\pi.$$

Problems (continued)

P 9.4.37. Let $\{c_n\}_{n=0}^{\infty}$ be the sequence of Ming–Catalan numbers. Use the ratio test for the convergence of series to show that the series

$$1 + 1 + \frac{1}{2} + \frac{1}{5} + \frac{1}{14} + \frac{1}{42} + \cdots + \frac{1}{c_n} + \cdots$$

converges.[12]

P 9.4.38. Let $f(x) \overset{\text{ops}}{\longleftrightarrow} \left\{\frac{1}{c_n}\right\}_{n=0}^{\infty}$, where c_n is the nth Ming–Catalan number. Using the recurrence relation $c_n = \frac{4n-2}{n+1} c_{n-1}$ of Proposition 5.18, prove that

$$x(x-4)f'(x) + 2(x+1)f(x) = 2.$$

P 9.4.39. As in Problem P 9.4.38, let $f(x) \overset{\text{ops}}{\longleftrightarrow} \left\{\frac{1}{c_n}\right\}_{n=0}^{\infty}$, where c_n is a Ming–Catalan number. Define

$$h(x) = 1 + \frac{x}{4-x} + \frac{6x}{(4-x)^2} + \frac{24\sqrt{x}\,\arcsin(\frac{\sqrt{x}}{2})}{(4-x)^{5/2}}.$$

(a) Use a computer algebra system to verify that $h(x)$ satisfies the same differential equation as $f(x)$. In other words, show that

$$x(x-4)h'(x) + 2(x+2)h(x) = 2.$$

(b) Show that $h(0) = f(0)$.

(c) Assuming that $h(x)$ is the closed form of the generating function $f(x)$ (see Koshy and Gao 2012 for the precise proof), use it to prove Proposition 9.56.

(d) Approximate $1 + 1 + \frac{1}{2} + \frac{1}{5} + \frac{1}{14} + \frac{1}{42} + \frac{1}{132} + \frac{1}{429}$ and compare it to $2 + \frac{4\sqrt{3}}{27}\pi$. Is the series converging quickly?

P 9.4.40. As in Problem P 9.4.38, let $f(x) \overset{\text{ops}}{\longleftrightarrow} \left\{\frac{1}{c_n}\right\}_{n=0}^{\infty}$, where c_n is a Ming–Catalan number.

(a) Use the differential equation satisfied by $f(x)$ (see Problem P 9.4.38) and Proposition 9.56 to find $f'(1)$.

(b) What is the sum of the series

$$\frac{1}{c_1} + \frac{2}{c_2} + \frac{3}{c_3} + \cdots + \frac{n}{c_n} + \cdots?$$

9.4.6 Generating Functions and Laplace Transforms

Given a sequence of numbers, we can construct both its ordinary and its exponential generating functions. How are these two functions related? Since we want to think of

---

[12] Problems P 9.4.37–P 9.4.40 are adapted from Koshy and Gao 2012.

these generating functions as functions, we will have to consider issues of convergence. In Problem P 9.4.42, you are asked to prove the following.

Proposition 9.57. *Let $a_0, a_1, \ldots, a_n, \ldots$ be a sequence of real numbers. Let*

$$A(x) \overset{\text{ops}}{\longleftrightarrow} \{a_n\}_{n=0}^{\infty}, \text{ and } E(x) \overset{\text{egf}}{\longleftrightarrow} \{a_n\}_{n=0}^{\infty}.$$

Assume that $a_0 + a_1 x + \cdots + \frac{a_n}{n!} x^n + \cdots$ converges to $E(x)$ for all x. Furthermore, assume that $a_0 + a_1 x + \cdots + a_n x^n + \cdots$ converges to $A(x)$ for $-R < x < R$ (where R is a positive real number). Then, for any real number s with $|s| < R$, we have

$$A(s) = \int_0^{\infty} E(st) e^{-t} \, dt.$$

Continuing with the integration theme, we consider Laplace transforms. Given a function $f(t)$ of a real variable t, the Laplace transform of f is defined to be

$$F(s) = \int_0^{\infty} f(t) e^{-st} \, dt,$$

if the improper integral converges. (The Laplace transform is very useful when s is allowed to be a complex number, but here we limit ourselves to real numbers s.) In Problem *P 9.4.43*, you are asked to explore the relation of the Laplace transform to the ordinary power series of a sequence.

Problems (continued)

P 9.4.41. Verify Proposition 9.57 for the infinite sequence of all 1's: $1, 1, \ldots, 1, \ldots$

P 9.4.42. Prove Proposition 9.57. You may find the following steps helpful:

STEP 1: Start with the definition of $A(s) = \sum_{n=0}^{\infty} a_n s^n$. Rewrite the nth term in the sum as $a_n \left(\dfrac{s^n}{n!} \right) n!$, and use the Gamma function of Problem P 1.3.18 to replace the second $n!$ by an integral.

STEP 2: Can you write $A(s)$ as an infinite sum of integrals? After doing so, switch the sum and the integral to write $A(s)$ as one (improper) integral of an infinite sum.

STEP 3: Every term of the infinite sum has an e^{-t} factor that can be factored out. Use the definition of $E(st)$ to complete the proof.

P 9.4.43. **The Laplace Transform.** Let $f(t)$ be a function of a real variable t, and assume that $F(s) = \int_0^{\infty} f(t) e^{-st} \, dt$, the Laplace transform of f, exists.

(a) $F(s)$ is the integral of a function, and we can approximate integrals by Riemann sums. Use rectangles of width 1 and appropriate height to approximate $F(s)$ by an infinite sum.

(b) In your approximation, do a change of variable, and let $s = -\ln(x)$. The result is the ordinary generating function of which sequence of numbers?

(c) Does it make sense to think of Laplace transforms as the continuous version of ordinary generating functions?

9.4.7 The Completed Counting Table

Now that we have completed "the counting table" (see the completed table following Chapter 9), the reader is advised to review the whole table.

Problems (continued)

P 9.4.44. Io, Europa, Ganymede, Callisto, Amalthea, and Himalia are six of my friends, and I have 10 copies each of Mary Shelley's *Frankenstein*, Jane Austen's *Pride and Prejudice*, Charlotte Brontë's *Jane Eyre*, and Virginia Woolf's *Mrs. Dalloway*, for a total of 40 books. Later, I clone Europa twice, so that there are three identical Europas if I need them (clones are identical to each other).

Find each of the following:

(a) the number of ways of giving a book to each of my six original friends so that among the six books distributed there is at least one copy of each type of book

(b) the number of ways of distributing the 10 copies of *Mrs. Dalloway* among my original six friends such that each of them gets at least one book

(c) the number of ways of organizing my six original friends into four teams (each team has to have at least one member)

(d) the total number of ways to seat my six friends around four round tables with no restriction on the number of persons at a given table; for example, they could all sit at one table and leave the other three tables empty, but even in this configuration there are a number of different seatings possible

(e) the number of ways of giving the 10 copies of *Mrs. Dalloway* to the 3 (cloned) Europas in a way that each gets at least one book

(f) the number of ways of arranging 10 of the books in a row so that the number of *Frankenstein*s is even and the number of *Jane Eyre*s is at least two

(g) the total number of ways of arranging 10 of the books in a row

(h) the number of ways of arranging three *Frankenstein*s, four *Jane Eyre*s, and three *Mrs. Dalloway*s in a row

(i) the number of ways of choosing 10 of the books so that the number of *Pride and Prejudice*s is even and the number of *Mrs. Dalloway*s is at least two

(j) the total number of ways of choosing 10 books

(k) the total number of ways of choosing 13 books.

Each of the following numbers is equal to one of the above 11. Which is equal to which?

(l) the number of ways of putting 10 identical balls into 3 identical boxes such that each box gets at least one ball

(m) the number of ways of placing 6 distinct balls into 4 distinct boxes such that each box gets at least one ball

(n) the number of non-negative integer solutions to $x_1 + x_2 + \cdots + x_6 = 4$

(o) the coefficient of x^{10} in $\dfrac{x^2}{(1-x)^4(1+x)}$

(p) the coefficient of $\dfrac{x^{10}}{10!}$ in $e^{2x}\dfrac{e^x + e^{-x}}{2}(e^x - 1 - x)$.

P 9.4.45. Diocletian, Constantine, Julian, Zeno, Justinian, Heraclius, and Irene are seven emperors of the Eastern Roman Empire, and each has minted coins with their own image on it. Assume that the coins of a particular ruler are all identical. Noosha, Shooka, and Borna are three of my friends.

(a) If you had exactly one coin from each ruler, in how many ways could you distribute the coins between my friends in such a way that each friend gets at least one coin?

(b) If you had 10 identical coins from Justinian, in how many ways could you distribute the coins between my friends in such a way that each friend gets at least 1 coin?

(c) If you had four coins from Julian and six from Constantine, in how many ways could you arrange them in a row in such a way that no two coins from Julian are next to each other?

(d) I have 10 coins from Heraclius and I want to put them into 3 piles (each pile should have at least 1 coin in it, otherwise it wouldn't be much of a pile). In how many ways can I do this?

(e) I have 10 coins from each of the 7 emperors. I want to arrange five of them in a row in such a way that the arrangement includes at least one coin from Diocletian and one coin from Zeno. How many different arrangements are possible?

(f) I have plenty of coins from each of the seven emperors. I want to pick 47 coins but I don't want any from Constantine or Justinian, at most 4 from Diocletian, at most 2 from Zeno, and a multiple of 5 from Irene. (I can have any number of coins from Julian or Heraclius). In how many ways can this be done?

(g) I have plenty of coins from each of the seven emperors. I want to arrange 11 coins in a row but I don't want any from Zeno, Julian, or Heraclius, and I want an even number from Irene, at most two from Diocletian, and any number from Constantine or Justinian. In how many ways can this be done?

P 9.4.46. For each entry of the completed counting table following Chapter 9, explain what kinds of problems fall in that category and how they could be solved.

Interregnum: Counting Table Completed

S & $T \rightarrow$ $u_i \downarrow$	T distinct S distinct	T distinct S identical	T identical S distinct	T identical S identical
$0 \le u_i \le 1, t \ge s$	$\boxed{1}$ $(t)_s$ Theorem 4.6	$\boxed{2}$ $\binom{t}{s}$ Theorem 4.21	$\boxed{3}$ 1 Theorem 4.23	$\boxed{4}$ 1 Theorem 4.23
$u_i \ge 0$	$\boxed{5}$ t^s Theorem 4.27	$\boxed{6}$ $\left(\!\binom{t}{s}\!\right)$ Theorem 4.40	$\boxed{7}$ $\sum_{k=0}^{t} \left\{{s \atop k}\right\}$ Theorem 6.10	$\boxed{8}$ $\sum_{k=0}^{t} p_k(s)$ Theorem 7.5
$u_i \ge 1$	$\boxed{9}$ $t!\left\{{s \atop t}\right\}$ Theorem 6.9	$\boxed{10}$ $\left(\!\binom{t}{s-t}\!\right)$ Theorem 4.45	$\boxed{11}$ $\left\{{s \atop t}\right\}$ Definition 6.4	$\boxed{12}$ $p_t(s)$ Theorem 7.5
$1 \le i \le t, 0 \le$ $u_i \le n_i, n_i \in \mathbb{Z}^{>0}$	$\boxed{13}$ Multinomials Exponential Generating Functions Theorem 4.33 Corollary 9.48	$\boxed{14}$ Inclusion– Exclusion Theorem 8.17		
$u_i \in N_i \subset \mathbb{Z}^{\ge 0}$ for $i = 1, \ldots, t$	$\boxed{15}$ Multinomials Exponential Generating Functions Theorem 4.34 Theorem 9.47	$\boxed{16}$ Ordinary Generating Functions Theorem 9.29		
	Permutations $\uparrow$	Combinations $\uparrow$		

Balls (S) and Boxes (T)
$|S| = s, |T| = t$
The ith box contains u_i balls.

Collaborative Mini-project 7: Ming–Catalan Numbers

In this mini-project, you will explore a remarkable sequence of integers, which appear in myriad combinatorial situations. This sequence is (almost universally) known as the *Catalan numbers*, and we already introduced them in Definition 5.17 of Section 5.1.5. See Chapter 5 footnote 2 for the rationale for calling them the *Ming–Catalan numbers*. (These numbers were also the subject of the optional Sections 5.4.3 and 9.4.5. This mini-project is not dependent on the material of those sections.)

Preliminary Investigations

Please complete this section before beginning collaboration with other students.

Three Sequences

In Problems MP7.1–MP7.3, three sequences of numbers are introduced. You are asked to understand the definitions and generate the first few terms of these sequences.

MP7.1 Consider an expression constructed using the subtraction operation, such as $4 - 3 - 2 - 1$. Such an expression is ambiguous; in fact, different answers may result depending on the order in which the subtractions are performed. For instance, $4 - (3 - (2 - 1)) = 2$, whereas $(4 - 3) - (2 - 1) = 0$. In general, let $*$ denote some abstract, non-associative operation and consider the ambiguous expression

$$x_1 * x_2 * x_3 * \cdots * x_n$$

with n operands. Let $d(n)$ denote the number of ways to parenthesize this expression to make it unambiguous.
   We define $d(0) = 0$. Find $d(1), d(2), d(3)$, and $d(4)$.

MP7.2 Consider walks in the two-dimensional coordinate plane where each step is $U : (x, y) \to (x+1, y+1)$ or $D : (x, y) \to (x+1, y-1)$. Thus, for example, if you are at the point $(3, 4)$, you can go either to $(4, 5)$ or to $(4, 3)$. We let u_n be the number of ways that we can start at $(0, 0)$ and end at $(2n, 0)$ by staying above the x axis, and, except for $(0, 0)$ and $(2n, 0)$, *never meeting or crossing* the x-axis. We have $u_0 = 0$. Find u_1, u_2, u_3, u_4.

MP7.3 Consider a sequence

$$a_1, a_2, \ldots, a_{2n}$$

with $2n$ terms, that has the following properties:
(i) n of the terms are $+1$, and n of the terms are -1
(ii) $a_1 + \cdots + a_j > 0$ for $j = 1, 2, \ldots, 2n - 1$.
Let the number of such sequences be $e(n)$, and define $e(0) = 0$. Find $e(1)$, $e(2)$, $e(3)$, and $e(4)$.

Collaborative Investigation

Work on the rest of the questions with your team. Before getting started on the new questions, you may want to compare notes about the preliminary problems.

The Sequence $d(n)$

We will start by exploring the sequence $d(n)$ defined in Problem MP7.1. You are asked to find a recurrence relation for this sequence and to compute its generating function. From the generating function, you will get an actual formula for this sequence of numbers.

MP7.4 By considering the operation that is performed last, find and prove a recurrence relation of the form

$$d(n) = \sum_{k=0}^{n} d(?)d(?) \quad \text{for } n \geq 2.$$

MP7.5 Let $g(x)$ denote the (ordinary) generating function for

$$d(0), d(1), d(2), \ldots$$

Find a closed formula for $g(x)$ and use it to find a formula for $d(n)$.
    You may find the following steps useful.
Step 1: Write down the first few terms of $g(x)$. Calculate the first few terms of $g(x)^2$ and find a relationship between $g(x)$ and $g(x)^2$. Using the recurrence relation found in Problem MP7.4, prove the relationship.

Step 2: Solve the equation for $g(x)$ that you found in the last step. You should now have a (closed) formula for $g(x)$ in terms of x.

Step 3: We now want to find an actual formula for $d(n)$. The integer $d(n)$ is the coefficient of x^n in the Taylor series expansion of the generating function $g(x)$. Thus

$$d(n) = \frac{g^{(n)}(0)}{n!},$$

where $g^{(n)}(x)$ is the nth derivative of g at x. Find the first few derivatives of (the closed form of) g and look for a pattern for $g^{(n)}(x)$. Prove your pattern using induction.

Step 4: By using your formula for $g^{(n)}(x)$ and the fact that $d(n) = g^{(n)}(0)/n!$, find a formula for $d(n)$ and simplify it.

Ming–Catalan Numbers

We call the sequence of integers $d(n)$ the *Ming–Catalan numbers*. These numbers arise in a substantial number of seemingly diverse combinatorial settings.

MP7.6 Using your formula from Problem MP7.5, find the first ten Ming–Catalan numbers. We have already defined a sequence c_n called the Ming–Catalan numbers in Definition 5.17. How is the sequence c_n related to the sequence $d(n)$? By doing Problem P 5.1.28 (which asks for a proof of Proposition 5.18) and reading Corollary 5.21, find alternative formulas for $d(n)$.

The Sequence u_n

Now we will consider the sequence u_n defined in Problem MP7.2. The steps U and D are defined in that problem.

MP7.7 Find a formula for the sequence u_n of Problem MP7.2.

You may find the following steps useful. Whenever we speak of paths, we are only considering paths consisting of steps of type U or D.

STEP 1: Let h, k, m, n be integers with $h < m$ and $k \leq n$. Ignore the issue of meeting or crossing the x-axis, and find the total number of paths from (h,k) to (m,n).

HINT: If the number of U-steps is a and the number of D-steps is b, then find a and b in terms of h, k, m, and n.

STEP 2: Assume that k and n are positive integers. Consider a path from $A = (h,k)$ to $B = (m,n)$ that *meets* or *crosses* the x-axis. Let C be the first point on the path that is on the x-axis. Reflect the portion of the path from A to C about the x-axis to get a path from $A' = (h, -k)$ to B. (See Figure MP7.1.)

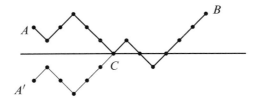

Figure MP7.1 Given a path from A to B, reflect the initial part of the path.

How are the following two numbers related? The number of paths from A to B that meet or cross the x-axis, and the number of all paths from A' to B. Write a proof for your conjecture.

STEP 3: Using the previous steps, find the number of paths from $(1, 1)$ to $(2n-1, 1)$ that cross or meet the x-axis. Use this to find the number of paths from $(1, 1)$ to $(2n - 1, 1)$ that do not cross or meet the x-axis.

STEP 4: How is your answer to the previous step related to the number of paths (using U and D steps) in the upper half plane from $(0,0)$ to $(2n,0)$ that do not cross or meet the x-axis (except of course at $(0,0)$ and $(2n,0)$)?

MP7.8 Is there a relationship between the sequence $e(n)$ defined in Problem MP7.3 and the sequence u_n? Can you come up with a proof for the relationship?

MP7.9 How are the sequence u_n and $e(n)$ of Problems MP7.2 and MP7.3 related to the Ming–Catalan numbers $d(n)$?

Mini-project Write-Up

In your write-up of the results, include at least the following:

- A definition of the Ming–Catalan numbers using a formula
- The definition of the sequences $d(n)$, u_n and $e(n)$, and a proof of their relationship with the Ming–Catalan numbers
- A recurrence relation, a generating function, and several equivalent formulas for the Ming–Catalan numbers (with proofs).

Collaborative Mini-project 8: Sperner's Theorem

In this mini-project, we consider the collection of all subsets of a finite set, and we focus on "inclusion" as the relation among these subsets. Given two subsets A and B of a set X, we may have that $A \subseteq B$ (i.e., elements of A are included among the elements of B), $B \subseteq A$, or neither. Studying the combinatorial properties of subsets of a finite set is an active area of research in combinatorics, and we will come back to this topic in Chapter 11.

Guiding Problem

Out of a group of 47 volunteers, you need to organize a large number of smaller groups – we will call them *committees* – for specific tasks. Your only requirement is that not only do you not want to use the same committee twice, but you don't want any one committee to be a subset of another committee. In other words, given any two committees A and B, you want at least one fresh voice who is a member of A but not B, and also at least one other person who is a member of B but not A. What is the largest number of such committees that you could create?

The Boolean Lattices

Let n be a non-negative integer. Recall that, for $n > 0$, $[n] = \{1, 2, \ldots, n\}$, is a set with n elements, and $[0] = \emptyset$. We denote the set of subsets of $[n]$ by $2^{[n]}$, and call it the *Boolean lattice of order n*.

For example, let $n = 3$. Then $[3]$ has eight subsets, and these are given in Figure MP8.1.

In Figure MP8.1, $2^{[3]}$ is drawn as a graph. In general, $2^{[n]}$ can be drawn as a graph where the vertices are the subsets of $[n]$, and we place larger subsets higher. Furthermore, we draw an edge from A to B, if $A \subset B$ and $|B| = |A| + 1$. Such a graph of $2^{[n]}$ is called the *Hasse diagram* of $2^{[n]}$ with the inclusion relation.

As another example, Figure 11.3 gives all the elements in $2^{[6]}$, the Boolean lattice of order 6. Within that figure, you can also find the Boolean lattices of order 2, 3, 4, and 5. Our guiding problem is a question about subsets of $[47]$, i.e., elements of $2^{[47]}$, the Boolean lattice of order 47.

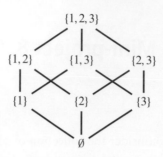

Figure MP8.1 The Boolean lattice of order 3, $2^{[3]}$.

Preliminary Investigations

Please complete this section before beginning collaboration with other students.

MP8.1 Draw a Hasse diagram akin to Figure MP8.1 for $2^{[5]}$. In other words, list all the subsets of [5], with $\emptyset$ at the bottom and $[5] = \{1, 2, 3, 4, 5\}$ at the top. Right above the empty set, and in the first "level," have all the subsets of size 1, followed by all the subsets of size 2 in the second level, and so on. Draw an edge between two subsets A and B if $A \subset B$ and $|B| = |A| + 1$.

The Levels of the Boolean lattice

Let $0 \leq k \leq n$ be non-negative integers. The *level k* of the Boolean lattice $2^{[n]}$ is the collection of all subsets of size k of [n].

The Profile of a Collection of Subsets

Let $\mathcal{A}$ be a collection of subsets of [n]. In other words, $\mathcal{A} \subseteq 2^{[n]}$. Such a collection is sometimes referred to as a *family* of subsets. The *profile* of $\mathcal{A}$ is the vector of integers

$$p(\mathcal{A}) = (p_0, p_1, \ldots, p_n),$$

where, for $0 \leq i \leq n$, p_i is the number of subsets of size i in $\mathcal{A}$.

MP8.2 Let $n = 6$, and let

$$\mathcal{A} = \{\{1, 2, 4, 5\}, \{1, 2, 4, 6\}, \{1, 2, 5, 6\}, \{1, 4, 5, 6\}, \{2, 4, 5, 6\}, \{3, 5, 6\}, \{1, 3\}, \{2, 3\}, \{3, 4\}\}.$$

What is $p(\mathcal{A})$, the profile of $\mathcal{A}$?

MP8.3 How many levels does $2^{[6]}$ have, and how many elements are on each of these levels?

We now name the kind of collections of subsets that our guiding problem called for.

Antichains

An *antichain* (or *Sperner family* or *clutter*) in $2^{[n]}$ is a family $\mathcal{A}$ of subsets of $[n]$ such that if A and B are in $\mathcal{A}$, then we have neither $A \subseteq B$ nor $B \subseteq A$.

Antichains get their name from chains, and we define these next.

Chains

A family $\mathcal{C} = \{C_1, C_2, \ldots, C_k\} \subseteq 2^{[n]}$ of subsets of $[n]$ is called a *chain of length $k - 1$* and *size k* if we have

$$C_1 \subset C_2 \subset C_3 \subset \cdots \subset C_k.$$

A chain of length n in $2^{[n]}$ is called a *maximal chain*. Such a chain, by necessity, starts with $\emptyset$, does not skip any levels, and ends at the full set $[n]$.

In $2^{[4]}$, as examples, $\{\{2, 3\}, \{1, 2, 4\}\}$ is an antichain, $\{\emptyset, \{2, 3\}, \{2, 3, 4\}\}$ is a chain of length 2, and $\{\emptyset, \{4\}, \{2, 4\}, \{1, 2, 4\}, \{1, 2, 3, 4\}\}$ is a maximal chain. Note that the length of a chain is the number of inclusions, and hence one less than the number of the subsets, in the chain. Our guiding problem can be translated to "What is the size of the largest antichain in $2^{[47]}$?"

MP8.4 Find a chain in $2^{[5]}$ consisting of three subsets of sizes 1, 3, and 4.

MP8.5 Find an antichain of size 10 in $2^{[5]}$.

MP8.6 Find an antichain of size 8 in $2^{[5]}$ that contains both a subset of size 2 and a subset of size 3. Do you think you can find such an antichain of size 9? (You don't have to give a proof.)

MP8.7 Let $\mathcal{A}$ be an antichain in $2^{[4]}$. Let $p(\mathcal{A}) = (p_0, \ldots, p_4)$ be the profile of $\mathcal{A}$. What is the maximum possible value for

$$p_0 + p_1 + p_2 + p_3 + p_4?$$

Make a conjecture.

Collaborative Investigation

Work on the rest of the questions with your team. Before getting started on the new questions, you may want to compare notes about the preliminary problems.

MP8.8 Let n be a positive integer. How many levels does $2^{[n]}$ have? How many elements are in each level?

MP8.9 Consider the set of all maximal chains in $2^{[n]}$. Show that there is a bijection between the maximal chains in $2^{[n]}$ and permutations of $[n]$. How many maximal chains does $2^{[n]}$ have?

MP8.10 Let $A = \{1, 2, 4, 6\} \subseteq [6]$. The subset A is an element of $2^{[6]}$. How many maximal chains of $2^{[6]}$ go through A? (We say that a chain *goes through* A if A is one of the sets in the chain.) More generally, let n be a positive integer and k an integer with $0 \leq k \leq n$. How many maximal chains of $2^{[n]}$ go through a particular subset $A \subseteq [n]$ if $|A| = k$?

MP8.11 Is the family of subsets in $\mathcal{A}$ in Problem MP8.2 an antichain? Can one maximal chain of $2^{[6]}$ go through more than one element of $\mathcal{A}$? What fraction of the maximal chains of $2^{[6]}$ go through some element of $\mathcal{A}$?

MP8.12 By looking at examples for small n, make a conjecture about the size of the largest antichain in $2^{[n]}$. (The correct conjecture is called *Sperner's theorem*.) Make a conjecture about the number of antichains of this maximum size. (Does it make a difference whether n is odd or even?)

MP8.13 Let $\mathcal{A}$ be an antichain in $2^{[n]}$. Let $p(\mathcal{A}) = (p_0, \ldots, p_n)$ be the profile of $\mathcal{A}$. By looking at some examples (for example, the family of subsets in $\mathcal{A}$ in Problem MP8.2), make a conjecture about

$$\sum_{i=0}^{n} \frac{p_i}{\binom{n}{i}}.$$

This result is known as the *LYM inequality* (or more accurately the *BLYM* inequality).[1]

REMARK: If $\mathcal{A}$ is a collection of subsets of $2^{[n]}$ and $p(\mathcal{A}) = (p_0, \ldots, p_n)$ is its profile, then for $0 \leq i \leq n$, the quantity

$$\frac{p_i}{\binom{n}{i}}$$

measures the fraction of elements at level i of the Boolean lattice that are in $\mathcal{A}$.

MP8.14 Generalize what you did in Problems MP8.10 and MP8.11. Let $\mathcal{A}$ be any antichain in $2^{[n]}$, and find an expression for the total number of maximal chains of $2^{[n]}$ that could go through members of $\mathcal{A}$. Your expression should be a sum and be a function of n p_0, p_1, ..., and p_n, where $p(\mathcal{A}) = (p_0, p_1, \ldots, p_n)$ is the profile of $\mathcal{A}$.

MP8.15 Prove your conjecture in Problem MP8.13, by first writing an inequality that compares the number of maximal chains through an antichain (as found in Problem MP8.14) with the total number of maximal chains in $2^{[n]}$. (Problem MP8.11 was a warm-up for this proof.)

MP8.16 Let $\mathcal{A}$ be an antichain. Using your conjecture of Problem MP8.13 – which you proved in Problem MP8.15 – add two inequalities to the expression below, and thus get a bound on the size of the largest antichain in $2^{[n]}$:

$$|\mathcal{A}| = \sum_{i=0}^{n} p_i = \binom{n}{\lfloor n/2 \rfloor} \sum_{i=0}^{n} \frac{p_i}{\binom{n}{\lfloor n/2 \rfloor}} \leq \text{??} \leq \text{??}$$

This result is called *Sperner's theorem*, and was possibly what you conjectured in Problem MP8.12.

MP8.17 We want the minimum number of chains in $2^{[n]}$ with the property that their union would include every subset of $[n]$. Does Sperner's theorem give a lower bound for this number? Make a guess (no proof needed) on whether the lower bound can always be achieved.

---

[1] Named after David Lubell (Lubell 1966), Kōichi Yamamoto (Yamamoto 1954), and Lev Mešalkin (Mešalkin 1963). Béla Bollobás (Bollobás 1965) also had a proof, but his initial is often left out.

Mini-project Write-Up

In your write-up of the results, include at least the following:

- The definition, with examples, of the Boolean lattices and their Hasse diagrams
- Definitions and examples of chains, antichains, and profile vectors
- A statement, an explanation, and a proof of the BLYM inequality for the Boolean lattices
- A theorem (with proof) that gives the size of the largest antichain in $2^{[n]}$
- A solution to the guiding problem on the number of subcommittees
- A conjecture about the number of antichains of maximum size
- A lower bound for the minimum number of chains needed to cover $2^{[n]}$.

10   Graph Theory

A soccer ball is often tiled with 12 pentagons and 20 hexagons. Is there anything special about those numbers? If you tile a soccer ball with pentagons and hexagons, what are the possibilities for the number of pentagons and the number of hexagons?

Graph theory is a vast subject – there are many texts devoted to it – and we have chosen a selection of topics to give you a flavor of elementary graph theory. After reviewing some of the basic vocabulary, we begin with graphic sequences.

We have given the basic definitions of graphs in Section 2.2. The reader is urged, before proceeding, to review the definitions of a graph, a subgraph, a vertex, an edge, a multigraph, a general graph, order, adjacency, incidence, degree, and graph isomorphism, as well as examples of paths, complete graphs, cycles, and complete bipartite graphs from Section 2.2 (which can be read independently of the rest of Chapter 2). In addition, the optional Mini-projects 5 and 6 are focused on graph theory and provide further experience with basic graph theory vocabulary and concepts. In lieu of repeating the basic definitions, we look at one example.

Example 10.1. Consider the graph G of Figure 10.1. A graph consists of a set of vertices and a set of edges. The set of vertices of G is the set $V = [10] = \{1, 2, \ldots, 10\}$. We say that G is a graph of *order* 10 since it has 10 vertices (see Definition 2.23). Intuitively (and visually), an edge is a direct connection between two vertices. More formally, an edge is a set consisting of a pair of vertices. The graph G has a total of 19 edges, and, for example, $\{1, 2\}$ is one edge, while $\{1, 6\}$ is not an edge. Note that $\{1, 2\}$ is a set, and so the order of the elements does not matter, and $\{1, 2\} = \{2, 1\}$. We denote the set of edges by E, write $G = (V, E)$, and we have

$$G = (V, E)$$
$$V = \{1, 2, 3, 4, 5, 6, 7, 8, 9, 10\}$$
$$E = \{\ \{1, 2\}, \{2, 3\}, \{3, 4\}, \{1, 5\}, \{4, 5\}, \{2, 6\}, \{5, 6\}, \{6, 7\}, \{2, 8\}, \{3, 8\},$$
$$\{4, 8\}, \{6, 8\}, \{7, 8\}, \{5, 9\}, \{6, 9\}, \{8, 9\}, \{5, 10\}, \{7, 10\}, \{8, 10\}\ \}\ .$$

This graph G is a *simple* graph (see Definition 2.17), since it does not have any double edges or loops. If G had repeated edges but no loops, then it would be called a *multigraph*, and if it had both repeated edges and repeated loops, then it would be a *general graph* (see Definition 2.20 and Remark 2.21). As an example of the vocabulary for graphs, $\{4, 8\}$ is an edge of G jointing 4 and 8, and so we say that 4 and 8 are *adjacent* vertices, while 4 and 9 are not adjacent. We also say that the vertex 4 is *incident* with the edge $\{4, 8\}$ (see Definition 2.23). The *degree* of a vertex is the number of edges incident with it (see Definition 2.24, and note that, in a

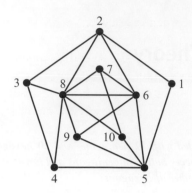

Figure 10.1 The graph $G = (V, E)$ of Example 10.1 with $|V| = 10$ and $|E| = 19$. The set of vertices $V = [10]$; three elements of E – the edges incident with the vertex 9 – are $\{5,9\}$, $\{6,9\}$; and $\{8,9\}$, and vertex 9 has degree 3.

general graph, a loop contributes 2 to the degree of its vertex). Vertex 4 is adjacent to vertices 3, 5, and 8, and so its degree is 3. For each of the vertices, we have recorded their degree in Table 10.1. In graph theory, we are not concerned with the labels of the vertices. So, if we keep the vertices and edges intact, but rearrange the names of the vertices (for example, switch 1 and 5 as well as replace 2 with 8, 7 with 2, and 8 with 7) we get a new graph that is basically the same graph as G. We say that these two graphs are *isomorphic* (see Definition 2.29).

Table 10.1 The degrees of the vertices of the graph
$G = (V, E)$ of Figure 10.1.

v	1	2	3	4	5	6	7	8	9	10
$\deg(v)$	2	4	3	3	5	5	3	7	3	3

For the record, we add a few oft-used terms to our vocabulary about graphs.

Definition 10.2 (Isolated Vertex, Leaf). Let G be a graph. A vertex with degree 0 is called an *isolated vertex*, while a vertex with degree 1 is called a *leaf*.

Definition 10.3 (Regular Graph). A graph is called *regular of degree d* or *d-regular* if each vertex has degree equal to d.

Definition 10.4 (Cubic Graph). A graph is called *cubic* if it is regular of degree 3 (i.e., if each vertex has degree equal to 3).

10.1 Graphic Sequences

Warm-Up 10.5. *Is there a simple graph with four vertices such that the degrees of the vertices are 3, 2, 1, and 1? What if the degrees were 2, 2, 1, 1?*

In Table 10.1, we recorded the degrees of all the vertices of the graph G of Figure 10.1. So, we know there exists a simple graph – namely, G – with 10 vertices where the degrees of those vertices are 7, 5, 5, 4, 3, 3, 3, 3, 3, and 2. This sequence is called the degree sequence of the graph G, and, in this section, we aim to prove a theorem that allows us to decide whether a given sequence of positive integers can be the sequence of the degrees of a graph without having to draw the graph.[1] First, we record the definition of a degree sequence, and the related graphic sequences.

Definition 10.6 (The Degree Sequence of a Graph; Graphic Sequences). The *degree sequence* of a graph is the list of the degrees of its vertices in non-increasing order.

A non-increasing sequence of non-negative integers is called *graphic* if there exists a simple graph whose degree sequence is precisely that sequence.

Example 10.7. The finite sequence 7, 5, 5, 4, 3, 3, 3, 3, 3, 2 is a graphic sequence since it is the degree sequence of the simple graph G of Figure 10.1. On the other hand, 1, 1, 1 is not a graphic sequence since it is not possible to construct a graph with three vertices so that all the three degrees are 1.

Question 10.8. Given a non-increasing sequence of non-negative integers, when is the sequence graphic? As an example, which of the following sequences are graphic?

(a) 5, 5, 4, 3, 3, 2, 2, 1
(b) 7, 5, 5, 4, 3, 2, 2, 0
(c) 6, 6, 6, 6, 4, 3, 3, 0
(d) 7, 6, 5, 5, 4, 4, 4, 2, 1.

In what follows, we will answer these questions with seemingly ad hoc arguments, but later we will present a general theorem that allows us to answer all such questions. First, we will record a useful but straightforward observation that was contained in the first paper on graph theory in the Western tradition, by Leonhard Euler (1741; translation in Biggs, Lloyd, and Wilson 1986). We can prove this result for general graphs, but note that a loop contributes 2 to the degree of its vertex.

Theorem 10.9. *Let $G = (V, E)$ be a general graph. Let $d_1, \ldots, d_p$ be the degrees of the vertices. Let $|E| = q$; then*

$$d_1 + d_2 + \cdots + d_p = 2q.$$

In particular, the number of vertices of G with odd degree is even.

Proof. Every edge contributes 2 to the sum of the degrees – 1 for each of its vertices – and hence the sum of the degrees is twice the number of edges.

---

[1] While it is not necessary, the reader may benefit from working through Problems MP5.1, MP5.2, and MP5.6 of Mini-project 5 before tackling this section.

If the number of vertices with odd degree was odd, then the sum of these odd degrees would be odd. The total sum of degrees would then be the sum of an odd and an even number, and hence odd. But the total sum of the degrees is twice the number of edges and hence even. The contradiction proves the claim.   □

Sequences of Question 10.8

(a) The sequence $5, 5, 4, 3, 3, 2, 2, 1$ is not graphic since the number of odd vertices is not even, and this contradicts Theorem 10.9.

(b) The sequence $7, 5, 5, 4, 3, 2, 2, 0$ does not contradict Theorem 10.9, and yet it is not graphic. It purportedly has eight vertices, and one of those vertices is not adjacent to any other vertex. Hence, the maximum degree of any of the other vertices can only be 6 (recall that we are not allowing double edges or loops).

(c) The sequence $6, 6, 6, 6, 4, 3, 3, 0$ has eight vertices, and one is not connected to any vertex, so the maximum degree of the others is 6. This means that four of the vertices are connected to all but the one isolated vertex. Thus the degree of every vertex (other than the isolated one) must be at least 4. Hence, the sequence is not graphic.

(d) A supposed graph with this degree sequence would have nine vertices. It is not clear how to proceed. No particular connection is forced a priori, and there seem to be many options. So, for the sake of argument, what if the vertex with degree 7 was *not* adjacent to the degree-1 vertex, but to all the others? If we take this vertex and all its edges out, we get the degree sequence $5, 4, 4, 3, 3, 3, 1, 1$. Again, there are many possibilities, and again, for the sake of argument, let us say that the degree 5 vertex is adjacent to the five other vertices with highest degree. Take this vertex and its edges out and we are left with the degree sequence $3, 3, 2, 2, 2, 1, 1$. Continuing in the same way, taking out the highest-degree vertex (and assuming that it is adjacent to the next-highest-degree vertices), we get $2, 1, 1, 2, 1, 1$, which we rearrange to get $2, 2, 1, 1, 1, 1$. Assuming that the two vertices of degree 2 are adjacent, and then removing one of them, we get $1, 1, 1, 1, 0$. The latter is certainly graphic. It is the disjoint union of two edges together with one isolated vertex. Now, we can work our way back, adding vertices one by one and, using our calculation as a guideline, connecting each new vertex to appropriate other vertices. We will get a graph with the original degree sequence. (See Figure 10.2.) This proves that the original sequence was graphic.

It is not at all clear that this method would always work. We made a lot of choices along the way. What if the final sequence that we arrived at was not graphic? Would we have to go back and make other choices (and, for example, assume that the high-degree vertex was connected to some of the vertices with smaller degree)?

We now prove that the algorithm used for the last sequence in Question 10.8 actually always works. Like many combinatorial theorems, one direction of the theorem is easy to prove, while the other direction requires some insight.

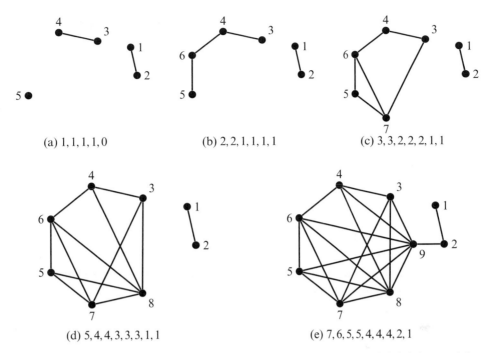

Figure 10.2 Starting with a simple graph with five vertices and degree sequence $1, 1, 1, 1, 0$, at each iteration, we add one vertex and judiciously connect it to other vertices to eventually get a simple graph with nine vertices and degree sequence $7, 6, 5, 5, 4, 4, 4, 2, 1$.

Theorem 10.10 (The Havel–Hakimi Algorithm).[2] *Consider the following two sequences of non-negative integers. Assume the first sequence is in non-increasing order and $t_s \geq 1$:*

(a) $s, t_1, \ldots, t_s, d_1, \ldots, d_n$
(b) $t_1 - 1, t_2 - 1, \ldots, t_s - 1, d_1, \ldots, d_n$.

Sequence (a) is graphic if and only if sequence (b) is graphic (after possibly rearranging it to make it non-decreasing).

Proof. ($\Leftarrow$) We first assume that sequence (b) is graphic. This means that there exists a graph G with this degree sequence. Just add a new vertex v to G, and connect v to the s vertices with degrees $t_1 - 1, \ldots, t_s - 1$ to get a new graph with degree sequence (a).

($\Rightarrow$) For the converse, assume that degree sequence (a) is graphic, and let H be a graph with this degree sequence. A vertex S in this graph has degree s, vertices $T_1, \ldots, T_s$ have degrees $t_1, \ldots, t_s$, respectively and vertices $D_1, \ldots, D_n$ have degrees $d_1, \ldots, d_n$ respectively. A priori, we do not know what the s vertices adjacent to S are. We will split the possibilities into two cases:

---

[2] Named after the Czech mathematician Václav J. Havel and the Iranian-American mathematician Seifollah L. Hakimi (1932–2005).

CASE 1: In the graph H, the vertex S is adjacent to all the vertices $T_1, \ldots, T_s$.

CASE 2: For some $1 \leq i \leq s$, in the graph H, the vertex S is not adjacent to the vertex T_i.

In the first case, the proof is straightforward. In the graph H, just remove the vertex S and all the edges incident with it, to get a graph with degree sequence (b).

For the second case, we will show that we can modify the graph H so that it will continue to have the same degree sequence as before, but with the vertex S adjacent to the vertex T_i. The vertex S has degree s and – since it is not adjacent to T_i – must be adjacent to a vertex D_j for some $1 \leq j \leq n$.

$$S \qquad\qquad T_i \qquad D_j$$

Figure 10.3 In H, the vertex S is not adjacent to T_i but is adjacent to D_j.

Since the degree sequence (a) is in descending order, we have that $t_i \geq d_j$. We split the possibilities in this case into two subcases:

SUBCASE 1: $t_i = d_j$.

SUBCASE 2: $t_i > d_j$.

For Subcase 1, we just switch the places of T_i and D_j, and we are back to the situation where S is connected to (the new) T_i.

$$S \qquad\qquad D_j \qquad T_i$$

Figure 10.4 For Subcase 1, we switch T_i and D_j.

In Subcase 2, we know that T_i has more neighbors than D_j and so there must be a vertex W that is adjacent to T_i but not adjacent to D_j. Hence – using dashed lines to show non-adjacency – we have the situation in Figure 10.5.

$$S \qquad T_i \qquad W \qquad D_j$$

Figure 10.5 In Subcase 2, W is adjacent to T_i but not to D_j.

Now, just remove the two edges $\{S, D_j\}$ and $\{T_i, W\}$, and add the two edges $\{S, T_i\}$ and $\{W, D_j\}$. The new graph – see Figure 10.6 – will have the same degree sequence as H and yet S is now connected to T_i.

If necessary, we repeat what we did for T_i for any other T that is not connected to S, and eventually we will be back to Case 1. This completes the proof. $\qquad\qquad\square$

Because of this theorem, if you are given a sequence of non-negative integers and want to know if it is graphic or not, you repeatedly use Theorem 10.10 to reduce your sequence to

$$S \qquad T_i \qquad W \qquad D_j$$

Figure 10.6 In H, remove $\{S, D_j\}$ and $\{T_i, W\}$, and add $\{S, T_i\}$ and $\{W, D_j\}$.

simpler sequences. If, at any point, the simpler sequence is graphic (or not graphic), then so is the original sequence.

Problems

P 10.1.1. A simple graph G has nine edges and the degree of each vertex is at least 3. What are the possibilities for the number of vertices? Give an example, for each possibility.

P 10.1.2. Is $7, 7, 6, 5, 4, 4, 4, 3, 2$ a graphic sequence?

P 10.1.3. Is there a simple regular graph (Definition 10.3) of degree 5 with eight vertices? Why?

P 10.1.4. Is there a simple graph on eight vertices where half of the degrees are 5 and the other half are 3?

P 10.1.5. The sequence $7, 5, 5, 4, 3, 3, 3, 3, 3, 2$ is graphic because it is the degree sequence of the simple graph of Figure 10.1. Apply the Havel–Hakimi algorithm of Theorem 10.10 to construct a graph with this degree sequence. Is the resulting graph isomorphic – see Definition 2.29 – to the graph of Figure 10.1?

P 10.1.6. Assume that you applied the Havel–Hakimi algorithm to a given sequence, and, at the end of the process, you arrived at a graphic sequence. You draw a simple graph corresponding to this final sequence, and work your way back to construct a simple graph with the original sequence as its degree sequence. As you work your way back, is it the case that, at every step, after adding a new vertex, you add edges between this new vertex and those existing vertices that have the highest degrees? Either prove that you do or provide an example where you don't.

P 10.1.7. Assume that you have a sequence d_1, d_2, d_3, d_4 (with $d_1 \geq d_2 \geq d_3 \geq d_4 \geq 0$) that you know is *not* graphic. You consider the sequence $d_1, d_2, d_3, d_4 + 1, 1$. Can this sequence be graphic (after possible rearranging so that it is in non-increasing order)? Either prove that it is never graphic or give an example where it becomes graphic.

P 10.1.8. A sequence is graphic if it is the degree sequence of a *simple* graph. Is there a sequence that is not graphic, and yet is the degree sequence of a multigraph? Either prove that there are no such sequences or give a specific example.

P 10.1.9. Can you find a sequence that is *not* the degree sequence of a multigraph, but is the degree sequence of a general graph?

P 10.1.10. Let $d_1, d_2, \ldots, d_n$ be a non-increasing sequence of n non-negative integers. By Theorem 10.9, if this sequence is the degree sequence of a general graph, then the sum $d_1 + d_2 + \cdots + d_n$ is even. What about the converse?

P 10.1.11. Find two non-isomorphic regular simple graphs of degree 3 and order 6.

P 10.1.12. Apply the Havel–Hakimi algorithm of Theorem 10.10 to the sequence $5, 4, 4, 2, 2, 1$. Is the sequence graphic? Can you use the Havel–Hakimi algorithm to find a general graph with this degree sequence that has exactly one loop?

P 10.1.13. (a) Prove that a sequence $d_1, d_2, \ldots, d_p$ is a graphic sequence if and only if the sequence
$$p - d_p - 1, \ldots, p - d_2 - 1, p - d_1 - 1 \text{ is graphic.}$$
(b) Is $9, 9, 9, 9, 9, 9, 9, 9, 8, 8, 8$ a graphic sequence?

P 10.1.14. I have a simple graph with six vertices. The degrees of five of the vertices are $5, 4, 4, 2$, and 2. What are the possibilities for the degree of the sixth vertex?

P 10.1.15. Can we have a simple graph where the degree sequence consists of all distinct integers? What about a multigraph?

P 10.1.16. Let G be a simple graph with 94 vertices. Assume that all the degrees of the vertices are odd integers. Prove that there are at least three vertices with the same degree.

P 10.1.17. Assume that $a_1 \geq a_2 \geq \cdots \geq a_n$ is a graphic sequence. Prove that, for $1 \leq k \leq n$,[3]

$$\sum_{i=1}^{k} a_i \leq k(k-1) + \sum_{i=k+1}^{p} \min(a_i, k).$$

P 10.1.18. Let p and t be positive integers with $2 \leq t < p$. Assume $a_1 \geq a_2 \geq \cdots \geq a_{t-1} > a_t \geq \cdots \geq a_{p-1} > a_p$ is the degree sequence of a simple graph. Prove that $a_1 \geq a_2 \geq \cdots \geq a_{t-1} \geq a_t + 1 \geq \cdots \geq a_{p-1} \geq a_p + 1$ is also the degree sequence of a simple graph.

P 10.1.19. (Kapoor, Polimeni, and Wall 1977) Does there exist a simple graph with 48 vertices where the set of the degrees of the vertices is $\{4, 7, 47\}$? In other words, we want all the degrees of the vertices to be either 4, or 7, or 47, and for there to be at least one vertex with each of these degrees. More generally, let $S = \{a_1, \ldots, a_k\}$ be any non-empty set of k positive integers with $a_1 < a_2 < \cdots < a_k$. Prove that there exists some simple graph G with $a_k + 1$ vertices, where the set of the degrees of the vertices is precisely S. (Note that we are not specifying the degree sequence of the graph, just the set of numbers that occur as degrees.) You may find the following steps helpful:

STEP 1: Assume $|S| = 1$, and $S = \{a_1\}$. Give an example of a simple graph where all the degrees are equal to a_1.

STEP 2: Construct a simple graph G with $p + q$ vertices as follows. Start with a complete graph K_p and q isolated vertices. Now add an edge between every isolated vertex and every vertex of K_p. What is the set of degrees of G?

STEP 3: Assume $|S| = 2$, and $S = \{a_1, a_2\}$ with $a_1 < a_2$. By a judicious choice of p and q in the previous step, give an example of a simple graph where the set of the degrees is precisely S.

---

[3] The Erdős–Gallai theorem asserts that this necessary condition, together with the (trivial) condition that the sum of the degrees be even, gives a sufficient condition for identifying graphic sequences. This theorem can be proved (see Choudum 1986) by induction on the sum of the degrees. Given a degree sequence satisfying the condition, you subtract 1 from the smallest degree and 1 from one of the other degrees (the smallest index t with $a_t > a_{t-1}$ or, if all the degrees are equal, then from a_{n-1}). After (tediously) showing that this new sequence also satisfies the Erdős–Gallai condition, you use the inductive hypothesis, and finish the proof using the result in Problem P 10.1.18.

STEP 4: Let p, q, r, and m be positive integers. Assume that H is a simple graph with r vertices, and that $S_1 = \{b_1, \ldots, b_m\}$ is the set of the degrees of H. Construct a simple graph G with $p + q + r$ vertices as follows. Start with a K_p (complete graph of order p), a copy of H, and q isolated vertices. Add an edge between every vertex in K_p and each of the other vertices. What is the set of degrees of the vertices of G?

STEP 5: To prove the general statement, induct on $|S|$. Using the inductive hypothesis, start with a graph H with degree set equal to $\{a_2 - a_1, a_3 - a_1, \ldots, a_{k-1} - a_1\}$, and judiciously choose p and q in Step 4.

10.2 Paths, Cycles, and Trees

Warm-Up 10.11. *Draw an example of a graph with degree sequence 3, 3, 2, 1, 1, 1, 1. How many vertices and how many edges does such a graph have? Do you think that it is possible for such a graph to have a triangle (three vertices with each adjacent to the other two)? Make a conjecture.*

In this section, after some preliminary but oft-used definitions, we give a number of equivalent characterizations of a special class of graphs called trees.[4]

Definition 10.12 (Walks, Trails, Paths, and Cycles). Let $G = (V, E)$ be a general graph. A sequence of m edges of the form

$$\{x_0, x_1\}, \{x_1, x_2\}, \ldots, \{x_{m-1}, x_m\}$$

is called a *walk of length m*, and this walk *joins the vertices* x_0 *and* x_m. The vertices $x_0, x_1, \ldots, x_m$ are called the vertices of the walk.

If $x_0 = x_m$, then the walk is *closed*. If $x_0 \neq x_m$, the walk is *open*.

If a walk has distinct edges, then it is a *trail*. A closed trail is called a *circuit*.

If a walk has distinct vertices (and hence also distinct edges), then it is a *path* or a *chain*.

If a walk has distinct vertices except $x_0 = x_m$ (and so it is a circuit with no repeated vertices), then it is called a *cycle*. (See Table 10.2 and Figure 10.7.)

Remark 10.13. The vocabulary in Definition 10.12 is widely used but is not standard. When you read a graph theory book, make sure that you know the appropriate definitions.

Remark 10.14. We defined a walk in terms of the edges in it, but we think of a walk as going from one vertex to another through an edge. In fact, in a general graph – where repeated edges are allowed – an expression of the form $\{x_2, x_3\}$ may be ambiguous since there may be more than one edge whose vertices are x_2 and x_3. In such cases, we have to give names such as e_5 and e_9 to these edges and use these names in any walk. We also often – and when there is no

---

[4] While not necessary, the reader may benefit from working through Problems MP5.3 and MP5.7 of Mini-project 5 on graphic sequences and planar graphs before reading this section.

ambiguity – just list the vertices of a walk. So instead of $\{x_0, x_1\}, \{x_1, x_2\}, \ldots, \{x_{m-1}, x_m\}$, we also sometimes write

$$x_0, x_1, x_2, \ldots, x_{m-1}, x_m, \text{ or } x_0 - x_1 - x_2 - \cdots - x_{m-1} - x_m$$

for the same walk. Finally, note that the *length* of a walk (or a trail, path, circuit, or cycle) is the number of *edges* in the walk. It is easy to be confused by the vocabulary in Definition 10.12. We summarize the expressions in Table 10.2 and give examples in Figure 10.7.

Table 10.2 Closed walks, circuits, and cycles begin and end at the same vertex. Walks and trails may or may not do so. Paths (or chains) do not repeat any vertices and so are always open.

	Repeated edges and vertices allowed	Repeated vertices allowed but no repeated edges	No repeated vertices or edges
Always open			path or chain
Could be open or closed	walk	trail	
Always closed	closed walk	circuit	cycle

Remark 10.15. In a general graph, a loop is a cycle of length 1, and a double edge is a cycle of length 2. For simple graphs, the length of the shortest cycle cannot be less than 3.

To enhance your familiarity with these concepts, we now prove a sequence of very useful results. In each case, the reader is urged to first construct some examples to illustrate the meaning of the statement, and then try to give a proof.

Lemma 10.16. *Let $G = (V, E)$ be a general graph, and let $x, y \in V$, with $x \neq y$. Assume that x and y are joined by a walk. Then they are joined by a path.*

Proof. Let $v_1, v_2, \ldots, v_n$ be among the vertices of G, and assume $x - v_1 - v_2 - \cdots - v_n - y$ is a walk joining x and y. Eliminate any portion between two occurrences of the same vertex (and one of the two offending vertices). You will obtain a path between x and y. □

Definition 10.17 (Connected Graph). A general graph is *connected* if, for each pair of distinct vertices, there is a walk joining them. Otherwise G is *disconnected*.

Let $G = (V, E)$ be a general graph, and $W \subseteq V$. If every pair of vertices in W have a walk joining them, and no vertex outside of W has a walk joining it to a vertex in W, then W is called a *connected component* of G.

Definition 10.18 (Bridge). Let G be a connected general graph, and e an edge of G. The edge e is a *bridge*, provided its removal from the graph leaves a disconnected graph. (See Figure 10.8.)

More generally in a (not necessarily connected) general graph, a bridge is an edge whose removal increases the number of connected components.

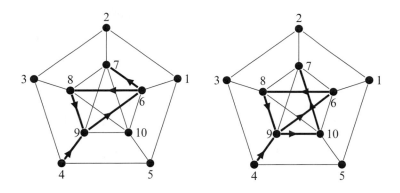

4–9–6–8–9–6–7 is a walk from 4 to 7   4–9–6–8–9–10–7 is a trail from 4 to 7

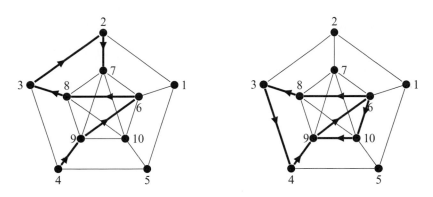

4–9–6–8–3–2–7 is a path or a chain from 4 to 7   4–9–6–10–9–6–8–3–4 is a closed walk

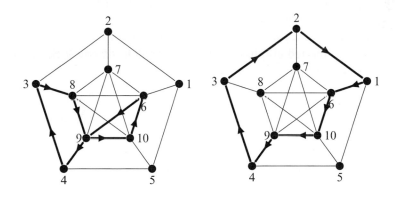

9–4–3–8–9–10–6–9 is a circuit   4–3–2–1–6–10–9–4 is a cycle

Figure 10.7 Every path is a trail and every trail is a walk, but not vice versa. Likewise, every cycle is a circuit and every circuit is a closed walk, but not vice versa.

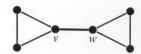

Figure 10.8 The edge $\{v, w\}$ is a bridge.

Lemma 10.19. *Let $G = (V, E)$ be a connected general graph, and $\alpha = \{x, y\} \in E$. Then α is a bridge if and only if no cycles of G contain α.*

Proof. ($\Rightarrow$) For one direction, assume that there was a cycle of G containing α. Then removing α will not make G disconnected since, in any walk using α, we can replace α with the rest of the cycle. Thus, if α is a bridge, then it cannot be part of a cycle.

($\Leftarrow$) For the other direction, assume α is not a bridge. Then $G - \{\alpha\}$ is connected, which means that there is a path from x to y (recall Lemma 10.16: if there is a walk, then there is a path). Add α to this path to get a cycle containing α. $\qquad\square$

Theorem 10.20. *Let $G = (V, E)$ be a connected general graph with $|V| = n$. Then $|E| \geq n - 1$. Also, there exist connected graphs $G = (V, E)$ with $|E| = |V| - 1$, and for such graphs, every edge is a bridge, and the graph is a simple graph.*

Proof. We know that G is connected and $|V| = n$. Start with n isolated vertices and then add edges one at a time. At the beginning, there are n connected components. Each edge decreases the number of connected components by at most 1. Hence, to get to one connected component, we need at least $n - 1$ edges.

The same argument shows that if a connected graph with n vertices has $n - 1$ edges, then, in the above process of adding edges one at a time – regardless of the order of inserting the edges – at every stage the number of connected components has to decrease by one. So, the deletion of each edge increases the number of connected components. Thus, every edge is a bridge. A loop or a double edge is not a bridge, and so a connected graph with n vertices and $n - 1$ edges must be a simple graph. A path of length $n - 1$ and a star with $n - 1$ edges are examples of connected graphs with n vertices and $n - 1$ edges. See Figure 10.9. $\qquad\square$

Figure 10.9 A path of length $n - 1$ and a star with $n - 1$ edges have n vertices.

Theorem 10.21. *Let $G = (V, E)$ be a connected simple graph. Then every edge of G is a bridge if and only if $|E| = |V| - 1$.*

Proof. ($\Leftarrow$) If we know that $|E| = |V| - 1$, then we already proved in Theorem 10.20 that every edge of G is a bridge.

($\Rightarrow$) For the other direction, assume that every edge of G is a bridge, and induct on $n = |V|$. For the base case, if $n = 1$, then we have one vertex and no edges, every edge is a bridge, and the number of edges is one less than $|V|$.

For the inductive step, let $n > 1$, and assume that the theorem is true for all simple graphs with less than n vertices. Now, let G be a simple graph with n vertices and m edges and assume that every edge is a bridge. We want to prove that $m = n - 1$. Let α be any edge of G, and let $G' = G \backslash \alpha$ be the graph that you get by removing the edge α. The graph G' has exactly the same vertices as G but one less edge. So G' has n vertices and $m - 1$ edges.

Since α was assumed to be a bridge, G' has two connected components G_1' and G_2'. Let k and ℓ be the number of vertices of G_1' and G_2', respectively. Then $k + \ell = n$ and both k, and ℓ are less than n. Every edge continues to be a bridge in G_1' and G_2', and each has less than n vertices. So, the inductive hypothesis applies to both graphs. As a result, G_1' and G_2' have $k - 1$ and $\ell - 1$ edges, respectively.

Thus, the number of edges of G – remembering α – is $m = (k-1)+(\ell-1)+1 = k+\ell-1 = n - 1$, and the proof is complete. $\qquad\square$

Definition 10.22 (Tree, Forest). Let $G = (V, E)$ be a simple graph. Then G is called a *forest* if G has no cycles. The graph G is called a *tree* if it has no cycles and is connected. (See Figure 10.10.)

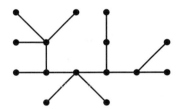

Figure 10.10 A tree is a connected graph with no cycles.

We are now ready for this section's main theorem, which gives various characterizations of trees. We have already proved many parts of this theorem, but bringing together the various statements could be helpful.

Theorem 10.23. *Let $G = (V, E)$ be a simple graph.*
  Then the following are equivalent:

(a) *G is a tree.*
(b) *Any two vertices of G are connected by a unique path.*
(c) *G is connected, and every edge is a bridge.*
(d) *G is connected, and $|E| = |V| - 1$.*
(e) *G has no cycles, and $|E| = |V| - 1$.*
(f) *G has no cycles, but if any edge is added to E, then the resulting graph will have a cycle.*

Proof. $(a \Rightarrow b)$ Let v and w be two vertices of G. Since G is connected, there is a walk from v to w. By Lemma 10.16, there is also a path from v to w. If there were two distinct paths from v

to w, then a cycle would exist. (As you go from v to w, the two paths possibly share a number of vertices before they part ways. Let x be the last vertex shared by the two paths such that all the previous vertices are shared by the two paths. The two paths will eventually come back together. Let y be the first time that happens. Then there is a cycle going from x to y on one path and coming back from y to x on the other.)

$(b \Rightarrow c)$ Since every two vertices are connected, the graph G is connected. It remains to show that every edge is a bridge. Let $\alpha = \{v, w\}$ be an edge, and assume α is not a bridge. Then $G\backslash\alpha$ is connected and hence there is a walk not containing α from v to w. So there is a path – not containing α – from v to w. We now have two distinct paths – one being just the edge α – that join v and w. This is a contradiction, and so α must be a bridge.

$(c \Leftrightarrow d)$ This was Theorem 10.21.

$(c \Rightarrow e)$ The fact that $|E| = |V| - 1$ follows from Theorem 10.21. The fact that there are no cycles follows from Lemma 10.19. (In a connected graph, an edge is a bridge if and only if there are no cycles through it.)

$(e \Rightarrow a)$ Since G has no cycles, every connected component is a tree. Let k be the number of connected components. We will be done when we show that $k = 1$. Now each connected component, being a tree, has one edge less than its number of vertices. So the total number of edges is $|V| - k$. But this is supposed to be $|V| - 1$. Hence $k = 1$.

$(b \Rightarrow f)$ Any two vertices on a cycle are connected by at least two distinct paths, and so G must have no cycles. If we add edge $\alpha = \{v, w\}$ to G, then α together with the unique path in G from v to w form a cycle.

$(f \Rightarrow a)$ We already know that G has no cycles and so we have to show that G is connected. Let v and w be two vertices in G. If there is an edge $\{v, w\}$, then there is certainly a walk from v to w. Otherwise, add this edge to the graph and get a cycle. v and w will be on the cycle and hence, in addition to this new edge, there is another path from v to w. The latter already existed before we added the edge and the proof is complete. □

Remark 10.24. In Mini-project 9, we will revisit trees to prove Cayley's formula on the number of labeled trees.

Problems

P 10.2.1. Let x and y be vertices of a general graph.
- (a) Suppose that there is a closed walk containing both x and y. Must there be a closed trail containing both x and y?
- (b) Suppose that there is a closed trail containing both x and y. Must there be a cycle containing both x and y?

P 10.2.2. If a vertex x of a graph is on a circuit, then is x also on a cycle?

P 10.2.3. Let $G = (V, E)$ be a general graph and let $x, y \in V$ with $x \neq y$. Assume that x and y are joined by a walk. In Lemma 10.16, we proved that x and y must be joined by a path. A friend objects to the proof given in the text. In that proof it was decreed: "Eliminate any portion between two occurrences of the same vertex (and one of the two offending vertices)." Your friend claims that this is an ambiguous proclamation since there may be multiple vertices occurring multiple times. It is not clear where to start and how to stop. Put your friend at ease by rewriting the proof using induction on the number of edges on the walk.

P 10.2.4. What is the maximum length of a cycle in the graph of Figure 10.1?

P 10.2.5. Consider the Petersen graph of Figure 2.1. What are the possible lengths of cycles in this graph? Make a conjecture. You don't have to prove your conjecture.

P 10.2.6. We defined a tree as a connected graph with no cycles (Definition 10.22). Could we have just as well defined a tree as a connected graph with no circuits? What about a connected graph with no closed walks?

P 10.2.7. I have a simple graph with five vertices and four edges. Could it be a tree? Does it have to be a tree?

P 10.2.8. Let G be a forest consisting of t trees. Assume G has n vertices. How many edges does G have?

P 10.2.9. Let G be a tree. Recall that a vertex v of G of degree 1 is called a leaf. Show that every finite tree with at least two vertices has at least two leaves.

P 10.2.10. Let G be a tree, and assume that G has one vertex of degree 4. What is the minimum number of leaves possible for G?

P 10.2.11. A mystery simple graph G has the degree sequence $3, 3, 3, 2, 1$. Prove that G has a cycle.

P 10.2.12. Assume that T is a tree with exactly two leaves. Prove that T is a path.

P 10.2.13. Assume that G is a tree with vertices $\{v_1, \ldots, v_{20}\}$. We know that there is not an edge between v_4 and v_{18}. The graph H is the same as the graph G except with one extra edge: $\{v_4, v_{18}\}$. What can you say about the number of cycles in H?

P 10.2.14. Let G be a connected simple graph with 47 vertices and 47 edges. What can you say about the number of cycles in G?

P 10.2.15. Let n be an integer greater than or equal to 2. Show that a simple graph (connected or not) with n vertices and at least n edges must have a cycle. What about general graphs?

P 10.2.16. We have a mystery simple graph $G = (V, E)$. We know that $|V| = 10$ and that the two largest degrees of the vertices of G are 5 and 4 (there may be other degree 4 vertices, but only one of degree 5). For each of the following scenarios, answer the question(s), and give the drawing of a possible graph satisfying the conditions.

(a) If G is a tree, then what are the possible degree sequences of the graph?

(b) Can G be a forest with two connected components? If so, what are the possibilities for $|E|$ and for the degree sequence of the graph?

(c) If G has two components and $|E| = 9$, then what are the possibilities for the number of cycles in G?

P 10.2.17. Let $n \geq 2$, and let $d_1 \geq d_2 \geq \cdots \geq d_n$ be a sequence of positive integers. Prove that this sequence is the degree sequence of a tree if and only if $\sum_{i=1}^{n} d_i = 2n - 2$.

10.2.1 Graph Vocabulary: Spanning Tree*

Not all graphs are trees, but each graph has subgraphs that are trees. A spanning tree of a connected graph G is a subgraph of G that is a tree and contains every vertex of G. In other words, a spanning tree of a graph G is a tree (and therefore connected) whose vertices are exactly the same as the vertices of G and whose edges are some of the edges of G. A spanning tree is a sort of skeleton for the graph G (see Figure 10.11). We record the formal definition.

Definition 10.25 (Spanning Tree). Let $G = (V, E)$ be a connected general graph. Let $E' \subseteq E$ be a set of edges such that the graph $T = (V, E')$ is a tree. The tree T is then called a *spanning tree* for G.

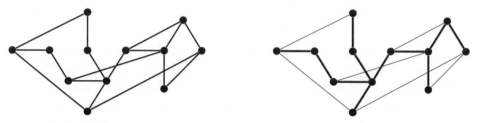

Figure 10.11 The dark edges on the right form a spanning tree for the graph on the left.

You are asked in Problem *P 10.2.21* (which uses Problem P 10.2.20) to prove that every connected general graph has at least one spanning tree.[5]

Problems (continued)

P 10.2.18. Let T be a tree. How many spanning trees can T have?

P 10.2.19. If T_1 and T_2 are both spanning trees of a graph G, then can T_1 and T_2 have different numbers of edges?

P 10.2.20. Let G be a connected general graph. Consider the set $\mathcal{S}$ consisting of all subgraphs of G that have all the vertices of G and are connected. (The set of edges of a graph in $\mathcal{S}$ is a subset of the edges of G.) Show that the following are equivalent:

    (a) T is a spanning tree of G.

    (b) $T \in \mathcal{S}$, and T has the smallest number of edges from among the elements of $\mathcal{S}$.

P 10.2.21. Let G be a connected general graph. Prove that G has a spanning tree.

P 10.2.22. Consider K_4, the complete graph of order 4.

    (a) How many non-isomorphic spanning trees does K_4 have?

---

[5] We will revisit spanning trees in Problems P 10.7.20 and P 10.7.21, where the tools of deletion and contraction (Definition 10.108, which can be read independently of the rest of that section) are used to help count the number of spanning trees in a graph.

(b) If the vertices of K_4 are labeled (so that they are distinguishable), then how many different spanning trees does K_4 have?

P 10.2.23. Find a spanning tree for the graph of Figure 10.1.

P 10.2.24. Find a spanning tree for the Petersen graph of Figure 2.1.

10.2.2 Graph Vocabulary: Adjacency Matrix, Distance, Diameter, Girth, and the Complement*

To discuss and compare graphs, it is helpful to have an appropriately rich vocabulary. Paths and cycles allow us to use geometrical notions of distance, diameter, and girth in graphs.

Definition 10.26 (Distance). Let G be a general graph, and u and v be two vertices of G. If there is a path in G from u to v, then the *distance* between u and v in G, denoted by $d(u,v)$, is the length of a shortest path from u to v in G. If there is no such path in G, then we define $d(u,v)$ to be infinite.

Definition 10.27 (Diameter of a Graph). Let G be a general graph. The *diameter* of G is the maximum distance between two vertices of G.

Definition 10.28 (Girth of a Graph). Let $G = (V, E)$ be a general graph. The *girth* of G is the number of edges in a shortest cycle of G. The girth of a forest is defined to be ∞.

Definition 10.29 (Complement of a Graph). Let G be a graph. We define a new graph called the *complement* of G (sometimes called the *inverse* of G) and denoted by $\overline{G}$. This new graph has the same vertices as G. Two vertices in $\overline{G}$ are adjacent if and only if they are not adjacent in G.

Definition 10.30 (Adjacency Matrix). Let G be a simple graph with n vertices, and number its vertices $1, 2, \ldots, n$ in some arbitrary manner. Let $A = [a_{ij}]$ be an $n \times n$ matrix such that the entry a_{ij}, in row i column j, is 1 if the vertices i and j are adjacent, and 0 otherwise. The matrix A is called the *adjacency matrix* of the graph G.

Remark 10.31. Note that the adjacency matrix of a simple graph is a symmetric (that is, $a_{ij} = a_{ji}$ for all $1 \leq i \leq j \leq n$) (0,1)-matrix with 0's on the diagonal. One can also define an adjacency matrix for a general graph. For such graphs, if $i \neq j$, then a_{ij} equals the number of edges joining the vertices i and j. For a general graph, the diagonal entries of the adjacency matrix are the number of loops at that vertex (sometimes, it is more convenient to have the diagonal entries be twice the number of loops).

Problems (continued)

P 10.2.25. In the graph of Figure 10.1, which vertex has the maximum distance from vertex 1.

P 10.2.26. What are the diameter and the girth of the graph of Figure 10.1?

P 10.2.27. What are the diameter and the girth of the Petersen graph of Figure 2.1?

P 10.2.28. Find the complement graphs of a path of length 3, a path of length 4, a cycle of length 4, and a cycle of length 5. Are any of these graphs "self-complementary"? In other words, if we denote the complement of a graph G by $\overline{G}$, is G isomorphic (see Definition 2.29) to $\overline{G}$ for any of these graphs?

P 10.2.29. Let G be a simple graph, and assume that the diameter of G is greater than 3. Can the diameter of $\overline{G}$, the complement of G, be greater than or equal to 3?

P 10.2.30. Give an example of a simple graph of girth 7 and diameter 3.

P 10.2.31. Let G be a simple graph with a cycle. Show that the girth of G is no more than two times the diameter of G plus 1.

P 10.2.32. Consider the graph of Figure 10.1. The adjacency matrix of this graph is a 10×10 matrix with its rows and columns indexed by the vertices of the graph. What is the second row of this matrix? What is its 8 column? Find the dot product of these two and explain its significance.

P 10.2.33. Let A be the adjacency matrix of a simple graph, G. Prove that the following integers are equal:

(a) The dot product of row i and row j of A

(b) The dot product of row i and column j of A

(c) The number of vertices in G that are adjacent to both vertices i and j

(d) The number of different paths of length 2 from vertex i to vertex j.

P 10.2.34. (a) Let A be the adjacency matrix of a simple graph. By looking at some examples, determine the significance of the entries in A^2.

(b) Generalize your result in the previous part.

P 10.2.35. Let G be a simple graph and A its adjacency matrix. Assume that I can easily find as many powers of A as I want: $A, A^2, A^3, \ldots, A^k, \ldots$ Can you find the diameter of G easily?

P 10.2.36. Let G be a simple graph, and let x and y be two vertices of G. Assume that $d(x, y)$, the distance between x and y, is 47. Further assume that $x = v_0 - v_1 - v_2 - \cdots - v_{47} = y$ is a path of length 47 from x to y. What can you say about $d(x, v_{23})$, the distance between x and v_{23}?

P 10.2.37. Let G be a simple connected graph of order n and diameter d. Let A be the $n \times n$ adjacency matrix of G. Prove that

$$\{I_n, A, A^2, \ldots, A^d\}$$

is a linearly independent set of matrices.[6]

---

[6] Using some more linear algebra, it follows that, if A has m distinct eigenvalues, then the diameter of G is less than or equal to $m - 1$. The linear algebra needed is beyond what we have been assuming, but, for the initiated, the argument goes as follows. The matrix A is a real symmetric matrix, and so it is diagonalizable. From this it follows that the minimum polynomial of A has distinct roots. But the degree of the minimum polynomial is the smallest exponent m with $I_n, A, \ldots, A^m$ linearly dependent. The result now follows from Problem P 10.2.37. These kinds of fascinating connections between linear algebra concepts and graph theoretic concepts are pursued under the rubric of "algebraic graph theory." See Biggs 1993 for much more.

10.2.3 Graph Vocabulary: Rooted Trees*

Let T be a tree. If we designate one vertex of T as the *root*, then the tree is called a *rooted tree*.

Note that any vertex can be the root of a tree. We just have to specify the root once and for all. The standard way to draw a rooted tree (see Figure 10.12) is to place the root at the top of the figure (you can think of this as a hanging tree). Then the vertices that are adjacent to the root are placed one level below the root, and the vertices (other than the root) that are adjacent to these are placed two levels below the root, and so on. The *level number* or *depth* of a vertex in the tree is the length of the (unique) path from the root to the vertex. Thus the root is at level 0, while the vertices adjacent to the root are at level 1, and so on. (See Figure 10.12.)

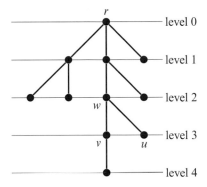

Figure 10.12 A rooted tree, with root r, of height 4. The nodes v and u are siblings and w is their parent.

Let T be a rooted tree with root r, and let v be a vertex of T other than r. Let w be the vertex such that $\{w, v\}$ is the last edge in the (unique) path from r to v. Then w is called the *parent* of v, and v is called a *child* of w.

Two vertices with the same parent are called *siblings*. Vertices with no children are called *leaves*, and vertices with children are called *internal* vertices (or internal nodes) of the tree. Let m be a positive integer. If each internal vertex of a rooted tree has at most m children, the tree is called an *m-ary tree*. 2-ary and 3-ary trees are more commonly called *binary* and *ternary* trees, respectively. (The tree of Figure 10.12 is an example of a ternary tree.) A *full m-ary tree* is an m-ary tree where each internal vertex has exactly m children.

The *height* of a rooted tree is the length of the longest path from the root to a leaf (or the largest level number of a vertex).

Problems (continued)

P 10.2.38. Redraw the tree of Figure 10.12 but with w as the root. What is the height of the new rooted tree?

P 10.2.39. We want to choose a different root s for the tree of Figure 10.12. Which choice of s would result in a rooted tree with the biggest height? Which choice would give the shortest height? For each of your choices, find the smallest integer m such that the (new) rooted tree is an m-ary tree.

P 10.2.40. Let T be a full ternary tree with n vertices. If $1 \le n \le 15$, then what are the possibilities for n?

P 10.2.41. Let T be a full ternary tree of height 4. What is the smallest and the largest number of vertices that T could have?

P 10.2.42. Let T be a full m-ary tree of height h. What is the minimum and maximum number of vertices possible for T?

P 10.2.43. Let T be a full binary tree.

(a) If T has 10 internal vertices, then what is the total number of vertices of T?
(b) If T has 11 vertices altogether, then how many of them are internal, and how many are leaves?

P 10.2.44. Let T be a full m-ary tree.

(a) If T has i internal vertices, then what is the total number of vertices of T?
(b) If T has a total of n vertices, then what is the number of internal vertices of T? What is the number of leaves of T?

P 10.2.45. Let T be a full binary tree with nine vertices. What are the possibilities for the height of T?

P 10.2.46. Let T be a full binary tree with n vertices. Let h be the height of T. Denote $\log_2(n)$ by $\lg(n)$, and show that

$$\lfloor \lg(n) \rfloor \le h \le \left\lfloor \frac{n}{2} \right\rfloor.$$

P 10.2.47. Let T be a (not necessarily full) binary tree with n vertices. Find a lower and an upper bound for the height of T in terms of n.

P 10.2.48. Let T be an m-ary tree of height h.

(a) Assume v is a leaf of T of depth less than h. If we add m children to v, then, compared to T, will the resulting tree have more or fewer leaves?
(b) What is the largest possible number of leaves for T?
(c) Assume T has ℓ leaves. Find a lower bound for h in terms of ℓ and m.

P 10.2.49. Let T be a full binary tree with five leaves. For each leaf, find its level number, and construct a multiset consisting of the five level numbers of the leaves. What are the possible multisets that you get? What is the average of the level numbers of the leaves for each possibility?

P 10.2.50. Let T be a binary tree with ℓ leaves. For each leaf, we find its level and then average these level numbers. Show that this average is at least $\lfloor \lg(\ell) \rfloor$.

10.3 Bipartite Graphs

Warm-Up 10.32. *Habib, Jamal, Kamal, Laila, Rana, and Sana are members of a club. You want to split them up into two groups of three, but you want to make sure that everyone will get to meet new people. You have asked each pair if they already know each other, and the information is organized in Figure 10.13. An edge in the graph means that the two vertices already know each other. What is the best way to split the group?*

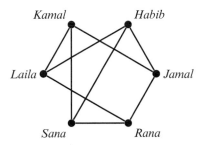

Figure 10.13 Friendship graph.

In this short section, we define and characterize bipartite graphs.

Definition 10.33 (Bipartite Graph). Let $G = (V, E)$ be a simple graph (or a multigraph). Then G is a *bipartite* graph (or multigraph) if we can partition the set of vertices into two sets X and Y (i.e., $V = X \cup Y$ and $X \cap Y = \emptyset$) such that all the edges of G have one end in X and one end in Y.

If G is a bipartite graph, then often we write $G = (X, E, Y)$, where $X \cup Y$ is the set of vertices, E is the set of edges, and all the edges have one end in X and one end in Y. The sets X and Y are called the *parts* of G.

Example 10.34. Let m and n be positive integers, and let L be a set of m vertices and R be a set of n vertices. Recall from Example 2.28 that, if we connect *every* vertex in L to every vertex in R, the resulting graph is called the *complete bipartite graph* on m and n vertices, and is denoted by $K_{m,n}$. There are no edges between vertices in R, and no edges between the vertices in L. (As examples, see $K_{2,3}$ and $K_{3,3}$ in Figure 10.14.)

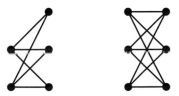

Figure 10.14 The complete bipartite graphs $K_{2,3}$ (on the left) and $K_{3,3}$ (on the right).

Example 10.35. The cycle of length 3, C_3, is not bipartite. This is because, if we partition the two sets of vertices into two parts, then two of the vertices will be in the same part and adjacent.

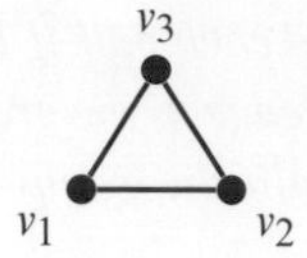

Figure 10.15 C_3 is not bipartite.

Example 10.36. The cycle of length 4, C_4, is bipartite. The usual depiction of C_4 – the graph on the left-hand side of Figure 10.16 – does not exhibit the parts of the graph. However, if we draw C_4 as in the right-hand side of Figure 10.16, then it is clear that it is bipartite.

Figure 10.16 Both graphs are C_4, but the right-hand figure presents C_4 as a bipartite graph.

We are now ready to give a characterization of bipartite graphs. Recall from Definition 10.26 that, if v and w are vertices of general graph, then the *distance* between v and w, denoted by $d(v, w)$, is the length of the shortest walk from v to w.

Theorem 10.37. *Let $G = (V, E)$ be a multigraph. Then G is bipartite if and only if each cycle of G has even length.*

Proof. ($\Rightarrow$) For one direction, assume that the graph $G = (X, E, Y)$ is bipartite with parts X and Y. Every cycle has to go back and forth between X and Y and has to come back to the starting point. Hence it will have even length.

($\Leftarrow$) For the other direction, assume that every cycle of G has even length. To show that G is bipartite, it is enough to prove that each connected component of G is bipartite. As a result, we can assume that G is connected. Pick an arbitrary element $u \in V$, and let X be the set of vertices of G whose distance from v is even, and let Y be the rest of the vertices, namely those whose distance from v is odd. In other words,

$$X = \{v \in V \mid d(u, v) \text{ is even}\}$$
$$Y = \{v \in V \mid d(u, v) \text{ is odd}\}.$$

Then, every vertex of G is in exactly one of X or Y, that is $V = X \cup Y$, and $X \cap Y = \emptyset$. We want to prove that X and Y are the parts of a bipartite graph. Let a, b be two vertices both in X (or both in Y), and we claim that $\{a, b\}$ is not an edge.

PROOF OF CLAIM: Assume $\{a, b\}$ was an edge.

Let $W_1 : u - \cdots - a$ be a shortest walk from u to a and $W_2 : u - \cdots - b$ be a shortest walk from u to b.

A shortest walk is necessarily a path, and so there cannot be any repeated vertices in W_1 or W_2. So if the edge $\{a, b\}$ was in either of these two walks, it would have to be the last edge (so as not to have a or b repeat). Assume by way of contradiction that the last edge of W_1 was $\{b, a\}$. Then the portion of W_1 without the last edge would be a shortest walk from u to b (since any portion of a shortest walk would itself have to be a shortest walk). But this would mean that the shortest distance from u to a differs by 1 from the shortest distance from u to b. This is not possible since a and b are both in X or in Y, and so their distances to u have the same parity. We conclude that the edge $\{a, b\}$ does not appear in W_1 or W_2.

Let z be the last common vertex of W_1 and W_2. So

$$u - \cdots - z - \cdots - a$$
$$u - \cdots - z - \cdots - b.$$

Note that the two walks from u to z do not have to be identical, but they do have to have the same length (otherwise one of the walks could be shortened). Also, the parts of the walks after z have no vertex in common.

Now the length of the walks $u - \cdots - a$ and $u - \cdots - b$ are both even or both odd, and so – since we are subtracting the same number from both – the lengths of $z - \cdots - a$ and $z - \cdots - b$ are both even or both odd as well. But this means that

$$z - \cdots - a - b - \cdots z$$

is a cycle (no repeated vertices) of odd length, which is a contradiction. Since there are no edges within X or Y, we conclude that G is bipartite.   □

Since forests do not have any cycles at all, we have:

Corollary 10.38. *Let G be a forest; then G is bipartite.*

Remark 10.39. We will revisit bipartite graphs in Mini-project 10, where we show that we can decide if a simple connected graph is bipartite by finding the rank of the incidence matrix of the graph.

Problems

P 10.3.1. Let $G = (X, \Delta, Y)$ be a bipartite graph. Assume that $|X| = 10$ and the degree of each vertex in X is 3. If the degree of each vertex in Y is 5, then what is $|Y|$?

P 10.3.2. Is the Petersen graph of Figure 2.1 bipartite? Either give a partition of the vertices that shows that it is bipartite or prove that it is not.

P 10.3.3. Among the graphs in Figure 2.11, which ones are bipartite?

P 10.3.4. Let $G = (X, \Delta, Y)$ be a bipartite graph. Suppose that there is a positive integer p such that each vertex in X is incident with at least p edges, and each vertex in Y is incident with at most p edges. Show that Y has at least as many vertices as X.

P 10.3.5. The graph G is in the form of an $n \times m$ grid. In other words, the number of vertices is nm and they are arranged as an $n \times m$ array. The edges are either horizontal or vertical and connect adjacent vertices. Is G bipartite? Why?

P 10.3.6. You have been investigating a mystery simple graph G. You know G has five vertices and seven edges. Prove that G has a cycle of length 3.

P 10.3.7. Consider K_6, the complete graph of order 6. A friend has asked you to remove five edges from the graph but in a such way that the resulting graph is bipartite. Is that possible? Why?

P 10.3.8. Let $G = (V, E)$ be a (simple) graph, and let $X \subseteq V$. The set X is called *independent* if no two vertices in X are adjacent in G (i.e., if $x, y \in X$, then $\{x, y\}$ is *not* in E). The size of the largest independent set for G is called the *independence number* of G, and often denoted by $\alpha(G)$.[7] Let G be a (simple) graph with 47 vertices, and assume that $\alpha(G) = 27$ and that G has no triangles.

(a) What is the largest possible degree for a vertex in G? Prove your assertion.

(b) Prove that the number of edges in G is no more than 540.

(c) Give an example of a simple graph with 540 edges that satisfies the hypotheses.

P 10.3.9. Show that any group of people can be partitioned into two subgroups such that at least half the friends of each person belong to the subgroup of which that person is *not* a member. Assume friendship is symmetric but not reflexive.

P 10.3.10. Let G be a multigraph with n vertices and m edges. Show that G has a bipartite subgraph that has n vertices and at least $m/2$ edges.

P 10.3.11. Prove that a simple graph of order n with no triangles has at most $n^2/4$ edges. For n even, give an example of a graph with n vertices, no triangles, and $n^2/4$ edges. You may find the following steps helpful:

STEP 1: Let v be a vertex in the graph with as large a degree as possible. Let S be the set of vertices that are adjacent to this vertex. Let T be the set of the vertices not in S. Can two elements in S be adjacent?

STEP 2: Argue that the sum of the degrees of the vertices in S is no more than $|S||T|$.

STEP 3: Argue that the sum of the degrees of the vertices in T is no more than $|S||T|$.

STEP 4: Argue that the number of edges in the graph is no more than $|S||T|$.

STEP 5: Write $|T| = n - |S|$ and decide which value of $|S|$ (in terms of n) maximizes $|S||T|$.

STEP 6: For an example, think bipartite.

P 10.3.12. What is the smallest value for n such that there exists a simple graph on n vertices with all degrees greater than or equal to 6, and yet no K_4 as a subgraph? (In other words, if you choose any four vertices of the graph, at least one pair of those vertices are not adjacent.)

---

[7] Independent sets and independence number will be revisited in subsection 10.7.1.

10.4 Eulerian Trails and Circuits

Warm-Up 10.40. *You want to draw the house of Figure 10.17 on a blackboard. You prefer to do so without taking your hand off the blackboard and without tracing any line twice (you are allowed to go through the vertices where the edges come together more than once). If that is not possible, you want to take your hand off the blackboard the minimum number of times necessary. How would you do it?*

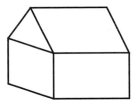

Figure 10.17 Draw this figure by starting at some point and taking your hand off the blackboard the smallest number of times possible.

In 1736, Euler (1741) solved the (relatively simple) Königsberg bridge problem, and with it ushered in the study of graphs in the Western tradition (see footnote 9 later in this chapter for earlier graph theory problems in the Islamic and Indian traditions). The River Pregel in Königsberg[8] had two islands and a total of seven bridges, as in Figure 10.18. The problem was whether it was possible to start from some point, go over every bridge exactly once, and return to the original point.

To translate this problem to graph theory, we put one vertex for each of the banks and each of the islands. For each bridge we put an edge between two of the vertices. Hence, we will have four vertices and seven edges. We get the multigraph in Figure 10.19. (If we wanted to have a simple graph, we would add one vertex in the middle of each of the two edges that are drawn as curved arcs.) We want to know whether it is possible to start at one vertex, go through every edge exactly once, and be back at the original vertex. The reader may want to review the vocabulary of walks, trails, etc. of Definition 10.12. (See also Table 10.2 and Figure 10.7.) First, we give a name to what we are looking for.

Definition 10.41 (Eulerian Trail, Eulerian Circuit). Let G be a general graph. An *Eulerian trail* in G is a trail – that is, a walk in G with no repeated edges – that contains every edge of G. An *Eulerian circuit* is a closed Eulerian trail.

---

[8] Königsberg was founded in the thirteenth century during the Northern Crusades, and is the birthplace of mathematicians Christian Goldbach (1690–1764) and David Hilbert (1862–1943). The philosopher Immanuel Kant (1724–1804) was first a student and later a professor at the University of Köningsberg, and, while there, he wrote his influential *Critique of Pure Reason*. In the eighteenth century the city was mostly part of Prussia, and before World War II it was one of the easternmost cities of Germany. The city was largely destroyed by allied bombing during the war, and was later annexed by the Soviet Union. Today, it is called Kaliningrad, and is an exclave of Russia surrounded by Poland and Lithuania.

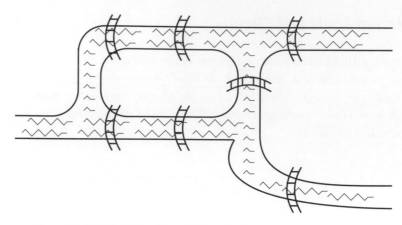

Figure 10.18 The River Pregel in Königsberg and its seven bridges.

Figure 10.19 The bridges of Königsberg as edges of a graph.

It may be a bit surprising to the reader that we will be able to completely characterize graphs that have an Eulerian circuit. We will do so after one lemma. Recall (Definition 2.24) that a loop contributes 2 to the degree of its vertex.

Lemma 10.42. *Let $G = (V, E)$ be a finite general graph. Assume that the degree of every vertex is positive and even. Then any given vertex lies on some circuit – that is, a closed trail.*

Proof. Suppose $x \in V$. Begin a walk at x and take any edge incident with x. Continue the walk, with the only condition being that we never repeat an edge.

CLAIM: Whenever you arrive at a vertex $y \neq x$, you can leave via a hitherto-unused edge.

PROOF OF CLAIM: Before your current visit to the vertex y, you may have visited y a total of h times and hence used $2h$ edges incident with y in your walk. You just entered y through a new edge, making the number of edges known to be incident with y equal to $2h + 1$. Since the degree of y is even, there has to be yet another edge incident with y. Hence you can leave using this edge.

Since the graph is finite, the walk has to stop sooner or later. It can only stop at x. Since we have not repeated any edges and we have started and ended at x, we have a closed trail containing x.

We are now ready to prove the main theorem on Eulerian circuits. The (easy) "only if" direction was proved by Leonhard Euler in 1736 (Euler 1741), while the "if" direction was only proved in 1871 by Carl Hierholzer (1840–1871) and published posthumously (Hierholzer 1873). □

Theorem 10.43. *Let G be a finite general graph with no isolated points. Then G has an Eulerian circuit if and only if G is connected and the degree of every vertex is even.*

Proof. ($\Rightarrow$) If G has an Eulerian circuit – that is, a closed walk that contains every edge of G exactly once – then every vertex appeared an even number of times (in and out). Since every edge is listed exactly once, the degree of each vertex is even.

($\Leftarrow$) For the converse, assume that G is connected and that the degree of every vertex is even. Let C be the longest circuit in G. If C contains every edge of G, then we are done. Hence assume that it does not.

Let H be the general graph consisting of the unused edges of G and their vertices (i.e., we have subtracted the edges of C from G and thrown away any isolated vertices to get H).

CLAIM: Every vertex of H has positive even degree.

PROOF OF CLAIM: Any vertex in H is there because it was incident to an edge in H. Hence, there are no isolated vertices in H, and so the degree of every vertex of H is positive. To construct H, for every vertex of G an even number of edges were taken away. Since the degrees of vertices of G were even, the degrees of vertices of H remain even as well.

CLAIM: The graphs C and H have a common vertex.

PROOF OF CLAIM: Since G is connected, there is some edge $e = \{x, y\}$ in G with x in H and y in C. Now, the edge e is either in C or in H. If it is in C, then x is in C – as well as in H – and if e is in H, then y is in H as well as in C. In either case, C and H have a common vertex.

Let z be a common vertex of C and H. By Lemma 10.42, z lies on some circuit D of H. Consider our two circuits:

$$C : \ldots \{x_1, z\}, \{z, x_2\}, \ldots$$
$$D : \ldots \{y_1, z\}, \{z, y_2\}, \ldots$$

We can extend C by inserting D at z to get

$$\ldots, \{x_1, z\}, \{z, y_2\}, \ldots, \{y_1, z\}, \{z, x_2\}, \ldots$$

This is a circuit – a closed walk with no repeated edges – and longer than C. The contradiction proves that C already contains every edge of G and is an Eulerian circuit. □

Remark 10.44. In graph theory – like many other areas of mathematics – while the results themselves are important, the techniques used in the proofs are even more important. In Theorem 10.43, we wanted to find a long circuit – in fact, we wanted a circuit that contained every edge of the graph – and we started by considering the *longest* circuit in the graph. Our hope was that this circuit would contain every edge of the graph, but initially we could not assume that it did. However, the fact that this was the longest circuit put restrictions on the rest of the graph. This idea – that is, start with considering the largest or smallest subgraph with some specific property – will be used repeatedly. The proof of the next theorem will also introduce a technique: sometimes it is helpful to add one or more edges to your graph. The moral of the story is that, to understand a mathematical topic deeply, it is not enough to know the statements of the theorems. The techniques used in the proofs are the tools that often can be widely applied.

Theorem 10.45. *Let G be a connected finite general graph, and let u and v be two distinct vertices of G. Then G has an (open) Eulerian trail joining u and v if and only if u and v are the only vertices of odd degree.*

Proof. Let $G' = G \cup \{e\}$, where e is a new edge joining u and v. The original graph G has an (open) Eulerian trail joining u and v if and only if G' has an Eulerian circuit. By the previous theorem, this happens if and only if the degree of every vertex of G' is even. And the latter happens if and only if u and v are the only vertices of odd degree in G.   □

The graphs of Theorem 10.43 were connected and had no isolated vertices. We close this section by relaxing these conditions.

Definition 10.46. Let G be a general graph, and let $H_1, H_2, \ldots, H_t$ be subgraphs of G. Assume that every edge of G is in exactly one of these subgraphs. Then we say that $H_1, \ldots, H_t$ is an *(edge) decomposition* of G.

Remark 10.47. Recall that if $G = (V, E)$ is a general graph, a subgraph of G is a general graph $H = (V', E')$ with $V' \subseteq V$ and $E' \subseteq E$ with the condition that if an edge is in E', then its vertices are included in V'. If G is decomposed into subgraphs $H_1, \ldots, H_t$, then each edge of G is in just one of $H_1, \ldots, H_t$, but each vertex may be in several of these subgraphs. Isolated vertices need not be in any of the subgraphs but, in fact, could be in as many of the subgraphs as they wish.

Lemma 10.48. *Let $G = (V, E)$ be a general graph, and let C be a circuit in G. Then C has a decomposition into cycles.*

Proof. Induct on the number of edges in the circuit. If the number of edges is one, then the circuit is a loop, and a loop is a cycle of length 1. (Figure 10.20 gives all the possible circuits with one or two edges.)

Figure 10.20 Possible circuits with one or two edges.

Assume that the result is true for all circuits with fewer edges than the circuit under consideration. If the circuit is a cycle already, then there is nothing to prove. If not, at least one vertex must be repeated at least twice in the circuit. Compare all the paths on the circuit that go from one occurrence of a vertex to another occurrence of the same vertex. Choose the shortest one. This part of the circuit will have to be a cycle – that is, a closed walk with no repeated vertex – since if a vertex was repeated inside this portion, then we would have had an even shorter cycle. Taking this shortest cycle out leaves us a shorter circuit. By induction, this circuit is a union of cycles. Together with the eliminated short cycle, we now have the original circuit as a union of cycles.   □

Corollary 10.49. *Let* $G = (V, E)$ *be a finite general graph. Then G has a decomposition into cycles if and only if every vertex of G has even degree.*

Proof. ($\Rightarrow$) Assume that $G = (V, E)$ has a decomposition into cycles, and let $x \in V$. If x belongs to h of the cycles in the edge decomposition of G, then $\deg(x) = 2h$, which is an even number.

($\Leftarrow$) Now assume every vertex of $G = (V, E)$ has even degree. Since a decomposition is about partitioning the edges, we can safely ignore isolated vertices. Now, by Theorem 10.43, every connected component of G has an Eulerian circuit, and, by Lemma 10.48, every circuit decomposes into cycles.   □

Problems

P 10.4.1. Does the graph of Figure 10.21 have an Eulerian circuit? What about an Eulerian trail? If the answer is yes to either question, is the Eulerian circuit/trail unique?

Figure 10.21 How many Eulerian circuits/trails?

P 10.4.2. Which complete graphs K_n have Eulerian circuits? Which have Eulerian trails?

P 10.4.3. I have a simple graph whose degree sequence is 4, 4, 4, 2, 2, 2, 2, 2, 2. Can this graph have a bridge? Prove your assertion.

P 10.4.4. Find all possible degree sequences of simple graphs with six vertices that have a closed Eulerian trail and no isolated vertices. For each such degree sequence, draw an actual graph.

P 10.4.5. Let G be a 5-regular simple graph of order 12. How many walks with six edges can you find?

P 10.4.6. We have a 2×5 block rectangular street network (see Figure 10.22). An activist wants to start and end at the same place and go through every street at least once and pass out

Figure 10.22 A 2×5 rectangular street network.

leaflets. Can she do this and go through every street exactly once? If the answer is no, then what is the minimum number of streets that she has to go through twice?

P 10.4.7. You want to draw Figure 10.23 by starting at some point and taking your hand off the paper the smallest number of times possible. What is the minimum number of times that you have to take your hand off the paper and move it elsewhere?

Figure 10.23 Draw this figure by starting at some point and taking your hand off the paper the smallest number of times possible.

P 10.4.8. Let G be a connected finite (simple) graph. Among the vertices of G there are exactly 10 vertices with odd degree. Prove that there exist five trails (i.e., walks with no repeated edges) in G such that, between them, they use each edge of G exactly once. (In other words, the five trails are edge-disjoint and the union of their edges is the set of edges of G.)

P 10.4.9. Let $G = K_8$, the complete graph of order 8. You want to partition the edges of G into open trails. In other words, you want to find a number of edge-disjoint open trails such that each edge of G is in exactly one of those trails. What is the minimum number of trails needed?

P 10.4.10. One one side of each domino piece, there are two sets of dots (see Figure 10.24). Each set of dots is from $[6] = \{1, \dots, 6\}$. We happen to have one domino for each pair of (not necessarily distinct) elements of $[6]$. So we have exactly one domino with three dots and four dots, exactly one domino with three and five dots, one domino with three and three dots, and so on. We want to arrange all our dominoes end to end with matching number of dots. (For example, follow the domino with one and two dots with the domino that has two and four dots, and so on).

(a) Is this possible?

(b) What if we first threw away all the dominoes that had a 6 on them?

(c) What if we kept all the dominoes and added a few duplicates? What is the minimum number of additional pieces (which would be duplicates of the one we have) needed?

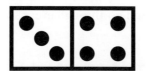 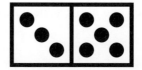

Figure 10.24 Two domino pieces.

P 10.4.11. Show that a general graph is 2-regular (Definition 10.3) if and only if each of its connected components is a cycle.

P 10.4.12. Assume that G is a connected graph, and the degree of every vertex is even. Theorem 10.43 assures us that G has an Eulerian circuit. If I want to actually find an Eulerian circuit for G, then how careful do I have to be? In particular, can I start the circuit at any of the vertices? After choosing the first vertex, can I continue along any of the edges incident with that vertex, or do I have to make a judicious choice? In fact, do I ever have to think ahead and plan the route of the Eulerian circuit, or, upon arriving at any of the vertices, can I proceed along one of the hitherto-unused edges without any planning? Either prove that no planning is necessary or give a methodical way (i.e., an algorithm) for constructing an Eulerian circuit.

P 10.4.13. Let G be a 4-regular connected general graph with no isolated vertices. Prove that we can find an edge decomposition of G into two graphs H_1 and H_2 such that every vertex of G is in both H_1 and H_2, and such that both H_1 and H_2 are 2-regular. (A 2-regular subgraph of G that contains all the vertices of G is called a *2-factor* of G.) Did we need to assume that G is connected or that it has no isolated vertices?

P 10.4.14. Let G be a given simple graph. All we know about G is that the degree of each vertex is at least 47. Let P be a longest (open) path in G. What is the smallest possible length for P? Why? Give an example to show that your answer is the best possible.

P 10.4.15. Just as in Problem P 10.4.14, all we know about the simple graph G is that the degree of each vertex is at least 47. I like long cycles. What is the length of the longest cycle that you can guarantee for G? How? Give an example to show that your answer is the best possible.

P 10.4.16. Let G be a finite simple graph. Let $\delta(G)$ denote the minimum degree among all the vertices of G (i.e., the degree of all vertices of G is no less than $\delta(G)$). Generalize your results of Problems P 10.4.14 and *P 10.4.15*. In other words, what can you say about the length of the longest path and the longest cycle in G if all you know is $\delta(G)$?

10.4.1 Graph Vocabulary: Line Graphs

Definition 10.50 (Line Graph). Let $G = (V, E)$ be a graph. We define a new graph called the *line graph of G* and denoted by $L(G)$. $L(G)$ has one vertex for each edge of G, and two vertices of $L(G)$ are adjacent if and only if the corresponding edges in G have a common vertex.

Example 10.51. Let G be a simple graph consisting of two triangles that share a vertex. G has five vertices and six edges, and so $L(G)$ has six vertices and an edge for each time two edges of G have a common vertex. (See Figure 10.25.)

Figure 10.25 A graph G (on the left) and its line graph $L(G)$ (on the right).

Problems (continued)

P 10.4.17. Let G be a path of length 4. What is $L(G)$, the line graph of G? What is $L(L(G))$, the line graph of the line graph of G?

P 10.4.18. Let G be a cycle of length n. What is $L(G)$, the line graph of G? What is $L(L(G))$, the line graph of the line graph of G?

P 10.4.19. What are the line graphs of K_3, the complete graph of order 3, and $K_{1,3}$, the complete bipartite graph on 1 and 3 vertices? Is the function that sends a simple graph to its line graph $L(G)$ a one-to-one function?

P 10.4.20. What is the line graph of K_4, the complete graph of order 4? What is the line graph of $K_{3,2}$, the complete bipartite graph on 3 and 2 vertices?

P 10.4.21. Let G be a simple connected graph, and let $L(G)$ denote its line graph.
   (a) Assume G has an Eulerian circuit, and show that $L(G)$ also has an Eulerian circuit.
   (b) Find a graph G that has no Eulerian circuit, but for which $L(G)$ has an Eulerian circuit.

10.5 Hamiltonian Paths and Cycles

Warm-Up 10.52. *What is the length of the longest cycle in the graph of Figure 10.26? Is there a cycle that contains every vertex of the graph?*

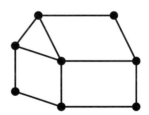

Figure 10.26 Start with one of the vertices, visit every other vertex exactly once, and end up back at the first vertex.

The Irish mathematician Sir William Hamilton (1805–1865) tried to market a puzzle that was basically the graph of the vertices and edges of a dodecahedron. The object of the puzzle was to start at some corner, move along the edges, and, after visiting every corner exactly once, to return to the original corner. The puzzle didn't sell that well, but the mathematical concept was named after Hamilton.[9]

---

[9] The study of "Hamiltonian cycles," and, in fact graph theory, has a much older history, and some of this history is intertwined with the game of chess. An older Persian version of the game called Shatranj was prevalent during the Sasanian Empire (224–651) of Iran. It is said that the game came to Iran from India. By the time of the Islamic Abbasid caliphate (750–1517), the players of the game were ranked, and chess problems and puzzles were popular.

Definition 10.53. Let G be a simple graph of order n. A cycle of length n is called a *Hamiltonian cycle*. A graph is *Hamiltonian* if it has a Hamiltonian cycle. Let a and b be two distinct vertices of G. A *Hamiltonian path* joining a and b is a path of length $n - 1$ in G that starts with a and ends at b.

Note that a Hamiltonian cycle has n vertices, and this means that the cycle starts at a vertex, goes through every other vertex exactly once, and returns to the original vertex. Likewise, a Hamiltonian path starts at a vertex a, goes through every vertex of the graph exactly once, and ends at vertex b.

On the face of it, Hamiltonian cycles and paths are very similar to Eulerian circuits and trails. Instead of going through every edge exactly once, you go through every vertex exactly once. However, the similarity is quite superficial. While we found a very clear criterion for graphs having Eulerian circuits and trails (Theorems 10.43 and 10.45), no such characterization of graphs with Hamiltonian cycles or paths exists. In fact, finding Hamiltonian cycles – in graphs that have them – can be difficult. For example, consider the following conjecture.

Conjecture 10.54 (Middle Levels Conjecture). *Let n be an odd integer, and let $[n] = \{1, \ldots, n\}$. Let the set V consist of subsets of size $\frac{n-1}{2}$ and subsets of size $\frac{n+1}{2}$ of $[n]$. Define a simple graph $G = (V, E)$ by letting V be its set of vertices. Two subsets $A, B \in V$ are adjacent in the graph if and only if $A \subsetneq B$.*
Then the graph G has a Hamiltonian cycle.

The conjecture goes back to the 1970s and has been attributed to many people, including Czech mathematician Ivan Havel, and is known as *Erdős's revolving door* or the *middle levels* conjecture. In the language of the upcoming Chapter 11, the graph G is the Hasse diagram (Definition 11.8) of the middle two levels of the Boolean lattice of order n (Definition 11.13). See Figure 10.27 for the case $n = 5$. As surprising as it is, this conjecture was open until very recently. A proof of the conjecture was announced only in 2014 by Torsten Mütze (Mütze 2016).

In this section, we give one obvious criterion that precludes the existence of a Hamiltonian cycle, and a more subtle criterion that ensures the existence of such a cycle.

Proposition 10.55. *Let G be a connected graph with a bridge. Then G does not have a Hamiltonian cycle.*

Proof. No matter where you start, you have to cross the bridge to go through the vertices on the other side. Later you will have to cross the bridge again to go back to the original

One question of interest was the existence of Knight's tours. Is it possible for a Knight to start at one of the squares, legally move to every other square exactly once and return to its original position? This question is the same as whether a certain graph is Hamiltonian. (See Problem P 10.5.5.) One early solution is attributed to al-Adli ar-Rumi (circa 800–870), a chess master who was originally from Anatolia but worked in Baghdad. He wrote a now-lost treatise on Shatranj, but later chess masters attributed a Knight's tour, on an 8×8 board from 840 CE to him. An earlier open Knight's tour, albeit on an 8×4 board, from circa 815 CE, is attributed to the Kashmiri poet Rudrata (ninth century). The tour was presented by a series of syllables on the squares of the board. The meaning of the text would only be revealed if read in the order of the tour. See Knuth 2020.

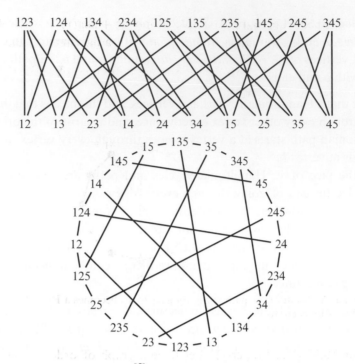

Figure 10.27 The middle two levels of $2^{[5]}$ (on the top) and a drawing of this graph exhibiting a Hamiltonian cycle (on the bottom).

starting point. Hence an edge and its two vertices will be on the walk twice. Hence, no Hamiltonian cycles.   □

Proposition 10.56 (Ore 1960). *Let $G = (V, E)$ be a simple graph of order $n \geq 3$. Assume $x, y \in V$ with $\{x, y\} \notin E$ and $\deg(x) + \deg(y) \geq n$. Define the simple graph $\tilde{G}$ by $\tilde{G} = (V, E \cup \{x, y\})$. In other words, $\tilde{G}$ has just one more edge – namely $\{x, y\}$ – than G. Then G is Hamiltonian if and only if $\tilde{G}$ is.*

Proof. Adding edges to a graph does not affect the already existing cycles, and so if G is Hamiltonian, so is $\tilde{G}$. Now assume that $\tilde{G}$ is Hamiltonian, and let $x = x_1, x_2, \ldots, x_n = y, x$ be a Hamiltonian cycle in this new graph. Thus, in G we have a Hamiltonian path $x = x_1, x_2, \ldots, x_n = y$ that goes from x to y and passes through every vertex exactly once. Now consider the sets

$$N_x = \{1 \leq i \leq n - 1 \mid x \text{ is adjacent to } x_{i+1} \text{ in } G\}$$
$$N_y = \{1 \leq i \leq n - 1 \mid y \text{ is adjacent to } x_i \text{ in } G\}.$$

So, for example, $4 \in N_x$ if and only if x is adjacent to x_5, and $4 \in N_y$ if and only if y is adjacent to x_4. Since $x_1, \ldots, x_n$ are all the vertices of G, we have $\deg(x) = |N_x|$ and $\deg(y) = |N_y|$. Since the members of N_x and N_y are from $1, \ldots, n-1$, if $N_x \cap N_y = \emptyset$ then $|N_x| + |N_y| \leq n-1$.

However, we are given that $|N_x| + |N_y| = \deg(x) + \deg(y) \geq n$. So $N_x \cap N_y \neq \emptyset$, and hence there is a j such that x is adjacent to x_{j+1} and y is adjacent to x_j. But then

$$x = x_1, x_2, \ldots, x_j, y = x_n, x_{n-1}, \ldots, x_{j+1}, x_1 = x$$

is a Hamiltonian cycle in G. See Figure 10.28. $\qquad \square$

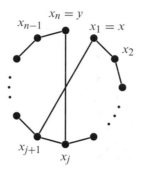

Figure 10.28 A Hamiltonian path connecting x_1 to x_n becomes a Hamiltonian cycle if x_1 is adjacent to x_{j+1} and x_n is adjacent to x_j.

Definition 10.57 (Ore Property). Let G be a graph of order n. We say that G satisfies the *Ore property* if, for each pair of distinct vertices x and y, either $\{x, y\}$ is an edge of G or $\deg(x) + \deg(y) \geq n$.

Example 10.58. Let G be a graph of order 50 with a vertex x of degree 4. If G has the Ore property, then the 45 vertices that are not adjacent to x must each have a degree of at least 46. In other words, the Ore property balances a vertex of low degree with a lot of other vertices of very large degree.

Corollary 10.59 (Ore 1960). *Let $G = (V, E)$ be a simple graph of order $n \geq 3$. Assume that G satisfies the Ore property. Then G has a Hamiltonian cycle.*

Proof. Let k be the number of pairs of non-adjacent vertices of G. Construct a sequence of graphs $G = G_0, G_1, \ldots, G_k = K_n$ where, for $1 \leq i \leq k$, G_i is the same as G_{i-1} except it has one more edge. In G_i an edge between two previously non-adjacent vertices has been added. By Proposition 10.56, for $0 \leq i \leq k - 1$, G_i is Hamiltonian if and only if G_{i+1} is. However, the complete graph $K_n = G_k$ is clearly Hamiltonian, and, as a result, so must be $G_0 = G$. $\quad \square$

Corollary 10.60 (Dirac 1952).[10] *Let G be a (simple) graph of order $n \geq 3$. If every vertex of a simple graph of order n has degree at least $n/2$ then the graph is Hamiltonian.*

[10] The Norwegian mathematician Øystein Ore (1899–1968) stated Corollary 10.59 as a consequence of a slightly different version of Proposition 10.56. The corollary generalized the earlier theorem (Corollary 10.60) proved by the Hungarian/British mathematician Gabriel Andrew Dirac (1925–1984). Ore got his mathematical training at the University of Oslo, and started to teach at Yale University in 1927. At Yale, he was the doctoral thesis advisor of Grace Hopper (1906–1992), a pioneer in computer science. G. A. Dirac was born Balázs Gábor in Budapest. His uncle on his mother's side was the physicist Eugene Wigner (1902–1995). In a second marriage, his mother married the physicist Paul Dirac (1902–1984). Gábor moved to England when he was 12 and changed his family name.

Problems

P 10.5.1. The cities $\{1, \ldots, 6\}$ are connected by highways as shown in Figure 10.29.

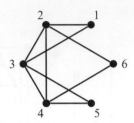

Figure 10.29 Cities $1, \ldots, 6$ are connected by highways as shown.

(a) Starting with city numbered 4, a salesperson wants to visit every vertex exactly once and return to 4. Can it be done?

(b) A garbage collection agency wants to start with vertex 4 and go through every edge exactly once and return to 4. Can it be done?

P 10.5.2. The graph in Figure 10.30 is a drawing in the plane of the vertices and edges of a dodecahedron. Does it have a Hamiltonian cycle?

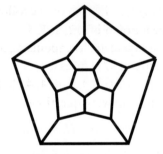

Figure 10.30 Does this graph have a Hamiltonian cycle?

P 10.5.3. (a) Does a cubic graph of order 6 have to be Hamiltonian?

(b) Does a simple graph with degree sequence $5, 4, 4, 4, 4, 4, 3$ have to be Hamiltonian?

P 10.5.4. How many Hamiltonian paths in K_7? How many Hamiltonian cycles in K_{11}?

P 10.5.5. In chess, a Knight makes an L-shaped move. From its current position, it moves two squares horizontally or vertically and then an additional square in a direction perpendicular to the first. See Figure 10.31. Consider a 3×4 chess board. We are interested in placing a Knight somewhere on the board, and then moving the Knight, by making legal moves, to other squares so as to visit every square on the board exactly once. Such a series of moves is called a *Knight's tour*. If, after visiting the last square, we can make a final move and go back to the starting square, then the Knight's tour is called a *closed* Knight's tour. Otherwise, it is an *open* Knight's tour.

(a) Translate the problem of finding Knight's tours to a problem of finding Hamiltonian paths or cycles on a graph. For the 3×4 board, draw the graph. What is the degree sequence of the graph?

(b) Can you find an open Knight's tour for the 3×4 board?

(c) Can you find a closed Knight's tour for the 3×4 board?

(d) If the answer to either of the above is yes, does the starting point matter?

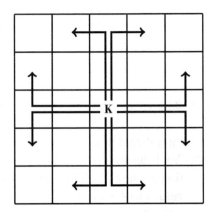

Figure 10.31 A Knight in chess makes an L-shaped move.

P 10.5.6. $G = (V, E)$ is a simple graph. In each of the following cases, all we know about G is given.

(A) The degree sequence of G is 3, 2, 1, 1, 1.

(B) G has exactly four connected components, and $|E| = |V| - 4 > 0$.

(C) The degree sequence of G is 5, 5, 3, 3, 3, 3.

(D) The degree sequence of G is 4, 4, 4, 4, 4, 2, 2, 2, 2.

In each case, answer the following questions:

(a) Is G a tree?

(b) Is G a forest?

(c) Is G connected?

(d) Is every edge of G a bridge?

(e) Does G have a bridge?

(f) Does G have a closed Eulerian trail?

(g) Does G have a Hamiltonian cycle?

Your answers could be: Yes, definitely; No, definitely not; or Maybe, sometimes yes and sometimes no. By giving adequate reasons and examples, justify your assertions.

P 10.5.7. (a) Let $G = (V, E)$ be a simple graph with a Hamiltonian cycle. Let S be a set of five vertices of G (i.e., $S \subseteq V$ and $|S| = 5$). H is the graph that we obtain from G by removing the vertices in S and their edges. Can H have six connected components?

(b) Generalize the previous part and state as a theorem.

P 10.5.8. (a) Find an example of a cubic simple graph with as few vertices as possible and with a Hamiltonian path.

(b) Find an example of a connected cubic simple graph that does not have a Hamiltonian path.

P 10.5.9. Let G be a graph with 47 vertices. The Dirac theorem, Corollary 10.60, says that if every vertex of G has degree at least 24, then G is Hamiltonian. Give an example of a non-Hamiltonian graph of order 47 with smallest degree 23. Does your graph have a Hamiltonian path?

P 10.5.10. Let G be a simple graph with vertices $[10] = \{1, \ldots, 10\}$. Assume that $1-2-\cdots-9-10-1$ is a Hamiltonian cycle in G, and that G has no cycles of length 3 or 4. Further assume that $\{1, 6\}$ is also an edge of G, and vertex numbered 2 has degree 3. Which vertices could vertex 2 be adjacent to?

P 10.5.11. Consider the sequence s: $7, 7, 6, 5, 4, 4, 3, 2$.

(a) Is s a graphic sequence?

(b) If s is the degree sequence of a simple graph G, must G be Hamiltonian? Is it possible to answer this question without reconstructing the graph G and by somehow utilizing Proposition 10.56?

P 10.5.12. Is the Petersen graph of Figure 2.1 a regular graph? A cubic graph? Does it have a Hamiltonian cycle? You may find the following steps helpful:

STEP 1: Let G be a regular cubic graph with 10 vertices, 15 edges, and no cycles of length 3 or 4.

STEP 2: Assume that the G had a Hamiltonian cycle. Draw the alleged Hamiltonian cycle as a circle, and arrange the 10 vertices around the circle: $1 - 2 - \cdots - 10 - 1$. How many other edges – chords of the circle – does the graph have?

STEP 3: Show that, if every vertex was adjacent to a vertex directly opposite it on the circle, then we would have a 4-cycle.

STEP 4: Assume that vertex 1 is not adjacent to 6. Which vertices could 1 be adjacent to without creating a 3- or a 4-cycle?

STEP 5: By considering the vertex numbered 6 and the vertices that it could be adjacent to, arrive at a contradiction.

P 10.5.13. Does the complement (see Definition 10.29) of the Petersen graph of Figure 2.1 have a Hamiltonian cycle? If so, draw this graph – the complement of the Petersen graph – by first putting the vertices on a circle in the order of the Hamiltonian cycle (akin to the graph at the bottom of Figure 10.27).

P 10.5.14. Let $G = L(K_5)$ be the line graph (Definition 10.50) of the complete graph of order 5, K_5.

(a) Is G Hamiltonian?

(b) Does G have an Eulerian circuit?

(c) Show that G is isomorphic to the complement of the Petersen graph. (See Problem P 10.5.13.)

(d) If we find the complement of the line graph of K_5, do we get the Petersen graph?

P 10.5.15. Let G be the line graph of the Petersen graph (see Definition 10.50). Is G Hamiltonian?

P 10.5.16. Is the Clebsch graph of Problem P 2.3.15 a regular graph? Does it have a Hamiltonian cycle? Answer by finding an appropriate drawing of this graph on the Internet.

P 10.5.17. I have a connected simple graph G sitting in my drawer. I am not sure how many vertices
or how many edges it has. I do know that

$$v_0 - v_1 - v_2 - \cdots - v_{212}$$

is a path of length 212 in G. I also know that there are no open paths of length 213 in G.
  (a) Given the above information, does G have to be Hamiltonian?
       For the next two questions, assume that, in addition to the above information, I also
       know that $\{v_0, v_{100}\}$ and $\{v_{99}, v_{212}\}$ are edges of G.
  (b) Does G have to have a cycle of length 213? Either give a counter-example, or prove
       that it must have such a cycle.
  (c) Does G have to be Hamiltonian? Either give an example where it is not, or prove that
       it has to be.

P 10.5.18. If everyone in a group is the friend of at least half the people in the group, then show that
the group can be seated around a table in such a way that everyone is seated between two
friends.

P 10.5.19. Suppose a classroom has 25 students seated in desks in a square 5×5 array. The students
want to confuse the teacher by each moving – when the teacher is not looking – to an
adjacent seat (just ahead, just behind, on the left, or on the right). Is this possible?

P 10.5.20. What if the classroom of Problem P 10.5.19 had 28 students and they were seated in a
rectangular array of 4×7?

P 10.5.21. Imagine a prison consisting of 64 cells arranged like the squares of an 8×8 chess board.
There are doors between all adjoining cells. A prisoner in one of the corner cells is told
that she will be released provided she can get into the diagonally opposite corner cell after
passing through every other cell exactly once. Can the prisoner obtain her freedom?

P 10.5.22. What if the prisoner of Problem *P 10.5.21* had to start at her corner cell, go through each
of the other 63 cells exactly once, and end up back in her own cell? Could it be done?

P 10.5.23. G is a (simple) connected graph with 20 vertices. Assume that, for any two non-adjacent
vertices x and y, we have $\deg(x) + \deg(y) \geq 12$.
  (a) Let $x_1 - x_2 - \cdots - x_\ell$ be the longest open path in G (recall that in a path we have
       distinct vertices and distinct edges). If $\ell < 20$, then could $\{x_1, x_\ell\}$ be an edge of G?
       Prove your assertion.
  (b) Prove that G has a cycle of length at least 7.

P 10.5.24. $G = (V, E)$ is a simple connected graph that does *not* have a Hamiltonian path. The longest
(open) path in G is

$$x_0 - x_1 - \cdots - x_k.$$

  (a) Can $\{x_0, x_k\} \in E$? Why?
  (b) Can x_0 be adjacent to a vertex not in the given path? Why?
  (c) For $1 \leq i \leq k - 1$, if $\{x_0, x_i\} \in E$, then can $\{x_{i-1}, x_k\} \in E$? Why?
  (d) If, in addition, we know that $\deg(x_0) + \deg(x_k) \geq 47$, then what can you say about k?
       Why?

P 10.5.25. Let $G = (V, E)$ be a simple graph. Assume that $|V|$, the number of vertices, is 47. Let x and y be two non-adjacent vertices of G. Assume that $\deg(x) + \deg(y) = 46$. What can you say about the distance (see Definition 10.26) of x from y? Prove your assertion.

P 10.5.26. Let G be a simple graph of order $n \geq 4$. Assume that G satisfies the Ore property. Prove that the diameter of G is no more than 2, the girth of G is no more than 4, and every vertex of G is on a cycle of length 4. Give an example of such a graph where the girth is exactly 4.

P 10.5.27. Let G be a simple graph of order $n \geq 5$, and let x and y be non-adjacent vertices of G with $\deg(x) + \deg(y) \geq 2n - 5$. Let $\tilde{G}$ be the graph obtained from G by adding the edge $\{x, y\}$. Prove that G has a cycle of length 5 if and only if $\tilde{G}$ does.[11]

P 10.5.28. Find a result similar to the one in Problem *P 10.5.27* for cycles of length 9. Anything special about 5 or 9?

P 10.5.29. We are trying to reconstruct a word (i.e., an ordered collection of letters) that is made from the letters A, B, C, D, and R. Table 10.3 gives a frequency table that shows the number of times a specific triple occurs in the word. For example, ABR occurs twice, ACA appears once, while RAB does not appear at all. We want to know *all* words with the same triples and with the same frequency table. The answer may be that there are no such words.

Table 10.3 The frequency of triples in a mystery word is given.

Triple	ABR	ACA	ADA	BRA	CAD	DAB	RAC
Frequency	2	1	1	2	1	1	1

(a) Construct a graph whose vertices are the triples. If a triple is supposed to occur two times, then we will have two different vertices for that triple. This graph will be a *directed graph*, which means that every edge will have a direction (usually denoted by putting an arrow on the edge). There is a directed edge between two vertices if the last two letters of the first vertex are the same as the first two letters of the second vertex. Can you use this directed graph to answer the question?

(a) Construct a different directed graph. This time the vertices are the pairs of letters that appear in the word, and so AB is a vertex and BR is another vertex. This time, each particular pair will be represented once. Instead, we may have multiple edges. There is a directed edge between a vertex labeled XY and a vertex labeled YZ every time XYZ occurs in the word. Thus, for our example, there are two directed edges from AB to BR since ABR occurs twice. Can you use this directed graph to answer the question?

P 10.5.30. Repeat Problem P 10.5.29 with the frequency table of Table 10.4.

Table 10.4 The frequency of triples in a mystery word is given.

Triple	ABR	ACA	ADA	BRA	CAD	DAB	RAB	RAC
Frequency	2	1	1	2	1	1	1	1

---

[11] Problems *P 10.5.27* and P 10.5.28 are adapted from Bondy and Chvátal 1976.

P 10.5.31.   Let $n \geq 3$. Take a complete graph on n vertices and color its edges using two colors. Prove that there will be a Hamiltonian cycle which either is monochromatic or consists of two monochromatic arcs.[12]

10.6   Planar Graphs and the Tiling of the Plane

Warm-Up 10.61. *The graph of Figure 10.32 has 6 vertices and 12 edges. Can you redraw it so that no edges are crossing?*

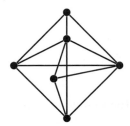

Figure 10.32   Can you redraw this graph such that no edges are crossing?

In this section, we consider graphs that can be drawn on a piece of paper without their edges crossing – the so-called planar graphs – and, in the process, we will also consider some combinatorial questions regarding possible tilings of the plane.[13] We begin with the opening problem for this chapter on tiling soccer balls.

Question 10.62. A typical soccer ball is tiled with pentagons and hexagons (see Figure 10.33 for a poor rendering of a soccer ball). Is there any constraint on the number of pentagons and hexagons that we can use?

On a traditional soccer ball, there are 12 pentagons. Each of these borders five hexagons. That gives $12 \times 5 = 60$ hexagons. But each hexagon is neighbors with three pentagons and hence is counted three times. Thus, the number of hexagons is $\frac{60}{3} = 20$. We also note that at any corner three shapes come together (see Figure 10.33). In another words, if we draw the hexagons and pentagons as a graph, every vertex of the graph has degree 3. Modifying our question, we will add a regularity condition. Tile the soccer ball with pentagons and hexagons in such a way that the degree of all vertices is equal (that is, at every "corner" the same number of shapes come together). What can you say about the number of hexagons and pentagons?

One further clarification before we continue. If you add an additional vertex on one of the edges of, say, a pentagon on the soccer ball, you may think that you have turned the pentagon

---

[12]   From Lovász 2007.

[13]   While not necessary, the reader may benefit from working through Problems MP5.4, MP5.5, and MP5.8–MP5.14 of Mini-project 5 on graphic sequences and planar graphs before tackling this section.

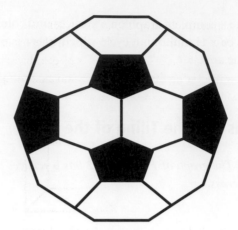

Figure 10.33 A typical soccer ball is tiled with pentagons and hexagons, and at every "corner" three shapes come together.

into a hexagon. This, of course, violates our rule that all vertices should have equal degree (the new vertex will have degree 2 while the others will have degree 3). However, to avoid silly configurations, we outlaw vertices of degree 1 or 2 altogether. In tiling a sphere, we will only consider vertices of degree 3 or higher.

Drawing a Tiling of a Soccer Ball

We want to reformulate the question further into a question about graphs. How can we draw the tiling of a soccer ball on a two-dimensional piece of paper? The tiling of the soccer ball creates regions on the ball. In the case of a traditional ball, these regions are pentagons and hexagons. Imagine cutting one of these regions out – hence puncturing the ball – and then opening the ball up by stretching the hole. At the end, the ball will be flattened and the tiling – albeit with the shapes distorted – will be a graph on the plane. We just have to remember that we are missing one of the tiles. In fact, the punched-out tile, which became the hole, is now the infinite region *outside* of the graph. This graph – the flattened soccer ball – has no edges crossing. Such graphs – as we shortly will define – are called *planar* graphs. Note that Figure 10.33 is *not* an example of a graph that depicts the traditional tiling of the soccer ball, since, in that figure, many of the tiles are on the "other" side and not visible.

From this point of view, any planar graph gives a tiling of a sphere (i.e., a soccer ball). We just have to remember that the sizes of the shapes and the straightness of the lines could have been distorted and that the infinite-area outside region turns into a finite region on the sphere. (In rendering a planar drawing of a tiling of a sphere, we have one more implicit rule/convention. All the vertices and edges have to be drawn in a bounded area of the plane, leaving exactly one "outside" region with infinite area.)

Example 10.63. Figure 10.34 depicts a tiling of the sphere with six quadrilaterals. (Remember to count the outside region, which is also bordered by four edges and hence is a quadrilateral.)

In our thought experiment, we stretched the soccer ball, and so the tiling of the plane given by Figure 10.34 gets distorted if you put it back on the sphere. The quadrilaterals may become squares, rectangles, trapezoids, or other kinds of quadrilaterals. In fact, this is exactly what we get if we "flatten" a three-dimensional cube. Such a cube – just like the graph of Figure 10.34 – has 6 square faces, 8 vertices, and 12 edges.

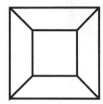

Figure 10.34 The tiling of the sphere (and the plane) with six quadrilaterals.

As another example, Figure 10.35 also depicts a tiling of the sphere, but into what shapes? You may first suspect that this is a tiling into 17 triangles. There are seven triangles in the bottom row, followed by five, then three, and finally one in the next rows. This comes to 16 triangles, and then there is the outside triangle, making for a total of 17.

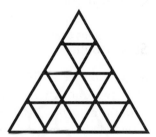

Figure 10.35 Another tiling of the sphere.

This is incorrect! Some of the shapes are not what they may seem at first. The top "triangle" is in fact not a triangle. The only vertices on the sphere are at the crossings of three or more edges, and so the very top of the triangle is not a vertex at all. In drawing a tiling of the sphere on the plane, lines get distorted. They can be bent and make corners, but this doesn't make for new vertices. What we have on the top – as well as the left and right corners – is a pair of edges going through two vertices (in other words, a multiple edge). It is not a triangle. It is a 2-gon. (We could have drawn it as ●━━━●) Likewise, the shape of the outside region is not a triangle. In fact, the outside has many more edges and it is a 9-gon. In other words, Figure 10.35 depicts a tiling of the sphere with three 2-gons, thirteen triangles, and a 9-gon.

Example 10.64. How about pentagons and hexagons? A pentagonal dodecahedron – one of the five Platonic solids – has twelve faces and each is a pentagon (see the left-hand diagram in Figure 10.36). This polyhedron naturally gives – imagine blowing up the polyhedron to make it into a sphere – a tiling of the sphere made up of just 12 pentagons and with all degrees equal to 3 (see the right-hand diagram in Figure 10.36 for a planar drawing of this tiling). Note that one of the pentagons is the outside region, and that the degree of each vertex is exactly 3. So, in addition to 12 pentagons and 20 hexagons, some other combinations of pentagons and hexagons are also possible.

 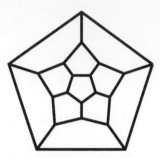

Figure 10.36 One side of a pentagonal dodecahedron, on the left, and the corresponding tiling of the sphere with 12 pentagons, on the right.

Figure 10.37 gives a tiling of the sphere with 2 hexagons and 12 pentagons. In this figure, the outside region and the shaded region in the center are both hexagons.

It is curious that, in the three examples, the number of pentagons is always 12. But maybe that is just an artifact of the examples that the author has chosen.

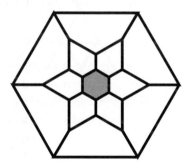

Figure 10.37 A tiling of the sphere with with 12 pentagons and 2 hexagons.

To further investigate the number of pentagons and hexagons in a tiling of the sphere, we turn to planar graphs and consider the relation between the number of vertices, the number of edges, and the number of regions in a planar graph.

Definition 10.65. A *planar graph* is a graph that can be drawn in the plane with no edges crossing. Such a drawing of a graph is called a *planar drawing* (or a *plane drawing* or a *planar embedding*) of the graph. A planar drawing of a graph partitions the plane into *regions*. In a planar drawing of a finite planar graph, one region (often referred to as the *outside* region) will have infinite area.

Remark 10.66. You can draw a particular graph in many different ways. It could be that in some drawings no edges cross and in other drawings some edges do cross. Such a graph, by our definition, is planar, although not all drawings of the graph are planar drawings. A *non-planar* graph is one where, no matter how you draw it, some edges will cross.

Theorem 10.67 (Jordan Curve Theorem). *A plane drawing of a cycle divides the plane into two regions.*

Proof. The most general version of this theorem is an important theorem in topology, and its rigorous proof actually – and surprisingly – is non-trivial. The proof for finite cycles is more manageable, but we have skipped it. (See Figure 10.38.)   □

Figure 10.38 A plane drawing of a cycle. While intuitively it makes sense that the plane is partitioned into two regions – the region inside the cycle and the region outside the cycle – it is not that straightforward to prove this fact rigorously.

Example 10.68. Let G be a finite tree. Since in drawing a graph there are no restrictions on the length of edges or the size of angles between different edges, intuitively you can always start with any vertex of G and draw all the other edges of a tree without any edge crossing (remember that trees don't have cycles and so you will never be stuck inside a cycle trying to get out). You can make this argument a bit more precise using induction on n, the number of vertices of G. If $n = 1$, then the tree is planar. For trees with more than one vertex, choose a leaf (a vertex of degree 1; finite trees always have them – see Problem *P 10.2.9*) and discard it and its corresponding edge. Now, by induction, you can draw the rest of the tree without any edges crossing. Go back and reattach that last edge and vertex. Regardless of how you had drawn the truncated graph, by making sure that the new edge is small enough, you can fit it without intersecting any other edge. We conclude that all trees – and therefore all forests – are planar.

Example 10.69. The complete graph of order 4, K_4, is planar. (See Figure 10.39.)

There is a very important (and beautiful) relationship between the number of vertices, edges, and regions of a planar graph (and, hence, of tilings of a sphere), called Euler's polyhedral formula. Our next task is to state and prove this theorem. We have chosen to

Figure 10.39 A planar drawing of K_4.

prove a slightly more general version of the theorem for a class of objects more general than planar graphs.[14]

Definition 10.70. Recall that a general graph is a graph where loops and double edges are allowed (Definition 2.20). We define a *squiggle* to be a planar drawing of a finite general graph with the proviso that we even allow a finite number of loops with no vertices. (Hence, we even allow the possibility of a squiggle having no vertices at all.) Squiggles, just like planar graphs, partition the plane into regions, and, because a squiggle has a finite number of vertices and edges, among the regions there will always be exactly one infinite region.

Remark 10.71. In a general graph, the edges can be ordinary edges (i.e, a pair of distinct vertices) or loops (i.e., a single vertex or a "pair of non-distinct vertices"). Moreover, these edges or loops can be repeated. For a squiggle, we are allowing a loop with no vertex (an empty set of vertices) as an edge. Note that a squiggle is not just a general graph augmented with loops. A squiggle is *a planar drawing* of such an object. If you just have a list of vertices and edges (including empty ones), you will not know what squiggle we are talking about. You have to have a drawing. In fact, two different planar drawings of the same set of vertices and edges may produce different squiggles. Since the drawing is decreed to be planar, whenever two edges cross we must have a vertex. A few examples will make the definition clear.

Example 10.72. A circle ◯ is a squiggle with one edge and no vertices. It splits the plane into two regions. A circle with a vertex either on or off it (◯ or ◉) is also a squiggle. In both cases, the squiggle has one vertex and one edge and the plane is split into two regions.

Figure 10.40 has two other squiggles. The squiggle on the left (which is just a planar graph, and, by definition, a planar drawing of a finite planar graph is a squiggle) has 24 vertices, 30 edges, and partitions the plane into 8 regions. The squiggle on the right has 9 vertices, 5 edges, and partitions the plane into 2 regions.

For every squiggle, we will be interested in the number of points, edges, regions, *and* the number of connected components of vertices. The latter is *not* the same as the number of connected components of a graph. This is because squiggles may have loops with no vertices on them. We just look at the vertices. The vertices that are connected to each other by walks through edges form a component, and we count the number of such components. So, for example, the number of connected components of vertices for the squiggle on the right in Figure 10.40 is five.

---

[14] I learned this approach to Euler's theorem from an expository talk on tilings of the soccer ball by Marty Isaacs.

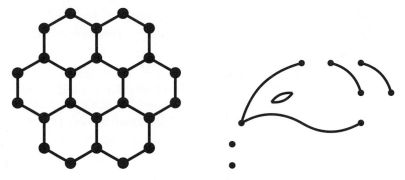

Figure 10.40 Two examples of squiggles.

We make a table, Table 10.5, of numbers of vertices, edges, regions, and connected components of vertices for various squiggles. We urge the reader to try to find a relationship between the various columns, before reading on.

Table 10.5 *v*, *e*, *r*, and *c* are, respectively, the number of vertices, edges, regions, and connected components of vertices of a squiggle.

Squiggle	v	e	r	c
● ● ● ● ●	5	0	1	5
◯	0	1	2	0
◯•	1	1	2	1
◎•	1	2	3	1
◯• △•	5	6	4	2
Figure 10.40 (left)	24	30	8	1
Figure 10.40 (right)	9	5	2	5

We are ready now to prove Euler's polyhedral formula for squiggles.

Theorem 10.73. *For an arbitrary squiggle, let v, e, r, and c denote the number of vertices, edges, regions, and connected components of vertices, respectively. Then*

$$v + r = e + c + 1.$$

Proof. Induct on the number of edges of the squiggle. For the base case, if the number of edges is zero, then we have n vertices and no edges. This means that $v = n$, $r = 1$, $e = 0$, and $c = n$, and the formula checks.

For the inductive step, let S be an arbitrary squiggle with a positive number of edges, and assume that we have already proved the theorem for any squiggle with fewer edges than S. Delete one edge (but no vertices) of S to get another squiggle, S'. (Eliminating just a vertex may turn a squiggle to a non-squiggle, but if you eliminate any edge, you will continue to have a squiggle.) Let v', e', r', and c', respectively, be the number of vertices, edges, regions, and connected components of vertices of S'. By construction, we already know that $e' = e - 1$ and $v' = v$. We can be less certain of how r and c have changed. However, since S' has fewer edges than S, the inductive hypothesis applies, and, by induction, we have $v' + r' = e' + c' + 1$. Replacing v' with v and e' with $e - 1$, we have

$$v + r' = e + c'. \tag{10.1}$$

How much could r' and c' have changed from their original values? When you eliminate an edge, the number of regions could have stayed the same. This happens when the parts of the plane on either side of the edge were already in the same region. Alternatively, eliminating an edge could combine two hitherto-separate regions and hence reduce the number of regions by one. What happens to the number of connected components of vertices when you eliminate an edge? When you eliminate an edge, the only change in c can come through vertices on that edge. (If there are not two vertices on an edge – and there cannot be more than two – and you eliminated that edge, then c could not have changed.) If a and b are the two vertices of the deleted edge, then either a and b remain in the same connected component as before, or each is now in a different connected component (and hence c has increased by 1). So far, we just know that r' is either r or $r - 1$, while c' is either c or $c + 1$.

If the deleted edge had no vertices on it, then it is a closed loop, and by the Jordan curve theorem, Theorem 10.67, this edge divides the plane into two separate regions, and deleting it combines the two regions into one. Hence, in this case, $r' = r - 1$ while $c' = c$ (since no vertices were disturbed).

If the deleted edge had one vertex on it, then, while it is possible for that vertex to be on other edges, the analysis stays the same as in the previous case. The deleted edge is still a closed loop, and, again by the Jordan curve theorem, originally the inside and the outside were two different regions. Removing the edge unites the two regions without affecting the number of connected components of vertices (since the deleted edge was not contributing in connecting that one vertex to other vertices and so deleting it does not change the connections of that vertex to other vertices). So, in this case also, $r' = r - 1$ and $c' = c$.

The only remaining case is if the deleted edge is an ordinary edge with two vertices a and b. There are two subcases here. Either the to-be-deleted edge is a bridge in S or it is not. In the latter case, by Lemma 10.19, our edge is part of cycle, and by the Jordan curve theorem, Theorem 10.67, in S, the inside and outside of the cycle cannot be part of the same region. In S', however, these two regions are combined. Moreover, since a and b are still connected in S' (by the rest of the cycle), the number of connected components of vertices has not changed.

So in this case, we continue to have $r' = r - 1$ while $c' = c$. We are left with the situation where, in S, the to-be-deleted edge is a bridge. Deleting the bridge results in separating one single connected component of vertices into two, and hence, in this case, $c' = c + 1$. Deleting a bridge in a squiggle does not reduce the number of regions. Intuitively, in a planar drawing of a finite general graph (and also a squiggle), the two sides of a bridge are in the same region. This is because the graph is finite, and if the bridge had different regions on its sides, then one of these would have been a finite region bounded by a cycle connecting a and b. This would contradict Lemma 10.19. (This part of the proof is admittedly hand-wavy since, to make it rigorous, we would have to get off-track and use the language and basic tools of topolgy.) Thus, in this final case, $r' = r$ and $c' = c + 1$.

We conclude that, in all cases, if $r' = r$, then $c' = c + 1$. On the other hand, if $r' = r - 1$, then $c' = c$. In the former case, Equation 10.1 becomes

$$v + r = e + (c + 1),$$

while in the latter case, Equation 10.1 becomes

$$v + (r - 1) = e + c.$$

In either case, $v + r = e + c + 1$ and the proof is complete. □

Corollary 10.74 (Euler's Polyhedral Formula).[15] *Let G be a connected planar general graph; then, in a planar drawing of G, we have*

$$v - e + r = 2.$$

Proof. For a connected planar graph, the number of connected components of vertices is one. The result follows from Theorem 10.73. □

What does all of this have to do with soccer balls? We can finally give the surprising answer to the opening problem of this chapter.

Corollary 10.75. *Let $d \geq 3$ be an integer. If you tile a soccer ball with pentagons and hexagons, as long as the degree of every vertex is d, then $d = 3$, and the number of pentagons is 12.*

Proof. Let P and H denote, respectively, the number of pentagons and hexagons in a tiling of the sphere (with the degree of each vertex equal to d). Then each pentagon has five vertices and each hexagon has six vertices. If we go through each of the pentagons and hexagons and count their vertices, we get $5P + 6H$. However, since the degree of each vertex is d, we have counted each vertex d times. Thus, v, the number of vertices in the tiling is $\frac{5P + 6H}{d}$. Similarly,

---

[15] Leonhard Euler (1707–1783) first proposed this formula in a letter dated November 14, 1750 to Christian Goldbach (1690–1764). He wrote. "6. In omni solido hedris planis incluso aggregatum ex numero hedrarum et numero angulorum solidorum binario superat numerum acierum, seu est $H + S = A + 2$, ..." The letter is in German but Euler drifts into Latin when giving mathematical statements. You can find this letter and much more at the Mathematical Association of America's online Euler Archive at http://eulerarchive.maa.org/. Euler attempted a proof two years later, but his proof was incomplete. The first correct proof was given by Adrien Marie Legendre (1752–1833) in 1794 (the year after he lost his family fortune because of the French Revolution). For a quick history of graph theory, see Wilson 2014.

each pentagon and hexagon has five and six edges, respectively. Each edge is in two of these shapes, and so e, the number of edges in the tiling, is $\frac{5P+6H}{2}$. The number of regions is just the total number of pentagons and hexagons and is equal to $P+H$. Hence, by Euler's polyhedral formula, Corollary 10.74, we have

$$\frac{5P+6H}{d} - \frac{5P+6H}{2} + P + H = 2.$$

Multiplying both sides by $2d$ and simplifying, we get

$$(10-3d)P + (12-4d)H = 4d.$$

If $d \geq 4$, then the left-hand side of this equation is a negative integer while the right-hand side is positive. This is impossible, and so $d=3$ is the only possibility. Plug in $d=3$ and we get $P=12$. The proof is complete. $\qquad\square$

Remark 10.76. If you tile a soccer ball with pentagons and hexagons and make sure that the degree of each vertex is 3, then we have just showed that the number of pentagons must be 12. We already have seen examples where the number of hexagons is 0 (Figure 10.36), 2 (Figure 10.37), or 20. In Problem P 10.6.4, you are asked to produce an example where the number of hexagons is 3. In fact, with more work, one can show that the number of hexagons can be any arbitrary non-negative integer except for 1.

Girth and Planar Graphs

Euler's polyhedral formula does much more than help us count the number of pentagons in a spherical tiling. We now use it to prove that certain graphs are non-planar. Recall (Definition 10.28) that the girth of a graph is the length of the shortest cycle in the graph. The girth of a forest (which does not have a cycle) is defined to be infinity.

Theorem 10.77. Let $G = (V,E)$ be a connected planar simple graph with finite girth. Let $v = |V|$ and $e = |E|$, and denote the girth of G by g. Then

$$e \leq \frac{g}{g-2}(v-2).$$

In particular, $e \leq 3(v-2)$, and, if G is triangle-free, then $e \leq 2(v-2)$.

Proof. Each region is bordered by a cycle, and each cycle has at least g edges. Hence, adding the number of edges in the cycles bordering all of the regions, we get at least rg. There may be other edges as well, and each counted edge is counted at most twice – once for each of the regions on either side of it – and so

$$e \geq \frac{rg}{2} \quad \Rightarrow \quad r \leq \frac{2e}{g}.$$

Now, we have $v - e + r = 2$ and hence

$$e = v + r - 2 \leq v + \frac{2e}{g} - 2.$$

Solving for e, we get

$$e\left[1 - \frac{2}{g}\right] \leq v - 2 \quad \Rightarrow \quad e \leq \frac{g}{g-2}(v-2).$$

A graph with a loop has girth 1 and a graph with no loops but with a double edge has girth 2. All simple graphs have girth $g \geq 3$. Now,

$$\frac{g}{g-2} = \frac{1}{1 - \frac{2}{g}},$$

and so a larger g will give a smaller value for $\frac{g}{g-2}$. Hence, since $g \geq 3$, the largest possible value for $\frac{g}{g-2}$ is $\frac{3}{3-2} = 3$. Thus, for all simple planar graphs, we have $e \leq \frac{g}{g-2}(v-2) \leq 3(v-2)$. If we know that the graph is triangle-free, then $g \geq 4$, and the maximum value for $\frac{g}{g-2}$ is $\frac{4}{4-2} = 2$. Hence, for triangle-free planar graphs, we have $e \leq \frac{g}{g-2}(v-2) \leq 2(v-2)$. $\square$

Corollary 10.78. *The complete graph of order 5, K_5, is not planar.*

Proof. The graph K_5 has 5 vertices and $\binom{5}{2} = 10$ edges. But $10 > 3(5-2)$, and this would contradict Theorem 10.77, if K_5 were planar. $\square$

Corollary 10.79. *The complete bipartite graph $K_{3,3}$ is not planar.*

Proof. The graph $K_{3,3}$ has 6 vertices, 9 edges, and is triangle-free (bipartite graphs do not have odd cycles). But $9 > 2(6-2)$ and this would contradict Theorem 10.77, if $K_{3,3}$ were planar. $\square$

An important theorem of the Polish mathematician Kazimierz Kuratowski (1896–1980) – whose proof is beyond the scope of this text – shows that in some sense K_5 and $K_{3,3}$ are the only obstructions to planarity.

Definition 10.80 (Subdivision). *Let $G = (V, E)$ be a graph and let $e = \{x, y\} \in E$. Inserting a vertex on the edge e means constructing a new graph that has one more vertex than G, and instead of e has two edges $\{x, u\}$ and $\{u, y\}$, where u is the new vertex.*

Given a graph G, a subdivision of G is a graph obtained by starting with G and, one by one, inserting a finite number of vertices.

If $G = (V, E)$ is a graph, recall (Definition 2.17) that a subgraph of G is any graph $H = (V', E')$ such that $V' \subseteq V$ and $E' \subseteq E$. In other words, starting with G, if you discard some of the edges and/or some of the vertices together with edges incident with them, then you get a subgraph of G. The following lemma is straightforward.

Lemma 10.81. *Let $G = (V, E)$ be a graph. Let H be a subgraph of G, and G' be a subdivision of G.*

(a) *If G is planar, then so is H.*
(b) *G is planar if and only if G' is planar.*

Proof. Assume G is a planar. Then a plane drawing of G will also contain a plane drawing of H. Also, inserting a vertex on any of the edges of a plane drawing of G will continue to be a plane drawing of a graph. Since G' is obtained from G by repeatedly inserting vertices, G' will be planar also.

When you insert a vertex, the new vertex will have degree 2, and so if you undo the insertion, you will not create any crossings. Hence, if G' is planar, then so is G.   □

The lemma can be used to show that certain graphs are not planar.

Example 10.82. Start with a cycle of length 8 and place the eight vertices $\{1, \ldots, 8\}$ around a circle. Then add four edges connecting vertices opposite each other on the circle (so, $\{1, 5\}$, $\{2, 6\}$, $\{3, 7\}$, and $\{4, 8\}$ are also edges). The resulting graph has 8 vertices and 12 edges. The graph has no triangles and its girth is 4. (See the graph on the left of Figure 10.41.) Since $12 = 2(8 - 2)$, Theorem 10.77 cannot be used to prove that this graph is not planar.

However, if we eliminate the edge $\{1, 5\}$, we get a subgraph (middle graph of Figure 10.41), that has two vertices (vertices numbered 1 and 5) of degree 2. We can think of these two vertices as having been inserted into some original graph. We undo these insertions. This means that we remove the vertex numbered 1 and we replace the two edges $\{1, 2\}$ and $\{1, 8\}$ with one edge $\{2, 8\}$ and similarly remove the vertex 5 and replace $\{4, 5\}$ and $\{5, 6\}$ with $\{4, 6\}$. The resulting graph has six vertices and is isomorphic to $K_{3,3}$. (See Figure 10.41.)

We conclude that the original graph had a subgraph that was a subdivision of $K_{3,3}$, and as a result is non-planar.

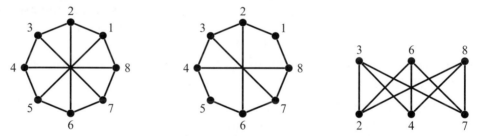

Figure 10.41 An 8-cycle with four chords joining opposite vertices (on the left) has a subgraph (in the middle) that is a subdivision of $K_{3,3}$ (on the right).

We can now state Kuratowski's theorem. The "only if" direction of this theorem follows directly from Lemma 10.81. The content is in the "if" direction, which we will state without proof. (For a proof, see West 2001, Theorem 6.2.2.)

Theorem 10.83 (Kuratowski's Theorem 1930). *Let $G = (V, E)$ be a simple graph. Then G is planar if and only if G does not have a subgraph that is a subdivision of K_5 or $K_{3,3}$.*

Remark 10.84. Just as in Example 10.82, we can use the easy direction of Kuratowski's theorem to prove that specific graphs are non-planar. If we suspect that a graph G is non-planar, we try to find a subgraph of G that is a subdivision of $K_{3,3}$ or K_5. The difficult direction of Kuratowski's theorem assures us that, if G is indeed non-planar, then we will succeed. In our quest to find a subgraph that is a subdivision of $K_{3,3}$ or K_5, we are allowed three types of moves. We can eliminate a vertex and all the edges incident with that vertex; we can eliminate any edge; and if v is a vertex of degree 2 with $\{v, x\}$ and $\{v, y\}$ as the edges incident with it, then we can eliminate v and replace these two edges with the one edge $\{x, y\}$. This last move amounts to undoing an insertion of a vertex. If some combination of these three moves results in $K_{3,3}$ or K_5, then the original graph is non-planar.

Another famous theorem about planar graphs is the four-color theorem. This theorem was proved in 1976 by American mathematician Kenneth Appel (1932–2013) and German-American mathematician Wolfgang Haken. The actual proof used computers in a non-trivial manner, and, in fact, to date a computer-free proof of this theorem is not available.

Definition 10.85. Let G be a simple graph. A *proper coloring* of G is an assignment of colors to the vertices of G in such a way that no two adjacent vertices receive the same color.

Let k be a positive integer. The graph G is *k-colorable* if there exists a proper coloring of G using k colors.

Theorem 10.86 (The Four-Color Theorem). *Every planar graph is 4-colorable.*

You will explore graph coloring in more detail in Section 10.7.

Problems

P 10.6.1. In Figure 10.36, a regular pentagon (i.e., a pentagon with equal internal angles and equal edge lengths) in the center is surrounded by five irregular pentagons. Could we have produced an alternative drawing (in the plane) where this second "circle" of pentagons were all regular as well?

REMARK: There does exist a tiling of the sphere into 12 regular pentagons.

P 10.6.2. Assume G is a simple planar graph with 12 edges and girth greater than 5. (See Definition 10.28.) What are the possibilities for the number of regions for a planar drawing of G? For each possibility, give an example.

P 10.6.3. You can continue the tiling in the figure on the left of Figure 10.40 to get a tiling of the entire plane by just hexagons. Does this mean that you can tile a sphere with just hexagons (and with the degree of every vertex equal to d, for some $d \geq 3$)? Why or why not? Comment.

P 10.6.4. Draw a tiling of a sphere (akin to Figures 10.36 and 10.37) with 12 pentagons and 3 hexagons.

P 10.6.5. I have a squiggle with no vertices but 46 (empty) loops. I don't remember how the loops were drawn (i.e., were they all separate or were some nested inside others?). In a planar drawing of such a squiggle, what are the possibilities for the number of regions? Why?

P 10.6.6. I have met a secretive connected planar graph. I have found out that every vertex has degree 4 and the number of regions is 10. How many vertices does the graph have?

P 10.6.7. Can you draw a connected planar graph that has five regions (including the infinite outside region), all quadrilaterals? If so, do so.

P 10.6.8. We want to tile a soccer ball with triangles and quadrilaterals in such a way that the degree of each vertex is 3. What are the possibilities for the number of triangles and quadrilaterals? For each possibility, draw an example.

P 10.6.9. Assume that we tile the surface of a soccer ball with triangles and quadrilaterals. Also assume that the degree of each vertex is 4. Can you do this? Give an example. Is there a restriction on the number of triangles? (We are allowing for the possibility that the number of either triangles or quadrilaterals, but not both, is zero.)

P 10.6.10. Is there a tiling of a soccer ball with triangles and quadrilaterals where the degree of each vertex is 4 and where the number of quadrilaterals is positive? If so, give an example. If not, prove why not.

P 10.6.11. Can we tile the surface of a soccer ball with hexagons and heptagons if the degree of each vertex is 3? Either give an example of such a tiling or prove that it cannot be done.

P 10.6.12. A friend of mine tiles a soccer ball using triangles and hexagons. She makes sure that at every corner three shapes come together.

(a) Is there any restriction on the number of triangles? Why?

(b) If she tells me that she used two hexagons (and some number of triangles), then what can you say about the number of vertices?

P 10.6.13. We want to tile the sphere with squares and hexagons such that the degree of each vertex is 3.

(a) Does Euler's polygonal formula give any restrictions on the number of squares or the number of hexagons?

(b) Give an example with the smallest number of regions possible.

(c) On the Internet, find a description of a "truncated octahedron." Is there a relation with this problem?

P 10.6.14. We want to tile the sphere with quadrilaterals and pentagons such that the degree of each vertex is 3.

(a) Are there any restrictions, due to Euler's formula, on the number of squares and/or the number of pentagons?

(b) Can you give an example of such a tiling where the number of pentagons is 2?

(c) Can you give an example of such a tiling where the number of regions is 9?

P 10.6.15. (a) Let G be a connected planar simple graph such that all regions (including the outside region of infinite area) are triangles, quadrilaterals, or pentagons. If every vertex has degree 3, show that G has at most 30 edges.

(b) Can you give an example of such a graph with 30 edges?

(c) What can you conclude if, in addition to triangles, quadrilaterals, and pentagons, hexagons are also allowed?

P 10.6.16. Can we have a planar simple graph with eight vertices such that the degree of every vertex is greater than or equal to 5? Why?

P 10.6.17. In Theorem 10.77, it was assumed that the connected planar simple graph G has finite girth. The final conclusions, that $e \leq 3(v - 2)$ and, in the case of triangle-free graphs, $e \leq 2(v - 2)$, do not mention girth. Do these conclusions remain true for planar graphs of infinite girth? Do we need any extra assumptions?

P 10.6.18. G is a simple planar graph with n vertices and $n+1$ edges. Assume that G has a Hamiltonian cycle. What can you say about the total number of distinct cycles of G?

NOTE: Here we consider two cycles to be the same if they have the same vertices in the same or in reverse (circular) order.

P 10.6.19. Let n be a positive integer. Consider the complete bipartite graph $K_{2,n}$. For which values of n is $K_{2,n}$ planar?

P 10.6.20. Find a planar graph whose line graph (Definition 10.50) is non-planar.

P 10.6.21. Is the Petersen graph of Figure 2.1 planar? Why?

10.6.1 Platonic Solids

A *Platonic solid* is a regular, convex polyhedron in $\mathbb{R}^3$. For our purposes, we can think of a Platonic solid as a tiling of the sphere – you can "blow up" a polyhedron into a sphere – with ℓ-gons for a specific positive integer $\ell \geq 3$, and such that all the vertices have the same degree $d \geq 3$. In other words, we want to tile the sphere with only triangles, only quadrilaterals, only pentagons, etc. For an actual Platonic solid, you also want to make sure that all the faces are congruent, but we are not concerned with that here. We just want to know what the possibilities are with the looser condition that you tile the sphere with one type of an ℓ-gon while keeping the condition that the degrees of all vertices are the same and greater than or equal to 3.

Problem P 10.6.22 asks you to pinpoint all the possibilities for the Platonic solids. Problem P 10.6.23 then asks you to show that one of the restrictions observed for the Platonic solids (that ℓ can only be 3, 4, or 5) is not limited to this particular configuration. In fact, surprisingly, the girth of any planar graph is always bounded by 5 as long as all the degrees are at least 3.

Problems (continued)

P 10.6.22. We tile a sphere with only ℓ-gons (i.e., each region in the tiling has exactly ℓ sides), and each vertex has degree equal to d. Assume $\ell, d \geq 3$, and let L be the number of ℓ-gons in the tiling.

(a) Find L in terms of ℓ and d.
(b) Argue that, for L to be positive, we need $\ell \leq 5$.
(c) For each possible ℓ, find the possible values for d.
(d) For each possible scenario, what is the value of L?

(e) For each of the possible tilings, if there isn't already a planar drawing of it in this section, then produce one.

P 10.6.23. Let G be a connected simple planar graph. Assume that the degree of each vertex of G is at least 3. Prove that the girth (Definition 10.28) of G is less than or equal to 5.

P 10.6.24. In Problem P 10.6.23, we insisted that the degree of each vertex be at least 3. Could we have allowed degree-1 vertices? What about degree-2 vertices?

10.6.2 Pick's Formula

In the ensuing problems, closely following Gaskell, Klamkin, and Watson (1976), modulo one geometric fact, you will use Euler's formula to prove the magical Pick's formula.[16] A planar drawing of a finite cycle with edges drawn as straight line segments is called a *simple polygon*. How do you find the area enclosed by such a polygon? You could split the enclosed area into triangles, find the area of each triangle, and then add them up. If you are lucky and all of the vertices of your polygon have integer coordinates, then, instead and more simply, you can use Pick's elegant formula. We start with a definition.

Definition 10.87. A point in the plane with integer coordinates will be called a *lattice point*,[17] and the set of lattice points in the plane will be denoted by $\mathbb{Z}^2$.

Theorem 10.88 (Pick's Formula). *Let C be a simple polygon whose vertices are lattice points. Let I and B be, respectively, the number of lattice points in the interior and on the boundary of C, and let A denote the area enclosed by C. Then*

$$A = I + \frac{B}{2} - 1.$$

Please note that Pick's formula is not an approximation. It is an exact formula!

Example 10.89. Consider the polygon of Figure 10.42. There are two lattice points in the interior, and so $I = 2$. There is a total of 11 lattice points on the boundary, and so the area enclosed by the polygon is

$$A = 2 + \frac{11}{2} - 1 = 6.5 \text{ units.}$$

To prove this theorem, we need one geometric fact (Lemma 10.91), which we will accept without proof. To state it, we make a definition.

---

[16] Named after Georg Alexander Pick (1859–1942) a Jewish-Austrian mathematician who discovered the formula in 1899. Pick got his doctorate at the University of Vienna but spent his career at the Charles-Ferdinand University in Prague. He helped Einstein get a job at the German University of Prague and introduced him to some of the mathematics that Einstein later found useful. On July 13, 1942, the Nazis sent Pick to the Theresienstadt concentration camp, where he died two weeks later.

[17] We use this term, as is customary, for the purposes of discussing Pick's formula. However, in other contexts, "lattice point" is defined much more broadly.

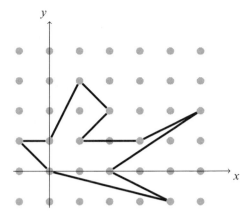

Figure 10.42 A simple polygon with vertices that are lattice points (i.e., have integer coordinates).

Definition 10.90 (Primitive Triangle, Primitive Triangulation). A *primitive triangle* is a triangle whose vertices are lattice points but that does not have any other lattice points on its boundary or interior. A *primitive triangulation* of the interior of a simple polygon is a partition of the area enclosed by the simple polygon with primitive triangles using every lattice point on the boundary and in the interior of the polygon as vertices.

Given a simple polygon with vertices in $\mathbb{Z}^2$, by using *every* lattice point on the boundary and in the interior of the polygon as vertices, we can certainly partition the interior of the polygon into triangles, and in fact we can make sure that each of these triangles is a primitive triangle. See Figure 10.43 for a primitive triangulation of the interior of the simple polygon of Figure 10.42.

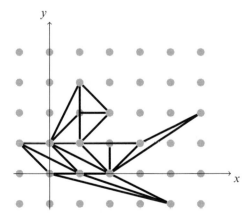

Figure 10.43 A triangulation of a polygon.

Lemma 10.91. *The area of every primitive triangle is* $1/2$.

If you look at some examples (Problem P 10.6.25), you can convince yourself that this must be true, but we will not take the time to prove it rigorously. For a proof see Honsberger (1970), Essay 5, or Gaskell et al. (1976).

Given Lemma 10.91, the area enclosed by a simple polygon will be one-half the number of triangles in a primitive triangulation of the interior of the simple polygon. So, to prove Pick's theorem, we now focus on the number of triangles. In fact, at this point, we need not worry about polygons or lattice points. Instead, we consider the more general situation of planar graphs where all the regions other than the infinite region are triangles, and the boundary of the infinite region is a cycle. To facilitate the discussion, we make a definition.

Definition 10.92 (Triangulation). Let C be the planar drawing of a finite cycle. By a *triangulation* of C, we mean the planar drawing of any finite simple graph $G = (V, E)$ such that

(a) G has no isolated vertices or leaves,
(b) the cycle C is the boundary of the outside infinite region, and
(c) other than the infinite region, all the other regions ("inside" C) are triangles.

Note that, for our purposes and in this general setting, in a triangulation, you can have any number of vertices in the interior of the cycle, and so the number of vertices, edges, or regions is not fixed.

Example 10.93. Figure 10.43 is a triangulation of the simple polygon of Figure 10.42. Figure 10.35 can be seen as a triangulation of a cycle of length 12. (Here, since we are not concerned about meaningful tilings of the sphere, we are allowing vertices of degree 2.)

In Problem P 10.6.26, you will use a nifty thought experiment and Euler's formula to prove the following:

Proposition 10.94. *Let $G = (V, E)$ be the triangulation of an n-cycle, and let t be the number of finite triangles in G (not counting the infinite region even if $n = 3$). Then*

$$t = 2|V| - n - 2.$$

Pick's formula is an immediate consequence (see Problem *P 10.6.27*.)

Problems (continued)

P 10.6.25. Consider the triangle whose vertices are $(0,0)$, $(1,1)$, and $(8,7)$. Is this a primitive triangle? What is its area?

P 10.6.26. **Proof of Proposition 10.94.** Let C be a cycle of length n drawn as a circle, and let $G = (V, E)$ be a triangulation of C. Let t be the number of triangles in the triangulation. Perform the following thought experiment. Draw C as the equator on a sphere. Draw two copies of G, one on the northern hemisphere and another on the southern hemisphere. The two copies of G share the vertices and edges of C but otherwise are disjoint.[18]

(a) Do you have a tiling of the sphere? Into what shapes? Do all the vertices necessarily have the same degree?

(b) Let v', e', and r' be the number of vertices, edges, and regions of the new (expanded) graph on the sphere. Find expressions for v' and r' in terms of n, $|V|$, and t.

(c) Find an expression for e' in terms of just t.

(d) Use Euler's formula to prove Proposition 10.94.

P 10.6.27. Proof of Pick's formula. Use Proposition 10.94 (proved in Problem P 10.6.26) and Lemma 10.91 to give a proof of Pick's formula of Theorem 10.88.

P 10.6.28. Let $G = (V, E)$ be the triangulation of an n-cycle, and let t be the number of finite triangles (not counting the infinite region).

(a) Show that $|E| = 3\,|V| - n - 3$.[19]

(b) Show that $t = |E| - |V| + 1$.

10.6.3 Counting the Regions in a Circle

Pick n points around a circle and draw a line segment from each of these points to each of the other $n - 1$ points, making sure that no three lines go through the same point in the interior of the circle. Some of these lines cross and hence the region inside the circle is cut up into smaller regions. Let s_n be the number of regions formed *inside* the circle. For example, $s_1 = 1$, $s_2 = 2$, and $s_3 = 4$. (See Figure 10.44.)

Figure 10.44 Place n points around the circle, join every pair, and count the number of regions inside the circle.

In Problems P 10.6.29–P 10.6.31, you are asked to investigate the sequence $\{s_n\}_{n=0}^{\infty}$.

Problems (continued)

P 10.6.29. (a) Consider the configuration for $n = 4$ in Figure 10.44, and add one new vertex at the intersection of the two diagonals to create a planar general graph. What is the number

---

[18] A similar thought experiment appeared in Stein 1969. Problems P 10.6.26 and P 10.6.27 were adpated from Gaskell et al. 1976.

[19] Adapted from Funkenbusch 1974.

of vertices, and what is the number of edges? Use Euler's polyhedral formula to find the number of regions. What is s_4?

(b) Based on the values of s_1, s_2, s_3, and s_4, do you have a conjecture about s_n?

(c) Check that your conjecture works for s_5.

(d) Now carefully (and tediously) check your conjecture for s_6.

(e) Comment.

P 10.6.30. Let n be a positive integer. Draw a line segment from each of n points on a circle to each of the other $n - 1$ points, making sure that no three lines go through the same point in the interior of the circle. Add additional vertices in the interior of the circle, at each of the intersections of the line segments, so that you have the planar drawing of a general graph.

(a) Prove that the number of vertices of this graph is $n + \binom{n}{4}$.

(b) If the number of edges is denoted by e_n, prove that

$$e_{n+1} = e_n + n + 1 + \sum_{i=2}^{n-1}(i - 1)(n - i).$$

(c) Use the previous parts and Euler's polyhedral formula to find s_7.

P 10.6.31. Let ℓ be a non-negative integer. Draw ℓ chords on a given circle (a chord is a straight line segment in the interior of a circle that starts and ends on the circle) in such a manner that no three chords intersect at the same point. Assume that the chords have m points of intersection in the interior of the circle.

(a) Show that the number of regions inside the circle is $1 + \ell + m$.

(b) Let n be a positive integer. Draw a line segment from each of n points on a circle to each of the other $n - 1$ points, making sure that no three lines go through the same point in the interior of the circle. Show that the number of regions inside the circle is

$$\binom{n}{0} + \binom{n}{2} + \binom{n}{4}.$$

(c) If $n = 7$, compare the answer in the previous part with that of Problem P 10.6.30.

10.6.4 Graph Vocabulary: Crossing Number

The *crossing number* $c(G)$ of a graph G is the minimum possible number of pairs of crossing edges among all the drawings of G on the plane. Note that if G is planar, then $c(G) = 0$.

In mathematics, whenever we are confronted with a question that has a binary "yes" or "no" answer, we strive to rephrase the question so that the answer can be a larger range of possibilities. Hence, instead of asking "Is the graph G planar?", we prefer to ask "What is the crossing number of G?" The answer to the latter not only tells us whether G is planar or not; it provides some way of measuring whether the graph is close to being planar.

Problems (continued)

P 10.6.32. We know that $K_{3,3}$ and K_5 are not planar. What is their crossing number?

P 10.6.33. Is it possible to draw $K_{3,3}$ on the plane with as few crossings as possible and with all the edges being straight line segments? If so, do so.

P 10.6.34. Consider the graph of Example 10.82 (a cycle of length 8 together with four more edges connecting opposite pairs of vertices). What is the crossing number of this graph?

10.6.5 Graphs on Other Surfaces

Planar graphs are graphs that we can draw on the plane without edges crossing (Definition 10.65). Finite planar graphs are also exactly the graphs that you can draw on a sphere without their edges crossing. We saw in the case of tilings of a soccer ball how to go from a drawing on a sphere to a planar graph and back. What about other surfaces? Is there a non-planar graph that we can draw on a torus without its edges crossing? A *torus* is the surface of a donut (or a tire).

For spheres, we figured out a way to draw, on the plane, a representation of a graph on a sphere (the discussion before Example 10.63). We will do the same for a torus. Consider a rectangular piece of paper. To make a torus from this, we first tape two opposing edges of the paper to make a cylinder, and then bend the cylinder so that the two circles, on either end, come together. Taping these circles together gives us a torus. We can reverse this process as well. You can cut the torus (think of a hollow donut) with a circular cross section (not in the direction that you would usually cut a bagel, but perpendicular to it), unwind it into a (hollow) cylinder, then cut along a straight line on the cylinder, and flatten it to get a rectangle.

So, instead of a torus, we can draw a rectangle. But we have to be careful. The opposing edges of the rectangle were glued together on the torus. We take care of this by "identifying" every point on the perimeter of the rectangle with the point directly across on the other side of the rectangle. Identifying two points means that the two points are really the same point. We have just drawn two copies of the same point. This allows edges that enter one of these points to exit their clones on the other side.

As an example, consider Figure 10.45. The rectangle denotes the torus but with the proviso that the points on opposite edges are identified. In other words, A and B may look like two different points, but they are not. A and B are two points on the plane, but on the torus they are the same point. Likewise, C and D are identified and are the same point on the torus. As a result, even though it may look like we have drawn two line segments – one from A to D and another from C to B – on the torus, this is really a loop! You go from A to D, which is the *same* point as C, and then you continue to B. But B is the point A and so you have completed a loop. You started from A and went around the torus and came back to A.

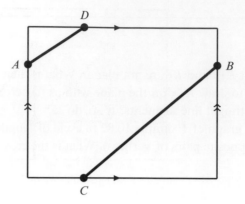

Figure 10.45 *A* and *B* are the same point, as are *C* and *D*. This is a drawing of a loop on a torus.

One convention for these drawings is the arrows on the edges. In Figure 10.45, you may have noticed a right arrow on each of the horizontal edges of the rectangle. This is meant to convey that the points on these two edges are identified. The double up arrows on each of the two vertical edges of the rectangles tell us that those two edges are also identified. We used a single arrow for one case and a double arrow for the other, so that there would be no confusion. For some other surfaces, the arrows on opposing edges may be in opposite directions (for example, a down arrow on one vertical edge and an up arrow on the other). This will mean that the two edges are first "twisted" and then identified. For example, Figure 10.46 is how we would depict a Möbius band . In this drawing, the point *A* is identified with point *B* but the point *C* has no clones (note that there are no arrows on the horizontal edges).

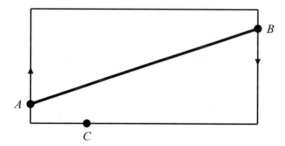

Figure 10.46 *A* and *B* are the same point, but *C* has no clones. This is a drawing of a loop on a Möbius band.

Example 10.95. We saw that the complete graph K_5 was not planar (Corollary 10.78) and so any drawing of it in the plane or on the sphere will have edges crossing. We can, however, draw K_5 on a torus without any edges crossing. See Figure 10.47.

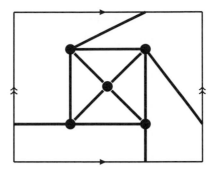

Figure 10.47 A drawing of the complete graph K_5 on a torus without any edges crossing.

Problems (continued)

P 10.6.35. We have four houses and three wells. We want to draw a path from each house to each well, in such a way that no two paths cross. Can we do this, if the houses and the wells are on a torus? I have already started the drawing (see Figure 10.48). If possible, complete it.

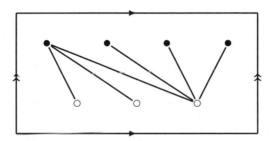

Figure 10.48 The vertices and some edges of $K_{4,3}$ (on a torus).

P 10.6.36. Can we draw K_5 on a Möbius band (see Figure 10.46) without the edges crossing?

P 10.6.37. Let n be a positive integer. For which n can you draw the complete bipartite graph $K_{3,n}$ on a torus (without any edges crossing)? Make a conjecture, and provide a drawing for the largest possible n.

P 10.6.38. Which of the complete graphs K_n can you draw on a torus (without any of the edges crossing)? Make a conjecture. Provide a drawing for the largest possible n.

P 10.6.39. Draw the Petersen graph of Figure 2.1 on a torus without its edges crossing. Your drawing of the Petersen graph partitions the torus into how many regions? If v, e, and r denote, as usual, the number of vertices, edges, and regions, respectively, then what is $v - e + r$ for your drawing?

P 10.6.40. Does the Jordan curve theorem, Theorem 10.67, remain valid on a torus? In other words, does every cycle on the torus divide the surface into two regions?

P 10.6.41. If we draw a simple graph on a sphere, then, by Euler's polyhedral formula, $v - e + r = 2$, where v, e, and r are the number of vertices, edges, and regions, respectively. Draw a single triangle on a torus in such a way that this formula also holds, and draw a different triangle for which this formula does not hold.

10.7 Graph Coloring and More*

Warm-Up 10.96. *You want to schedule six math classes in as few time slots as possible. The classes are Calc II (Calc), Linear Algebra (LA), Combinatorics (Comb), Abstract Algebra (ALG), Operations Research (OR), and Vector Calculus (VC). A student survey shows that there are students who want to take the following combinations of classes: {VC, LA}, {VC, OR}, {VC, Comb}, {LA, Comb}, {OR, ALG}, and {Comb, ALG}, and you want to allow them to do so. In addition, the same instructor is teaching both Calc and VC, and so these two classes have to be scheduled in different slots. What is the minimum number of time slots needed?*

In this optional section, by working through the problems, you will be introduced to the fertile area of graph coloring. This is an area of graph theory with many applications, as well as connections to a myriad of other areas. Early in the text, when doing Ramsey theory, we also dealt with graphs and colors. The emphasis here is a bit different, and we are more interested in grouping things – these could be vertices or edges – in a way that avoids certain conflicts. The most common scenario is when we are coloring the vertices of a graph, but we want adjacent vertices to have different colors. (Along these lines, we have already discussed the four-color theorem in the previous section.) But there are many other variants. For example, you may want to color the edges, or you may have a specific list of permissible colors for each vertex. In turn, we will discuss (1) the chromatic number, (2) the chromatic polynomial, (3) the edge chromatic number, and (4) list colorings and the choice number of a graph. We will close with a few tiling problems that tend to also employ certain kinds of colorings. Two highlights of this section are a proof of Richard Stanley's 1973 theorem (Stanley 1973) that connects the number of acyclic orientations of a graph to the chromatic polynomial of the graph (Problem P 10.7.23), and Carsten Thomassen's 1994 proof (Thomassen 1994) that every planar graph is 5-choosable and therefore 5-colorable (Problem *P 10.7.34*).

Each section begins with a few brief definitions. The actual material is developed through the problems.

10.7.1 Graph Vocabulary: Proper Coloring and Chromatic Number, Independent Set and Independence Number

Before stating the formal definitions, and as motivation, we give one example.

Example 10.97. In most computer programs, you define a number of variables. The value of these variables can be stored in CPU registers or in the computer's RAM. Accessing RAM is slower than accessing the CPU registers, but the number of CPU registers is limited. It is possible to store more than one variable in a register as long as two variables are not needed at the same time. One way to model this situation is to construct what is called the *interference graph*. The vertices of this graph are the variables, and two variables are adjacent if they are needed by the program at the same time at some point. You then try to assign colors to vertices, with the proviso that no two adjacent vertices can get the same color. All the vertices that are colored red, for example, can then be assigned to the same register. Since the number of registers is limited, you want to use the smallest possible number of colors for coloring the vertices. This number is the *chromatic number* of the graph.

We already defined proper coloring (Definition 10.85) when discussing the four-color theorem, but we will repeat the definition here for completeness.

Definition 10.98 (Proper Vertex Coloring, Chromatic Number). A coloring of the vertices of a simple graph G is called *proper* if no two adjacent vertices receive the same color. (See Figure 10.49.) The smallest number of colors needed for a proper coloring of a graph G is called the *chromatic number* of G and is denoted by $\chi(G)$.

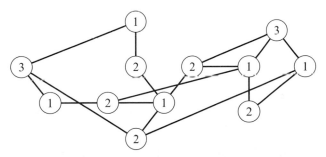

Figure 10.49 The 12 vertices of a graph have been colored (labeled) using the three colors 1, 2, and 3. The coloring is proper, since no two adjacent vertices have the same color. Since this graph cannot be properly colored using two colors, the chromatic number of the graph is 3.

Remark 10.99. In a general graph with a loop, there are no proper colorings, and having multiple edges versus having a single edge between two vertices makes no difference in properly coloring the vertices. As a result, when discussing proper vertex colorings and the chromatic number, there is no harm in limiting the discussion to simple graphs.

Definition 10.100 (Independent Set, Independence Number). Let G be a simple graph. An *independent set* (or a *stable set*) in G is a set of vertices of G, no two of which are adjacent. The size of the largest independent set in G is called the *independence number*, and is denoted by $\alpha(G)$.

Remark 10.101. When you properly color the vertices of a simple graph, none of the vertices with the same color share an edge, and so these vertices form an independent set of the graph. As a result, a proper coloring of a graph is equivalent to partitioning the set of vertices into independent sets.

Problems

P 10.7.1. What is the chromatic number of a cycle of length 5? What is the chromatic number of K_4, the complete graph of order 4?

P 10.7.2. What is the chromatic number of the Petersen graph of Figure 2.1?

P 10.7.3. A Sudoko board is constructed by arranging nine 3×3 boards in three rows and three columns to create a 9×9 board. Imagine constructing a simple graph $G = (V, E)$ where $|V| = 81$, and each vertex corresponds to a square on the Sudoko board. Two vertices of G are adjacent if they are in the same row or the same column, or if they are both in the same original 3×3 board. What can you say about $\chi(G)$, the chromatic number of this graph?

P 10.7.4. Let G be a simple graph with at least one edge. Prove that $\chi(G) = 2$ if and only if G is bipartite.

P 10.7.5. Let $G = (V, E)$ be a simple graph. As usual, $\alpha(G)$ and $\chi(G)$ denote the independence number and the chromatic number of G, respectively. Prove that

$$\alpha(G)\chi(G) \geq |V|.$$

P 10.7.6. Let G be a simple graph, and let $\Delta(G)$ denote the maximum degree of a vertex in G. Prove that

$$\chi(G) \leq \Delta(G) + 1.$$

Give examples of three simple graphs G such that $\chi(G) = \Delta(G) + 1$.

P 10.7.7. Let G be a simple graph with degree sequence $d_1 \geq d_2 \geq \cdots \geq d_n$. Make a table as follows:

i	1	2	...	n
degrees	d_1	d_2	...	d_n
$i - 1$	0	1	...	$n - 1$
$\min\{d_i, i - 1\}$	0	$\min\{d_2, 1\}$	...	$\min\{d_n, n - 1\}$

The last row of the table records $\min\{d_i, i - 1\}$, for $1 \leq i \leq n$. Let M_G be the maximum integer in the last row.

(a) What does the table look like and what is M_G for $G = K_n$, the complete graph on n vertices?

(b) What about for a cycle of length 5?

(c) And the Petersen graph of Figure 2.1?

(d) Prove that $\chi(G) \leq 1 + M_G$, for all simple graphs G.

P 10.7.8. Let G be a simple graph of order n. Order the vertices of G as follows. Let v_1 be a vertex of G of least degree. Now delete v_1 and all of its edges. Let v_2 be a vertex of least degree in the remaining graph. Continue eliminating the last chosen vertex and all of its edges, and letting the next chosen vertex be a vertex of minimum degree among the rest. Now that you have ordered the vertices $v_1, v_2, \ldots, v_n$, for each vertex count the number of neighbors that it has among only the vertices that come after it in the sequence. Let k be the maximum value of these counts. The integer k is called the *degeneracy* of the graph G, and is a measure of how "sparse" the graph is.

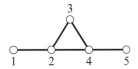

Figure 10.50   A graph with five vertices and five edges.

(a) What is the degeneracy of the graph of Figure 10.50?
(b) What is the degeneracy of the complete graph K_n?
(c) What is the degeneracy of a tree with at least one edge?
(d) Show that, for a simple graph G, $\chi(G) \leq 1 + \text{degeneracy}(G)$.

P 10.7.9. In Problem P 10.6.39, you drew the Petersen graph of Figure 2.1 on the torus without its edges crossing. In your drawing, give each region a color in such a way that no two neighboring regions have the same color. What is the minimum number of colors necessary?

10.7.2 Graph Vocabulary: Chromatic Polynomial

Remark 10.102. If $C = \{c_1, c_2, \ldots, c_k\}$ is a set of k colors, and if $G = (V, E)$ is a simple graph, then a coloring of G is a map $f : V \to C$. The function f assigns a color to each vertex, and two colorings are different if they assign different colors to any one of the (distinct) vertices.

Definition 10.103 (Chromatic Polynomial). Let G be a simple graph, and let k be a positive integer. Let $\chi(G; k)$ denote the number of ways that we can color the vertices of G properly using k colors. The function $\chi(G; k)$ is called the *chromatic polynomial* of G.

Remark 10.104. It turns out that $\chi(G; k)$ is a *polynomial* in k. This fact is not necessarily clear from the definition, and needs proof (see Corollary 10.112). It is clear, however, that $\chi(G; k) = 0$ for all $0 < k < \chi(G)$.

Example 10.105. Let G be the graph on the left of Figure 10.51. What is $\chi(G; k)$, the chromatic polynomial of G? We are counting the number of proper colorings of G using k colors. We have k choices of color for vertex 1. Regardless of what that choice is, we have $k - 1$ choices for vertex 2. At this point, vertex 1 is out of the picture. For vertex 3, we again

Figure 10.51 What are the chromatic polynomials of these graphs?

have $k-1$ choices of colors, since we cannot use whatever color we chose for vertex 2. Finally, for vertex 4, we can choose neither of the two different colors that we used for vertices 2 and 3. Hence, we have $k-2$ choices for this vertex. We conclude that the total number of proper colorings for G using k colors is $\chi(G;k) = k(k-1)^2(k-2)$. Note that plugging in $k = 0,1$, or 2 gives us 0, while $\chi(G;3) = 12$. This means that we cannot color G properly with 0, 1, or 2 colors, and using 3 colors we can properly color it in 12 ways. As a by-product, we know that $\chi(G)$, the chromatic number of G, is 3.

The polynomial $\chi(G;k)$, in addition to giving us $\chi(G)$, has quite a bit more information about the graph G. For example, if, in our case, you multiply the polynomial out, you get $\chi(G;k) = k^4 - 4k^3 + 5k^2 - 2k$. In this polynomial the absolute value of the coefficient of the second-highest power of k – in this case $|-4| = 4$ – is always the number of edges! Less surprisingly, the degree of the polynomial is the number of vertices.

And here is a remarkably strange fact. If you plug -1 into this polynomial – Why would you do that? That may seem like we are asking how many ways we can properly color the vertices of the graph using -1 colors – you get something worth knowing. In our case, we get 12. What does 12 count? Think of the edges as two-way streets. You may want to put an arrow on an edge to make it a one-way street. Twelve is the number of ways that you can make every edge on the graph a one-way street without creating a directed cycle (if you can start at some vertex and traverse the streets while observing the one-way signs and somehow get back to where you started, you will have a directed cycle). In more formal language (see Theorem 10.114), if G is a simple graph of order n, then $\chi(G, -1)$ is $(-1)^n$ times the number of acyclic orientations of the graph G! See Figure 10.52 for the six acyclic orientations of a triangle.

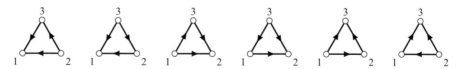

Figure 10.52 The six acyclic orientations of a triangle.

Example 10.106. Now consider a cycle of length 4. See the graph on the right of Figure 10.51. Can we proceed as we did in the previous example? We have k choices of colors for vertex 1, then $k-1$ choices of colors for vertex 2, and another $k-1$ choices for vertex 3. However, the situation is not as straightforward for vertex 4. For that vertex, we cannot use the colors that were used for vertices 1 or 3, but we don't know if we have used the same color or different colors for these two vertices, and that makes a difference for the number of our choices for vertex 4. For a complicated graph, finding the chromatic polynomial by asking for the number of choices of colors for each vertex does not work very well. We need other methods, which we will soon develop. However, a graph as small as a cycle of length 4 can be approached in many ways. One way to proceed is to use the principle of inclusion–exclusion. (See Example 10.113 for a much better method.) Let S be the set of all possible colorings of the vertices – proper or not – using k colors. Since there are k choices for each vertex, $|S| = k^4$. For $i = 1, 2, 3$, let A_i be the set of colorings where vertices i and $i + 1$ had the same color. Also let A_4 be the set of colorings where vertex 4 and vertex 1 had the same color. We want to find $\left| A_1^c \cap A_2^c \cap A_3^c \cap A_4^c \right|$. Now for $1 \leq i \leq 4$, $|A_i| = k^3$ (we have k choices for the two vertices whose colors are going to match and k choices for each of the other two). Similarly, for $1 \leq i < j \leq 4$, $\left| A_i \cap A_j \right| = k^2$. This is because, for $A_i \cap A_j$, we either want three consecutive vertices that have the same color, with total freedom for the fourth vertex, or we want two pairs of vertices, with the two vertices in each pair to get the same color. Regardless, the number of choices is k^2. For the triple intersections or the quadruple intersection, we have k choices of colors, since we have to use the same color for all the vertices. Hence, by the inclusion–exclusion principle,

$$\chi(C_4; k) = k^4 - \binom{4}{1}k^3 + \binom{4}{2}k^2 - \binom{4}{3}k + \binom{4}{4}k = k^4 - 4k^3 + 6k^2 - 3k$$
$$= k(k - 1)(k^2 - 3k + 3).$$

Remark 10.107. We claimed that the fact that $\chi(G; k)$ is a polynomial is not obvious from the definition, and needs proof. You may be tempted to think that, for every vertex of the graph, the number of choices is something like $k - a$ where a is some integer, and so $\chi(G; k)$ always factors as a product of such terms. If this were true, then roots of $\chi(G; k)$ would always be integers, and $\chi(G; k)$ would automatically be a polynomial. As Example 10.106 shows, this is false. A more striking example is the chromatic polynomial of the Petersen graph of Figure 2.1. It is

$$k(k - 1)(k - 2)\left(k^7 - 12k^6 + 67k^5 - 230k^4 + 529k^3 - 814k^2 + 775k - 352\right).$$

To have a robust way both to calculate specific chromatic polynomials and to prove general statements about them, we turn to a recursive formulation. For this, we first introduce two important operations, of deletion and contraction on a graph.

Definition 10.108. Let G be a general graph and $e = \{x, y\}$ an edge of G. We define two new graphs, $G - \{e\}$ (also denoted $G\backslash e$) and G/e, that are obtained from G by *deletion* and *contraction* of the edge e. The graph $G - \{e\}$ has the same vertices as the graph G, and all the edges of G except e.[20] The graph G/e is the same as the graph G except that the vertices x and y – the two ends of e – are identified and the edge e is discarded. In other words, the vertices

---

[20] The deletion of an edge may result in an isolated vertex. For some purposes, it is convenient to eliminate those, but for our current purpose we are keeping the isolated vertices.

x and y are fused together into one new vertex v_e, and any vertex that was adjacent to either x or y is now adjacent to v_e.[21]

Example 10.109. Let G be a 4-cycle, and let e be an edge of G. Then $G - \{e\}$ is a path of length 3 and G/e is a triangle. See Figure 10.53.

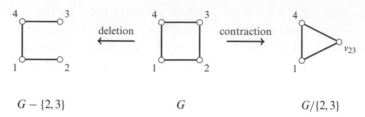

$$G - \{2, 3\} \qquad\qquad G \qquad\qquad G/\{2, 3\}$$

Figure 10.53 G is a cycle of length 4 (center). If we delete the edge $\{2, 3\}$, we get a path of length 3 (on the left), and if we contract the edge $\{2, 3\}$, we get a triangle (on the right).

Remark 10.110. Deletions and contractions are extremely versatile and important tools in advanced graph theory. Here, we are focusing on their use for finding chromatic polynomials recursively, and in Problem P 10.7.20 you will use them to count spanning trees. But there is much more than can be done with them. If G is a simple graph, any graph that you can obtain by a sequence (in any order) of deletions and contractions is called a *minor* of G. Much of structural graph theory is centered around the idea of classifying classes of graphs by focusing on minors that they can or cannot contain. For example, we have mentioned Kuratowski's theorem, Theorem 10.83, that a graph is planar if and only if it does not contain a subdivision of K_5 or $K_{3,3}$. Another version of this result, proved by the German mathematician Klaus Wagner (1910–2000), can be derived from Kuratowski's theorem, and says that a graph is planar if and only if it does not have K_5 or $K_{3,3}$ as a minor. A major result in structural graph theory is the *graph minor theorem* of Neil Robertson and Paul Seymour, which was proved in a series of papers from 1983 to 2004, and spans more than 500 journal pages. This theorem simply says that in *any* infinite family of finite simple graphs, you can find at least two graphs such that one is the minor of the other. Why is this important? If you are interested in a property of graphs (e.g., planarity) and you know that this property is minor-closed (a property of graphs is minor-closed if, whenever a graph has the property, all of its minors have the same property), then it follows from the graph minor theorem that there are a finite number of graphs that are obstructions to that property (just like K_5 and $K_{3,3}$ are obstructions to planarity). In other words, there is a finite list of graphs called the excluded minors for that property, and any graph has the property if and only if it does not contain, as a minor, one of the excluded minors. As a result, if you can find the list of the excluded minors, it

---

[21] The contraction of an edge, even in a simple graph, may create multiple edges between two vertices. For some purposes – e.g., counting spanning trees in Problem P 10.7.20 – it is important to keep the multiple edges. For the purposes of proper coloring of vertices, a double edge makes no difference, but if we want to end up with a simple graph after contracting an edge, we can discard the double edges. Deleting and contracting a loop result in identical graphs, and contracting an edge never results in a new loop.

is theoretically possible to classify all graphs with that property, the way the Kuratowski–Wagner theorem classifies planar graphs.

The following theorem, by using deletions and contractions, allows us to reduce the problem of finding chromatic polynomials to that of finding chromatic polynomials for graphs with fewer edges. As a corollary, we also get that the chromatic polynomial is a polynomial. You are asked to prove these in Problems P 10.7.14 and P 10.7.17.

Theorem 10.111 (Deletion–Contraction Theorem). *Let G be a simple graph and let e be an edge of G. Then*

$$\chi(G;k) = \chi(G - \{e\};k) - \chi(G/e;k).$$

Corollary 10.112. *Let G be a simple graph of order n. Then $\chi(G;k)$ is a polynomial of degree n with integer coefficients.*

Example 10.113. The chromatic polynomial of P_3, a path of length 2, is $k(k-1)^3$, and the chromatic polynomial of K_3, a triangle, is $k(k-1)(k-2)$. Hence, by Theorem 10.111, if C_4 is a cycle of length 4, then

$$\chi(C_4;k) = \chi(P_3;k) - \chi(K_3,k) = k(k-1)^3 - k(k-1)(k-2)$$
$$= k(k-1)(k^2 - 3k + 3).$$

The ease of calculation compares favorably with our method of Example 10.106.

Problems *P 10.7.22* and P 10.7.23 introduce acyclic orientations of graphs, and you will prove the following curious and delightful fact mentioned in Example 10.105.

Theorem 10.114 (Stanley 1973). *Let G be a simple graph of order n; then*

$$number\ of\ acyclic\ orientations\ of\ G = (-1)^n \chi(G;\ -1).$$

Problems (continued)

P 10.7.10. Let $G = K_n$ be the complete graph on n vertices. Find $\chi(G;k)$, the chromatic polynomial of G. If we expand this polynomial, what is the coefficient of k^t, for some non-negative integer t?

P 10.7.11. Find the chromatic polynomial of the graph in Figure 10.50.

P 10.7.12. Find the chromatic polynomial of the graph in Figure 10.54.

Figure 10.54 What is the chromatic polynomial of this graph?

P 10.7.13. Directly, and with no recourse to Theorem 10.111, find the chromatic polynomial of the graph in Figure 10.55.

Figure 10.55 A graph with four vertices and five edges.

P 10.7.14. **Proof of Theorem 10.111.** Let $G = (V, E)$ be a simple graph, and let $x, y \in V$ with $e = \{x, y\} \in E$. Let $G - \{e\}$ and G/e denote, as usual, the graphs obtained by, respectively, deleting and contracting e.

(a) Consider all proper vertex colorings of $G - \{e\}$ in which x and y have *different* colors. How do these compare with all the proper vertex colorings of G?

(b) Consider all proper vertex colorings of $G - \{e\}$ in which x and y have the *same* color. How do these compare with all the proper vertex colorings of G/e?

(c) Prove Theorem 10.111. In other words, show that

$$\chi(G; k) = \chi(G - \{e\}; k) - \chi(G/e; k).$$

P 10.7.15. Let G be the graph of Figure 10.55. Re-do Problem P 10.7.13, but this time use the deletion–contraction theorem, Theorem 10.111.

P 10.7.16. Let $n \geq 3$ be an integer, and let C_n denote a cycle of length n. Show that

$$\chi(C_n; k) = (k - 1)^n + (-1)^n (k - 1).$$

P 10.7.17. **Proof of Corollary 10.112.** Use Theorem 10.111 to prove that the chromatic polynomial of a simple graph of order n is indeed a polynomial of degree n with integer coefficients.

P 10.7.18. Let n be a positive integer. Let G be a simple graph of order n, and let $p(k) = \chi(G; k)$ be its chromatic polynomial (of degree n). Write

$$p(k) = a_n k^n + a_{n-1} k^{n-1} + \cdots a_1 k + a_0,$$

where $a_0, a_1, \ldots, a_n$ are integers. Prove that $a_0 = 0$, $a_n = 1$, and a_{n-1} is the negative of the number of edges of G.

P 10.7.19. Let G be a simple graph of order n. Prove that G is a tree if and only if $\chi(G; k) = k(k - 1)^{n-1}$. You may find the following steps helpful:

STEP 1: For the "only if" direction, use induction on n.

STEP 2: For the "if" direction, first use Problem P 10.7.18 to find the number of edges of G.

STEP 3: Argue that, if G is disconnected, then the chromatic polynomial of G is the product of two other chromatic polynomials.

STEP 4: Prove that, for any disconnected simple graph G, the chromatic polynomial $\chi(G; k)$ will have no linear term. That is, the coefficient of k will be zero.

STEP 5: Complete the proof.

P 10.7.20. Let G be a multigraph and let e be an edge of G. Let $G - \{e\}$ and G/e denote, as usual, the graphs obtained by, respectively, deleting and contracting e. However, if contracting e results in multiple edges, then keep those multiple edges. Denote the number of spanning trees (Definition 10.25) of G by $t(G)$. Show that

$$t(G) = t(G - \{e\}) + t(G/e).$$

P 10.7.21. Let G be the graph of Figure 10.55. What is $t(G)$, the number of spanning trees of G?

P 10.7.22. **Acyclic Orientations.** Let G be a simple graph of order n. You assign an *orientation* to the edges of G by giving each a direction (making them a one-way street as opposed to the original two-way street). A graph with a specific orientation for its edges is called a *directed graph*. A *directed cycle* in a directed graph is a cycle that respects the orientation of the edges (i.e., you can start at a vertex on the cycle, and, by going through the edges of the cycle in the designated direction, you can get back to the starting point). If the graph G has m edges, then there are 2^m ways to orient all the edges of G. From among all of these possibilities, let $a(G)$ be the number of *acyclic orientations*. These are the orientations that have produced no directed cycles. (See Figure 10.52 and the discussion in Example 10.105.)

Let $e = \{x, y\}$ be an edge of G, and give a combinatorial argument to show that

$$a(G) = a(G - \{e\}) + a(G/e).$$

You may find the following steps helpful:

STEP 1: Argue that every acyclic orientation of G is some acyclic orientation of $G - \{e\}$ with an appropriate orientation for e.

STEP 2: Argue that, in an acyclic orientation of $G - \{e\}$, if there is a directed path from x to y, then there cannot be a directed path from y to x. Hence, there are only two types of acyclic orientations of $G - \{e\}$. Type I orientations are those where there is no directed path in $G - \{e\}$ from x to y or from y to x. Type II orientations are those where, in $G - \{e\}$, there is a directed path from one of x or y to the other. Let $b_1(G - \{e\})$ and $b_2(G - \{e\})$ be, respectively, the number of type I and type II acyclic orientations of $G - \{e\}$.

STEP 3: Argue that $a(G) = 2b_1(G - \{e\}) + b_2(G - \{e\}) = a(G - \{e\}) + b_1(G - \{e\})$.

STEP 4: Argue that $b_1(G - \{e\}) = a(G/e)$.

STEP 5: Complete the proof and rewrite it concisely and clearly.

P 10.7.23. **Proof of Theorem 10.114.** Let G be a simple graph of order n, and let $\chi(G; k)$ be the chromatic polynomial of G. Define

$$c(G) = (-1)^n \chi(G; -1).$$

(a) If e is an edge of G, prove that $c(G) = c(G - \{e\}) + c(G/e)$.
(b) Adopting the definitions and notation of Problem P 10.7.22, prove that if G is a graph with no edges, then $a(G) = c(G)$.

(c) Prove that, for all simple graphs G of order n, we have

$$\text{number of acyclic orientations of } G = (-1)^n \chi(G; -1).$$

10.7.3 Graph Vocabulary: Edge Chromatic Number

Definition 10.115 (Edge Chromatic Number). A coloring of the edges of a graph is called a *proper edge coloring* if no two edges that share a vertex have been assigned the same color. The minimum number of colors needed for a proper edge coloring of a graph G is called the *edge chromatic number* of G, and is denoted by $\chi'(G)$.

We state a theorem of the Ukrainian mathematician Vadim Vizing (1937–2017) without proof (for a proof, see West 2001, Theorem 7.1.10).

Theorem 10.116 (Vizing's Theorem, 1964). *Let G be a simple graph, and let Δ be the maximum degree of the vertices of G. Then the edge chromatic number, $\chi'(G)$, is either Δ or $\Delta + 1$.*

We will revisit edge chromatic numbers in Theorem 11.57 of subsection 11.3.3, where we show that the edge chromatic number of a bipartite graph is always equal to the maximum degree of the graph.

Problems (continued)

P 10.7.24. Let G be a simple graph and $L(G)$ be its line graph (Definition 10.50). How is $\chi'(G)$, the edge-chromatic number of G, related to $\chi(L(G))$?

P 10.7.25. Recall that a graph is regular if the degrees of all the vertices are the same (Definition 10.3). Let G be a regular graph with an odd number of vertices. Using Vizing's theorem, prove that $\chi'(G) = \Delta + 1$.

10.7.4 Graph Vocabulary: List Coloring and the Choice Number; the Five-Color Theorem

The highlight of this section is the presentation of a proof that every planar graph is 5-colorable (Problem *P 10.7.34*). The best result in this direction is the celebrated four-color theorem (Theorem 10.86) that every planar graph is 4-colorable or, put another way, the chromatic number of any planar graph is no more than 4. That proof is difficult, and, in fact, all known proofs use a computer to check a non-trivial number of cases. In the problems, you will first find a proof of the more straightforward six-color theorem, and will then be guided through the magnificent proof by Danish mathematician Carsten Thomassen (Thomassen 1994) that every planar graph is 5-colorable.

In fact, both of these results prove more than 6-colorability or 5-colorability of planar graphs. They tackle a generalization of the usual vertex coloring problem. One way to generalize the concept of vertex coloring is through list colorings. If the vertices of a graph can be properly colored using k colors, then that means that, given colors $\{1, 2, \ldots, k\}$, we can choose one color for each vertex in such a way that no two adjacent vertices have the same color. But what if the lists of colors available for two different vertices are not the same?

Definition 10.117. Let $G = (V, E)$ be a graph. Assume that for every vertex $v \in V$ we have a list $L(v)$ of colors. A *list coloring* of a graph G is an assignment of colors to the vertices of G such that the color of vertex v is from the list $L(v)$, and such that no two adjacent vertices are assigned the same color. A graph G is *k-choosable* if there is a list coloring of G as long as $|L(v)| \geq k$ for all $v \in V$. The smallest k for which G is k-choosable is the *choice number* of G, denoted by ch(G).

In other words, let G be a graph and assume ch$(G) = k$. Then, given a list of k admissible colors for each vertex, we will be able to properly color G such that each vertex is assigned a color from its own list. For this same graph, it is possible to devise lists of size $k - 1$ in such a way that such a proper coloring is not possible.

In Problem P 10.7.28, you are asked to write down a proof that $\chi(G) \leq$ ch(G). Now, the four-color theorem (Theorem 10.86) says that $\chi(G) \leq 4$ for planar graphs G. There do exist, however, planar graphs for which ch$(G) > 4$. In Problem P 10.7.32, you are asked to prove the somewhat straightforward result that, for planar graphs, ch$(G) \leq 6$. We then turn in Problem P 10.7.34 to proving that, for a planar graph G, ch$(G) \leq 5$, which is the best we can hope for.

For that proof, we need a little bit of a setup. Let C be the planar drawing of a cycle. Recall from Definition 10.92 that by a *triangulation* of (the region enclosed by) C we mean a planar drawing of a finite simple planar graph $G = (V, E)$ such that C is the boundary of the outside region, G has no isolated vertices or leaves, and all the finite regions of G are triangles. In a triangulation of C, a *chord* is any edge of G that is not an edge of C and yet has both ends on the cycle C.

Given any graph, adding new edges makes the task of properly coloring the vertices harder, and, as a result, after you have added edges, the chromatic number or the choice number never goes down. Given a planar graph, we add enough edges to make all the finite regions into triangles. This allows us to limit ourselves, without loss of generality, to triangulations of cycles.

The proof that all planar graphs are 5-choosable (Problem P 10.7.34) is by induction on the number of vertices. But a straightforward inductive assumption that all smaller graphs are 5-choosable does not work. Instead, what is interesting about this proof is that we actually prove a stronger statement. Intuitively, proving a stronger statement should be more difficult. However, this is not necessarily the case when you are using a proof by induction. A stronger statement also give you a stronger inductive hypothesis, and that may be helpful. Here, for example, instead of providing lists of five permissible colors for each vertex, we limit the lists to just three colors for the vertices on the boundary and to just one permissible color for two of those vertices.

Problems (continued)

P 10.7.26. Let $G = K_{2,2}$ be the complete bipartite graph on 2 and 2 vertices. In Figure 10.56, two copies of G are given. For each copy of G, a list of admissible colors for each vertex is listed. For each copy, determine if, for the given lists, a list coloring of G is possible.

Figure 10.56 For each vertex, a list of permissible colors is given.

P 10.7.27. Let $G = K_{3,3}$. Show that $\chi(G) = 2$ but $\mathrm{ch}(G) > 2$.

P 10.7.28. Let G be a graph. What is the relationship between $\chi(G)$ and $\mathrm{ch}(G)$?

P 10.7.29. Show that $\mathrm{ch}(K_{2,2}) = 2$, and use it to find $\mathrm{ch}(K_{3,3})$.

P 10.7.30. Hiking Mount Damāvand, I run into an oracle who tells me that the choice number of the complete bipartite graph $K_{2,2}$ is 2. Continuing my hike, I consider a different graph G given in Figure 10.57.

Figure 10.57 Find the choice number of this graph.

(a) Does G have a subgraph isomorphic to $K_{2,2}$?

(b) What is $\mathrm{ch}(G)$? Prove your assertions.

P 10.7.31. Let G be a simple planar graph with 11 vertices. Show that both the chromatic number and the choice number of G are no greater than 5.

P 10.7.32. **The Six-Color Theorem.** G is a (simple) planar graph with n vertices.

(a) Can every vertex of G have degree greater than 5? Why?

(b) Prove that $\mathrm{ch}(G) \leq 6$.

P 10.7.33. Let $G = (V, E)$ be the triangulation of an n-cycle, and assume that G has no chords. Let u, v, and w be three consecutive vertices on C (in other words, $\{u, v\}$ and $\{v, w\}$ are both edges of C). Show that, in G, there is path $u - v_1 - \cdots - v_k - w$ such that, for $1 \leq i \leq k$, v_i is not a vertex of C (i.e., it is an internal vertex) and the set $S = \{u, v_1, \ldots, v_k, w\}$ is exactly the set of vertices of G adjacent to v. See Figure 10.58.

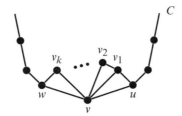

Figure 10.58 If the outer cycle C of a planar graph has no chords, then the neighbors of v form a path from one side of the outer cycle to the other. If there were chords, the path might contain more than two elements of C.

P 10.7.34. **Every Planar Graph is 5-choosable.**[22])Let $G = (V, E)$ be a triangulation of the interior of a cycle C. Assume that for each vertex $v \in V$ we are given a list L_v of admissible colors. Further assume that:

(a) For two adjacent vertices x and y on C, we have $|L_x| = |L_y| = 1$ and $L_x \cap L_y = \emptyset$,

(b) for all the other vertices z on C, we have $|L_z| = 3$, and

(c) for all internal vertices v of G, we have $|L_v| = 5$.

Then show that we can properly color the vertices of G using a permissible color for each of the vertices. In particular $ch(G)$, the choice number of G, and $\chi(G)$, the chromatic number of G, are both at most 5.

You may find the following steps useful.

STEP 1: Induct on $|V|$, and prove the result for the base case when $|V| = 3$.

STEP: 2: Assume $|V| > 3$, C is the cycle $v_1 - v_2 - \cdots - v_n - v_1$, and v_1 and v_n are the two vertices with distinct lists of size 1. Break up the proof into two cases: C has a chord or C does not have a chord.

STEP 3: If C has a chord $\{v_i, v_j\}$ (Figure 10.59 on the left) with $1 \leq i < j \leq n$, split the graph "along $\{v_i, v_j\}$" into two graphs that only share the vertices v_i and v_j and the edge $\{v_i, v_j\}$. Carefully, show how to apply the inductive hypothesis to the two graphs (and in which order) to get a proper coloring of G.

STEP 4: If C does not have a chord (Figure 10.59 on the right), consider the three consecutive vertices v_1, v_2, and v_3 and use the result of Problem P 10.7.33 to delete v_2 and its edges and replace the $v_1 - v_2 - v_3$ of the cycle C with another path from v_1 to v_3 consisting of internal vertices.

STEP 5: Before coloring the modified graph, change the admissible lists of color for the internal vertices that replaced v_2 on the path (but make no change to the lists of v_1 or v_3). Make sure that at least two of the colors in v_2's list are not on the lists of v_1 or on the lists of the internal neighbors of v_2.

---

[22] Adapted from Thomassen 1994.

STEP 6: Make sure that you can apply the inductive hypothesis to the remaining graph (G minus v_2 and its edges) and color all its vertices properly. Show that you can extend the coloring to v_2.

STEP 7: Rewrite your proof clearly and succinctly.

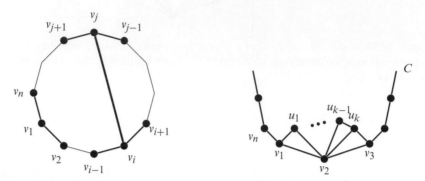

Figure 10.59 In Case 1 (on the left), there is a chord and you split the graph along the chord. In Case 2 (on the right), you eliminate v_2 and replace $v_1 - v_2 - v_3$ with $v_1 - u_1 - \cdots - u_{k-1} - u_k - v_3$.

P 10.7.35. Let $n \geq 3$. Let $G = (V, E)$ be the triangulation of an n-cycle C. In Problem P 10.6.28 you were asked to prove that

$$|E| = 3\,|V| - n - 3.$$

The proof outlined there ultimately was based on Euler's formula. Re-prove this result without using Euler's formula. You may find the following steps helpful:

STEP 1: Begin a proof by induction on $|V|$ by proving the base case.

STEP 2: Break the inductive step into two cases. Either G has a chord or it does not. If G has a chord, then clone the chord, and separate G into two smaller triangulations along the chord. Give names to the number of vertices and edges of each side and apply induction.

STEP 3: In the case where G does not have a chord, adopt the notation and result of Problem P 10.7.33. Delete the vertex v and all of its edges, keep track of the number of neighbors of v, and then apply induction.

P 10.7.36. Let $n \geq 3$. Let $G = (V, E)$ be the triangulation of an n-cycle C. Use the result of Problem P 10.7.35 to give a new proof of Euler's formula for G.

10.7.5 Tiling Rectangles with Tetris Pieces

The game Tetris uses seven flat pieces, each made up of four squares (see Figure 10.60). Each of these is called a *tetromino* or a *polyomino of order 4*. The problems in this section are about tiling rectangles using Tetris pieces. To answer these questions, it is often prudent to color the squares in the Tetris pieces in an appropriate way.

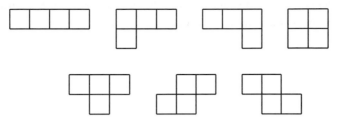

Figure 10.60 The seven Tetris pieces.

Problems (continued)

P 10.7.37. Assume that you have one each of the seven Tetris pieces. Can you put them together to form a rectangle without any holes?

P 10.7.38. Is it possible to cover an 8 × 8 chess board with 15 T tetrominoes (leftmost in the bottom row of Figure 10.60) and one square tetromino (rightmost in the top row of Figure 10.60)?[23]

P 10.7.39. Is it possible to cover an 8 × 8 chess board with 15 L tetrominoes (either of the two in the middle of the top row of Figure 10.60) and one square tetrominoe (rightmost in the top row of Figure 10.60)?

10.8   Open Problems and Conjectures

Graph theory is permeated with easy-to-state open problems and conjectures. There are profound and deep conjectures as well as more contorted ones. Sometimes, what was once thought of as a little puzzle will turn into a substantial area of research, with connections to other areas. There are whole papers and websites devoted to conjectures in graph theory (see Bondy 2014, for example). Just to whet your appetite, we will mention a few sample problems/conjectures that have not been resolved as of this writing.

Conjecture 10.118 (The Erdős–Gyárfás Conjecture, 1995). *Let G be a simple graph in which the degree of each vertex is at least* 3. *Then G contains at least one cycle whose length is a power of* 2.

Let $G = (V, E)$ be a graph, and $S \subseteq V$ be a subset of the vertices of G. Let $F \subseteq E$ be the set of all edges of G with both ends in S. The graph $H = (S, F)$ is called an *induced subgraph of G* (induced by S). In other words, to create an induced subgraph of G, you choose some of the

---

[23] Problems P 10.7.38 and P 10.7.39 are adapted from Golomb 1994.

vertices of G and *all* of the edges in G that have both ends in S. A graph G is 3-connected (see Mini-project 6) if removing any two vertices of G (and their edges) leaves a connected graph. The following theorem gives two special cases (there are others as well) of the Erdős–Gyárfás conjecture that are known to be true.

Theorem 10.119. *The Erdős–Gyárfás conjecture is true if one of the following additional conditions holds:*

(a) *(Daniel and Shauger 2001) G is a planar graph that does not have $K_{1,3}$ as an induced subgraph*

(b) *(Heckman and Krakovski 2013) G is a 3-connected cubic planar graph.*

Recall that a graph all of whose vertices have degree d is called *d-regular* (Definition 10.3).

Conjecture 10.120 (Barnette's Conjecture, 1969). *If G is a planar, 3-connected, 3-regular, bipartite graph, then G is Hamiltonian.*

Among other special cases, it is known that:

Theorem 10.121. *Barnette's conjecture is true if one of the following additional conditions holds:*

(a) *(Goodey 1975) The regions of the planar graph G are quadrilaterals and/or hexagons.*

(b) *(Holton, Manvel, and McKay 1985) If the number of vertices of G is no more than 66.*

Recall (Definition 10.46) that an *edge decomposition* of a graph G is a collection of subgraphs such that each edge of G is in exactly one of the subgraphs.

Conjecture 10.122 (Gallai's Conjecture). *Let G be any connected simple graph of order n. Then G has an edge decomposition consisting of at most $\lceil \frac{n}{2} \rceil$ paths.*

The following are among the known results:

Theorem 10.123. (a) *(Lovász 1968) Let G be any simple graph (connected or not) of order n. Then G has an edge decomposition consisting of at most $\lfloor \frac{n}{2} \rfloor$ paths and cycles. It follows that Gallai's conjecture is true if G has at most one vertex of even degree.*

(b) *(Pyber 1996) Gallai's conjecture is true if we make the additional assumption that every cycle of the graph has a vertex of odd degree.*

(c) *(Donald 1980) Let G be any connected simple graph of order n. Then G has an edge decomposition consisting of at most $\frac{3n}{4}$ paths.*

Recall that the edge coloring of a simple graph is proper if no two edges that share a vertex have been assigned the same color.

Conjecture 10.124 (Fiamčík's Acyclic Edge–Coloring Conjecture, 1978). *Let G be a simple graph with maximum degree Δ. Then the edges of G can be properly colored with $\Delta + 2$ colors in such a way that each cycle contain edges of at least three different colors.*

See Alon, Sudakov, and Zaks (2001) for some partial results. We next state the 1-2-3 conjecture.

Conjecture 10.125 (Karoński, Łuczak, and Thomason's 1-2-3 Conjecture, 2004). *Let G be a simple graph with the property that none of its connected components is just one edge. Then you can assign labels 1, 2, or 3 to the edges of G so that the sum of the labels of the edges incident to one vertex will not be the same as the sum of the labels of the edges incident to another, different vertex.*

As for partial results, Karoński, Łuczak, and Thomason (2004) proved that Conjecture 10.125 is true if the chromatic number of G is at most 3, and Kalkowski, Karoński, and Pfender (2010) proved that Conjecture 10.125 is true if, instead of just using 1, 2, or 3, you are allowed to use $1, \ldots, 5$ as labels.

We close with two seemingly hard problems on coloring graphs.

Problem 10.126. (a) *For which simple graphs G is the chromatic number $\chi(G)$ equal to the choice number $\mathrm{ch}(G)$?*

(b) *Which polynomials are the chromatic polynomial of some simple graph?*

Collaborative Mini-project 9: Cayley's Tree Formula

In this project, you are guided through two proofs of Cayley's formula[1] that the number of labeled trees on n vertices is n^{n-2}. The first proof uses multinomial coefficients and the multinomial theorem, and, in fact, also finds the number of labeled trees with specified degrees for each vertex. The second proof finds a way of cataloguing all labeled trees using something called the Prüfer code.

To become familiar with what we want to count, consider the following.

The Problem

Forugh, Simin, Sohrab, Siavash, Ahmad, Manouchehr, and Houshang are the first names of your favorite poets. You want to create a tree whose vertices are labeled by these. How many such trees are possible?

In this problem, we want to count trees of order 7 where each vertex is labeled with one of the names. More generally, we want to count the number of labeled trees with n vertices. Throughout this mini-project, the set of vertices of a tree with n vertices will be $[n] = \{1, \ldots, n\}$. Hence, each edge will be a subset of size 2 of $[n]$. Two trees will be different if they do not have the same set of edges, even if their graphical representations may otherwise look the same.

EXAMPLE. For $n = 3$, all trees of order 3 are isomorphic to each other, and each is a path of length 2. But how many labeled trees of order 3 are there? We are labeling the vertices with 1, 2, and 3, and we get that (1)—(2)—(3), (2)—(1)—(3), and (1)—(3)—(2) are all the different labeled trees of order 3. Hence, the number of labeled trees of order 3 is $3 = 3^{3-2}$, as predicted by Cayley's theorem. Note that the tree (3)—(2)—(1) is the same as the first tree listed, since, in both cases, $\{\{1,2\}, \{2,3\}\}$ is the set of edges.

---

[1] This result is almost universally attributed to Arthur Cayley (1821–1895), the prominent English mathematician, who published the result in 1889 (Cayley 1889). In that paper, he didn't give a proof but illustrated the result by showing that it worked for the case $n = 6$. However, Cayley himself ends the paper by saying "The forgoing theory presents itself in a paper by Borchardt ..." and goes on to cite the 1860 paper (Borchardt 1860) of German mathematician Carl Borchardt (1817–1880). A rigorous proof of the theorem was first given by Heinz Prüfer (1896–1934) in 1918 (Prüfer 1918).

Preliminary Investigations

Please complete this section before beginning collaboration with other students.

MP9.1 Let n be a positive integer. If a tree has n vertices, then what is the number of edges, and what is the sum of the degrees of the vertices?

MP9.2 Why does a tree with at least two vertices have at least two leaves? (Review Problem $P\ 10.2.9$.)

MP9.3 According to Cayley's formula, how many labeled trees of order 4 are there? If you ignore labels, how many non-isomorphic trees of order 4 are there?

MP9.4 Convince yourself that there is only one labeled tree with vertices $\{1, 2, 3, 4\}$ such that $\deg(1) = 3$ and $\deg(2) = \deg(3) = \deg(4) = 1$. Now, from this tree construct new graphs by attaching one new vertex labeled 5 to one of the old vertices (hence adding one more edge as well). What do the resulting graphs look like? Are they all trees? What are their degree sequences?

MP9.5 Review the definition of multinomial coefficients (Definition 4.28), the recurrence relation for multinomials (Theorem 5.28), and the multinomial theorem (Theorem 5.29).

Prüfer Codes

Let $T = (V, E)$ be a tree of order n whose vertices are labeled 1 through n. Use the following algorithm to assign to the tree an ordered sequence of $n - 2$ integers. Find the leaf with the smallest label. The first integer in the sequence is the label of the vertex adjacent to this leaf. Now eliminate the leaf and the edge incident with it, get a new smaller tree, and repeat. Namely, for the remaining tree find the leaf with the smallest label, and the next integer in the sequence will be the label of the vertex adjacent to this leaf. Stop when there is only one edge left (and you have recorded $n - 2$ integers for the sequence). The sequence constructed by this algorithm is called the *Prüfer code* for T.

MP9.6 Let T be the tree of order 9 in the top left corner of Figure MP9.1. What is the degree sequence of this tree? Among the leaves, which vertex has the smallest label? Why is 2 the first term in the Prüfer code for this tree?

MP9.7 Follow the steps in Figure MP9.1 and explain what is happening in each step.

MP9.8 How many seven-digit numbers can you make using the integers 1 through 9 (repetitions allowed)?

Collaborative Investigation

Work on the rest of the questions with your team. Before getting started on the new questions, you may want to compare notes about the preliminary problems.

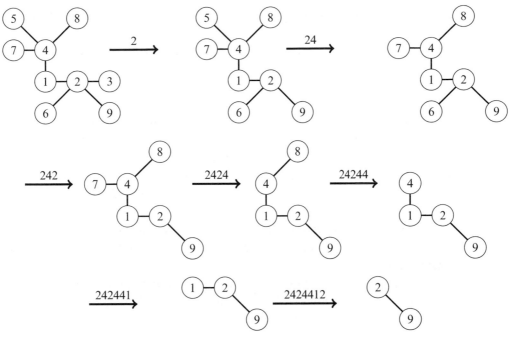

Figure MP9.1 Starting with the tree on the top left, if we sequentially delete the leaf with the lowest label and record the vertex that it is adjacent to, we get the Prüfer code 2424412. The vertices eliminated are in order 3, 5, 6, 7, 8, 4, and 1.

MP9.9 Find all labeled trees of order 5 where $\deg(1) = \deg(3) = \deg(5) = 1$, $\deg(2) = 2$, $\deg(4) = 3$. How many of them are there? What if we wanted the number of labeled trees of order 5 where $\deg(1) = 3$, $\deg(2) = 2$, and $\deg(3) = \deg(4) = \deg(5) = 1$? Are the two answers any different?

MP9.10 Let $n \geq 2$, and let $d_1, d_2, \ldots, d_n$ be a sequence of positive integers with the property that $d_1 + \ldots + d_n = 2(n-1)$. Let $T(n; d_1, d_2, \ldots, d_n)$ be the number of labeled trees $T = (V, E)$ such that $V = [n]$ and for $1 \leq i \leq n$, the degree of vertex i is equal to d_i. Prove that

$$T(n; d_1, d_2, \ldots, d_n) = \binom{n-2}{d_1 - 1 \ d_2 - 1 \ \ldots \ d_n - 1}.$$

You may find the following steps helpful:

STEP 1: Use induction on n. What is the base case?

STEP 2: For the inductive step, first argue (see Problem MP9.9) that you can assume $d_1 \geq d_2 \geq \cdots \geq d_n$. What is d_n?

STEP 3: Without making any prejudgements, argue that vertex n could be adjacent to vertex i with $1 \leq i \leq n-1$. If so, and if you eliminate vertex n and the edge $\{i, n\}$, what will the degrees of the vertices $\{1, \ldots, n-1\}$ in the remaining graph be? Is the remaining graph a tree? What is the sum of the degrees of the new graph? Can we use the inductive hypothesis to count the number of graphs that could result?

STEP 4: If $n > 2$, $d_n = 1$, and, for some $1 \leq i \leq n-1$, we have $d_i = 1$, then, in a tree, could vertex n be adjacent to vertex i? In such a case, does your count from Step 3 continue to work or do you have to make adjustments?

STEP 5: Using your count in Step 3, find an expression for $T(n; d_1, \ldots, d_n)$ as a sum of multinomial coefficients.

STEP 6: Use Theorem 5.28 to finish the proof.

MP9.11 Use the result of Problem MP9.10 to prove Cayley's formula that the number of labeled trees of order n is n^{n-2}. You may find the following steps helpful:

STEP 1: Write the total number of labeled trees of order n as a sum of multinomial coefficients, as given in Problem MP9.10. Be clear about what terms need to be included, and what you are summing over.

STEP 2: Use the multinomial theorem (Theorem 5.29) to expand $(x_1 + x_2 + \cdots x_n)^m$.

STEP 3: Specialize your identity from the previous step by letting $m = n-2$, and by choosing relevant parameters for the right-hand side of the expansion so that the resulting identity will include terms similar to what you had in Step 1.

STEP 4: Decide on what to substitute for $x_1, \ldots, x_n$, and complete the proof.

A Second Proof using Prüfer Codes

We now turn to a second and very different proof for Cayley's tree formula. The main point of this proof is to assign to each labeled tree a specific code – the Prüfer code described above – thereby having a complete catalogue of labeled trees. Then, instead of counting the trees, we count the codes. The latter is straightforward (see Problem MP9.12). What needs a bit of work is to show that there is a bijection between labeled trees and Prüfer codes. Since the algorithm for creating a Prüfer code is deterministic, for every labeled tree you will always get one code. So, we do have a function from the set of labeled trees to the set of Prüfer codes. What we have to do is to show that the map is one-to-one and onto. We do this by showing that the map has an inverse (see Proposition 3.25), and that there is an unambiguous way to reconstruct the tree from its Prüfer code.

MP9.12 What is the number of sequences of length $n - 2$ if each term is from $[n]$?

To show that every Prüfer code uniquely determines a tree, it is helpful to have some additional notation. (These notations are made up for the purposes of this mini-project, and are not commonly used.)

NOTATION: Let $n \geq 2$, and let $T = (V, E)$ be a tree with $V = [n] = \{1, \ldots, n\}$. Denote the Prüfer code of T by $\mathcal{P}(T)$. The process of constructing the code stopped when we reached one edge (and two vertices remained). Take the algorithm one more step, and call the resulting sequence the *extended Prüfer code* of T, denoted by $\mathcal{P}^{\star}(T)$. At each step, the algorithm for producing the Prüfer code deleted a vertex and one of its edges (the only edge incident with the vertex at that stage). The ordered list of the $n-2$ deleted vertices will be denoted by $\mathcal{D}(T)$, and, if we extend the algorithm by one more step, the extended ordered list of the $n-1$ deleted vertices will be denoted by $\mathcal{D}^{\star}(T)$.

MP9.13 What are $\mathcal{P}(T)$, $\mathcal{P}^{\star}(T)$, $\mathcal{D}(T)$, and $\mathcal{D}^{\star}(T)$ for the tree on the top left of Figure MP9.1?

For Problems MP9.14–MP9.17, let $T = (V, E)$ be a tree with $V = [n]$. Let $\mathcal{P}^{\star}(T) = (p_1, p_2, \ldots, p_{n-1})$ and $\mathcal{D}^{\star}(T) = (w_1, w_2, \ldots, w_{n-1})$ be, respectively, the extended Prüfer code and the extended ordered list of deleted vertices.

MP9.14 Argue that, under the extended Prüfer algorithm, the vertex n will always be the last vertex standing, and so $\mathcal{D}^{\star}(T)$ is some permutation of $[n - 1] = \{1, 2, \ldots, n - 1\}$.

MP9.15 Argue that $p_{n-1} = n$.

MP9.16 Prove that, for each $1 \le i \le n - 1$, $\{w_i, p_i\}$ is an edge of T, and that no two of these can be the same edge. As a result, these are all the edges of T.

MP9.17 Let $v \in V$. Use the fact that v is incident with $\deg(v)$ edges, and Problem MP9.14, to show that v appears in $\mathcal{P}(T)$ exactly $\deg(v) - 1$ times.

MP9.18 Assume that $n = 9$ and 9911128 claims to be the Prüfer code for a tree T. What is $\mathcal{P}^{\star}(T)$, the extended Prüfer code for T? Make a list of the (alleged) vertices of T, and write down the degree of each one. (Are Problems MP9.15 and MP9.17 relevant?)

MP9.19 Continuing with Problem MP9.18, let w_1 be the leaf of T with the smallest label. What is w_1, and what is the first element in $\mathcal{D}^{\star}(T)$, the extended sequence of deleted vertices? Do you know which vertex is adjacent to w_1? (Is Problem MP9.16 relevant?)

MP9.20 Continuing with Problem MP9.19, if you eliminate w_1 and its edge from T, then what are the degrees of the remaining vertices? Let w_2 be the leaf of the remaining tree with the smallest label. What is w_2, and what is the second element in $\mathcal{D}^{\star}(T)$? Do you know which vertex is adjacent to w_2?

MP9.21 Continuing with Problem MP9.20, reconstruct all of $\mathcal{D}^{\star}(T)$. Do you now know all of the vertices and edges of T? Can you produce a graph of T?

MP9.22 Let A be any sequence of length $n - 2$ whose terms are from $[n]$. By mimicking the procedure of Problems MP9.18–MP9.21, prove that there is a unique labeled tree T with $\mathcal{P}(T) = A$.

MP9.23 Can you give a complete proof of Cayley's tree formula using Prüfer codes?

Mini-project Write-Up

In your write-up of the results, include at least the following:

- An explanation, with examples, of what a labeled tree is
- A clear statement of Cayley's tree formula
- A statement and proof of the result in Problem MP9.10. Use it to give a first proof of Cayley's tree formula
- A definition, with example(s), of Prüfer codes

- A complete proof that there is a bijection between labeled trees and Prüfer codes. Use this to give a second proof of Cayley's tree formula.

(Extra Credit) Track down the paper by Avron and Dershowitz (2016), decipher it, and give a third proof of Cayley's tree formula.

Collaborative Mini-project 10: Incidence Matrices and Bipartite Graphs

In this mini-project – which uses some linear algebra – we will investigate the incidence matrices (defined below) of simple graphs and their rank. Somewhat surprisingly, there is a clear connection between the rank of these matrices and the structure of the underlying graphs. In fact, as we shall see, by just finding the rank of the incidence matrix, you can know whether the graph is bipartite or not. This use of linear algebra in graph theory is not unusual, and, in fact, in more advanced treatments of graph theory, probing the connections proves very fruitful.

Preliminary Investigations

Please complete this section before beginning collaboration with other students.

Incidence Matrices of a Graph

Let $n \geq 2$ and $m \geq 1$ be integers, and let $G = (V, E)$ be a simple graph with n vertices and m edges. Fix the order of the elements of $V = \{v_1, \ldots, v_n\}$ and $E = \{e_1, \ldots, e_m\}$. The *incidence matrix* of G is an $n \times m$ matrix whose rows are indexed by the vertices and whose columns are indexed by the edges. The (i, j) entry of the matrix (that is, the entry in the row corresponding to $v_i \in V$ and the column corresponding to $e_j \in E$) is 1 if v_i is incident (Definition 2.23) with e_j and 0 otherwise. For this mini-project, we will denote the incidence matrix of G by M_G.

Example 10.127. An incidence matrix for the graph of Figure MP10.1 is a 7×9 matrix. We have

$$
M_G = \begin{array}{c} \\ v_1 \\ v_2 \\ v_3 \\ v_4 \\ v_5 \\ v_6 \\ v_7 \end{array}
\begin{array}{c}
(v_1, v_2) \ (v_2, v_3) \ (v_3, v_4) \ (v_4, v_5) \ (v_5, v_6) \ (v_6, v_7) \ (v_5, v_7) \ (v_2, v_5) \ (v_1, v_4) \\
\left[\begin{array}{ccccccccc}
1 & 0 & 0 & 0 & 0 & 0 & 0 & 0 & 1 \\
1 & 1 & 0 & 0 & 0 & 0 & 0 & 1 & 0 \\
0 & 1 & 1 & 0 & 0 & 0 & 0 & 0 & 0 \\
0 & 0 & 1 & 1 & 0 & 0 & 0 & 0 & 1 \\
0 & 0 & 0 & 1 & 1 & 0 & 1 & 1 & 0 \\
0 & 0 & 0 & 0 & 1 & 1 & 0 & 0 & 0 \\
0 & 0 & 0 & 0 & 0 & 1 & 1 & 0 & 0
\end{array}\right].
\end{array}
$$

MP10.1 Let G be a path of length 2. By using three different orderings (for the vertices and edges), find three different incidence matrices for G. What are the ranks of these matrices?

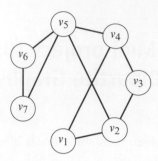

Figure MP10.1 A connected simple graph.

MP10.2 Let G be a cycle of length 3. Find one incidence matrix for G. What is the rank of the matrix? Find its determinant by a cofactor expansion.

MP10.3 Let G be a cycle of length 4. Find one incidence matrix of G. What is the rank of the matrix? Find its determinant by a cofactor expansion.

MP10.4 An incidence matrix of a mystery graph is

$$M_G = \begin{bmatrix} 1 & 0 & 0 & 1 & 1 & 0 \\ 0 & 0 & 1 & 1 & 0 & 0 \\ 0 & 0 & 0 & 0 & 0 & 1 \\ 0 & 1 & 1 & 0 & 1 & 0 \\ 1 & 1 & 0 & 0 & 0 & 1 \end{bmatrix}.$$

Draw the graph.

MP10.5 If M is an incidence matrix of a graph, then how many 1's are in each column? What does the number of 1's in each row tell you?

MP10.6 If A is some 53×57 matrix of rank 46, and $B = \left[\begin{array}{c|c} A & 0 \\ \hline 0 & 1 \end{array} \right]$, then what is the rank of B? Why?

MP10.7 If B is some 47×58 matrix of rank 46, and $C = \left[\ B \mid D\ \right]$, where D is a 47×13 matrix, then what are the possibilities for the rank of C? Why?

MP10.8 After reviewing the definition of a spanning tree (Definition 10.25), do Problem *P 10.2.21*.

Collaborative Investigation

Work on the rest of the questions with your team. Before getting started on the new questions, you may want to compare notes about the preliminary problems.

MP10.9 If G is a simple graph, and if you change the order of the vertices or the edges, then does the rank of the corresponding incidence matrix change?

The incidence matrix seems like an almost arbitrary matrix of 1's and 0's. It seems hard to believe that there would be connections between meaningful linear algebraic properties of matrices, and actual graph theoretical properties of graphs. On the other hand, the incidence matrix does contain all the information about a graph, and so maybe some structural properties of graphs can be gleaned from the well known parameters of matrices. In this mini-project, our focus is on the rank of the incidence matrix. Does it tell us anything about the graph?

MP10.10 Let G be a tree of order n. What size is an incidence matrix of G? By looking at some examples, make a conjecture about the rank of an incidence matrix for G.

MP10.11 Let G be a cycle of length n. By looking at some examples, make a conjecture about the rank and determinant of its incidence matrix. Does it make a difference whether n is even or odd?

MP10.12 Let G be a connected bipartite graph of order n and with at least one edge. By looking at some examples, make a conjecture about the rank of its incidence matrix.

MP10.13 Let $G = (V, E)$ be a simple graph with at least two edges. Assume that $v \in V$ is a leaf of G. Let $\widetilde{G}$ be the graph you obtain by eliminating v and the one edge e that v is incident to. Let M_G and $M_{\widetilde{G}}$, be respectively, be the incidence matrices of G and $\widetilde{G}$. Show that

$$\text{rank}(M_G) = \text{rank}(M_{\widetilde{G}}) + 1.$$

MP10.14 Use Problem MP10.13 and induction to prove your conjecture of Problem MP10.10.

MP10.15 Let G be a connected simple graph of order n with at least one edge, and let T be a spanning tree for G (Definition 10.25 and Problem P 10.2.21). When ordering the edges of G, put the edges of T first. Use your answers to Problem MP10.14 and Problem MP10.7 to prove that the rank of M_G is either $n - 1$ or n.

MP10.16 By showing that the rows of an incidence matrix of a connected bipartite graph are linearly dependent, prove your conjecture of Problem MP10.12.

MP10.17 Let G be a cycle of length n. Use cofactor expansion to find the determinant of M_G, an incidence matrix for G. Prove your conjecture of Problem MP10.11.

MP10.18 Assume G is an odd cycle with some trees attached to the vertices of the cycle. (The trees are disjoint from each other, and, other than that one original cycle, G does not have any cycles. See Figure MP10.2 for an example.) Let n be the order of the graph, and M_G be the incidence matrix of G. Then use your answers to Problems MP10.13 and MP10.17 to prove that $\text{rank}(M_G) = n$.

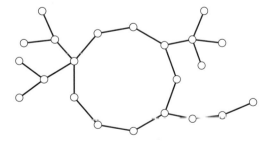

Figure MP10.2 An odd cycle with trees growing from some of its vertices.

MP10.19 Let G be a connected simple graph that is not bipartite, and let T be a spanning tree for G. Show that you can find at least one edge of G such that T together with this extra edge has exactly one odd cycle (see Problem P 10.2.13). Conclude that T together with this extra edge is a graph of the form described in Problem MP10.18.

MP10.20 Use Problem MP10.19 to prove that if G is a connected simple graph (with at least one edge) of order n, that is not bipartite, then $\mathrm{rank}(M_G) = n$.

MP10.21 If G is a simple graph with no isolated vertices, and its connected components are $G_1, \ldots, G_k$, show that $\mathrm{rank}(M_G) = \mathrm{rank}(M_{G_1}) + \cdots + \mathrm{rank}(M_{G_k})$.

MP10.22 Prove that if G is a simple graph of order n with at least one edge, then $\mathrm{rank}(M_G) = n - m$, where m is the number of connected components of G that are bipartite.

Mini-project Write-Up

In your write-up of the results, include at least the following:

- A definition, with examples, of incidence matrices of graphs
- An explanation of why it makes sense to talk about *the* rank of the incidence matrix of a graph, even though, for a specific graph, depending on the order of vertices and edges, there are a number of different incidence matrices
- An explanation of how, given a connected simple graph, the rank of the incidence matrix can be used to tell if the graph is bipartite
- Using the steps provided in the mini-project, a complete proof of the following theorem:

Theorem 10.128. *Let G be a simple graph of order n with at least one edge, and let M_G be an incidence matrix of G. Then*

$$\mathrm{rank}(M_G) = n - m,$$

where m is the number of the connected components of G that are bipartite. In particular, G has no bipartite connected components if and if only if $\mathrm{rank}(M_G) = n$.

(Extra credit) Let $G = (V, E)$ be the connected simple graph of Figure MP10.1. Assume that, for each vertex $v \in V$, we are given a real number as a label. Is it possible to assign a real number to each edge – we think of that as labeling the edges – in a way that the sum of the labels on the edges incident with any one of the vertices equals the label on the vertex? Translate the problem into one about the column space of M_G, and use Theorem 10.128 to answer it.

11   Posets, Matchings, and Boolean Lattices

Each of 450 first-year students is to be assigned to one of 30 themed activity groups during the orientation program. Each of the activity groups has a sign-up sheet with 45 slots and each student is asked to put their name on 3 of the sign-up sheets. The organizers, then, want to assign 15 students to each group in such a way that every student gets one of the groups that they signed up for. Will the organizers always succeed?

It is possible to organize much of elementary combinatorics around the language of graph theory. It is also possible to do the same using partially ordered sets. In the earlier chapters on enumerative combinatorics, we counted subsets, partitions of sets, partitions of integers, and many other things. Often it is possible to organize the objects that we are counting in a particular kind of a structure. The hope is that this organization will provide new insights. In this chapter, we introduce partially ordered sets (posets), and recast a number of (by now familiar) combinatorial objects in terms of them. We have also chosen to include a discussion of matchings and the celebrated "marriage theorem," Theorem 11.50, in this chapter as well. Our excuse is that we derive this theorem as a direct consequence of a general result – Dilworth's theorem, Theorem 11.42 – on partially ordered sets. Finally, we specialize to one particular class of partially ordered sets, namely the Boolean lattices, introduce the reader to some classical results from the combinatorics of finite sets, and argue that questions about Boolean lattices can often be thought of as generalizations of related questions in graph theory.

11.1   Posets, Total Orders, and Hasse Diagrams

Warm-Up 11.1. *Let X be the set of all subsets of $\{1,2,3\}$, and Y be the set of all positive integers that divide 30. List all the elements of X and all the elements of Y. Can you find a map $f : X \to Y$ such that f is 1-1 and onto, and, in addition, for all $x, y \in X$, $f(x)$ divides $f(y)$ if and only if x is a subset of y?*

Informally, a *partially ordered set* (or *poset* for short) is a set together with a way of ordering some of its elements. To define partial orders and partially ordered sets rigorously, it is helpful to first define what we mean by a relation on a set.

Definition 11.2 (Cartesian Product, Binary Relation). Let P be a set. Then $P \times P = \{(x,y) \mid x, y \in P\}$ is the Cartesian product of P with itself, and consists of ordered pairs (x,y) with $x, y \in P$. If R is a subset of $P \times P$, then R is called a *binary relation* on P. If the ordered pair $(x,y) \in R$, then we write xRy and we say that x is related to y.

Example 11.3. Let $P = \{1, 2, 4\}$. Then the elements of $P \times P$ are ordered pairs (a, b), where both a and b are elements of P. Since we have three choices for the first coordinate and, independent of that choice, three choices for the second coordinate, there are a total of $3 \times 3 = 9$ elements in $P \times P$. In fact,

$$P \times P = \{(1,1), (1,2), (1,4), (2,1), (2,2), (2,4), (4,1), (4,2), (4,4)\}.$$

Now, let $R = \{(1,1), (1,2), (1,4), (2,2), (2,4), (4,4)\}$. Evidently, R is a subset of $P \times P$, and so, according to Definition 11.2, it is a binary relation on P. If x and y are two elements of P, then saying that x is related to y – written xRy – is the same as saying that $(x, y) \in R$. So, for example, 1 is related to 1, 2, and 4, while 4 is only related to 4. In other words, the subset R just tells us which elements are related to which elements. Note that 2 is related to 4, while 4 is not related to 2, and so it is important that the elements of R are *ordered* pairs. The first element in the pair is related to the second element in the pair.

For this particular example, we could have defined the relation differently. We could have said that the set P is the set of positive integer divisors of 4, and, for $x, y \in P$, x is related to y if and only if the integer x divides the integer y. The integer 2 divides 4 and so $2R4$ – or $(2,4) \in R$ – while 4 does not divide 2 and so 4 is not related to 2, and $(4,2) \notin R$.

Next, we define partial orders. A partial order is a certain kind of a relation.

Definition 11.4 (Partial Order, Total Order). Let P be a set, and let R be a binary relation on P. The relation R is called a *partial order* on P if, for all $x, y, z \in P$, the following are satisfied:

- (reflexivity) xRx
- (antisymmetry) If xRy and yRx, then $x = y$
- (transitivity) If xRy and yRz, then xRz.

For a partial order, instead of R we often, but not always, use the notation $\leq$. A set P together with a partial order is called a *partially ordered set* or a *poset*.

To emphasize the relation on a set, we often write them as a pair. So if $\leq$ is a partial order on the set P, we say that $(P, \leq)$ is a poset.

Let $(P, \leq)$ be a poset, and $x, y \in P$. We say that x and y are *comparable* if $x \leq y$ or $y \leq x$. Otherwise, x and y are *incomparable*. We write $x < y$ if $x \leq y$ and $x \neq y$. We say that y *covers* x if $x < y$ and there is no z with $x < z < y$. A partial order is a *total order* if every two elements are comparable. A set with a total order is called a *totally ordered set*, a *linearly ordered set*, or a *chain*.

In this text, most of the time, we will be considering *finite* posets–that is, posets where the underlying set is a finite set.

Remark 11.5. When we say that $(P, \leq)$ is a poset, we mean that P is the underlying set and $\leq$ is the relation among the elements. We reiterate that the use of the $\leq$ symbol does not mean that the relation is the usual less than or equal to. We are substituting $\leq$ for the earlier R, and either one of these symbols could signify any relation. The use of $\leq$ for a partial order can be helpful since it makes it somewhat easier to remember the properties of a partial order.

The integers with the usual $\leq$ are also a poset (see Example 11.11), and so a statement such as $x \leq x$, or $x \leq y \leq z$ implies $x \leq z$, seems natural. The use of $\leq$ can sometimes be confusing, and so, for specific posets, depending on the actual relation, we may use a different symbol.

Example 11.6. Example 11.3 is an example of a partial order. It is reflexive since $1R1$, $2R2$, and $4R4$. It is antisymmetric because, except for when $x = y$, we don't have both xRy and yRx. Finally, the relation is transitive since, if x divides y and y divides z, then x divides z. In fact, this particular partial order is a total order, since $1R2$, $1R4$, and $2R4$, and so every pair of elements is comparable.

Example 11.7. Let P be the set of people in a family tree. For $x, y \in P$, define $x \leq y$ if either $x = y$ or if x is a descendant of y. Then $\leq$ is a binary relation on P. It is a transitive relation since, if x is a descendant of y and y is descendant of z, then x is also a descendant of z. The relation $\leq$ is also antisymmetric since, if x is a descendant of y and y is also descendant of x, then it must be that $x = y$. Finally, by definition, $\leq$ is reflexive. We conclude that $\leq$ is a partial order, and $(P, \leq)$ is a poset. This relation is not necessarily a total order, since it quite possible to have two people in a family tree who, while being relatives, are not descendants of each other.

Partial orders are best presented through their Hasse diagrams.

Definition 11.8. Let $(P, \leq)$ be a poset. The *Hasse diagram*[1] is a particular drawing of a graph associated with P. The vertices of this graph are the elements of the set P. We draw an edge between x and y if y covers x (i.e., $x < y$ and there is no $z \in X$ with $x < z < y$). In addition, if $x < y$ then y is drawn "above" x.

Note that in the Hasse diagram not all relations turn into edges. Even though $x \leq x$ for all $x \in P$, we do not draw any loops at the elements of P. More importantly, if $x < z < y$, we do not draw an edge between x and y. We only draw an edge if $x < y$ and there is *no z* with $x < z < y$. However, from the Hasse diagram, we can read *all* the relations, since all the relations that are not depicted follow from reflexivity and transitivity.

Example 11.9. Let $P = \{x, y, z, u\}$, and assume that, for all $v \in P$, we have $v \leq v$. We are also given that $x \leq y$, $y \leq z$, and $x \leq z$, and that there is no other relation among the elements of P. Then $(P, \leq)$ is a poset and its Hasse diagram is given on the left of Figure 11.1.

---

[1] Named after German mathematician Helmut Hasse (1898–1979). Hasse's mathematics was first-rate, and he made fundamental contributions to algebraic number theory. Politically, he was conservative and an ardent nationalist. German mathematics was unparalleled before the rise of the Nazis, and the effect of the 1933 anti-Jewish law was drastic. Many prominent mathematicians of Jewish origin, and those who refused to carry out the Nazi edicts, were dismissed. Among the ones who were allowed to stay, some were active and strong proponents of Nazism, while others seemed to be trying naively to save what they could of the German mathematical institutions. Hasse was originally supportive of the DNVP (Deutsche Nationale Volks Partei) and approved of some of Hitler's policies. In fact, he applied to join the Nazi party, but his application was not accepted since he had some Jewish ancestry. For more detailed context, see Segal (1980), who concludes that "Hasse was perhaps a typical example of the distinguished, conservatively apolitical and ideologically naive German professor, who, thrust into the ideologically charged situation of Nazi Germany, coped as best he could for the preservation of his discipline and himself."

Figure 11.1 Hasse diagrams for two different posets, each with four elements.

Example 11.10. Let $P = \{x, y, z, u\}$, and assume $x \leq y$, $z \leq y$, and $z \leq u$, and, for all $v \in P$, we also have $v \leq v$. If these are all the relations on P, then $(P, \leq)$ is a poset, and the Hasse diagram of P is given on the right of Figure 11.1.

Many particular posets play a natural role in combinatorics. We now go through a number of such examples.

Example 11.11. Let $P = \mathbb{Z}$ be the set of integers with the usual $\leq$ as the relation. In this poset, $5 \leq 7$, as you would expect, and all elements are comparable. In other words, $(\mathbb{Z}, \leq)$ is a totally ordered set. The Hasse diagram of $(\mathbb{Z}, \leq)$ is given in Figure 11.2. In fact, $\mathbb{Z}$ ordered by $\leq$ is our prototypical example of a (countably) infinite totally ordered set.

Figure 11.2 The Hasse diagram of $(\mathbb{Z}, \leq)$.

Example 11.12. Set inclusion is a relation for the set of all subsets of $\{1, 2, 3\}$. The *Boolean lattice of order* 3, denoted by $\mathbf{2}^{[3]}$, is the set of all subsets of $\{1, 2, 3\}$ with inclusion as the relation. A set with three elements has eight subsets, and so

$$\mathbf{2}^{[3]} = \{\emptyset, \{1\}, \{2\}, \{3\}, \{1, 2\}, \{1, 3\}, \{2, 3\}, \{1, 2, 3\}\}.$$

Which elements are related to which? If A and B are two subsets of $[3]$, then A is related to B if $A \subseteq B$. So, for example, $\{2\}$ is related to $\{2, 3\}$ since $\{2\}$ is a subset of $\{2, 3\}$. On the other hand, {2} and {1, 3} are incomparable. This relation is reflexive, antisymmetric, and transitive, and so $(2^{[3]}, \subseteq)$ is a poset. The Hasse diagram of $2^{[3]}$ is given in Figure MP8.1.

As the reader may suspect, there is nothing special about the choice of 3 in Example 11.12. Generalizing to arbitrary positive integers gives us one of the most ubiquitous posets in combinatorics.

Definition 11.13 (Boolean Lattice). As usual, let $[n] = \{1, \ldots, n\}$, and let $2^{[n]}$ denote the set of all the subsets of $[n]$. Set inclusion is a relation on $2^{[n]}$, and $(2^{[n]}, \subseteq)$ is called the *Boolean lattice of order n*. To emphasize the particular relation for this poset, we often refer to this poset as "$2^{[n]}$ ordered by inclusion."

Example 11.14. The number of subsets of [6] is $2^6 = 64$, and the Hasse diagram of $(2^{[6]}, \subseteq)$ is given in Figure 11.3. (To save space, curly brackets and commas are eliminated, and so, for example, 2356 stands for the subset $\{2, 3, 5, 6\}$.)

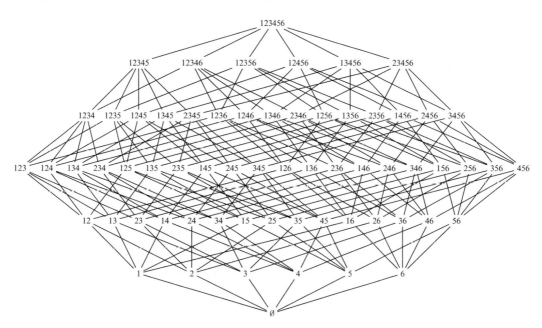

Figure 11.3 $2^{[6]}$, the Boolean lattice of order 6, ordered by inclusion.

Example 11.15. Let P be the positive integer divisors of 24 ordered by divisibility of integers. Then the elements of P are 1, 2, 3, 4, 6, 8, 12, and 24 – these are the divisors of 24 – and the partial order is divisibility. We use the symbol | to denote this relation. So, for example, 3 | 6 (this is read as 3 divides 6), while 4 ∤ 6. For this partial order, we did *not* use the customary ≤ notation to denote the relation. If we had, then we would have to write 3 ≤ 6 while 4 is not less than or equal to 6. This would be confusing, since, for integers, we are used to ≤ meaning the actual less than or equal to of numbers. The Hasse diagram of $(P, |)$ is given in Figure 11.4.

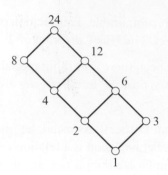

Figure 11.4 Divisors of 24 ordered by divisibility.

We generalize the example to get the divisor lattices.

Definition 11.16 (Divisor Lattice). Let n be a positive integer, and let $\mathcal{D}_n$ denote the poset of positive integer divisors of n ordered by divisibility. Then $(\mathcal{D}_n, \mid)$ is called a *divisor lattice*.

Next, we define the poset of partitions of a finite set ordered by refinement.

Definition 11.17. Let $[n] = \{1, \ldots, n\}$. A *partition* of $[n]$ is a set of disjoint non-empty subsets of $[n]$ whose union is $[n]$. In other words, $\mathcal{A} = \{A_1, \ldots, A_k\}$ is a partition of $[n]$ if, for $1 \leq i \leq k$, $A_i \neq \emptyset$, $\cup_{i=1}^{k} A_i = [n]$, and, for $1 \leq i < j \leq n$, $A_i \cap A_j = \emptyset$. The subsets $A_1, \ldots, A_k$ are called the *parts* of the partition. Given two partitions $\mathcal{A}$ and $\mathcal{B}$ of $[n]$, we define $\mathcal{A} \leq \mathcal{B}$ to mean that every part of $\mathcal{A}$ is a subset of one of the parts of $\mathcal{B}$. This relation is called *refinement*. The set of partitions of $[n]$ ordered by refinement is a poset.

Example 11.18. Let $n = 4$. Both $\{\{1, 2, 3\}, \{4\}\}$ and $\{\{1, 2\}, \{3, 4\}\}$ are partitions of $[4]$. For ease of reading, we write these two partitions as $123/4$ and $12/34$. Neither of these partitions is a refinement of the other, and so these two partitions are not comparable. On the other hand, $12/3/4$ is a partition of $[4]$ into three parts and is a refinement of $123/4$, $12/34$, and $124/3$. The Hasse diagram of partitions of $[4]$ ordered by refinement is given in Figure 11.5.

Let n be a positive integer. Recall – see Definition 7.2 – that a *partition* of n is a sequence $\lambda_1, \lambda_2, \ldots, \lambda_k$ of positive integers such that $\lambda_1 + \cdots + \lambda_k = n$, and $\lambda_1 \geq \lambda_2 \geq \cdots \geq \lambda_k > 0$. If $\lambda = (\lambda_1, \lambda_2, \ldots, \lambda_k)$ is a partition of n, then we write $\lambda \vdash n$, and the integers $\lambda_1, \ldots, \lambda_k$ are called the *parts* of the partition λ.

Definition 11.19. Let n be a positive integer, and let $\lambda = (\lambda_1, \lambda_2, \ldots, \lambda_k)$ and $\mu = (\mu_1, \mu_2, \ldots, \mu_\ell)$ be partitions of n. Append an infinite number of zeros at the end of both partitions, and say $\lambda \leq \mu$ if, for all $j > 0$, we have $\sum_{i=1}^{j} \lambda_i \leq \sum_{i=1}^{j} \mu_i$. This relation defines a partial order on partitions of n and is called the *dominance* order.

Example 11.20. Let $n = 6$. Then the poset of partitions of 6 ordered by dominance is given in Figure 11.6.

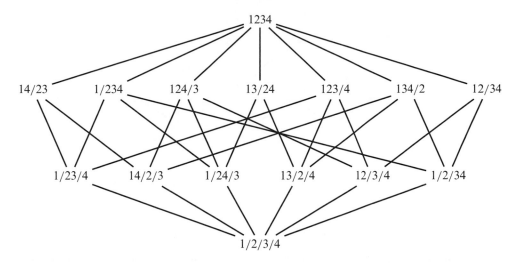

Figure 11.5 Partitions of [4] ordered by refinement.

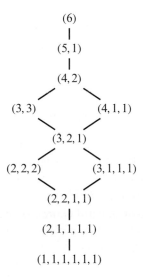

Figure 11.6 Partitions of the integer 6 ordered by dominance.

Definition 11.21. A *poset isomorphism* is a 1-1, onto map $\phi\colon P \to Q$ with the property that $x \leq y$ in P if and only if $\phi(x) \leq \phi(y)$ in Q. Two posets P and Q are *isomorphic* if there exists a poset isomorphism $\phi\colon P \to Q$.

Example 11.22. The poset of divisors of 30 ordered by divisibility is isomorphic to the Boolean lattice $\mathbf{2}^{[3]}$. This can be seen from their identical Hasse diagrams. See Figure 11.7.

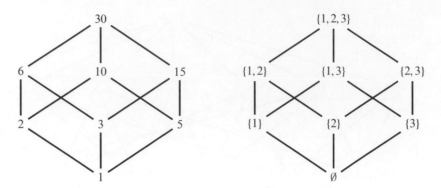

Figure 11.7 D_{30}, the poset of divisors of 30 ordered by divisibility, is isomorphic to the Boolean lattice of order 3.

Example 11.23. Figure 11.8 gives the Hasse diagrams of all posets (up to isomorphism) with four elements. This means that any poset with four elements is isomorphic to one of the posets listed.

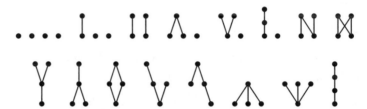

Figure 11.8 All possible Hasse diagrams of posets with four elements.

Remark 11.24. In studying combinatorial properties of ordered structures, there are (at least) two approaches. One approach would try to develop concepts and techniques applicable to *all* posets, or at least to very general classes of posets. As such, it would be natural to start studying posets with a small number of elements and work one's way up. The second approach would take as its prototypical posets the Boolean lattices, $2^{[n]}$, try to study their properties, and when possible generalize the results to wider classes of posets. In the second approach, the choice of questions and the choice of posets would be guided by what we know (or would like to know) about the Boolean lattices. My personal taste favors the second approach, and that will be evident in what follows.

Problems

P 11.1.1. In Definition 11.2, we defined a relation on P as a subset of $P \times P$. Let $P = [4] = \{1, 2, 3, 4\}$, and consider the following subset of $P \times P$:

$$R = \{(1,1), (2,2), (3,3), (4,4), (1,2), (1,3), (2,3), (4,2), (4,3)\}.$$

(a) Is R a relation on P?

(b) Is R a partial order on P? If so, draw its Hasse diagram.

(c) Is R a total order on P?

P 11.1.2. Let P be the set of the people in the world. For each of the relations below, decide if the relation is reflexive, antisymmetric, or transitive. Is the relation a partial order?

(a) For $x, y \in P$, x is related to y if x is y's sister.

(b) For $x, y \in P$, x is related to y if x has the same or fewer children than y.

(c) For $x, y \in P$, x is related to y if $x = y$ or if x has finished fewer years of school than y.

(d) For $x, y \in P$, x is related to y if $x = y$ or if x and y both live in New York City and x lives closer than y to Jacob's Soul Food Buffet on Frederick Douglass Blvd and 143rd Street.

P 11.1.3. Consider Definition 11.4 of a poset.

(a) Can we replace the "antisymmetry" condition with the following?

- If $x, y \in P$ with $x \neq y$ and xRy, then y is not related to x.

(b) Can we replace both the "reflexivity" and "antisymmetry" conditions with the following?

- If $x, y \in P$, then xRy and yRx if and only if $x = y$.

(c) Can we replace both the "reflexivity" and "antisymmetry" conditions with the following?

- If $x, y \in P$ with xRy, then yRx if and only if $x = y$.

P 11.1.4. Are the graphs of Figure 11.9 missing from Figure 11.8? If so, why?

Figure 11.9 Are these Hasse diagrams?

P 11.1.5. Can the Hasse diagram of a poset contain a triangle? Either give an example where it does, or prove that it cannot.

P 11.1.6. Let Q consist of those subsets of $[6]$ that contain $\{2, 6\}$. Is $(Q, \subseteq)$ a poset? If so, draw its Hasse diagram

P 11.1.7. Recall that $(\mathcal{D}_n, \mid)$ is the poset of divisors of the integer n ordered by divisibility (Definition 11.16). Draw the Hasse diagrams of $(\mathcal{D}_{16}, \mid)$ and $(\mathcal{D}_{36}, \mid)$.

P 11.1.8. Give examples of three integers n such that the Hasse diagram of $(\mathcal{D}_n, \mid)$, the poset of divisors of n ordered by divisibility, is a 5×5 grid, as in Figure 11.10.

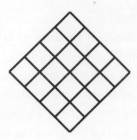

Figure 11.10 For which n is this the Hasse diagram of $(\mathcal{D}_n, \mid)$?

P 11.1.9. Let n be a positive integer. Do Stirling numbers and/or Bell numbers have anything to do with the poset of partitions of $[n]$ ordered by refinement (see Definition 11.17)? What?

P 11.1.10. Draw the Hasse diagram of the poset of partitions of the integers 5 and 7 ordered by dominance (see Definition 11.19).

P 11.1.11. If P and Q are posets, and $f \colon P \to Q$, then f is called an *order preserving map* if, for all $x, y \in P$, $x \le y$ (in P) implies $f(x) \le f(y)$ (in Q). Prove that a poset isomorphism is an order preserving bijection whose inverse is also order preserving.

P 11.1.12. Let $P = 2^{[7]}$ ordered by inclusion. Let $A = \{2, 4, 6\} \in P$. Let Q be the poset of subsets of A ordered by inclusion, and let R be the poset of subsets of $[7]$ that contain A again ordered by inclusion. Prove that Q and R are isomorphic to $2^{[3]}$ and $2^{[4]}$, respectively.

P 11.1.13. Let $P = 2^{[6]}$ ordered by inclusion.
  (a) Partition the elements of P into two disjoint sets Q and R such that, ordered by inclusion, each is isomorphic to $2^{[5]}$.
  (b) Can you explicitly partition the elements of P into four disjoint sets $P_1, \ldots, P_4$ such that, ordered by inclusion, each is isomorphic to $2^{[4]}$?

P 11.1.14. Find three distinct integers such that the poset of the divisors of each of them (ordered by divisibility) is isomorphic to $2^{[3]}$. The poset of divisors of which integers are isomorphic to a Boolean lattice?

P 11.1.15. Let $P = \mathcal{D}_{36}$ and $Q = \mathcal{D}_9$ be the posets of divisors of 36 and 9, respectively. Let $R = \{x \in P \mid x \text{ is divisible by } 4\}$.
  (a) What are the elements of R and Q?
  (b) Are the posets R and Q related? Are they isomorphic?

P 11.1.16. Let n and m be positive integers, and assume m divides n. Let $\mathcal{D}_n$ be the poset of divisors of n ordered by divisibility, and let $R = \{x \in \mathcal{D}_n \mid m \text{ divides } x\}$.
  (a) Is R (ordered by divisibility) a poset?
  (b) How is R related to $\mathcal{D}_{n/m}$, the poset of divisors of n/m? Are they isomorphic?

P 11.1.17. Let n be a positive integer and let λ and μ be two partitions of the integer n. Assume that λ and μ have k and ℓ parts, respectively, and that $k < \ell$. Is it possible that $\lambda \le \mu$ in the dominance order?

P 11.1.18. Let n be a positive integer, and let P_n be the set of all partitions of integers 1, 2, ..., n. If μ and ν are elements of P_n, then say $\mu \le \nu$ if the Ferrers diagram (Definition 7.9) of μ fits within the Ferrers diagram of ν. In other words, if $\mu = (\mu_1, \mu_2, \ldots, \mu_\ell)$ and $\nu = (\nu_1, \nu_2, \ldots, nu_k)$, then append an infinite number of zeros to each of them, and declare $\mu \leq \nu$ if, for all $j > 0$, we have $\mu_j \leq \nu_j$. Is $(P_n, \leq)$ a partially ordered set? If so, draw the Hasse diagram of P_4.

11.1.1 Products of Posets

Let $(P, \leq)$ and $(Q, \precsim)$ be posets. Recall that $P \times Q$ is the cartesian product of P and Q, and its elements are pairs (p, q), where $p \in P$ and $q \in Q$. We define a relation $\lesssim\kern-0.6em\lesssim$ on $P \times Q$ by

$$(p_1, q_1) \lesssim\kern-0.6em\lesssim (p_2, q_2) \text{ if and only if } p_1 \leq p_2, \text{ and } q_1 \precsim q_2.$$

Then $(P \times Q, \lesssim\kern-0.6em\lesssim)$ is a new poset called the *product of P and Q*.

Problems (continued)

P 11.1.19. Let $P = (\mathcal{D}_8, \mid)$ and $Q = (\mathcal{D}_9, \mid)$ be the posets of divisors of 8 and 9, respectively. Let $R = P \times Q$ be the product of the posets P and Q. How many elements does R have? What is the Hasse diagram of R? How is R related to $(\mathcal{D}_{72}, \mid)$?

P 11.1.20. Let $P = (\mathcal{D}_8, \mid)$ and $Q = (\mathcal{D}_4, \mid)$ be the posets of divisors of 8 and 4, respectively. Let $R = P \times Q$ be the product of the posets P and Q. Find a positive integer n such that the poset R is isomorphic to $(\mathcal{D}_n, \mid)$, the poset of divisors of n.

P 11.1.21. Let n and m be relatively prime integers. As usual, let $\mathcal{D}_n$ and $\mathcal{D}_m$ denote, respectively, the poset of divisors of n and m ordered by divisibility. Is $\mathcal{D}_n \times \mathcal{D}_m$ isomorphic to $\mathcal{D}_{nm}$? If so, prove it. If not, give a counter-example.

P 11.1.22. For a non-negative integer n, as usual, let $2^{[n]}$ denote the poset of subsets of $[n]$ ordered by inclusion. Find an integer m such that $2^{[3]} \times 2^{[2]}$ is isomorphic to $2^{[m]}$.

P 11.1.23. Let n be a positive integer. Prove that $2^{[n]}$ is isomorphic to $\underbrace{2^{[1]} \times 2^{[1]} \times \cdots \times 2^{[1]}}_{n}$.

11.2 Chains, Antichains, and Dilworth's Theorem

Warm-Up 11.25. *Figure 11.5 gives the Hasse diagram of partitions of $[4]$ ordered by refinement. (See Definition 11.17.) For each of the following, find the size of the largest collection $\mathcal{P}$ of partitions of $[4]$ that satisfies each of the following conditions:*

(a) *If $a, b \in \mathcal{P}$, then either $a \leq b$ or $b \leq a$.*
(b) *If $a, b \in \mathcal{P}$, then neither $a < b$ nor $b < a$.*

In a partially ordered set, two elements a and b could be comparable (that is, $a \leq b$ or $b \leq a$) or not. In this section, we consider subsets of a poset where every pair of elements is comparable, as well as subsets where no pair of elements is comparable. These dual concepts are related via the celebrated Dilworth theorem, which we will prove.

Definition 11.26. Let $(P, \leq)$ be a poset, and let $x_1, \ldots, x_k \in P$. Assume

$$x_1 < x_2 < \ldots < x_k.$$

Then $\{x_1, \ldots, x_k\}$ is called a *chain* of *size k* and *length $k - 1$*.

If, in addition, for $2 \leq i \leq k$, x_i covers x_{i-1}, then we say that the chain is *skipless*.

Example 11.27. In $2^{[6]}$, the Boolean lattice of order 6, the subsets

$$\{1, 3\} \subset \{1, 3, 5\} \subset \{1, 2, 3, 5, 6\}$$

form a chain of size 3 and length 2. This chain is not skipless, since both $\{1, 2, 3, 5\}$ and $\{1, 3, 5, 6\}$ are in between $\{1, 3, 5\}$ and $\{1, 2, 3, 5, 6\}$.

Definition 11.28. Let $(P, \leq)$ be a poset, and let $\mathcal{C}$ be a chain in P. The chain $\mathcal{C}$ is a *maximal chain* if $\mathcal{C} \cup \{x\}$ is *not* a chain for all $x \in P - \mathcal{C}$. The chain $\mathcal{C}$ is a chain of *maximum size* in P if there is no other chain with larger size than $\mathcal{C}$. The *height* of a poset is the maximum size (possibly infinite) of a chain in the poset.[2]

Example 11.29. Figure 11.11 is the Hasse diagram of a poset of height 6. This poset has six maximal chains. Three of these are of size 6 (and length 5) and three of them are of size 5 (and length 4).

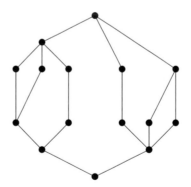

Figure 11.11 A poset of height 6 with maximal chains of sizes 5 and 6.

---

[2] The reader should be warned that some authors define the height to be one less than the maximum size of a chain in the poset.

Example 11.30. Consider the poset $(2^{[6]}, \subseteq)$, the Boolean lattice of order 6 (see Figure 11.3). One maximum-sized chain in the poset is

$$\emptyset \subset \{2\} \subset \{1,2\} \subset \{1,2,5\} \subset \{1,2,5,6\} \subset \{1,2,3,5,6\} \subset \{1,2,3,4,5,6\},$$

and as a result $2^{[6]}$ has height 7. Likewise,

$$1 \mid 2 \mid 4 \mid 12 \mid 24$$

is a maximum-sized (as well as a maximal) chain in the poset of divisors of 24 (see Figure 11.4), and so the height of this poset is 5.

Definition 11.31. Let $(P, \leq)$ be a poset, and let A be a subset of P. Then A is called an *antichain* if no two elements of A are comparable. The size of the largest antichain in P (possibly infinite) is called the *width* of P.

Example 11.32. In $2^{[6]}$, the Boolean lattice of order 6, the subsets

$$\{1,3\}, \{2,4\}, \{1,4,5\}, \{3,4,5\}, \{1,2,5,6\}$$

form an antichain of size 5, since none of these subsets contains any of the others.

Example 11.33. In the poset $(\mathbb{Z}, \leq)$ (see Figure 11.2) every two elements are comparable. Hence, no two elements form an antichain, and the size of the largest antichain is 1. So this poset has width 1.

Example 11.34. Consider the poset of divisors of 24 ordered by divisibility (Figure 11.4). All the divisors of 24 can be organized into two chains:

$$2 \mid 4 \mid 8 \mid 24$$
$$1 \mid 3 \mid 6 \mid 12.$$

If you pick three of these elements, then, by the pigeon-hole principle, at least two of the elements will be in the same chain. Two elements in the same chain are comparable and so cannot be in the same antichain. As a result, the size of an antichain in this poset cannot be more than 2. Now $\{2,3\}$ is an antichain in this poset and so the width of this poset is 2.

To illustrate a certain duality between chains and antichains, and guided by Example 11.34, we now ask a reasonably general question.

Problem 11.35. *Given a poset $(P, \leq)$, what is the minimum number of chains needed to partition P?*

Example 11.36. The Boolean lattice of order 3, $2^{[3]}$, has eight members. These are the subsets of $[3] = \{1,2,3\}$, and one way to partition these eight subsets into chains is as follows:

$$\emptyset \subset \{1\} \subset \{1,2\}$$
$$\{2\} \subset \{2,3\} \subset \{1,2,3\}$$
$$\{3\} \subset \{1,3\}.$$

This gives a partition of $2^{[3]}$ into three chains (of sizes 3, 3, and 2). Could we have partitioned $2^{[3]}$ into two chains? The answer is no. You can see this by considering the three subsets of size 2: $\{1,2\}$, $\{1,3\}$, and $\{2,3\}$. These three sets form an antichain, and, as a result, no two of them can be in the same chain. Thus to partition $2^{[3]}$, we need at least three chains, so that each of these subsets can be in one of those chains. Since we have already found three chains that partition $2^{[3]}$, we conclude that the minimum number of chains needed to partition $2^{[3]}$ is three. This argument also proves that the width of $2^{[3]}$ is three. We already saw an antichain of size 3, but why couldn't there be a bigger antichain? If there was an antichain of size 4, for example, then we would have needed at least four chains to cover the poset.

We distill the argument of Example 11.36 to create a tool for determining the width of a poset, as well as for finding the minimum number of chains needed to partition the poset.

Proposition 11.37. *Let P be a finite poset. Assume that*

(a) *P can be partitioned into m chains, and*
(b) *P has an antichain of size m.*

Then m is the minimum number of chains needed to partition P, and is also the width of P.

Proof. Two elements in the same antichain are, by definition, incomparable. As a result, they cannot be in the same chain. Thus, to partition P we need at least m chains. But since we already have a partition of P into m chains, the minimum number of chains needed to partition P is exactly m. The maximum size of an antichain in P is also m, since, if P had a bigger antichain, then there would not be a partition of P into m chains. □

Example 11.38. Consider the poset of subsets of [4] ordered by refinement (Figure 11.5). Here is an antichain of size 7 in this poset:

$$14/23, \ 1/234. \ 124/3, \ 13/24, \ 123/4, \ 134/2, \ 12/34.$$

On the other hand, here are seven chains that partition the poset:

$$1/2/3/4 \leq 1/23/4 \leq 14/23$$
$$14/2/3 \leq 124/3 \leq 1234$$
$$1/24/3 \leq 1/234$$
$$13/2/4 \leq 13/24$$
$$12/3/4 \leq 12/34$$
$$1/2/34 \leq 134/2$$
$$123/4.$$

Hence, by Proposition 11.37, this poset has width 7, and seven is the minimum number of chains needed to partition the poset. Also, the size of the largest chain in the poset is 4 and so the poset has height 4.

Remark 11.39. Let $(P, \leq)$ be a poset and let $C_1, C_2, \ldots, C_k$ be chains in P. If every element of P is in at least one of the chains, then we say that $\{C_1, C_2, \ldots, C_k\}$ *covers* P. If, for each pair C_i and C_j, we have $C_i \cap C_j = \emptyset$ – that is, the chains are pairwise disjoint – then we have a partition of P into chains.

Note that if you can cover P with k chains, then you can also partition P with k or fewer chains. This is because if you eliminate some of the elements of any of the chains, they continue to be chains. So, starting with a chain cover, you can go through the chains one by one and eliminate any element that has already appeared in previous chains to get a chain partition.

Remark 11.40. Proposition 11.37 can be thought of as a "weak duality" result – a concept that is formalized in linear programming and optimization courses. By constructing a certain chain partition and a certain antichain, as long as both have the same size, we deduce that the antichain is a maximum-sized antichain, and the chain partition has the minimum number of chains needed to cover it. This result is a *weak* duality result since it doesn't say that we can always find such an antichain and such a chain partition. It just says that *if* you find such a pair, then you know more. A stronger result – a "strong duality" result – would be if we could prove that, given a poset P, we can always find an antichain and a chain partition, both with the same size. For finite posets, the remarkably general Dilworth's theorem[3] gives us exactly this strong duality result. Unlike Proposition 11.37, the proof of this result is not straightforward, and will be presented next. In Problem P 11.2.10, you are asked to prove a dual result – with a surprisingly easier proof – that the minimum number of antichains needed to partition a poset is equal to the height of the poset.

Definition 11.41. Let P be a poset, and let $a \in P$. Then a is called a *maximal* element of P if there does not exist $b \in P$ with $b > a$. Likewise, a is called a *minimal* element of P if there is no $b \in P$ with $b < a$.

Theorem 11.42 (Dilworth's Theorem, Dilworth 1950). *Let P be any finite poset. Then the minimum number of chains needed to partition P is exactly equal to the width of P.*

Proof (Galvin 1994). Let $c(P)$ be the minimum number of chains needed to partition P, and let $m = \text{width}(P)$ be the size of the largest antichain in P. No two elements of an antichain can be on the same chain. Hence, $c(P)$ is at least as large as m. We need to show that we can partition P into m chains.

Induct on $|P|$. This is clearly true if $|P| = 1$. Hence, assume $|P| > 1$ and that the theorem has been proved for all posets of smaller size. Let a be a maximal element of P.

Consider the poset $P' = P - \{a\}$. Let n be the width of P'. By induction, P' can be partitioned into n disjoint chains. So assume

$$P' = C_1 \cup C_2 \cup \cdots \cup C_n,$$

[3] Due to the American mathematician Robert Dilworth (1914–1993).

where $C_1, \ldots, C_n$ are pairwise disjoint chains in P'. Now, if $m = n + 1$, then $P = C_1 \cup \cdots \cup C_n \cup \{a\}$ is a partition of P into $m = n + 1$ chains, and we are done.

The only remaining possibility is when $m = n$. In this case, we have to partition P into n chains. The reader may erroneously think that since a is a maximal element, we should always be able to extend one of the chains $C_1, \ldots, C_n$ by adding a to it. If this were possible, then P would have been partitioned into n chains, and we would be done. We can't quite make this argument, since maybe none of the elements "below" a are the top element of any of the chains $C_1, \ldots, C_n$. (See Figure 11.12.) A clever argument gets around this issue.

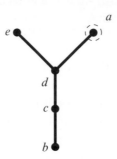

Figure 11.12 The point a can be a maximal element of a poset, and yet you may not be able to add it to the top of one of the chains – in this case the chain b-c-d-e – partitioning the rest of the poset.

To recap, we are in the case when P and $P - \{a\} = P' = C_1 \cup \cdots \cup C_n$ both have widths equal to n and we need to partition P into n chains. Since the width of P' is n, P' has an antichain of size n, and such an antichain must have exactly one element from each of C_i, for $1 \le i \le n$ (since these chains partition P' and none of the chains can have more than one element of any antichain). Let

$$a_i = \max\{x \in C_i \mid x \text{ is an element of an } n\text{-element antichain of } P'\}.$$

In other words, for every chain C_i, we start with the element at the top of the chain and ask "Do you belong to an antichain of P' of size n?" If the answer is yes, then that element is a_i. Otherwise, we pose the same question to the next element going down on C_i. The first time we get a "yes," we have found our element a_i. We already argued that each C_i must have at least one such element, and so we are assured of getting a "yes" sooner or later.

CLAIM: The set $\mathcal{A} = \{a_1, a_2, \ldots, a_n\}$ is an antichain.

PROOF OF CLAIM: Assume that, for $i < j$, we have $a_i \le a_j$. The chains $C_1, \ldots, C_n$ are disjoint chains and so $a_i \ne a_j$. We know that a_j is part of an n-element antichain. This antichain must have an element d in C_i. By definition of a_i, we have $d \le a_i$. But this means that $d \le a_j$, a contradiction which shows that a_i is not less than or equal to a_j. A similar argument shows that a_i cannot be bigger than a_j, and the proof of the claim is complete.

The set $\mathcal{A}$ is not only an antichain of P' but is also an antichain of size n in P. The width of P is also n, and so we can't have a larger antichain in P. Hence, the element a must be comparable to one of the elements of $\mathcal{A}$. (Otherwise, adding a to $\mathcal{A}$ would give an antichain of size $n + 1$ in P.) Say a is comparable to $a_i \in \mathcal{A}$, for some $1 \le i \le n$. Since a is a maximal element of P, we have $a_i < a$. This implies that

$$K = \{a\} \cup \{x \in C_i \mid x \le a_i\}$$

is a chain in P.

Now, from among the elements of C_i, $P - K$ only has those elements that are above a_i. As a result, $P - K$ cannot have any n-element antichains since otherwise a_i would not have been the largest element of C_i in an n-element antichain. Hence, by induction, $P - K$ can be partitioned into $n - 1$ chains. Now these $n - 1$ chains together with K partition P into n chains! □

You are guided through another proof of Dilworth's theorem, due to Micha Perles (Perles 1963), in Problem $P\ 11.2.19$.

Problems

P 11.2.1. Let P be a finite poset. Assume that you can partition P into m chains, and that you can find an antichain of size n. Prove, without using Dilworth's theorem, that $n \le m$. Give an example where $n < m$.

P 11.2.2. Consider $2^{[5]}$, the poset of subsets of [5] ordered by inclusion. Find a maximum-sized antichain and a partition of $2^{[5]}$ into as few chains as possible.

P 11.2.3. What is the width and height of the poset of partitions of the integer 6 ordered by domination? (See Figure 11.6.)

P 11.2.4. Let $\mathcal{D}_{36}$ be the poset of divisors of 36 ordered by divisibility. What is the height and width of this poset?

P 11.2.5. Find all maximum-sized antichains in the poset whose Hasse diagram is given in Figure 11.13.

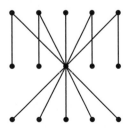

Figure 11.13 The Hasse diagram of a poset.

P 11.2.6. A poset P has width 7 and height 8. What can you say about $|P|$, the number of elements in P?

P 11.2.7. A mystery poset P has both height and width equal to 4. In addition, P has an element x which is only comparable to two other elements u and v, and we have $u \leq x \leq v$. What is the maximum possible size of P? What is the minimum possible size of P? Why?

P 11.2.8. **Dilworth's Theorem, Reworded.** Let s and t be positive integers, and let $n = (s-1)(t-1)+1$. Let P be a poset with n elements. Then show that P has either an antichain of size s or a chain of size t.

P 11.2.9. Let P be a poset and let $\mathcal{M}$ be the set of maximal elements of P. Let $Q = P - \mathcal{M}$, with the same partial order as in P.

    (a) Is $\mathcal{M}$ necessarily an antichain of P?

    (b) How does the height of Q compare to the height of P?

    (c) Let $\mathcal{A}$ be the set of maximal elements of Q. Is $\mathcal{A}$ necessarily an antichain of P?

P 11.2.10. **Mirsky's Theorem.**[4] Prove that the height of a finite poset equals the minimum number of antichains needed to partition the poset.

P 11.2.11. Let P be a poset of width w and height h. Give two proofs that $|P| \leq wh$. For one proof use Dilworth's theorem, Theorem 11.42, and for the second proof use Mirsky's theorem from Problem P 11.2.10.

P 11.2.12. Let $A_1, A_2, \ldots, A_{16}$ be an arbitrary set of 16 distinct sets. Let $P = \{A_1, \ldots, A_{16}\}$, and, as usual, let $\subseteq$ denote inclusion. Is $(P, \subseteq)$ a poset? Show that either both the width and the height of P are at least 4, or at least one of these (the width or the height) is at least 6.

P 11.2.13. Let $A_1, A_2, \ldots, A_{25}$ be an arbitrary set of 25 distinct sets. Prove that you can always choose five of these sets in such a way that none of the five chosen sets is the union of two from among the remaining four sets in the chosen collection. Can you generalize this result?

P 11.2.14. I have a secret partially ordered set (poset) that has 10 elements: $x_1, x_2, \ldots, x_{10}$. For each element x, I have found the size of the largest antichain that has x as a member. I have listed these sizes below:

Element	x_1	x_2	x_3	x_4	x_5	x_6	x_7	x_8	x_9	x_{10}
Max size of an antichain containing x	2	3	2	3	3	3	3	3	3	2

I am also letting you know that we have the following chains in our poset (there may exist many more chains):

$$x_1 \leq x_2 \leq x_4 \leq x_7,$$

$$x_3 \leq x_5 \leq x_8,$$

$$x_6 \leq x_9 \leq x_{10}.$$

Based on the given information, can you explicitly find an antichain of size 3? Prove your assertion.

---

[4] Named after the Russian-British mathematician Leonid Mirsky (1918–1983).

P 11.2.15. Let $x_1, x_2, \ldots, x_n$ be a sequence of n real numbers. Let $P = \{y_1, y_2, \ldots, y_n\}$ be a set with n elements. Define a relation $\precsim$ on P as follows: for $1 \leq i, j \leq n$, we define $y_i \precsim y_j$ if and only if $x_i \leq x_j$ and $i \leq j$.

    (a) Show that $(P, \precsim)$ is a partially ordered set.

    (b) Show that a chain in P corresponds to a non-decreasing subsequence of the original sequence.

    (c) Show that an antichain in P corresponds to a decreasing subsequence of the original sequence.

P 11.2.16. **Erdős–Szekeres Revisited.** Use either Dilworth's theorem (Theorem 11.42) or Mirsky's theorem (Problem P 11.2.10) to give a different proof of the Erdős–Szekeres theorem (Theorem 2.11 and Problem P 2.1.29). In other words, assume that, for positive integers r and s, a sequence of $(r-1)(s-1)+1$ real numbers is given. Prove that you can find either a non-decreasing subsequence of length r or a non-increasing subsequence of length s.

P 11.2.17. Let P be a given poset. The width of P is 5, the height of P is 7, and the Hasse diagram of P is connected.

    (a) If $\mathcal{A}$ is any antichain in P, we define

$$P^+(\mathcal{A}) = \{x \in P \mid x \geq y \text{ for some } y \in \mathcal{A}\}.$$

    Give an example of such a poset P with three antichains of maximum size: $\mathcal{A}_1, \mathcal{A}_2$, and $\mathcal{A}_3$ such that

$$P^+(\mathcal{A}_1) = P, \quad P^+(\mathcal{A}_2) = \mathcal{A}_2, \quad \mathcal{A}_3 \neq P^+(\mathcal{A}_3) \neq P.$$

    NOTE: It suffices to give the Hasse diagram of the poset and identify its properties.

    (b) Let $\mathcal{A}$ be an antichain with five elements in P. Let $a \in \mathcal{A}$. We construct Q from P by removing a and all elements of P comparable to a. Is Q a poset? What is the largest number of elements that Q could have? Why?

P 11.2.18. Let P be a finite poset of width w. Let $W \subseteq P$ be an antichain of size w. Define

$$P^+(W) = \{x \in P \mid x \geq y \text{ for some } y \in W\}$$
$$P^-(W) = \{x \in P \mid x \leq y \text{ for some } y \in W\}.$$

    (a) Show that $P = P^+(W) \cup P^-(W)$.

    (b) Show that $P^+(W) \cap P^-(W) = W$.

    (c) Show that, if each of $P^+(W)$ and $P^-(W)$ are partitioned into w chains, then these chains can be "glued" together to give a partition of P into w chains.

    (d) Assume that W is not the set of all maximal elements of P. Show that $P^-(W) \neq P$. Likewise, if W is not the set of all minimal elements of P, show that $P^+(W) \neq P$.

P 11.2.19. **Another Proof of Dilworth's Theorem.** Let P be a finite poset of width w. By following the steps outlined here,[5] give a different proof of Dilworth's theorem that P can be partitioned into w chains.

    STEP 1: For warm-up, you may want to do Problems P 11.2.9 and P 11.2.17.

---

[5] Adpated from Perles 1963.

STEP 2: Use induction on $|P|$. What is the base case?

STEP 3: For the inductive step, split the proof into two cases. For the first case, assume that P has an antichain of size w that is neither the set of maximal elements of W nor the set of minimal elements of W. In this case, use Problem P 11.2.18 and induction to complete the proof.

STEP 4: For the second case, assume that the only antichains of size w in P are the set of maximal elements of P, the set of minimal elements of P, or both. Can you find a and b, respectively a minimal and a maximal element of P with $a \leq b$? What is the width of $P - \{a, b\}$? Use induction and the chain $\{a, b\}$ to finish the proof.

11.2.1 Maximal Antichains, Fibres, and Cones

Let P be a poset. An antichain $\mathcal{A} \subseteq P$ is called a *maximal antichain* of P if $\mathcal{A}$ together with any other element of P would not be an antichain. In other words, an antichain is maximal if every element of P is comparable to some element already in the antichain.

If $(P, \leq)$ is a poset, then a *fibre* in P is a collection of elements of P that has a non-trivial intersection with each maximal antichain of P. A *minimal fibre* is a fibre P with the additional property that P without any of its elements is no longer a fibre.

If P is a poset, and $x \in P$, then the *cone* of x in P is the collection of all elements of P comparable to x.

Problems (continued)

P 11.2.20. Let P be a poset. Show that every cone in P is a fibre.

P 11.2.21. Let P be the poset whose Hasse diagram is given in Figure 11.14.

(a) Find all maximum-sized antichains in P.

(b) Find all maximal antichains in P.

(c) Is the cone of c a minimal fibre? What about the cone of d?[6]

(d) Find all the minimal fibres in P.

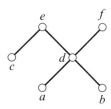

Figure 11.14

---

[6] See also Problem P 11.4.20.

11.3   Matchings and the Marriage Theorem

Warm-Up 11.43. *Batu, Bayarmaa, Bolormaa, Qacha, Qadan, and Xanadu are each to be assigned as discussion leaders for one of six books: (A) Isabel Allende's* The House of the Spirits, *(B) Octavio Butler's* Kindred, *(G) Nadine Gordimer's* Burger's Daughter, *(L) Jhumpa Lahiri's* Interpreter of Maladies, *(S) Gertrude Stein's* The Autobiography of Alice B. Toklas, *and (W) Alice Walker's* The Color Purple. *Each student is asked their preferences, and their responses are as follows: Batu* $= \{A, B\}$*, Bayarmaa* $= \{A, B, W\}$*, Bolormaa* $= \{A, G, L, W\}$*, Qacha* $= \{G, L, S\}$*, Qadan* $= \{B, W\}$*, and Xanadu* $= \{A, W\}$*. Assign one book per student in such a way that the largest possible number of students get a book from among their preferences.*

Warm-Up 11.43 is an example of a straightforward matching problem, where we want to match six students with six books. The opening problem of this chapter is also a matching problem – albeit a more complicated 15-to-1 matching. In mathematics, there is a useful maxim that says: "If there is a problem that you can't solve, there is an easier problem that you also can't solve. Solve the easier problem first."[7] So, before tackling the more difficult problem, we will start with a simpler problem – very similar to Warm-Up 11.43 – that we will use to guide our discussion of matchings and the marriage theorem.

Problem 11.44. *Each of the students* $S_1, S_2, \ldots, S_7$ *is to be assigned to one of seven different tasks:* $T_1, \ldots, T_7$*. Each student has different skill sets and, as a result, is best suited for certain tasks. For each student, here are the tasks that they are best qualified for:*

$$S_1 = \{T_3, T_4\} \qquad\qquad S_2 = \{T_1, T_2, T_4\}$$
$$S_3 = \{T_1, T_5, T_6, T_7\} \qquad\qquad S_4 = \{T_1, T_2, T_3\}$$
$$S_5 = \{T_5, T_6, T_7\} \qquad\qquad S_6 = \{T_2, T_3, T_4\}$$
$$S_7 = \{T_2, T_3, T_4\}.$$

We want to match each student with one task and each task with one student in such a way that we maximize the number of student–task pairs in which the student is qualified for the task.

For small problems, such as Warm-Up 11.43 and Problem 11.44, we can just eyeball an answer. Here, we first reformulate the problem into several equivalent forms, and, in the process, introduce some new vocabulary.

Transversals or Systems of Distinct Representatives

We have a collection of sets – in this case $S_1, \ldots, S_7$ – and ideally we want to choose one distinct element from each set. If we could do that, the set of distinct elements, one from

---

[7] A variation of this maxim is due to George Pólya (1887–1985), an influential mathematician and problem solver. Like any maxim, there are others that point in different directions. The brilliant mathematician Maryam Mirzakhani (1977–2017) – the first woman to win the Fields Medal – suggested in an interview (Klarreich 2014) that for doing deep mathematics "You have to ignore low-hanging fruit, which is a little tricky."

each set, is called a transversal or a system of distinct representatives for the sets. If this is not possible, we want to find the maximum number of sets for which we can find a transversal.

Definition 11.45 (Transversal, System of Distinct Representatives). Let $A_1, \ldots, A_r$ be a collection of r sets. A *transversal* or a *system of distinct representatives* (abbreviated *s.d.r.*) for $A_1, \ldots A_r$ is an ordered list $(a_1, \ldots, a_r)$ of r distinct elements such that, for $1 \leq i \leq r$, $a_i \in A_i$.

As an example, in Problem 11.44, the list $(T_3, T_2, T_5, T_1, T_7, T_4)$ is a transversal (or an s.d.r.) for $S_1, S_2, \ldots, S_6$, as is $(T_4, T_1, T_6, T_2, T_5, T_3)$. On the other hand, $(T_4, T_6, T_1, T_5, T_3, T_2)$ is a transversal for the sets $S_2, \ldots, S_7$. The five sets S_1, S_2, S_4, S_6, and S_7 – and, by extension, the seven sets $S_1, \ldots, S_7$ – do not have a transversal. These five sets, among themselves, contain only four elements, and so there is no way to find a transversal for them.

Rooks on Boards with Forbidden Positions

Problem 11.44 can be turned into a question about placing non-attacking rooks on a board with forbidden positions. Consider the 7×7 board with forbidden configurations of Figure 11.15, where, for each student, we have left open the positions that they are qualified for. Since each student is assigned to one task, and each task is assigned to one student, assigning students to tasks is the same as placing non-attacking rooks on this board while avoiding the forbidden positions. From the board we see, for example, that there are only two students qualified for the last three tasks, precluding any chance of placing seven non-attacking rooks. There are multiple ways of placing six rooks on the board, and each of these corresponds to assigning six of the students to tasks that they are qualified for (and presumably giving an unqualified student the final task).

	T_1	T_2	T_3	T_4	T_5	T_6	T_7
S_1	×	×			×	×	×
S_2			×		×	×	×
S_3		×	×	×			
S_4				×	×	×	×
S_5	×	×	×	×			
S_6	×				×	×	×
S_7	×				×	×	×

Figure 11.15 Finding a transversal (an s.d.r.) is the same as placing non-attacking rooks on a board with forbidden positions.

Matchings in Bipartite Graphs

Yet a third – and sometimes the best – way to think about the information of Problem 11.44 is to organize the data as a bipartite graph. The students and the tasks are the vertices, and there is an edge between a student and a task if the student is qualified for the task. See Figure 11.16. In this graph, any collection of disjoint edges – edges that do not share a vertex – corresponds to assigning some students to tasks for which they are qualified, while making sure that each student gets at most one task and each task gets at most one student. We can extend this interpretation to all simple graphs – as opposed to just bipartite graphs. We make some definitions.

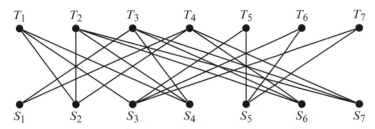

Figure 11.16 Finding a matching corresponds to finding a number of disjoint – in the sense of not sharing a vertex – edges.

Definition 11.46. Let $G = (V, E)$ be a simple graph. A *matching* (or an *independent edge set*) in G is a set of disjoint edges in G. In other words, $M \subseteq E$ is a matching in G if no two edges in M have a common vertex. The number of edges in M is the *size of the matching*. The *matching number* of G – often denoted as $\nu(G)$ – is the size of the largest matching in G. If a matching includes all vertices of G (i.e., if $\nu(G) = |V|/2$), then it is called a *perfect matching* or a *1-factor* for G. A *maximal matching* (as opposed to a maximum-sized matching) of G is a matching of G where you cannot add any more edges to the existing collection and keep it a matching.

Example 11.47. Figure 11.17 gives an example of a graph with matching number 2. A maximal matching of size 1 and a maximum-sized matching of size 2 are exhibited. A different graph with matching number 4 is depicted in Figure 11.18. In both figures, the thick edges on the left-hand diagram are a maximal matching, and the thick edges on the right-hand diagram give a maximum-sized matching. In the case of Figure 11.18, the maximum-sized matching is a perfect matching.

Figure 11.17 The thick edges constitute a maximal matching on the left, and a maximum-sized matching on the right.

Figure 11.18 The thick edges constitute a maximal matching on the left, and a perfect matching on the right.

The thick edges of Figure 11.19 give one example of a matching of size 6 for the bipartite graph of Figure 11.16. This is a maximum-sized matching for this graph, and so the matching number of this graph is 6.

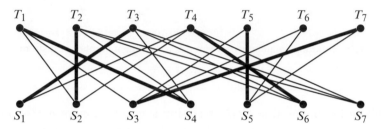

Figure 11.19 Finding a matching corresponds to finding a number of disjoint edges. The thick edges give a matching of size 6. The vertices S_7 and T_6 remain unmatched.

Chain Partitions of Posets

In Figure 11.16, we turned Problem 11.44 into a problem of matchings in a bipartite graph. What if we think of Figure 11.16 as the Hasse diagram of a poset of height 2? A maximum-sized matching then translates to a maximum set of disjoint chains of size 2. Furthermore, we can think of the remaining unmatched elements as chains of size 1, and we have a partition of the poset into chains.

Proposition 11.48. *Let $G = (X, \Delta, Y)$ be a bipartite graph, and let P be the poset of height 2 whose Hasse diagram is G. Then there is a natural bijection between partitions of P into chains and matchings in G. In this bijection, the number of chains of size 2 in a partition of P corresponds to the size of the matching in G. In particular, in this bijection, a partition of P into the minimum number of chains corresponds to a maximum-sized matching of G.*

Proof. A partition of P into chains consists of some chains of size 2 and some chains of size 1. Since, in a partition, the chains are disjoint, the chains of size 2 of a partition give a matching in G. Two different chain partitions of P give different matchings of G, and every matching in G gives rise to a chain partition in P (the edges in the matching will be the chains of size 2, and the remaining vertices are the chains of size 1). Hence, we have a bijection between chain partitions of P and matchings of G, and the number of chains of size 2 in the chain partition corresponds to the size of the matching. To get a partition of P into the minimum number of chains, we need to use as many chains of size 2 as possible. This results in a maximum-sized matching. □

Recall that the width of a poset is the size of the largest antichain in the poset (Definition 11.31), and that the matching number of a graph G (Definition 11.46) is denoted by $\nu(G)$. We then have:

Corollary 11.49. *Let $G = (X, \Delta, Y)$ be a bipartite graph, and let P be the poset of height 2 whose Hasse diagram is G. Then*

$$\nu(G) + \mathrm{width}(P) = |P|.$$

Proof. A partition of P into the minimum number of chains consists of k chains of size 2 and ℓ chains of size 1, with $2k + \ell = |P|$. The number of chains in this partition is $k + \ell$, and by Dilworth's theorem, Theorem 11.42, this number is equal to $\mathrm{width}(P)$. Hence, $|P| - \mathrm{width}(P) = k$.

By Proposition 11.48, a partition of P into the minimum number of chains corresponds to a maximum-sized matching of G of size k. (Basically, chains of size 2 give the matching and, to get a partition of P into as few chains as possible, we need the maximum number of chains of size 2 as possible.) The result now follows. □

We now present the celebrated marriage theorem, which gives a necessary and sufficient condition for the existence of a system of distinct representatives. Given a collection $A_1, \ldots, A_r$ of sets, when can we be assured of having a system of r distinct representatives? Consider a situation when $|A_1 \cup A_2 \cup A_3| = 2$. Then, the three sets A_1, A_2, and A_3 have two elements among themselves, and so there is no way that we can find three distinct elements with each one representing one of the sets. More generally, for there to be any chance of an s.d.r., any k of these sets must, among themselves, contain at least k different elements. The marriage theorem says that this necessary condition is also sufficient.

There are many proofs – including direct ones – of the marriage theorem, and, as the discussion above showed, the statement of the theorem can be reinterpreted in terms of matching in graphs (see Corollary 11.52), rooks on boards with forbidden positions (see Problem P 11.3.6), or chain partitions of posets of height 2. We have chosen to give a proof that uses Dilworth's theorem, Theorem 11.42, disguised as Corollary 11.49. For a direct proof, see Problems *P 11.3.21*–P 11.3.23 in subsection 11.3.1.

Theorem 11.50 (Marriage Theorem, Dénes Kőnig 1931, Philip Hall 1935).[8] *Let n be a positive integer, and let $A_1, \ldots, A_r$ be subsets of $[n]$. Then $A_1, \ldots, A_r$ possesses an s.d.r. if and only if, for each $1 \le m \le r$, the union of any m of the sets contains at least m elements.*

---

[8] This formulation of the marriage theorem is due to the influential British mathematician Phillip Hall (1904–1982), and is often called "Hall's marriage theorem." A number of other theorems – equivalent to the marriage theorem in some loose sense – have been independently proved by others. One is Kőnig's theorem (see Theorem 11.55), due to Dénes Kőnig (1884–1944). Kőnig was a Hungarian mathematician of Jewish descent. He wrote one of the first books on graph theory, and Paul Erdős (see footnote 16 of Chapter 2) took his graph theory class as a first-year college student. During World War II, Hungary signed a pact with Nazi Germany and handed over many Jews to Germany. Kőnig was brought up as a Christian, and so was exempt. However, towards the end of the war on October 15, 1944, the German-controlled Arrow Cross Party took over, and quickly started to persecute all Jews. Kőnig, apparently in order not to be captured by the fascists, committed suicide on October 19, 1944.

Proof. ($\Rightarrow$) Let m be a positive integer no larger than r. If we assume that the sets *have* an s.d.r. and take the union of any m of the sets, then this union will include at least the representatives for the m sets, and, since these representatives are distinct, there are exactly m representatives. Hence, the union will have at least m elements.

($\Leftarrow$) We are assuming that, for $1 \leq m \leq r$, the union of any m of the sets has at least m elements. We need to construct an s.d.r. To begin with, the union of all r sets must contain at least r elements, and these elements all come from $[n]$. So $r \leq n$.

We construct a bipartite graph G whose vertices are $\{1, 2, \ldots, n, A_1, \ldots, A_r\}$, and, for $1 \leq i \leq n$, $1 \leq j \leq r$, the vertex i is adjacent to the vertex A_j if and only if $i \in A_j$. (G has no other edges.) Let P be the poset whose Hasse diagram is G. The collection of sets has an s.d.r. if G has a matching of size r. By Corollary 11.49, this will be the case if the width of P is n. Clearly, P has an antichain of size n, namely $\{1, 2, \ldots, n\}$. So, we will be done when we show that there are no bigger antichains.

CLAIM: Every antichain of P has less than or equal to n elements.

PROOF: Assume $\{m_1, m_2, \ldots, m_k, A_1, \ldots, A_s\}$ is an antichain in P. (See Figure 11.20.) We want to show that $s + k \leq n$. By the hypothesis of the theorem, we have that the union of A_1 through A_s has at least s elements. But $m_1, \ldots, m_k$ cannot be members of $A_1, \ldots, A_s$ – otherwise, we wouldn't have an antichain – and so $A_1 \cup A_2 \cup \cdots \cup A_s$ has at most $n - k$ elements. Hence

$$s \leq |A_1 \cup A_2 \cup \cdots \cup A_s| \leq n - k \quad \Rightarrow \quad s + k \leq n. \qquad \square$$

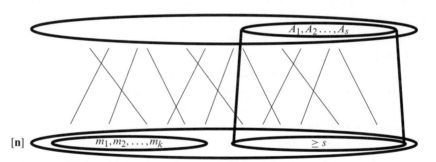

Figure 11.20 The sets $A_1, \ldots, A_s$ contain at least s elements, and yet contain none of $m_1, \ldots, m_k$. Hence, $k + s \leq n$.

Translating the marriage theorem to the language of bipartite graphs, we need one definition.

Definition 11.51 (Neighbor). Let $G = (V, E)$ be a general graph. For any set of vertices $S \subseteq V$, define the *neighbors* of S, denoted by $N(S)$, as the set of vertices of G that are adjacent to at least one vertex in S.

Corollary 11.52 (The Marriage Theorem for Bipartite Graphs). *Let $G = (X, \Delta, Y)$ be a bipartite graph with $|X| \leq |Y|$. Then G has a matching of size $|X|$ if and only if, for every $S \subseteq X$, we have $|N(S)| \geq |S|$.*

Proof. Replace every vertex in X with a set consisting of the neighbors of that vertex. Then, finding a matching of size $|X|$ is the same as finding an s.d.r. for the collection of sets in X. The statement of the corollary is then equivalent to the marriage theorem. □

We are finally ready to resolve the opening question of this chapter.

Problem 11.53. *Each of 450 first-year students is to be assigned to one of 30 themed activity groups during the orientation program. Each of the activity groups has a sign-up sheet with 45 slots, and each student is asked to put their name on three of the sign-up sheets. The organizers, then, want to assign 15 students to each group in such a way that every student gets one of the groups that they signed up for. Will the organizers always succeed?*

Solution. We have 450 students and 30 activity groups. Each student has chosen 3 activity groups and this has resulted in a list of 45 potential members for each group. We want to assign 15 students to each group. If you think of the students as elements and the lists as sets of elements, then each of the lists has 45 members. We don't want an s.d.r. but a "system of 15 distinct representatives." We employ a (beautiful) trick. Make copies of each list so that we have 15 copies of each list for an activity group. Now we have a total of $30 \times 15 = 450$ lists and we precisely want an s.d.r. An s.d.r. for these 450 sets would mean picking one distinct student for each set. This results in *every* student being assigned – since there are exactly 450 students – and *every* activity group getting 15 students – since there are fifteen copies of the list of students for each group. Does an s.d.r. exist? We have to check the marriage condition of Theorem 11.50. Let m be a positive integer no larger than 450. How many different students' names are on m arbitrarily chosen lists? Each list has 45 names, so there are a total of $45m$ names on these lists, but there are many duplicates. Each student wrote their name 3 times and we have 15 copies of each list. So a given student's name appears at most $3 \times 15 = 45$ times. Hence, the number of distinct names on the m lists is at least $\frac{45m}{45} = m$. The marriage condition is satisfied, hence an s.d.r. exists, and we conclude that no matter who wrote their name where, the administrators should be able to assign the students to activity groups properly! □

Problems

P 11.3.1. Let $S = \{a, b, c, d, e\}$, and let

$$A_1 = \{a, b, c\}, A_2 = \{a, b, c, d, e\}, A_3 = \{a, b\}$$

$$A_4 = \{b, c\}, A_5 = \{a\}, A_6 = \{a, c, e\}.$$

Does this collection of subsets of S have a system of distinct representatives ? If not, what is the largest number of sets in the collection with an s.d.r.?

P 11.3.2. Let $A_1 = \{1, 2, 3\}$, $A_2 = \{2, 4\}$, $A_3 = \{1, 4\}$, and $A_4 = \{2, 5\}$ be subsets of $[5]$. Does $\{A_1, A_2, A_3, A_4\}$ have an s.d.r.? Translate the problem of finding s.d.r.'s for this collection of subsets into a problem of placing non-attacking rooks on a board with forbidden positions. How many s.d.r.'s does this collection have? Translate the problem of finding an s.d.r., once again, but this time into a problem of finding a matching in a bipartite graph.

P 11.3.3. Let $A_1 = \{1,2\}$, $A_2 = \{1,4\}$, $A_3 = \{1,3,4,5\}$, $A_4 = \{2,4\}$, and $A_5 = \{1,4\}$. Does $\{A_1, A_2, A_3, A_4, A_5\}$ have an s.d.r.? Why? Construct a bipartite graph such that finding a maximum-sized matching of the graph corresponds to finding the maximum number of subsets in the collection with an s.d.r.

P 11.3.4. Let n be a positive integer, and let $A_1, A_2, \ldots, A_n$ be subsets of $[n]$ defined as follows:

$$A_i = \{1, 2, \ldots, n\} - \{i\} \text{ for } i = 1, 2, \ldots, n.$$

Prove that this family of subsets has an s.d.r., and find the number of s.d.r.'s.

P 11.3.5. Let $A = B = \{3,4,5\}$, $C = \{1,2,3,4\}$, $D = \{1,2,3\}$, and $E = \{1,2\}$. Find the number of systems of distinct representatives for the collection $\{A, B, C, D, E\}$.

P 11.3.6. Restate the marriage theorem, Theorem 11.50, in terms of non-attacking rooks on a board with forbidden positions.

P 11.3.7. Let n and r be positive integers, and let $A_1, A_2, \ldots, A_r$ be a family of subsets of $[n]$ with an s.d.r. Let x be an element of A_1. Does there always exist an s.d.r. which contains x? Does there always exist an s.d.r. in which x represents A_1?

P 11.3.8. Let n and r be positive integers, and let $A_1, A_2, \ldots, A_r$ be a family of subsets of $[n]$ that satisfy a condition stronger than the marriage condition. More precisely, suppose

$$\left| A_{i_1} \cup A_{i_2} \cup \cdots \cup A_{i_k} \right| \geq k + 1$$

for each $k = 1, 2, \ldots, r - 1$, and each choice of k distinct indices $i_1, i_2, \ldots, i_k$. Let x be an element of A_1. Prove that there is an s.d.r. for this family in which x represents A_1.

P 11.3.9. We have an airport with 20 terminals and we want to assign each terminal to one of 20 airlines. We ask each airline for their list of acceptable terminals and gather the 20 such lists. We notice that, for $1 \leq k \leq 20$, the union of the lists of any collection of k of the airlines contains at least k terminals. We then construct a new set of 20 lists, this time one for each terminal. For each terminal we list the airlines that have found it acceptable. Prove that, if you take the union of any 15 of these 20 new lists, then this union will have at least 15 airlines on it. Is there anything special about 15?

P 11.3.10. Let n be a positive integer, and let $\mathcal{S} = \{A_1, \ldots, A_n\}$ be a family of subsets of $[n]$. Assume that, for each subset $B \subseteq [n]$, the number of subsets in $\mathcal{S}$ that intersect B is at least $|B|$. Does $\mathcal{S}$ necessarily have an s.d.r.? Would the converse be true?

P 11.3.11. Does the marriage theorem work if the number of sets is infinite? Define $A_0 = \{1, 2, \ldots\}$, and, for every positive integer i, let $A_i = \{i\}$. Let $\mathcal{S} = \{A_0, A_1, A_2, \ldots\}$.
(a) Does $\mathcal{S}$ have an s.d.r.?
(b) Does $\mathcal{S}$ satisfy the marriage condition?
(c) Let m be a positive integer, and choose m arbitrary elements of $\mathcal{S}$. Does this collection have an s.d.r.?

P 11.3.12. I take k distinct books from my library to a holiday party. Eight of my friends are at the party and I want to give as many of them as possible a gift that they like. So I show them the books and ask each of them to give me a list of three of the books that they like. I collect the lists and inspect them. To my surprise, every one of my books appears on exactly two lists. Will I be able to give each of my friends one of the books on their list? If the answer is always yes, give a proof. Otherwise, give a specific example to show that the answer is sometimes no.

P 11.3.13. Recall (Definition 10.3) that a graph is regular if all of its vertices have the same degree. Let $G = (X, E, Y)$ be a regular bipartite graph. Show that $v(G) = |X| = |Y|$ and that, therefore, G has a perfect matching.

P 11.3.14. Let n and m be positive integers with n even, and $m \leq n/2$. Assume that there are n sports teams in a conference, and you are charged with creating a weekly game schedule for the teams. There are m weeks in the season, and each team is to play once every week. Over the course of the season, we want each team to play m other teams, never playing the same team twice. Can you come up with a schedule?

P 11.3.15. Let $G = (X, E, Y)$ be a bipartite graph, with $|X| \leq |Y|$. Assume that every vertex in X has degree u, while every vertex in Y has degree d. Show that $v(G) = |X|$.

P 11.3.16. $E = \{e_1, e_2, \ldots, e_m\}$ is the set of employees of your company. These employees are to be given tasks from the task list $J = \{t_1, t_2, \ldots, t_n\}$. Each employee is qualified for some of the tasks, and we want to assign each employee to *three* tasks. Each task will get one or zero employees. If $A \subseteq E$ is a collection of the employees, then denote by $N(A) \subseteq J$ the collection of all the tasks that can be performed by some member of A. Prove that we can accomplish our assignment if and only if

$$|N(A)| \geq 3|A| \text{ for all } A \subseteq E.$$

P 11.3.17. Let $G = (X, E, Y)$ be a bipartite graph. For any set of vertices A, as usual, the neighbors of A (Definition 11.51) are denoted by $N(A)$. Show that G has a matching of size t if and only if, for all $A \subseteq X$, we have

$$|N(A)| \geq |A| + t - |X| = t - |X - A|.$$

P 11.3.18. Let n be a positive integer, and assume that $2n$ teams are playing a round-robin tournament.[9] In each of $2n - 1$ consecutive days, each team plays one game, in such a way that, by the end of the tournament, each team has played every other team exactly once. There are no draws, and, in each game, one team loses and one team wins. Every day there are n games, and so every day there are n winners and n losers. After the games have ended, you are compiling a highlight video for the tournament. In your video, you want to have $2n - 1$ segments, one for each day of the tournament, and, for each segment, you want to highlight one new team. However, for each segment, you want to highlight a team that actually won their game on that day. Is that always possible? Either prove that you can always choose one winner per day for a total of $2n - 1$ different winners, or give an example to show that this may be an impossible task. What if you had the same restrictions but wanted to have $2n$ segments in your highlight video?

---

[9] Adapted from Problem B3 of the 2012 Putnam Competition.

P 11.3.19. Let n and r be positive integers, and let $A_1, A_2, \ldots, A_r$ be a family of subsets of $[n]$ with an s.d.r. Show that, for at least one of these subsets, you can decide a priori which of its elements should represent the subset in the s.d.r. In other words, show that there exists an integer i with $1 \le i \le r$ such that A_i has the property that, for each $x \in A_i$, you can find an s.d.r. for $A_1, A_2, \ldots, A_r$ such that x represents A_i. You may find the following steps helpful:

STEP 1: As a warm-up, do Problem P 11.3.7.

STEP 2: Do Problem P 11.3.8, and then argue that the only case remaining is if, for some $1 \le k \le r - 1$, the union of some k of the subsets had exactly k elements.

STEP 3: Among all candidates for k in the previous step, pick the smallest one. Without loss of generality, assume $|A_1 \cup \cdots \cup A_k| = k$.

STEP 4: Let $x \in A_1$. Make x the representative of A_1. Eliminate x from all other sets. Argue why the minimality of k implies that $A_2, \ldots, A_k$ satisfy the marriage condition and have an s.d.r.

STEP 5: Eliminate the k elements of $A_1 \cup \cdots \cup A_k$ from $A_{k+1}, \ldots, A_r$, and call the resulting sets $A'_{k+1}, \ldots, A'_r$. If, for $k + 1 \le i_1 < i_2 < \cdots < i_s \le r$, the union of $A'_{i_1}, \ldots, A'_{i_s}$ has fewer than s elements, then how many elements could $A_1 \cup \cdots A_k \cup A_{i_1} \cup \cdots A_{i_s}$ have? Is that possible? Do $A'_{k+1}, \ldots, A'_r$ satisfy the marriage condition?

STEP 6: Rewrite what you have to turn it into a concise and complete proof.

11.3.1 Two Other Proofs of the Marriage Theorem

In the text, we proved the marriage theorem (Theorem 11.50) as a consequence of Dilworth's theorem. There are many other proofs of the marriage theorem, including direct ones. The ensuing problems provide two other proofs. Problem P 11.3.20 guides you through an elegant proof by induction from 1950, due to Halmos and Vaughan (1950). Problems *P 11.3.21–P 11.3.23* guide you through a proof using "augmenting paths." One advantage of this proof is that it can be readily translated to an algorithm for finding maximum matchings in bipartite graphs.

Recall from Definition 11.46 that a matching is just a set of disjoint edges, and that, for a set of vertices S, $N(S)$ is the set of neighbors of S – see Definition 11.51.

Problems (continued)

P 11.3.20. Let n be a positive integer, and let $A_1, \ldots, A_r$ be a family of subsets of $[n]$. Assume that this family satisfies the marriage condition. In other words, for $1 \le k \le r$, the union of any k subsets in the family has at least k elements. Prove, by induction on r, that the collection possesses an s.d.r. You may find the following steps helpful:

STEP 1: What is the base case? What is the inductive hypothesis?

STEP 2: Consider the case when the family satisfies the stronger condition of Problem P 11.3.8. Show that, after picking a representative for A_1, the inductive hypothesis allows you to find an s.d.r. for the rest.

STEP 3: Argue that, without loss of generality, for some $1 \le k \le r - 1$ you can assume that $|A_1 \cup \cdots \cup A_k| = k$. Does the inductive hypothesis imply that we can find an s.d.r. for $A_1, \ldots, A_k$?

STEP 4: To complete the proof, show that if you eliminate the k elements of $A_1, \ldots, A_k$ from $A_{k+1}, \ldots, A_r$, the resulting sets still satisfy the marriage condition, and so you can apply the inductive hypothesis to them. (If you need more inspiration, do Problem P 11.3.19, or at least read the instructions for STEP 5 of that problem.)

P 11.3.21. Let $G = (X, E, Y)$ be a finite bipartite graph, and let M be a matching in G. Assume that, for all $S \subseteq X$, we have $|N(S)| \ge |S|$. Let x_0 be an unmatched vertex in X. Show that, for some integer $k \ge 0$, there is a sequence of vertices $x_0, y_0, x_1, y_1, \ldots, x_k, y_k$ in G with the following properties (see Figure 11.21):

(a) $x_0, x_1, \ldots, x_k \in X$
(b) $y_0, y_1, \ldots, y_k \in Y$
(c) For each $0 \le i \le k - 1$, there is one edge of G in M from y_i to x_{i+1}
(d) For each $0 \le i \le k$, there is at least one edge of G not in M from y_i to some vertex from among $x_0, \ldots, x_i$
(e) The vertex y_k is an unmatched vertex of Y.

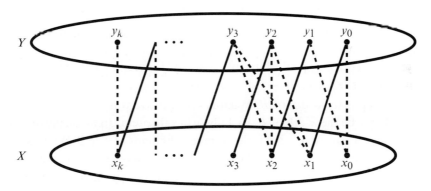

Figure 11.21 The solid lines are part of the matching. The dashed lines are edges in G not in the matching. The vertices x_0 and y_k are unmatched.

P 11.3.22. Let $G = (X, E, Y)$ be a finite bipartite graph, and let M be a matching in G. Assume that, for all $S \subseteq X$, we have $|N(S)| \ge |S|$. Let x be an unmatched vertex in X.

(a) Using Problem P 11.3.21, show that you can find an unmatched vertex $y \in Y$, and a path – called an *augmenting path* – from y to x in such a way that the first and the last edges of the path are not in M, and the edges alternate between being in M and not being in M.

(b) Start with M and the augmenting path that you constructed in the previous part. Delete from M any edge of M that is in that path, and, instead, add to M every edge on the path that was not already in M. Is the resulting set of edges a matching? How does its size compare to that of M?

P 11.3.23. Without recourse to Dilworth's theorem, and by using the results of Problems *P 11.3.21* and *P 11.3.22* give a different proof of the marriage theorem. In other words, let $G = (X, E, Y)$ be a bipartite graph with $|X| \le |Y|$, and show that G has a matching of size $|X|$ if and only if for every $S \subseteq X$, we have $|N(S)| \ge |S|$.

11.3.2 Vertex Covers

A concept dual to matchings in graphs is that of a *vertex cover*.

Definition 11.54. A *vertex cover* in a graph is a set of vertices such that each edge of the graph has at least one end in that set.

In the problems, you are asked to explore this notion, and, in the particular case of bipartite graphs, to prove the following.

Theorem 11.55 (Kőnig's Theorem). *Let G be a bipartite graph. Then $v(G)$, the matching number of G, is equal to the size of the smallest vertex cover of G.*

Problems (continued)

P 11.3.24. Let $G = (X, \Delta, Y)$ be the bipartite graph of Figure 11.22, and let P be a poset whose Hasse diagram is G.
(a) What is the width of P? What is $v(G)$? Prove your assertions.
(b) Let w be the width of P. By Dilworth's theorem, P can be partitioned into w chains. In such a partition, what is the number of chains of size 2? What is the number of chains of size 1? From among the chains of size 1, how many of them consist of an element of X? Do your answers depend on the particular partition of P into w chains?
(c) Find a vertex cover (Definition 11.54) for G that has $v(G)$ elements.

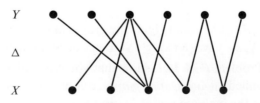

Figure 11.22 A bipartite graph and the Hasse diagram of a poset of height 2.

P 11.3.25. Let $G = (V, E)$ be a simple graph. Let $A \subseteq V$ be a set of vertices in G, and $B = V - A$ be the remaining set of vertices. If A is an independent set (Definition 10.100), then must B be a vertex cover? What about the converse?

P 11.3.26. Let $G = (V, E)$ be a simple graph. Recall that $\alpha(G)$ denotes the independence number of G (Definition 10.100). Let $\mu(G)$ denote the minimum size of a vertex cover in G. Prove that

$$\alpha(G) + \mu(G) = |V|.$$

P 11.3.27. Give an example of a graph where the minimum size of a vertex cover does not equal the matching number of the graph. Is the matching number of a simple graph always less than or equal to the minimum size of a vertex cover of the graph?

P 11.3.28. **Proof of Kőnig's Theorem.** Use Corollary 11.49 and Problem P 11.3.26 to prove Theorem 11.55. In other words, assume G is a bipartite graph, and prove that the minimum size of a vertex cover is equal to $\nu(G)$, the matching number of G.

11.3.3 Latin Squares and Edge Chromatic Number of Bipartite Graphs

In the ensuing problems, you are asked to tackle two seemingly unrelated problems. One question is about completing partially filled Latin squares, and the other is about the edge chromatic number of bipartite graphs. As we shall see, the solutions to both problems involve similar techniques, using matchings in bipartite graphs.

Definition 11.56 (Latin Square). Let L be an $n \times n$ matrix. Assume that

- the entries of the matrix are chosen from $\{1, 2, \ldots, n\}$,
- no two entries in a row are the same, and
- no two entries in a column are the same.

Then L is called a *Latin square*.

In general, if you have a partially filled Latin square (i.e., an $n \times n$ matrix where some of the entries are left blank, but those filled follow the rules above), it is not easy to complete it. However, surprisingly, in the case where some of the rows are completely filled, and other rows are completely empty, you will always be able to complete the square, one row at a time. Problem P 11.3.33 asks you for a proof.

Recall (Definition 10.3) that a graph is d-regular if all of its vertices have the same degree d. The key to the proof of the result on partially filled Latin squares is the very useful result of Problem *P 11.3.13*. It says that a regular bipartite graph will of necessity have a perfect matching. (Problem P 11.3.15 gives a more general version of this result.) The same technique can be used to find the edge chromatic number of any bipartite graph.

Let $\Delta(G)$ denote the maximum degree of G, and recall (Definition 10.115) that $\chi'(G)$, the edge chromatic number of G, is the smallest number of colors needed to color the edges of a graph in such a way that no two edges that share a vertex have the same color.

By first embedding an arbitrary bipartite graph in a regular bipartite graph, and then, using Problem *P 11.3.13*, Problems *P 11.3.34* and P 11.3.35 lead you to a proof of the following.

Theorem 11.57. *Let G be a bipartite graph. Then*

$$\chi'(G) = \Delta(G).$$

Problems (continued)

P 11.3.29. In how many ways can you complete the partially filled Latin square of Figure 11.23?

	1	2	
1			4
2			

Figure 11.23 A 4×4 partial Latin square.

P 11.3.30. Give an example of a partially filled 4×4 Latin square that *cannot* be completed.

P 11.3.31. Consider the incomplete Latin square of Figure 11.24. Denote the columns of the square by $c_1, c_2, \ldots, c_7$. Construct a bipartite graph $G = (X, \Delta, Y)$, where $X = \{c_1, \ldots, c_7\}$ and $Y = [7] = \{1, 2, \ldots, 7\}$. The vertex c_i is adjacent to vertex j if the symbol j does not appear in column c_i.

(a) Sketch G.
(b) What is the degree sequence of G?
(c) What is $\nu(G)$, the matching number of G?
(d) Explain how a perfect matching in G allows you to add a row to the incomplete Latin square while preserving the property that each row and each column contain at most one of the elements of [7].

6	4	7	5	2	3	1
3	5	1	2	6	7	4
2	1	6	7	4	5	3
1	3	2	6	7	4	5
7	2	3	4	5	1	6

Figure 11.24 A partial Latin square.

P 11.3.32. We continue with the assumptions and notation of Problem P 11.3.31.
(a) Find the number of perfect matchings of the bipartite graph $G = (X, \Delta, Y)$.

(b) Use one of these perfect matchings to add a sixth row to the square.

(c) Repeat and complete the 7×7 Latin square.

P 11.3.33. **Partially Filled Latin Squares.** Let k and n be positive integers with $k \leq n$. Assume that we are given a partially filled $n \times n$ Latin square where the first k rows are completely filled (and each element of $[n]$ appears at most once in every row and in every column), and the remaining $n - k$ rows are completely empty. Prove that you can always complete the Latin square. You may find the following steps helpful:

STEP 1: Argue that adding a row to the partially filled Latin square is equivalent to finding a perfect matching in the bipartite graph $G = (X, \Delta, Y)$ described in Problem P 11.3.31.

STEP 2: Find the degree sequence of G.

STEP 3: Use the result of Problem $P\ 11.3.13$ (or the more general Problem P 11.3.15) to show that G has a perfect matching, and complete the proof.

P 11.3.34. Let G be a d-regular bipartite graph. Prove that

$$\chi'(G) = d.$$

P 11.3.35. **Proof of Theorem 11.57.** Let G be a bipartite graph, and let $\Delta(G)$ and $\chi'(G)$ denote, respectively, the maximum degree and the edge chromatic number of G. Prove that $\chi'(G) = \Delta(G)$. You may find the following steps helpful:

STEP 1: Quickly argue that $\chi'(G) \geq \Delta(G)$, and so we need to show that $\Delta(G)$ colors suffice to properly color the edges of G.

STEP 2: Argue that, if we modify G by adding extra vertices and extra edges, as long as the old edges are in place it is sufficient to show that the edges of the modified G can be properly colored using $\Delta(G)$ colors.

STEP 3: In this and the next step, we will construct a (possibly much bigger) regular bipartite graph $\widetilde{G}$ that contains G as a subgraph. Let $G = (X, E, Y)$, where $X \cup Y$ is the set of vertices of G, E is the set of edges, and all edges have one end in X and the other in Y. If $|X| \neq |Y|$, add enough isolated vertices to X or Y so as to make the two parts equal. Let X_1 and Y_1 be, respectively, the new X and the new Y. Argue that there is an $x \in X_1$ with $\deg(x) \leq \Delta(G)$ if and only if there is also a $y \in Y_1$ with $\deg(y) \leq \Delta(G)$.

STEP 4: Assume that $x \in X_1$ and $y \in Y_1$, with $\deg(x) \leq \Delta(G)$ and $\deg(y) \leq \Delta(G)$. Let X_2 be X_1 together with $\Delta(G)$ new vertices. Likewise, Y_2 is Y_1 together with $\Delta(G)$ new vertices. Make every one of the new vertices in X_2 adjacent to every one of the new vertices in Y_2. (In other words, we have added a complete bipartite graph $K_{\Delta(G),\Delta(G)}$ to the graph (X_1, E, Y_1).) Let $a \in X_2$ and $b \in Y_2$ be two of the new vertices. Eliminate the (just added) edge $\{a, b\}$ and instead add in the edges $\{x, b\}$ and $\{a, y\}$. Let E_2 be the new E. In other words, E_2 is the set of edges of G together with $\{x, b\}$, $\{a, y\}$, and all the edges in the new $K_{\Delta(G),\Delta(G)}$ except $\{a, b\}$. Let $G_2 = (X_2, E_2, Y_2)$. Argue that, if you repeat this step enough times, you will arrive at a $\Delta(G)$-regular bipartite graph $\widetilde{G} = (\widetilde{X}, \widetilde{E}, \widetilde{Y})$ that contains all the vertices and edges of G (and possibly much more).

STEP 5: Use Problem $P\ 11.3.34$ to show that $\chi'(\widetilde{G}) = \Delta(G)$, and, using Step 2, complete the proof.

P 11.3.36. In doing Problem P 11.3.35, a student claims that Step 4 is unnecessarily complicated. She argues that, as argued in Step 3, if the graph (X_1, E, Y_1) is not regular, then there is a vertex $x \in X_1$ and a vertex $y \in Y_1$ such that the degree of both of them is less than $\Delta(G)$. Instead of adding a whole $K_{\Delta(G), \Delta(G)}$, as done in Step 4, just add an edge with these two vertices as endpoints. Comment.

11.4 Boolean Lattices: Symmetric Chains, Theorems of Sperner, and Erdős–Ko–Rado*

Warm-Up 11.58. *Let $2^{[4]}$ denote the poset of subsets of $[4]$ ordered by inclusion. What is the minimum number of chains needed to partition $2^{[4]}$? Why? Can you find a partition of $2^{[4]}$ into the minimum number of chains in such a way that all chains are of size 2 or 3?*

In this optional section, we focus on one set of posets, namely the Boolean lattices, the posets of subsets of finite sets ordered by inclusion. Such a poset is a very basic poset. After all, what can be more basic than the subsets of a finite set? What is surprising is that one can formulate many easy-to-state but hitherto-unsolved problems about the combinatorics of subsets of a finite set.[10] We first discuss a certain partition of the Boolean lattices into chains, and then use it to prove the celebrated Sperner's theorem, which gives the width of a Boolean lattice. We then discuss intersecting families of subsets and prove another celebrated theorem, the Erdős–Ko–Rado theorem. As usual, for n a positive integer, $[n] = \{1, 2, \ldots, n\}$, $[0] = \emptyset$, and $2^{[n]}$ denotes the set of subsets of $[n]$ ordered by inclusion (Definition 11.13).

Symmetric Chain Decomposition

Recall (Definition 11.31) that the width of a poset is the size of its largest antichain. Our next aim is to determine the width of $2^{[n]}$. Sperner's theorem, Theorem 11.67, provides the answer, and this influential little theorem has many different proofs. We have chosen to give a proof using symmetric chain decompositions, which we will first discuss.[11]

Recall (Definition 11.26) that a *chain* of size k and length $k - 1$ in $2^{[n]}$ is a collection $A_1, \ldots, A_k$ of subsets of $[n]$ with

$$A_1 \subsetneq A_2 \subsetneq \cdots \subsetneq A_k.$$

This chain is *skipless*[12] if $|A_i| = |A_{i-1}| + 1$ for $2 \leq i \leq k$.

---

[10] See Anderson 2002 for a wonderful introduction to the combinatorics of finite sets.
[11] In Mini-project 8, Sperner's theorem was discussed and a different proof using the BLYM inequality was given.
[12] Some authors use "saturated" instead of "skipless."

Definition 11.59 (Symmetric Chain Decomposition). Let m be a non-negative integer with $0 \le m \le \lfloor n/2 \rfloor$. A chain of elements of $2^{[n]}$,

$$A_m \subset A_{m+1} \subset \cdots \subset A_{n-m},$$

is a *symmetric chain* if $|A_i| = i$, for $i = m, m+1, \ldots, n-m$.

A partition of $2^{[n]}$ into symmetric chains is called a *symmetric chain decomposition* (SCD) of $2^{[n]}$.

Example 11.60. $\emptyset \subset \{1, 2\} \subset \{1, 2, 6, 8\}$ is a chain of size 3 and length 2 in $2^{[8]}$. This chain is neither maximal – you can insert $\{1, 2, 6\}$ and continue to have a chain – nor skipless.

$\{1, 3\} \subset \{1, 3, 4\} \subset \{1, 3, 4, 5\}$ is a symmetric chain in $2^{[6]}$. This chain is skipless, starts with a two-element set – and so, in the definition of a symmetric chain, $m = 2$ – and ends with a four-element set, and $n - m = 6 - 2 = 4$. This same chain is also a skipless chain in both $2^{[5]}$ and $2^{[7]}$, but it is not a symmetric chain in either of these posets.

Remark 11.61. If we draw the Hasse diagram of a Boolean lattice (see Figure 11.3), then the middle level(s) of this Hasse diagram are the biggest. For n even, there is a unique largest middle level, while, for n odd, there are two equally sized and large middle levels (see Theorem 5.8). A skipless chain is a symmetric chain if it is symmetric with respect to a horizontal line through the middle of the Hasse diagram. In other words, a symmetric chain starts as far below the middle as it goes above the middle.

Example 11.62. To construct a symmetric chain decomposition of $2^{[4]}$, for example, you have to place each of the $2^4 = 16$ subsets of $[4]$ into a symmetric chain. In other words, these symmetric chains will have to partition $2^{[4]}$. Here is one such possibility:

$$\emptyset \subset \{1\} \subset \{1, 2\} \subset \{1, 2, 3\} \subset \{1, 2, 3, 4\}$$
$$\{2\} \subset \{2, 3\} \subset \{2, 3, 4\}$$
$$\{3\} \subset \{1, 3\} \subset \{1, 3, 4\}$$
$$\{4\} \subset \{1, 4\} \subset \{1, 2, 4\}$$
$$\{3, 4\}$$
$$\{2, 4\}.$$

Defining a symmetric chain decomposition does not mean that it always exists. We next prove that the Boolean lattices do have a symmetric chain decomposition. The proof gives a pleasant inductive construction for SCDs.

Theorem 11.63. *For all positive integers n, $2^{[n]}$ has a symmetric chain decomposition.*

Proof. Induct on n. For $n = 1$, the one chain $\emptyset \subset 1$ provides a symmetric chain decomposition for $2^{[1]}$, and so assume $n \ge 2$. For the inductive step, we assume that we have a symmetric chain decomposition of $2^{[n]}$, and use it to construct a symmetric chain decomposition for $2^{[n+1]}$.

Let a symmetric chain in $2^{[n]}$ be given:

$$A_m \subset A_{m+1} \subset \cdots \subset A_{n-m-1} \subset A_{n-m},$$

where m is an integer with $1 \leq m \leq \lfloor n/2 \rfloor$, and $|A_i| = i$ for $m \leq i \leq n - m$. If $m = n - m$, then $m = n/2$, the chain consists of just one subset A_m, and $|A_m| = n/2$. Corresponding to this chain of size 1, we construct the following symmetric chain of size 2 in $2^{[n+1]}$:

$$A_m \subset A_m \cup \{n + 1\}.$$

If $m < n - m$, then, from the chain $A_m \subset \cdots \subset A_{n-m}$, we construct two symmetric chains in $2^{[n+1]}$:

$$A_m \subset A_{m+1} \subset \cdots \subset A_{n-m-1} \subset A_{n-m} \subset A_{n-m} \cup \{n + 1\}$$
$$A_m \cup \{n + 1\} \subset A_{m+1} \cup \{n + 1\} \subset \cdots \subset A_{n-m-1} \cup \{n + 1\}.$$

Both of these chains are skipless. The first one starts with a set of size m and ends with a set of size $(n + 1) - m$; the second one start with a set of size $m + 1$ and ends with a set of size $(n + 1) - (m + 1)$. Hence, both of the two new chains are symmetric chains in $2^{[n+1]}$.

Start with the symmetric chain decomposition of $2^{[n]}$ and apply the above construction to every chain in that SCD. You will now have a collection of symmetric chains in $2^{[n+1]}$. We claim that this is a symmetric chain decomposition of $2^{[n+1]}$.

Note that every element of $2^{[n+1]}$ is a subset of $[n + 1] = \{1, 2, \ldots, n, n + 1\}$. These subsets come in two varieties: those that contain $n+1$ and those that don't. Each of the latter subsets is already an element of $2^{[n]}$, and each of the former elements is just a subset of $[n]$ union $\{n + 1\}$. In our construction, we kept all the original sets (the subsets of $[n]$), and we also included the union of each of the original sets with $\{n + 1\}$. Hence, every subset of $[n + 1]$ was in one of our newly constructed symmetric chains. Moreover, every subset of $[n + 1]$ appeared exactly once, and so we did indeed construct a symmetric chain partition of $2^{[n+1]}$. The proof is complete. $\square$

Remark 11.64 (The SCD Construction: A Recap). The idea of the construction in Theorem 11.63 is straightforward. The subsets in $2^{[n+1]}$ come in two varieties, those that don't have $n + 1$ as a member and those that do. The subsets of $[n + 1]$ that do not have $n + 1$ as an element are exactly the subsets of $[n]$, and these form the Boolean lattice $2^{[n]}$. The rest of the subsets all have $n + 1$ as an element and, if we just ignore $n + 1$, these subsets are also exactly all the subsets of $[n]$. In other words, you can take all the subsets of $[n]$ and add the element $n+1$ to each of these sets, and you will get all the subsets in $2^{[n+1]}$ that have $n + 1$ in them. So this collection – subsets of $[n + 1]$ that have $n + 1$ as an element – also has the same poset structure as $2^{[n]}$. Hence, $2^{[n+1]}$ is made up of two copies of $2^{[n]}$. For concreteness, take $n = 3$. $\{1\} \subseteq \{1, 2\} \subseteq \{1, 2, 3\}$ is a symmetric chain in $2^{[3]}$. In $2^{[4]}$, we have two corresponding chains:

$$\{1\} \subseteq \{1, 2\} \subseteq \{1, 2, 3\} \quad \text{and} \quad \{1, 4\} \subseteq \{1, 2, 4\} \subseteq \{1, 2, 3, 4\}.$$

The first one is just the old chain in $2^{[3]}$, and the second one is also the original chain except that we have added in 4 to every subset in the chain. These chains are skipless but they are *not* symmetric. They start and end at the wrong places. We apply a simple fix. We move the top element of the second chain to the first chain and get two symmetric chains in $2^{[4]}$:

$$\{1\} \subseteq \{1,2\} \subseteq \{1,2,3\} \subseteq \{1,2,3,4\} \quad \text{and} \quad \{1,4\} \subseteq \{1,2,4\}.$$

The only time we don't get two chains is if our original symmetric chain consisted of just one set, A. This only happens if n is even and $|A| = n/2$. In this case, we just turn this chain into a chain of size 2: $A \subset A \cup \{n+1\}$. This is a symmetric chain and it ensures that we have taken care of the sets in our original chain as well as those sets union $\{n+1\}$. (You can also pretend that, even in this case, we did the same construction – of getting two chains from one – except that, after moving the top element of the second chain to the first chain, the second chain becomes the empty chain.)

Example 11.65. For $n = 0$, the SCD for $2^{[0]}$ is just $\emptyset$. Our inductive construction then gives $\emptyset \subset \{1\}$ for the SCD of $2^{[1]}$. This chain now becomes the two chains $\emptyset \subset \{1\} \subset \{1,2\}$ and $\{2\}$, and these two chains form an SCD for $2^{[2]}$. For the next step, the symmetric chain of size 3 in $2^{[2]}$ becomes two symmetric chains in $2^{[3]}$: $\emptyset \subset \{1\} \subset \{1,2\} \subset \{1,2,3\}$ and $\{3\} \subset \{1,3\}$. The symmetric chain of size 1 in $2^{[2]}$ results in just one symmetric chain in $2^{[3]}$: $\{2\} \subseteq \{2,3\}$, and we have the following symmetric chain decomposition of $2^{[3]}$:

$$\emptyset \subset \{1\} \subset \{1,2\} \subset \{1,2,3\}$$
$$\{2\} \subset \{2,3\}$$
$$\{3\} \subset \{1,3\}.$$

Repeating the inductive process one more time results in the SCD for $2^{[4]}$ given in Example 11.62.

Lemma 11.66. *Let n be a positive integer. The number of chains in a symmetric chain decomposition of $2^{[n]}$ is $\binom{n}{\lfloor n/2 \rfloor}$.*

Proof. The number of subsets of size $\lfloor n/2 \rfloor$ of $[n]$ is $\binom{n}{\lfloor n/2 \rfloor}$. By definition of a symmetric chain, every symmetric chain will have to contain exactly one element from the middle level if n is even and one from each of the two middle levels if n is odd. Hence, every symmetric chain contains exactly one subset of size $\lfloor n/2 \rfloor$, and, as a result, the number of chains in an SCD of $2^{[n]}$ is the same as the number of subsets of $[n]$ of size $\lfloor n/2 \rfloor$. $\square$

Sperner's Theorem[13]

We now turn to the width of $2^{[n]}$. What is the largest antichain in $2^{[n]}$? Let k be a non-negative integer no larger than n. Clearly, all subsets of $[n]$ of size k (the elements in level k of the

[13] Named after the German mathematician Emanuel Sperner (1905–1980).

Hasse diagram for $2^{[n]}$) form an antichain of size $\binom{n}{k}$. By Theorem 5.8, the largest of these is the middle level, with $\binom{n}{\lfloor n/2 \rfloor}$ elements. Can we find a larger antichain?

Theorem 11.67 (Sperner 1928). *Let A be an antichain in $2^{[n]}$. Then*

$$|A| \le \binom{n}{\lfloor n/2 \rfloor}.$$

In particular, the width of $2^{[n]}$ is $\binom{n}{\lfloor n/2 \rfloor}$.

Proof. Let A be an antichain in $2^{[n]}$. The poset $2^{[n]}$ has a symmetric chain decomposition, and A – being an antichain – has at most one element from each chain. Hence, the size of A is no more than the *number* of chains in an SCD decomposition of $2^{[n]}$. By Lemma 11.66, this number is $\binom{n}{\lfloor n/2 \rfloor}$. □

Intersecting Families

Definition 11.68 (Intersecting Family, Star). Let n be a positive integer, and let $A \subseteq 2^{[n]}$ be a collection of subsets of $[n]$. We call A an *intersecting family (in $2^{[n]}$)* if, for all $A, B \in A$, we have $A \cap B \neq \emptyset$. We call A a *star (in $2^{[n]}$)* if $\cap_{A \in A} A \neq \emptyset$. Let k be an integer with $1 \le k \le n$. An intersecting family or a star that consists solely of subsets of $[n]$ of size k is called a *k-uniform intersecting family* or a *k-uniform star*, respectively.

Remark 11.69 (Stars are Intersecting Families). Let A be a collection of subsets of $[n]$. This collection is an intersecting family if the intersection of every pair of subsets in A is non-trivial. This same collection is a star if the intersection of *all* subsets in A is non-trivial. Evidently, it is more difficult to be a star than to be intersecting, and so all stars are also intersecting families.

Example 11.70. The set $A = \{\{1,2\},\{2,3\},\{1,3\}\}$ is a 2-uniform intersecting family of subsets in $2^{[3]}$. Every element is a subset of size 2 of $[3]$ and every pair of subsets has a non-empty intersection. A is not a star since the intersection of all of the subsets in A is the empty set. The collection $S = \{\{1\},\{1,2\},\{1,3\},\{1,2,3\}\}$ is a star (and, hence, also an intersecting family) since the element 1 is an element of all the subsets in S.

It is easy enough to find small intersecting families, but what is the largest intersecting family that we can find? We pose two similar-sounding problems:

Problem 11.71. *Let n be a positive integer.*

(a) *What is the size of the largest intersecting family in $2^{[n]}$?*
(b) *Let k be an integer with $0 \le k \le n$. What is the size of the largest k-uniform intersecting family in $2^{[n]}$?*

These two problems are related but, as it turns out, the second one is a deeper question than the first. In a preliminary lemma, we answer the first question and the easy cases of the second question, and record the size of stars.

Lemma 11.72. *Let n be a positive integer, and let k be an integer with $1 \leq k \leq n$.*

(a) *Define $S = \{A \subseteq [n] \mid 1 \in A\}$. Then S is a star in $2^{[n]}$ with $|S| = 2^{n-1}$.*
(b) *The size of the largest intersecting family in $2^{[n]}$ is 2^{n-1}.*
(c) *Define $S^{\star} = \{A \subseteq [n] \mid 1 \in A, \text{ and } |A| = k\}$. Then $S^{\star}$ is a k-uniform star with*
$$|S^{\star}| = \binom{n-1}{k-1}.$$
(d) *If $k > n/2$, then the size of the largest k-uniform intersecting family is $\binom{n}{k}$.*

Proof. (a) How many subsets of $[n]$ include 1? If you take away 1 from each of the subsets in S, you will get all the subsets of $\{2, \ldots, n\}$. This is a set with $n - 1$ elements, and has 2^{n-1} subsets.

(b) Let $\mathcal{A}$ be an intersecting family in $2^{[n]}$. Simply observe that if $A \in \mathcal{A}$, then A^c, the complement of A, cannot be a member of $\mathcal{A}$ since $A \cap A^c = \emptyset$. This means that, for every subset of $[n]$ in $\mathcal{A}$, there is at least one subset of $[n]$ not in $\mathcal{A}$. Hence, $|\mathcal{A}| \leq \frac{1}{2} |2^{[n]}| = 2^{n-1}$. In part (a), we showed that a star is an intersecting family of size 2^{n-1}, and so the largest size of an intersecting family in $2^{[n]}$ is indeed 2^{n-1}.

(c) How many subsets of $[n]$ contain 1 and are of size k? If you eliminate 1 from all of these subsets, then you will have all subsets of size $k-1$ of $\{2, 3, \ldots, n\}$. A set with $n-1$ elements has $\binom{n-1}{k-1}$ subsets of size $k - 1$.

(d) If $k > n/2$, then every two subsets of $[n]$ of size k have a non-trivial intersection. As a result, the collection of all subsets of size k of $[n]$ is an intersecting family. A set with n elements has $\binom{n}{k}$ subsets of size k, proving the result. $\square$

We have yet to answer Problem 11.71b in the case when $1 \leq k \leq n/2$. We do already know (Lemma 11.72c) that, for $1 \leq k \leq n/2$, the maximum size of a k-uniform intersecting family in $2^{[n]}$ is *at least* $\binom{n-1}{k-1}$. The question is whether we can do better. Is there a k-uniform intersecting family with more elements than a k-uniform star? The answer is no, and the proof is given by the celebrated Erdös–Ko–Rado (EKR) theorem, which we will tackle next.

To begin with, instead of finding the largest intersecting family among *all* the subsets of size k of $[n]$, we first consider the largest intersecting families from among a particular collection of subsets.

Lemma 11.73. *Let n and k be positive integers with $k \leq n/2$. Arrange 1 through n around a circle, and let $\mathcal{F}$ be the collection of subsets of size k that consist of consecutive k elements on the circle. Then the maximum size of an intersecting family from among the n subsets of $\mathcal{F}$ is k.*

Proof. The elements of $\mathcal{F}$ are the following n subsets of size k of $[n]$:

$$\{1, 2, \ldots, k\}, \{2, \ldots, k, k+1\}, \ldots, \{k, k+1, \ldots, 2k-1\}, \ldots, \{n, 1, 2, \ldots, k-1\}.$$

Let $\mathcal{A}$ be an intersecting family consisting of elements of $\mathcal{F}$. There is really no difference among the different elements of $\mathcal{F}$, and so we can assume, without loss of generality, that $\{1, \ldots, k\}$ is in $\mathcal{A}$. But how many of the other elements of $\mathcal{F}$ could also be in $\mathcal{A}$? Since $\mathcal{A}$ is an intersecting family, we only need to worry about those elements of $\mathcal{F}$ that intersect $\{1, \ldots, k\}$. These are the subsets of $\mathcal{F}$ that either "start" or "end" at one of $1, 2, \ldots, k$. Ignoring $\{1, \ldots, k\}$ for now, we pair up the remaining subsets of $\mathcal{F}$ that have a non-trivial intersection with $[k]$ as follows:

$$\{2, 3, \ldots, k+1\} \quad \longleftrightarrow \quad \{n-k+2, \ldots, n-1, n, 1\}$$
$$\{3, 4, \ldots, k+2\} \quad \longleftrightarrow \quad \{n-k+3, \ldots, n, 1, 2\}$$
$$\vdots \qquad\qquad\qquad \vdots \quad \vdots$$
$$\{k-1, k, \ldots, 2k-2\} \quad \longleftrightarrow \quad \{n-1, n, 1, \ldots, k-2\}$$
$$\{k, k+1, \ldots, 2k-1\} \quad \longleftrightarrow \quad \{n, 1, 2, \ldots, k-1\}.$$

The paired subsets have an empty intersection, and, hence, at most one subset from each pair can be in $\mathcal{A}$. There are $k-1$ pairs, and $\mathcal{A}$ already contained $[k]$, and so we can conclude that $|\mathcal{A}| \leq (k-1) + 1 = k$. $\qquad\square$

In the preparation for the proof of the EKR theorem, we take a quick detour to say a few things about bijections from $[n]$ to $[n]$.

Remark 11.74 (Bijections as Relabeling). We defined bijections – the same thing as one-to-one, onto functions – back in Definition 3.23, and we have been working with them throughout the book. You can think of these functions as just permutations of $[n]$ (see Remark 6.26) or as relabelings. A 1-1, onto function from $[n]$ to $[n]$ just tells you how to switch around the numerical labels on (or in) an object. For example, if $f \colon [3] \to [3]$ and $f(1) = 3, f(2) = 1$, and $f(3) = 2$, then applying f is tantamount to changing all 1's to 3's, all 2's to 1's, and all 3's to 2's. This relabeling, when applied to a collection of subsets, changes the subsets, but does not change many combinatorial properties of the collection. The change is cosmetic.[14]

Definition 11.75 (Extending a Bijection to Sets and Families of Sets). Let $\pi \colon [n] \to [n]$ be a bijection. If $A \subseteq [n]$, then $\pi(A)$ is the subset of $[n]$ obtained by applying π to every element of A. In other words, $\pi(A) = \{\pi(a) \mid a \in A\}$. Likewise, if $\mathcal{F} \subseteq 2^{[n]}$ is a collection of subsets of $[n]$, then $\pi(\mathcal{F})$ is the collection of subsets of $[n]$ obtained by applying π to every subset in $\mathcal{F}$. In other words, $\pi(\mathcal{F}) = \{\pi(A) \mid A \in \mathcal{F}\}$.

If you think of bijections from $[n]$ to $[n]$ as relabeling, the following lemma is straightforward.

---

[14] This discussion is the tip of a very useful iceberg. In the parlance of group theory – see, for example, Chapter 4 of my *Algbera in Action* (Shahriari 2017) – the symmetric group S_n has a natural action on $[n]$ and this action extends to an action on families of subsets.

Lemma 11.76. *Let* $\pi : [n] \to [n]$ *be a bijection, and let* $\mathcal{A} \subseteq 2^{[n]}$ *be a collection of subsets of* $[n]$. *Then*

(a) $|\pi(\mathcal{A})| = |\mathcal{A}|$.
(b) *Let* $\mathcal{B} \subseteq \mathcal{A}$. *Then* $\mathcal{B}$ *is an intersecting family if and only if* $\pi(\mathcal{B})$ *is an intersecting family.*
(c) *The size of the largest intersecting family contained in* $\mathcal{A}$ *is the same as the size of the largest intersecting family in* $\pi(\mathcal{A})$.

Proof. You are asked to write down a proof in Problem P 11.4.15.   □

Lemma 11.77. *Let* n *and* k *be positive integers with* $k \leq n$. *Let* $A = \{a_1, a_2, \ldots, a_k\}$ *and* $B = \{b_1, b_2, \ldots, b_k\}$ *be two subsets of size* k *of* $[n]$. *Then there are* $k!\,(n-k)!$ *bijections* $\pi : [n] \to [n]$ *with the property that* $\pi(A) = B$. *In particular, there are* $n!$ *bijections from* $[n]$ *to* $[n]$.

Proof. To define a bijection $\pi : [n] \to [n]$, we have to decide on the image of every element of $[n]$. We first focus on elements of A. We have k choices – namely, $b_1, \ldots, b_k$ – for a_1. Given that choice, we have $k - 1$ choices for the image of a_2, and so on, for a total of $k!$ choices. We then decide on the images of elements of $[n]$ *not* in A. There are $n - k$ such elements, and we have $n - k$ choices for the image of the first one, $n - k - 1$ choices for the image of the second one, and so on, giving a total of $(n - k)!$ choices for the images of the elements of $[n]$ not in A. Thus, the total number of bijective maps from $[n]$ to $[n]$ that send A to B is $k!\,(n - k)!$.

In the particular case when $A = B = [n]$, we have $k = n$ and $k!\,(n - k)! = n!\,0! = n!$.   □

We are now ready to prove the Erdős–Ko–Rado theorem.

Theorem 11.78 (Erdös–Ko–Rado Theorem, 1938). *Let* n *be a positive integer, and let* $1 \leq k \leq n$. *Assume that* $\mathcal{A}$ *is a* k-*uniform intersecting family in* $2^{[n]}$. *Then*

$$|\mathcal{A}| \leq \binom{n-1}{k-1}.$$

In other words, in $2^{[n]}$, *the maximum size of a* k-*uniform intersecting family is equal to the size of a maximum-sized* k-*uniform star.*

Proof (G.O.H. Katona, 1972). Let S_n denote the set of all bijections from $[n]$ to $[n]$, and let $\mathcal{F}$ be defined as in Lemma 11.73. By Lemma 11.73, the size of the largest intersecting family in $\mathcal{F}$ is at most k. Let $\pi \in S_n$. Applying Lemma 11.76, we get that the size of the largest intersecting family in $\pi(\mathcal{F})$ is also at most k. Since $\mathcal{A}$ is an intersecting family, then so is $\pi(\mathcal{F}) \cap \mathcal{A}$. Hence, $|\pi(\mathcal{F}) \cap \mathcal{A}| \leq k$ for all $\pi \in S_n$. Since $|S_n| = n!$ (Lemma 11.77), we conclude that

$$\sum_{\pi \in S_n} |\pi(\mathcal{F}) \cap \mathcal{A}| \leq n!\,k. \tag{11.1}$$

We now use Lemma 11.77 to count $\sum_{\pi \in S_n} |\pi(\mathcal{F}) \cap \mathcal{A}|$ in a different way. In this sum, focus on a specific set $A \in \mathcal{A}$. How many times does A get counted? The set A gets counted every time a bijection π manages to relabel a set in $\mathcal{F}$ and turn it into A. There are n sets in $\mathcal{F}$, and by Lemma 11.77, for each of these there are $k!\,(n-k)!$ bijections that turn them into A. This happens for every set in $\mathcal{A}$. Hence,

$$\sum_{\pi \in S_n} |\pi(\mathcal{F}) \cap \mathcal{A}| = |\mathcal{A}|\, nk!\,(n-k)!. \tag{11.2}$$

Combining Equation 11.1 and Equation 11.2, we have

$$|\mathcal{A}|\, nk!\,(n-k)! \le n!\,k \quad \Rightarrow \quad |\mathcal{A}| \le \frac{(n-1)!}{(k-1)!\,(n-k)!} = \binom{n-1}{k-1}. \qquad \square$$

We close this section by pointing out the difference between "maximal" and "maximum-sized" intersecting families. A collection of elements is a *maximal* collection with some property P if it has that property *and* you cannot add any new elements without destroying property P. In contrast, a *maximum-sized* collection with some property P is a set of elements that has the largest possible size among the sets that have property P.

Example 11.79. Let $n = 6$ and consider $\mathcal{A} = \{\{1,2\}, \{1,3\}, \{2,3\}\}$. This collection of subsets is a 2-uniform intersecting family. By the EKR theorem, we know that the maximum size of a 2-uniform intersecting family in $\mathbf{2}^{[6]}$ is $\binom{5}{1} = 5$, and so $\mathcal{A}$ is *not* a maximum-sized 2-uniform intersecting family in $\mathbf{2}^{[6]}$. However, $\mathcal{A}$ is a *maximal* 2-uniform intersecting family. If you extend $\mathcal{A}$ by adding another subset of size 2 to it, the resulting collection will not be intersecting anymore.

In contrast to Example 11.79, all maximal intersecting families (as opposed to k-uniform intersecting families) are maximum-sized.

Proposition 11.80. *Let n be a positive integer and let $\mathcal{A}$ be an intersecting family in $\mathbf{2}^{[n]}$. If $|\mathcal{A}| < 2^{n-1}$, then $\mathcal{A}$ can be extended to an intersecting family of size 2^{n-1}.*

Proof. Suppose $\mathcal{A}$ is an intersecting family with $|\mathcal{A}| < 2^{n-1}$. This means that there is some subset A of $[n]$ such that neither A nor A^c is in $\mathcal{A}$. We want to show that we can add either A or A^c to $\mathcal{A}$ and keep it an intersecting family. By repeating this process, we would eventually get an intersecting family that contains $\mathcal{A}$ and, for every subset A of $[n]$, contains either A or A^c. Hence, we have extended $\mathcal{A}$ to an intersecting family of size 2^{n-1}.

Why can we add either A or A^c and be assured that the resulting bigger family continues to be an intersecting family? If A has a non-empty intersection with every element of $\mathcal{A}$, then, of course, we can add A to $\mathcal{A}$. On the other hand, if $A \cap B = \emptyset$ for some $B \in \mathcal{A}$, then $B \subseteq A^c$. (See Figure 11.25.) This does mean that A^c has a non-empty intersection with every element of $\mathcal{A}$, since if some element of $\mathcal{A}$ intersected A^c trivially, then its intersection with B would be trivial as well. Hence, we can add A^c to $\mathcal{A}$, and the new collection will continue to be intersecting. $\qquad \square$

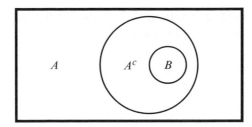

Figure 11.25 If $A \cap B = \emptyset$, then $B \subseteq A^c$, and every set intersecting B also intersects A^c.

Problems

P 11.4.1. What is the minimum number of chains needed to partition $2^{[11]}$?

P 11.4.2. Let $A_1 \subsetneq A_2 \subsetneq \cdots \subsetneq A_k$ be a chain in $2^{[n]}$. Show that this chain is symmetric if and only if it is skipless and $|A_1| + |A_k| = n$.

P 11.4.3. Give a maximal antichain in $2^{[4]}$ consisting of two subsets.

P 11.4.4. Assume that $\mathcal{A}$ is a maximal antichain in $2^{[n]}$. Let $\mathcal{B} = \{A^c \mid A \in \mathcal{A}\}$ be the collection of the complements of the sets in $\mathcal{A}$. Show that $\mathcal{B}$ is also a maximal antichain.

P 11.4.5. In $2^{[4]}$, is there an antichain of size 5? What about a maximal antichain of size 5?

P 11.4.6. Assume $\mathcal{A}$ is a maximal antichain in $2^{[4]}$. What are the possible sizes of $\mathcal{A}$?

P 11.4.7. Find all antichains of size 6 in $2^{[4]}$.

P 11.4.8. A talk show host has just bought 10 new jokes. Each night she tells some of the jokes. What is the largest number of nights on which you can tune in so that you never hear on one night at least all the jokes you heard on *one* of the other nights?

P 11.4.9. Let $\mathcal{A}$ be an antichain in $2^{[n]}$, and assume that $\mathcal{A}$ contains at least one subset of $[n]$ of size 1. What is the maximum size of $\mathcal{A}$?

P 11.4.10. Consider a symmetric chain decomposition of $2^{[6]}$. How many chains of what sizes does such a decomposition have?

P 11.4.11. Re-do Problem P 11.4.10 for $2^{[n]}$.

P 11.4.12. (Erdős 1945) Let $x_1, \ldots, x_n$ be real numbers, $x_i \geq 1$ for each i, and let I be any open unit interval on the real line. Show that the number of linear combinations $\sum_{i=1}^{n} \varepsilon_i x_i$ with $\varepsilon_i = 0$ or 1 lying inside I is at most $\binom{n}{\lfloor n/2 \rfloor}$.

P 11.4.13. What is the size of the largest intersecting family in $2^{[3]}$ that is *not* a star? What about $2^{[4]}$?

P 11.4.14. Let $n = 2k$ be a positive even integer, and let $\mathcal{A}$ be all the subsets of size k of $[n-1]$. What is $|\mathcal{A}|$? Is $\mathcal{A}$ a maximum-sized k-uniform intersecting family in $2^{[n]}$?

P 11.4.15. **Proof of Lemma 11.76.** Write down a proof of Lemma 11.76.

P 11.4.16. Let n be a positive integer, and consider the Hasse diagram of $2^{[n]}$ as a graph.
   (a) A subset of size k, for $0 \leq k \leq n$, is a vertex in the Hasse diagram of $2^{[n]}$. What is its degree?
   (b) What is the total number of edges in the Hasse diagram of $2^{[n]}$?

(c) In the Hasse diagram, for $0 \le k \le n - 1$, what is the total number of edges between subsets of size k and subsets of size $k + 1$?

(d) By counting the total number of edges in the Hasse diagram of $2^{[n]}$ in two different ways, get a combinatorial proof of the binomial coefficient identity in Proposition 5.26c.

P 11.4.17. Find $\sum_{B \subseteq [n]} \left(\sum_{A \subseteq [n]} |A \cap B| \right)$.

P 11.4.18. Let $[n] = \{1, \ldots, n\}$ be a set with n elements. Let $\mathcal{S} = \{(X, Y) \mid X, Y \subseteq [n], X \subsetneq Y\}$. Find $|\mathcal{S}|$ and give the answer in the form $?^n - ??^n$.

P 11.4.19. Let $\mathcal{A}$ be an antichain in $2^{[n]}$. Let $i(\mathcal{A})$ be all subsets of $[n]$ that have a non-empty intersection with every element of $\mathcal{A}$. Let $b(\mathcal{A})$ be the minimal elements of $i(\mathcal{A})$. In other words, $X \in b(\mathcal{A})$ if and only if $X \cap A \ne \emptyset$ for all $A \in \mathcal{A}$, and there is no other subset $Y \in i(\mathcal{A})$ with $Y \subsetneq X$.

(a) Show that $b(\mathcal{A})$ is an antichain.

(b) Let $n = 4$ and let $\mathcal{A}$ be the antichain consisting of all subsets of size 2 of $[4]$. Find $b(\mathcal{A})$ and $b(b(\mathcal{A}))$.

(c) Show that, in general, $b(b(\mathcal{A})) = \mathcal{A}$.

P 11.4.20. **Fibres and Cones in $2^{[n]}$.** Recall the definition of a fibre and a cone in subsection 11.2.1. Let A be a subset of $[n]$. Show that, in $2^{[n]}$, the cone of A is a minimal fibre.

P 11.4.21. **An Application of Linear Algebra.** Let n and k be positive integers with $k \le n$. Let $A_1, A_2, \ldots, A_b$ be a k-uniform intersecting family of subsets of $[n]$. Further assume that, for each $1 \le i < j \le b$, we have $|A_i \cap A_j| = 1$. Prove that $b \le n$. You may find the following steps helpful:[15]

STEP 1: Let N be the incidence matrix of elements versus subsets. In other words, N is an $n \times b$ matrix, its rows are indexed by elements of $[n]$, and its columns are indexed by $A_1, \ldots, A_b$. The (i, j) entry of this matrix is 1 if $i \in A_j$ and 0 otherwise. What is the dot product of column i of N with column j of N?

STEP 2: Let J be the $b \times b$ matrix of all 1's, and let I_b be the $b \times b$ identity matrix. Show that $N^t N = J + (k - 1)I_b$. (N^t is the transpose of N.)

STEP 3: Show that $N^t N$ does not have an eigenvector with eigenvalue 0. Conclude that $N^t N$ is an invertible matrix.

STEP 4: Use the fact that, for two matrices A and B, $\text{rank}(AB) \le \text{rank}(A)$ to show that $b \le n$.

---

[15] The suggested solution uses a fair amount of linear algebra.

11.5   Boolean Lattices and Graphs: Ramsey Theory Extended*

Warm-Up 11.81. *A total of* 4700 *briefcases are organized in* 47 *rows. In each row there are* 100 *briefcases, and each is identified with a tag of the form* (a, b), *where* $1 \leq a, b \leq 10$. *Some of these briefcases have cash in them, others don't. You are told that if a briefcase in row n has cash, then so does the briefcase with the same tag but in row* $n + 10$. *In addition, if a briefcase with tag* (a, b) *has cash in it, then so does a briefcase in the same row with tag* (c, d), *as long as* $c + d > a + b$. *You open the briefcase in row* 23 *and with tag* $(7, 7)$, *and it has cash in it. What can you conclude?*

In this optional section, we argue that Boolean lattices provide a context for generalizing the notion of a graph! As such, questions that can be asked about graphs can be generalized and investigated in a much more general setting. After explaining how we can think of a simple graph as a subset of a Boolean lattice, as proof of concept, we return to Ramsey theory and Ramsey numbers. In Chapter 2, we defined Ramsey numbers in the context of graphs, and by investigating the patterns that are constructed when you color the edges of a complete graph. Here, we prove a vast generalization of that idea, in the context of Boolean lattices.

Graphs as Subsets of $2^{[n]}$

When we specify a simple graph of order n, we can let $[n]$ be the set of vertices, and to specify the graph completely, we need to identify the set of edges. But each edge is a subset of size 2 of $[n]$, and hence an element of $2^{[n]}$. Thus a graph of order n is the same as a collection of subsets of size 2 of $[n]$. Hence, a graph of order n is a subset of the second level of the Boolean lattice $2^{[n]}$.

Example 11.82. The cycle of size 4, C_4, was depicted in Figure 2.5. This graph consists of four vertices and four edges, and the edges are pairs of vertices. So, if we denote the vertices as $[4] = \{1, 2, 3, 4\}$, then the edges are four elements in $2^{[4]}$, namely $\{1, 2\}$, $\{2, 3\}$, $\{3, 4\}$, $\{1, 4\}$. See Figure 11.26.

From this point of view, we can generalize the notion of a simple graph by allowing "edges" that continue to be subsets of $[n]$ but not necessarily of size 2. In other words, we can consider hyperedges that are subsets of any size of $[n]$. The resulting object is called a hypergraph.

Definition 11.83. A *hypergraph* consists of vertices and hyperedges. The *vertices* of a hypergraph are the elements of $[n]$, where n is a positive integer. The *hyperedges* of a hypergraph are then non-empty subsets of $[n]$. If all the hyperedges of a hypergraph are subsets of size k, for some $1 \leq k \leq n$, then the hypergraph is called a *k-uniform* hypergraph. A simple graph is just a 2-uniform hypergraph.

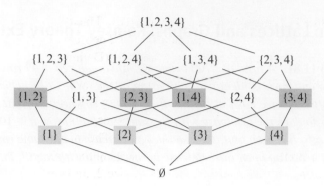

Figure 11.26 The cycle C_4 as a subset of $2^{[4]}$. The vertices of C_4 are the subsets of size 1, and the edges of C_4 are the identified subsets of size 2.

Ramsey's Theorem Generalized

Turning to Ramsey theory, we start by translating an arrow statement such as $K_6 \rightarrow K_3, K_3$ (Definition 2.36) to Boolean lattices. We can interpret this statement as follows: if we assign one of two colors to the subsets of size 2 of $2^{[6]}$, then we are guaranteed a blue subset of size 3 or a red subset of size 3. What do we mean by a blue subset of size 3? (Remember that we are only assigning colors to subsets of size 2.) A subset of size 3 is defined to be blue if all of its subsets of size 2 are blue.

How can we generalize this? Maybe instead of coloring subsets of size 2, we can color subsets of some other size!

Definition 11.84 (Monochromatic Set). Let n and t be positive integers with $t \leq n$. If we color the subsets of size t in $2^{[n]}$ with a number of colors including blue, then we say that a subset of size q (where $t \leq q \leq n$) is *monochromatic* and has color *blue* if all of its subsets of size t were assigned the color blue.

We are now ready to state Ramsey's theorem in the context of $2^{[n]}$. The previous versions of Ramsey's theorem (Theorem 2.38 and Theorem 2.52) are special cases of this.

Theorem 11.85 (Ramsey). *Let $k \geq 1$, $t \geq 1$, and $q_1, \ldots, q_k \geq t$ be all integers. Then there exists a positive integer s (dependent on $t, q_1, \ldots, q_k$) that has the following property. For $1 \leq i \leq k$, $s \geq q_i$, and, if we assign – in any way – one of k colors to each subset of size t in $2^{[s]}$, then, among the subsets in $2^{[s]}$, either a subset of size q_1 is monochromatic and has color 1, or a subset of size q_2 is monochromatic and has color 2, ..., or a subset of size q_k is monochromatic and has color k.*

Remark 11.86. We use the notation $K_s^t \rightarrow K_{q_1}^t, \ldots, K_{q_k}^t$ to denote the statement – which may be true or false – that, if we color each subset of size t of $2^{[s]}$ with one of k colors, then we will have either a monochromatic subset of size q_1 of the first color, or a monochromatic subset of size q_2 of the second color, ..., or a monochromatic subset of size q_k of the kth color.

Definition 11.87 (Ramsey Number, Generalized). The Ramsey number $r_t(q_1, q_2, \ldots, q_k)$ is the smallest number that satisfies the conclusion of Theorem 11.85. Note that $r_2(q_1, q_2, \ldots, q_k) = r(q_1, q_2, \ldots, q_k)$ are the earlier Ramsey numbers (Definition 2.51).

Remark 11.88. There are a number of parameters involved in the above definition, and that could be confusing. Here is a reprise. Everything is happening in a Boolean lattice $\mathbf{2}^{[n]}$. We are really reserving judgement on how big n needs to be, but for now assume n is some large finite number. Then choose a positive integer t. The integer t tells you which size subsets are going to be colored. Each subset of size t is going to be assigned one color. For example, if $t = 2$, then you will be coloring subsets of size 2, and this corresponds to coloring all the possible edges on a complete graph. On the other hand, if $t = 3$, then you will be coloring all subsets of size 3 of $[n]$. You always will be coloring the elements in one level of the Boolean lattice, and each subset gets one color, but how many colors do you use? The positive integer k is the number of colors used. The sequence $q_1, \ldots, q_k$ is your wish list. You are happy if you can find a subset of size q_1 that is monochromatic with color 1 (recall that for a set to be monochromatic means that all subsets of size t of this set have the same color). You will also be happy if, instead, you could find a set of size q_2 that is monochromatic but with color 2. And so on; you have a wish list of k different possibilities, one for each color, and you will be happy as long as at least one of your wishes comes true. Ramsey's theorem says that, if n is large enough, regardless of the values of t, k, and $q_1, \ldots, q_k$ (as long as the latter are no smaller than t), one of your wishes will come true. The smallest n that would work for a specific choice of parameters is a Ramsey number.

We will give a proof of this version of the Ramsey theorem for hypergraphs for the case of two colors. The more general version needs a bit more messy notation but is not much harder conceptually. The proof is by a double induction (somewhat akin to Warm-Up 11.81). We first induct on t, the size of the subsets that get the initial colorings. But to prove the inductive step, we use a second induction, this time inducting on $q_1 + q_2$. Now, we have already proved Ramsey's theorem for graphs, and that corresponds to $t = 2$. We could use this as our base case for induction, but the theorem is true even for $t = 1$. In fact, if we color subsets of size 1, then this version of Ramsey's theorem becomes the pigeonhole principle!

Proposition 11.89 (Pigeonhole Principle Revisited).

(a) We have $r_1(\underbrace{2, 2, \ldots, 2}_{k}) = k + 1$, and this is a restatement of the ordinary pigeonhole principle (Theorem 2.2).

(b) We have $r_1(q_1, q_2, \ldots, q_k) = q_1 + q_2 + \cdots + q_k - k + 1$, and this is a restatement of the strong form of the pigeonhole principle (Problem P 2.1.4).

Hence, we can think of Ramsey's theorem as a generalization of the pigeonhole principle.

Proof. In both cases, we are coloring the subsets of size 1, which are the vertices themselves.

(a) If we assign each of $k + 1$ vertices (or pigeons) one of k colors (or pigeonholes), then at least one pigeonhole (color) will have at least two vertices. Thus there will be a monochromatic subset of size 2. Clearly $k+1$ is the smallest number of vertices for which our argument works. Hence $r_1(2, \ldots, 2) = k + 1$.

(b) If we assign each of $q_1 + \cdots + q_k - k + 1$ pigeons to one of k pigeonholes, then either the first pigeonhole will have q_1 or more pigeons, or the second pigeonhole will have q_2 or more pigeons, ..., or the kth pigeonhole will have q_k or more pigeons. The result now follows.   □

We are now ready to prove Ramsey's theorem for hypergraphs and for two colors.

Theorem 11.90. *Let t be a positive integer, and let q_1 and q_2 be integers no less than t. Then there exists a positive integer N such that*

$$K_N^t \to K_{q_1}^t, K_{q_2}^t.$$

As a result, $r_t(q_1, q_2)$ is a finite number.

Proof. For the duration of this proof, let the first color be blue and the second color be red.

Induct on t. For $t = 1$, the result follows from Proposition 11.89.

For $t > 1$, assume that the theorem is already true for $t - 1$. To prove the theorem for the given t, we again use induction (so we are proving the inductive step using a different induction), but this time on $q_1 + q_2$. What is the base case for this induction? The smallest possible value for q_1 or for q_2 is t, and, in fact, as long as $p \geq t$, we have $r_t(p, t) = r_t(t, p) = p$. This is true, because if you assign blue or red to each of the t-subsets of $[p]$, then either you do not use red or you do. In the former case, all the t-subsets are assigned blue, and so $[p]$ itself is monochromatic and blue. In the latter case, one of the subsets of size t has been assigned red, and that subset is monochromatic and red. Hence, the base case for the induction (the second induction within the original induction) has been proved.

For general values of q_1 and q_2 – where both of these are greater than t – we let $n_1 = r_t(q_1 - 1, q_2)$ and $n_2 = r_t(q_1, q_2 - 1)$. Both n_1 and n_2 exist because of the (second) inductive hypothesis. Let $N = 1 + r_{t-1}(n_1, n_2)$. This number exists because of the first inductive hypothesis. The theorem will be proved when we show:

CLAIM: $K_N^t \to K_{q_1}^t, K_{q_2}^t.$

PROOF OF CLAIM: Arbitrarily assign blue or red to each subset of size t of $[N]$. We want to find a blue subset of size q_1 (i.e., a subset of size q_1 all of whose t-subsets are blue), or a red subset of size q_2.

We first use the given coloring to assign colors to subsets of size $t - 1$ of $[N - 1]$. If A is a subset of size $t - 1$ of $[N - 1]$, then assign to it the same color as $A \cup \{N\}$ (which is a subset of size t of $[N]$ and already colored). Since $N - 1 = r_{t-1}(n_1, n_2)$, and we just assigned blue or red to subsets of size $t - 1$ of $[N - 1]$, by the definition of Ramsey numbers we are assured that we can find either a blue subset of $[N - 1]$ of size n_1 or a red subset of $[N - 1]$ of size n_2. The two cases are going to be exactly similar. So, without loss of generality, let B be a subset of $[N - 1]$ of size n_1 such that all of its subsets of size $t - 1$ have been assigned blue.

Now B is itself a set of size n_1, its subsets of size $t - 1$ are all blue, but its subsets of size t are also colored, albeit with the original coloring, and they may have blue as well as red subsets.

Moreover, $n_1 = r_t(q_1 - 1, q_2)$. So, B either has a blue subset of size $q_1 - 1$ or a red subset of size q_2. In the latter case, we are already done, since we found the sought-after subset of size q_2, all of whose t-subsets are red. In the former case, we have a subset $C \subset B$ of size $q_1 - 1$ such that all of its t-subsets are blue. Now consider the set $D = C \cup \{N\}$. The subset D has q_1 elements and we claim that every subset of size t of D is blue. If this were true the proof would be complete.

What are the subsets of size t of D? They come in two varieties: those that include N and those that don't. The latter are really subsets of size t of C, and we know that these are all blue. But what about a subset E of size t of D that has N as one of its elements? If you go back to the way we colored subsets of size $t-1$ using the coloring for subsets of size t, you will notice that E has the same color as $E - \{N\}$. However, $E - \{N\}$ is a $(t-1)$-subset of $C \subset B$, and all $(t-1)$-subsets of B are blue. Hence, E is blue as well, and the proof is complete! $\square$

In Table 2.4, we summarized what is known for Ramsey numbers for $t = 2$. For $t > 2$, even less is known. The only (non-trivial) exact known value of $r_t(q_1, \ldots, q_k)$, as of this writing, is $r_3(4, 4)$, which was proved to be 13 by Brendan McKay and Stanisław Radziszowski in 1991. Other known bounds include $r_3(5, 4) \geq 35$, $r_3(6, 4) \geq 58$, $r_3(5, 5) \geq 82$, $r_3(4, 4, 4) \geq 56$, and $r_4(5, 5) \geq 34$. See the survey by Radziszowski (Radziszowski 2021) for more details about known Ramsey numbers.

Problems

P 11.5.1. Let $\mathcal{H} = (V, E)$ be a finite hypergraph. In other words, V is a finite set of vertices and E is a finite set of subsets of V. In Theorem 10.9, we proved the straightforward fact that, in a general graph, the sum of the degrees of the vertices is twice the number of edges. Make the appropriate definitions so as to prove a generalization of this statement for $\mathcal{H}$.

P 11.5.2. Let $\mathcal{H} = (V, E)$ be a (finite simple) hypergraph. We want to define proper coloring of vertices of $\mathcal{H}$.

(a) Assume that we define a proper coloring of vertices as an assignment of colors to the vertices in such a way that no two vertices that are elements of the same edge get the same color. Give an example of a hypergraph with four vertices and one edge that would need four colors for the proper coloring of the vertices.

(b) Assume we define a proper coloring of vertices as an assignment of colors to the vertices in such a way that no edge is monochromatic. Let $V = [4]$ and let E be all subsets of size 3 of V. How many colors would you need to properly color the vertices?

(c) Which of the two suggestions do you prefer for the definition of proper coloring?

P 11.5.3. Let $\mathcal{H} = (V, E)$ be a (finite simple) hypergraph with no edges of size 1. We say that a coloring of V is *proper* if there are no monochromatic edges. The chromatic number of $\mathcal{H}$ – denoted $\chi(\mathcal{H})$ – is the minimum number of colors necessary to color the vertices of $\mathcal{H}$ properly. The *degree* of a vertex is the number of edges that contain it, and the maximum degree among the degrees of vertices of $\mathcal{H}$ is denoted by $\Delta(\mathcal{H})$. Prove, in analogy with Problem P 10.7.6, that $\chi(\mathcal{H}) \leq \Delta(\mathcal{H}) + 1$.

P 11.5.4. Prove that $r_3(4,4) > 5$.

P 11.5.5. An oracle tells you that $r_3(4,4) = 13$. By coincidence, you are at a party with 13 of your friends. You make a list of all the possible subsets of size 3 from among your friends. Your list is a very long list of $\binom{13}{3}$ subsets. You then go and ask each group of three if the three of them have ever gone on a hike together, and you record their responses. What can you conclude from what the oracle told you?

P 11.5.6. Let m and t be positive integers with $m \geq t$. What can you say about the Ramsey number $r_t(m,t,t)$?

P 11.5.7. Let $q_1, q_2, \ldots, q_k$, and t be positive integers with each $q_i \geq t$. Let m be an integer with $m \geq q_i$ for $1 \leq i \leq k$. How does $r_t(\underbrace{m,m,\ldots,m}_{k})$ compare with $r_t(q_1, q_2, \ldots, q_k)$? Which one is bigger and why?

P 11.5.8. **Geometric Erdős–Szekeres Theorem.** A polygon in the plane is *convex* if every line segment connecting two points on the polygon remains inside the polygon. Accept the following two geometric facts: (1) given any five points in the plane with no three of them on a straight line, at least four of them form a convex polygon; and (2) given n points in the plane with no three on a straight line, if every four of them form a convex polygon, then so do all n. Fix an integer $n \geq 4$, and let m be the Ramsey number $r_4(5,n)$. Prove that *any* collection of m points in the plane with no three on a straight line contains n points that form a convex polygon.

P 11.5.9. **An Infinite Ramsey Theorem.** Let c and t be positive integers. Let X be an infinite set, and assign one of c colors to each subset of size t of X. Then prove that there exists $M \subseteq X$ with $|M| = \infty$ such that all subsets of size t of M have the same color.[16] You may find the following steps helpful:

STEP 1: Induct on t. What is the base case, and why is it true?

STEP 2: For the inductive case, choose $x_1 \in X$, and let $Y_1 = X - \{x_1\}$. Mimic proof of Theorem 11.90, and assign colors to $(t-1)$-subsets of Y_1 using the colors assigned to t-subsets of X. Apply the inductive hypothesis to Y_1 to get an infinite subset M_1 such that all subsets of size $t-1$ of M_1 have the same color.

STEP 3: Argue that all subsets of size t of X that consist of x_1 and $t-1$ elements of M_1 have the same color.

STEP 4: Choose $x_2 \in M_1$ and repeat Steps 2 and 3. Get an infinite set $M_2 \subseteq M_1$ such that all t-subsets of X that consist of x_2 and $t-1$ elements of M_2 have the same color (possibly a different color than the previous one).

STEP 5: Repeat the previous step and get an infinite sequence $x_1, x_2, x_3, \ldots$ of elements of X, as well as an infinite sequence $X = M_0 \supseteq M_1 \supseteq M_2 \supseteq M_3 \cdots$ of infinite sets such

---

[16] It is possible to get the finite Ramsey theorem as a corollary of this result.

that, for all i, $x_i \in M_{i-1}$, *and* all t-subsets of X that consist of x_i and $t-1$ elements of M_i are of the same color.

STEP 6: Use the fact that the number of colors is finite to get an infinite subsequence x_{i_1}, $x_{i_2}, \ldots$ of elements of X, with $i_1 < i_2 < \cdots$, such that the color associated with them from the previous step is the same color.

STEP 7: Let $M = \{x_{i_1}, x_{i_2}, \ldots\}$. Show that M is an infinite subset of X with the property that all of its t-subsets have the same color.

STEP 8: Rewrite what you have and give a complete and concise proof.

11.6 Möbius Inversion: Inclusion–Exclusion Extended*

Warm-Up 11.91. *Let P be a finite poset, and let f be a function of two variables. You feed f two elements of the poset, and f returns an integer. In other words, $f : P \times P \to \mathbb{Z}$. Let $a \in P$ be fixed. We are interested in the double sum $M = \sum_{x \le a} \left(\sum_{y \le x} f(x, y) \right)$, but we want to reverse the order of summation, and to write*

$$M = \sum_{y \le a} \left(\sum_{???} f(x, y) \right).$$

Is this possible? If so, what should be the range of variables in the (new) inner sum?

In this final section of the text, we generalize the principle of inclusion–exclusion. We will define, for any poset, something called the *Möbius function*,[17] and using it will prove a result known as *Möbius inversion*. We will then see that the inclusion–exclusion principle is just Möbius inversion applied to the Boolean lattices. By applying Möbius inversion to other posets – most notably, the posets of divisors of a positive integer – we will get a variety of other results. There is much more to be said about the Möbius function, and we will be only scratching the surface, with the hope of whetting your appetite. We begin with an example, and a motivating question.

Question 11.92. Let S be a finite set, and let A, B, C, and D be subsets of S. Assume that, for these particular subsets, we know (see Figure 11.27) that

$$D = A \cap B = A \cap C = B \cap C = A \cap B \cap C.$$

We are interested in $|A^c \cap B^c \cap C^c|$, and, using the principle of inclusion–exclusion (Theorem 8.6), we can write

$$\left| A^c \cap B^c \cap C^c \right| = |S| - |A| - |B| - |C| + |A \cap B| + |A \cap C| + |B \cap C| - |A \cap B \cap C|.$$

---

[17] Named after August Möbius (1790–1868). For a history of the Möbius function and earlier work of Euler on the topic, see Dunham 2018.

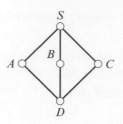

Figure 11.27 $D = A \cap B = A \cap C = B \cap C = A \cap B \cap C.$

However, in this particular situation, we know more about the sets A, B, and C and their various intersections. Exploiting what we are given, we have

$$\left| A^c \cap B^c \cap C^c \right| = |S| - |A| - |B| - |C| + 2\,|D|.$$

Given the configuration in Figure 11.27, it is not surprising that we can write $|A^c \cap B^c \cap C^c|$ in terms of the sizes of S, A, B, C, and D, and with no direct mention of $A \cap B$, $A \cap C$, etc. But could we have predicted this specific final result without first invoking the principle of inclusion–exclusion? In particular, could we have predicted the coefficient of 2 for $|D|$ in the final result?

After a preliminary definition, we will be ready to define the Möbius function of a poset.

Definition 11.93 (Intervals in Posets). Let $(P, \leq)$ be a poset, and let $x, y \in P$ with $x \leq y$. The *closed interval* $[x, y]$ is defined by

$$[x, y] = \{z \in P \mid x \leq z \leq y\}.$$

Likewise, the *open interval* (x, y) is defined by

$$(x, y) = \{z \in P \mid x < z < y\}.$$

Recall that $\mathbb{Z}$ denotes the set of integers, and, for a set P, $P \times P$ is the cartesian product of P with itself, and its elements are of the form (a, b) with $a, b \in P$ (Definition 11.2).

Definition 11.94 (The Möbius Function of a Poset). Let P be a finite poset. Define *the Möbius function* of P, $\mu_P \colon P \times P \to \mathbb{Z}$, by the following rules:

(a) $\mu_P(x, y) = 0$ if $x > y$ or if x and y are incomparable elements of P
(b) $\mu_P(x, x) = 1$ for all $x \in P$
(c) $\sum_{z \in [x, y]} \mu_P(x, z) = 0$ for any $x, y \in P$ with $x < y$.

Remark 11.95. The Möbius function of a poset is a function of two variables. You plug in two elements of your poset, and out comes an integer. The only time you could get a non-zero answer is if you plug in x and y with $x \leq y$. It is not obvious from the definition that we have properly defined the function for every choice of $x, y \in P$. In fact, this is a recursive definition (there are others). To use it, and to find $\mu_P(x, y)$, if $x < y$, you start by noting that $\mu_P(x, x) = 1$, and then work your way up to $\mu_P(x, y)$ by first finding $\mu_P(x, z)$ for all elements

z in the interval (x, y). Alternatively, as we shall see later, you can start with $\mu_P(y, y) = 1$, and work your way down. A few examples, will make this clearer.

It is also not necessary to assume that P is a finite poset, even though some kind of finiteness condition is needed. For most of what we do, it is enough for the poset to be locally finite. A poset is *locally finite* if, for all $x, y \in P$, $|[x, y]|$, the number of elements in the interval $[x, y]$, is finite.

Finally, we are most interested in the cases when the poset P has either a unique minimal element $\mathbf{0}$ or a unique maximal element $\mathbf{1}$. In such cases, for $x \in P$, we often focus entirely either on the value of $\mu_P(\mathbf{0}, x)$ or on the value of $\mu_P(x, \mathbf{1})$. Either choice results in a function of one variable.

Example 11.96. Let $P = 2^{[3]}$ be the Boolean lattice of order 3. To calculate $\mu_P(\emptyset, A)$ for all $A \in 2^{[3]}$, we start with $\mu_P(\emptyset, \emptyset) = 1$, and proceed to subsets of size 1. (For such a small poset, it is most convenient to write each known value of $\mu_P(\emptyset, A)$ next to A in the Hasse diagram of the poset, and calculate new values using the known ones. See Figure 11.28.) For $i \in [3]$, there are only two subsets in the interval $[\emptyset, \{i\}]$, and over any closed interval the sum of the values of the Möbius function – where the beginning of the interval is plugged in for the first variable – is equal to zero. So, we have

$$0 = \sum_{z \in [\emptyset, \{i\}]} \mu_P(\emptyset, z) = \underbrace{\mu_P(\emptyset, \emptyset)}_{1} + \mu_P(\emptyset, \{i\}) \Rightarrow \mu_P(\emptyset, \{i\}) = -1.$$

Moving one level up, for $1 \le i < j \le 3$, the interval $[\emptyset, \{i, j\}]$ is itself isomorphic to $2^{[2]}$, and has four elements. Since we know the value of the desired Möbius function for three of these, and we know that the sum of all of them has to be zero, we can calculate the fourth one:

$$0 = \sum_{z \in [\emptyset, \{i, j\}]} \mu_P(\emptyset, z) = \underbrace{\mu_P(\emptyset, \emptyset)}_{1} + \underbrace{\mu_P(\emptyset, \{i\})}_{-1} + \underbrace{\mu_P(\emptyset, \{j\})}_{-1} + \mu_P(\emptyset, \{i, j\})$$

$$\Rightarrow \mu_P(\emptyset, \{i, j\}) = 1.$$

Doing this one more time, we get $\mu_P(\emptyset, [3]) = -1$. See Figure 11.28.

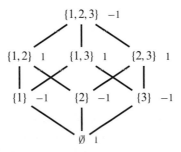

Figure 11.28 The values of $\mu_P(\emptyset, A)$ for all $A \in P = 2^{[3]}$.

We now generalize Example 11.96.

Proposition 11.97. *Let n be a non-negative integer, let $P = 2^{[n]}$, and let $A, B \in P$. Then*

$$\mu_P(A, B) = \begin{cases} (-1)^{|B|-|A|} & \text{if } A \subseteq B \\ 0 & \text{otherwise.} \end{cases}$$

Proof. If A is not a subset of B, then, by definition, $\mu_P(A, B) = 0$, and there is nothing to prove. So assume $A \subseteq B$, and let $r = |B| - |A|$. We prove the result by induction on r. For the base case, if $r = 0$, then $A = B$, and, by definition, $\mu_P(A, B) = 1 = (-1)^0$.

For the inductive step, assume $r \geq 1$, and assume that the result is true for smaller values of r. Now the interval (A, B) consists of all proper subsets of B (other than A) that contain A. If $B - A = \{x_1, \ldots, x_r\}$, then a typical element of (A, B) is $C = A \cup \{x_{i_1}, \ldots, x_{i_k}\}$ (with $k < r$). Now A and C are both subsets of $[n]$, and $|C| - |A| = k$, and we have $k < r$. Hence, by the inductive hypothesis, $\mu_P(C) = (-1)^k$. Now there are $\binom{r}{k}$ ways to choose k elements from $B - A$, and so, in the interval $[A, B]$, there are $\binom{r}{k}$ subsets C with $|C| - |A| = k$, and k can range from 0 to r. By definition of the Möbius function, we have $\sum_{C \in [A,B]} \mu_P(A, C) = 0$. Expanding this, one of the terms is $\mu_P(A, B)$, which we do not yet know. But all the others we know, and we get

$$\mu_P(A, A) + \sum_{\substack{C \in [A, B] \\ |C|-|A|=1}} \mu_P(A, C) + \sum_{\substack{C \in [A, B] \\ |C|-|A|=2}} \mu_P(A, C) + \cdots + \sum_{\substack{C \in [A, B] \\ |C|-|A|=r-1}} \mu_P(A, C)$$

$$+ \mu_P(A, B) = 0.$$

By the inductive hypothesis, the above becomes

$$1 + (-1)^1 \binom{r}{1} + (-1)^2 \binom{r}{2} + \cdots + (-1)^k \binom{r}{k} + \cdots + (-1)^{r-1} \binom{r}{r-1} + \mu_P(A, B) = 0.$$

Now substituting $x = -1$ in the Corollary 5.25 version of the binomial theorem, we have

$$0 = (1 - 1)^r = \sum_{k=0}^{r} \binom{r}{k}(-1)^k$$

$$= 1 + (-1)^1 \binom{r}{1} + \cdots + (-1)^k \binom{r}{k} + \cdots + (-1)^{r-1} \binom{r}{r-1} + (-1)^r \binom{r}{r}.$$

Comparing the two equations, we conclude that

$$\mu_P(A, B) = (-1)^r,$$

as desired. $\qquad\square$

Remark 11.98. In the proof of Proposition 11.97, we were essentially using the fact that if A, B are two finite sets with $A \subseteq B$, then the interval $[A, B]$ is isomorphic as a poset to $2^{[r]}$, where $r = |B| - |A|$. This is because, if you eliminate all the elements of A from each subset in $[A, B]$, then you get exactly all the subsets of $B - A$, a set with r elements. The set of subsets of a set with r elements – and, as a result, the interval $[A, B]$ – is isomorphic as a poset to $2^{[r]}$. In the poset $[A, B]$, the set A is playing the role of the empty set, and the subsets $A \cup \{x\}$, for $x \in B - A$, are playing the role of singletons, and so on.

Remark 11.99. While in the text we are investigating the Möbius function on posets of subsets, an even more fruitful path is to look at the Möbius function on the posets of divisors of an integer. You are asked to do this in the problems. If you have studied number theory, then it is interesting that what we are presenting as a generalization of the combinatorial inclusion–exclusion principle is, at the same time, related to well-studied topics in number theory.

The definition of the Möbius function privileged the "smaller" elements of a poset. To use it, it was most convenient to start at the bottom and work our way up. We next show that you can just as well start with elements at the top of the Hasse diagram of a poset.

Proposition 11.100. *Let P be a finite poset, and let x and y be elements of P with $x \leq y$. Then*

$$\sum_{d \in [x,y]} \mu_P(d,y) = \begin{cases} 0 & \text{if } x < y \\ 1 & \text{if } x = y. \end{cases}$$

Proof. We are thinking of x and y as two fixed elements of P, with $x \leq y$. We organize the proof as induction on $|[x,y]|$, the number of elements in the interval $[x,y]$. If $|[x,y]| = 1$, then $x = y$, and the result holds. For the inductive step, assume $|[x,y]| > 1$ (and in particular, $x \neq y$). Consider the double sum

$$\alpha = \sum_{d \in [x,y]} \left(\sum_{c \in [d,y]} \mu_P(d,c) \right).$$

We will calculate α in two different ways. First, as given, the inner sum is zero except for the case when $d = y$. So $\alpha = \sum_{c \in [y,y]} \mu_P(y,c) = \mu_P(y,y) = 1$. On the other hand, if we switch the order of summation, we get

$$\alpha = \sum_{d \in [x,y]} \left(\sum_{c \in [d,y]} \mu_P(d,c) \right) = \sum_{c \in [x,y]} \left(\sum_{d \in [x,c]} \mu_P(d,c) \right).$$

Now the inner sum is known by the inductive hypothesis as long as $c < y$ (since in those cases, $|[x,c]| < |[x,y]|$). So, again most of the terms are zero (except when $c = x$), and we have

$$\underbrace{\alpha}_{1} = \underbrace{\mu_P(x,x)}_{1} + \sum_{d \in [x,y]} \mu_P(d,y).$$

We conclude that $\sum_{d \in [x,y]} \mu_P(d,y) = 0$, and the proof is complete.   □

Example 11.101. Going back to Question 11.92, consider the poset $P = \{A, B, C, D, S\}$, whose Hasse diagram is given in Figure 11.27. For $x \in P$, we calculate the values of $\mu_P(x, S)$. (We will record the values on the Hasse diagram.)

We start at the top. We know $\mu_P(S, S) = 1$. Now, working our way down, we will use Proposition 11.100, which says, fixing the second coordinate, that the sum of the values of the Möbius function on every closed interval is equal to zero. Now the interval $[A, S]$ has only two elements, A and S. Hence, we have

$$0 = \sum_{z \in [A,S]} \mu_P(z,S) = \mu_P(A,S) + \underbrace{\mu_P(S,S)}_{1} \implies \mu_P(A,S) = -1.$$

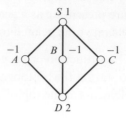

Likewise, $\mu_P(B,S) = \mu_P(C,S) = -1$. We can now tackle the interval $[D,S]$, and get

$$0 = \sum_{z \in [D,S]} \mu_P(z,S) = \mu_P(D,S) + \underbrace{\mu_P(A,S)}_{-1} + \underbrace{\mu_P(B,S)}_{-1} + \underbrace{\mu_P(C,S)}_{-1} + \underbrace{\mu_P(S,S)}_{1}$$

$$\implies \mu_P(D,S) = 2.$$

Curious minds want to know if there is a connection between this 2 and the coefficient of $|D|$ in the answer to Question 11.92.

We are now ready to prove Möbius inversion, the main result of this section. Recall that $\mathbb{C}$ stands for the set of complex numbers.

Theorem 11.102 (Möbius Inversion). *Let P be a finite poset. Let $f, g \colon P \to \mathbb{C}$. Then*

$$g(y) = \sum_{c \leq y} f(c) \text{ for all } y \in P \text{ if and only if } f(y) = \sum_{d \leq y} g(d) \mu_P(d,y) \text{ for all } y \in P.$$

Remark 11.103. Theorem 11.102 says that if you have a mystery function f on the poset, and the only information that you have is various sums of f (of the form $\sum_{c \leq y} f(c)$ for every $y \in P$), then you can recover f using the Möbius function. This should remind you of the fundamental theorem of calculus, where, given a mystery function $f(x)$, if you only know $g(x) = \int_a^x f(t)\, dt$ for a range of values of x, then you can recover f by differentiating g.

Proof. Both directions of the proof amount to using Definition 11.94 of the Möbius function as well as the alternative form given in Proposition 11.100, while carefully switching orders of summation.

For one direction, assume that $g(y) = \sum_{c \leq y} f(c)$ for all $y \in P$. Then, for $y \in P$, we have

$$\sum_{d \leq y} g(d)\mu_P(d,y) = \sum_{d \leq y} \left(\sum_{x \leq d} f(x) \right) \mu_P(d,y) = \sum_{d \leq y} \sum_{x \leq d} f(x)\mu_P(d,y)$$

$$= \sum_{x \leq y} \sum_{d \in [x,y]} f(x)\mu_P(d,y) = \sum_{x \leq y} f(x) \sum_{d \in [x,y]} \mu_P(d,y)$$

$$= f(y).$$

At one point in the above string of equalities, we switched the order of summation. The element y is fixed throughout. Originally, for each d with $d \leq y$, we would choose every x with $x \leq d$. Alternatively, to get the same terms in the summation, we can first choose an x with $x \leq y$, and then, for that x, choose every d with $x \leq d \leq y$. This explains the third step. For the last step, we note that $\sum_{d \in [x, y]} \mu_P(d, y) = 0$ if $x < y$, and so we only get a non-zero term when $x = y$.

We use very similar reasoning for the proof of the other direction. Assume that, for all $y \in P$, we have $f(y) = \sum_{d \leq y} g(d) \mu_P(d, y)$. For a fixed $y \in P$, we calculate

$$\sum_{c \leq y} f(c) = \sum_{c \leq y} \left(\sum_{d \leq c} g(d) \mu_P(d, c) \right) = \sum_{d \leq y} \sum_{c \in [d, y]} g(d) \mu_P(d, c)$$

$$= \sum_{d \leq y} g(d) \sum_{c \in [d, y]} \mu_P(d, c) = g(y),$$

and the proof is complete. $\qquad\qquad\qquad\qquad\qquad\qquad\qquad\qquad\qquad\qquad\qquad\square$

We will now consider a poset of subsets of a finite set, and apply Möbius inversion to one particular function. We will see that we get a new, more refined version of the principle of inclusion–exclusion! Said another way, Möbius inversion is a much more general version of the inclusion–exclusion principle. We first define the function of interest.

Definition 11.104. Let n be a positive integer, and let $P \subseteq 2^{[n]}$. In other words, P is a family of subsets of $[n]$, and $(P, \leq)$, where the relation $\leq$ is inclusion, is a finite poset itself. We say that P is a poset of subsets *closed under intersections* if, whenever $A, B \in P$, so is $A \cap B$.

For a poset of subsets P that is closed under intersections, we define a function $\chi_P \colon P \to \mathbb{Z}$ by

$$\chi_P(A) = |\{i \in A \mid i \notin B \text{ for any } B \in P \text{ with } B \subsetneq A\}| \quad \text{for } A \in P.$$

In other words, for $A \in P$, $\chi_P(A)$ records the number of elements of A that are not already a member of a smaller subset in P.

Remark 11.105. If P is a poset of subsets, and $A \in P$, then there is no reason why *every* subset of A should also be in P. Often in what follows, we are focused on the poset P, and interested only in subsets of A that are in P. To avoid confusion and yet be concise (especially as subscripts of summation signs), we will deliberately make use of the subtle difference between $\subseteq$ and $\leq$ in this context. The set $\{B \mid B \subseteq A\}$ is the family of all subsets of A (regardless of whether A or B are in a poset of subsets). On the other hand, if $(P, \leq)$ is a poset of subsets, then $\{B \mid B \leq A\}$ refers to elements of the poset that are less than or equal to A. Hence, these are $B \in P$ with $B \subseteq A$.

Example 11.106. Going back to Question 11.92, let P be the poset of subsets whose Hasse diagram is given in Figure 11.27. The poset P is a poset of subsets closed under intersections. For this poset, we have

$$\chi_P(D) = |D|$$
$$\chi_P(A) = |A| - |D|$$
$$\chi_P(S) = |S| - |A \cup B \cup C|.$$

Lemma 11.107. *Let $(P, \leq)$ be a poset of subsets closed under intersections, and let χ_P be defined as in Definition 11.104. Then*

(a) *If x is an element of any of the subsets in P, then among the subsets in P that contain x, there is a unique minimal one.*

(b) *Let $A \in P$; then*

$$\sum_{B \leq A} \chi_P(B) = |A|.$$

Proof. (a) For the element x, we want to show that there is a subset A in P with the property not only that $x \in A$ but that if $x \in B \in P$, then $A \subseteq B$. This is the case, because if x is an element of B and an element of C, then x is also an element of $B \cap C$, and we know, since P is closed under intersections, that $B \cap C$ is also in P. In fact, the unique subset that is minimal among all the subsets that contain x is $\bigcap_{x \in A} A$, the intersection of all the subsets in P that contain x.

(b) We want to argue that $\sum_{B \leq A} \chi_P(B)$ counts each element $x \in A$ exactly once, and, as a result, the sum is equal to $|A|$. So, let $x \in A$. For which $B \leq A$ does $\chi_P(B)$ count x? The value of $\chi_P(B)$ is increased by 1 on behalf of x if and only if $x \in B$ but x is not in any of the subsets of B that are elements of P. As a result, every x is counted exactly once by $\chi_P(B)$, where $B \in P$ is the subset that is minimal among the subsets in P that contain x. Since all elements of A are counted exactly once, the sum will be equal to $|A|$. □

We are now ready to use the Möbius function μ_P (Definition 11.94) and Möbius inversion (Theorem 11.102) to find an alternative expression for $\chi_P(A)$ of Definition 11.104.

Proposition 11.108. *Let $(P, \leq)$ be a poset of subsets closed under intersections. Then, for $A \in P$,*

$$\chi_P(A) = \sum_{B \leq A} |B| \, \mu_P(B, A).$$

Proof. We know that P is a finite poset, and that $\chi_P \colon P \to \mathbb{Z}$ is a well-defined function. Define a function $g \colon P \to \mathbb{Z}$ by $g(A) = |A|$.

By Lemma 11.107, we know that $g(A) = \sum_{B \leq A} \chi_P(B)$. Now by Möbius inversion, Theorem 11.102, we have

$$\chi_P(A) = \sum_{B \leq A} g(B) \mu_P(B, A) = \sum_{B \leq A} |B| \, \mu_P(B, A).$$

  □

Proposition 11.108 is a general form of the inclusion–exclusion principle in disguise. We are ready for the unveiling.

Theorem 11.109. *Let S be a finite set, and let $A_1, \ldots, A_n$ be a family of subsets of S. Let $U = A_1 \cup \cdots \cup A_n$, and let P be the poset of subsets that includes U, $A_1, \ldots, A_n$, and all of* *the various intersections of these sets, so that P is a poset of subsets closed under intersections. Then*

$$|A_1 \cup A_2 \cup \cdots \cup A_n| = - \sum_{A < U} |A| \, \mu_P(A, U).$$

Alternatively,

$$|A_1^c \cap A_2^c \cap \cdots \cap A_n^c| = |S| + \sum_{A < U} |A| \, \mu_P(A, U).$$

Proof. The second statement follows from the first, since

$$|A_1^c \cap A_2^c \cap \cdots \cap A_n^c| = |S| - |A_1 \cup A_2 \cup \cdots \cup A_n| \, .$$

To get the first statement, first note that every element of $U = A_1 \cup \cdots \cup A_n$ is an element of one of the A_i's, and so, by the definition of χ_P, we have $\chi_P(U) = 0$. Now apply Proposition 11.108 to $U \in P$ to get

$$0 = \chi_P(U) = \sum_{A \leq U} |A| \, \mu_P(A, U) = |U| \underbrace{\mu_P(U, U)}_{1} + \sum_{A < U} |A| \, \mu_P(A, U). \qquad \square$$

Remark 11.110. Theorem 11.109 can be turned into a new proof of the inclusion–exclusion principle of both Theorem 8.6 and Corollary 8.9.

We are finally ready to go back and fully answer Question 11.92.

Example 11.111. As in Question 11.92, let S be a finite set, and let A, B, C, and D be subsets of S. Assume that

$$D = A \cap B = A \cap C = B \cap C = A \cap B \cap C.$$

Let $P = \{A, B, C, D, S\}$, a poset of subsets closed under intersections. (See Figure 11.27 for the Hasse diagram of P.) In Example 11.101, for $x \in P$, we calculated the values of $\mu_P(x, S)$. Without using the original inclusion–exclusion principle, and by plugging the values of the Möbius function into the second identity of Theorem 11.109, we get

$$|A_1^c \cap A_2^c \cap \cdots \cap A_n^c| = |S| + \sum_{A < U} |A| \, \mu_P(A, U)$$

$$= |S| + \mu_P(A, S) \, |A| + \mu_P(B, S) \, |B| + \mu_P(C, S) \, |C| + \mu_P(D, S) \, |D|$$

$$= |S| - |A| - |B| - |C| + 2 \, |D| \, .$$

While at this point it may not be surprising that the Möbius function somehow precisely predicted the coefficients in the above identity, it is pretty amazing that all the seemingly endless formula crunching – not to mention the various switching of summation signs – was so productive.

Problems

P 11.6.1. Let $P = 2^{[3]}$. Find $\mu_P(\{2\}, A)$ for $A = \emptyset$, $A = \{2,3\}$, $A = \{1,3\}$, and $A = \{1,2,3\}$.

P 11.6.2. Let $n = 5$. I am interested in the following subsets of $[5]$: $\{1,2,3,4\}$, $\{1,3,4,5\}$, $\{1,4\}$, and $\{1,3\}$.

    (a) Is this collection a poset of subsets closed under intersections? If not, add to the collection as few subsets as possible, so that the resulting poset P is closed under intersections.

    (b) Draw the Hasse diagram of P.

    (c) For every subset $A \in P$, calculate $\chi_P(A)$.

    (d) Verify Lemma 11.107b for $A = \{1,3,4,5\}$.

P 11.6.3. Let $A = \{1,3,4,5\}$. For every element x of the poset P of Problem P 11.6.2, calculate $\mu_P(x, A)$. Then, verify Proposition 11.108 for A.

P 11.6.4. Let $A = \{1,2,3,4,5,7,8\}$, $B = \{1,2,3,4,6\}$, and $C = \{1,3,4,5,6\}$.

    (a) Find a poset of subsets P with as few elements as possible that includes A, B, and C, and is closed under intersections.

    (b) Draw the Hasse diagram of P.

    (c) Find $\mu_P(x, A)$ for all $x \in P$.

    (d) Verify Proposition 11.108 for A.

P 11.6.5. The sets A, B, C, D, E, and F are all subsets of a set S, and the poset $P = \{A, B, C, D, E, F\}$ is a poset of subsets closed under intersections. The Hasse diagram of P is given in Figure 11.29. Find an expression for $|A^c \cap B^c \cap C^c \cap D^c|$ in terms of $|S|, |A|, \ldots, |F|$.

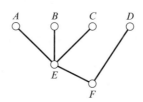

Figure 11.29 The Hasse diagram of a poset of subsets closed under intersections.

P 11.6.6. Recall that a poset P is locally finite if, for all $x, y \in P$, the interval $[x, y]$ has a finite number of elements. Let $\mathbb{R}$ denote the real numbers, and let P be the set of subspaces of $\mathbb{R}^3$. Is P ordered by inclusion a poset? Is it a finite poset? Is it locally finite?

P 11.6.7. **Discrete Version of the Fundamental Theorem of Calculus.** Let n be a positive integer, and let $P = [n]$ ordered by the ordinary $\leq$. (In other words, if $i, j \in [n]$, then we have $i \leq j$ if and only if the number i is actually less than or equal to the number j.)

    (a) For $i \in [n]$, what is $\mu_P(i, n)$?

(b) Let $f, g \in P \to \mathbb{C}$ be given. Apply Möbius inversion, Theorem 11.102, to get a statement of the form

$$g(m) = \sum_{i \leq m} f(i), \text{ for all } m \in [n] \text{ if and only if } f(1) = g(1) \text{ and } f(m) = ??.$$

Use your answer to part (a) to simplify this answer.

(c) The fundamental theorem of calculus says, in part, that if $g(x) = \int_0^x f(t)\,dt$, then $g'(x) = f(x)$. Explain why your result in part (b) could be called a "discrete version" of this theorem.

(d) Re-prove part (b) directly and with no recourse to Möbius inversion.

P 11.6.8. **Binomial Inversion Revisited.** In Problem P 5.4.12, for a sequence of integers $a_0, a_1, \ldots, a_n$, we defined another sequence of integers $b_0, b_1, \ldots, b_n$ by

$$b_k = \binom{k}{0} a_0 + \binom{k}{1} a_1 + \cdots + \binom{k}{k} a_k, \text{ for } 0 \leq k \leq n.$$

We then proceeded to use the inverse of the matrix of binomial coefficients to find the sequence $\{a_k\}_{k=0}^n$ in terms of $\{b_k\}_{k=0}^n$. Re-do this problem using Möbius inversion. You may find the following steps helpful:

STEP 1: Let $P = 2^{[n]}$. Define $f, g: 2^{[n]} \to \mathbb{Z}$ by $f(A) = a_{|A|}$ and $g(A) = b_{|A|}$. In other words (and for a specific example), every subset of $[n]$ of size 47 is mapped by f to a_{47}, while each subset of size 5 is mapped by g to b_5. Show that $g(A) = \sum_{B \leq A} f(B)$.

STEP 2: Use Möbius inversion to find f in terms of g.

STEP 3: Use Proposition 11.97 to finish the problem.

11.6.1 The Möbius Function for Divisor Lattices

Let n be a positive integer, and let $P = \mathcal{D}_n$ be the poset of the divisors of n ordered by divisibility (Definition 11.16). The Möbius function of this poset is used extensively in number theory.

Definition 11.112. Let $\mathbb{Z}^{>0}$ and $\mathbb{Z}$ denote the positive integers and the integers, respectively. Define a function $\mu: \mathbb{Z}^{>0} \to \mathbb{Z}$ by letting, for $n \in \mathbb{Z}^{>0}$, $\mu(n) = \mu_P(1, n)$ where $P = \mathcal{D}_n$ is the poset of divisors of n ordered by divisibility. The function μ is a function of one variable, and we call it the *Möbius function of number theory*.

In a sequence of problems culminating in Problem P 11.6.15, you are guided to prove:

Proposition 11.113. *Let $\mu: \mathbb{Z}^{>0} \to \mathbb{Z}$ be the Möbius function of number theory; then*

(a) *If $P = \mathcal{D}_n$ is the poset of divisors of n ordered by divisibility, and if m divides n, then*
$$\mu_P(m, n) = \mu(n/m).$$

(b) *If n is a positive integer, then*

$$\mu(n) = \begin{cases} 1 & \text{if } n = 1 \\ 0 & \text{if, for some prime } p, p^2 \text{ divides } n \\ (-1)^r & \text{if } n = p_1 \dots p_r \text{ is a product of } r \text{ distinct primes.} \end{cases}$$

As an application, we go a bit off track, and use Proposition 11.113 to give a formula for the number of integers between 1 and n that are relatively prime to n. Recall that two integers m and n are relatively prime if their greatest common divisor is 1.

Definition 11.114. For a positive integer n, let $\phi(n)$ denote the number of integers $m \in [n]$ that are relatively prime to n. The function $\phi(n)$ is called *Euler's totient function*, or *Euler's phi function*.

A sequence of problems, ending in Problem P 11.6.19, guide you through a proof of:

Corollary 11.115. *Let n be a positive integer, let μ be the Möbius function of number theory, and let $\phi(n)$ be the Euler phi function. Then*

$$\phi(n) = \sum_{d|n} \mu(d)\frac{n}{d} = n \prod_{\substack{p|n \\ p \text{ prime}}} \left(1 - \frac{1}{p}\right).$$

As a step in this proof, in Problem P 11.6.18 you prove the independently interesting fact that $\sum_{d|n} \phi(d) = n$. This particular proof of Corollary 11.115 is interesting since it not only gives an example of possible uses of Möbius functions in number theory, but also makes explicit the role of an inclusion–exclusion argument in the proof. However, it is also possible to get the result more directly. This is often done by first proving that $\phi(mn) = \phi(m)\phi(n)$ if m and n are relatively prime, and then using the fact, proved using a weak form of the inclusion–exclusion principle, that, for a prime p and a positive integer a, we have $\phi(p^a) = p^a - p^{a-1}$.[18]

Problems (continued)

P 11.6.9. Let $P = \mathcal{D}_{24}$ be the poset of divisors of 24 ordered by divisibility. See Figure 11.4 for the Hasse diagram of $\mathcal{D}_{24}$.
   (a) For each $d \in \mathcal{D}_{24}$, find $\mu_P(d, 24)$.
   (b) How is $\mu_P(1, 12)$ related to $\mu_P(2, 24)$? Can you answer this question without calculating the two quantities, but, instead, using the result of Problem P 11.1.16?
   (c) Define $f: P \to \mathbb{Z}$ by $f(x) = 1$ for all $x \in P$. Define $g: P \to \mathbb{Z}$ by $g(x) = \sum_{z \le x} f(z)$. What is g measuring? Give an alternative definition of g.

---

[18] The author's favorite, but very different, method for proving statements such as $\phi(mn) = \phi(m)\phi(n)$, if m and n are relatively prime, or $\sum_{d|n} \phi(d) = n$, is to use facts about the number of generators of finite cyclic groups from elementary group theory. See, for example, my *Algebra in Action* (Shahriari 2017), Problems 2.5.10 and 2.6.23.

(d) What does Möbius inversion (Theorem 11.102) say for the poset P and functions f and g? Using this, write down a specific mathematical statement about $n = 24$.

P 11.6.10. Let n be a positive integer, and let $f: \mathbb{Z}^{>0} \to \mathbb{C}$ be a function on positive integers.

(a) Show that $\sum_{d|n} f(d) = \sum_{d|n} f(n/d)$.

(b) Show that $\sum_{d|n} df(n/d) = \sum_{d|n} \frac{n}{d} f(d)$.

P 11.6.11. Let n be a positive integer, and let $P = \mathcal{D}_n$ be the poset of divisors of n ordered by divisibility. Let m be a divisor of n. Show that $\mu_P(m, n) = \mu_P(1, n/m) = \mu(n/m)$, where μ is the Möbius function of number theory (Definition 11.112).

P 11.6.12. Let p be a positive prime integer, and let $n = p^\alpha$ for some fixed positive integer α. If μ is the Möbius function of number theory, then what is $\mu(n)$?

P 11.6.13. Let n be a positive integer. Assume that n is a product of r distinct primes.

(a) Prove that the poset $\mathcal{D}_n$ of divisors of n is isomorphic to $\mathbf{2}^{[r]}$.

(b) Prove that $\mu(n) = (-1)^r$, where μ is the Möbius function of number theory.

P 11.6.14. Let $n = p^\alpha m$, where p is a prime, and $\alpha \geq 2$ is an integer. Prove that $\mu(n) = 0$. You may find the following steps helpful:

STEP 1: Use induction on m. Resolve the base case either directly or by appealing to Problem P 11.6.12.

STEP 2: For the inductive step, let $m > 1$. Partition the divisors of n into two sets: the divisors that are not divisible by p^α, and those that are. Call the former A and the latter B. Argue that, by the inductive hypothesis, if $k \in B$ and $k \neq n$, then $\mu(k) = 0$.

STEP 3: Let $s = p^{\alpha-1} m$, and let $Q = \mathcal{D}_s$, the poset of divisors of s ordered by divisibility. Argue that $A = Q$, and, by definition of the Möbius function, $\sum_{a \in A} \mu(a) = 0$.

STEP 4: Start by explaining why $\sum_{a \in A} \mu(a) + \sum_{b \in B} \mu(b) = 0$, and then, using previous steps, explain why this leads to $\mu(n) = 0$.

P 11.6.15. By combining Problems P 11.6.11–P 11.6.14, prove Proposition 11.113.

P 11.6.16. Let $n = p_1^{\alpha_1} p_2^{\alpha_2} \cdots p_k^{\alpha_k}$ be the factorization of the positive integer n into a product of primes, and let μ be the Möbius function from number theory. By expanding both sides, show that

$$\sum_{d|n} \frac{\mu(d)}{d} = \prod_{i=1}^{k} \left(1 - \frac{1}{p_i}\right).$$

P 11.6.17. **Möbius Inversion from Number Theory.** As usual, let $\mathbb{Z}$, $\mathbb{Z}^{>0}$, and $\mathbb{C}$ denote the integers, the positive integers, and the complex numbers, respectively, and let $\mu: \mathbb{Z}^{>0} \to \mathbb{Z}$ be the Möbius function of number theory (Definition 11.112). For functions $f, F: \mathbb{Z}^{>0} \to \mathbb{C}$, prove that

$$F(n) = \sum_{d|n} f(d) \text{ for all } n \in \mathbb{Z}^{>0} \text{ if and only if } f(n) = \sum_{d|n} F(d)\mu(n/d) \text{ for all } n \subset \mathbb{Z}^{>0}.$$

P 11.6.18. Let n be a fixed positive integer, and let $\phi(n)$ denote the Euler ϕ function (Definition 11.114).

(a) Find $\phi(7)$, $\phi(8)$, and $\phi(24)$.

(b) Find $\sum_{d|8} \phi(d)$ and $\sum_{d|8} \phi(8/d)$.

(c) Prove that $\sum_{d|n} \phi(d) = n$. You may find the following steps helpful:

STEP 1: Do Problem P 11.6.10(a), or directly argue that $\sum_{d|n} \phi(n) = \sum_{d|n} \phi(n/d)$.

STEP 2: Let d be a positive integer that divides n. Let B_d be the set of integers m, with $1 \le m \le n/d$, that are relatively prime to n/d. Let C_d be the set of the integers $m \in [n]$ such that the greatest common divisor of m and n is d. Define a map $g :: C_d \to B_d$ by $g(x) = x/d$. Show that g is a well-defined bijection. (If you are comfortable with elementary number theory, give a proof. Otherwise, give a reasonable argument.)

STEP 3: Consider the collection of sets $\{C_d \mid d \text{ divides } n\}$. Show that this collection partitions $[n]$, and as a result $\sum_{d|n} |C_d| = n$.

STEP 4: Combine Steps 1, 2, and 3 to prove that $\sum_{d|n} \phi(d) = n$.

P 11.6.19. (a) Let $f, F \colon \mathbb{Z}^{>0} \to \mathbb{Z}$ be defined by $f(n) = \phi(n)$, and $F(n) = \sum_{d|n} \phi(d)$. Use the result of Problem P 11.6.18 and the Möbius inversion from number theory of Problem P 11.6.17 (if you prefer, you can apply the original Möbius inversion of Theorem 11.102) to show that

$$\phi(n) = \sum_{d|n} d\mu(n/d).$$

(b) Use Problem P 11.6.10(b) to show that $\phi(n) = \sum_{d|n} \frac{n}{d}\mu(d)$.

(c) Use Problem P 11.6.16 to prove Corollary 11.115.

11.7 Open Problems and Conjectures

Ordered sets is an area of active research, with many interesting open problems and conjectures. The lay observer may be surprised to hear that there are lots of open problems even about the combinatorics of finite sets, and yet this is so. We will mention just a couple of conjectures here.

Conjecture 11.116 (Frankl's Conjecture, 1979). *Let n be a positive integer, and let A be a family of subsets of $[n]$ closed under intersections (Definition 11.104), with $|A| \ge 2$. Then there exists $i \in [n]$ such that i belongs to at most half of the subsets in the family.*

Example 11.117. The family $A = \{\emptyset, \{1\}, \{3\}, \{1,3\}, \{1,2,3\}\}$ of subsets of $[3]$ is closed under intersections and has five elements. From among the elements of $[3]$, 1 appears in $3/5$ of the sets, 2 appears in $1/5$ of the sets, and 3 appears in $3/5$ of the sets. So 2 does appear in less than half of the elements of A, as the conjecture claims. What if we delete $\{1,2,3\}$ so as not to use the problematic 2? But then 1 and 3 each appear in two out of the four remaining elements, and so both satisfy the conjecture.

Special cases of the conjecture have been proved. For example, the conjecture is true if $n \leq 11$, or if $|\mathcal{A}| < 47$. However, as of this writing, the general conjecture remains open. See Bruhn and Schaudt (2015) for different formulations of this conjecture and a survey of known results.

The Erdős–Ko–Rado theorem, Theorem 11.78, said that the size of the largest k-uniform intersecting family in $2^{[n]}$ is the same as the size of the largest k-uniform star in $2^{[n]}$ (see Definition 11.68 for the relevant definitions). Now a family of subsets $\mathcal{A} \subseteq 2^{[n]}$ is called a *downset* (or an *ideal* or a *simplicial complex*) if whenever $B \subseteq A$ and $A \in \mathcal{A}$, then B is also a member of $\mathcal{A}$.

Conjecture 11.118 (Chvátal's Conjecture, 1974 (Chvátal 1974)). *Let n be a positive integer, and let $\mathcal{A}$ be a family of subsets of $[n]$ that is a downset. Then the size of the largest intersecting family in $\mathcal{A}$ is the same as the size of the largest star in $\mathcal{A}$.*

Again, special cases of Chvátal's conjecture have been proved. For example, Daniel Kleitman and Thomas Magnanti (1974) proved that the conjecture is true if $\mathcal{A}$ itself is contained in two stars. However, the main conjecture remains open.

Many problems in graph theory can be and have been extended to hypergraphs. Often such problems become harder. For example, an active area of research is to choose your favorite small simple graph F with k vertices, and let n be an integer with $n \geq k$. Then the complete graph K_n certainly will have subgraphs isomorphic to F. But what is the largest number e of edges in a simple graph with n vertices that does *not* have a subgraph isomorphic to F? In other words, you want a graph on n vertices with as many edges as possible and yet no copy of F. Put another way, you want to find e such that *any* simple graph with n vertices and $e + 1$ or more edges will by necessity have a subgraph isomorphic to F. Problem P 10.3.11 (originally proved by Willem Mantel (1907)) was an example of this – probably the earliest theorem of this kind – where you showed that any graph on n vertices and more than $n^2/4$ edges must contain a triangle. Here is a similar, deceptively simple-sounding hypergraph problem that remains open despite much effort.

Problem 11.119 (Turán 1941). *Let $n \geq 4$ be a fixed integer. Let $\mathcal{F}$ be a family of subsets of size 3 of $[n]$. Assume that there are no subsets A of size 4 of $[n]$ with the property that all of the subsets of size 3 of A are in $\mathcal{F}$. What is the size of the largest such family $\mathcal{F}$? In other words, what is the size of the largest family of 3-subsets of $[n]$ that does not contain all the 3-subsets of some subset of size 4 of $[n]$?*

Problem 11.119 and the following related conjecture are both due to Paul Turán (1910–1976).

Conjecture 11.120 (Turán 1941). *Let $\mathrm{ex}_3(n, 4)$ denote the answer to Problem 11.119; then*

$$\lim_{n \to \infty} \frac{\mathrm{ex}_3(n, 4)}{\binom{n}{3}} = \frac{5}{9}.$$

We showed in Theorem 11.63 that the Boolean lattices have a symmetric chain decomposition. Symmetric chain decompositions are not unique, but the number of chains and their sizes are. Let $(c_1, \ldots, c_w)$ be the list of chain sizes in a symmetric chain decomposition of $2^{[n]}$. So, $w = \binom{n}{\lfloor n/2 \rfloor}$, and, for example, $c_1 = n + 1$, and $c_2 = \cdots = c_n = n - 1$, and so on (see Problem P 11.4.11). The sequence $(c_1, \ldots, c_w)$ is a partition of the integer 2^n, since $c_1 \geq c_2 \geq \cdots \geq c_w$ and $c_1 + \cdots + c_w = 2^n$. We call this specific integer partition the *SCD partition* of 2^n. We may ask which other partitions of 2^n can be realized by some chain partition of $2^{[n]}$. Do note that in a general chain partition of $2^{[n]}$, the chains are not required to be skipless (Definition 11.26). A beautiful conjecture of Jerrold R. Griggs[19] gives the answer.

Conjecture 11.121 (Griggs 1988). *Let σ be the SCD partition of 2^n, and let μ be an arbitrary integer partition of 2^n. Then there is a chain partition of $2^{[n]}$ with chain sizes given by μ if and only if $\mu \leq \sigma$ in the dominance order (Definition 11.19).*

A special case of this conjecture – that inspired Griggs to generalize it – is a question of Zoltán Füredi from 1985. Recall that $w = \binom{n}{\lfloor n/2 \rfloor}$ is the width of the Boolean lattice $2^{[n]}$ (Theorem 11.67), and that the minimum number of chains needed to partition a poset is equal to its width (Theorem 11.42).

Question 11.122 (Füredi 1985). Let $w = \binom{n}{\lfloor n/2 \rfloor}$. Can you partition $2^{[n]}$ into w chains in such a way that there are only two chain sizes?

Füredi is asking whether we can partition $2^{[n]}$ into the minimum number of chains possible with the chain sizes being as uniform as possible. Let $\alpha = 2^n / \binom{n}{\lfloor n/2 \rfloor} \approx \sqrt{\pi/2}\sqrt{n}$. Then we want a partition of $2^{[n]}$ into $\binom{n}{\lfloor n/2 \rfloor}$ chains such that all chain sizes are either $\lfloor \alpha \rfloor$ or $\lceil \alpha \rceil$. While Füredi's question and Griggs's more general conjecture remain open, some progress has been made. The author and his collaborators (Hsu et al. 2002, 2003) proved in 2003 that, for $n > 16$, the Boolean lattice $2^{[n]}$ can be partitioned into the minimum number of chains with all chain sizes between $\frac{1}{2}\sqrt{n}$ and $\frac{3}{2}\sqrt{n \log(n)}$. The bounds were improved in 2015 by István Tomon (2015) to $.8\sqrt{n}$ and $26\sqrt{n}$. Much more recently, Benny Sudakov, István Tomon, and Adam Wagner (Sudakov, Tomon, and Wagner 2019) have given a positive answer to Question 11.122 asymptotically. This roughly means that they constructed a chain decomposition with w chains and the property that, for very large n, except for a negligible set of chains, the sizes of the rest are off from α only by a negligible amount.

We close with a famous (maybe the most famous) conjecture from number theory. Let $\mu \colon \mathbb{Z}^{>0} \to \mathbb{Z}$ be the Möbius function of number theory (Definition 11.112). For $n > 1$, define $M(n) = \sum_{k=1}^{n} \mu(k)$. It follows from Proposition 11.113 that $M(n)$ is the number of square-free integers less than or equal to n with an even number of prime factors, minus the number of square-free integers less than or equal to n with an odd number of prime factors. Franz Mertenz (1840–1927) in 1897 conjectured that $|M(n)| \leq \sqrt{n}$ for all integers $n > 1$. By the 1960s, this was verified up to $n = 10$ billion. However, it was proved false in 1985,

---

[19] Jerry Griggs graduated from Pomona College in 1973, and is an Emeritus Professor at University of Southern Carolina.

by Andrew Odlyzko and Herman te Riele (Odlyzko and te Riele 1985). However, one of the major yet-unresolved conjectures in mathematics is the Riemann hypothesis, originally proposed by Bernhard Riemann (1826–1866) in 1859. This conjecture can be rephrased (see Wilf 1987) as

Conjecture 11.123 (Riemann Hypothesis, 1859). *Let $\epsilon > 0$; then there exists a positive integer N such that, for all $n > N$, we have*

$$|M(n)| < n^{\frac{1}{2}+\epsilon}.$$

APPENDIX A

Short Answers for Warm-Up Problems

1.1. For $k = 0, 1, \ldots, 18$, the briefcase numbered $1 + 3k$ has no cash.

1.6. In addition to briefcases numbered 1 and 8, every briefcase numbered ≥ 13 will have cash in it.

1.8. $f(1) = 1$ and $f(2) = 3$. If you buy Mustard on the first day, there are $f(n-1)$ ways to spend the rest of the money, while if you buy Mint or Marjoram on the first day, there are $f(n-2)$ ways to spend the rest of the money. Hence, $f(n) = f(n-1) + 2f(n-2)$.

1.12. Plug $\alpha n + \beta$ in the recurrence relation and solve for α and β to get $\alpha = \frac{1}{2}$ and $\beta = \frac{7}{4}$.

1.21. Yes. $a_n = 27a_{n-3} + 26$, and $a_n = 81a_{n-4} + 80$.

2.1. 45.

2.12. Create a graph by putting a vertex for each individual and connect two vertices if the two are friends.

2.31. With planning, you should be able to get zero monochromatic triangles.

2.53. No. Yes.

3.1. 74.

3.10. 1/2.

3.17. Let x_1 be the number of jars you give to Ananya, x_2 the number of jars to Anaya, and so on. We want the number of non-negative integer solutions to $x_1 + x_2 + x_3 + x_4 = 5$.

3.22. The domain of f is $\mathbb{Z}$ while the domain of g is $\mathbb{Z}^{>0}$. f is not 1-1, while g is.

4.1. $4 \times 3 \times 2 \times 1 = 24$.

4.18. 1/6.

4.24. 81, 12.

4.36. 10.

5.1. $\binom{n}{k}\binom{k}{b} = \binom{n}{b}\binom{n-b}{k-b}$.

5.3. All are alternatives for "Pascal's triangle."

5.5. For large n, an appropriately scaled normal distribution approximates $\left\{\binom{n}{k}\right\}_{k=0}^{n}$.

5.12. $(1, 1)$; $(4, 1)$.

5.16. 1, 1, 2, 5, 14.

5.22. $(10.1)^3 = 10^3 + 3(10^2)(.1) + 3(10)(.1)^2 + (.1)^3 = 1030.301$.

5.27. $\binom{10}{3\,3\,2\,2}$.

6.1. $2^7 - 2$.

6.24. 130.

6.34. $B = A^{-1}(D - C)$.

7.1. 2.

8.1. \$47.

8.16. $\left(\binom{7}{12}\right)$.

8.20. 9.

8.29. 4.

9.1. $g(x) = 1 - x^2$ is the Taylor polynomial of degree 3 at $x = 0$ for $f(x)$.

9.24. What is the number of positive integer solutions to $x_1 + x_2 + x_3 + x_4 = 47$ with x_2 even and x_1, x_3, and x_4 odd?

9.33. $5(n-1)a_{n-2}$.

9.38. $3\binom{7}{3\,2\,2} + 3\binom{7}{3\,3\,1}$.

10.5. No. Yes.

10.11. 7; 6; No.

10.32. $\{H, K, R\}$ and $\{J, L, S\}$.

10.40. You need to take your hand off once.

10.52. 8; Yes.

10.61. Yes.

10.96. 3.

11.1. Yes. For $x \in X$, define $f \colon X \to Y$ by $f(x) = 2^a 3^b 5^c$ where a, b, and c are either 1 or 0 depending on whether, respectively, 1, 2, or 3 are members of x or not.

11.25. 4; 7.

11.43. Batu (A); Bayarmaa (B); Bolarmaa (G); Qacha (S); Qadan (W); Xanadu (L).

11.58. 6; yes.

11.81. Every briefcase in rows 23, 33, and 43 whose tag is either $(7, 7)$ or (a, b) with $a + b > 14$ has cash.

11.91. The inner sum ranges over all $x \in P$ such that $y \le x \le a$.

APPENDIX B

Hints for Selected Problems

P 1.1.1. 2801 is 1 plus the sum of the first four terms.

P 1.1.5. Factor the resulting integers appropriately.

P 1.1.9. In the resulting fractions, make the numerators 1, 2, 3, . . .

P 1.1.11. At some point, by considering different cases when k is odd or even, prove that $k + 1 - \lceil \frac{k}{2} \rceil = \lceil \frac{k+1}{2} \rceil$.

P 1.1.23. Recall the trigonometry identities $\sin(\alpha + \beta) = \sin(\alpha)\cos(\beta) + \cos(\alpha)\sin(\beta)$ and $\cos(\alpha + \beta) = \cos(\alpha)\cos(\beta) - \sin(\alpha)\sin(\beta)$.

P 1.1.24. Do the cases $k = 1$ and $k = 2$ first.

P 1.2.10(b). The previous part may be relevant.

P 1.2.11. Guess the limit L and then subtract L from each term of the sequence.

P 1.3.2. Consider the data for $\frac{f(n)+5}{2}$.

P 1.3.3. Start covering the array from the left. Create two cases based on how you start.

P 1.3.9(b). How close is each term to twice the previous one? Use induction to prove your guess.

P 1.3.9(c). Try multiplying the sequence $f(1), f(2), \ldots, f(n), \ldots$ by various constants.

P 1.3.11. Let $f(n)$ be the number of integers whose digits are from among 1, 2, and 3, and the sum of the digits is n.

P 1.3.14(c). $3 \pm 2\sqrt{2} = (1 \pm \sqrt{2})^2$.

P 1.3.15(b). Consider different cases based on the first term of the sequence.

P 1.3.17. Consider smaller numbers of people and smaller amounts of money. Make a table and see if you can see any patterns.

P 1.4.1. Mimic Example 1.18.

P 1.4.5. Is $x = 1$ a root of the characteristic polynomial?

P 1.4.7. Do Problem P 1.4.6 first. What is $(x - 3)^3$?

P 1.4.8. Don't be deterred by complex numbers. Charge on.

P 1.4.12. Problem P 1.4.6 may be relevant.

P 2.1.9. The technique used in Example 2.6 may be relevant.

P 2.1.12. If you divide an integer by 100, what are the possible remainders?

P 2.1.14. List the numbers of candy bars that you gave to your friends, in increasing order.

P 2.1.16. Create 135 pigeonholes appropriately.

P 2.1.20. Find the remainders of the integers when divided by 100. One case is if two of these are the same, another is when they are all different. For the latter, put the remainders in 51 suitably chosen pigeonholes.

P 2.1.23. What are the possible remainders of $a_1, \ldots, a_m$ when divided by m? At some point, consider $a_j - a_i$.

P 2.1.25. Theorem 2.11 may be relevant.

P 2.1.28. Mimic the proof of Theorem 2.11.

P 2.1.29. First do Problem P 2.1.28.

P 2.1.31. For $1 \leq i \leq 10^{10} + 1$, let $s_i = ir - \lfloor ir \rfloor$ be the fractional part of ir. Find two of these that are close to each other.

P 2.2.2. Example 2.5 may be relevant.

P 2.2.9. If you got stuck, Google "Fano." After reading about the Italian town by that name, modify your search by adding the word "Math."

P 2.2.10. Consult with the advisors of the Queen of Problem P 2.2.9.

P 2.2.11. There is a solution with each team going to four meets and each meet having four teams present. After hitting your head against the wall for some time, Google "the projective plane over $\mathbb{F}_3$," and look under images.

P 2.3.8. Use a blue pen to draw an octagon and all of its four diagonals. Using a red pen, complete the K_8.

P 2.3.11. Recall that $r(3, 5) = 14$, and adapt the proof of Theorem 2.52.

P 2.3.12. Reread the proof of Theorem 2.52. Problem P 2.3.2 may also be relevant.

P 2.3.13. If you put 16 pigeons in 3 pigeonholes, then one of the holes will have at least 6 pigeons.

P 2.3.14. Problem P 2.3.13 is relevant, do it first.

P 2.3.15(g). First do Problem P 2.3.13.

P 2.3.19. Start with a blue cycle of length 17 and use it to relabel the vertices.

P 2.4.4(b). The previous part and Problem P 2.4.3 may be relevant.

P 2.4.9. Model the situation with a K_6. Color all the edges that are linked through the same facility with the same color. You are trying to avoid what kind of monochromatic graph? Problem P 2.4.8 may be relevant.

P 3.2.15. Focus on the first toss. If it is a tail, in how many ways can you complete the sequence. What if it is a head?

P 3.2.17. Use the recurrence relation of Problem P 3.2.16.

P 3.2.19(a). Problem P 3.2.18 may be relevant.

P 3.3.3. Place the fours first; the spaces before, between, and after the fours are the boxes.

P 3.4.8. If a committee doesn't include you, join it. If a committee does include you, leave it.

P 3.4.9. Do cases n odd or even separately. Problems P 3.4.3 and P 3.4.8 may be relevant.

MP1.6. Each edge is in how many triangles?

MP2.3. Right before you arrive at the box at (n, k), where could you have been? In how many ways could you get to these spots?

MP2.5. Generalize what you did in the last problem.

MP2.10. Are the sums of the squares also in the table?

MP3.7. Use the definition, or use induction and MP3.5.

P 4.1.17(a). Do Problem P 4.1.16 first.

P 4.2.23(b). Let $p(k)$ be the probability that 2 out of 20 fish have been marked. Note that this probability does depend on k. Maximizing $p(k)$ using calculus would be difficult, especially since we need to find an integer value for k. Thus instead find $\frac{p(k)}{p(k-1)}$. Find the values of k for which this ratio is greater than 1.

P 4.2.26. As a warm-up, do Problem P 4.2.25.

P 4.3.13. As a warm-up, do Problem P 4.3.12.

P 4.3.19. Do Problem P 4.3.18 first.

P 4.4.11. Define a slack variable x_8 so that $x_1 + \cdots + x_7 + x_8 = 47$.

P 4.4.12. As a warm-up do Problem P 4.4.11.

P 4.4.13. Consider $4x_1 + 4x_2 = 4k$ and $x_3 + x_4 = 24 - 4k$ separately.

P 4.5.11(c). Do Problem P 4.4.11 or P 4.4.12 first.

P 5.1.6. For a fixed k, use induction on n.

P 5.1.12. Use induction.

P 5.1.17. Try a combinatorial proof similar to the proof of Theorem 5.2(d). Consider a set of n green and 3 red balls. How many subsets of size k does it have?

P 5.1.18. For inspiration, go through the proof of Theorem 5.2(d) and do Problem P 5.1.17. Replace $\binom{m}{i}$ with $\binom{m}{?}$.

P 5.1.19. As a warm-up do Problem P 5.1.18.

P 5.1.21. First do Problem $P\ 5.1.19$.

P 5.1.25. What if the O comes after all the A's and all the B's?

P 5.1.26. Each term in the sum is a product of three binomial coefficients. Theorem 5.2(d) could be relevant.

P 5.1.27. Problem $P\ 5.1.7$ may be relevant.

P 5.1.31. Theorem 5.13 and Problem $P\ 5.1.29$ may be relevant.

P 5.2.7. An antiderivative for $(1+x)^n$ may be useful.

P 5.2.8. Proposition 5.26(b) may be relevant. Alternatively, see Problem P 3.4.9.

P 5.2.9. Proposition 5.26(c) and Problem $P\ 5.2.5$ may be relevant.

P 5.2.10. Do Problem P 5.2.8 first.

P 5.2.13. From among n people, choose a non-empty committee of unknown size and designate a chair and a vice-chair, allowing for the possibility that the same person holds both positions.

P 5.2.14. As a warm-up, do Problem $P\ 5.2.13$ first.

P 5.4.5(a). Use Rolle's theorem from calculus, and worry about roots with multiplicity separately.

P 5.4.6(b). Problem P 5.4.5 may be relevant.

P 5.4.7. Problems P 5.4.5 and P 5.4.6 may be relevant.

P 5.4.8(c). Problem P 5.1.18 may be relevant.

P 5.4.9. Make an appropriate substitution for x in Corollary 5.25.

P 5.4.10. Problem P 5.4.9 may be relevant.

P 5.4.11. Do Problems P 5.4.9 and P 5.4.10 first.

P 5.4.12. Problem $P\ 5.4.11$ may be relevant.

P 5.4.13(c). Problem $P\ 5.4.11$ may be relevant.

P 5.4.14. Use the "hockey-stick identity" of Problem P 5.1.6 twice.

P 5.4.16. Do Problem P 5.4.13 as a warm-up.

P 5.4.21(a). As a warm-up do Problem P 5.4.20.

P 5.4.22. Subtract the number of NE paths that are not Dyck paths from the total. Problem P 5.4.21 is relevant.

P 5.4.27. Can you reduce the problem to counting certain Dyck paths?

P 5.4.28. Problems P 5.4.22 and $P\ 5.4.27$ may be relevant.

P 5.4.31. Use induction on $a + b$, and focus on the last step.

P 5.4.32. For each part, the previous part is relevant.

P 5.4.34. Do Problem P 5.4.33 first.

P 5.4.39. Problem P 5.4.38 may be relevant.

P 5.4.43. Problems P 5.4.42 and P 5.1.6 may be relevant.

P 5.4.44. Problem $P\ 5.4.43$ may be relevant.

P 5.4.45. Problems P 5.4.42, $P\ 5.4.43$, and P 5.1.6 may be relevant.

P 6.1.16. Corollary 6.22 and Problem P 6.1.14 may be relevant.

P 6.1.17. Theorem 6.7 and Theorem 6.19 may be relevant.

P 6.1.23. Problem P 6.1.22 may be relevant.

P 6.1.24. Give a balls and boxes interpretation of E_n. Focus on one of the boxes.

P 6.1.25. Theorem 6.19 may be relevant.

P 6.1.26. Mimic the proof of Proposition 6.14.

P 6.1.27. You may get inspired by Problem P 6.1.25. Problem $P\ 6.1.19$ may also be relevant.

P 6.1.33. First do Problem P 6.1.30. You may want to work with one more matrix in addition to the matrices P_n, A_n, and C_n of Problem P 5.4.16. Problem $P\ 5.4.17$ may also be relevant.

P 6.2.5. Maybe try Problem P 6.2.4 first.

P 6.2.7. What are the possibilities for cycle sizes?

P 6.2.8. What are the possibilities for cycle sizes?

P 6.2.9. When you multiply the terms in $x^{(n)}$, how do you get an x^{n-2} term?

P 6.2.11. Theorem 6.31a may be relevant.

P 6.2.12. Theorem 6.31b may be relevant.

P 6.2.13. Theorem 6.31b may be relevant.

P 6.2.18. Problem P 6.2.10 may be relevant.

P 6.2.19. Use Theorem 6.33.

P 6.2.21. Problem *P 6.2.17* and Theorem 6.33 may be relevant.

P 6.2.25. Problems P 6.2.23 and P 6.2.24 may be relevant.

P 6.3.2. Problem *P 5.4.11* may be relevant.

P 6.3.3. Theorem 6.38 may be relevant. What happens if you multiply the matrix S_n by a column vector of all 1's?

P 6.3.5. Problem *P 6.1.19* may be relevant.

P 6.3.9. If you already have placed $1, \ldots, n-1$ into k ordered lists, then how many choices are there for placing n?

P 6.3.10. First choose k elements of $[n]$ to lead the ordered lists, then one by one place the remaining elements. At each step, how many choices do you have?

P 6.3.12. Do Problem P 6.2.5 as a warm-up.

P 6.3.13(b). Proposition 6.42c may be relevant.

P 6.3.13(c). Theorem 6.31 may be relevant.

P 6.3.14. Problem P 6.3.13 and Corollary 6.37 may be relevant.

P 6.3.15. Theorem 6.38, Problem P 6.3.4, and Proposition 6.42d may be relevant.

P 6.3.16. Proposition 6.42d may be relevant.

P 7.1.9. As a warm-up, do Problem P 7.1.6.

P 7.1.12. Do Problem *P 1.3.17* first.

P 7.1.13. Modify the proof of Theorem 7.6.

P 7.1.20. Use a Ferrers diagram.

P 7.1.22(d). Problem P 7.1.6c may be relevant.

P 7.1.23. Problem P 7.1.22 may be relevant.

P 7.1.24(d). The hypothesis that n is odd is important.

P 7.1.24(f). Problem P 7.1.23 may be relevant.

P 7.2.3. Problem P 5.1.14 may be relevant.

P 7.2.5. Do Problem P 7.2.4 first.

P 7.2.7. Do Problem P 7.2.6 first.

P 7.2.10(a). Use induction on k.

P 7.2.14. Problem P 7.2.11 may be relevant.

P 7.2.15. Only consider partitions with $|\ell(\lambda)| < |s(\lambda)|$.

P 7.2.16. First do Problems P 7.2.14 and *P 7.2.15*.

P 7.2.17. Problems P 7.2.14, *P 7.2.15*, and P 7.2.16 may be relevant.

P 8.1.7(a). Would your method give the right answer for integers between 7 and 14?

P 8.1.19. You have to cast out only the multiples of the first four primes. Why?

P 8.1.23. Problem P 5.1.12 may be relevant.

P 8.1.25. Do Problem *P 8.1.19* first.

P 8.3.9. Translate the problem into one about rooks.

P 8.3.11. You may want to change the order of the names.

P 8.3.14. Let A_j be the set of permutations of [7] that have j followed by $j + 1$. Use the inclusion–exclusion principle.

P 8.3.17(d). First, prove that $D_{n,k} = D_{n-1,k} + D_{n,k+1}$.

P 8.3.19. Replace $\binom{n}{k}$ with $\binom{n}{n-k}$, and count the total number of permutations of $[n]$ judiciously.

P 8.3.25. Do Problem P 8.3.24 first.

P 8.3.28. Do Problem P 8.3.27 first.

P 8.3.29. Do Problem *P 8.3.28* first.

P 9.1.3. Theorem 6.31 may be relevant.

P 9.1.7. Example 9.9 may be helpful.

P 9.1.12. Begin with $1/(1 - x)^6$.

P 9.1.13. If need be, use $\frac{x}{(1-3x)(1-5x)} = \frac{-1/2}{1-3x} + \frac{1/2}{1-5x}$.

P 9.1.14. Use partial fractions.

P 9.1.23. What is $f^{(m)}(x)$? Can induction help?

P 9.2.9. Let $y_2 = 3x_2$ and $y_3 = 5x_3$. What are the restrictions on y_2 and y_3?

P 9.2.10. As a warm-up do Problem *P 9.2.9*.

P 9.2.16. Problem P 9.1.6 may be relevant.

P 9.2.18. Example 9.21 may be relevant.

P 9.2.25. Do Problem *P 9.1.13* first.

P 9.2.26. Problem P 9.1.14 may be relevant.

P 9.2.31(a). Find two equations involving $A(x)$ and $B(x)$, and solve them together.

P 9.4.2. Problems P 9.4.1 and *P 9.2.9* may be good warm-ups.

P 9.4.5(a). Problem P 9.4.3 may be relevant.

P 9.4.6. Problem P 9.4.3 may be relevant.

P 9.4.7. Problem P 9.4.6 may be relevant.

P 9.4.9. Problems P 9.1.8 and P 9.4.8 may be relevant.

P 9.4.13. Corollary 9.32 may be relevant.

P 9.4.16. Re-do Corollary 9.32. Problem P 9.2.12 may be relevant. Problem P 9.4.13 is possibly a good warm-up.

P 9.4.21(a). Consult Section 3.4 and Problems P 3.4.14 and *P 3.4.15* for background.

P 9.4.21(d). Problem P 9.4.12 may be a good warm-up.

P 9.4.22. Review Example 9.23.

P 9.4.23(c). Use induction on k.

P 9.4.24(a). Theorem 6.9 may be relevant.

P 9.4.24(b). Theorem 9.47 may be relevant.

P 9.4.29(a). Problem P 9.4.28 may be relevant.

P 9.4.34. Problem P 9.4.32 may be relevant.

P 9.4.37. Proposition 5.18 may be relevant.

P 10.1.16. Problem P 2.1.15 could be relevant.

P 10.1.18. Let i be the vertex with degree a_i. If $\{t,p\}$ is already an edge, then find an edge $\{m,n\}$ with neither $\{t,m\}$ or $\{n,p\}$ as edges. Then perform a switcheroo akin to the proof of the Havel–Hakimi algorithm.

P 10.2.17. Induct on n. What can you say about d_n? Use the inductive hypothesis on the sequence $d_1 - 1, d_2, \ldots, d_{n-1}$.

P 10.2.21. Problem P 10.2.20 may be relevant.

P 10.2.33. As a warm-up, do Problem P 10.2.32.

P 10.2.37. Show that none of the matrices is a linear combination of the ones before by showing that each has a non-zero entry, where all the previous ones had a zero. Problems P 10.2.34 and P 10.2.36 may be relevant.

P 10.2.44. First do Problem *P 10.2.43*.

P 10.2.50. First do Problem P 10.2.49. Problem P 10.2.48 may be relevant.

P 10.3.9. Given a specific partition, can you improve it by moving just one person?

P 10.3.10. For inspiration, do Problem P 10.3.9 first.

P 10.3.12. Consider the set of pairs (e,q), where e is an edge of the graph and q is an induced subgraph with four vertices with e an edge. Count this set of pairs in two different ways and get an inequality that n must satisfy.

P 10.4.12. Think through and modify the proofs of Lemma 10.42 and Theorem 10.43.

P 10.4.13. Can you color every other edge of an Eulerian circuit red?

P 10.4.15. Do Problem P 10.4.14 first.

P 10.5.8(b). Start with a graph with degree sequence $3, 3, 3, 3, 2$. Use several copies and connect them appropriately.

P 10.5.11(b). Can you use Proposition 10.56 to put in some extra edges?

P 10.5.14(c). First do Problem P 10.5.13.

P 10.5.19. Is the relevant graph bipartite?

P 10.5.27. Either mimic the proof of Proposition 10.56 or focus on a small subgraph with five vertices and use Proposition 10.56.

P 10.5.28. Do Problem *P 10.5.27* first.

P 10.5.31. Use induction on n.

P 10.6.10. Problem *P 10.6.9* may be relevant.

P 10.6.18. Why is this problem in this section?

P 10.6.21. Problem P 10.2.27 may be relevant.

P 10.6.23. As a warm-up, do Problems P 10.1.1 and P 10.6.2 and review the proof of Theorem 10.77. Find a relationship between v and e, and between r and e.

P 10.6.28(a). Proposition 10.94 may be relevant.

P 10.6.31(a). Use Induction on ℓ.

P 10.7.7(d). As a warm-up, do Problem P 10.7.6.

P 10.7.10. Theorem 6.31 may be relevant.

P 10.7.14(c). The previous two parts may be relevant.

P 10.7.16. Induction and Theorem 10.111 may be helpful.

P 10.7.17. Induct on the number of edges.

P 10.7.20. The spanning trees of G come in two varieties.

P 10.7.21. Problem P 10.7.20 could be of help.

P 10.7.27. First do Problem P 10.7.26

P 10.7.31. Problem P 10.6.16 and an inductive argument may help.

P 10.7.32(b). Use induction on n, and the previous part.

P 11.1.16. As a warm-up, do Problem P 11.1.15.

P 11.1.21. As a warm-up, do Problem P 11.1.19.

P 11.2.10. Induct on the height of the poset. Problem P 11.2.9 may be relevant.

P 11.2.12. Use Dilworth's theorem.

P 11.2.13. Does it suffice to find a chain or an antichain of size 5? Problem P 11.2.12 may be relevant.

P 11.2.16. Use the construction of Problem P 11.2.15. Problem P 11.2.11 may be relevant.

P 11.3.10. Do Problem P 11.3.9 first.

P 11.3.12. As a first step, find k. Would it help to imagine a bipartite graph?

P 11.3.13. After showing $|X| = |Y|$, use the marriage theorem.

P 11.3.14. Let $n = 2t$, and start with the complete bipartite graph $K_{t,t}$. Problem P 11.3.13 may be relevant.

P 11.3.15. As a warm-up, do Problems P 11.3.12 and P 11.3.13.

P 11.3.16. Clone the employees so that you have three of each.

P 11.3.17. Add new vertices to Y and join each of these to every vertex in X.

P 11.3.18. Construct a bipartite graph for the $2n$ teams versus the $2n - 1$ days. Put an edge connecting a team and a day if the team won their game that day. Use the marriage theorem to show that there is a matching of size $2n - 1$. Note that if a team didn't win any games during m specific days, then during those same days m teams must have beaten them.

P 11.3.21. Construct the sequence one step at a time. If you have already constructed $x_0, y_0, \ldots, y_{m-1}, x_m$, how would you choose y_m and, if necessary, x_{m+1}?

P 11.3.23. Start with some matching and keep making it bigger.

P 11.3.26. Problem P 11.3.25 may be relevant.

P 11.3.33. Do Problems P 11.3.31 and P 11.3.32 as a warm-up.

P 11.3.34. Problem P 11.3.13 may be relevant.

P 11.4.5. If such an antichain contains $\{4\}$, then the rest of its elements are subsets of ??. You may be able to use Problem P 11.4.4.

P 11.4.6. Do Problems P 11.4.3 and P 11.4.5 first.

P 11.4.12. Associate with each sum the corresponding set of indices i for which $\varepsilon_i = 1$.

P 11.4.17. Focus on $i \in [n]$. How many subsets contain i, and, in the sum, how many times does i get counted?

P 11.4.18. First ignore the condition $X \neq Y$.

P 11.4.19(c). First show that if $A \in \mathcal{A}$, then A contains an element of $b(b(A))$, and then show the converse.

P 11.5.8. Give subsets of size 4 a color depending on whether they form a convex polygon or not.

P 11.6.11. Problem P 11.1.16 may be relevant.

P 11.6.13(a). As a warm-up, do Problem P 11.1.14.

P 11.6.13(b). Proposition 11.97 may be relevant.

P 11.6.16. Proposition 11.113 may be relevant.

APPENDIX C

Short Answers for Selected Problems

P 1.2.11(b). 2/3.

P 1.3.7. 297.

P 1.3.12(d). 19.

P 1.3.13. $a_7 = 34$.

P 1.3.16(d). 353.

P 1.4.6. $2^{n+2} - n2^{n-1}$

P 1.4.7.
$a_n = 3^n - 8n3^{n-2} + 2n^23^{n-2}$.

P 1.4.8. $a_n =$
$$\begin{cases} 3^n & \text{if } n = 4k \\ 3^{n-1} & \text{if } n = 4k + 1 \\ 3^{n-2} & \text{if } n = 4k + 2 \\ 7 \times 3^{n-2} & \text{if } n = 4k + 3. \end{cases}$$

P 1.4.12.
$-9 \times 2^n + 9n2^{n-1} + 3n + 13$.

P 1.4.15. $\frac{1}{2}(3^n + 5^n)$.

P 2.1.3. 37.

P 2.1.28. Yes.

P 2.2.3. $\lfloor n^2/4 \rfloor$.

P 2.2.4. 1081; 1035.

P 2.2.6. No. Yes.

P 2.2.14. Yes.

P 2.3.10. 14.

P 2.3.14. ≤ 66.

P 2.4.2. 5.

P 2.4.3. 5.

P 2.4.5. 6.

P 2.4.7. 7.

P 2.4.10. 6.

P 3.1.1. 125.

P 3.1.4. 72.

P 3.1.7. 37,512.

P 3.1.10. 200.

P 3.1.11. 560.

P 3.1.12. 676.

P 3.1.14. 30.

P 3.2.4. 0.0768.

P 3.2.13. $h(n) = F_{n+2}$.

P 3.2.14. $f(n, n - 2) = 5$, for $n \geq 4$.

P 3.2.15.
$g(n) = g(n-1) + g(n-2) + 1$.

P 3.2.17. For $k = 2$, the probability is 47/128.

P 3.2.18. $S(n, k) =$
$S(n - 1, k - 1) + S(n - 2, k) + S(n - 1, k) - S(n - 2, k - 1)$.

P 3.2.19(b). $\frac{F_{n-1}}{2^{n-1}}$.

P 3.3.2. Entry $\boxed{11}$.

P 3.3.5. Entry $\boxed{8}$.

P 3.4.5. Yes.

P 4.1.4. 94,620.

P 4.1.7. 0.0167.

P 4.1.10. 1/15.

P 4.1.11. 1,440.

P 4.1.13. 7! 6!.

P 4.1.15. 1152.

P 4.1.16(c). 0.294.

P 4.2.5. 4,651,200.

P 4.2.8. 0.014.

P 4.2.9. 945.

P 4.2.11. 35.

P 4.2.13. $7/2^7$.

P 4.2.14. 840.

P 4.2.15(a). 0.0565.

P 4.2.16. 108.

P 4.2.17. 0.0324.

P 4.2.18. 0.01.

P 4.2.25. 11; 440.

P 4.2.29. 2700.

P 4.2.30. 8990.

P 4.2.31. 5460.

P 4.3.5. 50,803,200.

P 4.3.6. 1,512,000.

P 4.3.7. 33,810.

P 4.3.9. $\frac{55}{252}$.

P 4.3.12. 15.

P 4.3.14. 90.

P 4.3.15. 17,047,279.

P 4.4.1. 56.

P 4.4.4. 324,632.

P 4.4.6. 118,755.

P 4.4.8. 3,276.

P 4.4.9. 14,950.

P 4.4.12. $\left(\binom{t+1}{s}\right) = \left(\binom{t}{s}\right) + \left(\binom{t}{s-1}\right) + \cdots + \left(\binom{t}{1}\right) + \left(\binom{t}{0}\right)$.

P 4.4.13. 252.

P 4.4.14. 126.

P 4.4.15. 126.

P 4.4.17. 4/143.

P 4.4.18. 0.2637.

P 5.1.17. $\binom{n+3}{k}$.

P 5.2.3. 47^{47}.

P 5.2.5. 0.

P 5.2.8. Both equal to 2^{n-1}

P 5.2.9. $n2^{n-2}$.

P 5.2.17. Yes; No.

P 5.3.2. 28; 3^6; 1.

P 5.3.4. 161; 2187.

P 5.3.5. 348,432.

P 5.3.6. 181,440.

P 5.3.7. 1,061,424.

P 5.4.16. $P_n = C_n A_n$.

P 5.4.21. $\binom{2n}{n-1}$.

P 5.4.26. Yes.

P 5.4.27. The Ming–Catalan number c_{n-1}.

P 5.4.42. $\binom{2n}{n}$, $\binom{2n-k-1}{n-1}$.

P 5.4.43. $\binom{2n}{n}$.

P 5.4.44. A Ming–Catalan number.

P 6.1.1. 729; 186.

P 6.1.2. 51.

P 6.1.3. 2094.

P 6.1.6. 150.

P 6.1.7. 40,824.

P 6.1.10. 302,400.

P 6.1.11. 1560.

P 6.1.14.
$\left\{ {n \atop 3} \right\} = \frac{1}{2}(3^{n-1} - 2^n + 1)$.

P 6.2.3. $44\binom{n}{5}$.

P 6.2.6. 13,402,620.

P 6.2.8. $\binom{n}{4}\binom{n}{2}$.

P 6.2.22. 20, 160.

P 7.1.1. 17.

P 7.1.4. 456; 3.

P 7.1.6. 2; 3; $\lfloor \frac{n}{2} \rfloor$.

P 7.1.7. $p(n - j)$.

P 7.1.17. $p(n) = p_n(2n)$.

P 7.2.3. F_{n-1}, a Piṅgala–Fibonacci number.

P 8.1.2. 275; 20.

P 8.1.3. 660.

P 8.1.5. 4,000.

P 8.1.6. 999,998,990,100.

P 8.1.7. 1429; 7194

P 8.1.9. 871.

P 8.1.10. 0.024.

P 8.1.12. 19,990.

P 8.1.13. ≈ 0.01.

P 8.1.17. 1,626,498.

P 8.1.19. 25.

P 8.1.22. 174.

P 8.2.1. 158.

P 8.2.2. 96.

P 8.2.3(a). 4.

P 8.2.6. 219,264.

P 8.3.1. 0.0153.

P 8.3.2. 0.264.

P 8.3.3. 630; 24024.

P 8.3.4. 1,625.

P 8.3.5. 2,880.

P 8.3.7. 240; 80; 166.

P 8.3.9. 134.

P 8.3.10. 9,120.

P 8.3.11. 20.

P 8.3.12. 5103.

P 8.3.14. 2119.

P 8.3.27. 6!.

P 9.1.5. $-1,176$.

P 9.1.9. No.

P 9.1.14. $4(4^k - 3^k)$.

P 9.1.19. $-5/6$; -100.

P 9.2.2. 1,128.

P 9.2.7. 1,364,480.

P 9.2.8(d). 112.

P 9.2.9. 88.

P 9.2.11. $h_{47} = 464$.

P 9.2.24. $h_6 = 43$.

P 9.2.28. 8,119.

P 9.2.30. 414,143.

P 9.3.11. 398,580.

P 9.3.12. $2^{93} + 2^{46}$.

P 9.3.19. $y'' = y' + y$.

P 9.4.11(c). $\frac{e^{-x}}{1-x}$.

P 9.4.12. Yes;
$\frac{1}{(1-x)(1-x^2)(1-x^5)}$.

P 9.4.21(d).
$f(x) = \frac{x^3}{(1-x^2)(1-x^3)(1-x^4)}$.

P 9.4.22. For $n \geq 9$,
$t(n) = t(n - 2) + t(n - 3) + t(n - 4) - t(n - 5) - t(n - 6) - t(n - 7) + t(n - 9)$.

MP8.10. $k! \ (n - k)!$.

P 10.2.6. Yes. No.

P 10.2.41. 13; 121.

P 10.3.8(a). 27.

P 10.3.12. 9.

P 10.5.11. Yes; Yes; Yes.

P 10.5.19. No.

P 10.5.25. 2.

P 10.6.2. $1 \leq r \leq 4$.

P 10.6.5. 47.

P 10.6.10. Yes.

P 10.6.30. 57.

P 10.6.33. Yes.

P 11.1.1. Yes; Yes; No.

P 11.3.5. 12.

P 11.3.11. No; Yes; Yes.

P 11.4.9. $1 + \binom{n-1}{\lfloor \frac{n-1}{2} \rfloor}$.

P 11.4.17. $n4^{n-1}$.

P 11.6.6. Yes; No; No.

APPENDIX D

Complete Solutions for Selected Problems

P 1.1.3. Let $S = 1 + a + \cdots + a^n$. Now $aS = a + a^2 + \cdots + a^{n+1}$ and $aS - S = a^{n+1} - 1$. Hence $S(a-1) = a^{n+1} - 1$ and thus, as long as $a \neq 1$, we have $S = \frac{a^{n+1}-1}{a-1}$. This is a non-inductive proof of the same identity as in Problem P 1.1.2.

P 1.1.7.

$$\frac{1}{1 \cdot 2} = \frac{1}{2}$$

$$\frac{1}{1 \cdot 2} + \frac{1}{2 \cdot 3} = \frac{2}{3}$$

$$\frac{1}{1 \cdot 2} + \frac{1}{2 \cdot 3} + \frac{1}{3 \cdot 4} = \frac{3}{4}.$$

There seems to be a pattern and it seems to continue. We will not know for sure without a proof. In general and in mathematical notation our informed guess could be

CLAIM: $\frac{1}{1 \cdot 2} + \frac{1}{2 \cdot 3} + \cdots + \frac{1}{n(n+1)} = \frac{n}{n+1}$, for $n = 1, 2, \ldots$

PROOF BY INDUCTION: For the base case, let $n = 1$. We have $\frac{1}{1 \cdot 2} = \frac{1}{1+1}$, which is true.

For the inductive step assume that our claim for $n = k$, that $\frac{1}{1 \cdot 2} + \frac{1}{2 \cdot 3} + \cdots + \frac{1}{k(k+1)} = \frac{k}{k+1}$, is true. We need to prove the claim for $n = k + 1$. In other words, we need to show that $\frac{1}{1 \cdot 2} + \frac{1}{2 \cdot 3} + \cdots + \frac{1}{(k+1)(k+2)} = \frac{k+1}{k+2}$. Now,

$$\frac{1}{1 \cdot 2} + \frac{1}{2 \cdot 3} + \cdots + \frac{1}{(k+1)(k+2)} = \left(\frac{1}{1 \cdot 2} + \cdots + \frac{1}{k(k+1)} \right) + \frac{1}{(k+1)(k+2)}$$

$$= \frac{k}{k+1} + \frac{1}{(k+1)(k+2)} = \frac{k(k+2)+1}{(k+1)(k+2)} = \frac{k^2 + 2k + 1}{(k+1)(k+2)}$$

$$= \frac{(k+1)^2}{(k+1)(k+2)} = \frac{k+1}{k+2},$$

and the proof is complete.

P 1.1.11. The first few sums are $1, -1, 2, -2, 3, -3$. So the sequence seems to be the sequence of integers and their negatives. Consider the sequence $\lceil k/2 \rceil$ for $k = 1, 2, 3, \ldots$ In other words, consider $\lceil 1/2 \rceil, \lceil 2/2 \rceil, \lceil 3/2 \rceil, \ldots, .$ This sequence is $1, 1, 2, 2, 3, 3, \ldots$, and this is very close to what we want. We thus conjecture that

$$1 - 2 + 3 - 4 + \cdots + (-1)^{n-1} n = (-1)^{n-1} \left\lceil \frac{n}{2} \right\rceil.$$

We prove this by induction on n. We have already checked several base cases. For the inductive step, assume that, for an arbitrary positive integer k, $1 - 2 + 3 - 4 + \cdots + (-1)^{k-1}k = (-1)^{k-1}\lceil\frac{k}{2}\rceil$, and we calculate the corresponding sum for $n = k + 1$:

$$1 - 2 + 3 - 4 + \cdots + (-1)^k(k+1) = 1 - 2 + 3 - 4 + \cdots + (-1)^{k-1}k + (-1)^k(k+1)$$

$$= (-1)^{k-1}\left\lceil\frac{k}{2}\right\rceil + (-1)^k(k+1) = (-1)^k\left[k + 1 - \left\lceil\frac{k}{2}\right\rceil\right].$$

The proof will be complete if we can show that $k + 1 - \lceil\frac{k}{2}\rceil = \lceil\frac{k+1}{2}\rceil$. We prove this statement by considering two cases. If k is even, then $\lceil k/2 \rceil = k/2$ and $k+1-\lceil\frac{k}{2}\rceil = k+1-k/2 = k/2+1$. Now $k/2+1$ is an integer, and $\lceil k/2+1/2 \rceil = k/2+1$. Hence, in this case, $k+1-\lceil\frac{k}{2}\rceil = \lceil\frac{k+1}{2}\rceil$. If k is odd, then $\lceil\frac{k}{2}\rceil = \frac{k+1}{2}$ and $k + 1 - \lceil\frac{k}{2}\rceil = k + 1 - (k+1)/2 = (k+1)/2$. The latter is an integer and so it is equal to its own ceiling. We have thus completed the proof.

P 1.1.17. We conjecture that 7 divides $5^n + 2^n$ if n is odd. We prove this by induction. The statement is true for $n = 1$. Assume that it is true for $n = k$, where k is an arbitrary odd integer, and prove it for $n = k+2$. (Note that since the statement is true only for odd integers, we need to jump from k to $k + 2$.) We have

$$5^{k+2} + 2^{k+2} = 25(5^k) + 4(2^k) = 21(5^k) + 4(5^k) + 4(2^k)$$
$$= 21(5^k) + 4(5^k + 2^k).$$

Now, by the inductive hypothesis, 7 divides $5^k + 2^k$, and it also divides $21(5^k)$ (because 7 divides 21). Hence, 7 divides their sum and the proof is complete.

P 1.1.19. We first prove that $2^n > 2n + 1$ for $n \geq 3$.

PROOF BY INDUCTION ON n: For the base case, let $n = 3$, and we have $2^3 = 8 > 7 = 2(3) + 1$.

For the inductive step assume that, for some $k \geq 3$, the statement $2^k > 2k + 1$ is true. We need to show that $2^{k+1} > 2(k + 1) + 1$ is also true. Now,

$$2^{k+1} = 2 \times 2^k > 2(2k + 1) = 4k + 2 = 2(k + 1) + 1 + (2k - 1) > 2(k + 1) + 1.$$

The last step follows since, for $k \geq 1$, we have that $2k - 1 > 0$.

Now we are ready to prove that $2^n > n^2$ for $n \geq 5$ (the inequality also holds for $n = 1$). The proof is by induction on n. For $n = 5$ we have $2^5 = 32 > 25 = 5^2$. Let $n > 5$, and assume that the theorem is true for k. We have to prove it for $k + 1$. We have

$$2^{k+1} = 2 \times 2^k = 2^k + 2^k > k^2 + 2k + 1 = (k + 1)^2.$$

Note that we used both the inductive hypothesis (that $2^k > k^2$) and the previous claim (that $2^k > 2k + 1$).

P 1.2.7. Recall that the Piṅgala–Fibonacci numbers satisfy the recurrence relation $F_{n+1} = F_{n-1} + F_n$, and the first few of these numbers are

$$F_0 = 0, \; F_1 = 1, \; F_2 = 1, \; F_3 = 2, \; F_4 = 3, \; F_5 = 5, \; F_6 = 8,$$
$$F_7 = 13, \; F_8 = 21, \; F_9 = 34, \; F_{10} = 55, \; \ldots$$

(a) When confronted with a sum such as the one in this problem, we could generate some data and make a guess or try to use the defining relation of the sequence to somehow find an alternative form for the given sum. We pursue the former approach here. The integer n can be $0, 1, 2, \ldots$ We generate some data for small n:

$n = 0$	$F_0 = 0$	$F_0 = 0$
$n = 1$	$F_1 = 1$	$F_0 + F_1 = 1$
$n = 2$	$F_2 = 1$	$F_0 + F_1 + F_2 = 2$
$n = 3$	$F_3 = 2$	$F_0 + \cdots + F_3 = 4$
$n = 4$	$F_4 = 3$	$F_0 + \cdots + F_4 = 7$
$n = 5$	$F_5 = 5$	$F_0 + \cdots + F_5 = 12$
$n = 6$	$F_6 = 8$	$F_0 + \cdots + F_6 = 20$
$n = 7$	$F_7 = 13$	$F_0 + \cdots + F_7 = 33$
$n = 8$	$F_8 = 21$	$F_0 + \cdots + F_8 = 54.$

We had to generate a fair amount of data since the first few terms (namely, 0, 1, 2, 4) are the starting integers in many sequences and thus we may be misled. Looking at the results, and comparing them to the original Piṅgala–Fibonacci sequence, we see a striking resemblance. If you start with F_2 and subtract 1 from the original sequence, you get the integers that we have generated. Writing this as a mathematical sentence, we have our claim (which needs proof since, at this point, it is just a guess and it may be wrong):

CLAIM: $F_0 + F_1 + \cdots + F_n = F_{n+2} - 1$ for $n \geq 0$.

PROOF: We use induction on n. For the base case we have $F_0 = 0 = 1 - 1 = F_2 - 1$. (We have actually already checked all the cases through $n = 8$.) Now we assume that $F_0 + F_1 + \cdots + F_n = F_{n+2} - 1$ and we try to prove that $F_0 + F_1 + \cdots + F_{n+1} = F_{n+3} - 1$. We have

$$F_0 + F_1 + \cdots + F_{n+1} = (F_0 + F_1 + \cdots + F_n) + F_{n+1}$$
$$= (F_{n+2} - 1) + F_{n+1}$$
$$= (F_{n+1} + F_{n+2}) - 1$$
$$= F_{n+3} - 1,$$

and the proof is complete. Note that, in the last step in our string of equalities, we use the fact that $F_{n+1} + F_{n+2} = F_{n+3}$. This is just the defining relation for the Piṅgala–Fibonacci sequence.

(b) We again generate some data:

$n = 0$	$F_0 = 0$	$F_0^2 = 0$
$n = 1$	$F_1 = 1$	$F_0^2 + F_1^2 = 1$
$n = 2$	$F_2 = 1$	$F_0^2 + F_1^2 + F_2^2 = 2$
$n = 3$	$F_3 = 2$	$F_0^2 + \cdots + F_3^2 = 6$
$n = 4$	$F_4 = 3$	$F_0^2 + \cdots + F_4^2 = 15$
$n = 5$	$F_5 = 5$	$F_0^2 + \cdots + F_5^2 = 40$
$n = 6$	$F_6 = 8$	$F_0^2 + \cdots + F_6^2 = 104$
$n = 7$	$F_7 = 13$	$F_0^2 + \cdots + F_7^2 = 273.$

This pattern is much harder to discern. You may realize that each resulting integer is the *product* of two consecutive Piṅgala–Fibonacci numbers, and so

CLAIM: $F_0^2 + F_1^2 + \cdots + F_n^2 = F_n F_{n+1}$, for $n = 0, 1, \ldots$

PROOF: We prove this by induction on n. As for the base case, we have already checked this for $n = 0$ through $n = 6$. So we assume that the statement is true for $n = k$ and prove it for $n = k + 1$:

$$F_0^2 + F_1^2 + \cdots + F_{k+1}^2 = \left(F_0^2 + F_1^2 + \cdots + F_k^2 \right) + F_{k+1}^2$$
$$= F_k F_{k+1} + F_{k+1}^2 = F_{k+1} \left(F_k + F_{k+1} \right) = F_{k+1} F_{k+2},$$

and the proof is complete.

P 1.2.9. For $n = 2$ and $n = 3$, we have $F_4 = 3 < 3.0625 = \left(\dfrac{7}{4} \right)^2$, and $F_5 = 5 < 5.359375 = \left(\dfrac{7}{4} \right)^3$. We now use strong induction and assume that the statement is true for $n = k$ and $n = k + 1$, and prove it for $n = k + 2$. Assuming that $F_{k+2} < \left(\dfrac{7}{4} \right)^k$ and $F_{k+3} < \left(\dfrac{7}{4} \right)^{k+1}$, and using the fact that $\dfrac{11}{4} < \left(\dfrac{7}{4} \right)^2$, we have

$$F_{k+4} = F_{k+2} + F_{k+3} > \left(\frac{7}{4} \right)^k + \left(\frac{7}{4} \right)^{k+1}$$
$$= \left(\frac{7}{4} \right)^k \left(1 + \frac{7}{4} \right) = \left(\frac{7}{4} \right)^k \frac{11}{4} < \left(\frac{7}{4} \right)^k \left(\frac{7}{4} \right)^2 = \left(\frac{7}{4} \right)^{k+2},$$

and the proof is complete.

P 1.3.3. Consider a rectangular array with two rows and n columns. On the left we can either put a vertical domino, in which case we can complete the rest of the tiling in h_{n-1} ways, or we can put two horizontal dominoes, in which case we can complete the task in h_{n-2} ways. Also, clearly, $h_1 = 1$ and $h_2 = 2$. Thus

$$h_n = h_{n-1} + h_{n-2}, \qquad h_1 = 1, h_2 = 2.$$

These are just the Piṅgala–Fibonacci numbers: $1, 2, 3, 5, 8, 13, 21, 34, \ldots$ The only difference is the index. The way we have defined the Piṅgala–Fibonacci numbers, we have $h_n = F_{n+1}$. Thus $h_8 = F_9$ is 34.

P 1.3.15.

(a) We can write down all the possible sequences and get $h(1) = 3$ (since 0, 4, 7 are the possible sequences), and $h(2) = 8$ (since 00, 04, 07, 40, 44, 47, 70, 77 are the possible sequences).

(b) We can start a sequence of length n with 0, 4, or 7. If we start with a 0, then we have $h(n-1)$ ways of completing the sequence. If we start with a 4, we have three choices for each entry afterwards, and thus we have 3^{n-1} ways of doing it. If we start with a 7, then we can never use a 4, and hence for each entry we have two choices, giving a total of 2^{n-1} ways of doing it. Thus, we have

$$h(n) = 3^{n-1} + 2^{n-1} + h(n-1).$$

We could have also considered the last entry. If we are going to add an entry at the end, then we have to start with a sequence of length $n - 1$ that satisfies the conditions. If we could add a 0, 4, or 7 at the end of each one, we would get $3h(n-1)$ sequences. However, some of these are not allowed. In particular, if the original sequence did not have a 4 and did have a 7, then we cannot add a 4 at the end. There are $2^{n-1} - 1$ such sequences (2^{n-1} sequences with no 4's, but one of them is the one with all 0's which does not have a 7). Thus, we also have

$$h(n) = 3h(n-1) - (2^{n-1} - 1).$$

(c) If we use the first recurrence relation, we get

$$h(3) = 3^2 + 2^2 + 8 = 21$$

$$h(4) = 3^3 + 2^3 + 21 = 56$$

$$h(5) = 3^4 + 2^4 + 56 = 153.$$

P 1.3.17. To make life easier, let \$25,000 be considered one unit of money. Thus we want to know the number of ways of dividing 11 identical units of money among 4 indistinguishable people. Let $l(m, k)$ denote the number of ways of dividing m identical units of money among k indistinguishable people. We make a table for small values of m and k:

$k \backslash m$	0	1	2	3	4	5	6	7	8	9	10	11
1	1	1	1	1	1	1	1	1	1	1	1	1
2	1	1	2	2	3	3	4	4	5	5	6	6
3	1	1	2	3	4	5	7	8				
4	1	1	2	3	5	6	9	11				
						$l(m,k)$							

CLAIM: $l(m,k) = l(m,k-1) + l(m-k,k)$ for $1 \le k \le m$.

PROOF: When you divide m units among k people, one of two things happens: either everyone gets something or else at least one person gets nothing. In the latter case, ignore the person who is getting nothing and you have to divide m units among $k-1$ people. This can be done in $l(m,k-1)$ ways. In the former case, everyone gets at least 1 unit. Distribute 1 unit to each person. You are left with $m-k$ units that you have to divide among k people. This can be done in $l(m-k,k)$ ways. Thus the claim follows.

Using the recurrence relation we can now finish the table:

$k \backslash m$	0	1	2	3	4	5	6	7	8	9	10	11
1	1	1	1	1	1	1	1	1	1	1	1	1
2	1	1	2	2	3	3	4	4	5	5	6	6
3	1	1	2	3	4	5	7	8	10	12	14	16
4	1	1	2	3	5	6	9	11	15	18	23	27
						$l(m,k)$							

Thus the number of ways of distributing 11 identical units among 4 indistinguishable individuals is 27.

NOTE: You were asked only to come up with a guess. I gave an actual proof. Sometimes it is hard to come up with a pattern without doing the proof. I came up with the pattern in the table by first making the above argument and then checking to see if it agrees with the table. In Chapter 7, we study the partitions of an integer. With the parlance of that chapter, $l(m,k)$ is the number of partitions of m into k or fewer parts.

P 1.4.3. If a function of the form a^n were to satisfy the recurrence relation $h_n = 2h_{n-1} + 2h_{n-2}$, then we would have (this is a repeat of the argument as in the text for finding the characteristic polynomial) $a^n = 2a^{n-1} + 2a^{n-2}$. Dividing by a^{n-2}, we have $a^2 - 2a - 2 = 0$. Hence, the characteristic equation is $x^2 - 2x - 2$. The roots of this polynomial are $1 \pm \sqrt{3}$. Hence, the general solution to the recurrence relation is

$$h_n = \alpha(1 + \sqrt{3})^n + \beta(1 - \sqrt{3})^n.$$

To find α and β we use the initial conditions. Since we are given $h_1 = 3$ and $h_2 = 8$, we could plug in 1 and 2 for n and solve for α and β. This does work, but we can make the calculations a bit easier if we start the sequence at $n = 0$. Using the same recurrence relation, we see that if we start with $h_0 = 1$ and $h_1 = 3$, we get the same sequence (albeit starting with h_0). We thus plug in $n = 0$ and $n = 1$, and get

$$\begin{cases} \alpha + \beta & = 1 \\ \alpha(1 + \sqrt{3}) + \beta(1 - \sqrt{3}) = 3 \end{cases} \Rightarrow \begin{cases} \alpha = \frac{3 + 2\sqrt{3}}{6} \\ \beta = \frac{3 - 2\sqrt{3}}{6}. \end{cases}$$

So, for $n \geq 0$,

$$h_n = \frac{3 + 2\sqrt{3}}{6}(1 + \sqrt{3})^n + \frac{3 - 2\sqrt{3}}{6}(1 - \sqrt{3})^n.$$

P 1.4.7.

(a) $a_3 = 9$, $a_4 = 81$, and $a_5 = 3^6 - 3^5 + 3^3 = 513$.
(b) The characteristic equation for this recurrence relation is $x^3 - 9x^2 + 27x - 27 = (x - 3)^3$. Hence 3^n, $n3^n$, and $n^2 3^n$ are all solutions to the recurrence relation (we can check each by plugging them into the recurrence relation). This means that $f(n) = \alpha 3^n + \beta n 3^n + \gamma n^2 3^n$, for scalars α, β, and γ, is also a solution to the recurrence relation. We need this to satisfy the initial conditions and so we have

$$\begin{cases} \alpha & = 1 \\ 3\alpha + 3\beta + 3\gamma & = 1 \\ 9\alpha + 18\beta + 36\gamma & = 1 \end{cases} \Rightarrow \begin{cases} \alpha = 1 \\ \beta = -\frac{8}{9} \\ \gamma = \frac{2}{9} \end{cases}.$$

So

$$a_n = 3^n - \frac{8n3^n}{9} + \frac{2n^2 3^n}{9} = 3^n - 8n3^{n-2} + 2n^2 3^{n-2}.$$

P 1.4.13.

$$a_2 = -2 - 9 + 4 = -7, \ a_3 = 7 + 12 - 27 + 16 = 8, \ a_4 = -8 - 42 - 81 + 64 = -67.$$

The characteristic polynomial of the corresponding homogeneous recurrence is $x^2 + x - 6 = (x - 2)(x + 3)$ and so the general solution to the homogeneous recurrence is $\alpha 2^n + \beta(-3)^n$. To find a particular solution to the non-homogeneous recurrence, we try an exponential of the form $a3^n + b2^{2n-2}$. We need to have

$$a3^n + b2^{2n-2} = -a3^{n-1} - b2^{2n-4} + 6a3^{n-2} + 6b2^{2n-6} - 3^n + 2^{2n-2}$$

$$\Rightarrow 3^{n-2} [9a + 3a - 6a + 9] + 2^{2n-6} [16b + 4b - 6b - 16] = 0$$

$$\Rightarrow (6a + 9)3^{n-2} + (14b - 16)2^{2n-6} = 0.$$

So if $a = -3/2$ and $b = 8/7$, then we have a particular solution. Hence,

$$a_n = \alpha 2^n + \beta(-3)^n - \frac{3^{n+1}}{2} + \frac{2^{2n+1}}{7}$$

is a general solution to the non-homogeneous recurrence. Using the initial condition, we get

$$\begin{cases} \alpha + \beta - 3/2 + 2/7 = 0 \\ 2\alpha - 3\beta - 9/2 + 8/7 = 2 \end{cases} \Rightarrow \begin{cases} \alpha = 9/5 \\ \beta = -41/70. \end{cases}$$

Thus, a closed formula for a_n is

$$a_n = \frac{9}{5}2^n - \frac{41}{70}(-3)^n - \frac{3^{n+1}}{2} + \frac{2^{2n+1}}{7}.$$

P 1.4.21. Before unwinding, let's just find some values of T. We are already given $T(0), \ldots, T(7)$, and

$$T(7) = 2$$
$$T(8) = T(1) + T(7) + 8 = 12$$
$$T(9) = T(2) + T(7) + 9 = 13$$
$$\vdots$$
$$T(13) = T(6) + T(7) + 13 = 17$$
$$T(14) = T(7) + T(7) + 14 = 18$$
$$T(15) = T(8) + T(7) + 15 = 29$$
$$T(16) = T(9) + T(7) + 16 = 31.$$

So the function T stays equal to 2 for a while, then makes a jump of ten to 12 and starts increasing by 1 for a while, then makes a jump of eleven to 29 and starts increasing by 2 for a while, then jumps, and so on. Now, $T(7) = 2$ means that $T(n) = T(n-7) + 2 + n$, and $T(n-7) = T(n-14) + 2 + n - 7$ as long as $n - 7 > 7$. So

$$T(n) = T(n-7) + 2 + n = [T(n-14) + 2 + n - 7] + 2 + n$$
$$= T(n-14) + 2(2+n) - 7 = [T(n-21) + 2 + n - 14] + 2(2+n) - 7$$
$$= T(n-21) + 3(2+n) - 21 = [T(n-28) + 2 + n - 21] + 3(2+n) - 21$$
$$= T(n-28) + 4(2+n) - 42 = [T(n-35) + 2 + n - 28] + 4(2+n) - 42$$
$$= T(n-35) + 5(2+n) - 70.$$

We write $7 = 1 \times 7, 21 = 3 \times 7, 42 = 6 \times 7, 70 = 10 \times 7$, and recognize 1, 3, 6, 10 as triangular numbers, and so, as long as $n - 7(k-1) > 7$, we have

$$T(n) = T(n-7k) + k(2+n) - \frac{7k(k-1)}{2}.$$

The condition $n - 7(k-1) > 7$ gives $n > 7k$. Thus, if we choose k such that $7k < n \le 7(k+1)$ (in other words, $k = \lfloor n/7 \rfloor$), then $n - 7k \le 7$. So, let $k = \lfloor n/7 \rfloor$ and

$$T(n) = T(n-7k) + k(2+n) - \frac{7k(k-1)}{2} = 2 + k(2+n) - \frac{7k(k-1)}{2}.$$

Hence,

$$T(n) = 2 + \lfloor n/7 \rfloor (2 + n) - \frac{7 \lfloor n/7 \rfloor (\lfloor n/7 \rfloor - 1)}{2}.$$

For example, if $n = 16$, then $k = \lfloor n/7 \rfloor = \lfloor 16/7 \rfloor = 2$ and so

$$T(16) = 2 + 2(2 + 16) - \frac{7 \times 2 \times 1}{2} = 38 - 7 = 31.$$

P 1.4.23. As an example, if $k = 3$, $V = \begin{bmatrix} 1 & a_1 & a_1^2 \\ 1 & a_2 & a_2^2 \\ 1 & a_3 & a_3^2 \end{bmatrix}$, and

$$\det(V) = a_2 a_3^2 - a_3 a_2^2 - a_1 a_3^2 + a_3 a_1^2 + a_1 a_2^2 - a_2 a_1^2.$$

Each term in $\det(V)$ is the product of one entry from each row and column of V (with a ± 1 coefficient). Hence, the degree of each term is $0 + 1 + \cdots + (k - 1) = \frac{k(k - 1)}{2}$.

If the rows or columns of a matrix are linearly dependent, then the determinant of the matrix is zero. Hence, if we replace a_j with a_i, then two columns of V are the same and hence the determinant of V becomes zero. This means that, for each $i \neq j$, $(a_j - a_i)$ is a factor of $\det(V)$. This is true for every choice of i and j and so $P = \prod_{1 \le i < j \le k} (a_j - a_i)$ is a factor of $\det(V)$. For each $j = 2, \ldots, k$, there are $j - 1$ parentheses of the form $(a_j - a_i)$ with $i < j$. So, the number of parentheses in P is $1 + 2 + \cdots + (k - 1) = \frac{k(k-1)}{2}$. This means that the degree of every term in P is $\frac{k(k-1)}{2}$ since every parenthesis contributes 1 to the degree of every term. Thus P and $\det(V)$ have terms of the same degree. So $\det(V)$ cannot have any non-constant factors other than P (since otherwise the degree of its terms would be higher than those of P). We conclude that $\det(V) = cP$. Now one term in $\det(P)$ is the product of the diagonals and this is $a_2 a_3^2 \cdots a_k^{k-1}$. This same term appears in P (by choosing the first term in each parenthesis) and so $c = 1$. We conclude that $\det(V) = P$ and $\det(V) \neq 0$.

P 2.1.13. If your sequence of nine integers is $a_1, a_2, \ldots, a_9$, then let $s_1 = a_1$, $s_2 = a_1 + a_2, \ldots, s_9 = a_1 + \cdots + a_9$ be the sequence of partial sums. Let $r_1, \ldots, r_9$ be the remainders of the partial sums when divided by 9. If any of these are zero, then we have an unbroken block of adjacent integers starting with the first one whose sum is divisible by 9, and we are done. Otherwise, all the remainders will have to be from among $1, \ldots,$ 8. We have nine r's and only eight possibilities. By the pigeonhole principle, two of them have to be equal to each other. So, in this case, two of the partial sums have the same reminder when divided by 9. This means that their difference – which is the sum of an unbroken block of adjacent integers – is divisible by 9. In general, the same proof – when you replace 9 by a positive integer n – shows that, if you write an arbitrary list of n integers, you will be able to find an unbroken block of adjacent integers whose sum is divisible by n.

P 2.1.17. Create 135 pigeonholes: $\{1, 2\}, \{3, 4\}, \ldots, \{269, 270\}$. We are choosing 136 numbers from $1, \ldots, 270$, and there are 135 pigeonholes. So at least two of the chosen integers must come from the same pigeonhole. Those two differ by 1.

P 2.1.19. There are 135 odd numbers between 1 and 270. These will be the pigeonholes. Any integer n can be written uniquely in the form $n = 2^k m$, where m is odd. Assign n to pigeonhole m. So, for example, if $20 = 2^2 \times 5$ and $40 = 2^3 \times 5$ are among the chosen integers, they will both be assigned to the pigeonhole with label 5. Given 136 integers between 1 and 270, two of them must fall in the same pigeonhole. The smaller of them divides the larger by a power of 2.

P 2.1.23.

(a) $2^{12} - 1 = 4095 = 13 \times 315$. For n smaller than 12, you can check that $2^n - 1$ is not divisible by 13, and so $n = 12$ is the smallest n for which a_n has a factor of 13.

(b) Let m be an arbitrary odd positive integer. We prove that at least one of $a_1, \ldots, a_m$ is divisible by m. First, find the remainders of $a_1, \ldots, a_m$ when divided by m. If one of these remainders is zero, then we have found an a_j that is divisible by m. By way of contradiction, assume that this is not the case and that none of the remainders is equal to zero. This means that the remainders are integers between 1 and $m - 1$. There are m integers (namely, $a_1, \ldots, a_m$) and $m - 1$ remainders (namely, $1, \ldots, m - 1$). By the pigeonhole principle, at least two of the remainders have to be the same. So assume that, for $i < j$, the remainder of a_i and a_j when divided by m are the same. This means that $a_i = km + r$ and $a_j = \ell m + r$ for some integers k, ℓ, and r. Consider the difference $a_j - a_i$. On the one hand, this difference is $(\ell m + r) - (km + r) = (\ell - k)m$, which is a multiple of m. On the other hand, this difference is

$$a_j - a_i = (2^j - 1) - (2^i - 1) = 2^i(2^{j-i} - 1) = 2^i a_{j-i}.$$

So it must be that m is a factor of $2^i a_{j-i}$. But m is odd and so it has no common factors with 2^i. Hence, $m \mid a_{j-i}$. But a_{j-i} is assumed to be an integer between 1 and $m - 1$ (recall that the a's were the remainders when dividing by m and we are assuming none of them is zero), and none of these are divisible by m. The contradiction proves the claim.

P 2.1.31. Let $s_1 = \text{frac}(r)$ be the fractional part of r (i.e., $s = r \mod 1$), $s_2 = \text{frac}(2r)$, and more generally, $s_i = \text{frac}(ir)$. Consider the real numbers

$$s_1, s_2, \ldots, s_{10^{10}+1}.$$

This is a list of $10^{10} + 1$ numbers and, for $1 \leq i \leq 10^{10} + 1$, we have $0 \leq s_i < 1$. If you divide the interval $[0, 1]$ into 10^{10} equal subintervals, then each will have a length of 10^{-10} and, by the pigeonhole principle, at least two of the s's have to be in the same subinterval. Hence, for some integers $j > i$, we have

$$|s_j - s_i| < 10^{-10}.$$

Now $s_j = jr - \lfloor jr \rfloor$ and so $s_j - s_i = (j - i)r - \lfloor jr \rfloor + \lfloor ir \rfloor$. Let $n = j - i$ and $m = \lfloor jr \rfloor - \lfloor ir \rfloor$. Then $|nr - m| = |s_j - s_i| < 10^{-10}$. Hence, $\left| r - \frac{m}{n} \right| < \frac{10^{-10}}{n}$.

P 2.2.5. The simple graph of order n with the greatest number of edges is K_n, and the number of edges of K_n is $M_1 = \frac{n(n-1)}{2}$. Now, if the graph is not connected, it has at least two parts, and the maximum number of edges will be achieved if each of the parts is a complete graph. If one part has m vertices, the other parts have $n - m$ vertices and, for $1 \leq m \leq n - 1$, the number of edges will be

$$\frac{m(m - 1)}{2} + \frac{(n - m)(n - m - 1)}{2} = m^2 - nm + \frac{n(n - 1)}{2}.$$

Now n is fixed, and $m^2 - nm + \frac{n(n-1)}{2}$ as a function of m is a concave-up parabola, and so its maximum is at one of the end points. At both end points the value of this function is $\frac{n(n-1)}{2} - (n - 1)$. Thus the number of edges of a disconnected graph with n vertices is at most $M_2 = \frac{n(n-1)}{2} - (n - 1)$. We conclude that $M_2 - M_1 = n - 1$.

P 2.2.11. The so-called *projective plane* over $\mathbb{F}_3$ has 13 points and 13 lines. Each line has 4 points and each point is on 4 lines. You are not expected to understand what this means. Suffice to say that there is an algebraic structure that provides the answer. One could also come up with the answer through trial and error. Let the teams be $X_1, X_2, X_3, X_4, A_1, A_2, A_3,$ $B_1, B_2, B_3, C_1, C_2,$ and C_3. Then here are the 13 meets (with the host for each meet boldfaced):

Meet 1	$\boldsymbol{X_1}$	X_2	X_3	X_4
Meet 2	X_1	$\boldsymbol{A_1}$	A_2	A_3
Meet 3	X_1	B_1	$\boldsymbol{B_2}$	B_3
Meet 4	X_1	C_1	C_2	$\boldsymbol{C_3}$
Meet 5	$\boldsymbol{X_2}$	A_1	B_1	C_1
Meet 6	X_2	$\boldsymbol{A_2}$	B_2	C_2
Meet 7	X_2	A_3	$\boldsymbol{B_3}$	C_3
Meet 8	X_3	A_2	B_3	$\boldsymbol{C_1}$
Meet 9	$\boldsymbol{X_3}$	A_1	B_2	C_3
Meet 10	X_3	$\boldsymbol{B_1}$	C_2	A_3
Meet 11	X_4	A_1	$\boldsymbol{C_2}$	B_3
Meet 12	X_4	C_1	B_2	$\boldsymbol{A_3}$
Meet 13	$\boldsymbol{X_4}$	B_1	A_2	C_3

If you are not satisfied with this seemingly ad hoc argument, here is a more sophisticated construction. Let $\mathbb{F}_3 = \{0, 1, 2\}$ and define addition and multiplication mod 3. In other words, when you want to add or multiply, instead of the usual answer, you put the remainder when divided by 3. So $2 + 2 = 1$, $2 + 1 = 0$, and $2 \times 2 = 1$. Now consider $(\mathbb{F}_3)^2 = \mathbb{F}_3 \times \mathbb{F}_3 = \{(a, b) \mid a, b \in \mathbb{F}_3\}$. The set $(\mathbb{F}_3)^2$ has nine points: $(0, 0), (1, 0), \ldots, (2, 2)$. Draw these, as usual, in the $x - y$ plane to get a 3×3 grid. See the drawing on the left in Figure D.1. Now the points satisfying $x = 0$ are three points on a vertical line. Likewise, $x = 1$ and $x = 2$ are also vertical lines parallel to $x = 0$. Similarly, the three lines $y = 0, y = 1$, and $y = 2$ are parallel horizontal lines, each with three points. Now, the equation $y = x + 1$ is also satisfied by three points:

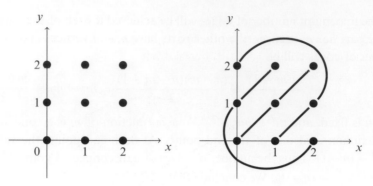

Figure D.1 On the left, 9 points with coordinates (i,j) for $0 \le i,j, \le 2$. On the right, 3 parallel lines with equations $y = x$, $y = x + 1$, and $y = x + 2$.

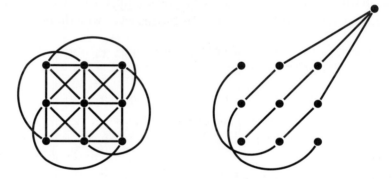

Figure D.2 On the left, 9 points and 12 lines. Each point is on 4 lines, and each line has 3 points. On the right, a point at infinity for the three parallel lines with equations $y = x$, $y = x + 1$, and $y = x + 2$.

$(0, 1)$, $(1, 2)$, and $(2, 0)$ (remember that $2 + 1 = 0$). We call these three points a line as well. In fact, the three lines $y = x$, $y = x + 1$, and $y = x + 2$ are also a set of parallel lines with three points on each (they are parallel because they don't intersect). See the drawing on the right of Figure D.1. Likewise, $y = 2x$, $y = 2x + 1$, and $y = 2x + 2$ are a fourth set of parallel lines with three points on each. So, altogether we have 9 points and 12 lines such that each point is on 4 lines (one horizontal, one vertical, one with "slope" equal to 1 and one with slope equal to $2 = -1$), and each line has 3 points. See the drawing on the left of Figure D.2. We have to modify this to get what we want. For each set of parallel lines, we add a point. We think of this as the point at infinity where the parallel lines meet. See the drawing on the right of Figure D.2. So we have one extra point that lies on each of $x = 0$, $x = 1$, and $x = 2$, another extra point that lies on $y = 2x$, $y = 2x + 1$, and $y = 2x + 2$, and so on. This makes the total number of points equal to $9 + 4 = 13$. Finally, we add one last line that only contains the four points at infinity. As a result, we have 13 points and 13 lines with the property that every line has 4 points and every point is on 4 lines. See Figure D.3. This is what we were after! Every team is a vertex, and every meet is a line. Hence, every team is in four meets and every meet has four participating teams. In the unexplained table that we first gave as a solution, $X_1, \ldots,$ X_4 are the points at infinity, while, for $1 \leq i \leq 3$, A_i, B_i, and C_i are points with coordinates $(i - 1, 0)$, $(i - 1, 1)$, and $(i - 1, 2)$, respectively.

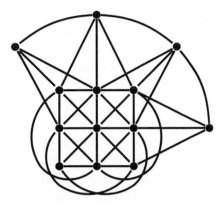

Figure D.3 By adding 4 points "at infinity" and a line through these 4 points, we have 13 points and 13 lines. Each point is on 4 lines and each line has 4 points.

P 2.2.13. The two graphs are isomorphic, and these are two depictions of a graph called the Petersen graph. To see the isomorphism, we have to map the vertices of the first graph to the vertices of the second graph in such a way that the map is bijective and two vertices that are adjacent in one graph are paired with vertices that are adjacent in the second graph. It turns out that it doesn't matter where you map vertex 1. Now, vertex 1 in the left diagram is adjacent to 2, 5, and 6. So the three vertices adjacent to the image of 1 need to be the images of 2, 5, and 6. You continue and, after deciding the labels of vertices adjacent to one of these, you will not have a choice and all labels will be determined. We have given a possible labeling of the second graph in Figure D.4 to show the isomorphism.

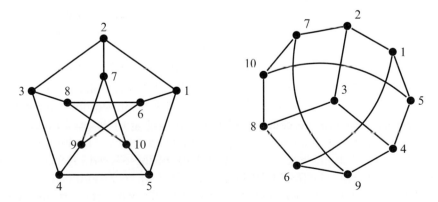

Figure D.4 The two versions of the Petersen graph are labeled so as to exhibit the isomorphism.

P 2.3.3. Consider the six vertices at the other ends of the six blue edges incident with v. These vertices form a K_6 and $K_6 \to K_3, K_3$. Thus, in that K_6 we have either a red K_3 – in which case we are done – or a blue K_3 which together with v form a blue K_4.

P 2.3.11. Let $n = r(14, 14)$. We need to show that $K_n \to K_3, K_3, K_5, K_5$. So, start with a K_n and arbitrarily color its edges using four colors. Think of the first and third colors as two hues of the same color. Likewise, think of the second and fourth colors as the hues of another color. As a result, the given coloring of K_n now uses only two colors. By definition of n (recall $n = r(14, 14)$), we must have a monochromatic K_{14} of one of the (new) colors. Without loss of generality, assume that we have a K_{14} whose edges have either the first or the third color. We know that $r(3, 5) = 14$, and so $K_{14} \to K_3, K_5$. Thus we have either a K_3 of the first color or a K_5 of the second color. The proof is now complete.

P 2.3.17. See Figure D.5 for the edges colored red and the ones colored blue. If $i < j < k$ were the vertices of a red triangle, then $j - i$, $k - j$, and $k - i$ would all be in S. We have $(j - i) + (k - j) = k - i$ but, by inspection, the sum of no two elements of S is in S. Thus, there are no red triangles in this coloring scheme.

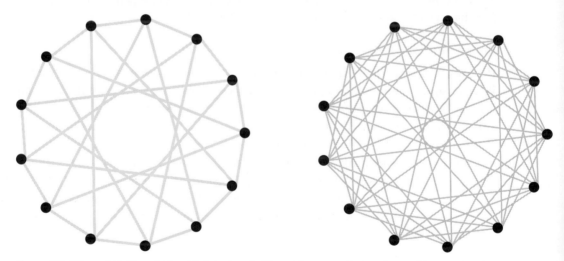

Figure D.5 The red (left) and blue (right) edges in K_{13} without a red triangle or a blue K_5 showing $r(3, 5) > 13$. A black and white version of this figure will appear in some formats. For the color version, please refer to the plate section.

If you put the vertices around a circle, each vertex has a blue edge to vertices that are 2, 3, 4, or 6 units away (in either direction). As such the graph is quite symmetric and you can't tell two vertices apart (this is called a *vertex-transitive* graph) except for their labels. Thus if there is a blue K_5, there will also be a blue K_5 with 0 as a vertex. This blue K_5 cannot have vertices 1, 5, 8, or 12 (since the edge between 0 and any of these is red). Hence, the other four vertices of the blue K_5 must be from the three subsets $\{2, 3, 4\}$, $\{6, 7\}$, and $\{9, 10, 11\}$. By the pigeonhole principle, two must come from the same subset. No two vertices of the blue K_5 can differ by 1 (since $1 \in S$ and so the edge between those two vertices would be red). So the two vertices from the subset must come from the subsets of size 3. In fact, they must be either 2 and 4, or 9 and 11. These two pairs are symmetrically located with respect to 0, and so the argument will be the same for both of them. Hence, without loss of generality, we can assume that, so far, our blue K_5 has 0, 2, and 4 as vertices. The edges between 2 and each of 3, 7, and 10 are all red, as are the edges between 4 and 9. So the only possibilities for the blue K_5 are 6 and 11. However, $11 - 6 = 5$ and so the edge between 6 and 11 is red as well. Thus there is no blue K_5.

We have constructed a 2-colored K_{13} with no red triangle and with no blue K_5. We conclude that $r(3, 5) > 13$.

P 2.4.7. Every vertex in K_6 has degree 5 and it is possible to color the edges of K_6 with red and blue so that every vertex is incident with two red edges and three blue edges (for a total of six red edges and nine blue edges). In this coloring there is no monochromatic $K_{1,4}$ since, for such a thing you would need a vertex that is incident with four edges of the same color. Hence, $r(K_{1,4}, K_{1,4}) > 6$. Now if you color the edges of K_7 with red and blue, then each vertex has degree 6, and the only way to avoid a monochromatic $K_{1,4}$ is if every vertex was incident with exactly three red and three blue edges. If this were possible, then summing up the number of red edges incident with the different vertices would give $3 \times 7 = 21$. But since each red edge is counted twice (once for each of its vertices), this number should be even. The contradiction proves that $r(K_{1,4}, K_{1,4}) \le 6$. We conclude that $r(K_{1,4}, K_{1,4}) = 6$.

P 2.4.9. Consider K_6, where the vertices are labeled x_1, x_2, x_3, x_4, x_5, and x_6. Each intermediate facility will be used to connect pairs of vertices. If you are going to use facility number 1 to connect vertices v and w, then color the edge (in K_6) between v and w with color 1. If a facility fails, then all the edges with that color become inoperative. The question then becomes what is the minimum number of colors needed, so that, if you eliminate the edges with one of the colors, then there is still some edge connecting each two distinct pairs. If $\{v, w\}$ and $\{y, z\}$ are two distinct pairs of vertices, then we don't want all the edges $\{v, y\}$, $\{v, z\}$, $\{w, y\}$, and $\{w, z\}$ to be the same color. Hence, we don't want a monochromatic $K_{2,2}$. Certainly, one color is not enough. Two colors is also not enough since, by Problem P 2.4.8, $r(K_{2,2}, K_{2,2}) = r(C_4, C_4) = 6$. (Recall that $K_{2,2}$ is isomorphic to C_4.) Hence, we need at least three colors. An example shows that three colors is enough. It also shows that $r(C_4, C_4, C_4) > 6$.

P 3.1.13. Every time you spell MATHEMATICS you have to use the middle M. Hence the number of ways of spelling MATHEMATICS is the same as the number of ways of spelling MATHEM times the number of ways of spelling MATICS. Now the number of ways to spell MATHEM is the same as the number of ways to spell MATICS because of symmetry. The number of ways of spelling MATHEM using the top A is 10 (by brute-force counting). The number of ways of spelling MATHEM using the bottom A is also 10 by symmetry, and thus the total number of ways of spelling MATHEM is 20. We conclude that the total number of ways of spelling MATHEMATICS is $20 \times 20 = 400$.

P 3.2.15. Let $n \ge 3$. The number of sequences of n tosses where every head is a single, there is at least one head, *and* the sequence starts with a T is $g(n - 1)$. If such a sequence starts with an H, then the second toss has to be a T (since otherwise every head would not be a single), and then you can complete the sequence in two ways. You could append a sequence with the longest run of heads equal to zero (i.e., the sequence of all tails) or one where the longest run of heads is 1. In other words, $g(n - 2) + 1$ ways. Hence, for $n \geq 3$,

$$g(n) = g(n - 1) + g(n - 2) + 1.$$

Now $g(1) = 1$ and $g(2) = 2$, hence $g(3) = 4$, $g(4) = 7$, $g(5) = 12$, $g(6) = 20$, $g(7) = 33$, and $g(8) = 54$.

P 3.2.19. We have $k(1) = 0$ since one coin toss is always a single. On the other hand, $k(2) = 2$ since both TT and HH have no singles. In Problem P 3.2.18, we showed that $S(n, k) = S(n - 1, k - 1) + S(n - 2, k) + S(n - 1, k) - S(n - 2, k - 1)$. If $k = 0$, two of the terms disappear, and we have

$$k(n) = k(n - 2) + k(n - 1).$$

Hence, the sequence $k(n)$ is $0, 2, 2, 4, 6, 10, \ldots$ These are exactly twice the Piṅgala–Fibonacci numbers (that is certainly the case for the first few, and the recurrence relation guarantees that this will continue throughout). Thus $k(n) = 2F_{n-1}$.

The total number of sequences of n coin tosses is 2^n and so the probability that there are no singles in a sequence of n tosses of a fair coin is $\frac{k(n)}{2^n} = \frac{F_{n-1}}{2^{n-1}}$.

P 3.3.5. Say you have 47 identical balls and 47 identical boxes (with no capacity restriction on the boxes). What are the different ways that you can put the balls into the boxes? Since the boxes and the balls are identical, the only thing that matters is the *number* of balls in the different boxes. Now, think of the balls as the blocks of wood. Every time you place the balls into the boxes, you get a sequence of positive integers adding up to 47. If you arrange these in non-increasing order, then you get one way of stacking the blocks of wood. Vice versa, each possibility for stacking the blocks of wood gives a non-increasing sequence of positive integers and these give you a recipe for placing balls into boxes (remember that the balls and boxes are identical and so it doesn't matter which balls go into which boxes; you just need to know how many balls should be together in a box, and some – maybe many – boxes can be empty). This is entry $\boxed{8}$ of the counting table.

P 3.4.11. We define a bijection from $\mathcal{B}$ to $\mathcal{A}$. Given a sequence of 19 coin tosses with exactly 7 heads, there are exactly 12 tails. (The tails are the donuts!) The number of tails before the first head is the number of donuts of type 1 that you choose. The number of tails between the first and second head is the number of donuts of the second type that you choose, and so on. Finally, the number of tails after the seventh head is the number of donuts of the eighth type that you choose. Every sequence of 19 coin tosses with exactly 7 heads encodes an unambiguous order, and, vice versa, each possible order for the donuts gives a corresponding sequence of coin tosses.

P 3.4.15. Assume that $g \colon \mathcal{B} \to \mathcal{A}$ is the inverse of f. We want to prove that f is both 1-1 and onto. Assume that $f(x_1) = f(x_2)$ for x_1 and x_2 elements of $\mathcal{A}$. Now apply g, and get $g(f(x_1)) = g(f(x_2))$. But g is the inverse of f and so $g(f(x_1)) = x_1$ and $g(f(x_2)) = x_2$.

We conclude that $x_1 = x_2$, and f is 1-1. To prove f is onto, let $y \in \mathcal{B}$, and let $x = g(y) \in \mathcal{A}$. Now $f(x) = f(g(y)) = y$ and so there is an element – namely $g(y)$ – that is mapped to y. We conclude that f is onto as well.

For the converse, assume that f is 1-1 and onto. We want to define a function $g \colon \mathcal{B} \to \mathcal{A}$. For $y \in \mathcal{B}$, since f is onto, there is at least one $x \in \mathcal{A}$ such that $f(x) = y$. But since f is also 1-1, there is exactly one such x. We now define $g(y) = x$. (In other words, we sent every element of $\mathcal{B}$ to the one element of $\mathcal{A}$ that is mapped to it by f). This defines a function $g \colon \mathcal{B} \to \mathcal{A}$, and because of the way we defined g, if we compose f with g (or g with f) we get the identity function. Hence, f has an inverse, namely the function g.

P 4.1.15. First seat Ms. Ashton and all the Shimashki reps around a circular table. The number of circular permutations of 5 objects is 4! and so you can finish your task in 4! ways. Given any such circular permutation, give a nickname to the Shimashki representatives starting with the one sitting to the right of Ms. Ashton. Call them in order A_1, A_2, A_3, and A_4 so that A_4 is the one sitting to the left of Ms. Ashton. (See Figure D.6.) Now you have 4 choices of Thrace reps to put between A_1 and A_2. Having made that choice, you have 3 choices of a Thrace rep to put between A_2 and A_3, and then 2 choices of a Thrace rep to put between A_3 and A_4. The final representative of Thrace region can sit on either side of Ms. Ashton. Hence, the total number of ways of making seating arrangements is

$$4! \times 4 \times 3 \times 2 \times 2 = 2(4!)^2 = 1152.$$

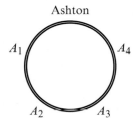

Figure D.6 First seat Ms. Ashton and the Shimashki A's.

P 4.1.17. Say the people in the room are numbered from 1 to 30 and you make a list of their birthdays. The total number of possibilities for such lists is $(365)^{30}$, and from among these $(365)_{30}$ have all distinct dates. Hence, the probability that a randomly chosen list has at least two occurrences of the same date is $1 - (365)_{30}/365^{30} \approx .706$. Hence, there is more than a 70% chance that two people will have the same birthday. For 20 people, the same probability is $1 - (365)_{20}/365^{20} \approx .411$ and for 40 people it is $1 - (365)_{40}/365^{40} \approx .891$.

P 4.2.9. Call the number that we want N. Let M be the number of ways to assign 10 police officers into 5 distinct squad cars. The numbers M and N are not the same since, in finding M, we are placing the officers in distinct cars, and so we not only pair the officers but permute the pairs among the cars. Hence, $M = 5! N$. To find M, first decide which two go in car 1, then from the remaining which two go in car 2, etc. We get $M = \binom{10}{2}\binom{8}{2}\binom{6}{2}\binom{4}{2}\binom{2}{2} = \frac{10!}{2^5}$. Thus

$$N = \frac{1}{5!} \frac{10!}{2^5} = 945.$$

ALTERNATIVE SOLUTION: The oldest police officer has 9 choices of partner. The oldest of the remaining eight has 7 choices of partner. Continuing in this way, the final answer is $9 \times 7 \times 5 \times 3 \times 1 = 945$.

P 4.2.17. The total number of ways to distribute 13 cards to each of 4 players is $\binom{52}{13}\binom{39}{13}\binom{26}{13}\binom{13}{13}$. If you want each to get 3 picture cards, then choose 3 of the 12 picture cards, and 10 of the 40 non-picture cards to give to the first player. Then choose another 3 picture cards from the remaining 9 and another 10 from the remaining 30 non-picture cards, and so on. Hence, the probability is

$$\frac{\binom{12}{3}\binom{40}{10}\binom{9}{3}\binom{30}{10}\binom{6}{3}\binom{20}{10}\binom{3}{3}\binom{10}{10}}{\binom{52}{13}\binom{39}{13}\binom{26}{13}\binom{13}{13}} = \frac{40!\,12!\,(13!)^4}{52!\,(10!)^4(3!)^4} = \frac{257,330,216}{7,937,669,495} \approx 0.0324.$$

P 4.2.23.

(a) Let $p(k)$ be the probability that 2 out of 20 fish have been marked if the pond has k fish. There are $\binom{k}{20}$ ways to choose 20 fish from the k fish. How many ways are there to pick 20 fish from the lake and get exactly 2 marked fish? There are $\binom{10}{2}$ ways to pick 2 marked fish and, regardless of our choice of marked fish, there are $\binom{k-10}{18}$ ways to pick the remaining 18 fish. Hence,

$$p(k) = \frac{\binom{10}{2}\binom{k-10}{18}}{\binom{k}{20}}.$$

(b) Consider $\frac{p(k)}{p(k-1)}$. This ratio is greater than 1 if and only if $p(k) > p(k-1)$. This will lead us to finding the value of k that maximizes $p(k)$. We have

$$\frac{p(k)}{p(k-1)} = \frac{\binom{10}{2}\binom{k-10}{18}}{\binom{k}{20}} \frac{\binom{k-1}{20}}{\binom{10}{2}\binom{k-11}{18}} = \frac{\binom{k-10}{18}}{\binom{k-11}{18}} \frac{\binom{k-1}{20}}{\binom{k}{20}}$$
$$= \frac{(k-10)!\,(k-29)!}{(k-28)!\,(k-11)!} \frac{(k-1)!\,(k-20)!}{(k-21)!\,k!} = \frac{(k-10)(k-20)}{(k-28)(k)}.$$

Now, this fraction is greater than 1 if the numerator is bigger than the denominator. This happens if $(k-10)(k-20) > (k-28)k$. Expanding both sides, we get $k^2 - 30k + 200 > k^2 - 28k$, which simplifies to $2k - 200 < 0$. Hence, $k < 100$. This means that for $k = 28, \ldots, 99, p(k) > p(k-1)$. To find when $\frac{p(k)}{p(k-1)} < 1$, we do the exact same calculation but with the sign reversed. So for $k > 100$, we have $p(k) < p(k-1)$. For $k = 100$, we get that $p(k) = p(k-1)$ and so we have $p(28) < p(29) < \ldots < p(99) = p(100) > p(101) > p(102) > \cdots$. This means that if we estimate that there are 99 or 100 fish in the lake, we have maximized the probability that if we catch 20 fish then 2 of the fish are marked.

We could have arrived at essentially the same estimate more directly by saying that we will assume that the proportion of the marked fish in the whole lake is the same as the proportion of marked fish in our catch. Hence, $\frac{10}{k} = \frac{2}{20}$ and so $k = 100$. Proceeding in this way, we would, however, not know that an estimate of $k = 99$ also gives exactly the same probability.

P 4.2.25. You have to walk 16 blocks, and you have to decide which 7 of these 16 blocks are going to be south-north blocks. Thus the answer is $\binom{16}{7} = \binom{16}{9} = 11,440$. The same argument says that the number of different NE lattice paths from $(0,0)$ to (n,k) is $\binom{n+k}{k} = \binom{n+k}{n}$.

P 4.2.29. Recall that, by Problem $P\ 4.2.25$, the number of NE paths from $(0,0)$ to (n,k) is $\binom{n+k}{k}$. Also, note that the number of paths from (a,b) to (n,k) is the same as the number of paths from $(0,0)$ to $(n-a, k-b)$. A path from $(0,0)$ to $(9,7)$ that goes through $(2,3)$ consists of two segments, the part from $(0,0)$ to $(2,3)$ and the part from $(2,3)$ to $(9,7)$. The number of the former is $\binom{5}{2}$, and the number of the latter – which is the same as the number of paths from $(0,0)$ to $(7,4)$ – is $\binom{11}{4}$. Hence, the number of NE paths from $(0,0)$ to $(9,7)$ that pass through $(2,3)$ is $\binom{5}{2}\binom{11}{4}$. From among these, how many also go through $(4,6)$? Such paths, must first go through $(2,3)$, then make their way to $(4,6)$ and from there continue to $(9,7)$. There are $\binom{5}{2}\binom{5}{2}\binom{6}{1}$ such NE paths. Hence, the total number of different NE lattice paths from $(0,0)$ to $(9,7)$ that pass through $(2,3)$ but avoid $(4,6)$ is

$$\binom{5}{2}\binom{11}{4} - \binom{5}{2}\binom{5}{2}\binom{6}{1} = 2700.$$

P 4.3.9. The total number of ways of arranging the letters is $\binom{9}{2\,2\,2\,1\,1\,1}$. To make sure that no two vowels appear consecutively, first arrange the six consonants W,S,C,N,S,N. There are $\binom{6}{2\,2\,1\,1}$ ways of doing this. There are seven gaps for the three vowels. You can put an I before, after, or both before and after the W. For the first two cases, you then have 5 choices for placing the other I and 5 choices for placing the O. For the last case, you have 5 choices for placing the O. Hence, you have a total of $2 \times 5 \times 5 + 5$ ways of placing the vowels. Hence, the total number of ways of arranging the letters with the constraints is $55\binom{6}{2\,2\,1\,1}$. The sought-after probability is

$$\frac{55\binom{6}{2\,2\,1\,1}}{\binom{9}{2\,2\,2\,1\,1\,1}} = \frac{55}{252}.$$

After placing the consonants, we could have counted the number of arrangements of vowels with the given constraints differently. Without the constraint on W,I the three vowels can be inserted in $\binom{7}{2} \cdot 5 = 105$ ways (insert the two I's in $\binom{7}{2}$ ways and then insert the O). If we insist that there be *no* adjacent W,I pair, then there are $\binom{5}{2} \cdot 5 = 50$ ways of inserting the vowels (since the W eliminates two of the possible places for the I's, but has no effect on the subsequent placement of the O). Hence, the number of ways to place the vowels and satisfy the constraints is $105 - 50 = 55$.

P 4.3.13. If you have k teams of size 3, then, to have the rest of the students in teams of size 4, $18 - 3k$ has to be divisible by 4. This happens only if $k = 2$ or $k = 6$. Thus we either have six teams of size 3 or two teams of size 3 and three teams of size 4. For the first case, if the teams were distinct, then, by Theorem 4.30, the number of possibilities would be $\binom{18}{3\,3\,3\,3\,3\,3}$. Since they are not distinct, we have counted every possibility 6! ways, and so the number of possibilities is $\frac{1}{6!}\binom{18}{3\,3\,3\,3\,3\,3}$. Similarly, the number of possibilities for the second case is $\frac{1}{2!\,3!}\binom{18}{4\,4\,4\,3\,3}$, and the final answer is

$$\frac{1}{6!}\binom{18}{3\,3\,3\,3\,3\,3} + \frac{1}{2!\,3!}\binom{18}{4\,4\,4\,3\,3} = 1{,}262{,}661{,}400.$$

P 4.3.19. There are 2^k possible sequences of k heads and tails. From among these, how many have the ith head on the kth toss? In such a case, we have a permutation of $\{(i-1)\cdot H, (k-i)\cdot T\}$ followed by a H. There are $\binom{k-1}{i-1\;k-i} = \binom{k-1}{i-1}$ such sequences. Hence, the probability that the ith head comes exactly on the kth toss is

$$\frac{1}{2^k}\binom{k-1}{i-1}.$$

P 4.4.7. Pair the donuts up so that I have 17 identical pairs. Give one pair of donuts to each individual (so that they each will get a *positive* number). Let $x_1, \ldots, x_5$ denote the number of additional pairs of donuts that Farzin, Farshad, Faranak, Fariborz, and Farzaneh get, respectively. I want the number of non-negative integer solutions to

$$x_1 + x_2 + x_3 + x_4 + x_5 = 12.$$

The answer is

$$\left(\binom{5}{12}\right) = \binom{5 + 12 - 1}{12} = \binom{16}{4} = \frac{16 \times 15 \times 14 \times 13}{4 \times 3 \times 2} = 1820.$$

P 4.4.11. If $x_1 + \cdots + x_7$ is less than or equal to 47, then the difference of 47 and $x_1 + \cdots + x_7$ is a non-negative integer. Let this difference be x_8. In other words, define a "slack variable" x_8, so that $x_1 + \cdots + x_7 + x_8 = 47$. We now want the number of non-negative integer solutions to the latter, and the number of such solutions is $\left(\binom{8}{47}\right) = \binom{54}{47} = 177{,}100{,}560$.

Alternatively, we could have said that the left-hand side of the equation can be equal to 47, $46, \ldots, 0$, and as a result the answer is $\left(\binom{7}{47}\right) + \left(\binom{7}{46}\right) + \cdots + \left(\binom{7}{0}\right)$. Incidentally, this gives a proof of the following identity:

$$\left(\binom{7}{47}\right) + \left(\binom{7}{46}\right) + \cdots + \left(\binom{7}{0}\right) = \left(\binom{8}{47}\right).$$

P 4.4.15. Seat the five Romulans and then place the empty chairs between them. There are six places to put the eight empty chairs (before, in between, and after the Romulans), and thus we want the number of integer solutions to

$$x_1 + x_2 + \cdots + x_6 = 8,$$

such that x_1 and x_6 are non-negative and the rest are positive. Let $y_i = x_i - 1$ for $i = 2, 3, 4, 5$ and let $y_j = x_j$ for $j = 1, 6$. Now the above is the same as the number of non-negative integer solutions to

$$y_1 + y_2 + \cdots + y_6 = 4.$$

We know that the answer is $\left(\binom{6}{4}\right) = \binom{4+6-1}{4} = 126$.

P 5.1.7. The resulting numbers are indeed the binomial coefficients right below where we stop (see Figure D.7). Hence, we conjecture that

$$\binom{n}{0} + \binom{n+1}{1} + \cdots + \binom{n+k}{k} = \binom{n+k+1}{k}.$$

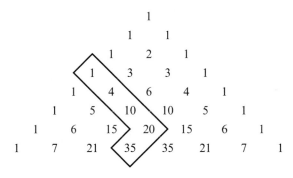

Figure D.7 Another hockey-stick theorem: $1 + 4 + 10 + 20 = 35$.

PROOF 1: Using the identity $\binom{n}{k} = \binom{n}{n-k}$ we can rewrite what we want to prove. The claim becomes

$$\binom{n}{n} + \binom{n+1}{n} + \cdots + \binom{n+k}{n} = \binom{n+k+1}{n+1},$$

and this is exactly the hockey stick identity of Problem P 5.1.6. In other words, if we take the old hockey stick and flip it over (using the fact that $\binom{n}{k} = \binom{n}{n-k}$), we get another hockey stick.

PROOF 2: We use the identity $\binom{n}{k} = \binom{n-1}{k} + \binom{n-1}{k-1}$ repeatedly:

$$\binom{n+k+1}{k} = \binom{n+k}{k} + \binom{n+k}{k-1} = \binom{n+k}{k} + \binom{n+k-1}{k-1} + \binom{n+k-1}{k-2}$$
$$= \binom{n+k}{k} + \binom{n+k-1}{k-1} + \binom{n+k-2}{k-2} + \binom{n+k-2}{k-3} = \cdots$$
$$= \binom{n+k}{k} + \binom{n+k-1}{k-1} + \binom{n+k-2}{k-2} + \cdots + \binom{n}{0}$$

PROOF 3: Fixing n, we use induction on k. For the base case, let $k = 0$, and we have $\binom{n}{0} = 1 = \binom{n+1}{0}$. For the inductive case, we assume that the identity is true for k, and prove it for $k + 1$:

$$\binom{n}{0} + \binom{n+1}{1} + \cdots + \binom{n+k}{k} + \binom{n+k+1}{k+1} = \binom{n+k+1}{k} + \binom{n+k+1}{k+1}$$

$$= \binom{n+k+2}{k+1}.$$

While the last two proofs were organized differently, they are basically the same proof, since the last step of the inductive proof also relied on the identity $\binom{n}{k} = \binom{n-1}{k} + \binom{n-1}{k-1}$.

P 5.1.11. Starting with $\binom{n}{m}$, we can write it as $\binom{n-1}{m-1} + \binom{n-1}{m}$. Now, unwind the second term, and replace it with the sum of two more binomial coefficients. Continue to unwind the term of the form $\binom{n-i}{m}$. We get

$$\binom{n}{m} = \binom{n-1}{m-1} + \binom{n-1}{m} = \binom{n-1}{m-1} + \binom{n-2}{m-1} + \binom{n-2}{m}$$

$$= \binom{n-1}{m-1} + \binom{n-2}{m-1} + \binom{n-3}{m-1} + \binom{n-3}{m} = \cdots$$

$$= \binom{n-1}{m-1} + \binom{n-2}{m-1} + \binom{n-3}{m-1} + \cdots + \binom{m}{m-1} + \binom{m}{m}$$

$$= \binom{n-1}{m-1} + \binom{n-2}{m-1} + \binom{n-3}{m-1} + \cdots + \binom{m}{m-1} + \binom{m-1}{m-1}.$$

This is the hockey-stick identity of Problem P 5.1.6. If we had stopped earlier, we would get

$$\binom{n}{m} = \binom{n-1}{m-1} + \binom{n-2}{m-1} + \cdots + \binom{n-j}{m-1} + \binom{n-j}{m},$$

for any $1 \leq j \leq n - m$. You could think of this as a generalization of the hockey-stick identity (maybe the "broken-hockey-stick" identity – see Figure D.8). Start with any term in the triangle ($\binom{n-j}{m}$ in our identity), add it to the term to its left, and then keep adding, going down this (new) column. Stop whenever you want, and the sum will be diagonally below and to the right of where you stopped. You can actually combine two hockey-stick identities to get the broken-hockey-stick identity. (Do you see how?)

P 5.1.19. Let S be a set with $m_1 + m_2$ elements. There are $\binom{m_1+m_2}{n}$ ways to choose n elements from S. Now, partition S into two disjoint subsets T and U such that T has m_1 elements and U has m_2 elements. Now to pick n elements from S, you can pick k elements from T and $n-k$ elements from U, and k can range from 0 to n. In this method, for a given k we get $\binom{m_1}{k}\binom{m_2}{n-k}$ subsets of S of size n. Thus

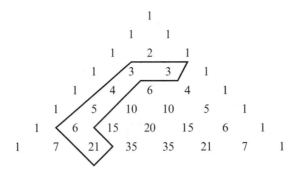

Figure D.8 The broken-hockey-stick theorem: $3 + 3 + 4 + 5 + 6 = 21$.

$$\binom{m_1 + m_2}{n} = \text{the total number of subsets of } S \text{ of size } n$$

$$= \binom{m_1}{0}\binom{m_2}{n} + \binom{m_1}{1}\binom{m_2}{n-1} + \cdots + \binom{m_1}{n}\binom{m_2}{0}$$

$$= \sum_{k=0}^{n} \binom{m_1}{k}\binom{m_2}{n-k}.$$

Recall that $\binom{n}{k}$ is defined to be zero if $k > n$, and so if n is bigger than m_1 (or m_2 or both), then some of the binomial coefficients in the identity will be zero. The identity will nevertheless be valid.

P 5.1.25. How many n-letter words can you make from r A's, s B's, one O, and $n - r - s - 1$ X's if the only condition is that O occurs somewhere to the right of all the A's and B's?

Pick the $r+s+1$ spots for the A's, B's, and the O. Place the O in the last spot, then permute the A's and the B's. The X's occupy all the remaining spots. Hence, the total number of ways of accomplishing the task is $\binom{n}{r+s+1}\binom{r+s}{s} = \frac{n!}{(r+s+1)\, r!\, s!}$.

Alternatively, if we assume O is in the kth position, then, from the previous $k - 1$ positions, we have to pick r spots for the A's, and, from what is left, s spots for the B's. Hence, in this case, the answer is $\binom{k-1}{r}\binom{k-r-1}{s}$. Varying k from 1 to n gives

$$\sum_{k=1}^{n} \binom{k-1}{r}\binom{k-r-1}{s} = \binom{n}{r+s+1}\binom{r+s}{s}.$$

(Note that if k is too small relative to r and s, the corresponding terms in the sum will be 0.)

P 5.1.29. Note that $\binom{8}{4} = 70 > 64 = 4^3$, and so the condition $n \geq 5$ is necessary. For the base case, we have $\binom{10}{5} = 252 < 256 = 4^4$. Now note that $\frac{2n+1}{n+1} < \frac{2n+2}{n+1} = 2$, and so, for the inductive step,

$$\binom{2n+2}{n+1} = \frac{(2n+2)(2n+1)}{(n+1)^2}\binom{2n}{n}$$

$$< \frac{2(2n+1)}{n+1} 4^{n-1} < 4 \times 4^{n-1} = 4^n.$$

P 5.2.5. Start with $(1 + x)^n = \sum_{k=0}^{n} \binom{n}{k} x^k$, differentiate both sides, and multiply both sides by x. You get

$$xn(1 + x)^{n-1} = \sum_{k=0}^{n} \binom{n}{k} k x^k,$$

at least for $n \geq 1$. Let $x = -1$, and get

$$\sum_{k=0}^{n} k(-1)^k \binom{n}{k} = 0.$$

P 5.2.9. Let $n \geq 2$ be an integer. We claim that

$$\binom{n}{1} + 3\binom{n}{3} + 5\binom{n}{5} + \cdots = 2\binom{n}{2} + 4\binom{n}{4} + 6\binom{n}{6} + \cdots.$$

In Proposition 5.26(c) and Problem P 5.2.5, we already showed that

$$0\binom{n}{0} + \binom{n}{1} + 2\binom{n}{2} + 3\binom{n}{3} + \cdots \qquad +n\binom{n}{n} = n2^{n-1}$$

$$0\binom{n}{0} - \binom{n}{1} + 2\binom{n}{2} - 3\binom{n}{3} + \cdots + (-1)^n n\binom{n}{n} = 0.$$

Averaging these two identities, we get

$$2\binom{n}{2} + 4\binom{n}{4} + \cdots = n2^{n-2}.$$

Since this is exactly one-half of the $\binom{n}{1} + 2\binom{n}{2} + 3\binom{n}{3} + \cdots + n\binom{n}{n}$, we conclude that the remaining terms add up to the same sum, and so

$$\binom{n}{1} + 3\binom{n}{3} + 5\binom{n}{5} + \cdots = n2^{n-2} = 2\binom{n}{2} + 4\binom{n}{4} + 6\binom{n}{6} + \cdots.$$

P 5.2.13. You have a room with n people and you are going to choose a committee and assign a chair and a vice-chair, allowing for the possibility that the same person may hold both positions. You could say that the committee is going to have k members for some $1 \leq k \leq n$. If the size of the committee is k, then we choose the committee in $\binom{n}{k}$ ways and then have k choices for the chair and k choices for the vice-chair. Hence, the total number of ways of accomplishing the task is

$$\binom{n}{1} + 2^2\binom{n}{2} + \cdots + k^2\binom{n}{k} + \cdots + n^2\binom{n}{n}.$$

On the other hand, we could decide on the chair and vice-chair first. This we break into two cases: all the cases when the chair and vice-chair are the same person and all the cases when the two are different people. In the first case, there are n choices for the chair/vice-chair person, and we then have to pick some additional committee members to join. So we just need the total number of subsets of the set with $n - 1$ members (everyone except the chosen chair/vice-chair), and this number is 2^{n-1}. In the second case, we have n choices for the chair, $n-1$ choices for the vice-chair, and 2^{n-2} choices for other folks to join the committee. Hence, the total number of ways of doing this is

$$n2^{n-1} + n(n-1)2^{n-2} = n2^{n-2}(2+n-1) = n(n+1)2^{n-2}.$$

Hence,

$$\binom{n}{1} + 2^2\binom{n}{2} + \cdots + k^2\binom{n}{k} + \cdots + n^2\binom{n}{n} = n(n+1)2^{n-2}.$$

P 5.2.15.

(a) The binomial theorem tells us that

$$(1+x)^n = \binom{n}{0} + \binom{n}{1}x + \binom{n}{2}x^2 + \cdots + \binom{n}{k}x^k + \cdots + \binom{n}{n}x^n.$$

If we take three derivatives in a row, we get

$$n(1+x)^{n-1} = \binom{n}{1} + 2\binom{n}{2}x + 3\binom{n}{3}x^2 + \cdots + k\binom{n}{k}x^{k-1} + \cdots$$
$$+ n\binom{n}{n}x^{n-1}$$

$$n(n-1)(1+x)^{n-2} = 2\binom{n}{2} + 3\cdot2\binom{n}{3}x + \cdots + k(k-1)\binom{n}{k}x^{k-2} + \cdots$$
$$+ n(n-1)\binom{n}{n}x^{n-2}$$

$$n(n-1)(n-2)(1+x)^{n-3} = 3\cdot2\binom{n}{3} + \cdots + k(k-1)(k-2)\binom{n}{k}x^{k-3} + \cdots$$
$$+ n(n-1)(n-2)\binom{n}{n}x^{n-3}.$$

Now plug in $x = 1$ in the last equation to get

$$\sum_{k=3}^{n} k(k-1)(k-2)\binom{n}{k} = n(n-1)(n-2)2^{n-3}.$$

Since the terms in the sum on the left would be zero for $k = 0, 1,$ or 2, we now have exactly what we found in Problem P 5.2.14:

$$\sum_{k=0}^{n} k(k-1)(k-2)\binom{n}{k} = n(n-1)(n-2)2^{n-3}.$$

(b) To be able to get k^3 – as opposed to $k(k-1)(k-2)$ – we have to (just as in the proof of Proposition 5.26(d)) multiply by x before taking the second and third derivatives.

We start by multiplying the first derivative, found in the previous part, by x, and then take a derivative, multiply by x, and take another derivative:

$$nx(1+x)^{n-1} = \binom{n}{1}x + 2\binom{n}{2}x^2 + \cdots + k\binom{n}{k}x^k + \cdots$$
$$+ n\binom{n}{n}x^n$$

$$n(1+x)^{n-1} + n(n-1)x(1+x)^{n-2} = \binom{n}{1} + 2^2\binom{n}{2}x + \cdots + k^2\binom{n}{k}x^{k-1} + \cdots$$
$$+ n^2\binom{n}{n}x^{n-1}$$

$$nx(1+x)^{n-1} + n(n-1)x^2(1+x)^{n-2} = \binom{n}{1}x + 2^2\binom{n}{2}x^2 + \cdots + k^2\binom{n}{k}x^k + \cdots$$
$$+ n^2\binom{n}{n}x^n$$

$$n(n^2x^2 + 3nx - x + 1)(1+x)^{n-3} = \binom{n}{1} + 2^3\binom{n}{2}x + \cdots + k^3\binom{n}{k}x^{k-1} + \cdots$$
$$+ n^3\binom{n}{n}x^{n-1}.$$

(I skipped a step factoring and simplifying the last expression on the left.) Now plug in $x = 1$ and get

$$\sum_{k=0}^{n} k^3\binom{n}{k} = n^2(n+3)2^{n-3},$$

which again is the same identity as the one for Problem P 5.2.14.

Combinatorial arguments (such as those in Problem P 5.2.14) are more conceptual and more fun. However, it is not always easy to come up with them. Algebraic methods (such as the ones in this problem) are more tedious but, in some ways, more predictable.

P 5.3.5. A typical term in the expansion of this expression is

$$x^i(x^{-1})^j(2x^{-4})^k.$$

where $0 \le i, j, k \le 17$, and $i + j + k = 17$. The coefficient of this term is $\binom{17}{i\,j\,k}$. We want the constant term and so we want – in addition to $i + j + k = 17$ – to have $i - j - 4k = 0$. You can find the common integer solutions to these equations by trial and error. You can also subtract the two equations and get

$$2j + 5k = 17.$$

We want integer solutions, and so k must be an odd integer. On the other hand, we have $17 - 5k = 2j \ge 0$. This implies that $5k \le 17$, which means that $0 \le k \le 3$. Thus the only possibilities for k are 1 and 3. For $k = 1$, we get $j = 6$ and $i = 10$, while for $k = 3$, we have $j = 1$ and $i = 13$. Thus, there are two constant terms, and adding them we get

$$2\begin{pmatrix} 17 \\ 10 \quad 6 \quad 1 \end{pmatrix} + 2^3\begin{pmatrix} 17 \\ 13 \quad 1 \quad 3 \end{pmatrix}.$$

P 5.4.7. By Problem P 5.4.5, $p'(x) = 4a_4x^3 + 3a_3x^2 + 2a_2x + a_1$ has real roots. By the result of Problem P 5.4.6, we have $9a_3^2 \geq 24a_2a_4$ and $4a_2^2 \geq 9a_1a_3$. As a result, $a_3^2 \geq \frac{8}{3}a_2a_4$ and $a_2^2 \geq \frac{9}{4}a_1a_3$. Thus, so far, we know that the sequence a_4, a_3, a_2, a_1 is log concave.

Invoking Problem P 5.4.5 again, we know that $x^4p(1/x) = a_4 + a_3x + a_2x^2 + a_1x^3 + a_0x^4$ also has real roots. Taking the derivative, we know that $a_3 + 2a_2x + 3a_1x^2 + 4a_0x^3$ also has real roots, and by Problem P 5.4.6, we have $9a_1^2 \geq 24a_2a_0$ and so $a_1^2 \geq \frac{8}{3}a_2a_0$. The proof is now complete.

P 5.4.11. The matrix of signed binomial coefficients is $(n+1)\times(n+1)$ and its rows and columns are indexed 0 through n. The (i,j) entry – the entry in row indexed i and column indexed j – is $(-1)^{i+j}\binom{i}{j}$. We mimic Problem P 5.4.10. Let V be the vector space of polynomials of degree less than or equal to n with real coefficients. Let $S = \{1, x, \ldots, x^n\}$ and $B = \{1, (x + 1), (x + 1)^2, \ldots, (x + 1)^n\}$. Both S and B are bases for V. We consider the change of basis matrices for these two bases. From Corollary 5.25, we have

$$(x + 1)^j = \sum_{k=0}^{j} \binom{j}{k} x^k.$$

Thus the jth column of the change of basis matrix from B to S is exactly the jth *row* of the Karaji–Jia triangle. This shows that the change of basis matrix from B to S is C_n^t, the transpose of the binomial coefficients. From Problem P 5.4.9, we know that

$$x^j = \sum_{k=0}^{j} (-1)^{j+k} \binom{j}{k} (x + 1)^k.$$

This shows that the change of basis matrix from S to B is the transpose of the matrix of signed binomial coefficients. If we let D_n stand for the $(n + 1) \times (n + 1)$ matrix of signed binomial coefficients, we now know that C_n^t and D_n^t are inverses of each other. (The change of basis matrix from B to S is always the inverse of the change of basis matrix from S to B.) But if we transpose both sides of $C_n^t D_n^t = I_{n+1}$, we get $D_n C_n = I_{n+1}$. As a result, we have proved that C_n and D_n are inverses of each other.

P 5.4.17. From Problem P 5.4.16 we have that $P_n = C_n A_n$. Thus $A_n = C_n^{-1} P_n$, and we know (from Problem P 5.4.11) that C_n^{-1} is the matrix of signed binomial coefficients. Now the (r, m) entry of A_n is $a_{r,m}$, and this will be equal to the dot product of the rth row of C_n^{-1} and the mth column of P_n. Hence

$$a_{r,m} = (-1)^r \binom{r}{0} 0^m + (-1)^{r+1} \binom{r}{1} 1^m + (-1)^{r+2} \binom{r}{2} 2^m + \cdots + (-1)^{2r} \binom{r}{r} r^m$$

$$= \sum_{k=0}^{r} (-1)^{r+k} \binom{r}{k} k^m.$$

We should mention a slightly subtle point. We have shown that if there are constants $a_{r,m}$ that give us identities of the form in Problem P 5.4.16, then they have to be the ones that we just found. However, we really showed that the above identities only work for $k = 0, \ldots, n$. How do we know that they are valid for all k? The above formula for $a_{r,m}$ does not depend on n, and we can make n as large as we want, and so each of those identities is valid for as large a range of values of k as we wish. Hence, they are true for all values of k. (In fact, both sides of the identity are polynomials, and if we bring them to one side, we have a polynomial equal to zero. Now a polynomial has only a finite number of roots unless it is the zero polynomial. We conclude that the two sides are equal for all values of k.)

P 5.4.25. Chung–Feller Theorem. Let n be a positive integer, and let i be an integer such that $0 \le i \le n$. Define S_i to be the collection all NE lattice paths from $(0,0)$ to (n,n) that have exactly i N steps above the line $y = x$. Then, regardless of the value of i,

$$|S_i| = \frac{1}{n+1} \binom{2n}{n}.$$

Proof. The total number of NE lattice paths from $(0,0)$ to (n,n) is $\binom{2n}{n}$, and each NE lattice path belongs to exactly one S_i. Hence, the theorem will be proved once we show that the $n+1$ sets $S_0, S_1, \ldots, S_n$ all have the same size.

We show that all these sets have the same size by showing that each two consecutive ones have the same size. We do this by providing a bijection between the paths in S_i and the paths in S_{i+1} for $0 \le i \le n-1$.

For this purpose, fix an i with $0 \le i \le n-1$, and let $P \in S_i$. Since $i \ne n$, P has some N step that is below the diagonal, and so the path visits the area below $y = x$. Let h be the smallest non-negative integer such that $(h+1, h)$ is on P. After $(h+1, h)$, the path will eventually get to (n,n), and so it visits the line $y = x$ at (n,n), but maybe it does so earlier. Let k be the least positive integer greater than h with (k,k) on the path. Now, we split P into three parts: P_1 is the part from $(0,0)$ to $(h+1, h)$, P_2 is the part from $(h+1, h)$ to (k,k), and P_3 is the part from (k,k) to (n,n).

Define the NE path $P^\star$ as the path that you get by switching the place of P_1 and P_2 and appropriately translating them. In other words, we translate P_2 to start at $(0,0)$. Since P_2 itself went from $(h+1, h)$ to (k,k), the translation of P_2 will go from $(0,0)$ to $(k-h-1, k-h)$. We then translate P_1 to start at $(k-h-1, k-h)$ and end at (k,k). We then continue on P_3 to (n,n).

The path P_2 went from $(h+1, h)$ to (k,k) and all of its steps were under the line $y = x$, and its last N step went from $(k, k-1)$ to (k,k). After translating it, almost all steps of the translated P_2 will continue to be under the line $y = x$ except the last N step, which now goes from $(k-h-1, k-h-1)$ to $(k-h-1, k-h)$. So the translated P_2 has one N step above $y = x$ while the original P_2 had none.

The path P_1 originally had all of its steps except the last one – which went from (h, h) to $(h+1, h)$ – above $y = x$ (it is possible that its first step was from $(0,0)$ to $(1,0)$, but even then P_1 has just one step, and the statement stands that all of its steps except the last one are above $y = x$.). We translated P_1 to start not on $y = x$ but one step north of $y = x$. Hence, the translated P_1 has all of its steps above $y = x$. In particular, it has as many N steps above $y = x$ as the original P_1 did.

The path P_3 has not been moved, and so we conclude that $P^\star$ has one more N step than P and so $P^\star \in S_{i+1}$. Thus the map sending P to $P^\star$ is a map from S_i to S_{i+1}. The map is 1-1 since, if you change the original P, the resulting $P^\star$ will change as well. To show that the map is onto, let $Q \in S_{i+1}$. We can now reverse-engineer what we did above to find a path P with $P^\star = Q$. More specifically, let Q have a first N step above $y = x$, and so let h be the least non-negative integer with the point $(h, h+1)$ on the path Q. Likewise, let k be the smallest positive integer greater than h such that (k, k) is on Q. Again split Q into three parts and switch the first two parts (after translating them appropriately). The resulting path P has the property that $P^\star = Q$, proving that our map is onto. The proof is now complete. □

P 5.4.27. A lattice path of the type we are interested in has to have an N step right at the beginning and an E step right at the end. Thus all such paths are NE lattice paths from $(0, 1)$ to $(n-1, n)$, and they never go below the diagonal connecting $(0, 1)$ and $(n-1, n)$. Eliminate the first and the last step (which are the same for all such paths) and translate the path to start at $(0,0)$. We get a NE lattice path from $(0,0)$ to $(n-1, n-1)$ that always stays above the diagonal but may touch it. These are exactly the Dyck paths from $(0,0)$ to $(n-1, n-1)$, and their number, by Problem P 5.4.22, is $\frac{1}{n}\binom{2n-1}{n-1} = c_{n-1}$, the $(n-1)$th Ming–Catalan number.

P 5.4.37.

(a) The acts E and S do not affect each other and are independent events. Hence, the order of performing them makes no difference. You can perform them simultaneously or in either order. This is the same as in Problem P 5.4.35, and the probability is

$$\frac{47\binom{46}{6}}{47\binom{47}{7}} = \frac{7}{47}.$$

(b) The probability now is

$$\frac{\binom{46}{6}}{\binom{47}{7}} = \frac{7}{47}.$$

If we do E first and then S, then replacing E with E' doesn't change anything, since 23 is one of the possibilities for the outcome of E and all outcomes of E are created equal and would lead to the same conclusion. Again, the order of E' and S does not matter, since they do not affect each other.

(c) The probability in this case is

$$\frac{7}{47},$$

and the order does not matter either.

(d) The probability here is 1, since 23 is a member of the chosen set.

(e) Consider the question "What is the probability that the element that you picked is in the subset that you picked?" If you have done E' and S, then the answer is p. If you have done E and S', then the answer is q. But we have seen that the answer in both cases is the same as if you had actually done E and S and then asked the same question. Hence, $p = q$.

P 5.4.43. The integer k can be $0, 1, \ldots, n$. Hence, adding the answers to Problem P 5.4.42, and using the hockey-stick identity of Problem P 5.1.6, we get

$$\sum_{k=0}^{n} \binom{2n-k-1}{n-1} = \binom{2n-1}{n-1} + \binom{2n-2}{n-1} + \cdots + \binom{n-1}{n-1} = \binom{2n}{n}.$$

P 6.1.7. Consider the ordered list 12345678. For each of the numbers 1 through 8, we have to replace them with one of the names of the Ostrogothic Kings in a way that all four names are used. First partition [8] into four non-empty subsets. Each of these would then be given one of the names. There are $\begin{Bmatrix} 8 \\ 4 \end{Bmatrix}$ ways of partitioning [8] into four non-empty parts, and 4! factorial ways of assigning the four names to the parts. Hence, the answer is $4! \begin{Bmatrix} 8 \\ 4 \end{Bmatrix} = 24(1701) = 40{,}824$.

P 6.1.15. The number of ways of distributing n distinct objects in three distinct boxes is $3! \begin{Bmatrix} n \\ 3 \end{Bmatrix}$ (entry $\boxed{9}$). We do this count in a different way. In how many ways can I place the n objects in exactly one of the boxes? The answer is clearly 3, since you just have to pick which box you are going to use. In how many ways can I place the n objects in exactly two of the boxes? Pick the two boxes in $\binom{3}{2} = 3$ ways, and then you have $2! \begin{Bmatrix} n \\ 2 \end{Bmatrix} = 2(2^{n-1} - 1) = 2^n - 2$ choices. Now, if you allow empty parts, there are 3^n ways to distribute the n objects in the three boxes (you have three choices for each object). But some of these have empty boxes. So we have to subtract the number of configurations where we use exactly one or exactly two boxes. Hence, we have $3^n - 3(2^n) - 2) - 3$. This gives the same count as $3! \begin{Bmatrix} n \\ 3 \end{Bmatrix}$, and so

$$\begin{Bmatrix} n \\ 3 \end{Bmatrix} = \frac{1}{3!} [3^n - 3(2^n - 2) - 3]$$

$$= \frac{1}{2} [3^{n-1} - 2^n + 1].$$

The above argument is a combinatorial proof of this identity. Alternatively, you could try to look at some data, guess a formula, and then prove it by induction. The proof by induction can also be a second check to convince us that we didn't miss anything in our argument. We give the induction proof next.

CLAIM: $\begin{Bmatrix} n \\ 3 \end{Bmatrix} = \frac{1}{2}(3^{n-1} - 2^n + 1)$.

Complete Solutions for Selected Problems 563

PROOF: We prove the claim by induction on n. For $n = 3$ we have $\left\{{3 \atop 3}\right\} = 1 = \frac{1}{2}(3^{3-1} - 2^3 + 1)$. Now assume that the claim is true for all $n < m$. We prove that the claim is true for $n = m$. We have

$$\left\{{m \atop 3}\right\} = 3\left\{{m-1 \atop 3}\right\} + \left\{{m-1 \atop 2}\right\} = \frac{3}{2}(3^{m-2} - 2^{m-1} + 1) + (2^{m-2} - 1)$$

$$= \frac{1}{2}(3 \cdot 3^{m-2} - 3 \cdot 2^{m-1} + 3 + 2 \cdot 2^{m-2} - 2) = \frac{1}{2}[3^{m-1} - 2^{m-1}(3 - 1) + 1]$$

$$= \frac{1}{2}(3^{m-1} - 2^m + 1).$$

P 6.1.17. We use the recurrence relation for the Stirling numbers of the second kind and then apply Theorem 6.19.

$$\left\{{n \atop 1}\right\} + 4\left\{{n \atop 2}\right\} + 12\left\{{n \atop 3}\right\} + 24\left\{{n \atop 4}\right\} + 24\left\{{n \atop 5}\right\}$$

$$= \left\{{n-1 \atop 1}\right\} + \left\{{n-1 \atop 0}\right\} + 8\left\{{n-1 \atop 2}\right\} + 4\left\{{n-1 \atop 1}\right\} + 36\left\{{n-1 \atop 3}\right\} + 12\left\{{n-1 \atop 2}\right\}$$

$$+ 96\left\{{n-1 \atop 4}\right\} + 24\left\{{n-1 \atop 3}\right\} + 120\left\{{n \atop 5}\right\} + 24\left\{{n \atop 4}\right\}$$

$$= \left\{{n-1 \atop 0}\right\} + 5\left\{{n-1 \atop 1}\right\} + 20\left\{{n-1 \atop 2}\right\} + 60\left\{{n-1 \atop 3}\right\} + 120\left\{{n-1 \atop 4}\right\} + 120\left\{{n-1 \atop 5}\right\}$$

$$= (5)_0\left\{{n-1 \atop 0}\right\} + (5)_1\left\{{n-1 \atop 1}\right\} + (5)_2\left\{{n-1 \atop 2}\right\} + (5)_3\left\{{n-1 \atop 3}\right\} + (5)_4\left\{{n-1 \atop 4}\right\} + (5)_5\left\{{n-1 \atop 5}\right\}$$

$$= 5^{n-1}.$$

P 6.1.19. Consider two cases: $t \geq s$ and $t < s$. In the former case, for all values of i with $s < i \leq t$, we have $\left\{{s \atop i}\right\} = 0$, and so we can rewrite the identity as

$$t^s = \sum_{i=0}^{s}(t)_i\left\{{s \atop i}\right\} = \sum_{i=0}^{s}\left\{{s \atop i}\right\}i!\binom{t}{i}.$$

In the latter case ($t < s$), for all values of i with $t < i \leq s$, both $(t)_i$ and $\binom{t}{i}$ are equal to zero. (Recall that, for $k > 0$, $(x)_k = x(x-1)(x-2)\cdots(x-k+1)$, and so if x is a non-negative integer less than k, then $(x)_k = 0$.) Thus, there is no harm in adding terms to the sum and writing

$$t^s = \sum_{i=0}^{s}(t)_i\left\{{s \atop i}\right\} = \sum_{i=0}^{s}\left\{{s \atop i}\right\}i!\binom{t}{i}.$$

P 6.1.27. Let $f(x) = x^t$. If we define $f_0(x) = f(x)$ and, for $n > 0$, $f_n(x) = xf'_{n-1}(x)$, then $f_1(x) = xf'(x) = tx^t$, $f_2(x) = t^2x^t$, and, more generally, $f_s(x) = t^sx^t$. On the other hand,

$f'(x) = tx^{t-1}$, $f''(x) = (t)_2 x^{t-2}$, and, more generally, for $1 \le k \le t$, $f^{(k)}(x) = (t)_k x^{t-k}$. Hence, the result of Problem P 6.1.26 becomes

$$t^s x^t = \sum_{k=0}^{s} (t)_k {s \brace k} x^k x^{t-k} = x^t \sum_{k=0}^{s} (t)_k {s \brace k}.$$

Canceling x^t from both sides, we get

$$t^s = {s \brace 0} + (t)_1 {s \brace 1} + (t)_2 {s \brace 2} + \cdots + (t)_s {s \brace s}.$$

In Problem *P 6.1.19*, we argued that the latter is the same as the result of Theorem 6.19.

P 6.1.31. We have

$$f_n(x) = x + \frac{2^n x^2}{2!} + \frac{3^n x^3}{3!} + \cdots + \frac{k^n x^k}{k!} + \cdots$$

$$e^{-x} = 1 - \frac{x}{1!} + \frac{x^2}{2!} - \frac{x^3}{3!} + \cdots + (-1)^k \frac{x^k}{k!} + \cdots$$

When we multiply these two series, we get an x^k term whenever we multiply an x^i term from the first series with an x^{k-i} term from the second series. Hence, the coefficient of x^k in the product is

$$\frac{k^n}{k!} - \frac{(k-1)^n}{1! \, (k-1)!} + \frac{(k-2)^n}{2! \, (k-2)!} + \cdots + (-1)^{k-2} \frac{2^n}{(k-2)! \, 2!} + (-1)^{k-1} \frac{1^n}{(k-1)! \, 1!}$$

$$= \sum_{i=0}^{k} \frac{(-1)^i (k-i)^n}{i! \, (k-i)!}.$$

According to Proposition 6.14, $e^{-x} f_n(x) = \sum_{k=0}^{n} {n \brace k} x^k$. Thus, we must have

$${n \brace k} = \sum_{i=0}^{k} \frac{(-1)^i (k-i)^n}{i! \, (k-i)!}.$$

Now change variables and let $r = k - i$. We have $i = k - r$, and we get

$${n \brace k} = \sum_{r=0}^{k} \frac{(-1)^{k-r} r^n}{r! \, (k-r)!},$$

which is the statement of Theorem 6.23b.

P 6.1.33. Recall, from Problem P 5.4.16, that C_n, A_n, and P_n are all $(n+1) \times (n+1)$ matrices and their rows and columns are numbered 0 through n. The (i,j) entries of C_n, A_n, and P_n, respectively, are $\binom{i}{j}$, $a_{i,j}$, and i^j (we stipulated that $0^0 = 1$). We showed in Problem P 5.4.16 that $P_n = C_n A_n$. Now let F_n be a diagonal $(n+1) \times (n+1)$ matrix whose ith diagonal entry is $1/i!$ (again the matrix's rows and columns are indexed 0 to n). In our solution to Problem P 6.1.30, we showed that

$${s \brace t} = \frac{a_{t,s}}{t!},$$

and so $(S_n)^t = F_n A_n$. Now (this was Problem P 5.4.17) $A_n = C_n^{-1} P_n$, and so

$$(S_n)^t = F_n A_n = F_n C_n^{-1} P_n.$$

Thus the entry in row t and column s of $F_n C_n^{-1} P_n$ is $\{^s_t\}$.

As an example, let $n = 4$; then

$$C_4 = \begin{bmatrix} 1 & 0 & 0 & 0 & 0 \\ 1 & 1 & 0 & 0 & 0 \\ 1 & 2 & 1 & 0 & 0 \\ 1 & 3 & 3 & 1 & 0 \\ 1 & 4 & 6 & 4 & 1 \end{bmatrix}, P_4 = \begin{bmatrix} 1 & 0 & 0 & 0 & 0 \\ 1 & 1 & 1 & 1 & 1 \\ 1 & 2 & 4 & 8 & 16 \\ 1 & 3 & 9 & 27 & 81 \\ 1 & 4 & 16 & 64 & 256 \end{bmatrix}, F_4 = \begin{bmatrix} 1 & 0 & 0 & 0 & 0 \\ 0 & 1 & 0 & 0 & 0 \\ 0 & 0 & \frac{1}{2} & 0 & 0 \\ 0 & 0 & 0 & \frac{1}{6} & 0 \\ 0 & 0 & 0 & 0 & \frac{1}{24} \end{bmatrix},$$

and

$$F_4 C_4^{-1} P_4 = \begin{bmatrix} 1 & 0 & 0 & 0 & 0 \\ 0 & 1 & 1 & 1 & 1 \\ 0 & 0 & 1 & 3 & 7 \\ 0 & 0 & 0 & 1 & 6 \\ 0 & 0 & 0 & 0 & 1 \end{bmatrix}.$$

This last matrix is the transpose of the table of Stirling numbers of the second kind!

P 6.2.7. $\left[^n_{n-2}\right]$ counts the number of permutations with $n - 2$ cycles. There are two types of such permutations. The first type has one cycle of size 3 and $n-3$ cycles of size 1. The second type has two cycles of size 2 and $n - 4$ cycles of size 1. For the first type, you first pick three elements, and then arrange them in a circle. There is no other choice, and so the number of such permutations is $2\binom{n}{3}$. For the second type, you pick the four elements that are going to be in cycles of size 2, and then arrange them in two cycles. There are $\frac{1}{2}\binom{4}{2} = 3$ ways of arranging four elements in cycles of size 2, and so the number of permutations of the second type is $3\binom{n}{4}$. Hence

$$\left[^n_{n-2}\right] = 2\binom{n}{3} + 3\binom{n}{4} = \binom{n}{3}\left[2 + \frac{3(n-3)}{4}\right] = \frac{1}{4}(3n-1)\binom{n}{3}.$$

P 6.2.17. If $f(x) = x(x+1)(x+2)\cdots(x+n-1)$, then $f'(x) = (x+1)\cdots(x+n-1) + x(x+2)\cdots(x+n-1) + x(x+1)(x+3)\cdots(x+n-1) + \cdots x(x+1)\cdots(x+n-2)$, where each term in the sum is just missing one of the factors of $f(x)$. Hence,

$$f'(1) = \frac{n!}{1} + \frac{n!}{2} + \cdots \frac{n!}{n} = n!\left[1 + \frac{1}{2} + \cdots + \frac{1}{n}\right] = \left[^{n+1}_2\right],$$

by Theorem 6.33. On the other hand, by Theorem 6.31a,

$$f'(x) = \left[^n_1\right] + 2\left[^n_2\right]x + 3\left[^n_3\right]x^2 + \cdots + n\left[^n_n\right]x^{n-1},$$

and so

$$\left[^{n+1}_2\right] = f'(1) = \left[^n_1\right] + 2\left[^n_2\right] + \cdots + n\left[^n_n\right].$$

P 6.2.21. There are a total of $n!$ permutations of $[n]$. What is the total number of cycles of these? $\begin{bmatrix} n \\ 1 \end{bmatrix}$ of these permutations have 1 cycle, $\begin{bmatrix} n \\ 2 \end{bmatrix}$ of the permutations have 2 cycles, and so on. So the total number of cycles of all the permutations of $[n]$ is

$$\begin{bmatrix} n \\ 1 \end{bmatrix} + 2\begin{bmatrix} n \\ 2 \end{bmatrix} + \cdots + n\begin{bmatrix} n \\ n \end{bmatrix}.$$

By Problem $P\ 6.2.17$, this is equal to $\begin{bmatrix} n+1 \\ 2 \end{bmatrix}$. Hence, the average number of cycles of a permutation of $[n]$ is

$$\frac{\begin{bmatrix} n+1 \\ 2 \end{bmatrix}}{n!} = 1 + \frac{1}{2} + \frac{1}{3} + \cdots + \frac{1}{n},$$

by Theorem 6.33.

P 6.2.31. From Fermat's little theorem (Problem P 5.3.8), we know that if $\alpha = 0, 1, 2, \ldots, p-1$, then the remainder of α^p is the same as the remainder of α when divided by p. This means that $\alpha^p - \alpha$ has remainder 0 when divided by p. In other words, α is a root of $q(x) = x^p - x$ when we think of $q(x)$ as a polynomial over $\mathbb{F}_p$ (this means that the coefficients are in $\mathbb{F}_p$ and we do all arithmetical operations in $\mathbb{F}_p$). Thus we have found p different roots for $q(x)$ and so q cannot have any other roots. Moreover, each of these roots gives one factor of $q(x)$. We conclude that

$$x^p - x = q(x) = x(x-1)(x-2)\cdots(x-p+1) = (x)_p = s(p,0) + s(p,1)x + \cdots + s(p,p)x^p.$$

(The last equality was Theorem 6.31b.) But this means that $s(p,0) = 0$, $s(p,1) = -1$, $s(p,p) = 1$ (all of which we already knew), and $s(p,k) = 0$ for $2 \le k \le p-1$. Now remember that we are in the world of $\mathbb{F}_p$, and so $s(p,k) = 0$ doesn't mean that this number is actually zero. Rather it means that it is the zero of $\mathbb{F}_p$, and that means that it is divisible by p. Now $\begin{bmatrix} n \\ k \end{bmatrix}$ is $\pm s(p,k)$, and so, for $2 \le k \le p-1$, $\begin{bmatrix} n \\ k \end{bmatrix}$ is divisible by p.

P 6.3.5. Let n and k be positive integers. Let S_n be the $(n+1) \times (n+1)$ matrix of the Stirling numbers of the second kind, and let s_n be the $(n+1) \times (n+1)$ matrix of the signed Stirling numbers of the first kind. Theorem 6.38 says that these two matrices are inverses of each other. Also, recall that both of these matrices are lower triangular. Theorem 6.19 says

$$n^k = (n)_0 \begin{Bmatrix} k \\ 0 \end{Bmatrix} + (n)_1 \begin{Bmatrix} k \\ 1 \end{Bmatrix} + (n)_2 \begin{Bmatrix} k \\ 2 \end{Bmatrix} + \cdots + (n)_n \begin{Bmatrix} k \\ n \end{Bmatrix}.$$

We actually need to rewrite this slightly. You were asked to do this in Problem $P\ 6.1.19$ – basically, depending on whether $k \le n$ or $k > n$, an appropriate number of the final terms are zero, making the two expression the same. This version is

$$n^k = (n)_0 \begin{Bmatrix} k \\ 0 \end{Bmatrix} + (n)_1 \begin{Bmatrix} k \\ 1 \end{Bmatrix} + (n)_2 \begin{Bmatrix} k \\ 2 \end{Bmatrix} + \cdots + (n)_k \begin{Bmatrix} k \\ k \end{Bmatrix}.$$

Define two column vectors $U_n = \begin{bmatrix} (n)_0 \\ (n)_1 \\ \vdots \\ (n)_n \end{bmatrix}$ and $H_n = \begin{bmatrix} n^0 \\ n^1 \\ \vdots \\ n^n \end{bmatrix}$, respectively, of falling factorials

and powers of n. Then,

$$S_n U_n = \begin{bmatrix} \{{}^0_0\}(n)_0 \\ \{{}^1_0\}(n)_0 + \{{}^1_1\}(n)_1 \\ \vdots \\ \{{}^k_0\}(n)_0 + \{{}^k_1\}(n)_1 + \cdots + \{{}^k_k\}(n)_k \\ \vdots \\ \{{}^n_0\}(n)_0 + \{{}^n_1\}(n)_1 + \cdots + \{{}^n_n\}(n)_n \end{bmatrix}$$

$$s_n H_n = \begin{bmatrix} s(0,0)n^0 \\ s(1,0)n^0 + s(1,1)n^1 \\ \vdots \\ s(k,0)n^0 + s(k,1)n^1 + \cdots + s(k,k)n^k \\ \vdots \\ s(n,0)n^0 + s(n,1)n^1 + \cdots + s(n,n)n^n \end{bmatrix}.$$

We now can summarize Theorem 6.19 and Theorem 6.31b as

$$S_n U_n = H_n \quad \text{and} \quad s_n H_n = U_n.$$

P 6.3.9. The function $L(n,k)$ counts the number of partitions of $[n]$ into k non-empty ordered lists. To partition $[n]$ into k non-empty ordered lists, you first partition $[n-1]$ into either $k-1$ or k ordered lists, and then place n. You can partition $[n]$ into $k-1$ ordered lists in $L(n-1,k-1)$ ways, and, if that is what you have done, then you have to place n in a list all by itself in order to get k lists. There is only one way to do the latter. On the other hand, if you already have placed $1, \ldots, n-1$ into k ordered lists, then you need to place n in one of existing lists. You can either place n to the left of any of the $n-1$ elements already in lists, or you can place n at the end of any of the k lists. Hence, in this case, you can place n in $(n-1)+k$ ways. As a result,

$$L(n,k) = L(n-1,k-1) + (n+k-1)L(n-1,k).$$

P 6.3.15. As in Problem P 6.3.4, let D_n be the $(n+1) \times (n+1)$ diagonal matrix where, for $0 \le i \le n$, the (i,i) entry of D_n is $(-1)^i$. Also, as usual, s_n and S_n are the $(n+1) \times (n+1)$ matrices of signed Stirling numbers of the first kind and unsigned Stirling numbers of the second kind, respectively. Also let $|s|_n$ be the $(n+1) \times (n+1)$ matrix of unsigned Stirling numbers of the first kind. In all of our matrices, the rows and columns are numbered 0 through n. In Problem P 6.3.4, you were asked to show that $D_n^{-1} = D_n$, and that, for any $(n+1) \times (n+1)$ matrix A, $D_n A D_n$ is the "signed version" of A. Hence, $D_n |s|_n D_n = s_n$ and $D_n s_n D_n = |s|_n$.

We also know (Proposition 6.42d) that $L_n = |s|_n S_n = D_n s_n D_n S_n$, and (Theorem 6.38) that S_n and s_n are inverses of each other. As a result,

$$L_n^{-1} = S_n^{-1} D_n s_n^{-1} D_n = s_n D_n S_n D_n = D_n D_n s_n D_n S_n D_n = D_n |s|_n S_n D_n = D_n L_n D_n.$$

Hence, L_n^{-1} is the matrix of "signed" Lah numbers.

P 7.1.11. We first give a combinatorial argument. The function $p_k(n)$ counts the number of ways you can put n identical balls into k identical (non-empty) boxes. So that none of the boxes end up empty, we first put one ball in each box. (There is only one way to do that.) Next you have to place the remaining $n-k$ balls into any number of boxes (as long as you use no more than k boxes). The number of ways of putting the remaining $n - k$ balls in j boxes is $p_j(n - k)$, and j can go from 0 to k. Hence,

$$p_k(n) = p_k(n - k) + p_{k-1}(n - k) + \cdots + p_1(n - k) + p_0(n - k).$$

For an alternative approach, we start with the recurrence relation for $p_k(n)$ and, each time, apply the same recurrence relation to one of the terms. In other words,

$$p_k(n) = p_k(n - k) + p_{k-1}(n - 1) = p_k(n - k) + p_{k-1}(n - k) + p_{k-2}(n - 2)$$
$$= p_k(n - k) + p_{k-1}(n - k) + p_{k-2}(n - k) + p_{k-3}(n - 3)$$
$$\vdots$$
$$= p_k(n - k) + p_{k-1}(n - k) + \cdots + p_1(n - k) + p_0(n - k).$$

P 7.1.15. Let $\mathcal{B}$ be the set of all partitions of n that have at least one part equal to 1. Given the Ferrers diagram of an element of $\mathcal{B}$, take the last row which consists of one dot off to get a partition of $n - 1$. This function has an inverse – namely add a dot as a new last row to the Ferrers diagram of any partition of $n - 1$ – and so is a bijection. We conclude that $|\mathcal{B}| = p(n - 1)$. The result follows since $|\mathcal{A}| + |\mathcal{B}| = p(n)$.

P 7.1.19. The partitions of 6 into distinct parts are:

The partitions of 6 into parts such that the smallest part is 1 and each part is either the same or one more than the next one are:

Each of the partitions in the second set is the conjugate of a partition in the first set. Hence:

CLAIM: Let n be a positive integer. The number of partitions of n into distinct parts is equal to the number of partitions of n into parts $\lambda_1 \geq \lambda_2 \geq \ldots \geq \lambda_k > 0$ such that the difference between two consecutive parts is 0 or 1, and $\lambda_k = 1$.

PROOF: Consider a partition of n into distinct part. This means that the rows of the Ferrers diagram are never equal. Look at the columns. The last column will have to be of size 1 (otherwise the two first parts would be equal). The penultimate column will have to be of size 1 or 2, since if it were 3 or more, then various rows would have the same size. In fact, every column will have to be the same size or 1 more than the one after, since otherwise two rows would have the same size. Hence, the conjugate of this partition will have smallest part equal to 1 and each part is at most 1 less than the previous part. Conversely, if the rows of a partition don't change drastically (their consecutive size differences are 0 or 1), then the columns cannot be of the same size. (If two columns were the same size, then the difference between the row corresponding to the bottom of these columns and the next row would be at least 2.) We conclude that conjugation gives a bijection between the two sets of partitions of n, and so their number is equal.

P 7.2.7.

(a) $m = n + 1 + 2 + \cdots + (k-1) = n + \frac{k(k-1)}{2}$.

(b) The original parts were non-decreasing. Starting with the smallest part, we are adding 0, 1, 2, ... This ensures that the resulting partition always has distinct parts.

(c) Since μ has distinct parts, it has $k!$ different permutations and each gives a positive integer solution to $x_1 + \cdots + x_k = m$.

(d) For each partition of n into k parts, we get $k!$ positive integer solutions to the equation. But the equation may have more such solutions. Hence, $k! \, p_n(k)$ is possibly smaller but never bigger.

(e) The number of positive integer solutions to $x_1 + \ldots x_k = m$ is $\left(\!\binom{k}{m-k}\!\right)$ and so

$$p_n(k) \leq \frac{1}{k!}\left(\!\binom{k}{m-k}\!\right).$$

(f)

$$\left(\!\binom{k}{m-k}\!\right) = \binom{m-1}{m-k} = \binom{m-1}{k-1} = \binom{n + \frac{(k+1)(k-2)}{2}}{k-1}.$$

The result follows.

P 7.2.15. Let n be a positive integer. As with all the problems in this section, $\mathcal{O}$ and $\mathcal{E}$, respectively, are the sets of partitions of n into an odd or even number of distinct parts. The operator D of Problem P 7.2.12 can be applied to any partition $\lambda \vdash n$ as follows: in the Ferrers diagram for λ, take the dots in $\ell(\lambda)$ (Definition 7.25) and create a new row (appropriately placed) with them. Most often – but not always – D sends elements of $\mathcal{O}$ to $\mathcal{E}$, and sends elements of $\mathcal{E}$ to $\mathcal{O}$.

For convenience, we call *problematic* any case when there exists an element of $\mathcal{E}$ that is not mapped by D to $\mathcal{O}$ or, vice versa, if there is an element of $\mathcal{O}$ that is not mapped by D to $\mathcal{E}$. We will only be considering partitions with $|\ell(\lambda)| < |s(\lambda)|$.

If $|\ell(\lambda)| < |s(\lambda)|$ then D turns the dots in $\ell(\lambda)$ into the last row of the Ferrers diagram (see Problem P 7.2.12). Thus, the operator D always increases the number of rows by one. So an even number of parts becomes an odd number and vice versa. The only problem could be if we ended up with identical rows, and this could only happen if the newly created row has as many dots as the row above it. If $\ell(\lambda) \cap s(\lambda) = \emptyset$, then this would not happen since the final two rows would have $s(\lambda)$ and $\ell(\lambda)$ dots and these are assumed not to be equal. So in the problematic case, $\ell(\lambda) \cap s(\lambda) \neq \emptyset$. This means that the sizes of all rows were consecutive and $\ell(\lambda) = k$, where k is the number of parts of the partition. Moreover, $s(\lambda) = \ell(\lambda) + 1 = k + 1$ (so that after we take a dot off it and create a new row, both rows will have $\ell(\lambda)$ dots). Hence, for the unique problematic case, we have

$$n = (k+1) + (k+2) + \cdots + (k+k) = \frac{k(3k+1)}{2}.$$

These are exactly the generalized pentagonal numbers that are not pentagonal. (Plugging in positive integers in $\frac{k(3k+1)}{2}$ is the same as plugging in negative integers in $\frac{k(3k-1)}{2}$ – this was part of Problem P 7.2.10).

Hence, the unique problematic partition is $\lambda = (2k, 2k-1, \ldots, k+1)$ with $D(\lambda) = (2k-1, 2k-2, \ldots, k, k)$. If k is even, $\lambda \in \mathcal{E}$, and if k is odd, $\lambda \in \mathcal{O}$. Hence, our result is:

Proposition D.1. *Let n be a positive integer, and the map D be as defined in Problem P 7.2.12. We call a partition $\lambda \vdash n$ problematic if $|s(\lambda)| > |\ell(\lambda)|$, and either $\lambda \in \mathcal{E}$ and $U(\lambda) \notin \mathcal{O}$, or $\lambda \in \mathcal{O}$ and $U(\lambda) \notin \mathcal{E}$.*

(a) *A problematic $\lambda \in \mathcal{E}$ exists if and only if n is a generalized pentagonal number $\frac{k(3k+1)}{2}$ with k even.*

(b) *A problematic $\lambda \in \mathcal{O}$ exists if and only if n is a generalized pentagonal number $\frac{k(3k+1)}{2}$ with k odd.*

Moreover, in either case, the problematic λ is unique.

P 8.1.13. Let S denote the set of 10 card hands from a standard deck. Let 11, 12, and 13 denote Jack, Queen, and King, respectively, and, for $1 \le i \le 13$, let A_i denote an element of S that contains four of a kind of the card numbered i. We want to find $|A_1 \cup A_2 \cup \cdots \cup A_{13}|$. For $1 \le i \le 13$, we have $|A_i| = \binom{48}{6}$ (the hand will consist of the 4 cards numbered i and 6 other cards from among the remaining 48). For $1 \le i < j \le 13$, we have $|A_i \cap A_j| = \binom{44}{2}$ (you have to choose 2 more cards from the remaining 44 cards to augment the 4 cards numbered i and the 4 cards numbered j), and the intersection of 3 or more of the sets is empty (no room for 3 four of a kinds). By the inclusion–exclusion principle,

$$|A_1 \cup A_2 \cup \cdots \cup A_{13}| = \sum |A_i| - \sum |A_i \cap A_j| = \binom{13}{1}\binom{48}{6} - \binom{13}{2}\binom{44}{2}.$$

The probability that a 10-card hand will have at least 1 four of a kind is

$$\frac{|A_1 \cup A_2 \cup \cdots \cup A_{13}|}{|S|} = \frac{\binom{13}{1}\binom{48}{6} - \binom{13}{2}\binom{44}{2}}{\binom{52}{10}} = \frac{6483}{643195} \approx .01.$$

There is about a 1% chance that you will have at least 1 four of a kind. Note that, in this problem, if you ignored the corrective term $\binom{13}{2}\binom{44}{2}$, the probability would not change discernibly.

P 8.1.19. Let $S = \{2, \ldots, 100\}$. Note that if an integer $n \in S$ is composite, then it has at least one prime factor less than $\sqrt{100} = 10$. This is because otherwise, n would be the product of at least two integers bigger than 10. This would mean $n > 100$, a contradiction.

Now the prime numbers up to 10 are 2, 3, 5, and 7. For $1 \le i \le 4$, let A_i be the integers in S that are multiples of the ith prime. Hence A_2 is all the even numbers in S, and $A_4 = \{7, 14, 21, 28, \ldots, 98\}$ consists of all multiples of 7 in S.

If we find $|A_1^c \cap A_2^c \cap A_3^c \cap A_4^c|$, then the number of primes up to 100 will be this number plus 4 (since 2, 3, 5, and 7 are primes and not counted). The number of integers in $\{2, \ldots, 100\}$ that are multiples of an integer m is $\left\lfloor \frac{100}{m} \right\rfloor$. Also, as an example, if a number is a multiple of 2, 5, and 7, then it is a multiple of their least common multiple, namely 70. Thus by the inclusion–exclusion principle,

$$
\begin{aligned}
&|A_1^c \cap A_2^c \cap A_3^c \cap A_4^c| \\
&= 99 - \left\lfloor \frac{100}{2} \right\rfloor - \left\lfloor \frac{100}{3} \right\rfloor - \left\lfloor \frac{100}{5} \right\rfloor - \left\lfloor \frac{100}{7} \right\rfloor \\
&\quad + \left\lfloor \frac{100}{6} \right\rfloor + \left\lfloor \frac{100}{10} \right\rfloor + \left\lfloor \frac{100}{14} \right\rfloor + \left\lfloor \frac{100}{15} \right\rfloor + \left\lfloor \frac{100}{21} \right\rfloor + \left\lfloor \frac{100}{35} \right\rfloor \\
&\quad - \left\lfloor \frac{100}{30} \right\rfloor - \left\lfloor \frac{100}{42} \right\rfloor - \left\lfloor \frac{100}{70} \right\rfloor - \left\lfloor \frac{100}{105} \right\rfloor + \left\lfloor \frac{100}{210} \right\rfloor \\
&= 99 - 50 - 33 - 20 - 14 + 16 + 10 + 7 + 6 + 4 + 2 - 3 - 2 - 1 - 0 + 0 = 21.
\end{aligned}
$$

Hence, the number of primes less than 100 is $21 + 4 = 25$.

P 8.1.21. We have k choices for the ceiling, and after choosing a color for the ceiling, we cannot use that color for any of the four walls. Thus, we have to color the four walls using $k - 1$ colors and then multiply the answer by k. We find the number of ways of coloring the walls by using the principle of inclusion–exclusion.

Let S be the set of different ways of coloring four walls with $k - 1$ colors (and with no restriction). Let A_1 be the set of coloring methods in S for which the north and the west walls are colored the same. Likewise A_2 is the set of colorings in S in which the west and the south walls are colored the same, A_3 is the set of colorings in which the south wall and the east wall get the same color, and finally A_4 is the set of colorings in which the east wall and the north wall get the same color. We want $|A_1^c \cap A_2^c \cap A_3^c \cap A_4^c|$. For $1 \le i < j < k \le 4$, we have

$$
\begin{aligned}
\alpha_0 &= |S| = (k - 1)^4 \\
\alpha_1 &= |A_i| = (k - 1)^3 \\
\alpha_2 &= |A_i \cap A_j| = (k - 1)^2 \\
\alpha_3 &= |A_i \cap A_j \cap A_k| = k - 1 \\
\alpha_4 &= |A_1 \cap A_2 \cap A_3 \cap A_4| = k - 1.
\end{aligned}
$$

Note that, in calculating α_2, it may seem that $|A_1 \cap A_2|$ could be different from $|A_1 \cap A_3|$. But in both cases we can choose colors twice and for each we have $k - 1$ choices. For both α_3 and α_4 all the walls have to be the same color and hence we have $k - 1$ choices. Now, by the inclusion–exclusion principle, we have that the total number of ways of coloring the room is

$$k \left[(k-1)^4 - \binom{4}{1}(k-1)^3 + \binom{4}{2}(k-1)^2 - \binom{4}{1}(k-1) + (k-1) \right].$$

If we expand this we get $k^5 - 8k^4 + 24k^2 - 31k^2 + 14k = k(k-1)(k-2)(k^2 - 5k + 7)$. This fifth-degree polynomial has three real roots at 0, 1, and 2, and somehow manages to give the right answers when we plug in non-negative integers. For $k = 0, 1$, and 2, it gives 0. For $k = 3$, it gives 6, for $k = 4$ it gives 72, and for $k = 5$ it gives 420.

REMARK. Put a vertex for each of the four walls and the ceiling and connect two vertices if they share a border. We then get the graph in Figure D.9.

Figure D.9 Graph for a ceiling and four walls.

Now what we want to do is to assign a color to each vertex such that no two adjacent vertices have the same color. Such a coloring is called a *proper vertex coloring* of the graph, and the minimum number of colors needed for such a coloring is called the *chromatic number* of the graph. In the case of our graph, the chromatic number is 3 since we need at least 3 colors. The problem asked for a polynomial in k that gives us the *number* of different proper colorings if we have k colors. This is called the *chromatic polynomial* of the graph. Note that when we plug in positive integers in the chromatic polynomial, we will get zeros until we reach the chromatic number. Starting with the chromatic number, every value of k will give a positive integer. Very surprisingly, if you plug in -1 for k and take the absolute value, you will get the number of acyclic orientations of the graph. An *acyclic orientation* of a graph is any way of assigning directions to the edges such that you do not get any directed cycles (if you think of the directed edges as one-way streets, then no directed cycles means that there is no way to start driving from a vertex and come back to the same vertex). See Section 10.7.2.

P 8.1.24. In Problem P 8.1.23, you were asked to prove that, if $s \in S$ and if s belongs to $n > 0$ of the sets $A_1, A_2, \ldots, A_m$, then s contributes $(-1)^t \binom{n-1}{t-1}$ to e_t. If s belongs to none of $A_1, \ldots, A_m$, then s contributes 0 to e_t.

We just have to add the contributions of every element of S to the right-hand side of the equation for e_t to get e_t. If $s \in S$ belongs to none of the subsets, then its contribution is zero. Now, if t is even, then each element of S that belongs to some of the A's makes a *non-negative* contribution. As a result, $e_t \geq 0$. Similarly, if t is odd, the contribution of each of these elements is non-positive and so $e_t \leq 0$.

By the definition of e_t, we have $e_{t+1} = e_t - (-1)^t N_t$. If t is even, then $e_{t+1} = e_t - N_t$. But $t+1$ is odd, and so $e_{t+1} \leq 0$. Thus, $e_t - N_t \leq 0$ and so $e_t \leq N_t$. On the other hand, if t is odd, then $e_{t+1} = e_t + N_t$, and this quantity is greater than or equal to zero since $t+1$ is even. Thus, in this case, $e_t \geq -N_t$. We conclude that if t is even, then $0 \leq e_t \leq N_t$, and if t is odd, then $-N_t \leq e_t \leq 0$, and the proof is complete.

COROLLARY. We always have $|e_t| \leq N_t$.

P 8.2.5. For $1 \leq i \leq 5$, let x_i denote the number of balls that we put in box number i. We need the number of non-negative integer solutions to

$$x_1 + x_2 + x_3 + x_4 + x_5 = 24,$$

with the conditions that $0 \leq x_1, x_2, x_3 \leq 5$ and $x_4, x_5 \geq 0$. We let S be the set of non-negative integer solutions to $x_1 + x_2 + x_3 + x_4 + x_5 = 24$ (with no upper bounds on the variables). For $i = 1, 2, 3$, let A_i be the set of those solutions in S for which $x_i \geq 6$. Thus, we want $|A_1^c \cap A_2^c \cap A_3^c|$.

We have $|S| = \left(\binom{5}{24}\right) = \binom{28}{4}$. The sizes of $A_1, A_2,$ and A_3 are all the same. To find $|A_1|$, put 6 balls into the first box to begin with, and then decide how to distribute the rest of the 18 balls. Hence, for $1 \leq i \leq 3$, $|A_i|$ is the same as the number of non-negative integer solutions to $x_1 + x_2 + \cdots + x_5 = 18$, and is equal to $\left(\binom{5}{18}\right) = \binom{22}{4}$. Similarly, for $1 \leq i < j \leq 3$, $|A_i \cap A_j| = \left(\binom{5}{12}\right) = \binom{16}{4}$. Finally, to find elements in $A_1 \cap A_2 \cap A_3$, first put 6 balls in the first three boxes, and then distribute the remaining 6 balls. This amounts to the number of non-negative integer solutions to $x_1 + \cdots + x_5 = 6$, and so $|A_1 \cap A_2 \cap A_3| = \left(\binom{5}{6}\right) = \binom{10}{4}$. Thus, by the inclusion–exclusion principle, the answer is

$$\binom{28}{4} - \binom{3}{1}\binom{22}{4} + \binom{3}{2}\binom{16}{4} - \binom{10}{4} = 3{,}780.$$

NOTE: This problem can also be solved using the generating function techniques of Chapter 9.

P 8.3.5. Let the original seating be $\{1,2,3,4,5,6,7,8\}$ so that the pairs sitting opposite each other are $\{1,5\}, \{2,6\}, \{3,7\},$ and $\{4,8\}$. Let T be all the circular permutations of $\{1,\ldots,8\}$. Now for $i = 1,2,3,4$, let A_i consist of those elements of T where i and $i+4$ are sitting opposite each other. We want to find $|A_1^c \cap A_2^c \cap A_3^c \cap A_4^c|$. Now, $\alpha_0 = |T| = 7!$, $\alpha_1 = |A_i| = 6!$ (seat 1 and 5 opposite each other anywhere, and you have 6! possibilities for the rest), $\alpha_2 = |A_i \cap A_j| = 6 \times 4!$ (place 1 and 5 opposite each other anywhere, and you have 6 choices for placing 2 and 6 facing each other and 4! for the rest), $\alpha_3 = |A_i \cap A_j \cap A_k| = 6 \times 4 \times 2!$, and $\alpha_4 = |A_1 \cap A_2 \cap A_3 \cap A_4| = 6 \times 4 \times 2$. Thus the answer using the inclusion–exclusion principle is

$$7! - 4 \times 6! + 6 \times 6 \times 4! - 4 \times 6 \times 4 \times 2! + 6 \times 4 \times 2 = 2{,}880.$$

P 8.3.11. We can change the problem into one about placing non-attacking rooks on a board with forbidden positions. If we use the order of names and dialogues given, then the forbidden positions will be scattered around the board. However, after some experimentation, we can find a better ordering for the names and the dialogues, and we have:

	Meno	Crito	Protagoras	Euthyphro	Parmenides
Faranak	✕✕				
Fereydoon		✕✕	✕✕		
Rostam		✕✕	✕✕		
Tahmineh				✕✕	
Roudabeh				✕✕	✕✕

We let r_k be the total number of ways of placing k non-attacking rooks in the forbidden positions. We get (by inspection) that $r_1 = 8$, $r_2 = 2+1+4+3+12 = 22$, $r_3 = 12+6+2+1+4 = 25$, $r_4 = 2+6+4 = 12$, and $r_5 = 2$. Hence the number of ways to assign the dialogues is

$$5! - 8 \times 4! + 22 \times 3! - 25 \times 2! + 12 \times 1! - 2 \times 0! = 20.$$

P 8.3.19. For every permutation of $[n]$, count the number of elements of $[n]$ that are in their natural position. You will get $0, 1, \ldots, n$. For $1 \le i \le n$, how many permutations have exactly i elements of $[n]$ in their natural position? First choose the i elements, place them in their own position, then find a derangement of the remaining $n - i$ elements. Hence, there are $\binom{n}{i}D_{n-i}$ such permutations. Since i ranges from 0 to n, and the total number of permutations of $[n]$ is $n!$, we have

$$\binom{n}{0}D_n + \binom{n}{1}D_{n-1} + \cdots + \binom{n}{n}D_0 = \sum_{i=0}^{n}\binom{n}{i}D_{n-i} = n!.$$

Now, just replace $\binom{n}{i}$ with $\binom{n}{n-i}$ to get

$$\binom{n}{0}D_0 + \binom{n}{1}D_1 + \cdots + \binom{n}{n}D_n = \sum_{i=0}^{n}\binom{n}{i}D_i = n!.$$

P 8.3.28.

(a) Without the cardinal rule, every artisan has n choices for each tile, and so each artisan can color the tiles in n^s ways. There are $n!$ artisans and so the total number of colorings is $|X| = n! \, n^s$.

(b) Assume that T_1 is colored with color 1 and this is a violation. You can color all the other tiles using any of the n colors. So there are n^{s-1} sets of colorings where T_1 is colored with color 1. But for how many artisans would such a choice of colors constitute a violation? The artisan would have to be associated with a permutation $\mu \in S_n$ where 1 is in its natural position. There are $(n-1)!$ such permutations. Hence, there are $(n-1)! \, n^{s-1}$ colorings where T_1 is colored with color 1 and this is a violation.

(c) To violate the cardinal rule for the coloring of T_1, the tile T_1 could be colored with any one of the n colors. Since there is a choice of n colors, the total number of elements of X for which T_1 is colored in a way that violates the cardinal rule is $n (n-1)! \ n^{s-1} = n! \ n^{s-1}$.

(d) There are two cases: the two tiles are colored either the same or differently. If T_1 and T_2 are colored the same, first pick the color (n choices), then pick the color of every other tile (n^{s-2} choices), then decide the number of artisans for whom such a coloring would be a violation. For such an artisan, the chosen color has to be in its natural position, and the rest of the colors can be permuted at will. So there are $(n-1)!$ choices. So, in the case of T_1 and T_2 colored with one color, we have $n \, n^{s-2} \, (n-1)! = n! \ n^{s-2}$ possibilities. In the second case, we proceed similarly, but we have to choose one color for T_1, then another for T_2 ($n(n-1)$ choices), then color the rest of the tiles (n^{s-2} choices), and finally decide for the number of artisans for whom such a choice of colors would be two violations of the cardinal rule. The answer to the latter is $(n-2)!$, and so, in this case, the total number of elements of X that violate the cardinal rule for both T_1 and T_2 is $n(n-1)n^{s-2} \, (n-2)! = n! \, n^{s-2}$. Thus the final answer is $2n! \ n^{s-2}$.

(e) You could be using $1, 2, \ldots, j$ colors for $T_1, \ldots, T_j$. If for some ℓ, with $1 \leq \ell \leq j$, you are using ℓ colors to color $T_1, \ldots, T_j$, then you have to (1) choose the ℓ colors, (2) partition $T_1, \ldots, T_j$ into ℓ distinct non-empty parts (in order to know which is getting which color), (3) assign a color to each of the remaining tiles, and (4) decide the number of artisans for whom such a choice of colors would be a violation of the cardinal rule for each of $T_1, \ldots, T_j$. The number of choices is $\binom{n}{\ell}$ for (1), $\ell! \, \left\{ {j \atop \ell} \right\}$ for (2), n^{s-j} for (3), and $(n-\ell)!$ for (4). So if you are using ℓ colors, the number we are looking for is $\binom{n}{\ell} \ell! \, \left\{ {j \atop \ell} \right\} n^{s-j} (n-\ell)! = n! \, \left\{ {j \atop \ell} \right\} n^{s-j}$. Since $1 \leq \ell \leq j$, the total number of elements of X for which the coloring for each of $T_1, \ldots, T_j$ constitutes a violation of the cardinal rule is

$$\sum_{\ell=1}^{j} n! \, \left\{ {j \atop \ell} \right\} n^{s-j} = n! \, n^{s-j} \sum_{\ell=1}^{j} \left\{ {j \atop \ell} \right\} = n! \, n^{s-j} B(j).$$

P 9.1.13. Since $\frac{x}{(1-3x)(1-5x)} = \frac{-1/2}{1-3x} + \frac{1/2}{1-5x}$, we have

$$g(x) = \frac{1}{1-3x} \left[\frac{x}{1-5x} + 1 \right] = \frac{x}{(1-3x)(1-5x)} + \frac{1}{1-3x} = \frac{-1/2}{1-3x} + \frac{1/2}{1-5x} + \frac{1}{1-3x}$$

$$= \frac{1/2}{1-3x} + \frac{1/2}{1-5x}.$$

Now $\frac{1}{1-ax}$ is the generating function for $1, a, a^2, \ldots, a^n, \ldots$ Hence $g(x)$ is the generating function for

$$h_n = \frac{1}{2}(3^n + 5^n).$$

P 9.1.19. We know the power series for $\ln(1+x)$:

$$\ln(1+x) = x - \frac{x^2}{2} + \frac{x^3}{3} - \cdots .$$

So $\ln(1+x)$ is the ordinary generating function for $0, 1, -\frac{1}{2}, \frac{1}{3}, \ldots$ Hence, $f(x)$ is the generating function of the convolution of this sequence with itself. The coefficient of x^5 is therefore

$$-\frac{1}{1 \cdot 4} - \frac{1}{2 \cdot 3} - \frac{1}{3 \cdot 2} - \frac{1}{4 \cdot 1} = -\frac{5}{6}.$$

The coefficient of x^5 in the power series expansion of $f(x)$ is also $\frac{f^{(5)}(0)}{5!}$, and so

$$f^{(5)}(0) = -5! \, \frac{5}{6} = -100.$$

P 9.1.23. Let $f^{(0)}(x) = f(x)$ and $f^{(m)}(x)$ denote the mth derivative of f. We prove by induction on m that

$$f^{(m)}(x) = m! \, a_m + \frac{(m+1)!}{1} a_{m+1} x + \frac{(m+2)!}{2!} a_{m+2} x^2 + \cdots + \frac{(m+n)!}{n!} a_{m+n} x^n + \cdots .$$

Check it for the base case $m = 0$, and get that $f(x) = a_0 + a_1 x + a_2 x^2 + \cdots + a_n x^n + \cdots$, as expected. Now, assuming the conclusion to be true for $m = k$, we have

$$f^{(k)}(x) = k! \, a_k + \frac{(k+1)!}{1} a_{k+1} x + \frac{(k+2)!}{2!} a_{k+2} x^2 + \cdots + \frac{(k+n)!}{n!} a_{k+n} x^n + \cdots .$$

Taking one derivative, we get

$$f^{(k+1)}(x) = (k+1)! \, a_{k+1} + \frac{(k+2)!}{1!} a_{k+2} x + \cdots + \frac{(k+n)!}{(n-1)!} a_{k+n} x^{n-1}$$
$$+ \frac{(k+n+1)!}{n!} a_{k+n+1} x^n + \cdots ,$$

which is what we want for the $(k+1)$th derivative. Hence, the result follows by induction.

Now, for $m = n$, plug in $x = 0$, to get $f^{(n)}(0) = n! \, a_n$, and so

$$a_n = \frac{f^{(n)}(0)}{n!}.$$

P 9.2.9. You can use Corollary 9.31 or make a direct argument. We do the latter here. Let $y_2 = 3x_2$ and $y_3 = 5x_3$. Consider the number of non-negative integer solutions to

$$x_1 + y_2 + y_3 = n,$$

where y_2 is a multiple of 3 and y_3 is a multiple of 5. Let the answer be h_n, and let $g(x)$ be the ordinary generating function for h_n. Then

$$g(x) = (1 + x + x^2 + \cdots)(1 + x^3 + x^6 + \cdots)(1 + x^5 + x^{10} + \cdots)$$
$$= \frac{1}{(1-x)(1-x^3)(1-x^5)}.$$

Using a symbolic algebra software, we find the coefficient of x^{47} in the power series expansion of $g(x)$, and get $h_{47} = 88$.

P 9.2.21.

(a) If $f(x) = \sum_{n \geq 0} h_n x^n$, then

$$xf(x) = h_0 x + h_1 x^2 + \cdots + h_{n-1} x^n + \cdots$$

$$\frac{3x}{(1-x)^2} = 3x + 3(2x^2) + \cdots + 3(nx^n) + \cdots$$

$$\frac{-2}{1-x} = -2 - 2x - 2x^2 - \cdots - 2x^n - \cdots.$$

Adding these up, we get

$$xf(x) + \frac{3x}{(1-x)^2} - \frac{2}{1-x} = -2 + (h_0 + 3 - 2)x + (h_1 + 3(2) - 2)x^2 + \cdots$$

$$+ \underbrace{(h_{n-1} + 3n - 2)}_{h_n} x^n + \cdots$$

$$= -2 + h_1 x + h_2 x^2 + \cdots + h_n x^n + \cdots = -2 + f(x) - h_0$$

$$= f(x) - 5.$$

Solving for $f(x)$, we have

$$f(x) = \frac{5}{1-x} - \frac{2}{(1-x)^2} + \frac{3x}{(1-x)^3},$$

and therefore the coefficient of x^n is

$$h_n = 5 - 2\left(\binom{2}{n}\right) + 3\left(\binom{3}{n-1}\right) = 5 - 2\binom{n+1}{n} + 3\binom{n+1}{n-1}$$

$$= 5 - 2(n+1) + \frac{3(n+1)n}{2}$$

$$= (6 - n + 3n^2)/2.$$

(b) First, we solve the corresponding homogeneous recurrence relation $h_n = h_{n-1}$. The solution is $h_n = c$, where c is some constant. We also look for a particular solution for the non-homogeneous equation. We try $h_n = an^2 + bn + c$ (the more obvious choice of $h_n = an + b$ does not work), and get

$$an^2 + bn + c = a(n-1)^2 + b(n-1) + c + 3n - 2 \implies (2a-3)n - a + b + 2 = 0$$

$$\implies a = \frac{3}{2}, \; b = -\frac{1}{2}.$$

Thus $h_n = \frac{3}{2}n^2 - \frac{1}{2}n$ is a particular solution to the non-homogeneous equation. We conclude that the general solution to our non-homogeneous equation is $\frac{3}{2}n^2 - \frac{1}{2}n + c$. Turning to the initial condition, we get $c = 3$, and our final solution is

$$h_n = \frac{3}{2}n^2 - \frac{1}{2}n + 3.$$

P 9.2.31.

(a) We have

$$A(x) = a_0 + a_1 x + a_2 x^2 + \cdots + a_n x^n + \cdots$$
$$B(x) = b_0 + b_1 x + b_2 x^2 + \cdots + b_n x^n + \cdots$$
$$xA(x) = a_0 x + a_1 x^2 + a_2 x^3 + \cdots + a_{n-1} x^n + \cdots$$
$$xB(x) = b_0 x + b_1 x^2 + b_2 x^3 + \cdots + b_{n-1} x^n + \cdots$$
$$xA(x) + xB(x) = a_1 x + a_2 x^2 + a_3 x^3 + \cdots + a_n x^n + \cdots = A(x) - a_0 = A(x)$$
$$xA(x) + A(x) = a_0 + b_1 x + b_2 x^2 + \cdots + b_n x^n + \cdots = B(x) - b_0 + a_0 = B(x) - 1,$$

Thus, we have

$$\begin{cases} (x-1)A(x) + xB(x) = 0 \\ (x+1)A(x) - B(x) = -1. \end{cases}$$

Multiply the second equation by x, add the two resulting equations, and solve for $A(x)$ to get

$$A(x) = \frac{-x}{x^2 + 2x - 1}.$$

We can now substitute for $A(x)$ to find $B(x)$:

$$B(x) = \frac{-x(x+1)}{x^2 + 2x - 1} + 1 = \frac{x-1}{x^2 + 2x - 1}.$$

(b) If $x^2 + 2x - 1 = 0$, then $x = -1 \pm \sqrt{2}$. Let $\alpha = -1 + \sqrt{2}$, $\beta = -1 - \sqrt{2}$, and write

$$\frac{x-1}{x^2 + 2x - 1} = \frac{C}{x - \alpha} + \frac{D}{x - \beta},$$

where C and D are yet-to-be-determined constants. Combining the fractions on the right (note that $(x - \alpha)(x - \beta) = x^2 + 2x - 1$), we see that we must have

$$x - 1 = (C + D)x - (\beta C + \alpha D).$$

Hence,

$$\begin{cases} C + D = 1 \\ \beta C + \alpha D = 1 \end{cases} \Rightarrow \begin{cases} C = \frac{1-\alpha}{\beta - \alpha} = -\alpha/2 \\ D = 1 - \frac{1-\alpha}{\beta - \alpha} = -\beta/2. \end{cases}$$

Thus

$$B(x) = \frac{-\alpha/2}{x - \alpha} + \frac{-\beta/2}{x - \beta} = \frac{1/2}{1 - \frac{x}{\alpha}} + \frac{1/2}{1 - \frac{x}{\beta}}.$$

Now, substituting x/α for x in the power seres expansion of $\frac{1}{1-x}$, we get that the coefficient of x^n in the power series expansion of $\frac{1}{1-x/\alpha}$ is $1/\alpha^n$. We conclude that

$$b_n = \frac{1}{2}\left(\frac{1}{\alpha^n} + \frac{1}{\beta^n}\right) = \frac{(1+\sqrt{2})^n + (1-\sqrt{2})^n}{2}.$$

(c) Proceeding as before, we have

$$A(x) = \frac{-x}{x^2 + 2x - 1} = \frac{\frac{\sqrt{2}-2}{4}}{x - \alpha} + \frac{\frac{-\sqrt{2}-2}{4}}{x - \beta},$$

and (arguing as before),

$$a_n = \frac{2-\sqrt{2}}{4}(1+\sqrt{2})^{n+1} + \frac{-\sqrt{2}-2}{4}(1-\sqrt{2})^{n+1}.$$

(d) We prove the claim by induction on n. You can check the claim for small values of n, and, for the inductive step, we have

$$\begin{aligned}
b_{n+1}^2 - 2a_{n+1}^2 &= (2a_n + b_n)^2 - 2(a_n + b_n)^2 \\
&= 4a_n^2 + 4a_n b_n + b_n^2 - 2a_n^2 - 4a_n b_n - 2b_n^2 \\
&= 2a_n^2 - b_n^2 = -(b_n^2 - 2a_n^2) \\
&= -(-1)^n = (-1)^{n+1}.
\end{aligned}$$

We now can argue that

$$\frac{b_n^2}{a_n^2} = 2 + \frac{(-1)^n}{a_n^2} \Rightarrow \lim_{n\to\infty}\frac{b_n}{a_n} = \lim_{n\to\infty}\sqrt{2 + \frac{(-1)^n}{a_n^2}} = \sqrt{2}.$$

P 9.3.17. We have

$$G(x) = a_0 + \frac{a_1 x}{1} + \frac{a_2 x^2}{2!} + \cdots + \frac{a_r x^r}{r!} + \cdots$$

$$H(x) = b_0 + \frac{b_1 x}{1} + \frac{b_2 x^2}{2!} + \cdots + \frac{b_r x^r}{r!} + \cdots .$$

Now the coefficient of x^n in the product $G(x)H(x)$ is

$$\sum_{r+s=n} \frac{a_r b_s}{r!\, s!},$$

and so the coefficient of $\frac{x^n}{n!}$ in $G(x)H(x)$ is

$$n! \sum_{r+s=n} \frac{a_r b_s}{r!\, s!} = \sum_{r+s=n} \frac{n!\, a_r b_s}{r!\, s!} = \sum_{r=0}^{n} \binom{n}{r} a_r b_{n-r}.$$

Thus, the product $G(x)H(x)$ is the exponential generating function for $\{c_n\}$, where

$$c_n = \sum_{r=0}^{n} \binom{n}{r} a_r b_{n-r}.$$

P 9.4.11. Let n be a positive integer, and let D_n denote the number of derangements of $[n]$. By Theorem 8.24,

$$\frac{D_n}{n!} = 1 - \frac{1}{1!} + \frac{1}{2!} - \frac{1}{3!} + \cdots + (-1)^n \frac{1}{n!}. \qquad (D.1)$$

Now let $a_n = (-1)^n \frac{1}{n!}$, and consider the sequence $\{a_n\}_{n=0}^{\infty}$: $1, -\frac{1}{1!}, \frac{1}{2!}, \ldots, (-1)^n \frac{1}{n!}, \ldots$ What is the ordinary power series of $\{a_n\}_{n=0}^{\infty}$? It is

$$1 - \frac{1}{1!}x + \frac{1}{2!}x^2 - \frac{1}{3!}x^3 + \cdots + (-1)^n \frac{1}{n!}x^n + \cdots = e^{-x}.$$

Now, by Problem P 9.4.8, $\dfrac{e^{-x}}{1-x}$ is the ordinary generating function for the sequence of partial sums of $\{a_n\}_{n=0}^{\infty}$. But, by Equation D.1, the sequence of partial sums of $\{a_n\}_{n=0}^{\infty}$ is the sequence $\left\{ \dfrac{D_n}{n!} \right\}_{n=0}^{\infty}$. Hence,

$$D_0 + \frac{D_1}{1!}x + \frac{D_2}{2!}x^2 + \cdots + \frac{D_n}{n!}x^n + \cdots = \frac{e^{-x}}{1-x}.$$

The power series to the left is the ordinary power series for $\left\{ \dfrac{D_n}{n!} \right\}$, but it is also the exponential generating function for D_n. Hence,

$$\frac{e^{-x}}{1-x} \overset{\text{egf}}{\longleftrightarrow} \{D_n\}_{n=0}^{\infty}.$$

P 9.4.15. (a) = (b). A 0,1 solution to $x_1 + 2x_2 + \cdots + 47x_{47} = 47$ is the same as a number of distinct integers adding up to 47. This is exactly the same as partitioning 47 into distinct parts.

(a) = (c). Every term of the product will have one term from each factor. From the ith factor, we can take either 1 or x^i. Now, any 0, 1 solution to $x_1 + 2x_2 + \cdots + 47x_{47} = 47$ will tell us which to pick from each term. Conversely, given a way of multiplying one term from each factor to get x^{47}, we have a 0, 1 solution to $x_1 + 2x_2 + \cdots + 47x_{47} = 47$.

(c) = (d). In finding the infinite product, the only terms that are relevant in finding x^{47} are the terms in the finite product.

P 9.4.27. We have

$$F_{k-1}(x) = \frac{\left[{k-1 \atop k-1} \right]}{(k-1)!}x^{k-1} + \frac{\left[{k \atop k-1} \right]}{k!}x^k + \cdots + \frac{\left[{n \atop k-1} \right]}{n!}x^n + \cdots,$$

and

$$F'_k(x) = \frac{\left[{k \atop k} \right]}{(k-1)!}x^{k-1} + \frac{\left[{k+1 \atop k} \right]}{k!}x^k + \cdots + \frac{\left[{n+1 \atop k} \right]}{n!}x^n + \cdots.$$

According to Problem P 9.4.8, if f is the ordinary power series for a sequence, then $f/(1-x)$ is the ordinary power series for the partial sums of that sequence. Now $F_{k-1}(x)$ is the ordinary power series for the sequence $\{\frac{\left[\begin{smallmatrix} n \\ k-1 \end{smallmatrix}\right]}{n!}\}_{n=k-1}^{\infty}$, and we know $F_k'(x)$ is the ordinary power series of $\{\frac{\left[\begin{smallmatrix} n+1 \\ k \end{smallmatrix}\right]}{n!}\}_{n=k-1}^{\infty}$. So each term of the latter must be the partial sums of the former, and so we have

$$\frac{\left[\begin{smallmatrix} n+1 \\ k \end{smallmatrix}\right]}{n!} = \frac{\left[\begin{smallmatrix} k-1 \\ k-1 \end{smallmatrix}\right]}{(k-1)!} + \frac{\left[\begin{smallmatrix} k \\ k-1 \end{smallmatrix}\right]}{k!} + \cdots + \frac{\left[\begin{smallmatrix} n \\ k-1 \end{smallmatrix}\right]}{n!}.$$

If you add extra terms equal to zero to the beginning of the sum, you will get the identity of Problem P 6.2.15:

$$\frac{1}{n!}\begin{bmatrix} n+1 \\ k \end{bmatrix} = \frac{1}{0!}\begin{bmatrix} 0 \\ k-1 \end{bmatrix} + \frac{1}{1!}\begin{bmatrix} 1 \\ k-1 \end{bmatrix} + \frac{1}{2!}\begin{bmatrix} 2 \\ k-1 \end{bmatrix} + \frac{1}{3!}\begin{bmatrix} 3 \\ k-1 \end{bmatrix} + \cdots + \frac{1}{n!}\begin{bmatrix} n \\ k-1 \end{bmatrix}.$$

P 9.4.43. The height of the function being integrated at $0, 1, \ldots$ is $f(0), f(1)e^{-s}, f(2)e^{-2s}, \ldots$ So

$$F(s) = \int_0^\infty f(t)e^{-st}\, dt \approx f(0) + f(1)e^{-s} + f(2)e^{-2s} + \cdots + f(n)e^{-ns} + \cdots.$$

Letting $s = -\ln(x)$, we have $e^{-ns} = e^{n\ln(x)} = x^n$, and so

$$F(s) \approx f(0) + f(1)e^{-s} + f(2)e^{-2s} + \cdots + f(n)e^{-ns} + \cdots$$

$$f(0) + f(1)x + f(2)x^2 + \cdots + f(n)x^n + \cdots.$$

Hence, the resulting power series is the ordinary generating function of $f(0)$, $f(1)$, $\ldots, f(n), \ldots$

When we find the ordinary generating function of $f(0), f(1), \ldots, f(n), \ldots$, we are concerned with the values of the function f only at $0, 1, \ldots, n, \ldots$ However, this was a crude approximation of the integral. To get the actual integral, the values of f at all non-negative real numbers comes into play. In this sense, the ordinary generating function is a discrete version of the Laplace transform and, vice versa, the Laplace transform is the continuous version of the ordinary generating function.

P 10.1.11. Applying the Havel–Hakimi algorithm of Theorem 10.10, we get

$$3, 3, 3, 3, 3, 3$$
$$3, 3, 2, 2, 2$$
$$2, 2, 1, 1$$
$$1, 1, 0.$$

Reconstructing the graph by starting from the last sequence and working our way back, we get the graph on the left of Figure D.10. On the other hand, if after the first step of the algorithm we say that maybe a vertex of degree 3 is adjacent to the three vertices of degree 2 (instead of to one vertex of degree 3 and two vertices of degree 2), then we get the following sequence of graphic sequences:

$$3,3,3,3,3,3$$
$$3,3,2,2,2$$
$$3,1,1,1$$
$$0,0,0.$$

Starting with three isolated vertices and working our way back, we get the figure on the right of Figure D.10. The two graphs are not isomorphic since the first one has triangles but the second one doesn't.

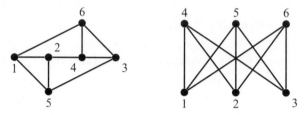

Figure D.10 Two non-isomorphic graphs with degree sequence $3,3,3,3,3,3$.

P 10.1.19. Note that K_{a_1+1}, the complete graph of order $a_1 + 1$, has $a_1 + 1$ vertices, and the degree of each vertex is a_1. Hence, the set of degrees for this graph is $\{a_1\}$.

Take K_p, a complete graph on p vertices, and q other vertices. Add an edge between each of the q vertices and each of the vertices of K_p. Then the vertices of (the old) K_p will now have degree $p + q$ while the (old) isolated vertices will now have degree p. So if $p = a_1$ and $q = a_2 - a_1$, then the set of degrees of the vertices will be $\{a_1, a_2\}$.

Assume that H is a simple graph with r vertices, and that $S_1 = \{b_1, \ldots, b_m\}$ is the set of the degrees of H. Construct a simple graph G with $p + q + r$ vertices as follows. Start with a K_p (complete graph of order p), a copy of H, and q isolated vertices. Add an edge between every vertex in K_p and all the other vertices (the isolated vertices and those in H). The set of degrees of G is $\{(p - 1) + q + r, b_1 + p, b_2 + p, \ldots, b_k + p, p\}$. At this level of generality, there may or may not be any duplicates in this list. In other words, it is possible for two of the numbers in the set to be the same.

Let $S = \{a_1, \ldots, a_k\}$ be any non-empty set of k positive integers with $a_1 < a_2 < \cdots < a_k$. We claim that there exists some simple graph G with $a_k + 1$ vertices, where the set of the degrees of the vertices is precisely S.

We prove this by induction on $|S|$. For the base case, we have already proved the cases when $|S| = 1$ or $|S| = 2$. So assume $S = \{a_1, \ldots, a_k\}$ with $k \geq 3$, and $a_1 < a_2 < \cdots < a_k$.

By the inductive hypothesis, we know there exists a graph H with $a_{k-1} - a_1 + 1$ vertices and with degree set $\{a_2 - a_1, a_3 - a_1, \ldots, a_{k-1} - a_1\}$.

Construct a graph G by starting with a K_{a_1}, a copy of H, and $a_k - a_{k-1}$ isolated vertices. (In our previous construction we are letting $r = k - 2$, $p = a_1$, and $q = a_k - a_{k-1}$.) Connect every vertex in K_{a_1} to each of the vertices in H as well as to the isolated vertices. The resulting graph G will have $a_1 + (a_{k-1} - a_1 + 1) + (a_k - a_{k-1}) = a_k + 1$ vertices. What are the degrees? Each of the originally isolated vertices will now have degree a_1. The vertices in the original K_{a_1} will now have degree $(a_1 - 1) + (a_{k-1} - a_1 + 1) + (a_k - a_{k-1}) = a_k$. The degrees in the original graph H will now become $\{a_2, a_3, \dots, a_{k-1}\}$, and we have a graph with the desired properties.

P 10.2.9. Assume G has n vertices. Since it is a tree, it must have $n - 1$ edges, and so the sum of the degrees of the vertices of G is $2(n - 1) = 2n - 2$. On the other hand, assume that there was at most one leaf. Then the sum of the degrees would be at least $2(n - 1) + 1 = 2n - 1$ which is bigger than $2n - 2$. The contradiction proves that there are at least two leaves.

P 10.2.17. ($\Rightarrow$) A tree with n vertices has $n - 1$ edges, and since the sum of the degrees is twice the number of edges, it will have to be $2(n - 1)$.

($\Leftarrow$) We claim that if, for positive integers $d_1 \geq d_2 \geq \cdots \geq d_n$, we have $\sum_{i=1}^{n} d_i = 2n - 2$, then there exists a tree with degree sequence $d_1, d_2, \dots, d_n$.

We prove the claim by induction on n. If $n = 2$, then the condition on the sum of the degrees becomes $d_1 + d_2 = 2$, and the degree sequence becomes $1, 1$. A path of length 1 will be the desired tree.

For $n > 2$, assume the claim is true for smaller values of n, and let the sequence of positive integers $d_1 \geq d_2 \geq \cdots \geq d_n$ be given. Every tree has a leaf (Problem P 10.2.9) and so $d_n = 1$ (the direct proof is straightforward: if all the n integers were ≥ 2, then the sum would be at least $2n$). Likewise, $d_1 \geq 2$, since if all the degrees were 1, then the sum of the degrees would be n and not $2n - 2$ ($n \neq 2n - 2$ since $n > 2$). Hence, consider the sequence $d_1 - 1, d_2, \dots, d_{n-1}$, and appropriately rearrange it to make it non-increasing. The terms of this sequence are $n - 1$ positive integers that add up to $2n - 4$. So, by the inductive hypothesis, there is a tree with degree sequence $d_1 - 1, d_2, \dots, d_{n-1}$ (appropriately rearranged). Now add a new vertex, and make it a leaf adjacent to the vertex of degree $d_1 - 1$. The resulting graph will have the right degree sequence and will be a tree, since you did not create any cycles.

P 10.2.21. Let the set S consist of all subgraphs of G that have all the vertices of G and are connected. Since $G \in S$, the set S is not empty. From among the elements of S, pick a subgraph T with the smallest number of edges. By Problem P 10.2.20, T is a spanning tree of G. (The argument is as follows. Since $T \in S$, it is connected and does have all the vertices of G, and so the only issue might be that T may have a cycle and not be a tree. But if T did have a cycle, you could remove one edge of it, and the resulting subgraph would still be in S, but would have one less edge than T. This contradicts our assumption that T has the smallest number of edges of all graphs in S.)

P 10.2.31. Let d and g be the diameter and the girth of G, respectively. We are assuming that G has a cycle. Let C be the cycle in G with the shortest length. The girth is the length of the shortest cycle and so the length of C is g. This means that there are g vertices and edges on C. Let x be an arbitrary vertex on this cycle, and let y be the vertex on the cycle furthest away from x. Other than x and y, there are $g - 2$ vertices on the cycle, and so, other than x or y, one of the two paths from x to y along C has $\left\lfloor \frac{g-2}{2} \right\rfloor$ vertices, and the other has $\left\lceil \frac{g-2}{2} \right\rceil$ vertices (if g is even, then these two are equal). So the shorter of the two paths on C from x to y has a total (including x and y) of $\left\lfloor \frac{g-2}{2} \right\rfloor + 2$ vertices and is of length $\left\lfloor \frac{g-2}{2} \right\rfloor + 1$. (Remember that, if there are k vertices on a path, then the length of the path is $k - 1$.)

Could there be a path possibly not on C from x to y that was shorter? The answer is no, since if there were such a path, we could combine it with a part of C (possibly the whole shorter path from x to y on C, or some part of that path if the new path intersects C again before reaching y) and get a cycle that is even shorter than C. The upshot is that the distance between x and y is $\left\lfloor \frac{g-2}{2} \right\rfloor + 1$ (since we have a path of this length from x to y and there cannot be a shorter path.)

Now the distance between any two points cannot be more than the diameter, and so $\left\lfloor \frac{g-2}{2} \right\rfloor + 1 \le d$. We conclude that

$$\left\lfloor \frac{g-2}{2} \right\rfloor \le d-1 \Rightarrow \begin{cases} \frac{g-2}{2} \le d-1 & \text{if } g \text{ is even} \\ \frac{g-3}{2} \le d-1 & \text{if } g \text{ is odd} \end{cases} \Rightarrow g \le 2d+1.$$

P 10.2.43.

(a) Every internal vertex has two children. This gives 20 vertices, all of whom have a parent. The only vertex without a parent is the root. So the total number of vertices is 21.

(b) Assume that i of the vertices are internal; then the total number of vertices is $2i + 1$. Hence, if $2i + 1 = 11$, then $i = 5$. Vertices are either leaves or internal, and so the number of leaves is 6.

P 10.3.7. When you are done, you want to have a bipartite graph with 6 vertices. The number of vertices of the two parts of the bipartite graph will be 5, 1, or 4, 2, or 3, 3. The greatest number of edges that each of these possibilities could have would be if they were a complete bipartite graph. In other words, to get to a bipartite graph with many edges, you want to get to one of $K_{5,1}$, $K_{2,4}$, or $K_{3,3}$. Which one has the greatest number of edges? Well, $K_{5,1}$ has 5 edges, $K_{2,4}$ has 8 edges, and $K_{3,3}$ has 9 edges. The original K_6 has $\binom{6}{2} = 15$ edges. So the smallest number of edges to take out before getting to a bipartite graph is 6. Thus, it is impossible to arrive at a bipartite graph by taking out only 5 edges.

P 10.4.9. The graph K_8 has eight vertices of degree 7. Pair these vertices into four pairs of vertices. Add an edge between the two vertices in each pair. The resulting graph is a multigraph since some pairs of vertices now have two edges connecting them. Also, the degree of each of the vertices is now 8. By Theorem 10.43, the resulting multigraph has an Eulerian circuit. This Eulerian circuit contains each edge of G (as well as each of the four new edges) exactly once. Take one of the new edges off and you will have one Eulerian trail starting from one vertex, ending in another, and containing each of G (and the three remaining new edges) exactly once. Now take off each of the three remaining new edges. The result will be four open trails such that each edge of G will be in exactly one of those trails.

Can we possibly cover G with fewer than four open trails? Assume that the union of three open trails contained every edge of G exactly once. This would mean that G has at most six vertices of odd degree, which is not enough. Hence, the minimum number of trails needed to cover K_8 is four.

P 10.4.13. If the number of vertices of G is v, then the sum of the degrees of G is $4v$. Hence, the number of edges is $2v$ and is even. G has all even degrees and is connected. So it has an Eulerian circuit that contains an even number of edges. Color the edges of the circuit alternately red and blue. Now each vertex has degree 4 and there are three possibilities: (1) the vertex is incident with two blue edges and two red edges, or (2) the vertex is incident with two edges of the same color and a loop of the other color, or (3) the vertex is incident with two loops, each of a different color. In any case, every vertex will be incident with some red edge and some blue edge. Moreover, the blue edges (as well as the red edges) form a 2-regular subgraph. So let H_1 be the set of blue edges and their vertices, and let H_2 be the set of red edges and their vertices. H_1 and H_2 give an edge decomposition of G.

Isolated vertices pose no issue at all since they could be included in either, both, or none of H_1 and H_2. The graph G need also not be connected. Our argument applies to each connected component, and the union of the H_1's from each component would be one of the subgraphs for all of G. Likewise for H_2. Hence, a 4-regular general graph has an edge decomposition into two 2-factors.

P 10.4.15. Recall that a cycle has no repeated vertices or edges except that the first and the last vertex are the same. Let P be a longest path in G, and let $P: x_1 - x_2 - \cdots - x_n$. All the neighbors of x_1 are on the path P, since otherwise we could have lengthened P but P is a longest path. Since x_1 has at least 47 neighbors and they are all on this path, the length of P is at least 47 (i.e., $n \geq 48$). Let x_i be the vertex on P adjacent to x_1 that has the largest index i. Since there are at least 47 neighbors for x_1 on P, i has to be at least 48 (it would be equal to 48 if $x_2, \ldots, x_{48}$ were all the neighbors of x_1). Now the cycle $x_1 - x_2 - \cdots - x_i - x_1$ will be a cycle of length 48. We conclude that, as long as every degree is at least 47, we are guaranteed a cycle of length at least 48. (See Figure D.11.)

Figure D.11 All neighbors of x_1 are on the path P.

How do we know we couldn't have done better and with some other more clever argument showed that the graph is guaranteed to have a longer cycle? To show that our answer is the best possible, we need an example of a graph that convinces us that we could not have proved a stronger statement. One example is the graph K_{48}, the complete graph with 48 vertices. In this graph the minimum degree is 47 and so the graph does satisfy the given condition. The length of the longest cycle in K_{48} is 48 (since the number of vertices is 48, there is can't be any longer cycles).

P 10.5.15. The line graph of a graph G has as many vertices as G has edges, and two of these vertices are adjacent if the corresponding edges in the original graph have a vertex in common. The vertices of the Petersen graph are $[10] = \{1, \ldots, 10\}$. We will use x instead of the vertex numbered 10. This graph has 15 edges, and we denote each edge as a pair of vertices. So, for example, 12 is an edge, as is $7x$. The line graph of the Petersen graph has one vertex for each edge, and so, using the labels of Figure 2.1, the vertices of the complement are

$$\{12, 23, 34, 45, 15, 16, 27, 38, 49, 5x, 68, 8x, 7x, 79, 69\}.$$

With this labeling of the vertices of the line graph, it is easier to know which vertex is adjacent to which. The vertex 12, for example, is adjacent to any vertex that has 1 or 2 in its label. Hence the neighbors of 12 are $\{23, 15, 16, 27\}$. Figure D.12 shows the graph with the Hamiltonian cycle pronounced.

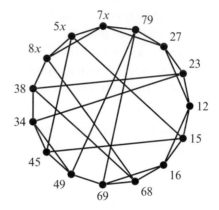

Figure D.12 The line graph of the Petersen graph does have a Hamiltonian cycle.

P 10.5.21. We will prove that it is impossible to fulfill this quest. If we put a vertex for each square, and connect each pair of vertices if there is a door between them, then we get an 8×8 grid. (See Figure D.13.) The problem is asking if this graph has a Hamiltonian path starting from one corner and ending up at the other corner.

Color the vertices of the graph alternately black and white. (See Figure D.13.) Then the two vertices at the opposite corners will have the same color. If you are at a black vertex then after one move you will be at a white vertex and vice versa. Hence, to go from any vertex to another vertex of the same color, you have to make an even number of moves. Thus any path from one corner to the other corner will have an even number of edges. Hence any such path will have an odd number of vertices on it since the number of vertices is one more than the number of edges. This walk cannot be a Hamiltonian path – that is, it cannot go through every vertex – since the total number of vertices is even!

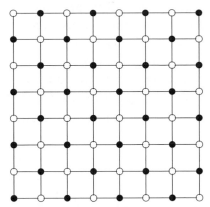

Figure D.13 To go from one corner to another, you need an even number of edges. If no vertex is repeated, this gives an odd number of vertices. You could not have gone through all the vertices since the total number of vertices is even.

P 10.5.27. If you add an edge, a cycle does not go away, and so if G has a cycle of length 5, then so does $\tilde{G}$. Now assume that $\tilde{G}$ has a cycle of length 5. If this cycle does not include the edge $\{x, y\}$, then it is, of course, also a cycle of G, and we are done. Hence, we can assume that $\{x, y\}$ is an edge in the cycle. As a result, we now know that in the original graph G, there is a path of length 4 from x to y:

$$x = v_1 - v_2 - v_3 - v_4 - v_5 = y.$$

There are n vertices in the graph and five of them are represented in this path. Thus there are $n - 5$ vertices of the graph remaining. The vertices x and y could theoretically be adjacent to *all* the other $n - 5$ vertices. This gives a total of $(n - 5) + (n - 5) = 2n - 10$ adjacencies. But we know that $\deg(x) + \deg(y) \geq 2n - 5$. So, between them, x and y are adjacent to five of the vertices on the path. The vertices of x and y are evidently adjacent to v_2 and v_4, respectively. So, other than these two adjacencies, there must be at least three more; and note that we know that $\{x, y\}$ is not an edge of G. Let $N_x = \{2 \leq i \leq 3 \mid \{x, v_{i+1}\}$ is an edge of $G\}$ and $N_y = \{2 \leq i \leq 3 \mid \{y, v_i\}$ is an edge of $G\}$. We have decided that $|N_x| + |N_y| \geq 3$. But $N_x \cup N_y \subseteq \{2, 3\}$, and so it must be that $N_x \cap N_y \neq \emptyset$. If $i \in N_x \cap N_y$ (we are using i but really i is either 2 or 3), then

$$x - v_{i+1} - \cdots - y - v_i - \cdots - x$$

is a cycle of length 5 in G.

ALTERNATIVE PROOF. Consider the graph H consisting of the vertices $x = v_1, \ldots, v_5 = y$ and all their edges in G. The graph H has five vertices and in it the sum of the degrees of x and y

is 5. Hence, by Proposition 10.56, since adding the edge $\{x, y\}$ results in a Hamiltonian cycle (for $\tilde{H}$), then H must have a Hamiltonian cycle as well. So G has a cycle of length 5.

P 10.6.7. We want the number of regions, r, to be 5. Each region is surrounded by four edges, and so summing the number of edges for all the regions gives $4r = 20$. This counts every edge twice, and so the number of edges, e, is 10. By Euler's theorem, $v - e + r = 2$, and so the number of vertices is $v = e - r + 2 = 10 - 5 + 2 = 7$. Figure D.14 gives a connected planar graph with 7 vertices, 10 edges, 5 regions, and with every region bounded by a 4-cycle.

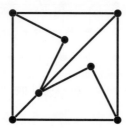

Figure D.14 A connected planar graph with 7 vertices, 10 edges, 5 regions, and every region bounded by a four cycle.

P 10.6.9. Let T be the number of triangles and S the number of quadrilaterals. Now the number of vertices is $\frac{3T+4S}{4}$ and the number of edges is $\frac{3T+4S}{2}$. Also the number of regions is $T + S$. Hence by Euler's formula we have

$$\frac{3T + 4S}{4} - \frac{3T + 4S}{2} + T + S = 2 \quad \Rightarrow \quad 3T + 4S - 6T - 8S + 4T + 4S = 8 \quad \Rightarrow \quad T = 8.$$

Thus the number of triangles has to be eight. An example, with $T = 8$ and $S = 0$, is given in Figure D.15.

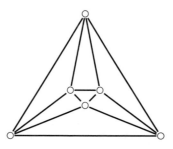

Figure D.15 A tiling of a soccer ball with eight triangles and with all vertices of degree 4.

P 10.6.27. Let C be a simple polygon with n lattice points as vertices. Using all lattice points in the interior and on the boundary (including but not limited to the vertices), triangulate C into primitive triangles. By Lemma 10.91, the area of each such triangle is $1/2$, and so, if t is the number of triangles in the triangulation, then the area enclosed by C is $t/2$. The triangulation of C is exactly a triangulation of a cycle as in Definition 10.92. Hence, by Problem P 10.6.26 – which used Euler's formula on a corresponding tiling of the sphere into all triangles – we have

$$t = 2|V| - n - 2.$$

Now the vertices of the graph are all lattice points, and they are either in the interior or on the boundary. If B and I are the number of vertices on the boundary and in the interior, respectively, then $n = |B|$ and $|V| = |B| + |I|$. Hence we have

$$t = 2|I| + |B| - 2.$$

If A denotes the area enclosed by C, then

$$A = \frac{1}{2}t = |I| + \frac{|B|}{2} - 1,$$

which is Pick's formula.

P 10.6.31.

(a) Induct on ℓ. If $\ell = 0$, then $m = 0$ also, and there are $1 + \ell + m = 1$ regions inside the circle. For the inductive step, assume that the result is true for ℓ chords. Given $\ell + 1$ chords, take one chord out. By induction, the number of regions is $1 + \ell + m'$ where m' is the number of points of intersection of the ℓ remaining chords. Now the new chord starts at a point p_0 on the circle. Follow the chord along. Let its points of intersection with the other chords and the circle in order be $p_1, p_2, \ldots, p_k$ (where p_k is the endpoint of the chord, also on the circle). This means that this chord intersected $k - 1$ other chords, and so m, the number of points of intersection of the $\ell + 1$ chords, is equal to $m' + (k - 1)$. On the other hand, for $0 \le i \le k - 1$, each of the line segments $p_i p_{i+1}$ split one region in the interior of the circle into two regions. Hence, we have an additional k regions. So the total number of regions is

$$1 + \ell + m' + k = 1 + (\ell + 1) + (m' + k - 1) = 1 + (\ell + 1) + m,$$

as desired.

(b) With n points around a circle, the number of chords is $\binom{n}{2}$. Likewise, given any four of the n points, we construct six chords, and two of these intersect in the interior of the circle. So every choice of four points results in one intersection point in the interior of the circle. Thus the number of points of intersection in the interior is $\binom{n}{4}$. The result now follows from the previous part.

(c) If $n = 7$, then $\binom{7}{0} + \binom{7}{2} + \binom{7}{4} = 1 + 21 + 35 = 57$.

P 10.6.37. By inspection, we conjecture that we can draw $K_{3,n}$ on a torus with no edges crossing for $1 \le n \le 6$, and we cannot do so for $n \ge 7$. See Figure D.16 for a drawing of $K_{3,6}$ on the torus without any edges crossing.

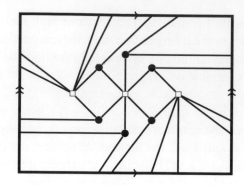

Figure D.16 An embedding of $K_{3,6}$ on the torus.

P 10.7.7.

(a) For the complete graph K_n, all the degrees are $n-1$ and so the table becomes

i	1	2	$\ldots$	n
degrees	$n-1$	$n-1$	$\ldots$	$n-1$
$i-1$	0	1	$\ldots$	$n-1$
$\min\{d_i, i-1\}$	0	$\min\{n-1,1\}$	$\ldots$	$\min\{n-1,n-1\}$

and so $M_{K_n} = n-1$.

(b) For a cycle of length 5, the table is

i	1	2	3	4	5
degrees	2	2	2	2	2
$i-1$	0	1	2	3	4
$\min\{d_i, i-1\}$	0	1	2	2	2

and so $M_{C_5} = 2$.

(c) The Petersen graph is a cubic graph, and so we have

i	1	2	3	4	5	6	7	8	9	10
degrees	3	3	3	3	3	3	3	3	3	3
$i-1$	0	1	2	3	4	5	6	7	8	9
$\min\{d_i, i-1\}$	0	1	2	3	3	3	3	3	3	3

We conclude that $M_G = 3$ for the Petersen graph.

(d) Let G be an arbitrary simple graph. Assume that you have $1 + M_G$ colors. You have to show that it is possible to properly color the vertices of G with these colors. Proceed using the greedy algorithm, starting by assigning color 1 to a vertex with the largest degree. At each stage, pick a vertex that has not yet been colored and that has the largest degree from among the yet-to-be-colored vertices. Color that vertex with a color that is not used by one of its colored neighbors. For $1 \le i \le n$, when you get to vertex numbered i, you have already colored $i - 1$ vertices. If you are unlucky – and you have to assume that you are – all the neighbors of vertex i have different colors, and you have to choose a color other than those. Will there be a color available? The number of colors used by the neighbors cannot be more than d_i since vertex i has d_i neighbors. This number also cannot be more than $i - 1$ since, when you get to vertex i, you have colored only $i - 1$ vertices altogether. Hence, the total number of colors used on the neighbors of vertex i is less than or equal to $\min\{d_i, i - 1\}$. M_G is at least as big as this number since M_G is the maximum of all such numbers. Since the number of colors is one more than M_G, we are safe, and a suitable color is always available. We conclude that $1 + M_G$ colors always suffice and so $\chi(G) \le 1 + M_G$.

For the complete graphs and odd cycles this bound was perfect as $\chi(K_n) = n$ and $\chi(C_5) = 5$. In Problem P 10.7.2, you were asked to find the chromatic number of the Petersen graph. It is 3 but the bound in this problem gave 4, which was not too bad but not the best possible.

P 10.7.9. By the four-color theorem, for a planar graph, four colors suffice to color the regions of the graph properly. This is not so for the graphs on a torus. The corresponding theorem to the four-color theorem, but for a torus, says that to properly color the regions of a graph embedded on a torus, we need at most seven colors. However, for the special case of the Petersen graph, only five colors suffice, as seen in Figure D.17.

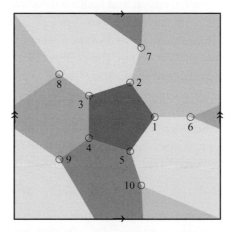

Figure D.17 The Petersen graph embedded on a torus, and its regions colored. A black and white version of this figure will appear in some formats. For the color version, please refer to the plate section.

P 10.7.19. ($\Rightarrow$) Let G be a tree of order n. We prove that $\chi(T;k) = k(k-1)^{n-1}$ by induction on n. If $n = 1$, we have k choices of colors for that one vertex and so the chromatic polynomial is k. If the theorem is true for smaller values of n, then, given a tree of order n, eliminate a leaf and its edge. The remaining tree, by induction, will have a chromatic polynomial equal to $k(k-1)^{n-2}$. Now put the leaf back in. After having colored the rest of the graph, we have $(k-1)$ choices for the leaf, since the only restriction is that it cannot have the same color as the one vertex adjacent to it. Hence, the total number of ways of coloring the tree using k colors is $k(k-1)^{n-1}$ as claimed.

($\Leftarrow$) If $\chi(G;k) = k(k-1)^{n-1}$, then what is the coefficient of k^{n-1} in the expansion of this polynomial? By the binomial theorem, the coefficient of k^{n-2} in the expansion of $(k-1)^{n-1}$ is $n-1$. Hence, the coefficient of k^{n-1} in $\chi(G;k)$ is $n-1$. By Problem P 10.7.18, this is the number of edges of G. So G has $n-1$ edges. To show that G is a tree, all we need to know is that G is connected, since a connected simple graph of order n with $n-1$ vertices is a tree.

Assume $G = (V,E)$ was disconnected. Write $V = V_1 \cup V_2$ with no edges between vertices of V_1 and vertices of V_2, and write $E = E_1 \cup E_2$ with E_1 being all the edges of G between the vertices in V_1, and E_2 the rest of the edges. Hence, $G = G_1 \cup G_2$, where $G_1 = (V_1, E_1)$ and $G_2 = (V_2, E_2)$. Now when you color the vertices of G, since there are no edges between G_1 and G_2, the coloring of G_1 is completely independent of the coloring of G_2, and hence $\chi(G;k) = \chi(G_1;k)\chi(G_2;k)$. Now neither $\chi(G_1;k)$ nor $\chi(G_2;k)$ has a constant term, and so their product will not have a linear term.

We now notice that $k(k-1)^{n-1}$ does have a linear term and so G cannot be disconnected. Since G is a connected graph of order n with $n-1$ edges, so it must be a tree.

P 10.7.22. Note that if you have an acyclic orientation of G, and you eliminate e, you still have an acyclic orientation of $G - \{e\}$. But this is not a 1-1 correspondence since sometimes an acyclic orientation of $G - \{e\}$ leads to two acyclic orientations of G.

Now, let an acyclic orientation of $G - \{e\}$ be given. This implies that there cannot be both a directed path from x to y and one from y to x in $G - \{e\}$, since otherwise the two directed paths would make a directed circuit and hence a directed cycle (somewhere along the directed circuit). Hence, for $G - \{e\}$, there are two types of acyclic orientation. Type I: there is no directed path from x to y or from y to x; Type 2: there is one such directed path from one of x or y to the other. Let $b_1(G - \{e\})$ be the number of type I acyclic orientations of $G - \{e\}$, and let $b_2(G - \{e\})$ be the number of type II acyclic orientations of $G - \{e\}$. We have

$$a(G - \{e\}) = b_1(G - \{e\}) + b_2(G - \{e\}).$$

Now each type I acyclic orientation of $G - \{e\}$ results in two acyclic orientations for G and each type II results in just one acyclic orientation for G, and, as we first noted, every acyclic orientation of G does come from some acyclic orientation of $G - \{e\}$. Hence,

$$a(G) = 2b_1(G - \{e\}) + b_2(G - \{e\}) = a(G - \{e\}) + b_1(G - \{e\}).$$

The proof will be complete once we show that $b_1(G - \{e\}) = a(G/e)$.

Consider an acyclic orientation of G/e. In that orientation there cannot be a directed path from x to y or from y to x since, in G/e, x and y are the same vertex and such a directed path would be a directed cycle. So the acyclic orientations of G/e are in 1-1 correspondence with the acyclic orientations of $G - \{e\}$ of type I, and the proof is complete.

P 10.7.34. We induct on $|V|$. For the base case, let $|V| = 3$. In this case, all vertices are on C, and two of them have distinct lists of size 1. We color them with that color. The third vertex has a list of size 3 and so, no matter what we did with the first two, there will be a color left for them.

For the inductive step, assume $|V| > 3$. Assume that the vertices of C are listed in order $v_1, v_2, \ldots, v_n$. Without loss of generality, assume that v_1 and v_n have the distinct lists of size 1. (For these two vertices the coloring is forced.)

There are two cases: C has a chord or it doesn't.

If C has a chord $\{v_i, v_j\}$ with $1 \le i \le j \le n$, then, since this is a chord, v_i and v_j cannot be consecutive and so $j - i \ge 2$. Create two planar graphs, both triangulations of a cycle and both with fewer vertices than G, basically by cutting G along the chord (but you have to clone the chord so that both subgraphs have it). The outer cycle of the first one is v_1, $\ldots, v_i - v_j - v_{j+1} - \cdots - v_n - v_1 - \ldots - v_{i-1} - v_i$. The outer cycle of the second one is $v_i - v_{i+1} - \cdots - v_j - v_i$. The inner vertices and edges of G are split between the two graphs. The only vertices shared by the two graphs are v_i and v_j (and one, but not both, of these could be v_1 or v_n). The only edge shared is the chord $\{v_i, v_j\}$. Since $\{v_i, v_j\}$ is a chord, there is no edge with one end in one graph and the other end in the other graph.

The first graph satisfies the conditions of the theorem, and we can color its vertices properly by induction. We then turn to the second graph. At this point, the vertices v_i and v_j have already been assigned a color. We will make their assigned color their list of size 1, and every other vertex continues to have their lists intact. We can again color every vertex in the second graph properly by induction. There is no conflict between the two colorings, and so we are done in this case.

If C has no chord, then consider the three consecutive vertices $v_1 - v_2 - v_3$ on C. By Problem P 10.7.33, there is a path of internal vertices from v_1 to v_3 consisting of the entirety of the neighbors of v_2. Eliminate v_2 and its edges, and let H be the remaining graph. The graph H continues to be the triangulation of a cycle; it is just that the cycle has changed and the path from v_1 to v_3 consisting of previously internal vertices has replaced $v_1 - v_2 - v_3$. Also make some adjustments to the permissible color lists of the vertices. Eliminate the one color that is in L_{v_1} from the list of L_{v_2}, to leave v_2 with two colors, say colors α and β. Don't do anything to L_{v_3} but eliminate α and β from each of the lists of the internal vertices on the path from v_1 to v_3 (these were all neighbors of v_2). These internal neighbors of v_2 used to have five available colors but now may have as few as three. On the outside cycle of the new graph H, each of the lists for the vertices v_1 and v_n continues to consist of just one color, and the two lists are disjoint. All the other vertices on this cycle have at least three colors on their lists. By induction, we can properly color all of the vertices of H. At this point none of the neighbors of v_2 other than v_3 could have used the colors α or β. So, depending on what color v_3 gets (it could be one of α or β), there is still a color left for v_2. The coloring is now complete.

P 11.1.13.

(a) Let Q be all the subsets of [5], and let R be all the subsets of [6] that contain 6. The poset Q is evidently the same as (and hence isomorphic to) $2^{[5]}$. On the other hand, define a map $\phi: Q \to R$ by $\phi(A) = A \cup \{6\}$. In other words, given a subset in Q, add the element 6 and get a subset in R. ϕ is one-to-one and onto, and preserves inclusion. Hence, R is also isomorphic to $2^{[5]}$. Now, each element of $2^{[6]}$ either contains 6 or does not contain 6, and hence is either in R or in Q. Also, every element of R has a 6 and hence cannot be in Q. We conclude that $2^{[6]} = R \cup Q$ is a partition of $2^{[6]}$ into two copies of $2^{[5]}$.

(b) Let Q be all the subsets of [4], R be those subsets of [6] that contain 5 but not 6, S be those subsets of [6] that contain 6 but not 5, and T be those subsets of [6] that contain both 5 and 6. It is clear that $2^{[6]}$ is the disjoint union of Q, R, S, and T. The poset Q is $2^{[4]}$ itself. The poset R is obtained from Q by adding 5 to every subset in Q. Likewise, you add 6 to every subset in Q to get S. Finally, adding 5 *and* 6 to every element of Q gives T. Hence, each of R, S, T is isomorphic to $2^{[4]}$.

P 11.1.19. The elements of P are 1, 2, 4, and 8, the divisors of 8. Likewise, the elements of Q are 1, 3, and 9, the divisors of 9. The elements of R are of the form (a, b) where $a \in P$ and $b \in Q$. So, there are $4 \times 3 = 12$ elements in R. The Hasse diagram of R is given in Figure D.18. The map sending (a, b) to ab gives an isomorphism from R to $\mathcal{D}_{72}$. The two posets are isomorphic, and so we can think of $\mathcal{D}_{72}$ as $\mathcal{D}_8 \times \mathcal{D}_9$.

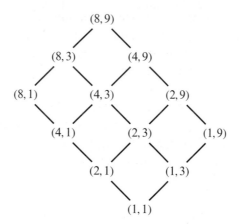

Figure D.18 The Hasse diagram of $\mathcal{D}_9 \times \mathcal{D}_8$.

P 11.2.7. By Dilworth's theorem, P is covered by four chains. What are the chains that could cover x? The possible such chains are $u - x - v$, $u - x$, $x - v$, and just x. Hence, at least one of these must be a chain in a chain partition of P. Consider the poset $Q = P - \{u, x, v\}$. In the poset Q, every chain of P that contained x has been removed, and so Q can be partitioned by one less chain than P. The maximum size of each chain in Q is 4, and so $|Q| \leq 3 \times 4 = 12$. This means that the maximum size of P is 15. An example (see the poset on the left of Figure D.19) shows that 15 is possible.

To find the minimum possible size for P, note that P has at least one maximum-sized chain of size 4. This chain does not include x, and so x together with the elements on the chain of size 4 make for five elements in P (these five elements could include or not include u or v or both). These five elements are partitioned into two chains (the chain consisting of x alone and the chain of size 4) and they cannot be partitioned into one chain. So the maximum size of an antichain from among these five elements is 2. P has an antichain of size 4, and so it must have at least two other elements. So the minimum size of P is 7. But maybe there is an argument that convinces us that the size of P has to be even bigger. An example (the poset on the right of Figure D.19) shows that 7 is indeed the minimum possible size for P.

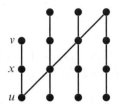 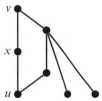

Figure D.19 Two examples of a poset P with both width and height equal to 4, as well as an element x that is only comparable to elements u and v.

P 11.2.15.

(a) For $1 \leq i \leq n$, we have $x_i \leq x_i$ and $i \leq i$, and so $y_i \precsim y_i$. Hence $\precsim$ is reflexive. If $y_i \precsim y_j$ and $y_j \precsim y_k$, then $x_i \leq x_j \leq x_k$ and $i \leq j \leq k$. As a result, $x_i \leq x_k$ and $i \leq k$. So $y_i \precsim y_k$ and $\precsim$ is transitive. Finally, if $y_i \precsim y_j$ and $y_j \precsim y_i$, we have $i \leq j$ and $j \leq i$. So $i = j$ and $y_i = y_j$, proving that $\precsim$ is antisymmetric.

(b) If $y_{i_1} \precsim y_{i_2} \precsim \cdots \precsim y_{i_k}$ is a chain in P, then $i_1 \leq i_2 \leq \cdots i_k$ and $x_{i_1} \leq x_{i_2} \leq \cdots \leq x_{i_k}$. Hence, $x_{i_1}, x_{i_2}, \ldots, x_{i_k}$ is a non-decreasing subsequence of the original sequence.

(c) If $\mathcal{A} = \{y_{i_1}, y_{i_2}, \ldots, y_{i_k}\}$ is an antichain in P, we can rearrange them if necessary so that $i_1 < i_2 < \cdots < i_k$. If $x_{i_1} \leq x_{i_2}$, then y_{i_1} and y_{i_2} would be comparable and $\mathcal{A}$ would not be an antichain. So $x_{i_1} > x_{i_2}$. Likewise, $x_{i_2} > x_{i_3}$, and so on. We conclude that $x_{i_1}, x_{i_2}, \ldots, x_{i_k}$ is a decreasing subsequence of the original sequence.

P 11.2.19. Let P be a finite poset of width w. We use induction on $|P|$ to prove that P can be partitioned into w chains.

If $|P| = 1$, then its width is also 1 and the one element of P is a chain that single-handedly partitions the poset.

So assume that $|P| > 1$ and that the result holds for smaller posets. First assume that P has an antichain W of size w such that W is neither the maximal elements of P nor the minimal elements of P. Next, as in Problem P 11.2.18, define

$$P^+(W) = \{x \in P \mid x \geq y \text{ for some } y \in W\}$$
$$P^-(W) = \{x \in P \mid x \leq y \text{ for some } y \in W\}.$$

We proved in that problem that $P^+(W)$ and $P^-(W)$ are smaller posets of width w (they both include W). By induction, each can be partitioned into w chains, and we showed that we can glue these chains together to get a partition of P into w chains. So we are done in this case.

For the second case, assume that no such W exists. This means that the only antichains of size w are either the set of all maximal elements of P, or the set of all minimal elements of P, or both. Let a be a minimal element of P. Every element of P is less than or equal to some maximal element of P. So let $b \in P$ be a maximal element with $a \leq b$. (It is possible that $a = b$.) Let $Q = P - \{a, b\}$. By our hypothesis for this case, Q does not have an antichain of size w but it must have an antichain of size $w - 1$ (one of the antichains of P with a or b thrown out). Hence, the width of Q is $w - 1$. Hence, by the inductive hypothesis, Q can be partitioned into $w - 1$ chains. These chains together with the chain $\{a, b\}$ give a partition of P into w chains.

P 11.3.13. Let d be the degree of each vertex of G. The total number of edges in G is $d\,|X|$ as well as $d\,|Y|$, which means that $|X| = |Y|$. We want to show that this graph has a matching of size (cardinality) $|X|$. In other words, we want to show that we can match every element of X to a distinct element of Y. We will use the marriage theorem for bipartite graphs, Corollary 11.52.

If A is a subset of X, then how many elements of Y are neighbors of A? There are $|A|\,d$ edges incident with elements of A. The other end of these edges are vertices of Y, and, since every element of Y has d edges, we conclude that these edges are incident with at least $\frac{|A|d}{d} = |A|$ vertices in Y. Hence the number of neighbors of A is at least as big as $|A|$. Hence, $|N_Y(A)| \geq |A|$ for all $A \subseteq X$, the marriage theorem applies, and we have a matching of size $|X|$.

P 11.3.17. First of all, since A is some subset of X, we have $|X - A| = |X| - |A|$, and thus we do have $t - |X - A| = t - (|X| - |A|) = |A| + t - |X|$.

($\Rightarrow$) Assume G has a matching of size t, and A is some arbitrary subset of X. This means that t of the vertices of X are matched with t of the vertices in Y. At least how many of these are in A? There are $|X| - |A|$ vertices of $|X|$ that are not in A, and maybe all of them were used in the matching. This leaves $t - (|X| - |A|)$ vertices of X that must be used in the matching. These vertices must be in A, and each of them is adjacent to a distinct vertex of Y. Thus, the number of vertices of Y that are adjacent to some vertex of A is at least $t - (|X| - |A|)$. We conclude that $N(A) \geq t - (|X| - |A|)$.

($\Leftarrow$) We are assuming that for every $A \subseteq X$ we have $|N(A)| \geq |A| + t - |X|$. We want to show that there exists a matching of size t.

Construct a new bipartite graph G' from G by adding $|X| - t$ vertices to Y. Add an edge from *each* vertex of X to each of the new vertices in Y. We claim that this bipartite graph has a matching of size $|X|$. This would mean that every element of X can be matched with some element of Y. By the marriage theorem, this happens if and only if $|N(A)| \geq |A|$ for every $A \subseteq X$. For the graph G', for every $A \subseteq X$, we have $|N(A)| \geq |A| + t - |X| + (|X| - t) = |A|$ since every one of the new vertices in Y is adjacent to the vertices in A. Thus, the marriage theorem applies to G', and G' has a matching of size $|X|$. Now from among the vertices in X, at most $|X| - t$ vertices may be matched with the new vertices of Y. This leaves t vertices of X that must be matched with the old vertices in Y. Thus, G must have a matching of size at least t.

P 11.3.21. To keep track of the different edges, whenever we run into an edge, if it is in the matching M we color it blue, and if it is not in the matching M, we color it red.

We know that x_0 has at least one neighbor $y_0 \in Y$. Color the edge $\{x_0, y_0\}$ red. If y_0 was also unmatched in our matching, then let $y_k = y_0$, and x_0, y_0 will be our sequence. But if y_0 is already matched to $x_1 \in X$, then color the edge $\{y_0, x_1\}$ blue. Now consider $S_1 = \{x_0, x_1\}$. We know that $y_0 \in N(S_1)$ but, by our assumption, $|N(S_1)| \geq |S_1| = 2$. Let y_1 be an element of $N(S_1)$ distinct from y_0. Since y_1 is in $N(S_1)$, there is at least one edge from one element of S_1 to y_1. Color all such edges red. If y_1 is unmatched, then we stop, and our sequence will be x_0, y_0, x_1, y_1. On the other hand, if y_1 is already matched to some element $x_2 \in X$, then color the edge $\{y_1, x_2\}$ blue, and let $S_2 = \{x_0, x_1, x_2\}$. Again $N(S_2)$ will have at least three elements, and so let y_2 be an element of $N(S_2)$ distinct from y_0 and y_1. The vertex y_2 is adjacent to at least one of $\{x_0, x_1, x_2\}$. Color all such edges red, and continue.

In general, assume that we have defined $x_0, y_0, \ldots, x_{m-1}, y_{m-1}, x_m$ in such a way that all the conditions of the claim except the last one have been satisfied. Consider $S_m = \{x_0, \ldots, x_m\}$. By assumption, $|N(S_m)| \geq |S_m| \geq m + 1$, and so let y_m be an element of $N(S_m)$ distinct from $y_0, \ldots, y_{m-1}$, and color all edges from y_m to elements of S_m red. If y_m is unmatched, stop. Otherwise, let x_{m+1} be the element of X matched to it, and color the edge $\{y_m, x_{m+1}\}$ blue. The new sequence will again have all the needed properties (except the last condition), since each of $y_0, y_1, \ldots, y_{k-1}$ are incident with one blue edge and a number of red edges. For each y, the blue edge is with the "next" x while the red edges are all with the previous x's. This process can continue unless we reach a y that is unmatched. Since the graph is finite, the process will have to end, and so we will eventually arrive at an unmatched vertex y_k and can stop.

P 11.3.25. We prove that A is an independent set if and only if B is a vertex cover.

($\Rightarrow$) Assume A is an independent set, and assume that B is not a vertex cover. This means that there is some edge $e \in E$ with neither end in B. This means that both ends of e are in A, and as a result two vertices in A are adjacent. This means A is not an independent set. The contradiction proves that B is a vertex cover.

($\Leftarrow$) Assume B is a vertex cover and that A is not an independent set. This means that two vertices in A are adjacent. If e is the edge connecting these two vertices, then e has neither end in B. Hence, B is not a vertex cover. The contradiction proves that A is an independent set.

P 11.3.34. We use induction on d. If $d = 0$, then there are no edges, and no colors are needed. For the inductive step, assume that $d > 0$, and that the result has been proved for all d-regular bipartite graphs with a smaller d. By Problem $P\ 11.3.13$, every regular bipartite graph has a perfect matching. The edges in a perfect matching are disjoint, and so can be colored with one color. Assign the color numbered d to all the edges in the perfect matching, and then eliminate them. The resulting graph will be a $(d - 1)$-regular bipartite graph, and so, by induction, its edges can be colored using $d-1$ colors. Color them using the first $d-1$ colors. You have now colored all the edges of G properly using d colors. Since the degree of each vertex is d, you need at least d colors to color the edges properly. Hence, $\chi'(G) = d$.

P 11.4.7. Let $[n] = \{1, 2, \ldots, n\}$. Sperner's theorem says that the size of the largest antichain in $2^{[n]}$ is $\binom{n}{\lfloor n/2 \rfloor}$. A companion to Sperner's theorem says that the only antichains of maximum size in $2^{[n]}$ are the middle levels (i.e., the collection of subsets of size $\lfloor n/2 \rfloor$ or the collection of subsets of size $\lceil n/2 \rceil$). For the special case requested in this problem, we can prove the result directly:

Let $[4] = \{1, 2, 3, 4\}$. We claim that $\mathcal{M} = \{\{1, 2\}, \{1, 3\}, \{2, 3\}, \{1, 4\}, \{2, 4\}, \{3, 4\}\}$ is the only antichain of size 6 in $2^{[4]}$.

Assume that $\mathcal{A}$ was an antichain of size 6 in $2^{[4]}$ and that $\{4\} \in \mathcal{A}$. In this case, $\mathcal{A}$ has at most $\{4\}$ together with an antichain of subsets of $\{1, 2, 3\}$. By Sperner's theorem, the largest antichain in $[3]$ has $\binom{3}{1} = 3$ elements, and so the size of $\mathcal{A}$ could not have been 6. We conclude that $\mathcal{A}$ does not include any singletons. Now assume that $\{1, 2, 3\} \in \mathcal{A}$. If we take the complements of every element in $\mathcal{A}$, we get another antichain (see Problem P 11.4.4) that will include $\{4\}$, and we just showed that this means that $\mathcal{A}$ has at most four elements. We conclude that $\mathcal{A}$ must consist of only elements of size 2.

P 11.4.17. Let $i \in [n]$. How many times does i get counted in this sum? This is the same as asking how many of the intersections contain i. An intersection will contain an i exactly when we intersect two subsets A and B that both contain i. You get a subset with i if you add i to a subset that does not have i. There are 2^{n-1} of the latter, and so there are 2^{n-1} choices for A and 2^{n-1} choices for B. This means that there are $2^{n-1} \times 2^{n-1} = 2^{2n-2} = 4^{n-1}$ intersections in the above sum that will contain i. We conclude that each element of $[n]$ gets counted 4^{n-1} times in the above sum, and so the final answer is $n4^{n-1}$.

P 11.5.3. We want to prove that $\Delta(\mathcal{H}) + 1$ colors are enough to properly color the vertices of $\mathcal{H}$. We prove this statement by induction on the number of vertices. If the number of vertices is 1, then one color suffices, and the maximum degree of that one vertex is 0 (since here we are not allowing singletons as edges). Hence, the result holds for the base case. For the inductive case, let the number of vertices be greater than one. Let v be one vertex. Eliminate v and all of its edges. By induction, you can find a proper coloring of the rest of the vertices using $\Delta(\mathcal{H}) + 1$ colors. Now we want to assign a color to v. The edges that contain v and two or more other vertices are already multicolored. As far as these edges are concerned, v could have any color that it wants. This leaves the edges consisting of just two vertices: v and one other vertex. To keep the coloring proper, we need to give a color to v that is different from the color of the other vertex. But how many such vertices could there be? Each problematic edge has just one such vertex and so the number of such vertices is at most $\Delta(\mathcal{H})$. So in the worst-case scenario, $\Delta(\mathcal{H})$ colors have already been used for these vertices. But we have one more color, and we can use it to color v.

P 11.6.7.

(a) The Hasse diagram of the poset $(P, \leq)$ is a chain. By definition, $\mu_P(n,n) = 1$, and we have $\mu_P(n,n) + \mu_P(n-1,n) = 0$, and so $\mu_P(n-1,n) = -1$. Now $\underbrace{\mu_P(n,n)}_{1} + \underbrace{\mu_P(n-1,n)}_{-1} +$

$\mu_P(n-2,n) = 0$, and so $\mu_P(n-2,n) = 0$. Continuing in the same fashion, we conclude that, for $n \geq 2$, $\mu_P(i,n) = \begin{cases} 1 & \text{if } i = n \\ -1 & \text{if } i = n-1 \\ 0 & \text{for } 1 \leq i \leq n-2. \end{cases}$ If $n = 1$, then we have $\mu_P(1,1) = 1$.

(b) By Theorem 11.102, $g(m) = \sum_{i \leq m} f(i)$ for all $m \in [n]$ if and only if $f(m) = \sum_{i \leq m} g(i) \mu_P(i,m)$ for all $m \in [n]$. Now $\mu_P(i,m)$ is zero except when $i = m$ and $i = m-1$. Hence, the statement becomes $g(m) = f(1) + \cdots + f(m)$ for all $m \in [n]$ if and only if $f(m) = g(m) - g(m-1)$ for all $2 \leq m \leq n$, and $f(1) = g(1)$.

(c) One part of the fundamental theorem of calculus says that if $g(x) = \int_0^x f(t)\, dt$, then $g'(x) = f(x)$. In our context, are we taking some kind of derivative of g? Since g is only defined on the positive integers, we cannot find the derivative of g. The best we can do is to approximate the instantaneous rate of change of g at $x = m$ with the average rate of change of g: $\frac{g(m)-g(m-1)}{m-(m-1)} = g(m) - g(m-1)$, and this is exactly what f is. Hence, f is a "discrete derivative of g." But is g the "discrete integral" of f? Now integration is about finding the limit of Riemann sums. Again, here we cannot find a limit, since our function f is only defined on integers. So again the best we can do is to find a Riemann sum. Using subintervals of length 1 from 0 to m, and using a right-hand Riemann sum, we get $f(1) \times 1 + f(2) \times 1 + \cdots + f(m) \times 1$. This was our definition of g. Thus, our result says that if g is the "discrete integral" of f, then the "discrete derivative" of g is f – the same idea as the fundamental theorem of calculus.

(d) We want to prove that

$$g(m) = f(1) + \cdots + f(m) \text{ for all } m \in [n] \text{ if and only if}$$
$$f(m) = g(m) - g(m-1) \text{ for all } 2 \leq m \leq n, \, \& f(1) = g(1).$$

($\Rightarrow$) Assume that $g(m) = f(1) + \cdots + f(m)$ for all $m \in [n]$. Then $f(1) = g(1)$, and, for $2 \leq m \leq n$,

$$g(m) - g(m-1) = (f(1) + \cdots + f(m)) - (f(1) + \cdots + f(m-1)) = f(m),$$

as desired.

($\Leftarrow$) Assume that $f(1) = g(1)$ and, for $2 \leq m \leq n$, we have $f(m) = g(m) - g(m-1)$. Then, for $m \in [n]$, we have

$$f(1) + f(2) + \cdots + f(m) = g(1) + (g(2) - g(1)) + (g(3) - g(2)) + \cdots + (g(m) - g(m-1))$$
$$= g(m),$$

as desired.

P 11.6.16. From Proposition 11.113, we know that $\mu(d) = 0$ if the square of a prime divides d. So the only relevant terms of the sum are those where d is square-free. In other words, to get a non-zero term, d has to be a product of distinct primes (and still divide n). So,

$$\sum_{d|n} \frac{\mu(d)}{d} = \frac{\mu(1)}{1} + \frac{\mu(p_1)}{p_1} + \cdots + \frac{\mu(p_k)}{p_k}$$

$$+ \frac{\mu(p_1 p_2)}{p_1 p_2} + \frac{\mu(p_1 p_3)}{p_1 p_3} + \cdots + \frac{\mu(p_{k-1} p_k)}{p_{k-1} p_k}$$
$$+ \frac{\mu(p_1 p_2 p_3)}{p_1 p_2 p_3} + \cdots + \frac{\mu(p_{k-2} p_{k-1} p_k)}{p_{k-2} p_{k-1} p_k}$$

$$\vdots$$

$$+ \frac{\mu(p_1 p_2 \cdots p_k)}{p_1 p_2 \cdots p_k}.$$

Now $\mu(p_1 \ldots p_r) = (-1)^r$ (by Proposition 11.113), and so

$$\sum_{d|n} \frac{\mu(d)}{d} = \frac{1}{1} - \frac{1}{p_1} - \cdots - \frac{1}{p_k} + \frac{1}{p_1 p_2} + \cdots + \frac{1}{p_{k-1} p_k}$$

$$- \frac{1}{p_1 p_2 p_3} - \cdots - \frac{1}{p_{k-2} p_{k-1} p_k} + \cdots + (-1)^k \frac{1}{p_1 p_2 \cdots p_k}.$$

On the other hand, if we expand

$$\left(1 - \frac{1}{p_1}\right)\left(1 - \frac{1}{p_2}\right)\cdots\left(1 - \frac{1}{p_k}\right),$$

a typical term will be a product of k terms, and each of these terms will be either a 1 or a term of the form $-\frac{1}{p_i}$. So the typical terms is

$$(-1)^s \frac{1}{p_{i_1} p_{i_2} \cdots p_{i_s}},$$

and these are exactly the terms in the expansion above. Hence the two sides are equal as declared.

Bibliography

Many of the references below are followed by an MR number. "MR" stands for *Mathematical Reviews*, a journal published by the American Mathematical Society (AMS) that contains brief synopses, and in some cases evaluations, of many articles in mathematics, statistics, and theoretical computer science. Reading the review may be a good prelude to reading the actual work. In addition, MathSciNet, the online database associated with MR, gives you the list of references in a work, as well as a list of articles that have followed a given paper and have cited that paper. As a result, you can follow a mathematical paper's journey both backwards and forwards through *Mathematical Reviews*.

Alon, Noga, Benny Sudakov, and Ayal Zaks, *Acyclic edge colorings of graphs*, J. Graph Theory **37** (2001), no. 3, 157–167. MR 1837021

Anderson, Ian, *Combinatorics of Finite Sets*, Dover Publications, Mineola, NY, 2002. Corrected reprint of the 1989 edn. published by Oxford University Press. MR 1902962

Andrews, George E., *Euler's pentagonal number theorem*, Math. Mag. **56** (1983), no. 5, 279–284. MR 720648

———, *The Theory of Partitions*, Cambridge Mathematical Library, Cambridge University Press, Cambridge, UK, 1998. Reprint of the 1976 original. MR 1634067

Andrews, George E. and Jordan Bell, *Euler's pentagonal number theorem and the Rogers-Fine identity*, Ann. Comb. **16** (2012), no. 3, 411–420. MR 2960013

Andrews, George E. and Kimmo Eriksson, *Integer Partitions*, Cambridge University Press, Cambridge, UK, 2004. MR 2122332

Angeltveit, Vigleik and Brendan D. McKay, $R(5,5) \leq 48$, J. Graph Theory **89** (2017), no. 1, 5–13. MR 3828124

Avron, Arnon and Nachum Dershowitz, *Cayley's formula: A page from the book*, Amer. Math. Monthly **123** (2016), no. 7, 699–700. MR 3539855

Barnier, William and James Jantosciak, *Duality and symmetry in the hypergeometric distribution*, Math. Mag. **75** (2002), no. 2, 135–143.

Bajri, Sanaa, John Hannah, and Clemency Montelle, *Revisiting Al-Samaw'al's table of binomial coefficients: Greek inspiration, diagrammatic reasoning and mathematical induction*, Arch. Hist. Exact Sci. **69** (2015), no. 6, 537–576. MR 3413617

Bayer, Margaret and Keith Brandt, *The pill problem, lattice paths, and Catalan numbers*, Math. Mag. **87** (2014), no. 5, 388–394. MR 3324707

Benjamin, Arthur, *Combinatorics and campus security*, UMAP J. **17** (1996), no. 2, 111–116. MR 1799735

Benjamin, Arthur, Gregory Preston, and Jennifer Quinn, *A Stirling encounter with harmonic numbers*, Math. Mag. **75** (2002), no. 2, 95–103.

Bennet, Mark (https://math.stackexchange.com/users/2906/markbennet), *Recurrence relations with multiple roots of auxiliary equation*, Mathematics Stack Exchange, https://math.stackexchange.com/q/816393 (version: 2014-05-31).

Berggren, John L., *Episodes in the Mathematics of Medieval Islam*, 2nd edn., Springer, New York, 2016. MR 3617353

Berkovich, Alexander and Ali Kemal Uncu, *Some elementary partition inequalities and their implications*, Ann. Comb. **23** (2019), no. 2, 263–284. MR 3962857

Biggs, Norman. *Algebraic Graph Theory*, 2nd edn., Cambridge Mathematical Library, Cambridge University Press, Cambridge, UK, 1993. MR 1271140

Biggs, Norman L., E. Keith Lloyd, and Robin J. Wilson, *Graph Theory, 1736–1936*, 2nd edn., The Clarendon Press, Oxford University Press, New York, 1986. MR 879117

Bindner, Donald J. and Martin Erickson, *Alcuin's sequence*, Amer. Math. Monthly **119** (2012), no. 2, 115–121. MR 2892424

Bloom, David M., *Singles in a sequence of coin tosses*, College Math. J. **29** (1998), no. 2, 120–127. MR 1615510

Bollobás, Belá, *On generalized graphs*, Acta Math. Acad. Sci. Hungar. **16** (1965), 447–452. MR 183653

Bondy, J. Adrian, *Beautiful conjectures in graph theory*, European J. Combin. **37** (2014), 4–23. MR 3138588

Bondy, J. Adrian and Václav Chvátal, *A method in graph theory*, Discrete Math. **15** (1976), no. 2, 111–135. MR 414429

Borchardt, Carl Wilhelm, *Ueber eine der interpolation entsprechende darstellung der elliminations-resultante*, Crelle **LVII** (1860), 111–121.

Bose, Raj Chandra and Bennet Manvel, *Introduction to Combinatorial Theory*, Wiley Series in Probability and Mathematical Statistics: Probability and Mathematical Statistics, John Wiley & Sons, New York, 1984. MR 734898

Boyadzhiev, Khristo N., *Close encounters with the Stirling numbers of the second kind*, Math. Mag. **85** (2012), no. 4, 252–266. MR 2993818

Bréard, Andrea, *China*. In Robin J. Wilson and John J. Watkins (eds.), *Combinatorics: Ancient and Modern*, Oxford University Press, Oxford, 2013, pp. 65–81. MR 3204727

Brualdi, Richard A., *Introductory Combinatorics*, 5th edn., Pearson Prentice Hall, Upper Saddle River, NJ, 2010. MR 2655770

Bruhn, Henning and Oliver Schaudt, *The journey of the union-closed sets conjecture*, Graphs Combin. **31** (2015), no. 6, 2043–2074. MR 3417215

Callen, David, *Pair them up! A visual approach to the Chung-Feller theorem*, College Math. J. **26** (1995), no. 3, 196–198.

Carroll, Lewis, *A Tangled Tale*, Macmillan & Co., London, 1885.

Cayley, Arthur, *A theorem on trees*, Quart. J. Pure Appl. Math. **23** (1889), 376–378. Collected Mathematical Papers, vol. 13, Cambridge University Press, Cambridge, UK, 1897, 26–28.

Choudum, Sheshayya A., *A simple proof of the Erdős–Gallai theorem on graph sequences*, Bull. Austral. Math. Soc. **33** (1986), no. 1, 67–70. MR 823853

Chung, Kai Lai and W. Feller, *On fluctuations in coin-tossing*, Proc. Natl. Acad. Sci. U. S. A. **35** (1949), 605–608. MR 0033459

Chvátal, Václav, *Intersecting families of edges in hypergraphs having the hereditary property*. In *Hypergraph Seminar: Ohio State University, 1972*, Lecture Notes in Mathematics, vol. 411, Springer-Verlag, Berlin, 1974, pp. 61–66. MR 0369170

Chvátal, Václav and Frank Harary, *Generalized Ramsey theory for graphs. II: Small diagonal numbers*, Proc. Amer. Math. Soc. **32** (1972), 389–394. MR 0332559

Clarke, Robert J. and Marta Sved, *Derangements and Bell numbers*, Math. Mag. **66** (1993), no. 5, 299–303. MR 1251443

Conlon, David, Jacob Fox, and Benny Sudakov, *Recent developments in graph Ramsey theory*. In *Surveys in Combinatorics 2015*, London Mathematical Society Lecture Note Series, vol. 424, Cambridge University Press, Cambridge, UK, 2015, pp. 49–118. MR 3497267

Daboul, Siad, Jan Mangaldan, Michael Z. Spivey, and Peter J. Taylor, *The Lah numbers and the nth derivative of $e^{1/x}$*, Math. Mag. **86** (2013), no. 1, 39–47. MR 3030333

Daniel, Dale and Stephen E. Shauger, *A result on the Erdős–Gyárfás conjecture in planar graphs*.

In *Proceedings of the Thirty-second Southeastern International Conference on Combinatorics, Graph Theory and Computing* (Baton Rouge, LA, 2001), vol. 153, 2001, pp. 129–139. MR 1887475

Dasef, Martha and Steven Kautz, *Some sums of some significance*, College Math. J. **28** (1997), no. 1, 52–55.

Davidson, Roger R. and Bruce R. Johnson, *Interchanging parameters of the hypergeometric distribution*, Math. Mag. **66** (1993), no. 5, 328–329.

DeTemple, Duane W., *Combinatorial proofs via flagpole arrangements*, College Math. J. **35** (2004), no. 2, 129–133.

Dilworth, Robert P., *A decomposition theorem for partially ordered sets*, Ann. of Math. (2) **51** (1950), 161–166. MR 32578

Donald, Alan, *An upper bound for the path number of a graph*, J. Graph Theory **4** (1980), no. 2, 189–201. MR 570353

Djebbar, Ahmed, *Islamic combinatorics*. In Robin J. Wilson and John J. Watkins (eds.), *Combinatorics: Ancient and Modern*, Oxford University Press, Oxford, 2013, pp. 83–107. MR 3204727

Dunham, William, *Euler: The master of us all*, The Dolciani Mathematical Expositions, vol. 22, Mathematical Association of America, Washington, DC, 1999. MR 1669154

———, *The early (and peculiar) history of the Möbius function*, Math. Mag. **91** (2018), no. 2, 83–91. MR 3777904

Edelman, Alan and Gilbert Strang, *Pascal matrices*, Amer. Math. Monthly **111** (2004), no. 3, 189–197. MR 2042124

Edwards, Anthony W. F., *Pascal's Arithmetical Triangle*, Charles Griffin & Co., London; The Clarendon Press, Oxford University Press, New York, 1987. MR 930876

Encyclopedia Britannica, *Gamma function*, www.britannica.com/science/gamma-function (version 2019-09-06).

Erdős, Paul, *On a lemma of Littlewood and Offord*, Bull. Amer. Math. Soc. **51** (1945), 898–902. MR 14608

———, *Some remarks on the theory of graphs*, Bull. Amer. Math. Soc. **53** (1947), 292–294. MR 19911

———, *On the combinatorial problems which I would most like to see solved*, Combinatorica **1** (1981), no. 1, 25–42. MR 602413

Erdős, Paul and Richard Rado, *A partition calculus in set theory*, Bull. Amer. Math. Soc. **62** (1956), 427–489. MR 81864

Erdős, Paul and George Szekeres, *A combinatorial problem in geometry*, Compositio Math. **2** (1935), 463–470. MR 1556929

Euler, Leonhard, *Solutio problematis ad geometriam situs pertinentis*, Commentarii academiae scientiarum Petropolitanae **8** (1741), 128–140 (Latin). English translation in Norman L. Biggs, E. Keith Lloyd, and Robin J. Wilson, *Graph Theory, 1736–1936*, 2nd edn., The Clarendon Press, Oxford University Press, New York, 1986. MR 879117

Fiamčík, Iosif, *The acyclic chromatic class of a graph*, Math. Slovaca **28** (1978), no. 2, 139–145. MR 526851

Franklin, Fabian, *Sur le développement du produit infini* $(1 - x)(1 - x^2)(1 - x^3) \cdots$, Comptes Rendus **82** (1881), 448–450.

Funkenbusch, W. W., *Classroom notes: From Euler's formula to Pick's formula using an edge theorem*, Amer. Math. Monthly **81** (1974), no. 6, 647–648. MR 1537447

Füredi, Zoltán Füredi, *Problem session*, Kombinatorik geordneter Mengen, Oberwolfach, B. R. D., 1985.

Galvin, Fred, *A proof of Dilworth's chain decomposition theorem*, Amer. Math. Monthly **101** (1994), no. 4, 352–353. MR 1270960

Gaskell, R. W., Murray S. Klamkin, and Peter Watson, *Triangulations and Pick's theorem*, Math. Mag. **49** (1976), no. 1, 35–37. MR 400046

Golomb, Solomon W., *Polyominoes: Puzzles, Patterns, Problems, and Packings*, 2nd edn., Princeton University Press, Princeton, NJ, 1994. MR 1291821

Golomb, Solomon W. and Leonard D. Baumert, *Backtrack programming*, J. Assoc. Comput. Mach. **12** (1965), 516–524. MR 0195585

Goodey, Paul R., *Hamiltonian circuits in polytopes with even sided faces*, Israel J. Math. **22** (1975), no. 1, 52–56. MR 410565

Goodman, Adolph W., *On sets of acquaintances and strangers at any party*, Amer. Math. Monthly **66** (1959), 778–783. MR 0107610

Graham, Ronald L., Bruce L. Rothschild, and Joel H. Spencer, *Ramsey Theory*, Wiley Series in Discrete Mathematics and Optimization, John Wiley & Sons, Hoboken, NJ, 2013. Paperback edition of the 2nd (1990) edition [MR1044995]. MR 3288500

Granville, Andrew, *Zaphod Beeblebrox's brain and the fifty-ninth row of Pascal's triangle*, Amer. Math. Monthly **99** (1992), no. 4, 318–331. MR 1157222

———, *Arithmetic properties of binomial coefficients. I: Binomial coefficients modulo prime powers*. In *Organic Mathematics: Proceedings of the Organic Mathematics Workshop, December 12–14, 1995*, Simon Fraser University, Burnaby, BC, Canadian Mathematical Society Conference Proceedings, vol. 20, American Mathematical Society, Providence, RI, 1997, pp. 253–276. MR 1483922

Greenwood, Robert E. and Andrew M. Gleason, *Combinatorial relations and chromatic graphs*, Canad. J. Math. **7** (1955), 1–7. MR 0067467

Griggs, Jerrold R., *Problems on chain partitions*, Discrete Math. **72** (1988), no. 1-3, 157–162, Proceedings of the First Japan Conference on Graph Theory and Applications (Hakone, 1986). MR 975534

Halmos, Paul R. and Herbert E. Vaughan, *The marriage problem*, Amer. J. Math. **72** (1950), 214–215. MR 33330

Harper, Larry H., *Stirling behavior is asymptotically normal*, Ann. Math. Statist. **38** (1967), 410–414. MR 211432

Hathout, Heba, *The old hats problem revisited*, College Math. J. **35** (2004), no. 2, 97–102.

Heckman, Christopher Carl and Roi Krakovski, *Erdős–Gyárfás conjecture for cubic planar graphs*, Electron. J. Combin. **20** (2013), no. 2, Paper 7, 43. MR 3066346

Heule, Marijn J. H., *Schur number five*. In *Proceedings of the Thirty-Second AAAI Conference on Artificial Intelligence (AAAI-18), New Orleans, LA, February 2–7, 2018*, AAAI Press, 2018, pp. 6598–6606.

Hierholzer, Carl, *Über die möglichkeit, einen Linienzug ohne Wiederholung und ohne Unterbrechung zu umfahren*, Mathematische Annalen **VI** (1873), 30–32 (German).

Hoffman, Paul, *The Man Who Loved Only Numbers: The Story of Paul Erdős and the Search for Mathematical Truth*, Hyperion Books, New York, 1998. MR 1666054

Holton, Derek A., Bennet Manvel, and Brendan D. McKay, *Hamiltonian cycles in cubic 3-connected bipartite planar graphs*, J. Combin. Theory Ser. B **38** (1985), no. 3, 279–297. MR 796604

Honsberger, Ross, *Ingenuity in Mathematics*, New Mathematical Library, vol. 23, Random House, New York, 1970. MR 3155264

Hsu, Tim, Mark J. Logan, Shahriar Shahriari, and Christopher Towse, *Partitioning the Boolean lattice into chains of large minimum size*, J. Combin. Theory Ser. A **97** (2002), no. 1, 62–84. MR 1879127

———, *Partitioning the Boolean lattice into a minimal number of chains of relatively uniform size*, European J. Combin. **24** (2003), no. 2, 219–228. MR 1961561

Jameson, Graham J. O., *Fibonacci periods and multiples*, Math. Gaz. **102** (2018), no. 553, 63–76. MR 3766792

Jewett, Robert I. and Kenneth A. Ross, *Random walks on* **Z**, College Math. J. **19** (1988), no. 4, 330–342. MR 962485

Joseph, Anthony, Anna Melnikov, and Rudolf Rentschler (eds.), *Studies in Memory of Issai Schur*, Progress in Mathematics, vol. 210, Birkhäuser, Boston, MA, 2003. Papers from the Paris Midterm Workshop of the European Community Training and Mobility of Researchers (TMR) Network held in Chevaleret, France, May 21–25, 2000, and the

Schur Memoriam Workshop held in Rehovot, Israel, December 27–31, 2000. MR 1985184

Kalkowski, Maciej, Michał Karoński, and Florian Pfender, *Vertex-coloring edge-weightings: Towards the 1-2-3-conjecture*, J. Combin. Theory Ser. B **100** (2010), no. 3, 347–349. MR 2595676

Kapoor, Shashichand F., Albert D. Polimeni, and Curtiss E. Wall, *Degree sets for graphs*, Fund. Math. **95** (1977), no. 3, 189–194. MR 480200

Katz, Victor J., *Combinatorics and induction in medieval Hebrew and Islamic mathematics*, Vita mathematica (Toronto, ON, 1992; Quebec City, PQ, 1992), MAA Notes, vol. 40, Math. Assoc. America, Washington, DC, 1996, pp. 99–106. MR 1391737

Karoński, Michał, Tomasz Łuczak, and Andrew Thomason, *Edge weights and vertex colours*, J. Combin. Theory Ser. B **91** (2004), no. 1, 151–157. MR 2047539

Klarreich, Erica, *A tenacious explorer of abstract surfaces*, Quanta Magazine (2014), www.quantamagazine.org/maryam-mirzakhani-is-first-woman-fields-medalist-20140812/.

Kleitman, Daniel J., and Thomas L. Magnanti, *On the number of latent subsets of intersecting collections*, J. Combin. Theory Ser. A **16** (1974), 215–220. MR 329908

Knuth, Donald E., *Two thousand years of combinatorics*. In Robin J. Wilson and John J. Watkins (eds.), *Combinatorics: Ancient and Modern*, Oxford University Press, Oxford, 2013, pp. 3–37. MR 3204727

———, *Hamiltonian paths and cycles*. In *The Art of Computer Programming*, vol. 4, https://www-cs-faculty.stanford.edu/~knuth/taocp.html, online draft as Pre-Fascicle 8a. To be published by Addison-Wesley.

Koshy, Thomas and Zhenguang Gao, *Convergence of a Catalan series*, College Math. J. **43** (2012), no. 2, 141–146. MR 2897477

Kulenović, Mustafa R. S. and Gerasimos Ladas, *Dynamics of Second Order Rational Difference Equations*, Chapman & Hall/CRC, Boca Raton, FL, 2002. With open problems and conjectures. MR 1935074

Kusuba, Takanori and Kim Plofker, *Indian combinatorics*. In Robin J. Wilson and John J. Watkins (eds.), *Combinatorics: Ancient and Modern*, Oxford University Press, Oxford, 2013, pp. 41–62. MR 3204727

Lagarias, Jeffrey C. (ed.), *The Ultimate Challenge: The $3x + 1$ Problem*, American Mathematical Society, Providence, RI, 2010. MR 2663745

Lah, Ivo, *Eine neue Art von Zahlen, ihre Eigenschaften und Anwendung in der mathematischen Statistik*, Mitteilungsbl. Math. Statist. **7** (1955), 203–212. MR 74435

Lam, Lay Yong, *The Chinese connection between the Pascal triangle and the solution of numerical equations of any degree*, Historia Math. **7** (1980), no. 4, 407–424. MR 601289

Larcombe, Peter J., *The 18th century chinese discovery of the Catalan numbers*, Mathematical Spectrum **32** (1999/2000), no. 1, 5–7.

Levin, Oscar and Gerri M. Roberts, *Counting knights and knaves*, College Math. J. **44** (2013), no. 4, 300–306. MR 3096494

Lovász, László, *On covering of graphs*. In *Theory of Graphs (Proceedings of the Colloquium Held at Tihany, Hungary, September, 1966)*, Academic Press, New York, 1968, pp. 231–236. MR 0233723

———, *Combinatorial Problems and Exercises*, 2nd edn., AMS Chelsea Publishing, Providence, RI, 2007. MR 2321240

Lubell, David, *A short proof of Sperner's lemma*, J. Combin. Theory **1** (1966), 299. MR 194348

Mann, Henry B. and Daniel Shanks, *A necessary and sufficient condition for primality, and its source*, J. Combin. Theory Ser. A **13** (1972), 131–134. MR 0306098

Mantel, Willem, *Problem 28*, Wiskundige Opgaven **10** (1907), 60–61 (Solution by H. Gouwentak, W. Mantel, J. Teixeira de Mattes, F. Schuh and W. A. Wythoff).

Mazur, David R., *Combinatorics: A Guided Tour*, MAA Textbooks, Mathematical Association of America, Washington, DC, 2010. MR 2572113

Merris, Russell, *Combinatorics*, PWS Publishing Company, Boston, MA, 1996. MR 1650711

Meshalkin, Lev D., *Generalization of Sperner's theorem on the number of subsets of a finite set*, Theory Probab. Appl. **8** (1963), no. 2, 203–204.

Moser, Leo and Max Wyman, *Problem B-6*, Fibonacci Quart. **1** (1963), no. 1, 74.

Mütze, Torsten, *Proof of the middle levels conjecture*, Proc. Lond. Math. Soc. (3) **112** (2016), no. 4, 677–713. MR 3483129

Needham, Joseph, *Science and Civilization in China*, vol. 3, Cambridge University Press, New York, 1959.

Nunemacher, Jeffrey and Robert M. Young, *On the sums of consecutive kth powers*, Math. Mag. **60** (1987), no. 4, 237–238.

Odlyzko, Andrew and Herman J. J. te Riele, *Disproof of the Mertens conjecture*, J. Reine Angew. Math. **357** (1985), 138–160. MR 783538

Oro, Ken, *Parity of the partition function*, Electron. Res. Announc. Amer. Math. Soc. **1** (1995), no. 1, 35–42. MR 1336698

Ore, Oystein, *Note on Hamilton circuits*, Amer. Math. Monthly **67** (1960), 55. MR 118683

Osler, Thomas J., *Fermat's little theorem from the multinomial theorem*, College Math. J. **33** (2002), no. 3, 239.

Pak, Igor, *History of Catalan numbers*. In Richard P. Stanley, *Catalan Numbers*, Cambridge University Press, New York, 2015, pp. 177–190. MR 3467982

Perles, Micha A., *A proof of Dilworth's decomposition theorem for partially ordered sets*, Israel J. Math. **1** (1963), 105–107. MR 168496

Petkovšek, Marko and Tomaž Pisanski, *Combinatorial interpretation of unsigned Stirling and Lah numbers*, Pi Mu Epsilon J. **12** (2007), no. 7, 417–424.

Prüfer, Heinz, *Neuer beweis eines satzes über permutationen*, Arch. Math. Phys. **27** (1918), no. 3, 142–4.

Pyber, László, *Covering the edges of a connected graph by paths*, J. Combin. Theory Ser. B **66** (1996), no. 1, 152–159. MR 1368522

Radziszowski, Stanislaw, *Small Ramsey numbers*, Electron. J. Combin. Dynamic Survey (2021), DS 1.16, 1–116.

Ramras, Mark, *A codeword proof of the binomial theorem*, College Math. J. **34** (2003), no. 2, 144.

Ramsey, Frank P., *On a problem of formal logic*, Proc. Lond. Math. Soc. (2) **30** (1929), no. 4, 264–286. MR 1576401

Renault, Marc, *Four proofs of the ballot theorem*, Math. Mag. **80** (2007), no. 5, 345–352. MR 2362634

Rowland, Eric, *Two binomial coefficient conjectures*, arXiv:1102.1464v1 [math.NT] (2011), 1–10.

Schur, Issai, *Gesammelte Abhandlungen: Bände I, II, & III*, Springer-Verlag, Berlin-New York, 1973. MR 0462891, 0462892, 0462893

Segal, Sanford L., *Helmut Hasse in 1934*, Historia Math. **7** (1980), no. 1, 46–56. MR 559835

Seidenberg, Abraham, *A simple proof of a theorem of Erdős and Szekeres*, J. Lond. Math. Soc. **34** (1959), 352. MR 0106189

Shahriari, Shahriar, *On the structure of maximum 2-part Sperner families*, Discrete Math. **162** (1996), no. 1-3, 229–238. MR 1425790

———, *Approximately Calculus*, American Mathematical Society, Providence, RI, 2006. MR 2275512

———, *Algebra in Action: A Course in Groups, Rings, and Fields*, Pure and Applied Undergraduate Texts, vol. 27, American Mathematical Society, Providence, RI, 2017. MR 3675414

Singh, Parmanand, *The so-called Fibonacci numbers in ancient and medieval India*, Historia Math. **12** (1985), no. 3, 229–244.

Singmaster, David, *How often does an integer occur as a binomial coefficient?*, Amer. Math. Monthly **78** (1971), no. 4, 385–386.

Smullyan, Raymond M., *What is the Name of this Book?: The Riddle of Dracula and Other Logical Puzzles*, Prentice Hall, Englewood Cliffs, NJ, 1978.

Spencer, Joel, *Ramsey's theorem: A new lower bound*, J. Combin. Theory Ser. A **18** (1975), 108–115. MR 0366726

Spivey, Michael Z., *Deranged exams*, College Math. J. **41** (2010), no. 3, 197–202. MR 2656317

Stanley, Richard P., *Acyclic orientations of graphs*, Discrete Math. **5** (1973), 171–178. MR 317988

———, *Enumerative Combinatorics*, vol. 1, 2nd edn., Cambridge Studies in Advanced Mathematics, vol. 49, Cambridge University Press, Cambridge, 2012. MR 2868112

———, *Catalan Numbers*, Cambridge University Press, New York, 2015. MR 3467982

Stein, Sherman K., *Mathematics: The Man-Made Universe*, Freeman, San Francisco, CA, 1969.

Stigler, Stephen M., *Isaac Newton as a probabilist*, Statist. Sci. **21** (2006), no. 3, 400–403. MR 2329703

Sudakov, Benny, István Tomon, and Adam Zsolt Wagner, *Uniform chain decompositions and applications*, arXiv:1911.09533v1 [math.CO] (2019), 1–22.

Thomassen, Carsten, *Every planar graph is 5-choosable*, J. Combin. Theory Ser. B **62** (1994), no. 1, 180–181. MR 1290638

Tomon, István, *On a conjecture of Füredi*, European J. Combin. **49** (2015), 1–12. MR 3349520

Tucker, Alan, *Applied Combinatorics*, 3rd edn., John Wiley & Sons, Inc., New York, 1995. MR 1287611

Turán, Paul, *Eine Extremalaufgabe aus der Graphentheorie*, Mat. Fiz. Lapok **48** (1941), 436–452. MR 18405

van Lint, Jacobus H. and Richard M. Wilson, *A Course in Combinatorics*, 2nd edn., Cambridge University Press, Cambridge, 2001. MR 1871828

West, Douglas B., *Introduction to Graph Theory*, 2nd edn., Prentice Hall, Upper Saddle River, NJ, 2001. MR 1367739

Wilf, Herbert S., *A greeting: And a view of Riemann's hypothesis*, Amer. Math. Monthly **94** (1987), no. 1, 3–6. MR 873600.

———, *Generating Functionology*, 3rd edn., A K Peters, Wellesley, MA, 2006. MR 2172781

Wilson, Robin J., *History of graph theory*. In Jonathan L. Gross, Jay Yellen, and Ping Zhang (eds.), *Handbook of Graph Theory*, 2nd edn., Discrete Mathematics and its Applications, CRC Press, Boca Raton, 2014, pp. 31–51. MR 3185588

Wilson, Robin J. and John J. Watkins (eds.), *Combinatorics: Ancient and Modern*, Oxford University Press, Oxford, 2013. MR 3204727

Yamamoto, Koichi, *Logarithmic order of free distributive lattice*, J. Math. Soc. Japan **6** (1954), 343–353. MR 67086

Index